LABORATORY MANUAL—
with Cat and Pig Dissections

Anatomy & Physiology: An Integrative Approach

Fourth Edition

Kyla Turpin Ross
Georgia Institute of Technology

Leslie Day
Texas A&M University

Joseph Comber
Villanova University

Christine M. Eckel
Indiana University School of Medicine, Northwest

LABORATORY MANUAL—ANATOMY & PHYSIOLOGY: AN INTEGRATIVE APPROACH, FOURTH EDITION

Published by McGraw Hill LLC, 1325 Avenue of the Americas, New York, NY 10121. Copyright ©2022 by McGraw Hill LLC. All rights reserved. Printed in the United States of America. Previous editions ©2016 and 2014. No part of this publication may be reproduced or distributed in any form or by any means, or stored in a database or retrieval system, without the prior written consent of McGraw Hill LLC, including, but not limited to, in any network or other electronic storage or transmission, or broadcast for distance learning.

Some ancillaries, including electronic and print components, may not be available to customers outside the United States.

This book is printed on acid-free paper.

3 4 5 6 7 8 9 LMN 26 25 24 23 22 21

ISBN 978-1-264-26544-2 (bound edition)
MHID 1-264-26544-1 (bound edition)

Portfolio Manager: *Matthew Garcia*
Product Developer: *Melisa Seegmiller*
Marketing Manager: *Valerie Kramer*
Content Project Managers: *Laura Bies & Brent dela Cruz*
Buyer: Sandy *Ludovissy*
Designer: *Daivd W. Hash*
Content Licensing Specialist: *Lori Hancock*
Cover photo: *Erin O'Loughlin*
Cover illustration: *Libby Wagner; MPS North America, LLC*
Compositor: *MPS Limited*

The Internet addresses listed in the text were accurate at the time of publication. The inclusion of a website does not indicate an endorsement by the authors or McGraw Hill LLC, and McGraw Hill LLC does not guarantee the accuracy of the information presented at these sites.

mheducation.com/highered

brief contents

about the authors

KYLA TURPIN ROSS received her undergraduate degree from Louisiana State University in biological and agricultural engineering and her Ph.D. in biomedical engineering from Georgia Institute of Technology and Emory University. Kyla then served as a postdoctoral fellow in the Fellowships in Research and Science Teaching (FIRST) program at Emory University, a National Institutes of Health (NIH)-funded program that provides training in both research and teaching. In 2008, she joined the Department of Biology at Georgia State University, where she mentored faculty and students, and worked closely with administrators to improve instructional effectiveness in anatomy and physiology courses. She directed teaching assistant training, ensuring quality training in ethics, instructional effectiveness, and risk management. She also taught in a wide range of courses and classroom environments at both the undergraduate and graduate levels.

In 2016, Kyla joined the Wallace H. Coulter Department of Biomedical Engineering (BME) at Georgia Institute of Technology and Emory University as the Director of Graduate Training. In 2019, Kyla transitioned to the role of Assistant Vice Provost for Advocacy and Conflict Resolution at Georgia Institute of Technology, where she works closely with administrators, faculty, staff, and students to resolve conflicts in accordance with Institute policies and procedures. Kyla has more than 13 years of experience in physiology education and program development. In addition to teaching undergraduate physiology courses, she provides training opportunities for faculty and students that promote positive lab, work, and class environments. She has been an active member of the Human Anatomy and Physiology Society (HAPS) since 2012, hosted an annual conference, served as HAPS steering committee chair, and will assume the role of HAPS president in July 2021. In addition to academic endeavors, Kyla enjoys traveling and spending quality time with her family and friends. She views life experiences through an optimist's lens, and always looks for opportunities to reflect and grow.

LESLIE DAY earned her B.S. in Exercise Physiology from University of Massachusetts at Lowell, an M.S. in Applied Anatomy & Physiology from Boston University, and a Ph.D. in Biology from Northeastern University with her research on the kinematics of locomotion. Starting in 2002, she worked as a Lecturer in the Biology Department of Northeastern University in Boston, Massachusetts, teaching several sections of anatomy and physiology. In 2008, she transferred to the Department of Physical Therapy, Movement and Rehabilitation Sciences at Northeastern University to run the Cadaver Laboratory and teach gross anatomy and neuroscience to undergraduate and graduate students in a wide range of majors. In addition, she coordinated and taught postgraduate continuing education courses. During her time, she was promoted to Associate Clinical Professor and served as the Associate Department Chair. In 2019, she moved to Texas A & M University as Instructional Associate Professor to join a new program in the College of Medicine called Engineering Medicine (ENMED). The program combines medicine and engineering degrees utilizing nontraditional, innovative teaching methods. In her role, she is the course director for medical gross anatomy, teaches in the neuroscience course, and aids in the innovative curricular development.

She has received Northeastern University's teaching with technology award three times and in 2009 was awarded the Excellence in Teaching Award. In 2017, she received national recognition for her teaching by being the recipient of the ADInstruments Sam Drogo Technology in the Classroom Award from the Human Anatomy and Physiology Society (HAPS). She has been an active part of HAPS for several years, and currently serves as Treasurer on the Board of Directors. She is also a member of the American Association for Anatomy and the American Association of Clinical Anatomists. Her current research focuses on the effectiveness of different teaching pedagogies, including the flipped-classroom and TBL, and its effect on students' motivation and learning.

JOSEPH COMBER received his B.S. in Biology from Neumann University, his M.S. in Biology from Villanova University, and his Ph.D. in Immunology and Microbial Pathogenesis from Thomas Jefferson University. While completing his graduate degrees, he served as a teaching assistant in the anatomy and physiology course at Villanova University and as an instructor in anatomy and physiology laboratories at Neumann University. After finishing his Ph.D., he completed a postdoctoral fellowship in vaccine immunology, where his research focused on identifying T-cell epitopes generated during infection. He joined the faculty at Villanova University in 2014 and is currently an Associate Teaching Professor in the Biology Department. In this role he teaches anatomy and physiology for pre-health profession students, histology and human anatomy for Biology majors, and a course in vaccines and public perception for nonscience majors.

Joseph is a two-time semifinalist for the Lindback Award for Distinguished Teaching at Villanova University and serves on several departmental and college committees. He is a member of the Human Anatomy and Physiology Society (HAPS), the National Science Teaching Association, and the Society for College Science Teachers. When not in the classroom, Joseph enjoys coaching Little League baseball and spending time with his wife and son.

With love and thanks to my entire family, including the dogs.

©Christine Eckel

CHRISTINE MARIE ECKEL received her B.A. in Integrative Biology and M.A. in Human Biodynamics from the University of California, Berkeley, and her Ph.D. in Neurobiology and Anatomy at the University of Utah School of Medicine. She has taught a two-semester anatomy and physiology course for pre-nursing and pre-health-science majors, and an advanced dissection course for premedical students at Carroll College; stand-alone general biology, human anatomy, and human physiology courses at the University of California at Berkeley and Salt Lake Community College (SLCC); human gross anatomy and medical histology for medical students, and anatomy review courses for residents in orthopedic surgery and pathology both at the University of Utah School of Medicine (U of USOM) and the West Virginia School of Osteopathic Medicine (WVSOM). She has also advised pre-med, pre-PA, pre-nursing, and pre-PT students. Christine also headed the Body Donor Program at WVSOM.

Christine is the author of Human Anatomy Laboratory Manual, 3e (McGraw-Hill Education). She has also authored several supplements and individual chapters for textbooks in human anatomy and human physiology.

Christine is currently serving on the Board of Directors as Secretary for the Human Anatomy and Physiology Society (HAPS). Her previous service to HAPS includes two terms serving as Western Regional Director, and many years serving as Chair and member of the Cadaver Use Committee. Christine has also served on other committees for both HAPS and the American Association of Anatomists (AAA). She is an ad hoc reviewer for the journals *Anatomical Sciences Education* and *Medical Education*. Her research is in the field of teaching innovation and educational outcomes research.

With over 25 years of experience engaging with community college students, medical students and medical residents in orthopedic surgery, pathology, and gynecologic surgery, Christine has a unique appreciation for the learning challenges experienced by students at all levels. Christine's passions for human anatomy, classroom and laboratory teaching, biological dissection, and photography are evident throughout the pages of this laboratory manual.

In her spare time, Christine loves to take her English Setter, Zelda, hiking, mountain biking, cross-country skiing, and exploring the great outdoors—always with her camera in hand.

contents

Chapter 13
The Muscular System: Appendicular Muscles   305

PART IV | COMMUNICATION AND CONTROL   339

Chapter 14
Nervous Tissues   339

Chapter 15
The Brain and Cranial Nerves   365

PART V | MAINTENANCE AND REGULATION   517

Chapter 20
The Cardiovascular System: Blood   517

Chapter 21
The Cardiovascular System: The Heart   537

Chapter 22
The Cardiovascular System: Vessels and Circulation   571

Chapter 23
The Lymphatic System and Immunity   619

Chapter 24
The Respiratory System   643

Chapter 25
The Urinary System   677

Chapter 28
Cat Dissection Exercises   783

Chapter 29
Fetal Pig Dissection Exercises   849

Human anatomy and physiology is a complex yet fascinating subject, and is perhaps one of the most personal subjects a student will encounter during his or her education. It is also a subject that can create concern for students because of the sheer volume of material, and the misconception that "it is all about memorization."

The study of human anatomy and physiology really comes to life in the anatomy and physiology laboratory, where students get hands-on experience with human cadavers and bones, classroom models, preserved and fresh animal organs, and histology slides of human tissues, and explore the process of scientific discovery through physiology experimentation. Yet most students are at a loss regarding how to approach the anatomy and physiology laboratory. For example, students are often given numerous lists of structures to identify, histology slides to view, and "wet labs" to conduct, but are given comparatively little direction regarding how to recognize structures, or how to relate what they encounter in the laboratory to the material presented in the lecture. In addition, most laboratory manuals on the market contain little more than material repeated from anatomy and physiology textbooks, which provides no real benefit to a student.

This laboratory manual takes a very focused approach to the laboratory experience and provides students with tools to make the subject matter more relevant to their own bodies and to the world around them. Rather than providing a recap of material from classroom lectures and the main textbook for the course, this laboratory manual is much more of an interactive workbook for students: a "how-to" guide to learning human anatomy and physiology through touch, dissection, observation, experimentation, and critical thinking exercises. Students are guided to formulate a hypothesis about each experiment before beginning physiology exercises. Diagrams direct students in how to perform experiments and don't just show the end results. The text is written in a friendly, conversational tone to put students at ease as they discover, organize, and understand the material presented in each chapter.

Organization

Because observation of histology slides, human cadavers or classroom models, and "wet lab" experiments are usually performed in separate physical spaces or at specific times within each laboratory classroom, chapters in this laboratory manual are similarly separated into three sections: Gross Anatomy, Histology, and Physiology. Each exercise within these chapter sections has been designed with the student's actual experience in the anatomy and physiology laboratory in mind. Thus, each exercise covers only a single region of the human body, classroom model, histology slide, or wet lab experiment. At the same time, within-chapter Concept Connection and Clinical View boxes provide an opportunity to integrate the material from all three sections of each chapter. Learning Strategies boxes provide mnemonics, study tips, and other helpful hints to assist students in recall of pertinent information. In addition, Can You Apply What You've Learned? and Can You Synthesize What You've Learned? questions in Post-Laboratory Worksheets provide further opportunities for students to integrate the information and apply it to clinically relevant and practical situations. Organization of each chapter into a series of discrete exercises makes the laboratory manual easily customizable to any anatomy and physiology classroom, allowing an instructor

to assign certain exercises while telling students to ignore other exercises. Post-Laboratory Worksheets are also organized by exercise and are coded to Learning Objectives within the chapter, which makes it easy for an instructor to assign questions that relate only to the exercises and/or Learning Objectives covered in their classroom.

Changes to the Fourth Edition

Anatomy & Physiology: An Integrative Approach Laboratory Manual, fourth edition, continues to serve as a resource for students both in and out of the lab, providing a "how-to" guide for learning anatomy and physiology. The interactive pages within serve as a stand-alone manual, while also complementing the textbook, McKinley/O'Loughlin/Bidle: *Anatomy & Physiology: An Integrative Approach*, fourth edition.

The fourth edition includes a consolidation of the MAIN, CAT, and PIG versions of the lab manual into a single, stand-alone resource. Having only one version of the lab manual makes textbook selection more straightforward for instructors and students. In this edition, chapter 28 covers the dissection of the cat (*Felis domesticus*), and chapter 29 covers the dissection of the fetal pig (*Sus scrofa domesticus*). Dissection provides the student an opportunity to view, feel, and explore the spatial relationships of anatomical structures. The exercises included in this combined lab manual contain detailed images, which are beneficial for relating content covered in earlier chapters to the anatomy of cats and fetal pigs. This comparative approach allows students to make connections and note similarities in vertebrate anatomy.

- Gross Anatomy has been moved prior to Histology in each chapter to allow students and instructors to explore macroscale structures before microscale structures.

- Exercises and labeled figures have been reviewed and revised for accuracy.

- The introductory text in each chapter has been revised to be more engaging and relatable for students and instructors.

- Content in Clinical Views has been revised to be more easily accessible and generally applicable for students and instructors.

- Concept Connections have been revised to review content covered in previous chapters rather than to preview content in upcoming chapters, so as to reduce confusion.

- Reference tables have been revised and reorganized for coverage and ease of learning.

- Content throughout the manual has been revised to align with rather than to duplicate coverage in the complementary textbook.

- Exercises throughout the manual have been revised to be more interactive with the addition of question prompts and fill-in tables.

- Pre-Laboratory Worksheets and Post-Laboratory Worksheets have been revised to include broader coverage of topics and to eliminate the use of true/false questions.

- BIOPAC exercises have been revised to include updated instructions that reflect updates in the software.

- Ph.I.L.S. exercises have been revised to include new screen captures that illustrate updates in the software.

- The number of labels in labeling activities has been reduced to decrease cognitive overload.

The fourth edition of this lab manual contains improvements and updates to text and figures throughout, and updated figures are added into Connect.

Changes by Chapter

The following is a list of the most significant changes made in each chapter in the fourth edition of this lab manual.

Chapter 1

- Revised Learning Strategy on learning techniques
- New Clinical View on Common Safety Training by Healthcare Professionals
- Revised Figure 1.7 Scalpel Blade Handles and Blades

Chapter 2

- Moved Exercise 2.2 Regional Terms before Exercise 2.3 Directional Terms
- Revised Table 2.2 Directional Terms to include more terms
- Revised Exercise 2.3 to include integration of directional and regional terms
- Deleted Table 2.3 on Selected Regional Terms
- Incorporated Learning Strategy on directional terms into text
- Revised Figure 2.4 Regional Terms for accuracy
- Revised Figure 2.7 Body Cavities for accuracy
- Revised Exercise 2.4 on Body Cavities to include activity on membranes
- Incorporated Concept Connection on Serous Membranes into text
- New Figure 2.6 on Serous Membranes

Chapter 3

- Revised and moved to earlier chapter Concept Connection on viewing histological slides
- Revised Exercise 3.1 Parts of a Compound Microscope to include a table
- Revised Exercise 3.3 Measuring the Diameter of the Field of View to be more interactive
- Revised Figure 3.5 Estimating Specimen Size

Chapter 4

- Revised Introduction to better explain the overall structure and function of cells and tissues
- Reorganized Table 4.1 Parts of a Generalized Animal Cell to sort terms by membrane-bound and non-membrane-bound organelles
- Replaced Concept Connection on excitable cells with Concept Connection on plasma membrane proteins and membrane transport due to relevance to chapter content
- Deleted Concept Connection on molecular weight calculation to avoid duplicated content
- Replaced Clinical View on Hemodialysis with Clinical View on Dehydration due to relevance to chapter content

Chapter 5

- Revised Introduction to explain the process of preparing tissue samples for histological staining

- Revised Clinical View: Histopathology to move information about tissue preparation to the introduction
- Revised Learning Strategy on identifying epithelial tissue on a slide to help make this identification easier for the student
- Revised Table 5.3 Cell Surface Modifications and Specialized Cells of Epithelial Tissues to more clearly differentiate between surface modifications and specialized cells

Chapter 6

- Revised Introduction to include several important functions of skin
- Revised Gross Anatomy introductory text: Integument Model to discuss the layers of skin in more depth
- Revised Clinical View: Fingerprinting to highlight the contribution of dermal papillae to fingerprint formation
- Revised Clinical View: Melanoma to include evolution of a mole as a characteristic of melanoma clinical presentation
- Revised Figure 6.11 Longitudinal Section of a Nail
- Revised Table 6.5 Parts of a Nail

Chapter 7

- Revised Introduction to cite the relevance of chapter content to real-life scenarios
- Replaced the term *bone marrow cavity* with *medullary cavity* to align with the textbook
- Revised Exercise 7.3 Cow Bone Dissection for brevity
- Revised Figure 7.4 The Human Skeleton
- Revised Table 7.2 Types of Bone Cells
- Revised Exercise 7.5 Compact Bone for brevity
- Revised Figure 7.9 Spongy Bone
- Revised Concept Connection on the composition of bone for brevity
- Revised Concept Connection on hormones that influence bone growth to better relate to chapter content
- Revised Learning Strategy on five functional layers of the epiphyseal plate

Chapter 8

- Incorporated Learning Strategy on learning bony projections into Introduction
- Revised all tables for accuracy and brevity
- Revised Figure 8.1 Anterior View of the Skull
- Eliminated table on The Axial Skeleton: Anterior View of the Skull for redundancy
- Revised Exercise 8.1B The Orbit to be more interactive
- Revised Figure 8.2 The Orbit
- Revised Concept Connection on the nasal cavity
- Revised Figure 8.4 The Mandible
- New Table 8.3 Sutures and Craniometric Points
- Revised Figure 8.5 Lateral View of the Skull
- Revised Figure 8.6 Posterior View of the Skull
- Revised Figure 8.8 Inferior View of the Skull
- Revised Figure 8.11 Superior View of the Cranial Floor
- Revised Clinical View on Spondylolisthesis
- Revised Figure 8.21 Sacrum and Coccyx

- New Exercise 8.11 Radiographs
- New Figure 8.24 A Radiograph of Skull
- New Figure 8.25 A Chest X-Ray
- Revised Post-Laboratory Worksheet to include a labeling figure and some new questions

Chapter 9

- Incorporated Concept Connection on learning bony features into the introduction
- Moved Learning Strategy on identifying individual bones to earlier in the chapter
- Revised all tables for accuracy and brevity
- New Figure 9.1 Right Pectoral Girdle
- Revised Figure 9.3 The Right Scapula
- Revised Exercise 9.1B The Scapula due to redundancy
- Refined Clinical View on Clavicular Fracture
- Revised Exercise 9.2A The Humerus
- Revised Figure 9.4 The Humerus
- Revised Figure 9.5 The Radius
- Revised Exercise 9.2D The Carpals to be more interactive
- Refined Learning Strategy on learning carpal bone names
- New Exercise 9.2F on Articulated Bones in the Upper Extremity
- Revised Figure 9.9 The Elbow
- New Figure 9.10 X-Ray of the Wrist
- Revised Figure 9.12 The Right Ox Coxae
- Revised Exercise 9.4B Male and Female Pelves
- Refined Clinical View on Pregnancy and Childbirth
- Revised Figure 9.14 The Proximal and Distal Femur
- Revised Figure 9.15 The Right Femur
- Revised Figure 9.16 The Tibia
- Revised Exercise 9.5D The Tarsals to be more interactive
- New Exercise 9.5F Articulations in the Lower Extremity
- New Figure 9.20 Pelvis X-Ray, Anterior View
- New Figure 9.21 X-ray of the Knee, Medial View
- Refined Concept Connection on red bone marrow
- Revised Post-Laboratory Worksheet to include some new questions and decrease length

Chapter 10

- Revised Introduction to cite the relevance of chapter content to real-life scenarios
- Replaced the term *synovial cavity* with *joint (articular) cavity* to align with the textbook
- Reorganized Table 10.4 Components of Synovial Joints
- Revised Figure 10.3 Diagram of a Representative Synovial Joint
- Revised Table 10.5 Classification of Synovial Joints
- Revised Figure 10.4 Classification of Synovial Joints
- Revised Table 10.6 Movements of Synovial Joints to convert a reference table to an exercise activity
- Revised Figure 10.6 A Representative Synovial Joint: The Right Knee Joint

- Revised Concept Connection on movement of synovial joints to better relate to chapter content
- Revised Clinical View: Low Back Pain for brevity

Chapter 11

- Revised Introduction to better present an overview of chapter content
- Revised the terms for skeletal muscle fiber types to align with the textbook
- Revised Table 11.4 Fascial Compartments of the Limbs and Their General Muscle Actions to convert a reference table to an exercise activity
- Revised Concept Connection on the three types of muscle tissue for brevity
- Simplified Figure 11.7 The Neuromuscular Junction
- Moved Figure 11.8 Cardiac Muscle Tissue within Exercise 11.7 Cardiac Muscle Tissue for chapter consistency
- Moved Figure 11.9 Smooth Muscle Tissue within Exercise 11.8 Smooth Muscle Tissue for chapter consistency
- Revised Table 11.6 Properties of Skeletal Muscle Fiber Types
- Revised Figure 11.10 Recruitment of Motor Units

Chapter 12

- Revised all tables for accuracy and brevity
- Revised Figure 12.1 Muscles of Facial Expression
- New Table 12.6 Muscles that Move the Head and Neck
- Deleted Clinical View on Stroke due to relevance to chapter content
- Simplified Table 12.7 Muscles of the Vertebral Column
- New Clinical View on Strains, Sprains, and Spasms
- Incorporated Concept Connection on pulmonary ventilation into the text
- New Table 12.10 Structures Related to Abdominal Musculature
- Revised Post-Laboratory Worksheet to include a labeling figure and some new questions

Chapter 13

- Revised all tables for accuracy and brevity
- Added new figure to Clinical View on Winged Scapula
- Reorganized Table 13.6 Posterior (Extensor) Compartment of the Forearm
- Revised Figure 13.6 Posterior (Extensor) Compartment of the Forearm
- Reorganized Table 13.8 Muscles of the Gluteal Region
- Reorganized Exercise 13.5 Muscles That Act About the Hip Joint/Thigh for clarity
- Revised Figure 13.8 Actions of Gluteal Muscles During Locomotion
- New Table 13.10 Medial Compartment of the Thigh
- Reorganized Table 13.13 Posterior Compartment of the Leg
- Revised Figure 13.15 Lateral View of the Leg
- Revised Post-Laboratory Worksheet to include labeling figures and some new questions

Chapter 14

- Revised Introduction to discuss the two major branches of the nervous system
- New Figure 14.1 Organization of the Nervous System
- Revised Concept Connection on the excitability and conductivity of nervous tissue
- New Figure 14.2 Gray Matter of the Spinal Cord
- Revised Physiology introductory text: Resting Membrane Potential to describe this potential and what contributes to the generation of the potential
- Revised Learning Strategy on potential and kinetic energy and resting membrane potential

Chapter 15

- New Concept Connection on neuron control of skeletal muscle
- New Table 15.1 Brain Structures Visible from the Superior View
- New Table 15.2 Brain Structures Visible from the Lateral View
- New Table 15.3 Brain Structures Visible from the Inferior View
- New Table 15.4 Brain Structures Visible from the Midsagittal View
- New Table 15.5 General Meningeal Layers
- New Table 15.6 Associated Meningeal Structures
- Revised Figure 15.6 Meningeal Structures
- New Table 15.7 Secondary Brain Vesicles and Associated Structures of the Brain

Chapter 16

- Revised text in The Spinal Cord section for clarity and breadth
- Added new content to Table 16.1 Gross Anatomy of the Spinal Cord
- Revised Figure 16.1 Regional Gross Anatomy of the Spinal Cord
- Revised Exercise 16.3 The Brachial Plexus for clarity
- Revised Table 16.2 Organization of the Brachial Plexus for brevity
- Revised Figure 16.4 for accuracy
- New Clinical View on Additional Nerves of the Brachial Plexus and Spinal Cord Injury
- Revised Table 16.5 for brevity
- Revised Exercise 16.4 The Lumbar and Sacral Plexus
- Refined Learning Strategy on pudendal nerve
- Revised Exercise 16.5 Histological Cross Sections of the Spinal Cord to make more interactive
- Refined Concept Connection on reflexes
- Revised Exercise 16.7 Patellar Reflex
- Revised Post-Laboratory Worksheet to include some new questions

Chapter 17

- Revised Introduction to better explain an overview of chapter content
- Revised the terms for cranial nerves to align with the textbook
- New Figure 17.1 Lower Motor Neurons of the Autonomic Nervous System

- Revised Figure 17.2 Comparison of the Parasympathetic and Sympathetic Divisions of the ANS
- Reorganized Table 17.1 Comparison of Parasympathetic and Sympathetic Divisions to emphasize structure and function
- New Table 17.2 Parasympathetic Division Outflow for content coverage and activity
- Incorporated Learning Strategy on parasympathetic innervation of visceral organs into Exercise 17.1 Parasympathetic Division
- Revised Learning Strategy on parasympathetic innervation of visceral organs to sympathetic innervation of the adrenal gland
- New Table 17.3 Sympathetic Division Pathways for content coverage and activity
- Revised Clinical View on Pheochromocytoma for brevity
- Revised Post-Laboratory Worksheet to include new application questions

Chapter 18

- Revised Figure 18.1 Skin
- Revised Table 18.2 External and Accessory Structures of the Eye
- Revised Table 18.3 Internal Structures of the Eye
- Revised Table 18.4 Extrinsic Eye Muscles
- Revised Figure 18.8 Classroom Model of the Ear
- Revised Table 18.5 Structures of the External, Middle, and Inner Ear
- Revised Table 18.6 Sensory Receptors in Thick Skin
- Revised Table 18.8 Olfactory Epithelium
- Revised Table 18.9 The Retina
- Revised Figure 18.17 The Cochlea
- Revised Table 18.10 The Cochlea
- New Table 18.11 The Spiral Organ
- New Clinical View: Tinnitus

Chapter 19

- Revised Table 19.1 The Hormone-Secreting Cells of the Pituitary Gland
- New Concept Connection on growth hormone effects on skeletal tissue
- Revised Figure 19.8 Endocrine Portion of the Pancreas
- Revised Concept Connection on hormones and transport proteins

Chapter 20

- Revised Introduction for relevance of chapter content and brevity
- Revised Table 20.1 The Four Greek Humors
- Revised Caution to include the hazards of Wright's stain
- Combined and reorganized Table 20.2 Characteristics of the Formed Elements of Blood and Table 20.3 Leukocyte Characteristics
- Removed Exercise 20.3 Identification of Formed Elements of the Blood on Classroom Models or Charts to avoid duplicated content
- Revised Table 20.3 Normal Ranges for Laboratory Blood Tests

- Revised Concept Connection for relevance of chapter content

Chapter 21

- Added new questions to Pre-Laboratory Worksheet to cover entire chapter
- New content on the layers of the heart wall in Exercise 21.2 Gross Anatomy of the Human Heart
- Revised Exercise 21.4 Superficial Structures of the Sheep Heart to make more interactive
- Moved Exercise 21.9 Auscultation of Heart Sounds earlier in chapter for better flow of content
- New Learning Strategy on auscultation of the heart
- Revised Figure 21.20 Normal Components of an ECG for accuracy
- Revised Clinical View on ECG to be more interactive
- Revised Post-Laboratory Worksheet to include a labeling figure and some new questions

Chapter 22

- Revised opening Gross Anatomy text
- New Exercise 22.2 Great Vessels of the Heart
- New Figure 22.2 Great Vessels of the Heart
- Revised Figure 22.7 Circulation to the Thoracic and Abdominal Walls
- Revised Exercise 22.6 Circulation to the Abdominal Cavity to be more interactive
- Revised Figure 22.12 Circulation from the Abdominal Aorta to the Sigmoid Colon and Back to the Right Atrium of the Heart.
- New Concept Connection on blood vessel wall
- Revised Figure 22.21 Elastic Artery
- Revamped Exercise 22.15 Capillaries to be more interactive and concise
- Revised Concept Connection on blood flow
- New Figure 22.27 Blood Pressure Measurement
- Revised figures in Post-Laboratory Worksheet for accuracy

Chapter 23

- Revised Introduction to more clearly explain the major roles of the lymphatic system
- Revised Histology introductory text: Lymphatic Vessels to more clearly describe the histology of lymphatic vessels
- New Figure 23.6 Lymphatic Nodule
- Revised Clinical View: Appendicitis
- Revised Exercise 23.6 The Thymus to more clearly describe the role of selection during T-cell development
- Revised Table 23.6 Cells of the Immune System

Chapter 24

- New opening text in Gross Anatomy describing the respiratory system
- New Figure 24.1 Parts of Respiratory System
- Revised Figure 24.5 The Pleural Cavities for clarity
- Revised Figure 24.6 The Right Lung
- Revised Figure 24.7 The Left Lung

- Revised Figure 24.10 Histology of the Trachea
- Refined Learning Strategy on the trachea for brevity
- Revised Exercise 24.9 The Lungs to be more interactive
- New Table 24.7 Microscopic Structures Within the Alveoli

Chapter 25

- Revised Introduction to more clearly explain the major roles of the urinary system and the functional components that carry out these roles
- Revised Figure 25.1 Coronal Section Through the Right Kidney
- Revised Figure 25.2 Blood Supply to the Kidney
- Revised Figure 25.3 Model of the Kidney Demonstrating the Blood Supply to the Kidney
- Revised Exercise 25.3 Urine-Draining Structures Within the Kidney to allow students to trace the flow of filtrate
- Revised Table 25.1 Histological Features of the Kidney
- Revised Learning Strategy on glomerular filtration rate
- Revised Concept Connection on glomerular filtration

Chapter 26

- Revised Introduction to more clearly explain how modifications of the GI tract impact function
- Revised Figure 26.2 Oral Cavity
- Reorganized Table 26.2 Gross Anatomic Features of the Liver, Gallbladder, Pancreas, and their Associated Ducts
- Revised Histology introductory text: The Stomach
- Revised Exercise 26.8 Histology of the Stomach
- Revised Figure 26.12 Gastric Pits and Gastric Glands
- Revised Figure 26.13 The Small Intestine

Chapter 27

- Revised Table 27.4 Developmental Stages of Ovarian Follicles
- Revised Table 27.8 Phases of the Menstrual Cycle
- New Concept Connection on smooth muscle in the reproductive system
- New Clinical View: Testicular Cancer and Testicular Self Exam
- Revised Figure 27.17 Oogenesis

Chapter 28

- Revised Figure 28.5 Muscles of the Head, Neck, and Thorax (Anterior View)
- Revised Figure 28.6 Muscles of the Thorax and Abdomen
- Revised Figure 28.8 Superficial Back Muscles
- Revised Figure 28.21 The Thoracic Cavity
- Revised Figure 28.22 The Respiratory System
- Revised Figure 28.24 The Abdominal Cavity

Chapter 29

- Revised Figure 29.4 The Pig Skeleton
- Revised Exercise 29.13 The Heart, Lungs, and Mediastinum

the learning system

Features

The Ross/Day/Comber/Eckel: *Anatomy & Physiology Laboratory Manual* works well as a complement to the McKinley/O'Loughlin/Bidle: *Anatomy & Physiology: An Integrative Approach* textbook, or to accompany any other anatomy and physiology text. Each chapter opener includes an **outline** that lists a set of **learning objectives** for the chapter.

- A chapter **Introduction** opens with a real-life scenario that emphasizes the section of the body covered in the chapter, to connect the anatomy of our bodies with the physiology that helps us to perform day-to-day activities.

- The laboratory manual exhibits the highest-quality **photographs and illustrations** of any laboratory manual on the market.

The Muscular System: Muscle Structure and Function

chapter 11

OUTLINE AND LEARNING OBJECTIVES

GROSS ANATOMY 248

Gross Anatomy of Skeletal Muscles 248

EXERCISE 11.1: NAMING SKELETAL MUSCLES 248
1. Explain some of the logic behind the naming of skeletal muscles

EXERCISE 11.2: ARCHITECTURE OF SKELETAL MUSCLES 250
2. Use anatomical terminology to describe the architecture of skeletal muscles, and describe how the architecture of a skeletal muscle is related to its action

Organization of the Human Musculoskeletal System 252

EXERCISE 11.3: MAJOR MUSCLE GROUPS AND FASCIAL COMPARTMENTS OF THE LIMBS 253
3. Describe the location and major actions of the major muscle groups of the body
4. Describe the fascial compartments of the limbs, and explain the major actions associated with each fascial compartment

HISTOLOGY 255

Skeletal Muscle Tissue 255

EXERCISE 11.4: HISTOLOGY OF SKELETAL MUSCLE FIBERS 257
5. Identify skeletal muscle tissue through the microscope, and describe the features unique to skeletal muscle tissue
6. Name the visible bands that form the striations in skeletal muscle tissue

EXERCISE 11.5: CONNECTIVE TISSUE COVERINGS OF SKELETAL MUSCLE 258
7. Describe the layers of connective tissue that surround skeletal muscle tissue

The Neuromuscular Junction 258

EXERCISE 11.6: THE NEUROMUSCULAR JUNCTION 259
8. Define motor unit, and describe how the concept of a motor unit applies to neuromuscular junctions

Cardiac Muscle Tissue 259

EXERCISE 11.7: CARDIAC MUSCLE TISSUE 260
9. Identify cardiac muscle tissue through the microscope, and describe the features unique to cardiac muscle tissue
10. Compare and contrast the structure of skeletal, smooth, and cardiac muscle tissues

Smooth Muscle Tissue 260

EXERCISE 11.8: SMOOTH MUSCLE TISSUE 261
11. Identify smooth muscle tissue through the microscope, and describe the features unique to smooth muscle tissue
12. Describe how two layers of smooth muscle tissue act as antagonists

PHYSIOLOGY 262

Force Generation of Skeletal Muscle 262

EXERCISE 11.9: MOTOR UNITS AND MUSCLE FATIGUE (HUMAN SUBJECT) 263
13. Compare and contrast the characteristics of slow oxidative, fast oxidative, and fast glycolytic muscle fibers
14. Describe the sequence of motor unit recruitment
15. Explain the relationship between motor unit recruitment, muscle force, and fatigue

EXERCISE 11.10: CONTRACTION OF SKELETAL MUSCLE (WET LAB) 264
16. Describe the effect of adding ATP alone, ATP + salts, or salts only on contraction of glycerinated muscle
17. Explain the relationship between ATP supply to skeletal muscle and rigor mortis

EXERCISE 11.11: PhILS LESSON 5: STIMULUS-DEPENDENT FORCE GENERATION 268
18. Describe the events of a muscle twitch
19. Describe the relationship between stimulus intensity and maximum isometric twitch force
20. Describe the threshold voltage, and explain what happens when suprathreshold stimuli are applied to skeletal muscle

EXERCISE 11.12: PhILS LESSON 7: THE LENGTH-TENSION RELATIONSHIP 269
21. Describe the relationship between muscle length and tension for a maximal isometric contraction
22. Describe what is happening at the sarcomere level when a muscle contracts isometrically at its optimal length

EXERCISE 11.13: PhILS LESSON 8: PRINCIPLES OF SUMMATION AND TETANUS 272
23. Describe the relationship between stimulus frequency and muscle tension
24. Demonstrate the concepts of summation, incomplete tetanus, and complete tetanus

EXERCISE 11.14: PhILS LESSON 9: EMG AND TWITCH AMPLITUDE 273
25. Describe the relationship between muscle tension and EMG amplitude

EXERCISE 11.15: BIOPAC ELECTROMYOGRAPHY (EMG) 274
26. Describe the relationship between motor unit recruitment and EMG amplitude

Anatomy & Physiology Revealed 4.0

Module 6: Muscular System

245

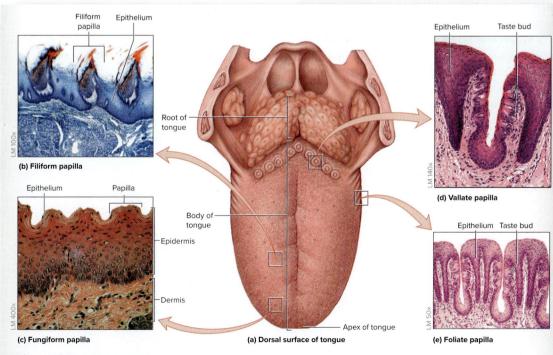

Figure 18.12 Taste Buds. Gustation (taste) requires taste buds, which are associated with tongue papillae. (*a*) Dorsal surface of tongue, (*b*) Filiform papilla, (*c*) Fungiform papilla, (*d*) Vallate papilla, (*e*) Foliate papilla.

(*b*) ©CNRI/Science Photo Library/Corbis; (*c*) McGraw-Hill Education/Christine Eckel, photographer; (*d*) ©McGraw-Hill Education/Alvin Telser; (*e*) ©Jose Luis Calvo/Shutterstock.com

- The content of the laboratory manual is informed by the textbook, and both the textbook and the laboratory manual share similar pedagogic elements: **Concept Connection**, **Learning Strategy**, and **Clinical View** features from the text are also employed in the laboratory manual.

  - **Integrate: Concept Connection** boxes draw concepts from the classroom into the laboratory for a real-time review of how previously covered concepts relate to body systems.

  - **Integrate: Learning Strategy** boxes offer tried-and-tested learning strategies that consist of everyday analogies, mnemonics, and useful tips to aid understanding and memory.

  - **Integrate: Clinical View** sidebars reinforce facts through a clinical discussion of what happens when the body doesn't perform normally.

- **Pre-Laboratory Worksheets** at the start of each chapter consist of important refresher points to provide students with a "warm-up" before entering the laboratory classroom. Some questions pertain to previous activities that are relevant to upcoming exercises, while others are basic questions that students should be able to answer if they have read the chapter from their lecture text before coming into the laboratory classroom. The goal of completing these worksheets is to have students arrive at the laboratory prepared to deal with the material they will be covering, so valuable laboratory time isn't lost in reviewing necessary information. All Pre-Laboratory Worksheet questions are assignable within Connect.

INTEGRATE

CONCEPT CONNECTION

Two basic cell types make up nervous tissue: neurons and neuroglia. Although these cell types vary in structure, function, and location, they share similarities with one another as well as with most other cell types found throughout the body. Like other cells, neurons are surrounded by a selectively permeable plasma membrane (the neurolemma) that regulates the movement of molecules into

The neuron can be divided into segments based on structural features and function. For example, the *receptive segment* contains the dendrites and the soma of a neuron. The neurolemma of these structures contains many ligand-gated ion channels, which are opened or closed by the binding of neurotransmitters ("ligands") released by the axon terminals of neighboring neurons. Therefore, the structures that make up the receptive segment *receive* the signal. Because these structures do not contain voltage-gated sodium channels, the membrane can only

---

INTEGRATE

LEARNING STRATEGY

To identify superficial muscles and tendons, place your left palm on the medial epicondyle of your right humerus. In this position, the order of the muscles on your right forearm, from lateral to medial, is:

Index finger—pronator teres (PT)

Middle finger—flexor carpi radialis (FCR)

Ring finger—palmaris longus (PL)

Pinky finger—flexor carpi ulnaris (FCU)

While performing this exercise, flex your wrist and digits to identify the tendons, from lateral to medial, of the flexor carpi radialis, palmaris longus, and flexor carpi ulnaris.

Pronator teres
Flexor carpi radialis
Palmaris longus
Flexor carpi ulnaris

(Left hand covers medial epicondyle)

©McGraw-Hill Education/JW Ramsey

INTEGRATE

CLINICAL VIEW
Piriformis Syndrome

The piriformis muscle is a "pear-shaped" muscle that lies in close proximity to important structures within the gluteal region, such as the sciatic nerve, and the gluteal arteries and nerves. Piriformis syndrome is a painful condition that results from inflammation or overuse of the piriformis muscle. The incidence of piriformis syndrome is relatively common in athletes such as runners and cyclists, who may develop an imbalance in the strength of the piriformis muscle

as compared to the gluteal muscles. As the piriformis muscle becomes inflamed or spasms, it may also compress the underlying sciatic nerve, resulting in sciatica. **Sciatica** is a tingling, painful, or even numbing sensation that travels down the path of the sciatic nerve. Patients complain of shooting pain that runs from the gluteal region down the lateral aspect of the thigh and toward the leg. Often the pain may be exacerbated when the body is held in certain positions, such as prolonged sitting or standing. Many health professionals, including physical therapists and chiropractors, can help treat patients with piriformis syndrome.

Chapter 14: Nervous Tissues

Name: _____
Date: _____ Section: _____

PRE-LABORATORY WORKSHEET

These Pre-Laboratory Worksheet questions may be assigned by instructors through their ■connect· course.

1. For each structure listed, write the corresponding letter as labeled in the diagram.

___ Axon   ___ Dendrite
___ Axon hillock   ___ Synaptic knobs
___ Cell body (soma)

A ____   D ____

B ____

C ____

E ____

2. Classify the following parts of the neuron as receptive or conductive:

a. axon: _____

b. cell body (soma): _____

c. dendrite: _____

3. Match the description listed in column A with the corresponding cell type listed in column B.

Column A	Column B
___ 1. cells that myelinate axons in the central nervous system	a. astrocytes
___ 2. cells that help reinforce the blood-brain barrier	b. ependymal cells
___ 3. glial cells found within peripheral nerve ganglia	c. microglia
___ 4. cells that myelinate axons in the peripheral nervous system	d. neurolemmocytes (Schwann cells)
___ 5. cells that engage in phagocytosis in response to tissue injury	e. oligodendrocytes
___ 6. cells that line the ventricles of the brain	f. satellite cells

4. Match the description listed in column A with the corresponding connective tissue structure listed in column B.

Column A	Column B
___ 1. surrounds a fascicle of axons	a. endoneurium
___ 2. surrounds an individual axon	b. epineurium
___ 3. surrounds the entire nerve	c. perineurium

■ In-chapter **activities** offer a mixture of labeling exercises, sketching activities, table completion exercises, data recording and analysis, palpation of surface anatomy, and other sources of learning. In the gross anatomy exercises of this manual, structures such as cranial bones and muscles of the body are *not* always presented as labeled photos, since students already have labeled photos provided in their anatomy and physiology textbook. Instead, images are presented as labeling activities with a checklist of structures. The checklists serve two purposes: (1) they guide students to items they should be able to identify on classroom models, fresh specimens, or cadavers (if the laboratory uses human cadavers), and (2) they double as a list of terms students can use to complete the labeling activities. Answers to the labeling activities are provided in the Appendix. Thus, if a student does not know what a leader line is pointing to, or cannot remember the correct term, the Appendix serves as a resource for locating the correct answer.

■ **Anatomy & Physiology REVEALED® 4.0 (APR)** correlations, indicated by the APR logo, direct students to related content in this cutting-edge software.

■ Each chapter contains numerous **tables,** which concisely summarize critical information and key structures and serve as important points of reference while in the laboratory classroom. Most tables contain a column that provides word origins for each structure listed within the table. These word origins are intended to give students continual exposure to the origins of the language of anatomy and physiology, which is critical for learning and retention.

■ Numerous **Physiology Interactive Lab Simulations©** **(Ph.I.L.S.) 4.0** exercises throughout the laboratory manual make otherwise difficult and expensive experiments a breeze, and offer additional opportunities to aid student understanding of physiology.

■ **BIOPAC©** exercises are included in chapters 11, 15, 17, 21, 22, and 24.

■ **Post-Laboratory Worksheets** at the end of each chapter serve as a review of the materials just covered and challenge students to apply knowledge gained in the laboratory. The Post-Laboratory Worksheets contain in-depth critical thinking questions and are perforated so they can be torn out and handed in to the instructor, if so desired. Assessment questions are organized by exercise and are keyed to the Learning Objectives from the chapter opener outline.

  ■ **Do You Know the Basics?** questions quiz students on the material they have just learned in the chapter, using a variety of question formats including labeling, table completion, matching exercises, and fill-in-the-blank.

  ■ **Can You Apply What You've Learned?** questions are often clinically oriented and expose health-sciences students to problem solving in clinical contexts.

  ■ **Can You Synthesize What You've Learned?** questions combine concepts learned in the chapter to ensure student understanding of each chapter's objectives.

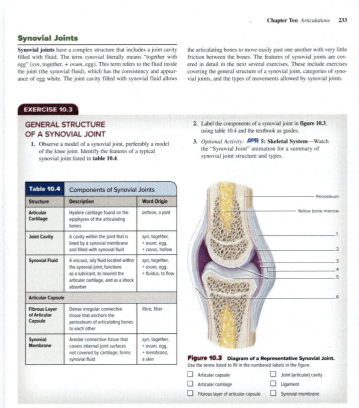

Teaching Supplements

Answers to the Pre-Laboratory and Post-Laboratory Worksheets can be found within the **Instructor's Manual** for this Laboratory Manual within Connect, by accessing the McKinley/O'Loughlin/Bidle: *Anatomy & Physiology,* 4th edition Instructor Resources. **Image files** for use in presentations and teaching materials are also provided for instructor use at this location.

Anatomy & Physiology Revealed® 4.0: An Interactive Cadaver Dissection Experience

Available in Connect and online at aprevealed.com, Anatomy & Physiology Revealed (APR) is an interactive human cadaver dissection tool built to enhance both lecture and lab. APR contains all the systems covered in A&P and Human Anatomy courses, including Body Orientation, Cells and Chemistry, and Tissues. Detailed cadaver photographs blended with a state-of-the-art layering technique provide a uniquely interactive dissection experience.

With a new streamlined, user- and mobile-friendly interface, increased accessibility, updated animations and 3D interactive models, APR was built to increase the success of your A&P laboratory course.

- Dissection: Peel away layers of the human body to reveal structures. Structures can be pinned and labeled, just as in a real dissection lab. Each labeled structure is accompanied by detailed information, audio pronunciation, and alternate views. Dissection images can be captured and saved. A direct link tool also can be used to bring students to the exact view you specify.

- Animations: Modern, updated animations demonstrate muscle action, show detailed attachments, clarify anatomical relationships, and explain difficult concepts.

- Histology: Labeled light micrographs presented with each body system allow students to study the cellular detail of tissues.

- Imaging: Labeled x-ray, magnetic resonance imaging (MRI), and computed tomography (CT) images familiarize students with the appearance of key anatomical structures as seen through different medical imaging techniques.

- Self-Quizzing: Challenging exercises allow students to test their ability to identify anatomical structures in a timed practical exam format or with traditional multiple choice questions. A results page provides an analysis of test scores and links back to all incorrectly identified structures for review.

- Rotatable 3D Models: Interactive, rotatable 3D models enhance the learning experience and allow students to see the spatial relationship of structures in the human body. Side-by-side corresponding cadaver images provide perspective.

- My Course Content: Instructors may customize APR 4.0 to their course by selecting the specific structures they require in their course. Once the structure list is generated, APR highlights these selected structures for students.

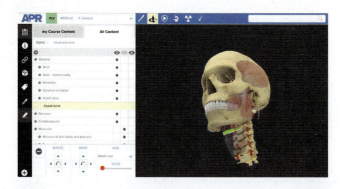

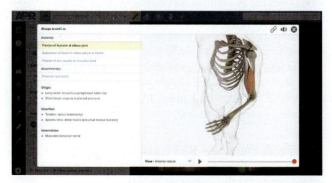

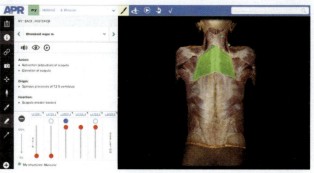

McGraw-Hill Education

Connect Virtual Labs is a fully online lab solution that can be used in conjunction with the lab manual as a preparation, supplement, or make-up lab. These simulations help students learn the practical and conceptual skills needed, then check for understanding and provide feedback. With adaptive pre-lab and post-lab assessment available, instructors can customize each assignment.

ꟿPhILS

Physiology Interactive Lab Simulations© 4.0 (Ph.I.L.S.) offers 42 lab simulations that may be used to supplement or substitute for wet labs. Users may adjust variables, view outcomes, make predictions, draw conclusions, and print lab reports. Ph.I.L.S is now more readily available than ever with a new user- and mobile-friendly interface, along with increased accessibility features.

McGraw-Hill empowers students to learn and succeed in the Anatomy and Physiology course.

SMARTBOOK®

SmartBook provides personalized learning to individual student needs, continually adapting to pinpoint knowledge gaps and focus learning on concepts requiring additional study. The result? Students are highly engaged in the content and better prepared for lecture.

LEARNSMART PREP®

LearnSmart Prep helps students thrive in college-level A&P by helping solidify knowledge in the key areas of cell biology, chemistry, study skills, and math. The result? Students are better prepared for the A&P course.

PhILS

Ph.I.L.S. 4.0 (Physiology Interactive Lab Simulations) software is the perfect way to reinforce key physiology concepts with powerful lab experiments. The result? Students gain critical thinking skills and are better prepared for lab.

Concept Overview Interactives are ground-breaking interactive animations that encourage students to explore key physiological processes and difficult concepts. The result? Students are engaged and able to apply what they've learned while tackling difficult A&P concepts.

Practice ATLAS

Practice Atlas for A&P is an interactive tool that pairs images of common anatomical models with stunning cadaver photography, allowing students to practice naming structures on both models and human bodies, anytime, anywhere. The result? Students are better prepared, engaged, and move beyond basic memorization.

*Statistic courtesy of The New England Journal of Higher Education

Virtual Labs

Connect Virtual Labs helps connect the dots between lab and lecture, boosts student confidence and knowledge, and improves student success rates. The result? Students are engaged, prepared, and utilize critical thinking skills.

Anatomy & Physiology Revealed® 4.0

Anatomy & Physiology Revealed® (APR) 4.0 is an interactive cadaver dissection tool to enhance lecture and lab that students can use anytime, anywhere. The result? Students are prepared for lab, engaged in the material, and utilize critical thinking.

INTRODUCTION

Welcome to the human anatomy and physiology laboratory! You are about to embark on a fascinating journey. The human body is one of the most amazing machines in existence. Provide it with fuel (i.e., food and water), and it can accomplish feats that put most mechanical machines to shame. Are you aware of any other machine that has an engine capable of sustaining it for 100 years without ever stopping? Yet, the human heart—the body's "engine"—can keep the body running for decades upon decades without ceasing. In this course, you will learn the many wonderful ways the body is able to accomplish this. It is well known that this course can be challenging. Students often experience both excitement and anxiety about it. These mixed emotions are normal and to be expected. It is our hope that the exercises in this laboratory manual will make your study both enjoyable and rewarding.

This laboratory manual is designed for an integrated, systems-based course that combines human gross anatomy, histology, and physiology. **Gross anatomy** is the study of structures that can be seen with the naked eye. This includes any structure that can be seen without the use of a microscope. **Histology** is the study of tissues and requires the use of a microscope. **Physiology** is the study of body functions. After completing this course, it is our hope that you will have developed an understanding and appreciation for how gross anatomical structures relate to tissue structures, and how all levels of structure relate to function. That said, in the laboratory itself, you will often be studying the three somewhat separately. That is, laboratory studies in gross anatomy will likely involve observing classroom models, dissecting animal specimens, or making observations of human bones and/or human cadavers; laboratory studies in histology will likely involve observing histology slides with a microscope or using some sort of virtual microscopy system; and laboratory studies in physiology will involve performing wet lab or virtual (computer software–based) experiments. To assist you in these endeavors, the exercises in this manual are divided into three types of activities: gross anatomy, histology, and physiology activities. Where applicable, each chapter will begin with a section on gross anatomy and will end with a section on physiology. Although you will perform the activities somewhat separately, the goal is to integrate what you learn in each exercise and to associate structure with function. "Concept Connection" boxes and questions within exercises in each chapter will assist with this task.

The purpose of this introductory chapter is to familiarize you with the process of science, systems of measurement, common equipment and dissection techniques encountered in the anatomy and physiology laboratory, proper disposal of laboratory waste materials, and common dissection techniques.

List of Reference Tables

Table 1.1	Body Temperature Data for Five Study Subjects	p. 5
Table 1.2	Distance Traveled by a Bowling Ball	p. 7
Table 1.3	Metrics	p. 8
Table 1.4	Common English-Metric Conversions	p. 9
Table 1.5	Common Dissection Instruments	p. 11
Table 1.6	Preservative Chemicals Encountered in the Human Anatomy & Physiology Laboratory	p. 13

INTEGRATE

CLINICAL VIEW
Use of Human Cadavers in the Anatomy and Physiology Laboratory

Where did that body lying on a table in the human anatomy and physiology laboratory come from? Typically, the body was donated by a person who made special arrangements before the time of death to donate his or her body to a body donor program so it could be used for education or research. Individuals who donate their bodies for these purposes made a conscious decision to do so. Such individuals have given us an incredible gift—the opportunity to learn human anatomy and physiology from an actual human body. It is important to remember that what that person has given is, indeed, a gift. The cadaver deserves the utmost respect at all times. Making jokes about any part of the cadaver or intentionally damaging or "poking" at parts of the cadaver is unacceptable behavior.

The idea that one will be learning anatomy and physiology by observing structures on what was, at one time, a living, breathing human being might make a person feel very uncomfortable at first. It is quite normal to have an emotional response to the cadaver upon first inspection. It takes time and experience to become comfortable around the cadaver. Even if you think you will be just fine around the cadaver when you are to observe it for the first time, it is important to be aware of your initial response and of the responses of fellow classmates. If at any time you feel faint or light-headed, sit down immediately. Fainting, though rare, is a possibility, and can lead to injuries if a fainting person falls down unexpectedly. Be aware of fellow students. If they appear to lose color in their faces or start to look sick—they might need your assistance.

Typically the part of the body that evokes the most emotional response is the face, because it is most indicative of the person that the cadaver once was. Because of this, the face of the cadaver should remain covered most of the time. This does not mean you are not allowed to view it. However, when you wish to do so, make sure that other students in the room know that you will be uncovering the face. If you have a particularly strong emotional response to the cadaver, take a break and come back to it later when you are feeling better.

Individuals with a great deal of experience around cadavers had a similar emotional response during their first time as well. In time one learns to disconnect one's emotions from the experience. Certainly at one time the body that is the cadaver in the laboratory was the home of a living human being. However, now it is just a body. Eventually students do become comfortable using the cadaver and find that it is an invaluable learning tool that is far more useful than any model or picture could ever be. There is nothing quite like the real thing to help students truly understand the structure of the human body. Make the most of this unique opportunity—and give thanks to those who selflessly donated their bodies to provide students with the ultimate learning experience in anatomy and physiology.

Students who are curious about the uses of cadavers in science and research are encouraged to check out the following book from the library: Mary Roach, *Stiff: The Curious Lives of Human Cadavers* (New York: W.W. Norton, 2003).

These Pre-Laboratory Worksheet questions may be assigned by instructors through their Mc Graw connect course.

1. The study of structures is called _____ (anatomy/physiology), whereas the study of functions is called _____ (anatomy/physiology).

2. Which of the following metric unit(s) is/are used to report mass? (Check all that apply.)

    _____ a. centimeter

    _____ b. decigram

    _____ c. kiloliter

    _____ d. microgram

    _____ e. millimeter

3. Number the following steps involved in the scientific method in the correct order.

    _____ a. conclusions

    _____ b. data analysis

    _____ c. data collection

    _____ d. experiment

    _____ e. hypothesis

4. When presenting data, the _____ (dependent/independent) variable is plotted on the X-axis, whereas the _____ (dependent/independent) variable is plotted on the Y-axis.

5. For each of the following metric prefixes, write the corresponding power of ten in the space provided.

    a. centi-:   _____

    b. deci-:   _____

    c. kilo-:   _____

    d. milli-:   _____

    e. micro-:   _____

6. Which of the following chemical(s) require(s) the use of personal protective equipment? (Check all that apply.)

    _____ a. ethanol

    _____ b. formalin

    _____ c. methylene blue

    _____ d. phenol

    _____ e. potassium permanganate

7. When removing a scalpel blade, be sure to point the blade _____ (toward/away from) you to avoid injury.

8. Which of the following dissecting tools is the most beneficial for attempting to loosen the hold between a specimen's skin and underlying fascia? (Circle one.)

    a. dissecting probe

    b. finger

    c. scalpel

    d. scissors

LEARNING STRATEGY

The volume of material students cover in an anatomy and physiology course is likely to be greater than that of any other college course. This is the reason it is considered by most to be a very challenging course. Learning how to approach a subject with such a vast amount of material can be quite difficult at first and might require a change of study habits. The list below provides several suggestions that you may want to incorporate into your study habits to make you more effective and efficient. Everyone learns in their own way, so all tips may not be suitable for all students.

1. Study a little bit *every day*. "Cramming" simply does not work in a course such as human anatomy and physiology in which there are a large number of terms to learn. Instead, break down the topics that are to be covered into manageable chunks. For example, if a student sits down to study with the goal of learning the details of every bone in the human body, that student may feel so overwhelmed that he or she has no idea where to start. On the other hand, if the student studies with the goal of learning the details of *one* bone at a time (e.g., the humerus), then the task becomes much simpler and the student is more likely to complete it.

2. Only study with a group when you are *reviewing* material that you have already covered on your own. Because all individuals learn differently, group study when first learning material is very inefficient. On the other hand, having someone ask you questions to review what you know is an excellent way to prepare for exams.

3. Use *active* study methods whenever possible. Active methods include writing, speaking, labeling, and the like. *Passive* methods include reading the book, listening to a lecture, and the like. Many students mistakenly believe they have mastered the material once they feel as if they understand what the instructor lectured about or what they read in the book. Do not make this mistake! Force yourself to recall information by writing it down, drawing it, or telling it to a friend or family member (even the family cat or dog will do as an audience). You will quickly find out what you do or don't know.

4. Use spaced repetition. Studies show that reviewing material at regular intervals is an effective way to help solidify the information in your head. As you plan your daily (yes, *daily*) and weekly study schedule, be sure to include time to review material you studied previously. It does not have to be a lot of time, but it is absolutely necessary to keep the material fresh in your head.

5. Learning anything new can sometimes be a struggle. The more you struggle to learn something, the better you actually learn the material. Most of us feel better when we review things we already know well because it gives us confidence. However, this approach may not be effective. Think of it like physical exercise. To gain a benefit, you must apply stress to your body and then allow your body to adapt to that stress. The stress may not feel good while you are exercising—in fact, it may feel difficult! However, with time your body adapts, and the stress/overload becomes easier to deal with. Your brain works much the same way.

Be patient and persistent. Have confidence in your ability to learn the material. Think of how AMAZING the human body is, and feel fortunate that you have this opportunity to learn about it. You *can* do this!

GROSS ANATOMY

The Scientific Process of Discovery

What is science? **Science** is a way of acquiring knowledge of the natural world through observation and experimentation. A **scientist** is an individual who engages in research using the scientific method to learn facts about the world. The **scientific method** is a systematic approach to inquiry that assumes that the answer to a question can be explained by phenomena that are *observable* and *measurable*. In the anatomy and physiology laboratory, the scientific method will be used to make observations and conclusions about the structure and function of the human body.

The scientific method is a rigorous and systematic approach to inquiry that requires certain steps be taken. However, the steps need not be followed precisely. That is, there is some flexibility in the order in which the steps take place. Very often several steps take place concurrently as the process of discovery evolves. The steps involved in the scientific method follow the general pattern shown below:

Observation → Hypothesis → Experiment → Data Collection →
Data Analysis → Conclusions

Observation

The first step of the scientific method is **observation**. When an unknown phenomenon is observed, the observer often makes a tentative explanation as to the cause of that phenomenon. This explanation is called a **hypothesis**. For example, consider the observation that body temperature changes during the day, and an observer of this fact is interested in knowing if body temperature also changes during the course of the night while a person is asleep. An observation has been made (body temperature changes during the day) that was followed up with a question: Does body temperature change during the night? The next step is to formulate a hypothesis.

Hypothesis

The second step of the scientific method is to **formulate a hypothesis**. One of the key features of a good hypothesis is that it must be testable. That is, some aspect of the variable of interest must be measurable. In the current example, body temperature is the variable of interest, and it can be measured using a thermometer. Another feature of a good hypothesis is that it must be specific, yet limited in scope. This is not to say that the explanation for a phenomenon is limited, nor that the questions asked are limited. Instead, it means that the testable hypothesis must be limited to something that is measurable while all other conditions are controlled. An example of a simple, testable hypothesis is the following: Body temperature changes over time during the night. Once the hypothesis has been formulated, the next step is to design an experiment to test the hypothesis.

Experiment

One of the most creative and interesting aspects of the scientific method is to **design an experiment** to test a hypothesis. Designing a good experiment to test a hypothesis is a challenging task. The key feature of good experimental design is to attempt to predict any variables that may have an influence on the variables of interest and control for them. In the current example, the variables of interest are body temperature and time during the night. To determine if body temperature changes during the night, an experiment must be designed to measure body temperature at given time intervals during the night.

Variables

A **variable** is a characteristic that may or may not influence the outcome of an experiment. In the current experiment, two variables of interest have been identified: body temperature and time. Based on convention, the variables in the experiment must be categorized as independent, dependent, or confounding variables.

The **independent variable** is a variable that is set at the outset of the experiment. It does not change as a result of the experimental procedure. Thus, it is said to be *independent* of the experimental procedure. In this example, time is the independent variable. Note that it is impossible for time to change as a result of any experimental procedure. The **dependent variable** is the unknown variable that is going to be measured and is often expected to change as a result of the experimental procedure. Thus, it is said to be *dependent* on the experimental procedure. In the current example, body temperature is the dependent variable. **Confounding variables** are any variables that may affect the variable of interest (dependent variable). Examples of some confounding variables are the amount and type of clothing the subject is wearing, the type of bedding the subject uses, and how much and what the subject eats or drinks before going to sleep. In setting up an experiment, great efforts are made to control as many confounding variables as possible. Finally, a **control value** with which to compare the measured values of body temperature is required. In this example, a "normal" body temperature of 37 °C is the obvious control value.

Data Collection

Once a controlled experiment has been designed to test the hypothesis, data collection can begin. When a scientist conducts an experiment, he or she begins by performing a statistical test to determine the sample size necessary to get a meaningful result from the experiment. Why? In this example, if body temperature was measured for only one subject, it would be neither reasonable nor appropriate to extrapolate that data to include all individuals, because the person measured may not be typical of most individuals. For experiments conducted in the anatomy and physiology laboratory, the study subjects will likely consist of the students in the class. Thus, the data set will be limited in scope compared to the ideal situation. However, in most cases enough data will have been collected to obtain reasonable results.

Once the sample size has been determined, the next step is to **collect the data**. In this example, body temperature would likely be measured for each of the study subjects at specific time intervals during the night. At the same time, confounding variables would be controlled for by having all study subjects wear the same clothing, use the same bedding, refrain from eating within a certain number of hours before bed, and go to bed at a prescribed time. After data collection comes the fun part: data analysis!

Data Analysis

Once the data have been collected, the experimenter must **analyze the data** in a way that makes sense both to the experimenter(s) and to the rest of the scientific community. There are several ways to present data, and presentation depends somewhat on the variable in question.

For any given data set, a mean, median, and standard deviation must be calculated for the value to ensure the value represents the data for the group of study subjects as a whole. The **mean** is the average of all the data points and is calculated by taking the sum of all the data points and dividing by the number of study subjects. The **median** is the middle value of all the data points. The **standard deviation** is a measure of the variability of the individual data points as compared to the mean. When the standard deviation is small, it means that individual data points are all very close to the mean. When the standard deviation is large, it means there is much variability between individual data points and the mean.

For the purposes of the exercises in this manual, standard deviations will not be reported for experimental data because this generally requires using a computer program to perform the calculations. An easier way of determining variability is to calculate and report the *range* of values. The **range** is simply the difference between the highest and lowest values, and it is calculated by subtracting the lowest value from the highest value. If the range is small, it indicates that there is very little variability of individual data points as compared to the mean. If the range is large, it indicates that there was quite a bit of variability of individual data points as compared to the mean. **Table 1.1** shows hypothetical temperature data for five study subjects taken at two different times (12 a.m. and 6 a.m.). The mean, standard deviation, and range of the data are also shown in the table. Notice that even though the standard deviation and range are different for each time, they have the same pattern. That is, for the data taken at 6 a.m., both the standard deviation and the range are higher than for the data taken at 12 a.m. Thus, both of these measures demonstrate there was greater variability in the individual temperature values at 6 a.m. than there was at 12 a.m.

Conclusions

The final step in the scientific method involves **drawing conclusions** based on the results of the experiment. This requires reviewing the hypothesis in light of the data collected. That is, the data will either support or refute the hypothesis. In this example, the hypothesis was that body temperature would change over the course of the night. **Figure 1.1** is a sample graph that is based on data from actual studies that looked at the variation in body temperature of a large number

Table 1.1	Body Temperature Data for Five Study Subjects	
Study Subject	**Body Temperature at 12 a.m. (°C)**	**Body Temperature at 6 a.m. (°C)**
Subject A	36.2	35.0
Subject B	36.8	34.0
Subject C	37.2	34.3
Subject D	37.0	35.9
Subject E	36.5	36.0
Mean	**36.7**	**35.0**
Standard Deviation	**0.4**	**0.9**
Range	**1.0**	**2.0**

of study subjects during a daily cycle. Thus, it can be used as an estimate of data that might have been obtained had the hypothetical experiment been performed. In this example, the data support the hypothesis that body temperature does change as a result of time of the night. Furthermore, it is possible to be more specific and say *how* body temperature changes during the night. That is, body temperature appears to decrease steadily throughout the night and reaches its lowest point just before waking (5 a.m.).

The scientific method is a continual process. Often a hypothesis and related experiment will result in a few answers, but even more questions. For example, after concluding that body temperature changes during the night, a follow-up question may be asked: *why* does body temperature change during the night? Using the scientific method, the scientific process can continue through to the formulation of a new hypothesis. The new hypothesis will lead to further experimentation, data collection, and data analysis. As the process continues, more and more details concerning the area of interest emerge.

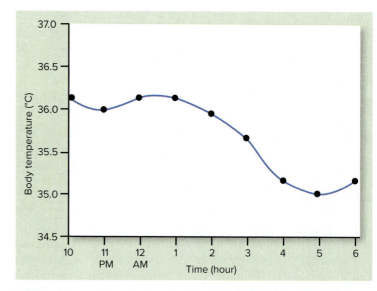

Figure 1.1 **Temperature Variance During the Course of the Night.** Each data point represents the mean for a number of study subjects.

EXERCISE 1.1

THE SCIENTIFIC METHOD

Let's say that you have observed that heart rate varies with activity level. That is, when an individual exercises, heart rate is higher, and when an individual is at rest, heart rate is lower. Now, let's say that because you have observed that heart rate is lower when an individual is in a "rested" state, you are interested in knowing if meditation, an activity that is designed to put the body in a calm/rested state, causes an individual's heart rate to decrease. You decide to perform a scientific experiment to figure out if this is accurate.

1. State your hypothesis: _____
   _____
   _____
   _____
   _____
   _____

2. What is the independent variable? _____
   _____

3. What is the dependent variable? _____
   _____

4. Design an experiment to test your hypothesis. In the explanation of your experimental design, describe what variable(s) you will control and what variable(s) you will measure.
   _____
   _____

5. Finally, describe any confounding variables that may affect the results of the experiment, and explain how you might control for such variables.

EXERCISE 1.2

PRESENTING DATA

Once experimental data have been collected, they must be presented in a meaningful way so that others who view the data will be able to draw their own conclusions about the data. In the previous description of the scientific method, data were presented using a data table (table 1.1) and a graph (figure 1.1). These are two of the most common modes of data presentation and are the modes most often used to present data for the experiments performed in this manual. When presenting data using a **table**, it is common to include mean, number of study subjects, and standard deviation (or range) for each data point (see table 1.1 for an example). When presenting data using a **graph**, there are conventions to follow. Namely, the *independent* variable is plotted on the X-axis (horizontal) and the *dependent* variable is plotted on the Y-axis (vertical). Graphing data this way allows the individual reading the graph to determine what effect, if any, variable X has on variable Y. For the sample experiment looking at the effect of time of the night on body temperature, the data have been plotted in the graph shown in figure 1.1.

1. Why was time plotted on the X-axis? _____

   _____

2. Why was temperature plotted on the Y-axis? _____

   _____

Other items to note when creating a graph are the following:

1. Each axis must be labeled with the appropriate numbers indicating the value of the measurement. For example, in the graph (figure 1.1), these are the measured values for "hour" and "degrees Celsius."

2. Each axis must have a title and must provide the units that are used. In the graph (figure 1.1), the X-axis is labeled "Time" and the unit is "hour"; the Y-axis is labeled "Body temperature" and the units are "degrees Celsius."

It is very important to always place labels on graphs. A common mistake many students make is to assume that their instructor—or whoever is grading their laboratory report—already knows what is supposed to be graphed, therefore the student does not need to put the units on the graph. Do not make this mistake! Always assume that the person who is going to read a graph has no idea what the graph is trying to present. Failing to put units on a graph results in a graph that means nothing. For example, let's say the label "Time" is on the X-axis, but there are no units provided. The reader of the graph is left to wonder if the intervals indicate time in seconds, minutes, hours, or some other unit. Likewise, failing to indicate the units for "Body temperature" on the Y-axis leaves the reader wondering if the temperature is given in Celsius (°C) or Fahrenheit (°F).

Sample Graphing Exercise

1. **Table 1.2** represents experimental data obtained by rolling a bowling ball on a track and measuring how far the ball rolls over time.

Table 1.2	Distance Traveled by a Bowling Ball
Time	**Distance Traveled**
(sec)	(meters)
0	0
1	6
2	8
3	9.5
4	10.5
5	11
6	11.5
7	11.75
8	11.8
9	11.9
10	12

Questions:

a. What is the independent variable? _____

   _____

b. What is the dependent variable? _____

   _____

2. Using the data in table 1.2, graph the data on the grid provided. Be sure to label the axes and provide the appropriate units.

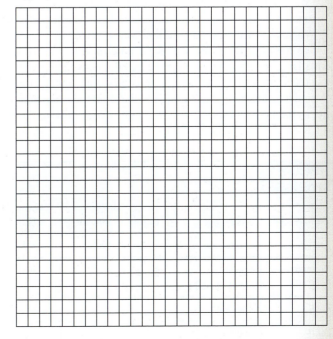

(continued on next page)

(continued from previous page)

3. Now that the data have been graphed, write a brief paragraph explaining the results in the space provided.

_____

_____

_____

_____

_____

_____

_____

_____

Measurement in Science

Systems of Measurement

If one has ever baked a cake or done another activity that required making measurements, common units of measurement such as cups, gallons, teaspoons, liters, and the like should be familiar. There are two systems of measurement that are most commonly used in the world: English and metric systems. Because of its uniformity and ease of use, the **metric system** is the system most widely used throughout the world. It is also the system of measurement that scientists use when reporting data. Thus, when performing the laboratory activities in this manual, students will be reporting results using metric units.

The metric system is relatively simple to use because all units are given as powers (multiples) of ten. To calculate a power of ten, simply take the superscript (i.e., 10^x, where x is the superscript) and multiply ten by ten that many times. Thus, $10^2 = 10 \times 10 = 100$; $10^3 = 10 \times 10 \times 10 = 1000$. The only unusual numbers are 10^0 and 10^1. When you see these numbers, just remember that ten to the zero power is 1 and ten to the 1st power is ten. In the metric system, specific prefixes denote the power of ten that is used in a measurement. For example, a kilometer is 10^3 meters and thus represents 1000 meters. **Table 1.3** lists common metric prefixes and the power of ten that each represents.

When converting one metric unit to another, move the decimal point to the right or to the left, depending on if the conversion is from a larger unit to a smaller unit (e.g., kilograms to grams) or vice versa (e.g., grams to kilograms). When converting from a larger unit to a smaller unit, move the decimal to the *right* the same number of units as the power of ten. This is the same as *multiplying* the larger measurement by the appropriate factor of ten. For example, to convert kilometers to meters (1 km = 1000 meters), move the decimal point to the right *three* positions because a kilometer is one meter times ten to the *third* power (10^3). Conversely, when converting from a smaller unit to a larger unit, move the decimal to the *left* the same number of units as the power of ten. This is the same as *dividing* the smaller measurement by the appropriate factor of ten. For example, to convert meters to millimeters (1 mm = 1/1000 meter), move the decimal point to the left *three* positions because a millimeter is one meter times ten to the negative *third* power (10^{-3}). Another way to remember this is that a power of ten that is positive moves the decimal to the right (forward) and a power of ten that is negative moves the decimal to the left (backward).

larger unit → smaller unit   MULTIPLY by appropriate power of ten

smaller unit → larger unit   DIVIDE by appropriate power of ten

Practice:

Convert 457 milligrams to grams

_____

Convert 5698 centimeters to decimeters

_____

Convert 4.3 kilometers to meters

_____

Convert 0.5 liter to milliliters

_____

Table 1.3	Metrics							
Metric Prefix	**Meaning**	**Symbol**	**Amount**	**Power of Ten**	**Length Measure**	**Mass Measure**	**Volume Measure**	**Word Origin**
kilo	one thousand	k	1000	10^3	kilometer (km)	kilogram (kg)	kiloliter (kL)	*chlioi*, a thousand
No prefix/ base unit	Use the standard unit	–	1	10^0	meter (m)	gram (g)	liter (L)	NA
deci-	one-tenth	d	0.1	10^{-1}	decimeter (dm)	decigram (dg)	deciliter (dL)	*decimus*, tenth
centi-	one-hundredth	c	0.01	10^{-2}	centimeter (cm)	centigram (cg)	centiliter (cL)	*centum*, hundred
milli	one-thousandth	m	0.001	10^{-3}	millimeter (mm)	milligram (mg)	milliliter (mL)	*mille*, thousand
micro-	one-millionth	μg	0.000001	10^{-6}	micrometer (μm)	microgram (μg)	microliter (μL)	*mikros*, small

Metric Conversions

If measurements are made using the English system, the measurements will need to be converted from English units to metric units for laboratory reports. **Table 1.4** lists some common conversions between English and metric measurements. Some of these conversions are conversions that you should be able to make in your head easily. For example, 1 inch = 2.54 centimeters and 1 pound = 0.45 kilogram.

For example, to convert 16 ounces (English measurement) into milliliters (metric measurement), multiply the number of ounces (16) by 30 (see table 1.4) to get milliliters:

$$16 \text{ ounces} \times 30 = 480 \text{ milliliters}$$

Notice that the conversion in table 1.4 tells how to convert ounces to milliliters. What if we want to convert ounces to liters instead? Recall from table 1.3 that one milliliter is one one-thousandth of a liter, and its power of ten is 10^{-3}. Thus, to convert ounces to liters, first convert to milliliters using the multiplier listed in table 1.4, then convert milliliters to liters by moving the decimal point three places to the left: 480.0 milliliters = 0.48 liter.

Temperature Scales

Just as there are two systems of measurement for volume, length, and mass, there are also two systems of measurement for temperature. In the United States most of our temperatures (such as air temperatures reported by weather stations) are reported as degrees Fahrenheit (°F). However, often temperatures will instead be reported as degrees Celsius (°C, also known as centigrade). In science, it is appropriate to report temperature readings (such as body temperature) in degrees Celsius. To make conversions between these two temperature readings, use the following equations:

To convert degrees Celsius to degrees Fahrenheit:

$$°F = ((°C \times 9) / 5) + 32$$

To convert degrees Fahrenheit to degrees Celsius:

$$°C = ((°F - 32) \times 5) / 9$$

Table 1.4	Common English-Metric Conversions			
To Convert From:	**To:**	**Multiply:**	**By:**	**Conversion**
centimeters	inches	centimeters	0.39	1 cm = 0.39 inch
feet	centimeters	feet	30.48	1 foot = 30.48 cm
feet	meters	feet	0.3	1 foot = 0.3 m
fluid ounces	milliliters	fluid ounces	30	1 oz = 30 mL
gallons	liters	gallons	3.78	1 gallon = 3.78 L
grams	ounces	grams	0.035	1 g = 0.035 oz
inches	centimeters	inches	2.54	1 inch = 2.54 cm
kilograms	pounds	kilograms	2.2	1 kg = 2.2 lb
kilometers	miles	kilometers	0.62	1 km = 0.62 mile
liters	quarts	liters	1.06	1 L = 1.06 qt
liters	gallons	liters	0.26	1 L = 0.26 gallon
meters	yards	meters	0.9	1 m = 1.1 yd
meters	feet	meters	3.3	1 m = 3.3 feet
miles	kilometers	miles	1.61	1 mi = 1.61 km
milliliters	fluid ounces	milliliters	0.03	1 mL = 0.03 oz
millimeters	inches	millimeters	0.039	1 mm = 0.039 inch
fluid ounces	grams	fluid ounces	28.3	1 oz = 28.3 g
fluid ounces	milliliters	fluid ounces	29.6	1 oz = 29.6 mL
pounds	grams	pounds	453.6	1 lb = 453.6 g
pounds	kilograms	pounds	0.45	1 lb = 0.45 kg
quarts	liters	quarts	0.95	1 qt = 0.95 L
yards	meters	yards	0.9	1 yd = 0.9 m

UNITS OF MEASUREMENT

1. Make each of the following conversions. Use tables 1.3 and 1.4 and the temperature conversion equations given in this chapter as references:

    a. 5 millimeters = _____ centimeters

    b. 21 grams = _____ ounces

    c. 500 mL = _____ liters

    d. 5 inches = _____ centimeters

    e. 3 meters = _____ feet

    f. 3 gallons = _____ liters

    g. 3 liters = _____ mL

2. The set point for human body temperature is 98.6°F. This means that 98.6°F is the temperature set point that homeostatic mechanisms in the body strive to maintain. It is appropriate in science to report temperature values in degrees Celsius. Convert 98.6°F to degrees Celsius in the space provided. Be sure to show all of your work.

Laboratory Equipment

The typical human anatomy and physiology laboratory classroom consists of laboratory tables or benches that provide ample room for use of microscopes, laboratory equipment, and dissection materials. If human cadavers are used in the classroom, there will also be a space dedicated to the tables where the cadavers are stored. When entering the classroom, look around and familiarize yourself with the environment. Pay particular attention to the location of the sinks, eyewash stations, and safety equipment such as first-aid kits and fire extinguishers. The instructor will provide a detailed introduction specific to the laboratory classroom, safety procedures, and accepted protocols. The main purpose of this chapter is to introduce common safety devices and dissection equipment. *Do not* use the information in this chapter as the sole source of information on laboratory safety. This laboratory manual is not intended to be a safety manual for the laboratory.

Protective Equipment

The human anatomy and physiology laboratory poses few risks, although it is important to be aware of what these are. The main risks are damage to skin or eyes from exposure to laboratory chemicals (covered in the next section) and cuts from dissection tools. As a general precaution, whenever you are working with fresh or preserved specimens (animal or human), wear protective gloves to keep any potentially infectious or caustic agents from contacting your skin. If there is a risk of squirting fluid, then wear protective eyewear (safety glasses or safety goggles). When wearing gloves, be sure to wear the correct size for your hands. If the gloves are too small, they may tear easily. If they are too big, it may be difficult to handle instruments and tissues. When gloves become excessively dirty, remove them and put on a new pair. When removing a glove, start at the wrist and pull toward your fingers, turning the glove inside-out as it is removed. This will prevent any potentially damaging fluids from contacting your skin during removal of the glove.

When using dissecting tools there is always a risk of cutting yourself or others. First and foremost—*never* wear open-toed shoes to the laboratory. Dissecting tools are sometimes dropped, and they will cut your feet if you are not wearing protective footwear. When using sharp tools such as scalpels, always be aware of where the scalpel blades are pointing. Scalpel blades should always be pointed away from you *and* away from others in the laboratory. When dissecting, be aware of where others are standing or sitting, and consider the risk posed to yourself and others if your hand were to slip. *Never* put your hands in the dissecting field when someone else is dissecting. If another person asks for assistance holding tissues while dissecting, use forceps or some other device to hold the tissue so your hands are not within reach of the scalpel blade. Always be aware of the location of any scalpels, particularly when they are not in use. Scalpels left on dissecting trays or tables can cut people who are not expecting them to be there. Remove dissecting pins from the specimen once the dissection is complete so that the pins will not poke unsuspecting individuals.

IDENTIFICATION OF COMMON DISSECTION INSTRUMENTS

Several dissection instruments are commonly found in the human anatomy and physiology laboratory classroom. **Table 1.5** describes each of these instruments and their uses.

1. Obtain a dissection kit from the laboratory instructor, or use your own dissection kit if you were required to purchase your own materials.

2. Identify the instruments listed in **figure 1.2**, using table 1.5 as a guide. Then label figure 1.2.

Table 1.5	Common Dissection Instruments		
Tool	**Description and Use**	**Photo**	**Word Origin**
Blunt Probe	An instrument with a blunt (not sharp) end on it. It is used to pry and poke at tissues without causing damage. Some probes come with a sharper point on the opposite end that can be used for "picking" at tissues.	©Christine Eckell	*proba*, examination
Dissecting Needles	Long, thick needles that have a handle made of wood, plastic, or metal. These needles are used to pick at tissues and to pry small pieces of tissue apart.	©Christine Eckell	*dissectus*, to cut up
Dissecting Pins	"T" shaped pins that are used to pin tissues to a dissecting tray, thus allowing a particular area to be seen more easily	©Christine Eckell	*dissectus*, to cut up
Dissecting Tray	Metal or plastic tray used to hold a specimen. The tray is filled with wax or plastic. The wax and/or plastic is soft enough to pin tissues to.	©Christine Eckell	*dissectus*, to cut up
Forceps	Resemble tweezers, and are used for holding objects. Some are large and have tongs on the ends that assist with grabbing tough tissues. Some are small and fine (needle-nose) for picking up small objects. Forceps may also be straight-tipped or curve-tipped.	©Christine Eckell	*formus*, form + ceps, taker
Hemostat	In surgery these are used to compress blood vessels and stop bleeding (hence the name). For dissection they are useful as "grabbing" tools. The handle locks in place, which allows you to pull on tissues without causing hand and forearm muscles to fatigue.	©Christine Eckell	*haimo-*, blood + statikos, causing to stop

(continued on next page)

(continued from previous page)

Table 1.5	Common Dissection Instruments (*continued*)		
Scalpel	A sharp cutting tool. Generally the blade and the blade handle will be separate, except when using a disposable scalpel. See specific directions in the text regarding proper use of a scalpel, as they can be dangerous!	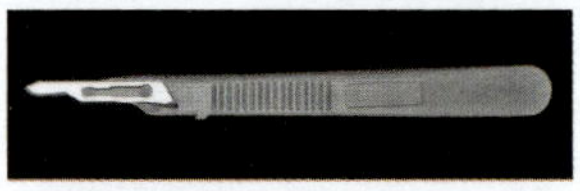 ©Christine Eckel	*scalpere,* to scratch
Scalpel Blade	Both the cutting part and the disposable part of a scalpel. The number on the blade indicates the size of the blade, and it must be matched with an appropriately numbered blade handle. When a blade becomes dull, it may be removed and replaced with a new blade. Used blades must be disposed of in a sharps container.	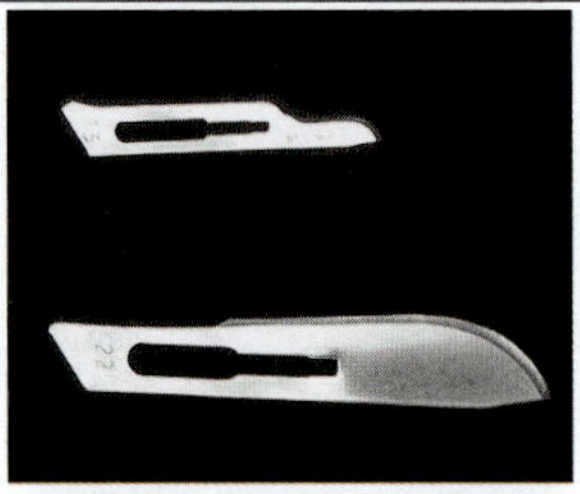 ©Christine Eckel	*scalpere,* to scratch
Scalpel Blade Handle	The nondisposable part of a scalpel that is used to hold the blade. The number on the handle indicates the size of the handle and is used to match it with a particular blade size. A scalpel blade handle can be a very useful tool for blunt dissection when used *without* a blade attached.	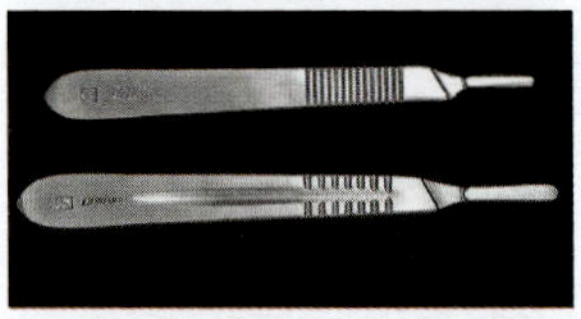 ©Christine Eckel	*scalpere,* to scratch
Scissors	Some scissors come with pointed blades and some have one curved (blunt) and one pointed blade. Scissors with the curved/blunt edge are used when you need to be careful not to damage structures. To use them, direct the curved blade toward the structures you do not want to damage. Pointed-blade scissors are particularly helpful for using "open scissors" technique (see exercise 1.8).	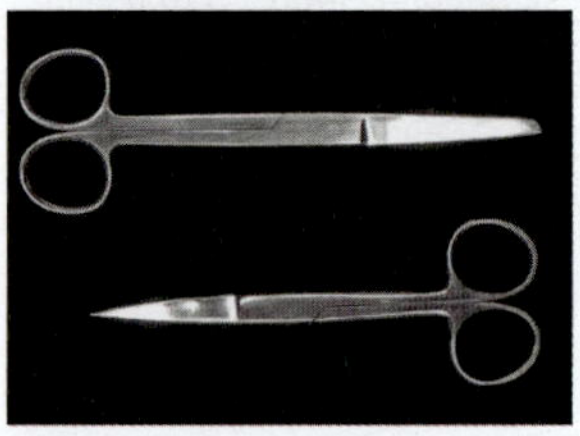 ©Christine Eckel	*scindere,* to cut

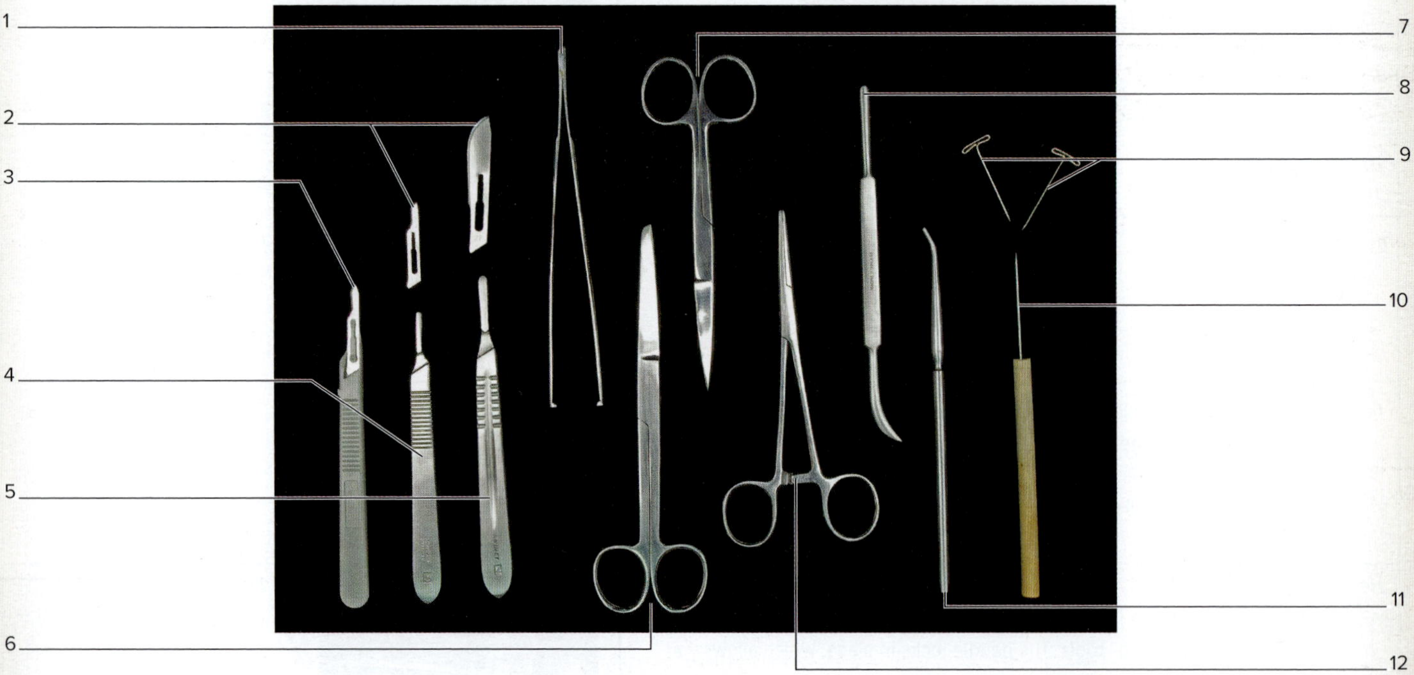

Figure 1.2   Identification of Common Dissection Instruments.
Use the terms below to fill in the numbered labels in the figure. Answers may be used more than once.

- ☐ Blunt probe
- ☐ Dissecting needle
- ☐ Dissecting pins
- ☐ Forceps
- ☐ Hemostat
- ☐ Scalpel (disposable)
- ☐ Scalpel blade handle (#3)
- ☐ Scalpel blade handle (#4)
- ☐ Scalpel blades
- ☐ Scissors (curved)
- ☐ Scissors (pointed)

©McGraw-Hill Education/Christine Eckel, photographer

Hazardous Chemicals

Various chemicals are used in the human anatomy and physiology laboratory. The discussion in this chapter will address chemicals that are used to preserve, or "embalm," animal specimens or human cadavers. Safety precautions for chemicals used in physiology "wet lab" experiments will be covered when these chemicals are encountered within specific laboratory exercises. Most embalming chemicals are not used in their full-strength form in the laboratory. Instead, most tissues and specimens encountered in the laboratory will have been previously injected with solutions containing these chemicals. Thus, safety measures in the laboratory are designed to protect users from the forms of these chemicals that are most likely to be encountered. The most common chemicals used for embalming purposes are formalin, ethanol, phenol, and glycerol.

Table 1.6 summarizes the uses and hazards of these chemicals. The majority of these chemicals are used to fix tissues and prevent the growth of harmful microorganisms, such as bacteria, viruses, and fungi. **Fixation** refers to the ability of the chemical to stabilize proteins, thus preventing their breakdown. **Preservatives** both fix tissues and inhibit the growth of harmful microorganisms. Because most preservatives also dehydrate tissues, humectants are added to embalming solutions. **Humectants**, such as glycerol, attract water. When humectants act alongside preservatives, they help keep tissues moist. Other chemicals that may be added to embalming solutions are pigments, which either make the tissues look more natural or mask the odors of the preservative chemicals. **Formalin** and **phenol** are the most toxic and odoriferous preservative chemicals. Luckily,

exposure to them in the anatomy and physiology laboratory will be very low. Although it may smell as if the concentrations of these chemicals are high, the odor is often misleading because these chemicals can be detected by odor in extremely small quantities. Although the concentrations of formalin and phenol that users are exposed to may be very low, if these chemicals have been used to preserve specimens, then protective clothing is required to prevent the chemicals from contacting the user's skin or eyes. Use gloves whenever handling specimens, and use protective eyewear whenever there is a risk of chemicals getting into your eyes. If your skin is exposed, rinse it immediately. If your eyes are exposed, use the eyewash station in the laboratory to rinse your eyes thoroughly. If you have experienced contact exposure to these chemicals and your skin or eyes continue to feel irritated after rinsing, consult a medical doctor.

Proper Disposal of Laboratory Waste

There are several types of waste that must be disposed of in the human anatomy laboratory. Much of this waste is "normal" waste, such as tissues, paper towels, or gloves. Such waste should be disposed of in the regular garbage/waste container found in the classroom. However, any potentially **hazardous waste** must be disposed of in a special container. The general rule for determining if something is potentially hazardous or not is this: If you think someone else may be injured *in any way* from handling this waste, it is hazardous. Follow this rule, and be sure to ask the instructor how to properly

Table 1.6	Preservative Chemicals Encountered in the Human Anatomy & Physiology Laboratory					
Chemical	**Description**	**Use**	**Hazard**	**Preventing Exposure**	**Disposal**	
Ethanol	Inhibits growth of bacteria and fungi	Preservative	Flammable, so requires storage in a fire-safe cabinet. Generally safe in small quantities.	Gloves and eye protection. Rinse body part immediately if exposed, particularly eyes. Seek medical attention if irritation persists.	Small amounts may be flushed down the sink along with plenty of water to dilute the solution.	
Formalin	Fixes tissues by causing proteins to cross-link (stabilize). Destroys autolytic enzymes, which initiate tissue decomposition. Inhibits growth of bacteria, yeast, and mold.	Preservative	Flammable, so requires storage in a fire-safe cabinet. Toxic at full strength. Penetrates skin. Corrosive. Burns skin. Damages lungs if inhaled. May be carcinogenic.	Gloves and eye protection. Rinse body part immediately if exposed, particularly eyes. Seek medical attention if irritation persists.	Do not pour into sinks.	
Glycerine (glycerol)	Helps control moisture balance in tissues. Counteracts the dehydrating effects of formalin.	Humectant	Flammable, so requires storage in a fire-safe cabinet. Generally safe. Can pose a slipping hazard if spilled on the floor.	Gloves and eye protection. Rinse body part immediately if exposed, particularly eyes. Seek medical attention if irritation persists.	Small amounts may be flushed down the sink along with plenty of water to dilute the solution.	
Phenol	Assists formalin in fixing tissues through protein solidification. Inhibits growth of bacteria, yeast, and mold.	Preservative	Flammable, so requires storage in a fire-safe cabinet. Extremely toxic at full strength. Rapidly penetrates the skin. Corrosive. Burns skin. Damages lungs if inhaled. NOTE: When used as embalming preservative, concentration (and thus toxicity) is extremely low.	Gloves and eye protection. Rinse body part immediately if exposed. Use an eyewash station if solution gets in the eyes. Seek medical attention if irritation persists.	Do not pour into sinks.	

dispose of something any time there is a question as to whether it is hazardous or not. It is always better to err on the side of caution.

What is hazardous waste?

1. Any sort of fresh tissue and/or blood
2. Laboratory chemicals
3. Broken glass, scalpel blades, or any other sharp item that may cut an individual who handles the waste

Sharps Containers

Sharps containers (figure 1.3) are plastic containers (often red or orange) that are used to dispose of anything "sharp," such as needles, scalpel blades, broken glass, pins, or anything else that has the potential to cut or puncture a person who handles it. Such items should NEVER go in the garbage, because they may injure anyone who handles the garbage thereafter. When in doubt, put it in the sharps container.

Biohazard Bags

Special **biohazard bags** may be available in the laboratory. These are used for biological material such as blood or other fresh animal tissue that requires special disposal. When it comes to human blood, an item containing a small amount of blood (such as a Band-aid™) can be disposed of in a normal wastebasket. However, if a towel is soaked with blood, then it must be disposed of in a biohazard bag. A biohazard bag is usually red or clear and has the symbol shown in **figure 1.4** on it. When dealing with tissues that must be disposed of in a biohazard bag, the instructor generally will provide disposal instructions. Again, when in doubt, always ask before disposing of something potentially hazardous. Important note: Human cadaveric tissues do *not* go into biohazard bags. They must be kept with the cadaver. Any piece of human tissue removed from a cadaver must eventually be returned to the cadaver to be cremated with the entire body.

Figure 1.3   Sharps Containers.   Samples of two different models of sharps containers. Such containers allow one to place sharp objects into the container, but the objects cannot be removed once placed inside. Note the biohazard warning symbol on the containers. ©Christine Eckel

Figure 1.4   **Biohazard Waste Symbol.**

INTEGRATE

CLINICAL VIEW
Common Safety Training by Health-Care Professionals

In the health care field, medical professionals are usually required to complete different types of training before working with patients. Training often includes: working with and proper disposal of hazardous waste; requirements for personal protective equipment (PPE) such as gloves and masks; safe use and disposal of sharps; and working with blood-borne pathogens. The training is designed not only to create a safe environment for employees and patients, but also to help prevent the spread of infection. The Occupational Safety and Health Administration (OSHA) is a U.S. federal agency that sets and enforces standards to ensure that employees are trained to work in a safe environment.

EXERCISE 1.5

PROPER DISPOSAL OF LABORATORY WASTE

Circle the letter (a, b, or c) of the correct waste receptacle (shown in **figure 1.5**) for each item listed below.

1.	broken scissors	A	B	C	**6.**	gloves	A	B	C
2.	cotton swab	A	B	C	**7.**	scalpel blades	A	B	C
3.	dissecting pins	A	B	C	**8.**	fresh tissue	A	B	C
4.	glass slide	A	B	C	**9.**	hypodermic needle	A	B	C
5.	paper towel	A	B	C					

(a)　　　　　　　　　　(b)　　　　　　　　　　(c)

Figure 1.5　**Common Waste Receptacles in the Laboratory.**　(*a*) Sharps container. (*b*) Wastebasket. (*c*) Hazardous waste bag. ©Christine Eckel

Dissection Techniques

The word *dissect* literally means to cut something up. Most people have been led to think that the first thing a surgeon or anatomist does when planning to dissect is to pick up a scalpel and cut. However, skilled dissection does not always involve actually cutting tissues. In fact, the dissector's best friend is a technique called "blunt dissection." Blunt dissection specifically involves separation of tissues *without* using sharp instruments (hence the term *blunt*). When dissecting tissues, always try using blunt dissection before picking up sharp instruments such as scissors and scalpels. Sharp instruments are very handy for cutting things. However, often students will end up cutting things they do not wish to cut, purely by accident. Thus, being sparing and prudent in the use of sharp tools is one of the most important tips for performing a good dissection.

For this exercise, the demonstration of techniques will be shown using a fresh chicken purchased from a grocery store. However, the instructor may choose another specimen for you to practice on. For now, the goal of the dissection is to separate the skin from the underlying tissues such as bones and muscle (the "meat") of the specimen.

Sharp Dissection Techniques

Sharp dissection techniques are the techniques most familiar to most people. These techniques involve the use of sharp instruments such as scissors and scalpels. They are "cutting" techniques. They are advantageous in that tough tissues may be easily separated from each other, or tissue pieces may be removed from a dissection specimen. The danger in using sharp techniques is that novice and experienced dissectors alike will often end up cutting things they do not wish to cut, such as blood vessels and nerves. Thus, sharp dissection techniques should be used with care.

PLACING A SCALPEL BLADE ON A SCALPEL BLADE HANDLE

Scalpels come in many forms. Some are of the disposable type, which typically means that the handle and blade come as one unit and the handle is made out of plastic **(figure 1.6)**. Often the blades and handles are separate items. Such items allow the blade to be replaced whenever it becomes dull from use. This exercise covers how to properly place a scalpel blade on a scalpel blade handle and how to properly remove the blade once finished.

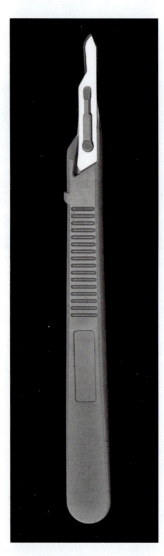

Figure 1.6 **Disposable Scalpel.** The scalpel blade and scalpel blade handle are both disposable. The entire unit must be disposed of in a sharps container. ©Christine Eckel

1. Obtain a **scalpel blade** and **scalpel blade handle** from the instructor. Scalpel blades and handles come in various sizes, and it is important to match the size of the blade to the size of the blade handle. Observe the scalpel handle and look for a number stamped on it, which will be a 3 or a 4 **(figure 1.7a)**. Next, observe the blade packet and note the number on it (figure 1.7b). A number 3 handle is used to fit number 10, 10A, 11, 12, 12D, and 15 blades. A number 4 handle is used to fit number 18, 20, 21, 22, 23, 24, 24D, and 25 blades. Larger handles and blades are generally used for making bigger, deeper cuts, whereas the smaller handles and blades are generally used for finer dissection. One of the most commonly used combinations in anatomy laboratories is a number 4 handle matched with a number 22 blade.

2. Once the scalpel handle and blade size are properly paired, carefully open the scalpel blade packet halfway **(figure 1.8a-1)**. Note the bevel on the blade. This bevel matches the bevel on the blade handle, so that there is only one way to properly place the scalpel blade on the handle. The blade handle has a bayonet fitting that is matched to the opening on the scalpel blade (figure 1.8a-2). This locks the blade in place on the handle. The safest way to place the blade on the handle is to first grasp the end of the blade using **hemostats** (figure 1.8a-2; table 1.5). Then, while matching the bevel on the blade to the bevel on the handle, slide the blade onto the handle until it clicks, indicating it is locked in place (figure 1.8a-3, a-4). If the blade does not go on easily, make sure that the blade has not been placed on the handle incorrectly (example: figure 1.8b). Now it is ready for use!

3. The safest way to remove a blade from a handle is to use a device that is both a **blade remover** and a sharps container all in one (an example is shown in **figure 1.9**).

4. If a blade remover is not available, remove the blade using hemostats. Obtain a pair of hemostats. Pointing the blade away from you (but not toward someone else), clamp the part of the blade nearest the handle with the hemostats **(figure 1.10-1)**. Once you have a firm grip on the blade, slide it over the bayonet on the handle and away from yourself until the blade comes off the handle (figure 1.10-2). Using the hemostats, transport the blade to a sharps container and dispose of it in the sharps container (figure 1.10-3).

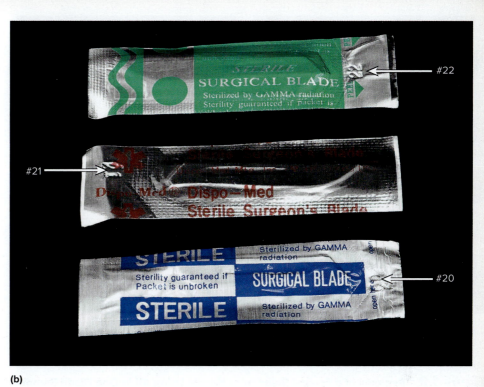

(a)

(b)

Figure 1.7 **Scalpel Blade Handles and Blades.** (*a*) The number on the scalpel blade handle indicates what size blades will fit on the handle. (*b*) The number on the blade wrapper indicates the size of the blade. See text for description of what size blades fit on what size blade handles. ©Christine Eckel

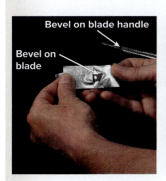

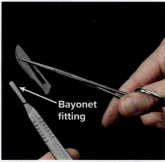

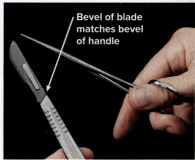

1 Open the foil packet and note the bevel on the blade.

2 Grasp the blade firmly using hemostat and line the blade up so that it matches the bevel on the blade handle.

3 Slide the blade onto the bayonet of the blade handle.

4 The blade should "click" as it locks in place on the blade handle.

(a)

(b)

Figure 1.8 **Scalpel Blade Placement.** (*a*) Correct procedure. (*b*) Incorrect placement of a blade on a blade handle. Notice that the bevel on the blade does not match up with the bevel on the blade handle. If placed in this fashion, the blade will not be secure on the handle and may slip off the handle and injure someone. ©Christine Eckel

(continued on next page)

(continued from previous page)

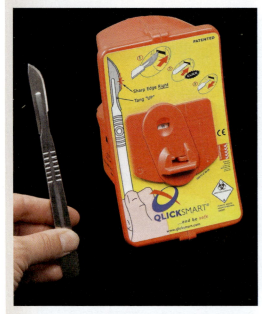
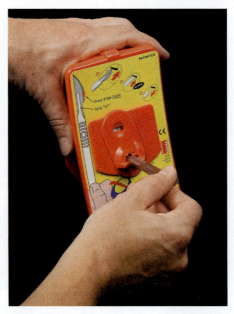
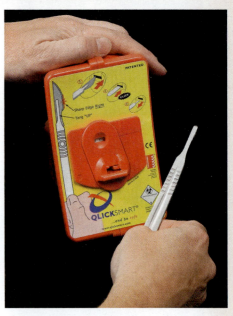

(1) Orient the blade and blade handle with sharp edge of the blade pointed to the right, as shown on the front of the device.

(2) Push the blade into the slot on the device until a distinct "click" is both heard and felt.

(3) While holding the removal device firmly with your free hand, pull the blade handle out of the device.

Figure 1.9 **Removal of a Scalpel Blade from Handle Using All-in-One Blade Remover/Sharps Container.** ©Christine Eckel

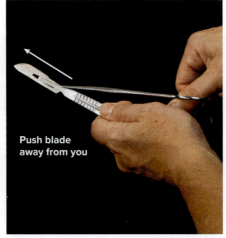
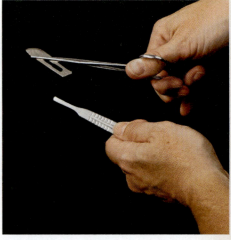

(1) With the blade pointed away from your body and the bayonet surface of the handle also directed away from your body, grasp the base of the blade with hemostats and lock the hemostats firmly to the blade.

(2) Slide the blade off the bayonet on the blade handle. Again, push it away from your body (and away from others in your vicinity as well).

(3) Once the blade has been removed from the handle, continue to grasp it firmly with the hemostats. Dispose of the scalpel blade in a sharps container.

Figure 1.10 **Removal of a Scalpel Blade from Handle Using Hemostats.** ©Christine Eckel

DISSECTING WITH A SCALPEL

1. Obtain a dissection specimen and place it on a dissecting tray.

2. Obtain a scalpel with a blade (see exercise 1.4) and some **tissue forceps** (see table 1.5). If the tissue is difficult to grasp, use **hemostats** instead of forceps. Hemostats allow you to "lock" on to the tissue so the tissue is not dropped when you release your grip on the handle.

3. Using the forceps or hemostats, pull the skin away from the muscle on the dissection specimen (**figure 1.11-1**). Carefully cut into the skin, using the tip of the scalpel blade (figure 1.11-2). Note how easily a new blade cuts into the tissue. When cutting with a scalpel, take care not to cut too deep, or too aggressively, or underlying tissues will be damaged. Once a small slit has been cut in the skin, observe the stringy tissue that lies between the skin and the muscle. This tissue is a loose connective tissue called fascia (figure 1.11-3), which is discussed further in chapter 5. Because the goal is to separate the skin from the muscle, it is necessary to loosen the "grip" of the fascia that holds the skin and muscle together. One way to do this is to cut into the fascia using the scalpel.

4. Next, *without* holding the skin away from the muscle with forceps, cut into the skin using a considerable amount of pressure. Note how easy it is to cut through the skin directly into the muscle. This is not desirable. To avoid damaging the underlying tissues, push a blunt probe or scalpel handle (*without* blade attached) into the space between the skin and muscle, thus protecting the underlying tissues. Then cut with the scalpel superficial to the probe (**figure 1.12**). This way the probe limits the depth at which the scalpel blade can cut, thus protecting the underlying tissues.

5. Once enough skin has been pulled back for it to be easily grasped with forceps or hemostats, put as much tension on the skin as possible, thus stretching out the fibers in the fascia (**figure 1.13**). Once the fascia is stretched, use the scalpel to cut the fascia and remove the skin from the specimen. When cutting with the scalpel, always point the sharp end of the blade toward the skin, not toward the underlying tissues, so as to protect those underlying tissues.

6. Practice using the forceps, hemostats, blunt probe, and scalpel to remove the skin from part of the specimen. Note areas where this is more difficult than others. When practicing, consider carefully whether the scalpel is the best instrument for the job, or if using it is causing damage to tissues.

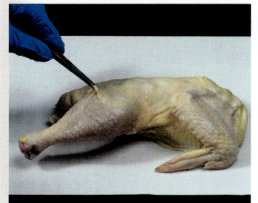

① Pull the skin away from the underlying tissues using tissue forceps.

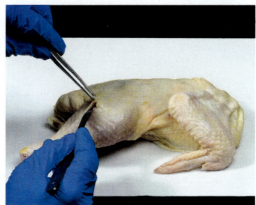

② Begin cutting the skin with the scalpel, taking care not to cut delicate tissues deep to the skin.

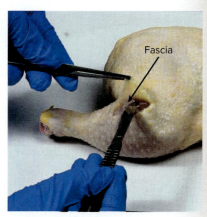

③ To assist with removal of the skin, use the scalpel to gently cut away the fascia that loosely holds the skin to the muscle. Maintain as much tension on the skin as possible and always keep the sharp end of the blade pointed toward the skin, not the underlying tissues.

Figure 1.11   Dissecting with a Scalpel. ©McGraw-Hill Education/Christine Eckel, photographer

(continued on next page)

(continued from previous page)

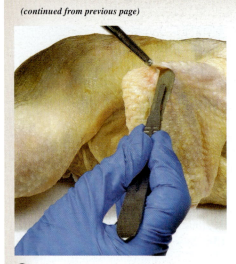

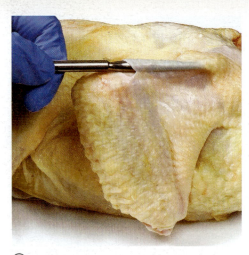

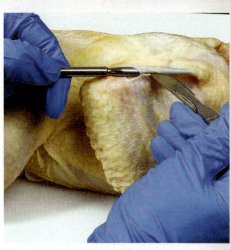

1 Pull the skin away from the underlying tissues using tissue forceps and cut a small slit in the skin with the scalpel, taking care not to cut delicate tissues deep to the skin.

2 Push the probe under the skin along the line where the cut will be made.

3 Cut the skin superficial to the probe with the scalpel. Notice how the blunt probe limits the depth at which the scalpel can cut, thus protecting underlying tissues.

Figure 1.12 **Protecting Underlying Tissues with a Probe When Dissecting with a Scalpel.** ©McGraw-Hill Education/Christine Eckel, photographer

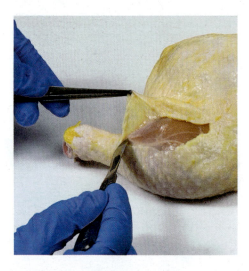

Figure 1.13 **Removing the Rest of the Skin with the Scalpel.** Use the forceps to pull the skin away from the underlying tissues and keep as much tension as possible on the fascia. Cut the fascia with the scalpel, always keeping the sharp part of the scalpel blade pointed toward the skin. This way, if the blade slips accidentally, it will cut the skin, not the underlying tissues. ©McGraw-Hill Education/Christine Eckel, photographer

EXERCISE 1.8

DISSECTING WITH SCISSORS

1. Using the same dissection specimen used in exercise 1.7, practice using scissors to cut tissues.

2. Obtain a pair of **pointed scissors** and **forceps** (table 1.5). Using the forceps, grasp part of the skin covering part of the specimen that has not already been dissected and pull it away from the muscle. Next, cut a small slit into the skin until the fascia beneath it is visible. Continue to lengthen the cut until it is about 2 inches long (**figure 1.14**).

3. **Open scissors technique:** There are a lot of tissues within the fascia that may need to be preserved, such as nerves and blood vessels. When using "sharp" techniques, these structures may accidentally get cut. For this reason, "blunt" dissection technique is preferred to preserve important structures. One blunt dissection technique is called an "open scissors" technique, so named because the dissecting action of the scissors is performed by starting with the scissors closed and then actively opening them. This is the opposite of how most scissors are used.

4. With the scissors closed, push the tip of the scissors into the space between the skin and the muscle so that it pierces the fascia **(figure 1.15-1)**. Once the tip of the scissors is within the fascia, open the scissors (figure 1.15-2). Notice how this action causes the fibers within the fascia to separate from each other and loosens the hold between the skin and the fascia. When using the open scissors technique, be sure to keep the scissors open within the specimen. Remove the scissors completely before closing the scissors to prevent any unwanted damage to surrounding tissues and structures.

5. Continue to loosen the fascia using the open scissors technique. While doing this, observe small structures such as blood vessels and nerves that may run in the space between the skin and the underlying tissues. Notice how the fibers in the fascia easily separate from each other without damaging the vessels and nerves when using the open scissors technique. At times, the hold of the fascia will be too tight, and the open scissors technique will no longer work effectively. At those times, switch to "normal" scissors technique and simply cut away the tough tissue.

6. Practice using both open and normal scissors techniques to continue to remove skin from underlying tissues.

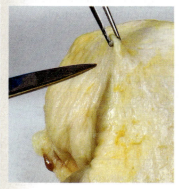 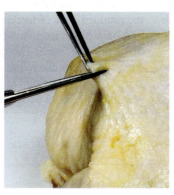 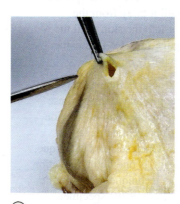 

(1) Use tissue forceps to pull the skin away from underlying tissues.

(2) Make a small cut in the skin with the scissors.

(3) A small hole has now been created in the skin. Insert the tip of the scissors into this hole and begin cutting directionally along the skin.

(4) Continue the cut along the skin. Continue to use the tissue forceps to pull the skin away from underlying tissues before making each cut to avoid damaging underlying structures.

Figure 1.14 **Dissecting with Scissors.** ©McGraw-Hill Education/Christine Eckel, photographer

 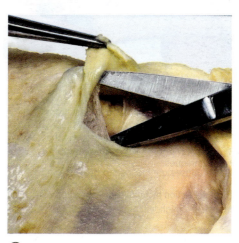

(1) Make a small cut in the skin. Using forceps, pull the skin away from the underlying fascia. Push the tip of the *closed* scissors into the fascia that lies deep to the skin.

(2) Open the scissors, thus separating the fibers of the fascia and loosening the skin from the underlying tissues.

Figure 1.15 **Open Scissors Technique.** ©McGraw-Hill Education/Christine Eckel, photographer

6. The following data set was obtained by measuring the distance traveled by a golf ball over time after it was hit with a putter. Graph the data using the grid provided. Be sure to include all units and labels on the graph. **6**

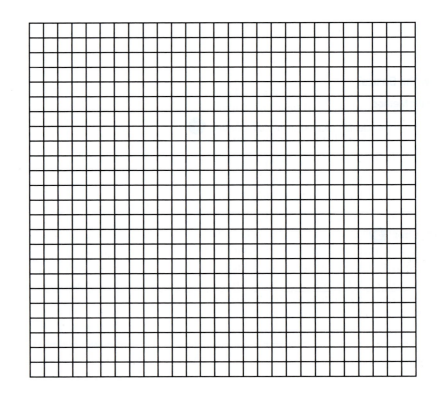

Time (min)	Distance (cm)
1	2
2	3.75
3	5.5
4	7
5	8.25
6	9.25
7	10
8	10.5
9	11
10	11.25

7. Observe the graph that was created in question 6. Describe the relationship between distance traveled and time in the space provided (e.g., is it linear/curved, does it increase/decrease?). **7**

_____

_____

_____

Exercise 1.3: Units of Measurement

8. Match each of the following prefixes listed in column A with the correct power of ten that it represents, listed in column B. **8**

Column A

_____ 1. centi-

_____ 2. deci-

_____ 3. kilo-

_____ 4. micro-

_____ 5. milli-

_____ 6. no prefix/base unit

Column B

a. 10^3

b. 10^0

c. 10^{-1}

d. 10^{-2}

e. 10^{-3}

f. 10^{-6}

9. List the metric unit used to measure each of the following: **9**

a. Length: _____

b. Mass: _____

c. Temperature: _____

d. Volume: _____

10. Make the appropriate English-to-metric conversions described: **10**

    a. You have a piece of dialysis tubing that is 8 inches long. How many centimeters long is the piece of dialysis tubing? _____cm

    Show your calculation in the space provided.

    b. You are weighing a patient, a 160-lb male. How many kilograms does the patient weigh? _____kg

    Show your calculation in the space provided.

    c. The boiling point of water is 100°C. What is the boiling point of water in Fahrenheit? _____°F

    Show your calculation in the space provided.

    d. The freezing point of water is 0°C. What is the freezing point of water in Fahrenheit? _____ °F

    Show your calculation in the space provided.

Exercise 1.4: Identification of Common Dissection Instruments

11. An instrument that resembles tweezers and is used to grasp objects is a _____. (Circle one.) **11**

    a. blunt probe

    b. forceps

    c. hemostat

    d. scalpel

    e. scissors

Exercise 1.5: Proper Disposal of Laboratory Waste

12. Which of the following items are described in the text as "hazardous waste"? (Check all that apply.) **12**

    _____ a. broken glass

    _____ b. cotton swab

    _____ c. fresh tissue

    _____ d. laboratory chemicals

    _____ e. paper towels

13. Which of the following is the correct waste receptacle for scalpel blades? (Circle one.) **12**

    a. hazardous waste bag

    b. sharps container

    c. wastebasket

14. Which of the following statement(s) is/are true of formalin and phenol? (Check all that apply.) **12**

    _____ a. Formalin and phenol are potentially hazardous chemicals.

    _____ b. It is not necessary to wear gloves and eye protection when working with formalin and phenol.

    _____ c. Seek medical attention if irritation persists following exposure to formalin and phenol.

    _____ d. Skin and eyes should be rinsed immediately if exposed to formalin and phenol.

Exercise 1.6: Placing a Scalpel Blade on a Scalpel Blade Handle

15. Describe the proper techniques for putting a scalpel blade on a scalpel handle and for removing it from the scalpel handle. **13** **14**

    _____

    _____

    _____

Exercise 1.7: Dissecting with a Scalpel

16. List an example of a case where a scalpel is the preferred tool for dissection. **15**

_____

_____

17. Describe a technique used to prevent damage to underlying tissues when dissecting with a scalpel. **16**

_____

_____

Exercise 1.8: Dissecting with Scissors

18. List an example of a situation where scissors are the preferred tool for dissection. **17**

_____

_____

19. Describe the *open scissors* technique, and list an example of a situation where the open scissors technique is the preferred method for dissection. **18**

_____

_____

_____

Exercise 1.9: Blunt Dissection Techniques

20. Define *blunt dissection,* and list an example of a situation where a blunt dissection technique is the preferred method for dissection. **19** **20** **21**

_____

_____

_____

Can You Apply What You've Learned?

21. A scientist designs an experiment to test the effect of body weight on the risk of developing type II diabetes mellitus.

    a. What is the independent variable in this experiment? _____

    b. What is the dependent variable in this experiment? _____

22. You have just finished dissecting a fresh cow bone as part of the day's laboratory activities. What is the most appropriate way to dispose of this waste?

_____

_____

Can You Synthesize What You've Learned?

23. You observe a classmate wearing open-toed shoes during a human cadaver dissection. The student accidentally spills some unknown fluid from "inside" the human cadaver directly onto his skin. Discuss possible harmful chemicals that could be in the unknown fluid. What steps would you recommend that the student take to address the chemical exposure?

_____

_____

_____

24. You are dissecting the wing of a chicken, and the skin is held tight to the bones beneath it. You would like to remove the skin, but in doing so you would like to preserve the bone and muscle beneath the skin. What tools might you use, and how will you use them, to complete this task? (There is more than one correct answer to this question.)

_____

_____

_____

chapter 2

Orientation to the Human Body

OUTLINE AND LEARNING OBJECTIVES

Anatomy & Physiology Revealed® 4.0

Module 1: Body Orientation

INTRODUCTION

The human body is both beautiful and complex. A course in anatomy and physiology allows students to develop a deeper understanding of that beauty and complexity. To be successful in this venture, a great deal of work is required. Success will be achieved with time and persistence. The study of anatomy and physiology requires dedicated time and commitment.

To put things in perspective, consider this: A beginning student in human anatomy and physiology is typically asked to learn more new words in a one-semester anatomy and physiology course than a beginning student learns in the first semester of a foreign language class. In fact, it *is* a new language. It is the common language used by all medical professionals that allows for effective communication while caring for patients. This language has its origins principally in Latin and Greek. To successfully learn this language, it is important to establish a firm understanding of the meanings of common word origins. Lists of common word origins are located in the "Word Origin" columns of tables and in the context of exercises in this manual. Students are encouraged to look up the origins of all words that are new or unfamiliar. The rich knowledge base that is developed early on allows students to interpret the meanings of new words encountered later in the course and to develop an impressive vocabulary of anatomical/medical terms.

Practice this by analyzing the origins of the words *anatomy* and *physiology*. The word *anatomy* can be broken down into two parts, *ana-* and *-tomé*. The word part *ana-* means "apart." The word part *-tomé* means "to cut."

Thus, the word **anatomy** literally means "to cut apart." Though students might not literally be cutting up human bodies as medical students do and early anatomists did, they will at the very least be *conceptually* "cutting up" the body to understand its component parts. The word **physiology** consists of two parts, *physis-,* nature and *-logos,* study. Physiology means to study the nature of how the body functions.

The laboratory exercises in this chapter reinforce the use of anatomically correct directional, regional, and sectional terms to describe the body and its parts. Major body organs will be located on a human torso model. If the laboratory uses human cadavers, the organs may also be located on the cadaver. Additional exercises are designed to familiarize students with the organ systems of the body and to allow them to practice using anatomic terminology to describe locations of the organs that compose these organ systems.

List of Reference Tables

These Pre-Laboratory Worksheet questions may be assigned by instructors through their ■ connect° course.

1. Describe the anatomic position in your own words. _____

   _____

   _____

   _____

2. The plane that separates the body into anterior and posterior portions is the _____ (coronal/sagittal) plane.

3. Match the definition listed in column A with the appropriate directional term listed in column B.

   Column A

   _____ 1. closer to the attachment point; near the beginning

   _____ 2. in front of; toward the front surface

   _____ 3. away from the midline of the body

   _____ 4. in back of; toward the back surface

   _____ 5. toward the midline of the body

   _____ 6. farther from the attachment point; closer to the end

   Column B

   a. anterior (ventral)

   b. distal

   c. lateral

   d. medial

   e. posterior (dorsal)

   f. proximal

4. Match the regional name listed in column A with the description listed in column B.

   Column A

   _____ 1. brachial

   _____ 2. cephalic

   _____ 3. cervical

   _____ 4. femoral

   _____ 5. lumbar

   Column B

   a. arm

   b. head

   c. lower back; loin

   d. neck

   e. thigh

5. What separates the thoracic cavity from the abdominopelvic cavity? (Circle one.)

   a. bone

   b. mediastinum

   c. thoracic diaphragm

   d. pericardium

6. The mediastinum is located within the _____ (abdominal/thoracic) cavity of the body.

7. The abdominopelvic cavity can be divided into a total of _____ quadrants or _____ regions. The central point of reference

   for dividing the abdominopelvic cavity into quadrants or regions is the _____.

8. The liver is located in which abdominopelvic quadrant? (Circle one.)

   a. right upper quadrant

   b. left upper quadrant

   c. right lower quadrant

   d. left lower quadrant

9. Identify which of the following are abdominopelvic regions. (Check all that apply.)

   _____ a. epigastric region

   _____ b. hypergastric region

   _____ c. left cervical region

   _____ d. right hypochondriac region

   _____ e. umbilical region

(continued from previous page)

EXERCISE 2.1B   Sectioning a Specimen

Obtain the Following:

- dissecting tray
- knife and/or scalpel
- specimen (e.g., sheep brain, sheep heart, sheep kidney)

1. **Figure 2.3** demonstrates a sheep heart that has been sectioned along coronal, sagittal, and transverse planes. Using figure 2.3 as a guide, section the specimen along the following planes:

   ☐ **Coronal**   ☐ **Sagittal**   ☐ **Transverse**

2. Sketch the appearance of each of the resulting portions of the specimen in the spaces provided.

Coronal section	Sagittal section	Transverse section

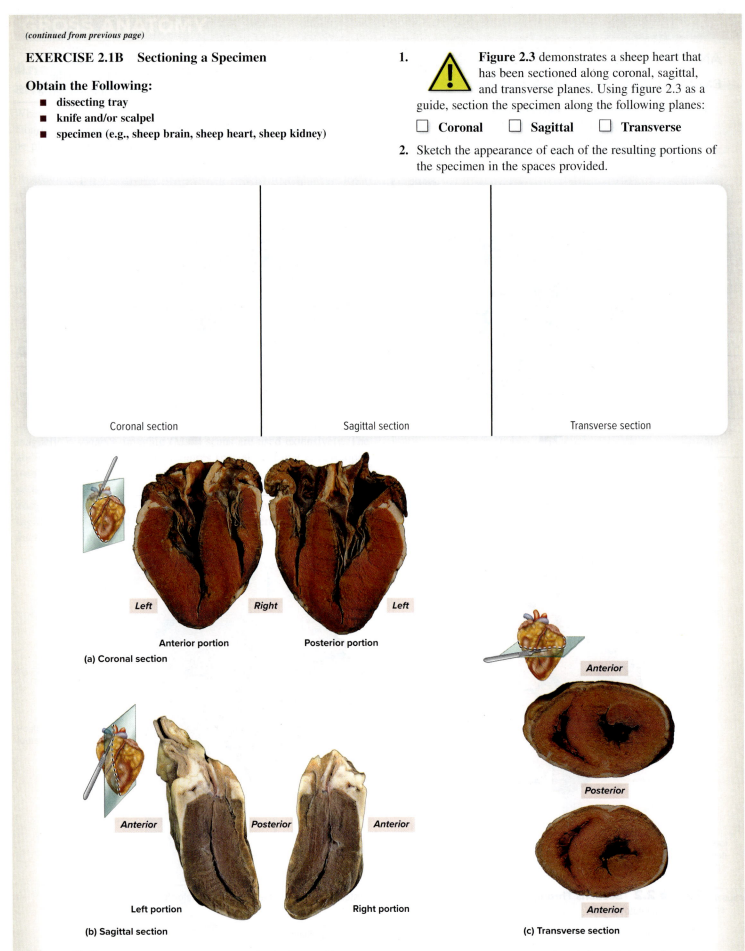

(a) Coronal section

Left | Right | Left
Anterior portion | Posterior portion

(b) Sagittal section

Anterior | Posterior | Anterior
Left portion | Right portion

(c) Transverse section

Anterior
Posterior
Anterior

Figure 2.3   **Sections Through a Sheep Heart.**   This figure demonstrates a sheep heart that has been sectioned along (*a*) coronal, (*b*) sagittal, and (*c*) transverse planes. (*a*) and (*c*) ©Christine Eckel (*b*) ©McGraw-Hill Education/Photo and Dissection by Christine Eckel

Regional Terms

There are common, everyday terms that are often used to describe regions of the body such as *arm* or *back*. The correct anatomic terms to describe regions of the body are basically synonyms for these terms and are closer to the Latin or Greek derivatives of the terms. Refer to the textbook for regional anatomy terms paired with words commonly used to describe the same region in everyday language. Use the textbook as a reference when completing exercise 2.2.

EXERCISE 2.2

REGIONAL TERMS

1. Identify the body regions listed in **figure 2.4** on your own body.

2. Label figure 2.4 with the appropriate regional terms, using a textbook as a guide.

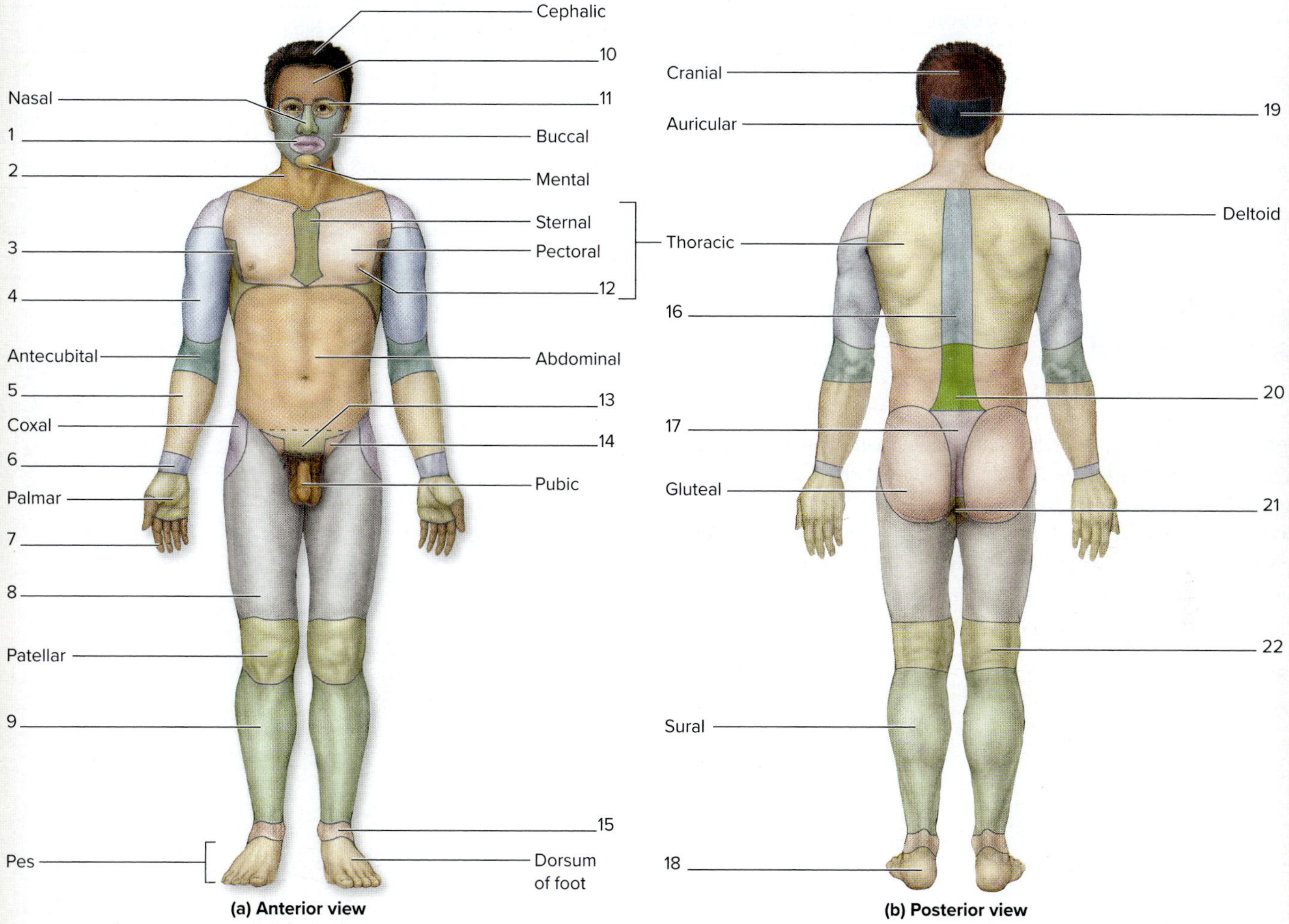

(a) Anterior view

Cephalic
10
11
Nasal
1
Buccal
2
Mental
Sternal
3
Pectoral
4
12
Antecubital
Abdominal
5
13
Coxal
14
6
Palmar
Pubic
7
8
Patellar
9
15
Pes
Dorsum of foot

(b) Posterior view

Cranial
Auricular
19
Deltoid
Thoracic
16
17
20
Gluteal
21
Sural
18
22

Figure 2.4 **Regional Terms.** Use the terms listed below to fill in the numbered labels in the figure.

☐ Antebrachial	☐ Carpal	☐ Femoral	☐ Mammary	☐ Pelvic	☐ Tarsal
☐ Axillary	☐ Cervical	☐ Frontal	☐ Occipital	☐ Perineal	☐ Vertebral
☐ Brachial	☐ Crural	☐ Inguinal	☐ Oral	☐ Popliteal	
☐ Calcaneal	☐ Digital	☐ Lumbar	☐ Orbital	☐ Sacral	

Directional Terms

Everyday terms like "front" and "back," and "on top of" or "on the bottom of," are often used to give directions. While this is perfectly appropriate in everyday language, it can cause confusion when referring to directions in the human body. For instance, when saying "on top of," we need to know to what part of the body "on top of" refers. Different people might use the term in different ways or in reference to different structures. Furthermore, an individual may be thinking that the direction "on top of" can change, and is relative to different body positions. This becomes problematic in a medical setting because it leads to confusion, which has the potential to create severe consequences for a patient. Thus, an agreed-upon set of directional terms is used in anatomy and medicine to be as specific as possible when describing directions, but also to ensure that everyone is speaking the same language. **Table 2.2** lists these directional terms and gives definitions of each of them. Exercise 2.3 involves practicing the use of these directional terms. The directional terms *superior* and *inferior* are used when describing one structure with respect to another structure in the trunk of the body. The directional terms *proximal* and *distal* are used when describing the position of one structure with respect to another structure on the limbs. Thus, it is more appropriate to say the elbow is located *proximal* to the wrist, rather than to say it is superior to the wrist.

Table 2.2	Directional Terms
Directional Term	**Definition**
Anterior (ventral)	Toward the front of the body (the belly side)
Posterior (dorsal)	Toward the back of the body (the back side)
Superior	Above; closer to the head
Inferior	Below; closer to the feet
Cranial (cephalic)	At the head end of the body
Caudal	At the tail end of the body
Medial	Toward the midline of the body
Lateral	Away from the midline of the body
Superficial	Toward the surface of the body; on the outside
Deep	Beneath the surface of the body; on the inside
Proximal	Near; closer to the point of attachment to the trunk
Distal	Far; farther from the point of attachment to the trunk
Contralateral	On the opposite side
Ipsilateral	On the same side

EXERCISE 2.3

DIRECTIONAL TERMS

Figure 2.5 shows a posterior view of a human. Three locations on the body are marked with the circled numbers 1–3. Describe the location of the markings 1–3 as specifically as possible using correct anatomic regional and directional terminology. When finished, compare your answers to those of other students in the class to see how similar the answers are. Note that there is more than one correct answer for each of these.

Location 1

_____

_____

_____

Location 2

_____

_____

_____

Location 3

_____

_____

_____

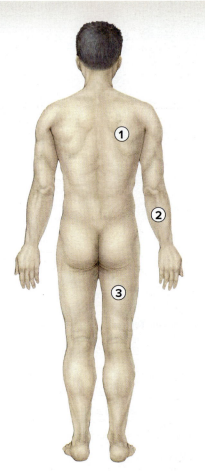

Figure 2.5 **Posterior View of an Individual with Three Reference Locations (1–3) Marked.**

Chapter 2: (

The **1** corresponds to the Lear

▲ **Do You Know the**

Exercise 2.1: Anatomic Planes

1. Which of the following corr

_____ a. one can be si

_____ b. the feet are d

_____ c. no two bones

_____ d. the palms of t

2. A _____ (pla
dimensional flat surfaces.

3. To divide this image of a Va

4. Match each definition liste

   Column A

   _____ 1. separates the

   _____ 2. separates ant

   _____ 3. separates sup

   _____ 4. runs at an an

5. Which of the following sha
transverse, and coronal? ((

   a. pyramid

   b. rectangular box

   c. egg

   d. square box

Exercise 2.2: Regional Terms

6. Which of the following stat
(Circle one.) **5**

   a. The *brachial* artery is i

   b. The *carpal* bones are i

   c. The *cervical* vertebrae

   d. The *femoral* nerve is in

   e. The *popliteal* artery is i

Exercise 2.3: Directional Term

7. For each of the following, i

   a. The elbow is located __

   b. The mouth is located __

   c. The lungs are located _

   d. The umbilicus is located

   e. The nose is located ___

   f. A scratch wound, which
the skin, is referred to a

Body Cavities and Membranes

Many organs within the body are compartmentalized and separated from each other by a *body cavity.* Compartmentalizing the organs this way allows the separate organs to perform their functions without interfering with the functioning of other organs. For example, the pumping action of the heart does not interfere with the expansion and recoil of the lungs because each organ is enclosed in its own cavity. In addition, the encasement of organs within separate cavities helps to prevent the spread of infection from one cavity to another.

Serous (*serosus,* watery fluid) **membranes** cover organs within the body cavities (**figure 2.6**). These membranes are composed of two layers: visceral and parietal. The **visceral** (*viscus,* internal organ) layer lies adjacent to the organ, whereas the **parietal** (*paries,* wall) layer attaches to the wall of the body cavity. The cells that compose the serous membrane secrete a slippery substance called **serous fluid** into the space between the two layers of membrane to act as a lubricant. When organs such as the heart and lungs expand and contract, serous fluid reduces friction between the organs and surrounding structures. Note that a specific serous membrane that surrounds an organ generally has a unique name that relates to the organ it encases (e.g., the pericardium surrounds the heart).

INTEGRATE

CLINICAL VIEW
Inflammation of Serous Membranes

Serous membranes are membranes that line body cavities, such as the pericardial cavity (which encases the heart). Inflammation of serous membranes can be a serious health risk. Clinical terms related to inflammation of serous membranes reference the specific serous membrane affected. For example, *pericarditis* (*peri-*, around + *cardio*, heart + *itis*, inflammation) is inflammation of the pericardium surrounding the heart; *pleurisy* is inflammation of the pleurae surrounding the lungs; *peritonitis* is inflammation of the peritoneum surrounding the abdominopelvic cavity. One consequence of inflammation of a serous membrane is swelling, which can impede the function of surrounding organs within the body cavity. For example, pericarditis may interfere with the heart's ability to pump blood. Pleurisy may prevent adequate gas exchange in the lungs.

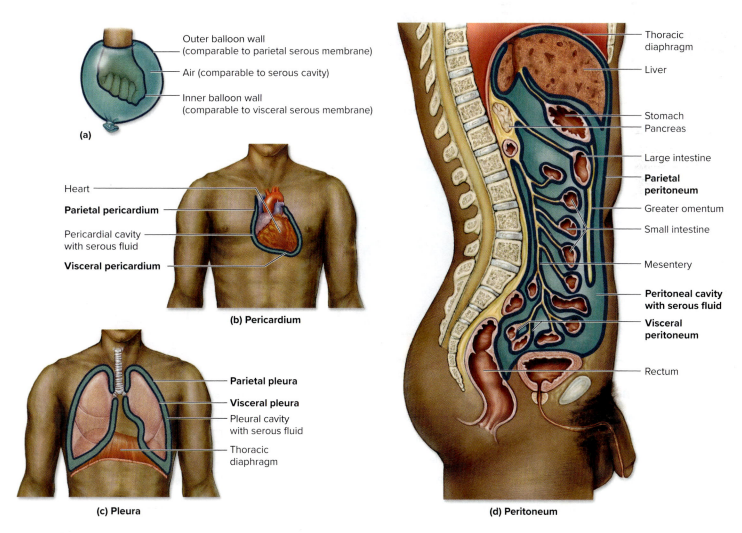

Figure 2.6 Serous Membranes. Serous membranes are found within the thoracic and abdominal cavities. (*a*) The membrane is continuous, with the visceral layer in contact with the organ and the parietal layer lining the cavity. This layout is similar to a fist pressing into a balloon. (*b*) The parietal and visceral pericardium surrounds the heart. (*c*) The parietal and visceral pleura surrounds the lungs. (*d*) The parietal and visceral peritoneum surrounds the abdominal viscera.

(continued from previous pa

Organ
Left kidney
Liver
Pancreas
Small intestine
Spleen
Stomach
Urinary bladder

The Microscope

OUTLINE AND LEARNING OBJECTIVES

Anatomy & Physiology Revealed® 4.0

Module 1: Body Orientation

INTRODUCTION

The study of anatomy involves observation of both gross (large) and microscopic (small) structures. A **microscope** is used to view the detailed structure of body tissues and cells. Both light microscopes and electron microscopes allow detailed observation of the ultrastructure of tissues. Specifically, microscopes allow the user to view specific cellular components of the various tissues that make up the body. Such cellular components include the shapes of the cells, intracellular components such as the nucleus, and modifications of the plasma membrane (e.g., cilia or microvilli). Observing specific cellular features provides a visual image that assists with the process of integrating structure and function.

To get the most from the experience of observing microscopic structures, it is important to know how to use a microscope properly. Although you may have used a microscope before, pay very careful attention to the instructions in this chapter. Poor technique when using microscopes causes frustration and is an impediment to learning. The exercises in this chapter provide an opportunity to refine and improve upon your technique.

This chapter's exercises cover both how to care for and how to use a compound microscope. A **compound microscope** is used to view detailed cell and tissue structures that are not visible with the naked eye. The proper use and care of a dissecting microscope (also called a stereomicroscope) is also covered in this chapter. A **dissecting microscope** is used to view larger specimens such as hair and muscle fibers. Performing the exercises in this chapter will help prepare you to successfully navigate the process of making observations of histology (tissue-level anatomy) in later chapters.

List of Reference Tables

INTEGRATE

CONCEPT CONNECTION

In each chapter of this laboratory manual, there are photographs of slides demonstrating cells, tissues, and organs. These slides will be used to determine both structure and function of the viewed specimens. Keep in mind that the body is a three-dimensional structure, even though what you view on the slides or in the images is two-dimensional. When viewing these images, look for consistent structural patterns among tissue types, and relate these patterns to the function of the cells, tissues, or organs. For example, the tissue in the small intestine and kidney might have similar characteristics because both organs play a role in absorption. When viewing histological slides, make every attempt to make connections among similar structures, their associated functions, and the body systems in which the structures are located.

These Pre-Laboratory Worksheet questions may be assigned by instructors through their ■ connect course.

1. Which of the following are proper procedures for holding, transporting, and storing a microscope? (Check all that apply.)

_____ a. Always use both hands when carrying the microscope.

_____ b. Place one hand on the base of the microscope and the other on the arm of the microscope.

_____ c. Use care and move deliberately when carrying the microscope.

_____ d. You may leave slides on the stage when storing a microscope.

_____ e. Move the stage to its highest position when storing a microscope.

_____ f. Wrap the power cord around the base and replace the dust cover after using the microscope.

2. Broken slides should be placed in the _____ (sharps container/trash).

3. Match the definition listed in column A with the microscope part listed in column B.

Column A

_____ 1. everything that is visible when looking through the eyepiece

_____ 2. the part of the microscope that connects the head to the base

_____ 3. the platform that a slide is placed upon

_____ 4. lens that is attached to the nosepiece

_____ 5. a knob that moves the mechanical stage up and down in small increments

_____ 6. a knob that moves the mechanical stage up and down in large increments

Column B

a. arm

b. coarse focus adjustment knob

c. field of view

d. fine focus adjustment knob

e. mechanical stage

f. objective lens

4. When viewing structures using the compound microscope, which of the following statements describes an appropriate action to take when light is visible in the field of view, but there is no clear specimen in the field of view? (Circle one.)

a. Check the power switch and make sure it is turned on.

b. Adjust the condenser.

c. Set the microscope to the scanning power objective and lower the stage; adjust the coarse focus adjustment knob until the specimen comes into view.

d. Make sure the microscope is plugged into a working power outlet.

e. Move the eyepieces either closer together or farther apart until a clear image is visible.

5. Which of the following correctly describes how to properly calculate the total magnification of a microscope? (Circle one.)

a. Add the ocular lens magnification and objective lens magnification.

b. Subtract the ocular lens magnification from the objective lens magnification.

c. Multiply the ocular lens magnification by the objective lens magnification.

d. Divide the objective lens magnification by the ocular lens magnification.

6. As total magnification power increases, depth of field _____ (increases/decreases).

7. As total magnification power increases, diameter of the field of view _____ (increases/decreases).

8. A dissecting microscope is used to view _____ (microscopic/macroscopic) structures.

9. Which of the following statements is an accurate comparison between a dissecting microscope and a compound microscope? (Check all that apply.)

_____ a. A dissecting microscope can view smaller structures than a compound microscope.

_____ b. An image viewed with a dissecting microscope is not inverted, as it is with a compound microscope.

_____ c. Tissues must be prepared prior to viewing with a compound microscope, but no such preparation is needed for a dissecting microscope.

_____ d. Cells and tissues can be viewed with either a dissecting microscope or a compound microscope.

EXERCISE 3.6

PARTS OF A DISSECTING MICROSCOPE

1. Obtain a dissecting microscope.

2. Identify the following structures on the dissecting microscope. Use figure 3.7 and table 3.1 as guides.

☐ Arm
☐ Base
☐ Diopter adjustment ring
☐ Eyepieces
☐ Focusing knob
☐ Head
☐ Head locking screw
☐ Illuminator lens adjustment

☐ Mirror
☐ Mirror tilt adjustment
☐ Nosepiece with objective lens
☐ Objective lens control knob
☐ Stage plate

3. Record the magnifications of the lenses on the microscope in the spaces below.

Magnification of ocular lenses: _____

Magnification of objective lenses:

Lowest _____   Highest (if applicable) _____

4. What is the total magnification of the microscope? If there is more than one objective lens, list the lowest and highest total magnifications possible.

Total magnification:
Lowest _____   Highest (if applicable) _____

5. Place a pen or pencil on the microscope stage and adjust the controls (focus, zoom, illuminator, etc.) while observing the object through the ocular lenses. Place a piece of paper with text written on it on the microscope stage and view that through the ocular lenses. Is there any change in the orientation of the letters as seen through the microscope as compared to viewing the letters without the microscope? _____
If so, describe what differences are observed. _____
_____
_____
_____
_____
_____
_____
_____

INTEGRATE

CLINICAL VIEW

Nail Fungus

Knowing how to view structures using a microscope is an essential skill for physicians to possess. Many diseases are accurately diagnosed only after viewing structures using a microscope. For example, a dermatologist may use a microscope to determine if a patient is suffering from nail fungus. The dermatologist first scrapes under the patient's nail, then places the debris on a slide. After adding a drop or two of an isotonic wetting solution, the dermatologist looks for the presence of fungi under the microscope before the patient even leaves the examination room. All types of these microorganisms (i.e., bacteria and fungi) require a microscope for viewing because none of them are visible with the naked eye. Proper and rapid diagnosis of the microorganism causing the infection allows the dermatologist to determine the best course of treatment for the patient. This saves the patient and the physician both expense and time as compared to the alternative of having to send the tissue sample out to an external laboratory for testing.

Chapter 3: The Microscope

The ❶ corresponds to the Learning Objective(s) listed in the chapter opener outline.

Do You Know the Basics?

Exercise 3.1: Parts of a Compound Microscope

1. Label the parts of the compound microscope. ❶

©McGraw-Hill Education/Christine Eckel, photographer

2. Describe how to perform each of the following tasks: ❶

a. Transport the microscope. _____

b. Position the microscope on a laboratory workstation. _____

c. Clean the microscope lenses. _____

d. Prepare the microscope for storage. _____

Exercise 3.2: Viewing a Slide of the Letter *e*

3. What is the total magnification of a microscope set up with an ocular lens magnification of 10× and an objective lens magnification of 43×? _____ ❷

Exercise 3.3: Measuring the Diameter of the Field of View

4. Complete the following table with the numbers observed or calculated in exercises 3.1, 3.2, and 3.3. ❸ ❹ ❺

Power	Ocular Magnification	Objective Magnification	Total Magnification	Working Distance	Diameter of the Field of View
Scanning					
Low					
High					

Exercise 3.4: Estimating the Size of a Specimen

5. Refer back to the calculations on the sample "specimens" in figure 3.5. Enter information from those calculations in the spaces provided. Pay attention to the units specified next to the answers, because they are not all the same. ❻

A.

Total magnification = __40×__

Diameter of field = __4.8__ mm

Length of object = _____ mm

B.

Total magnification = __200×__

Diameter of field = _____ mm

Length of object = _____ mm

C.

Total magnification = __500×__

Diameter of field = _____ μm

Length of object = _____ μm

Exercise 3.5: Determining Depth of Field

6. Referring back to exercise 3.5, record the answers to the questions about colored threads in the spaces provided. **7**

   Color of top thread: _____   Color of middle thread: _____   Color of bottom thread: _____

7. Explain why proper microscope technique requires always viewing a slide with the scanning objective first before moving to higher-power objectives. Use the concept of *depth of field* in the explanation. **7**

   _____

   _____

   _____

   _____

   _____

Exercise 3.6: Parts of a Dissecting Microscope

8. Identify two differences between a dissecting microscope and a compound microscope. **8**

   a. _____

   b. _____

9. List three types of specimens a dissecting microscope might be used to view: **9**

   a. _____

   b. _____

   c. _____

Can You Apply What You've Learned?

10. What microscope structures are used to control the amount of light illuminating the specimen?

    _____

11. What happened to the light intensity when switching from low to high power?

    _____

12. What adjustment will typically have to be made to the light after changing from the low-power to the high-power objective?

    _____

13. a. How does working distance change as total magnification increases?

    _____

    b. What are the practical consequences of this change in working distance?

    _____

14. If four cells are visible within the field of view at the field's maximum diameter, and the total magnification is 200×, how many cells will be visible at a total magnification of 500×?

    _____

Can You Synthesize What You've Learned?

15. Describe why images viewed with a compound microscope are two-dimensional, whereas images viewed with a dissecting microscope are three-dimensional.

    _____

    _____

16. A patient presented to his physician complaining of an unusual growth on the skin of his upper back. The physician was unable to identify the growth, so she decided to perform a biopsy (take a tissue sample). After obtaining a sample of the unusual growth on the patient's back, the physician sent the sample to the pathology lab. Discuss how the pathologist would make use of a compound microscope to correctly diagnose the identity of the unusual growth.

    _____

    _____

Design Elements: Integrate: Clinical View icon (clipboard): ©Laia Design Studio/Shutterstock.com; Integrate Learning Strategy (pencil): ©Slavoljub Pantelic/Shutterstock.com

Cell Structure and Membrane Transport

chapter

4

OUTLINE AND LEARNING OBJECTIVES

Anatomy & Physiology Revealed® 4.0

Module 2: Cells & Chemistry

INTRODUCTION

The cell is the basic unit of life. Organisms can be unicellular or multicellular, but they must be composed of cells to be considered living entities. Cells are compartmentalized due to a plasma membrane that separates the intracellular and extracellular environments. The structural components of the cell determine its function, and cells must be functioning properly to sustain life for the organism. Human beings are, of course, multicellular organisms. Groups of cells with a common function and their associated extracellular materials come together to function as units known as **tissues**. **Histology** [*histos*, web (tissue) + *logos*, study] is the study of tissues that first requires identifying cells and cellular organelles under the microscope. Most cells are easily seen with a light microscope, but most cellular organelles are too small to be seen without the use of a more powerful electron microscope.

Most animal cells are transparent and would not provide valuable information upon viewing without special tissue preparation. When tissue samples are prepared for use in the anatomy and physiology laboratory, the slides are stained so cellular details will be visible when viewed under a microscope. Different parts of a cell attract biological stains to different degrees, which makes some parts of the cell appear darker in color, and others appear lighter in color, or even transparent. The nucleus of the cell has a high attraction for most biological stains, so it is often the most recognizable part of a cell.

Most of the slides that are viewed in the anatomy and physiology laboratory contain cells that have been stained with hematoxylin and eosin, and they are labeled "H and E." This stain makes the **cytoplasm** of the cell appear pink and makes visible the outline of the cell where the **plasma membrane** (the boundary of the cell) is located. The **nucleus** of the cell appears dark purple, and the **nucleolus** often appears as a dark spot within the nucleus. Note that these structures have been described using references to colors that result from the use of hematoxylin and eosin stains (i.e., pink and purple). It is important to remember that other types of stains may be used on slides prepared for the laboratory. Use of stains other than H and E will cause the same structures to have different colors. For this reason, do not use color alone as an identifying feature when viewing slides. Instead, learn to recognize cells and cellular organelles based on *shape*.

Once cells and organelles have been identified, it can be helpful to relate specific structures to their respective functions. Think about the size, shape, and definable features of cells and organelles. Make associations with what action that cell and/or organelle performs, and remember that a cell is an independent, functioning entity that contains machinery needed to perform functions. For example, some muscle cells contain a high density of mitochondria that produce the adenosine triphosphate (ATP) needed to sustain the generation of muscular force.

The exercises in the Gross Anatomy and Histology sections of this chapter are designed to aid in the identification of cellular organelles that are visible using a light microscope and to aid in the identification of the stages of mitosis in a whitefish embryo. The exercises in the Physiology section of this chapter explore the selective permeability of the plasma membrane, which allows the cell to control both its internal and external environments through the physiological processes of membrane transport. Wet lab activities in this chapter demonstrate the processes of diffusion, osmosis, and filtration. These exercises explore the effect of factors such as temperature on the rate of diffusion and the process of osmosis across an artificial membrane. A simulation activity (Ph.I.L.S.) covers observation of the effect of placing erythrocytes (red blood cells) in solutions that mimic changes in extracellular concentration. Although filtration is not a cellular process, the process of filtration will be observed. An additional activity covers the effect of changing fluid (hydrostatic) pressure on filtration rate.

List of Reference Tables

These Pre-Laboratory Worksheet questions may be assigned by instructors through their connect **course.**

1. Match the function listed in column A with the cell structure listed in column B.

   Column A

   ____ 1. provides a selectively permeable barrier between the intracellular and extracellular environment of the celll

   ____ 2. contains the cell's genetic material (DNA)

   ____ 3. synthesizes new proteins destined for the plasma membrane, for lysosomes, or for secretion from the cell

   ____ 4. site where proteins are modified, packaged, and sorted for delivery to other organelles or to the plasma membrane of the cell

   ____ 5. site of cellular respiration

   Column B

   a. Golgi apparatus

   b. mitochondria

   c. nucleus

   d. plasma membrane

   e. rough endoplasmic reticulum (RER)

2. Number the following stages of mitosis in the correct order.

   _____ a. anaphase

   _____ b. metaphase

   _____ c. prophase

   _____ d. telophase

3. During interphase, individual chromosomes _____ (are/are not) visible within the nucleus of the cell.

4. Which of the following factors increase the rate of diffusion of a substance? (Check all that apply.)

   _____ a. decreased temperature

   _____ b. decreased viscosity of the solvent

   _____ c. increased molecular weight of the substance

   _____ d. increased permeability of the membrane

5. When a red blood cell is placed in a hypertonic solution, which of the following may occur? (Check all that apply.)

   _____ a. crenation

   _____ b. intracellular volume will decrease

   _____ c. intracellular volume will increase

   _____ d. lysis

6. Match the definition listed in column A with the term listed in column B.

   Column A

   _____ 1. the dissolved substance in a solution

   _____ 2. a solution into which another substance dissolves

   _____ 3. a solution that has a lower osmotic pressure than intracellular fluid

   _____ 4. a solution that has the same osmotic pressure as intracellular fluid

   _____ 5. the movement of water across a semipermeable membrane

   Column B

   a. hypotonic

   b. isotonic

   c. osmosis

   d. solute

   e. solvent

7. A _____ (concentration/pressure) gradient drives the movement of solutes in diffusion, whereas a _____ (concentration/pressure) gradient drives the movement of fluid in filtration.

GROSS ANATOMY

Models of a Generalized Animal Cell

Exercises in this section involve observing classroom models demonstrating a generalized animal cell and observing classroom models demonstrating the stages of mitosis. **Figure 4.1** is a photograph of a classroom model of a generalized animal cell, which has the organelles labeled for reference.

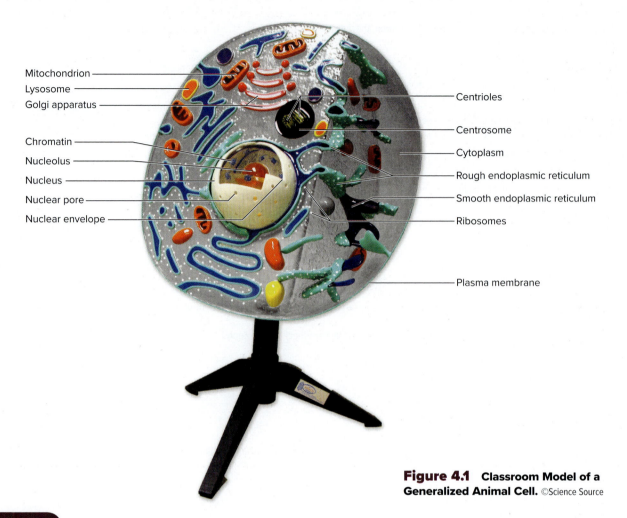

Mitochondrion
Lysosome
Golgi apparatus
Chromatin
Nucleolus
Nucleus
Nuclear pore
Nuclear envelope

Centrioles
Centrosome
Cytoplasm
Rough endoplasmic reticulum
Smooth endoplasmic reticulum
Ribosomes
Plasma membrane

Figure 4.1 **Classroom Model of a Generalized Animal Cell.** ©Science Source

EXERCISE 4.1

OBSERVING CLASSROOM MODELS OF CELLULAR ANATOMY

1. Obtain a classroom model demonstrating a generalized animal cell.

2. Identify the listed structures on the classroom model of a generalized animal cell. Use figure 4.1, table 4.1, and the textbook as guides.

☐ **Centrioles**
☐ **Chromatin**
☐ **Cytoplasm**
☐ **Golgi apparatus**
☐ **Lysosome**
☐ **Mitochondria**
☐ **Nuclear envelope**
☐ **Nucleolus**

☐ **Nucleus**
☐ **Peroxisome**
☐ **Plasma membrane**
☐ **Nuclear pore**
☐ **Ribosomes**
☐ **Rough ER**
☐ **Smooth ER**

3. Sketch the appearance of a generalized cell, including all visible organelles, in the space provided or on a separate sheet of paper.

HISTOLOGY

Structure and Function of a Generalized Animal Cell

Table 4.1 lists the parts of an animal cell and gives descriptions of the functions and microscopic features of each. The parts of an animal cell that are most readily visible under a light microscope are the nucleus, nucleolus, and plasma membrane (which is the boundary of the cell). While observing animal cells under the light microscope, focus on finding these parts of a typical animal cell. Learning to recognize what parts of an animal cell are typically visible under the light microscope serves as preparation for observing different cell types that will be presented in future laboratory exercises.

Table 4.1	Parts of a Generalized Animal Cell			
Organelle/ Structure	**Function**	**Microscopic Features**	**Word Origin**	**Appearance**
Cytoplasm	Includes cytosol, membrane-bound organelles, and non-membrane-bound organelles	Clear and homogeneous in appearance; may contain granular substances (e.g., glycogen)	*kytos*, a hollow (cell), + *plasma*, something formed	
Membrane-Bound Organelles				
Endoplasmic Reticulum (ER)	Site of lipid synthesis and detoxification of drugs and alcohol (smooth ER). Site of protein synthesis (rough ER).	Visible under the light microscope. In neurons, the rough ER stains very dark and is called chromatophilic substance (Nissl bodies).	*endon*, within, + *plasma*, something formed, + *rete*, a net	
Golgi Apparatus	Receives proteins from the rough ER and then modifies, packages, and sorts them for delivery to other organelles or the plasma membrane of the cell	Not generally visible under a light microscope	*Golgi*, Camillo, Italian histologist and Nobel laureate, 1843–1926	
Lysosomes	Membrane-enclosed sacs that contain digestive enzymes and extracellular components; function in the breakdown of intracellular debris	Not generally visible under a light microscope	*lysis*, a loosening, + *soma*, body	
Mitochondria	Site of cellular respiration; the metabolic pathway that utilizes oxygen in the breakdown of food molecules to produce ATP	Visible under a microscope	*mitos*, thread, + *chondros*, granule	

(continued on next page)

Table 4.1	Parts of a Generalized Animal Cell *(continued)*			
Organelle/Structure	Function	Microscopic Features	Word Origin	Appearance
Peroxisomes	Membrane-enclosed sacs that contain catalase and other oxidative enzymes that break down lipids and toxic substances	Not generally visible under a light microscope	*peroxi*, relating to hydrogen peroxide, + *soma*, body	
Ribosomes	Sites of protein synthesis; may be bound to the ER (bound) or within the cytoplasm (free)	Not generally visible under a light microscope	*ribose*, the sugar in RNA, + *soma*, body	Free ribosomes / Fixed ribosomes
Non-Membrane-Bound Organelles				
Centrosome	Contains paired organelles that are used to organize the spindle microtubules that attach to chromosomes during mitosis	Visible when a cell is undergoing mitosis	*kentron*, center	Centrosome / Centriole
Cytoskeleton	Provides structural support for the cell; composed of microtubules, intermediate filaments, and microfilaments	Not generally visible under a light microscope	*kytos*, a hollow (cell), + *plasma*, something formed	Cytoskeleton / Intermediate filament / Microfilament / Microtubule
Nucleus	Contains the cell's genetic material (DNA)	Typically stains very dark	*nucleus*, a little nut	Nucleus / Nuclear pores / Nucleolus / Nuclear envelope / Chromatin
Nucleolus	Synthesizes rRNA and assembles ribosomes in the nucleus	Small, dark circular structure within the nucleus		
Plasma Membrane	Provides a selectively permeable barrier between the intracellular and extracellular environments of the cell	Visible only using an electron microscope. Plasma membrane is often visible under the light microscope.	*plasma*, something formed, + *membrane*, a membrane	Plasma membrane

EXERCISE 4.2

OBSERVING CELLULAR ANATOMY WITH A COMPOUND MICROSCOPE

Preparing a Wet Mount of Human Cheek Cells

This exercise involves taking a sample of cells from the inside of the cheek and preparing a wet mount. A wet mount is a procedure that involves placing a tissue sample in a wet medium onto a microscope slide. The "wet" medium is typically an isotonic saline solution. Why is it important for the solution to be isotonic? _____

_____

The cells on the inside of a human cheek are squamous cells, which are flattened cells. The inside of the cheek is lined with multiple layers of these cells. Therefore, a few cells may be gently scraped off with a toothpick without causing damage to the entire epithelium (lining) of the inside of the mouth. Details regarding the structure of cheek cells are described in the histology section covering epithelial tissues in chapter 5.

After obtaining cheek cells, they will be placed on a microscope slide, a stain (methylene blue) will be applied, and they will be covered with a coverslip. The stain is necessary to visualize the cells because normal cells are nearly transparent. Methylene blue is basophilic (base-loving) and is attracted to eosinophilic (acid-loving) components of the cell. The most eosinophilic part of the cell is the nucleus, which contains the nucleic acids DNA and RNA. Thus, the nucleus of the cell stains more intensely than other parts of the cell.

Obtain the Following:
- **compound microscope**
- **microscope slide and coverslip**
- **toothpick or wood applicator stick**
- **methylene blue solution with eyedropper**
- **fine tissue paper or KimWipes®**

1. Place the microscope slide on a piece of white paper on the lab bench. The piece of white paper will make observations easier.

2. Place a small drop of normal saline on the microscope slide (**figure 4.2a**). Very *gently* scrape the toothpick along the inside of your mouth to pick up a few cells. This process should not be painful, and most definitely should not draw blood!

3. Next, place the tip of the toothpick in the drop of saline on the slide (figure 4.2b). Roll the toothpick around gently so the cells detach from the toothpick and fall into the drop of saline.

(a) Place a drop of normal saline on the slide.

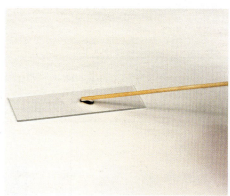

(b) After collecting cheek cells, gently roll the tip of the toothpick in the drop of saline.

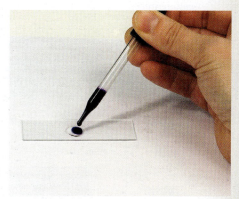

(c) Place a drop of methylene blue on the slide.

(d) Slowly lower a coverslip over the drop of liquid containing cheek cells.

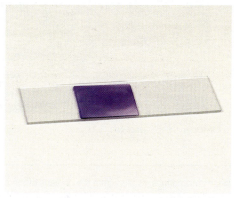

(e) Make sure there are no air bubbles between the coverslip and the slide. The slide is now ready to view with the microscope.

Figure 4.2 **Preparing a Wet Mount of Human Cheek Cells.** ©Christine Eckel

(continued on next page)

(continued from previous page)

4. Obtain a vial of methylene blue solution. Place a single drop of methylene blue on the drop of saline containing cheek cells (figure 4.2*c*).

5. Obtain a coverslip and place it on the edge of the liquid on the microscope slide as shown in figure 4.2*d*. Carefully and slowly lower the coverslip onto the drop of liquid. The goal is to place the coverslip over the drop of saline without introducing air bubbles. Simply dropping the coverslip on the slide may cause large air bubbles to form between the slide and the coverslip, which will interfere with the ability to see the cells on the slide. To prevent air bubbles from forming, carefully and slowly lower the coverslip, starting on one side and lowering it down at an angle (figure 4.2*d*). Most air bubbles will be pushed out of the way as the coverslip is placed. Obtain a piece of tissue paper or a KimWipe® and use it to dab any excess liquid on the sides of the coverslip (if necessary). A successful wet mount of cheek cells should resemble that shown in figure 4.2*e*.

6. Observe the slide with your naked eye. Can you see anything on the slide? In particular, can you see any cheek cells? _____

7. Arrange the objective lens on the microscope so it is set to use the scanning objective. Place the slide containing the cheek cells on the microscope stage and bring the tissue sample into focus using the scanning objective. Next,

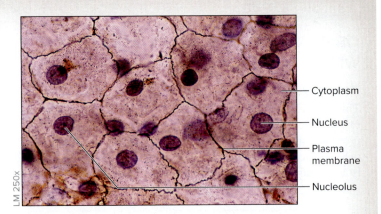

LM 250x

Figure 4.3 **Human Cheek Cells.** ©Ed Reschke/Getty Images

change to a higher power and bring the tissue sample into focus once again. The cells observed should somewhat resemble those in **figure 4.3**, although they will be isolated from one another rather than being in a sheet as in figure 4.3.

8. Scan the slide until cheek cells are visible in the field of view. Sketch the cheek cells as seen through the microscope in the space provided, or on a separate sheet of paper. Label the following on the sketch:

☐ **Cytoplasm**   ☐ **Nucleolus**

☐ **Nucleus**   ☐ **Plasma membrane**

Table 4.2	Appearance of Whitefish Embryo Cells Undergoing Mitosis During Phases of the Cell Cycle		
Stage	**Interphase**	**Prophase**	
Histological View	Nucleus with chromatin LM 450x ©Michael Abbey/Science Source	Nucleus with dispersed chromosomes LM 450x ©Carolina Biological Supply Company/Phototake	
Recognizable Features	Loose chromatin is visible within the nucleus of the cell. No chromosomes are visible because they are uncoiled (i.e., chromatin) at this stage.	Chromosomes become visible as the chromatin coils. If the cell is in early prophase, the fibers of the mitotic spindle may be visible adjacent to the nucleus.	
Word Origins	*inter*, between, + *phasis*, an appearance	*pro*, before + *phasis*, an appearance	

Troubleshooting:

If the slide appears to consist mostly of cellular debris instead of intact cells, it is likely that distilled water was mistakenly used in place of saline when making the slide. Distilled water is a hypotonic solution. When cells are placed in a hypotonic solution, they will lyse, leaving only cellular debris on the slide.

Sometimes the view through the microscope appears to vibrate or shake, which makes it impossible to visualize the cells. If this happens, there is too much fluid between the coverslip and the slide. Use a KimWipe® to draw some of the excess fluid out from under the coverslip. Then, observe the slide again.

9. *Optional Activity:* **APR 2: Cells & Chemistry—** Examine the "Generalized cell" dissection and test yourself on cell structures in the Quiz area.

Mitosis

The cell cycle describes the events that occur during the process of forming a new cell. As a cell passes through the stages of the cell cycle, two identical daughter cells are formed from one original parent cell. The cell cycle is divided into two main phases: interphase and mitotic phase. Interphase is the time in which the genetic material is uncoiled as chromatin. Mitotic phase (*mitos*, thread) includes the processes by which cells reproduce and includes mitosis (nuclear division, which is divided into four stages) and cytokinesis (*cyto-*, cell, + *kinesis*, movement), which is division of the cytoplasm. Note that casual usage of the term "mitosis" usually implies that the cytoplasm and cellular organelles have also divided. **Table 4.2** describes the microscopic appearance of cells in interphase and each of the four stages of mitosis. The four stages are (in order): prophase, metaphase, anaphase, and telophase. (The stages of mitosis can be remembered with the acronym P-MAT.) In this laboratory exercise the goal is to locate cells in interphase and each of the stages of mitosis by observing whitefish embryos (blastulas).

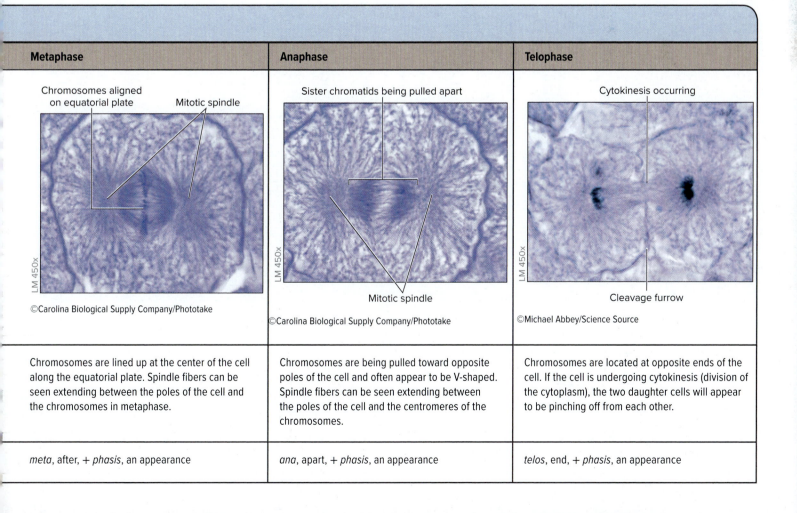

Metaphase	Anaphase	Telophase
Chromosomes aligned on equatorial plate — Mitotic spindle	Sister chromatids being pulled apart — Mitotic spindle	Cytokinesis occurring — Cleavage furrow
LM 450x	LM 450x	LM 450x
©Carolina Biological Supply Company/Phototake	©Carolina Biological Supply Company/Phototake	©Michael Abbey/Science Source
Chromosomes are lined up at the center of the cell along the equatorial plate. Spindle fibers can be seen extending between the poles of the cell and the chromosomes in metaphase.	Chromosomes are being pulled toward opposite poles of the cell and often appear to be V-shaped. Spindle fibers can be seen extending between the poles of the cell and the centromeres of the chromosomes.	Chromosomes are located at opposite ends of the cell. If the cell is undergoing cytokinesis (division of the cytoplasm), the two daughter cells will appear to be pinching off from each other.
meta, after, + *phasis*, an appearance	*ana*, apart, + *phasis*, an appearance	*telos*, end, + *phasis*, an appearance

OBSERVING MITOSIS IN A WHITEFISH EMBRYO

1. Obtain a compound microscope and a prepared slide of a whitefish embryo (blastula) or a slide of a different type of cell undergoing mitosis.

2. Scan the slide and locate cells in interphase and the four stages of mitosis that are listed in table 4.2. Once a cell in a particular phase has been located, switch to a higher-power objective to see the cell more clearly. Note that not all cells undergo mitosis simultaneously. Therefore, cells in various stages of mitosis and interphase may all be observed on the same slide.

☐ Anaphase   ☐ Prophase

☐ Interphase   ☐ Telophase

☐ Metaphase   ☐

3. Sketch the appearance of cells in each of the phases listed in step 2 in the spaces provided or on a separate sheet of paper.

Stage: Interphase

Stage: Anaphase

Stage: Prophase

Stage: Telophase

Stage: Metaphase

PHYSIOLOGY

Mechanisms of Passive Membrane Transport

Passive transport mechanisms are mechanisms substances use to cross the plasma membrane that do not require an input of energy. Because of this, such mechanisms can be studied using artificial membranes and/or simulations of membranes. The following exercises explore several of the factors that affect the rates of passive transport across plasma membranes. Before beginning these exercises, read the textbook to become familiar with the following concepts: kinetic energy, passive transport, molecular weight, simple diffusion, osmosis, hypertonic solution, hypotonic solution, isotonic solution, and membrane impermeable solute.

EXERCISE 4.4

DIFFUSION (WET LAB)

Diffusion (*diffundo*, to pour in different directions) refers to the net movement of solute particles from an area of high concentration to an area of low concentration. Diffusion occurs when a concentration gradient (difference in concentration between two areas) exists. The substance dissolved in the solution is the **solute**, and the substance in which the solute is dissolved is the **solvent**. When working with solutions in the anatomy and physiology laboratory, the solvent will typically be water.

The **second law of thermodynamics** (also called the **law of entropy**) is the law governing the natural behavior of molecules, including molecules in solution. This law states that, over time, all molecules move at random toward a state of increasing disorder. **Entropy** (*entropia*, a turning toward) is a measure of the disorder of a system. Thus, for a system to become more disorderly, no energy input is required (as the saying goes, "entropy happens"). Similarly, energy must be put into a system for the system to become more orderly. How does the law of entropy apply to diffusion? Consider the situation in **figure 4.4**. The purple dots represent solute particles in solution. On the left of figure 4.4*a* the solution is more concentrated, and on the right of figure 4.4*a* the solution is more dilute. Observe each particle in relation to the other solute particles. Notice that where the solution is more concentrated, the particles are packed close together, and thus are more orderly. Where the solution is more dilute, the solute particles are farther away from each other and thus are more disorderly. When following the movement of the solute particles over time, the random movement of solute particles in solution is expected to result in a situation where the particles become as disorderly as possible. This means each particle will be as far away from another particle as is possible. Figure 4.4*b* shows what happens after diffusion has taken place.

When two solutions of differing concentrations are separated from each other by a membrane that is permeable to the solute in question, solute particles will move, at random, both with and against the concentration gradient because of their inherent kinetic energy (kinetic energy is energy of *motion*). However, the *net* movement of particles will be from an area of high concentration to an area of low concentration (**figure 4.5**). Once the concentration of solute is the same on both sides of the membrane, the solution is said to be at *equilibrium*. It is important to note that although equilibrium has been reached at that point, this does not mean that solute particles have stopped moving across the membrane. On the contrary, particles continue to move at random and will still cross the membrane. However, the rate at which particles move from container A to container B will be equal in magnitude but opposite in direction to the rate at which particles move from container B to container A. Thus, *no net movement* of solute will occur. This is how the point of equilibrium is defined.

Some factors that affect the rate at which a substance diffuses include

1. viscosity of the solvent

2. temperature of the solvent

3. molecular weight of the solute

4. permeability of the membrane

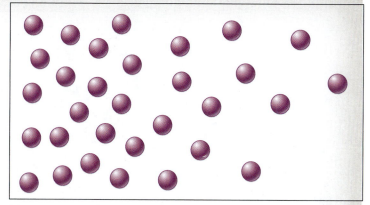

(a)

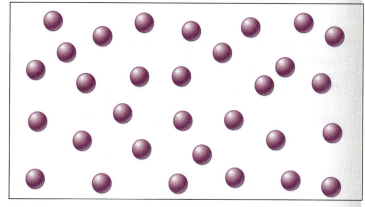

(b)

Figure 4.4 **Diffusion.** (*a*) Initially concentrated molecules. (*b*) Molecule distribution after diffusion takes place.

(continued on next page)

(continued from previous page)

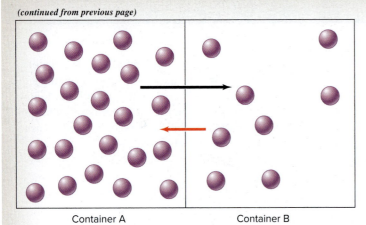

Container A   Container B

(a)

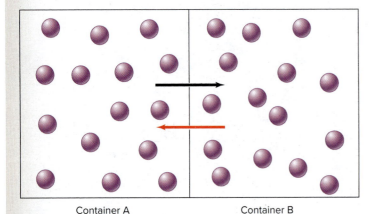

Container A   Container B

(b)

Figure 4.5 **Diffusion Through a Semipermeable Membrane.** (*a*) Diffusion of solute from A → B is greater than the diffusion of solute from B → A. Thus, the net movement is from A → B. (*b*) Once equilibrium has been reached, the movement from A → B is equal in magnitude but opposite in direction to the movement from B → A. Thus, there is no net movement at equilibrium, and the concentration of solute in both containers is the same.

INTEGRATE

CONCEPT CONNECTION

The plasma membrane contains proteins such as channels, carrier proteins, and pumps. These proteins make the membrane selectively permeable to the substances they transport. Water and small, mainly lipid-soluble substances pass freely into and out of the cell through the plasma membrane by passive transport. Ions can flow down their electrochemical gradients (passive transport) when channels are open, and ions can move against their electrochemical gradients (active transport) when pumps are active.

Plasma membrane proteins are made by the cell, for the cell. Bound ribosomes attached to the ER synthesize proteins that are destined for the plasma membrane. Once membrane proteins are synthesized by the rough ER, they are packaged in vesicles, moved to the Golgi apparatus for modification, and released as secretory vesicles for insertion into the membrane.

In the following exercises, the effect of each of these factors on rates of diffusion will be observed.

EXERCISE 4.4A Effect of Viscosity on Rate of Diffusion

This exercise involves observing the diffusion of a solute (potassium permanganate) in media of differing viscosities (water and agar). Agar is a gel-like substance that is made from algae. Agar is 98% water. Thus, molecules will diffuse through it. However, agar is more viscous (thicker) than water. In this experiment one crystal of potassium permanganate will be placed in a petri dish containing agar. Another crystal of potassium permanganate will be placed in a petri dish containing water. The rate of diffusion will then be observed in each dish. Before beginning the experiment, state a hypothesis regarding the effect of solvent viscosity on the rate of diffusion.

Hypothesis: _____

_____

_____

_____

Obtain the Following:

- **letter-size (8.5" × 11") piece of white paper**
- **petri dish containing distilled water**
- **petri dish containing agar**
- **fine-tissue forceps**
- **potassium permanganate crystals**
- **one large (~30 cm) and 2 small (~15 cm) metric rulers**
- **a stopwatch or other device for recording the time**

1. Obtain a small weigh boat containing potassium permanganate crystals (**figure 4.6**). Only two moderate-sized crystals are needed for this experiment, one for the agar dish and one for the water dish. Thus, only

Pour a small amount of potassium permanganate into a weigh boat or other small container. Obtain a single moderate-sized crystal using tissue forceps.

Figure 4.6 **Obtaining Potassium Permanganate Crystals.**
©Christine Eckel

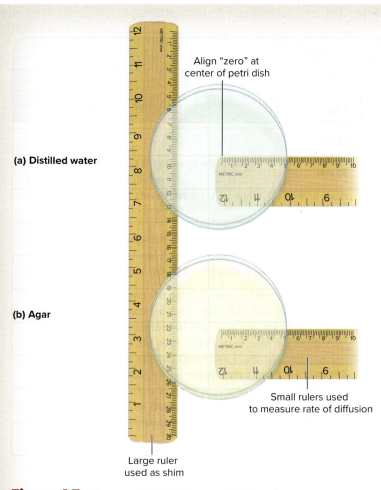

Align "zero" at
center of petri dish

(a) Distilled water

(b) Agar

Small rulers used
to measure rate of diffusion

Large ruler
used as shim

Figure 4.7 **Setup of Petri Dishes and Rulers for Exercise 4.4A.**

obtain a small number of crystals. Be aware! Potassium permanganate is a dye. Wear gloves when handling the crystals, and take care not to spill the crystals on the lab table, books, or clothing because they will permanently dye those items. If a spill occurs, first sweep the crystals up with a dry brush or cloth. Try not to use a wet towel until most of the crystals have been cleaned up because the water will create a liquid dye that can spread more easily (and thus stain more items) than the powdered form.

2. Place the rulers and the petri dishes on the white paper, as shown in **figure 4.7**. The large ruler will be used as a shim to keep the petri dishes level, whereas the small rulers will be used to measure rates of diffusion. If using very thin rulers, a shim will not be necessary. Place the large ruler on the left side of the paper and orient it vertically. Place the small rulers at right angles to the large ruler, approximately 8 cm apart from each other, and with the zero point approximately at the center of the petri dish.

3. Label the petri dishes "A – distilled water" and "B – agar" on the paper in the space next to each dish (see figure 4.7).

4. The diffusion of dye crystals in **water** will be observed first (**figure 4.8**). This process happens fairly quickly, so be prepared before beginning the experiment. One lab partner will place the crystal and record the distance diffused. Another student will start the stopwatch and record time intervals. To begin: Carefully place a single potassium permanganate crystal in the center of the petri dish containing water (dish A) at the zero mark on the ruler. The moment the crystal is placed in the water, start the stopwatch and observe diffusion of the dye. Record the radius of the dye spot in 30-second intervals for a total of at least 5 minutes (see **figure 4.9** for a description of

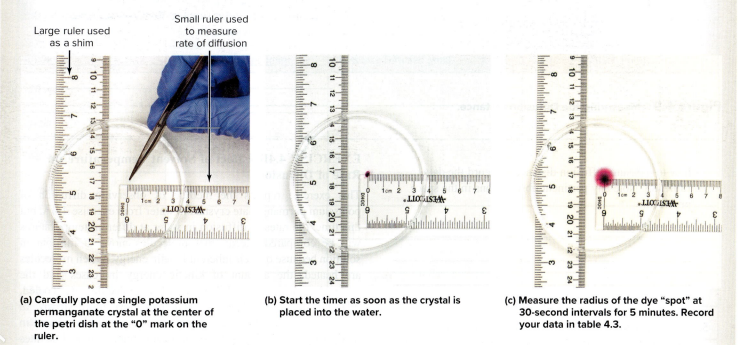

Large ruler used
as a shim

Small ruler used
to measure
rate of diffusion

(a) Carefully place a single potassium permanganate crystal at the center of the petri dish at the "0" mark on the ruler.

(b) Start the timer as soon as the crystal is placed into the water.

(c) Measure the radius of the dye "spot" at 30-second intervals for 5 minutes. Record your data in table 4.3.

Figure 4.8 **Observing Diffusion of Potassium Permanganate Crystals in Water.** ©Christine Eckel

(continued on next page)

Filtration

Filtration (*filtro*, to strain through) is a process that is likely familiar if a person has ever brewed coffee using a filtration brewing system. Filtration is a process by which solutes are separated from a solvent by being passed through a filter. The size of pores in the filter prevents solute particles that are larger than the size of the pores from passing through, while allowing any solute particles that are smaller than the size of the pores to pass. At times, the charge on a filter can also prevent substances from passing through. For example, if the substance is small and negatively charged and the filter is also negatively charged, even if the substance is small enough to fit through the pores in the filter, it will likely be prevented from passing because of the repellent forces of the negative charges. Notably, if enough fluid pressure, or **hydrostatic pressure** (*hydro*, water + *statikos*, causing to stand), is exerted on the fluid, the pressure can overcome the influence of like charges repelling each other. In the case of making coffee using a filtration system, coffee grounds are placed in a filter and hot water is passed over the grounds. The extracts of the coffee, such as caffeine, readily pass through the filter, but the coffee grounds do not pass because they are too large. The resulting beverage that ends up in the coffee pot is a **filtrate**.

EXERCISE 4.7

FILTRATION (WET LAB)

Although there are multiple factors that can affect filtration, this exercise explores only two of those factors:

1. *Particle size*—particles smaller than the size of the pores in the filter will pass through. Particles larger than the size of the pores in the filter will remain in the filter.

2. *Hydrostatic pressure*—a higher column of fluid exerts more pressure and thus increases filtration rate.

   In this exercise a solution containing substances of varying sizes will be made. The solution is poured into a funnel containing a paper filter. The rate at which fluid passes through the filter and fills up a graduated cylinder will be measured. The rate of filtration when the filter is completely full will then be compared to the rate of filtration when the filter is only 50% full. The solution will contain the following:

   a. distilled water

   b. 10% uncooked starch solution

   c. 10% dark corn syrup solution

   d. charcoal or ground black pepper

Before beginning the experiment, state a hypothesis regarding the effect of particle size and hydrostatic pressure on filtration rate. Will there be a difference in filtration rate when the filter is 100% vs. 50% full?

Hypothesis: _____

_____

_____

_____

_____

Obtain the Following:

- a ring stand and ring clamp to hold a glass funnel
- one piece of filter paper (18.5 cm diameter)
- glass funnel (10 cm diameter)
- 100 mL glass beaker
- 100 mL graduated cylinder
- uncooked dry starch
- dark corn syrup
- charcoal or black pepper
- stopwatch or other timing device

- Lugol's iodine (to test for starch)
- glucose test strips (to test for glucose)

1. Set up the ring stand with the funnel and graduated cylinder, as shown in **figure 4.16**. To prepare the filter paper, first fold it in half. Then fold it in half once again and open it up into a cone that will fit into the glass funnel. Place the filter in the funnel.

2. Mix the following solution:

   a. 100 mL distilled water

   b. 1 teaspoon charcoal or black pepper

   c. 1 tablespoon dark corn syrup

   d. 1 tablespoon dry uncooked corn starch

3. One lab partner will be in charge of timing; the other will be in charge of pouring the solution into the funnel

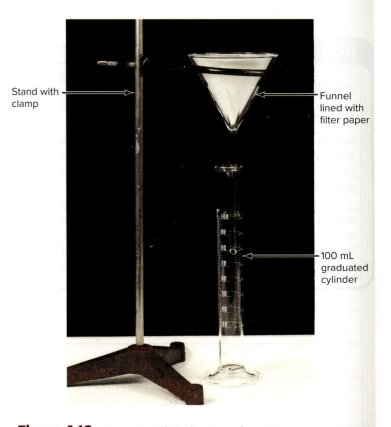

Stand with clamp

Funnel lined with filter paper

100 mL graduated cylinder

Figure 4.16 **Setup for Filtration Exercise.** ©Christine Eckel

and recording the times in **table 4.8**. Before pouring the solution into the funnel, give it a quick stir so that solids are not settled at the bottom of the beaker. Then, quickly pour the solution into the filter-lined funnel. The liquid should come up to just below the lip of the filter paper. Start the timer. Note the time it takes to fill the cylinder

Table 4.8	Filtration Rate
Volume (mL)	Time (sec)
5	
10	
15	
20	
25	
30	
35	
40	
45	
50	
55	
60	
65	
70	
75	
80	
85	
90	

in increments of 5 mL, and record these times in table 4.8. NOTE THE TIME WHEN THE COLUMN OF FLUID IN THE FUNNEL IS AT 50 mL. When that happens, record the time here: _____. This will allow comparison of filtration rates.

4. Allow filtration to continue until the level of filtrate in the cylinder reaches 80 or 90 mL. When filtration is complete, remove the funnel containing the filter. Note any substances retained by the filter in the space provided:

_____

_____

_____

_____

5. Next, perform the following tests on the filtrate in the graduated cylinder:

   a. Observe: Are there **charcoal** or black pepper flecks present in the filtrate? _____

   b. Dip a glucose test strip into the filtrate and look for a color change. Is **glucose** present in the filtrate? _____

   c. Place a couple of drops of Lugol's iodine into the filtrate and look for a color change. Is there **starch** in the filtrate? _____

6. Make a note regarding any conclusions of this experiment (i.e., did it support or refute your hypothesis?) in the space provided.

_____

_____

_____

_____

HISTOLOGY

Epithelial Tissue

Epithelial tissues are tissues that cover body surfaces, line body cavities, and form the majority of glands. As such, they will have a free surface (see Learning Strategy in this chapter). They are characteristically highly **cellular** (mostly composed of cells, with little extracellular material) and **avascular** (no blood vessels). Epithelial cells exhibit **polarity**; they have a distinct *basal* (bottom) and *apical* (top) surface. On their basal surface, they have a specialized extracellular structure called a **basement membrane**, which anchors the epithelium to the underlying tissues. The characteristics used to classify epithelial tissue include (a) the number of layers of cells (simple, stratified, or pseudostratified) **(table 5.1)**, (b) the shape of the cells on the *apical* surface of the epithelium **(table 5.2)**, and (c) the presence of any surface modifications **(table 5.3)**.

Figure 5.2 is a flowchart for classification of epithelial tissues that can be used as a tool when attempting to identify an unknown slide containing epithelial tissue. Following the flowchart will assist in the process of deciding how to classify an epithelial tissue.

Table 5.1	**Classification of Epithelial Tissue by Number of Cell Layers**		
Cell Layers	**Simple Epithelium**	**Stratified Epithelium**	**Pseudostratified Epithelium**
Micrograph	Lumen — Simple columnar epithelium — LM 70x ©McGraw-Hill Education/Al Telser	Lumen — Stratified squamous epithelium — LM 50x ©McGraw-Hill Education/Al Telser	Lumen — Pseudostratified columnar epithelium — LM 600x ©McGraw-Hill Education/Al Telser
Description	One cell layer thick; all cells make direct contact with the basement membrane	Contains two or more layers of cells; only the deepest layer of cells makes direct contact with the basement membrane	Appears stratified because all cell nuclei are not located the same distance from the basal surface; all cells make direct contact with the basement membrane
Generalized Functions	Absorption, diffusion, filtration, or secretion	Protection or to resist abrasion	Protection, secretion, and movement of particles (when epithelium is ciliated)

INTEGRATE

CLINICAL VIEW
Histopathology

Knowledge of the microscopic structure of tissues is critical for health professionals so that they are able to communicate with other medical professionals about tissue-level structures. Although most health professionals will rarely view slides of tissues in practice, nearly all will need to be able to interpret histopathology reports that are pertinent to their patients' diagnoses. A *histopathologist* is a physician and/or scientist who analyzes tissue samples that have been taken from a patient via biopsy (*bi*, two + *-opsy*, inspection). Once the tissue sample is received in the laboratory, a histopathologist goes through the process of creating a microscopic slide containing a slice of the tissue. This process is very similar to the process used to create the slides that are viewed in the anatomy & physiology laboratory, which is described in the introduction to this chapter.

Once a slide of a patient's tissue has been made **(figure 5.1)**, a histopathologist analyzes the sample to determine if the tissue appears as expected. Any variations from expected structure are then characterized and described. These observations are analyzed in conjunction with lab tests to aid in making a clinical diagnosis of the patient's condition.

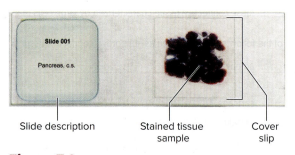

Slide description   Stained tissue sample   Cover slip

Figure 5.1 **Life-Size Histology Slide.** ©Christine Eckel

Table 5.2	Classification of Epithelial Tissue by Cell Shapes			
Cell Shape	**Squamous**	**Cuboidal**	**Columnar**	**Transitional**
Micrograph	Squamous cells / Lumen / LM 205x / ©Victor P. Eroschenko	Cuboidal cells / Lumen / LM 165x / ©Victor P. Eroschenko	Columnar cells / Lumen / LM 500x / ©McGraw-Hill Education/Al Telser	Transitional cell / Lumen / LM 180x / ©McGraw-Hill Education/Al Telser
Description	Cells are flattened and have irregular borders	Cells are as tall as they are wide	Cells are taller than they are wide	Cells change between flattened and dome-shaped, depending on the stretch of the epithelial tissue.
Generalized Functions	If only one cell layer thick, provides a very thin barrier for *diffusion*. If several layers thick, provides *protection*.	Cuboidal cells generally function in *secretion* and/or *absorption*.	Columnar cells generally function in *secretion* and/or *absorption*.	The fact that these cells change shape means that they are good at *stretching and relaxing* without being torn apart from each other.
Identifying Characteristics	In cross section, the nucleus is the most visible structure. The nucleus will be very flattened. In a surface view of the epithelium, the cell borders will be irregular in shape.	Generally, cuboidal cells are identified by their very round, plump nucleus, and by equal amounts of cytoplasm on all sides of the nucleus.	The nuclei of columnar cells are either oval or round, and they generally line up in a row close to the basal surface of the epithelium.	Transitional cells are located on the apical surface of the epithelium. Transitional cells appear much more rounded or dome-shaped than typical cuboidal cells, and they are sometimes binucleate.

Table 5.3	Cell Surface Modifications and Specialized Cells of Epithelial Tissues			
	Surface Modifications			**Specialized Cells**
	Cilia	**Microvilli**	**Keratinization**	**Goblet Cells**
Micrograph	Cilia / Lumen / LM 1000x / ©McGraw-Hill Education/Dennis Strete	Brush border of microvilli / Lumen / LM 130x / ©Victor P. Eroschenko	Keratinization / Lumen / LM 25x / ©Ed Reschke/Photolibrary/Getty Images	Columnar epithelial cell / Goblet cell / Mucin within goblet cell / Lumen / Goblet cell nucleus / Location of basement membrane / LM 250x / ©Ed Reschke/Getty Images
Description and Function	Cilia are small, hairlike structures that extend from the apical surface of epithelial cells. Cilia actively move to *propel substances along the apical surface of an epithelial sheet.*	Microvilli are extremely small extensions of the plasma membrane of the apical surface of cells. Microvilli *increase the surface area* of the cell and enhance *absorption.*	Stratified squamous epithelial cells of the skin contain keratin, which makes the cells appear to be a single homogeneous unit. Keratin imparts *strength* and *protection* to dead skin epithelial cells.	Goblet cells are named for their shape. They are rounded near the apical surface and they narrow toward their basal surface. Goblet cells *produce mucus* that assists in *transport* of substances along the epithelial sheet, *provides a protective barrier*, or *provides lubrication.*
Identifying Characteristics	When cilia are present, and the slide is viewed at sufficient magnification, what appear to be individual "hairs" are visible on the apical surface of the epithelial cells.	Individual microvilli will *not* be visible with a light microscope. Instead, the apical surface of the epithelium will appear "fuzzy."	Keratinization is recognized as a homogeneous, acellular-looking portion of a stratified squamous epithelium.	The shape of goblet cells is similar to the shape of a wineglass. The cells often appear white or "empty" in traditional H&E staining.

(continued from previous page)

EXERCISE 5.3D Dense Regular Connective Tissue

1. Obtain a slide of a tendon or ligament **(figure 5.15)**.

2. Place the slide on the microscope stage and bring the tissue sample into focus on low power. Then change to high power.

3. Identify the following structures, using figure 5.15 and tables 5.4 and 5.5 as guides:

   ☐ **Collagen fibers**   ☐ **Fibroblast nuclei**

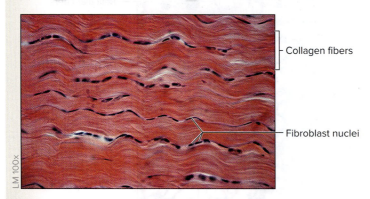

Figure 5.15 **Dense Regular Connective Tissue.** Dense regular connective tissue consists of bundles of collagen fibers all oriented in the same direction, with fibroblasts located between the bundles of collagen fibers.
©Ed Reschke/Getty Images

4. Sketch dense regular connective tissue as seen through the microscope in the space provided or on a separate sheet of paper. Be sure to label the structures listed in step 3 in the drawing.

_____ ×

EXERCISE 5.3E Dense Irregular Connective Tissue

1. Obtain a slide of skin **(figure 5.16)**. Skin consists of two major layers, an outer epidermis, composed of stratified squamous epithelial tissue, and an inner dermis, composed of connective tissue (areolar and dense irregular).

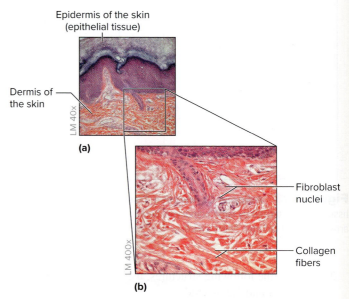

Figure 5.16 **Dense Irregular Connective Tissue.** Dense irregular connective tissue is located in the dermis of the skin (*a*). Dense irregular connective tissue consists of bundles of collagen fibers oriented in many different directions, with fibroblasts located between the bundles of collagen fibers (*b*). ©McGraw-Hill Education/Christine Eckel

2. Place the slide on the microscope stage and bring the tissue sample into focus using the scanning objective. Identify the epithelial tissue in the epidermis (figure 5.16*a*), then move the stage so the lens focuses on the underlying connective tissue in the dermis (figure 5.16*b*). Switch to a higher-power objective.

3. Identify the following structures, using figure 5.16 and tables 5.4 and 5.5 as guides:

   ☐ **Collagen fibers**

   ☐ **Fibroblast nucleus**

4. Sketch dense irregular connective tissue as seen through the microscope in the space provided or on a separate sheet of paper. Be sure to label the structures listed in step 3 in the drawing.

_____ ×

EXERCISE 5.3F Elastic Connective Tissue

1. Obtain a slide of the aorta or an elastic artery **(figure 5.17)**.

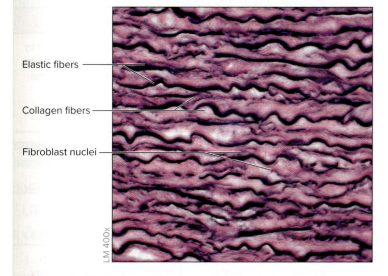

Elastic fibers

Collagen fibers

Fibroblast nuclei

LM 400x

Figure 5.17   **Elastic Connective Tissue.** The wall of the aorta consists of dense regular elastic connective tissue, which contains both collagen and elastic fibers (which stain black). All the fibers are oriented in the same direction. In this slide, the collagen fibers are dark pink and the elastic fibers are black and wavy. Some fibroblast nuclei are visible between the bundles of fibers. ©McGraw-Hill Education/Christine Eckel

2. Place the slide on the microscope stage and bring the tissue sample into focus on low power. Then switch to high power.

3. Identify the following structures, using figure 5.17 and tables 5.4 and 5.5 as guides:

   ☐ **Collagen fibers**

   ☐ **Elastic fibers**

   ☐ **Fibroblasts**

4. Sketch elastic connective tissue as seen through the microscope in the space provided. Be sure to label the structures listed in step 3 in the drawing.

_____ ×

IDENTIFICATION AND CLASSIFICATION OF FLUID CONNECTIVE TISSUE

1. Obtain a slide of a blood smear (**figure 5.23**).

2. Place the slide on the microscope stage and bring the tissue sample into focus on low power. Change to medium power, and then bring the sample into focus using the oil immersion lens. Consult the instructor if assistance is needed using the oil immersion lens.

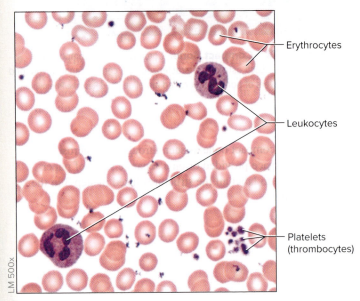

LM 500x

Figure 5.23   Blood. Blood is a fluid connective tissue containing erythrocytes (red blood cells), leukocytes (white blood cells), platelets (thrombocytes), and an extracellular matrix called plasma. ©McGraw-Hill Education/Alvin Telser

3. Identify the following structures, using figure 5.23 as a guide:

☐ **Erythrocytes**   ☐ **Platelets**
☐ **Leukocytes**   **(thrombocytes)**

4. Sketch blood as seen through the microscope in the space provided or on a separate sheet of paper. Be sure to label all the structures listed in step 3 in the drawing.

_____ ×

5. *Optional Activity:* **APR** **3: Tissues**—Watch the "Connective Tissue Overview" animation.

CLINICAL VIEW
Carcinomas and Sarcomas

Clinically, a tumor derived from epithelial tissues is called a *carcinoma* (*karkinos*, crab + *-oma*, tumor), and a tumor derived from connective tissues is called a *sarcoma* (*sarc-*, flesh + *-oma*, tumor). Carcinomas are considered noninvasive when they do not penetrate the basement membrane that lies between the epithelial and connective tissue layers. Once rapidly dividing cells penetrate the basement membrane, the cancer is considered invasive. While epithelial tissue is avascular, the underlying connective tissue is not. Thus, invading cancer cells can easily metastasize (*meta-*, change + *-ize,* an action) to other locations of the body through blood and lymphatic vessels. Sarcomas, which arise from connective tissues, pose a similar risk. However, they tend to grow and metastasize much more readily than carcinomas because of the highly vascular nature of connective tissues.

Muscle Tissue

Muscle tissue is both excitable and contractile. Excitable tissues are able to generate and propagate electrical signals called action potentials. As a contractile tissue, muscle has the ability to actively shorten and produce force. There are three types of muscle tissue: skeletal muscle, cardiac muscle, and smooth (visceral) muscle. The three types of muscle tissue are distinguished by the presence or absence of visible striations, the shape of the cells, and the number and location of nuclei. Skeletal muscle is found in the voluntary muscles that move the skeleton and the facial skin. Cardiac muscle is found in the heart, and smooth muscle is found in the walls of soft viscera, such as the blood vessels, stomach, urinary bladder, intestines, and uterus. **Table 5.7** compares the three types of muscle tissue, and the flowchart in **figure 5.24** explains steps that can be used to identify muscle tissues.

Table 5.7	Muscle Tissue		
Type of Muscle	**Description**	**Generalized Functions**	**Identifying Characteristics**
Skeletal	Elongate, cylindrical cells with multiple nuclei. Nuclei are peripherally located. Tissue appears striated (light and dark bands along the length of the cell).	Provides voluntary movement	Length of cells (extremely long), striations, multiple peripheral nuclei
Cardiac	Short, branched cells with 1–2 nuclei. Nuclei are centrally located. Dark bands (intercalated discs) are seen where two cells join. Tissue appears striated (light and dark bands along the length of each cell).	Responsible for pumping action of the heart	Branched, 1–2 nuclei/cell, striations, intercalated discs
Smooth	Elongate, spindle-shaped cells (fatter in the center, narrowing at the ends) with single, cigar-shaped or spiral nuclei. Nuclei are centrally located. No striations are apparent.	Creates movement within viscera such as intestines, bladder, uterus, and stomach. Regulates diameter of certain blood vessels.	Spindle-shaped cells, no striations, cigar-shaped nuclei that are centrally located

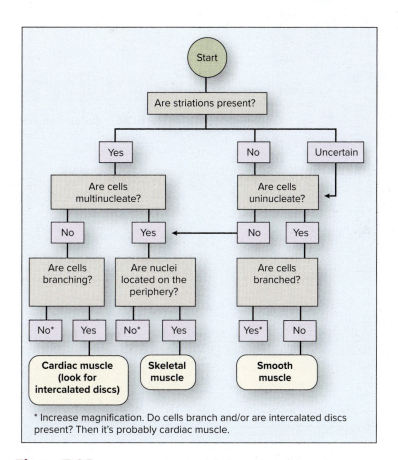

Figure 5.24   **Flowchart for Classifying Muscle Tissues.**

EXERCISE 5.6

IDENTIFICATION AND CLASSIFICATION OF MUSCLE TISSUE

EXERCISE 5.6A Skeletal Muscle Tissue

1. Obtain a slide of skeletal muscle (**figure 5.25**).

2. Place the slide on the microscope stage and bring the tissue sample into focus on low power. Then change to high power.

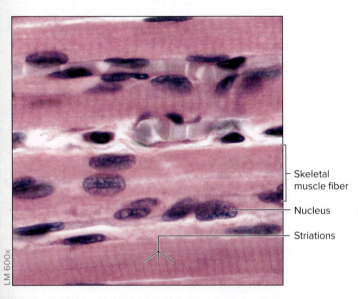

LM 600x

— Skeletal muscle fiber

— Nucleus

— Striations

Figure 5.25   **Skeletal Muscle.** Skeletal muscle fibers are elongated and striated and contain multiple peripherally located nuclei. ©McGraw-Hill Education/Al Telser

3. Identify the following structures, using figure 5.25 and table 5.7 as guides:

☐ **Nucleus**   ☐ **Striations**

☐ **Skeletal muscle fiber**

4. Sketch skeletal muscle as seen through the microscope in the space provided or on a separate sheet of paper. Be sure to label all the structures listed in step 3 in the drawing.

_____ ×

EXERCISE 5.6B Cardiac Muscle Tissue

1. Obtain a slide of cardiac muscle (**figure 5.26**).

2. Place the slide on the microscope stage and bring the tissue sample into focus on low power. Then change to high power.

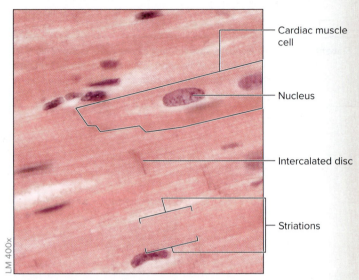

LM 400x

— Cardiac muscle cell

— Nucleus

— Intercalated disc

— Striations

Figure 5.26   **Cardiac Muscle.** Cardiac muscle cells are short and branched, striated, and contain 1-2 centrally located nuclei. ©McGraw-Hill Education/Al Telser

3. Identify the following structures, using figure 5.26 and table 5.7 as guides:

☐ **Cardiac muscle cell**   ☐ **Nucleus**

☐ **Intercalated disc**   ☐ **Striations**

4. Sketch cardiac muscle as seen through the microscope in the space provided or on a separate sheet of paper. Be sure to label all the structures listed in step 3 in the drawing.

_____ ×

EXERCISE 5.6C Smooth Muscle Tissue

1. Obtain a slide of smooth muscle (**figure 5.27**).

2. Place the slide on the microscope stage and bring the tissue sample into focus on low power. Then change to high power.

3. Identify the following structures, using figure 5.27 and table 5.7 as guides:

   ☐ **Smooth muscle cell**   ☐ **Nucleus**

4. Sketch smooth muscle as seen through the microscope in the space provided or on a separate sheet of paper. Be sure to label all the structures listed in step 3 in the drawing.

_____ ×

5. *Optional Activity:* **APR** 3: **Tissues**—Watch the "Muscle Tissue Overview" animation.

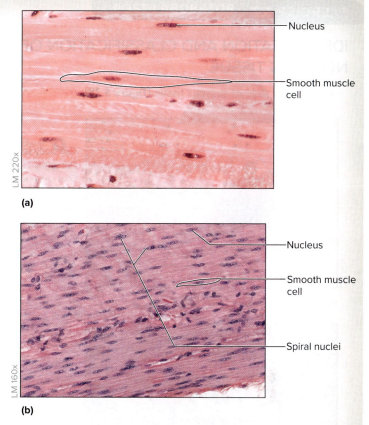

(a)

(b)

Figure 5.27   Smooth Muscle. Smooth muscle fibers are short and spindle-shaped, not striated, and contain only one centrally located nucleus (*a*). In figure 5.27*b*, some nuclei that have taken on a spiral shape are visible. This happens when the muscle fibers contract. As the fibers shorten, the nuclei start to coil up, or spiral. (*a*) ©McGraw-Hill Education/Al Telser; (*b*) ©Christine Eckel

Nervous Tissue

Nervous tissue is characterized by its excitability: the ability to generate and propagate electrical signals called action potentials. Nervous tissue is composed of two basic cell types (**table 5.8**). **Neurons** are excitable cells that send and receive electrical signals. They have a limited ability to divide and multiply in the adult brain. **Glial cells** are supporting cells that support and protect neurons. Glial cells maintain the ability to divide and multiply in the adult brain. Glial cells constitute over 60% of the cells found in nervous tissue.

Table 5.8	Nervous Tissue	
Cell Type	**Description**	**Generalized Functions**
Neurons	Though varied in shape, most neurons appear to have numerous branches coming off the cell body (soma). Neurons contain large amounts of rough endoplasmic reticulum (ER). The ribosomes and rough ER stain very dark and are collectively called chromatophilic substance.	These cells are responsible for *generating and transmitting information via electrical impulses* within the nervous system.
Glial Cells	Even more varied in shape than neurons, glial cells are generally much smaller than neurons with fewer (if any) branching processes.	These cells are the *general supporting cells* of the nervous system. They protect, nourish, and support neurons.

10. Which of the following is/are a type of bone tissue? (Check all that apply.) **12**

    ___ a. areolar   ___ d. reticular

    ___ b. compact   ___ e. spongy

    ___ c. elastic

Exercise 5.5: Identification and Classification of Fluid Connective Tissue

11. Which of the following is a type of fluid connective tissue? (Check all that apply.) **13**

    ___ a. blood   ___ d. lymph

    ___ b. bone   ___ e. mesenchyme

    ___ c. cartilage

12. Which of the following are possible cells present in fluid connective tissue? (Check all that apply.) **14**

    ___ a. chrondrocytes   ___ d. megakaryocytes

    ___ b. erythrocytes   ___ e. osteocytes

    ___ c. leukocytes

Exercise 5.6: Identification and Classification of Muscle Tissue

13. In the following table, compare and contrast the characteristics of the three types of muscle tissue. **15**

Characteristic	Skeletal Muscle	Cardiac Muscle	Smooth Muscle
Location and Number of Nuclei			
Cell Shape			
Presence or Absence of Striations			

14. One location where smooth muscle might be found is in _____ (the heart/blood vessels). **16**

Exercise 5.7: Identification and Classification of Nervous Tissue

15. The two main cell types found in nervous tissue are neurons and _____ (chondrocytes/glial cells). **17**

16. The processes branching from the cell body of the neurons that are responsible for receiving electrical signals are _____ (axons/dendrites).

▲ Can You Apply What You've Learned?

17. In the following table, compare and contrast epithelial and connective tissues with respect to the following:

Characteristic	Epithelial Tissues	Connective Tissues
Cell Number and Arrangement		
Polarity		
Extracellular Matrix		
Vascularity		

18. Match the description listed in column A with the specific type of connective tissue listed in column B.

    Column A

    ____ 1. collagen fibers are not visible; ground substance appears clear and glassy

    ____ 2. collagen fibers arranged in concentric rings; hard ground substance; osteocytes in lacunae

    ____ 3. fluid ground substance; many biconcave disc-shaped pink cells and larger, dark-staining cells

    ____ 4. chondrocytes lined up in rows in between bundles of collagen fibers

    ____ 5. densely packed elastic fibers; chondrocytes in lacunae

    Column B

    a. blood

    b. bone

    c. elastic cartilage

    d. fibrocartilage

    e. hyaline cartilage

19. What does the presence of cilia indicate about the function of an epithelial tissue?

    _____

    _____

20. Identify which statements about connective tissue are accurate. (Check all that apply.)

    ____ a. conducts neural impulses

    ____ b. contains bundles of nerve fibers and muscle fibers

    ____ c. forms bones, cartilage, ligaments, and tendons

    ____ d. forms deep layers of skin (dermis)

    ____ e. forms glands (e.g., salivary, pancreas)

    ____ f. forms a structural framework for organs

    ____ g. highly vascular (with some exceptions)

    ____ h. includes blood and components of blood vessel walls

21. Match the description listed in column A with the name of the connective tissue listed in column B.

    Column A

    ____ 1. makes up the cartilage discs between the vertebrae and between the pubic bones

    ____ 2. transports materials within blood vessels

    ____ 3. inner supporting framework of spleen, liver, and lymph nodes

    ____ 4. tough tissue that connects bone to bone and muscle to bone

    ____ 5. cartilage that retains its original shape after being deformed, such as in the cartilage of the ear

    ____ 6. allows the aorta (a large blood vessel) to stretch with each pulse of blood pumped into it from the heart and then return to its original size

    ____ 7. helps keep the body warm and stores excess fuel (energy)

    ____ 8. forms the ends of long bones, the larynx, the costal cartilages, and the embryonic skeleton

    Column B

    a. adipose connective tissue

    b. blood

    c. dense regular connective tissue

    d. elastic cartilage

    e. elastic connective tissue

    f. fibrocartilage

    g. hyaline cartilage

    h. reticular connective tissue

22. Explain the characteristics that make blood a connective tissue. (Hints: From what embryonic connective tissue is blood derived? What is the composition of the extracellular matrix? What is the cellular component? What are the fibers?)

    _____

    _____

    _____

    _____

    _____

    _____

    _____

EXERCISE 6.4

PIGMENTED SKIN

The color of a person's skin is partially due to the pigment **melanin** (*melas*, black). Melanin is produced by cells called **melanocytes**, which are found within the stratum basale of the epidermis. Melanocytes have long dendritic (*dendrites*, relating to a tree) processes that extend between keratinocytes. Melanocytes transfer the melanin granules they produce to adjacent keratinocytes. The melanin then accumulates on the apical side of the nuclei of the keratinocytes, which protects keratinocytes from harmful UV light that can damage their DNA. A single melanocyte produces melanin for several adjacent keratinocytes (the melanocyte plus the keratinocytes served by it are referred to as an *epidermal-melanin unit*). Some areas of the body have a higher density of melanocytes than others (e.g., the skin of the areola of the breast has more melanocytes than the skin on the rest of the breast). However, the *total number of melanocytes in light- and dark-skinned individuals is the same*. Differences in skin color have to do with the *amount of melanin* produced by melanocytes and *how quickly* the melanin is broken down. The melanocytes of dark-skinned individuals simply produce greater amounts of melanin that is broken down more slowly than those of light-skinned individuals.

1. Obtain a slide of pigmented skin and place it on the microscope stage. Observe at low power to locate the epidermis, and then change to a higher-power lens.

2. Focus on the stratum basale and stratum spinosum of the pigmented skin. Look for distinct black/brown melanin granules within the keratinocytes (**figure 6.3**). Melanocytes themselves are located in the stratum basale and are characterized by small, round nuclei and pale-staining cytoplasm. Try to locate a melanocyte on the slide, keeping in mind that melanocytes are not easily identified on the standard slides found in the laboratory.

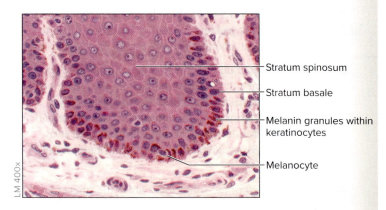

LM 400×

Stratum spinosum
Stratum basale
Melanin granules within keratinocytes
Melanocyte

Figure 6.3 Pigmented Skin. Pigmented skin containing melanin granules (brown) within keratinocytes, particularly in the stratum basale of the epidermis. Melanocytes themselves are distinguished by a halo of clear cytoplasm that surrounds the nucleus. ©Tom Caceci

INTEGRATE

CLINICAL VIEW
Melanoma

Melanoma (*melano-*, black + *-oma*, tumor) is an aggressive and sometimes deadly skin cancer that involves uncontrolled proliferation of melanocytes in the stratum basale layer of the epidermis. Melanoma usually presents as a mole on the surface of the skin that exhibits the following ABCDE characteristics: **asymmetry** (A), **irregular borders** (B), **varied colors** (C), a **diameter of greater than 6 mm** (D), and has **evolved in size, color, etc. over time** (E) **(figure 6.4)**. Clinicians test the suspicious lesion by taking a biopsy of the affected tissue to confirm the diagnosis (figure 6.4*b*). Once the lesion has been identified as melanoma, clinicians classify the severity of the case by identifying the "stage" of cancer. Tumors that remain in the outer portions of the epidermis are classified as stage I. These melanomas are the most responsive to treatment. Stage II occurs when the tumor invades the underlying dermis, but there are no signs of spread to local lymph nodes. Because the dermis contains both lymph vessels and blood vessels, there is the risk of metastasis, or spreading of the cancer cells from one part of the body to another. When the cancer spreads to nearby lymph nodes or tissues, it is classified as stage III. Once there is evidence of metastasis to distant organs, it is classified as stage IV.

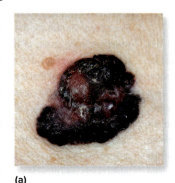

(a)

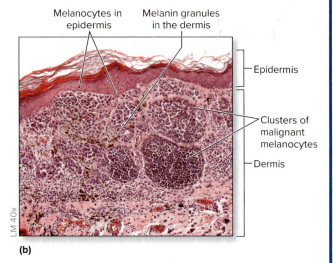

Melanocytes in epidermis   Melanin granules in the dermis

Epidermis

Clusters of malignant melanocytes

Dermis

LM 40×

(b)

Figure 6.4 Malignant Melanoma. (*a*) This mole demonstrates the A, B, C, Ds of melanoma: A = asymmetry, B = irregular borders, C = color, D = diameter >6 mm. (*b*) Histopathology of a malignant melanoma. Note the large clusters of melanocytes in the dermis and the brownish melanin pigment granules scattered throughout. (*a*) ©James Stevenson/Science Source; (*b*) ©McGraw-Hill Education/Christine Eckel, photographer

The Dermis

The dermis consists of two layers: an outer **papillary layer**, and an inner **reticular layer (figure 6.5)**. The papillary layer is the part of the dermis that contains "nipplelike" extensions that project upward into the epidermis (the dermal papillae). The papillary layer is generally quite thin and is composed of areolar connective tissue. The reticular layer is named for its "networked" appearance (*rete*, a net), not because it contains reticular fibers. In fact, the major fiber type found in this layer is the collagen that composes the dense irregular connective tissue. These collagen fibers are interwoven into a meshwork that surrounds structures of the dermis such as hair follicles, sweat glands, sebaceous glands, blood vessels, and nerves. The slide descriptions in this section of the laboratory manual serve as a guide both for general observation of the structure and function of the dermis and for identification of the various skin appendages found within the dermis.

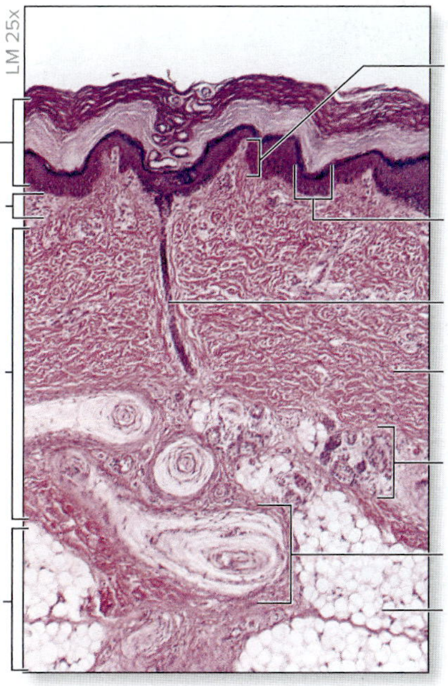

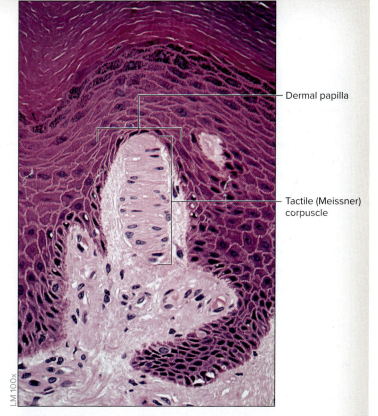

Figure 6.5 **Dermis.** Thick skin demonstrating several skin appendages, such as lamellated corpuscles and merocrine sweat glands in the dermis.
©Ed Reschke

INTEGRATE

LEARNING STRATEGY

The next time you see a piece of leather, look at it carefully and think about your own skin. The very tough tissue that composes most of the leather is dense irregular connective tissue (collagen fibers) from the dermis of the animal's skin.

EXERCISE 6.5

LAYERS OF THE DERMIS

1. Obtain a slide of thick skin or thin skin and place it on the microscope stage.

2. Scan the slide at low power to identify the skin tissue, and then bring the dermal layer of the skin into the center of the field of view.

3. Switch to high power and distinguish the papillary dermis from the reticular dermis (figure 6.5). Locate a **tactile (Meissner) corpuscle** within a dermal papilla **(figure 6.6).** The cells of the tactile corpuscle are oriented horizontally, parallel to the skin surface. Tactile corpuscles are sensory receptors for fine touch (see table 6.3 for a description). Note that tactile corpuscles will not be visible within every dermal papilla on the slide.

4. Identify the following structures of the dermis on the slide, using figures 6.5 and 6.6 as guides:

   ☐ **Dense irregular**
   connective tissue

   ☐ **Dermal papillae**

   ☐ **Papillary dermis**

   ☐ **Reticular dermis**

   ☐ **Tactile (Meissner)**
   corpuscle

Figure 6.6 **Tactile Corpuscle.** High-magnification view of a dermal papilla containing a tactile corpuscle. ©Ed Reschke/Photolibrary/Getty Images

(continued on next page)

(continued from previous page)

5. Sketch the dermis as seen through the microscope in the space provided or on a separate sheet of paper. Be sure to label all the structures listed in step 4 in the drawing.

_____ ×

EXERCISE 6.6

MEROCRINE (ECCRINE) SWEAT GLANDS AND SENSORY RECEPTORS

1. Obtain a slide of thick skin or thin skin and place it on the microscope stage. Begin by observing the epidermal and dermal layers of the skin. When viewing the current slide, focus on the numerous skin appendages found within the dermis.

2. Locate the coiled, tubular glands located deep within the reticular dermis (these glands open to the skin surface, and a duct traveling to the surface of the skin may be visible). These are **merocrine (eccrine) sweat glands** (*meros*, to share, + *krino*, to separate) (figure 6.5). Merocrine sweat glands are located in skin covering nearly the entire body (as opposed to apocrine sweat glands, which are located only in the axillary region, pubic region, and areola). **Table 6.2** lists the types of glands located in the dermis and describes their locations and functions.

3. Focus on sensory receptors found in the dermis. **Table 6.3** summarizes the characteristics of sensory receptors in the

Table 6.2	Glands in the Dermis				
Gland	**Location**	**Description**	**Mode of Secretion**	**Function**	**Word Origin**
Apocrine Sweat Glands	Axilla, areola of the breast, pubic, and anal regions	Coiled tubular glands located next to hair follicles. Ducts open into the hair follicle.	*Merocrine*— exocytosis of vesicles containing product into the duct of the gland	Produce a thick, slightly oily sweat	apo-, away from, + *krino*, to separate or secrete
Merocrine (Eccrine) Sweat Glands	Most of the surface of the body	Coiled tubular glands with main secretory portions found deep within the reticular layer of the dermis. Ducts open to the surface of the skin.	*Merocrine*— exocytosis of vesicles containing product into the duct of the gland	Produce the thin, watery sweat that cools the body	*meros*, share, + *krino*, to separate or secrete
Sebaceous Glands	Wherever hair follicles are found. Particularly abundant on the scalp.	Glands located next to hair follicles. Ducts commonly open into the hair follicle.	*Holocrine*— disintegrated whole cells filled with product are discharged into the duct of the gland	Produce sebum, an oily substance that lubricates the skin surface, keeps it from drying out, and inhibits the growth of bacteria	*sebaceous*, relating to sebum; oily; holos, whole, + *krino*, to separate or secrete

Table 6.3	Sensory Receptors in the Dermis		
Sensory Receptor	**Location**	**Structure/Appearance**	**Function**
Free Nerve Ending	At epidermal/dermal junction	Dendritic endings of sensory neurons. There is no special structure at the end of the nerve (hence "free" nerve ending).	Light touch, temperature, pain, and pressure
Lamellated (Pacinian) Corpuscle	Deep in the dermis and hypodermis	Dendritic endings of sensory neurons are ensheathed with an inner core of neurolemmocytes and outer concentric layers of connective tissue. Lamellated corpuscles resemble an onion in cross section.	Sensation of deep pressure and high-frequency vibration
Tactile (Merkel) Cell	At epidermal/dermal junction	Tactile cells are round cells located in the stratum basale of the epidermis. A tactile cell associates with a sensory nerve ending in the dermis (a tactile disc).	Sensation of fine touch, textures, and shapes
Tactile (Meissner) Corpuscle	In dermal papillae	Highly intertwined dendritic endings enclosed by modified neurolemmocytes and connective tissue. Oval-shaped structure with cells that appear almost layered on top of each other.	Sensation of fine, light touch and texture

dermis. Many of the receptors are difficult to identify histologically, so refer to the gross anatomy section of this manual to identify them on classroom models of integument. Observe the many dermal papillae found in the slide. Center one or more papillae in the field of view and increase the magnification.

4. Using figure 6.6 as a guide, once again try to locate the tiny sensory receptors called **tactile (Meissner) corpuscles** (*tactus,* to touch, + *corpus,* body) within the papillae. These sensory receptors, appropriately located near the surface of the skin, are responsible for sensing fine touch.

5. After identifying the tactile corpuscles, change back to a lower magnification and scan the lower reticular dermis and subcutaneous regions of the skin. Look for a large onion-shaped organ. This is a sensory receptor called a **lamellated (Pacinian) corpuscle** (*lamina,* plate, + *corpus,* body) (figure 6.5). Such sensory organs, located deep within the dermis, are responsible for sensing deep pressure applied to the skin. The other main sensory receptors found at the junction between the dermis and epidermis, the **tactile (Merkel) cells** and **free nerve**

endings, are not easily identifiable through the light microscope, so classroom models may be necessary for their identification.

6. Sketch the skin appendages observed through the microscope in the space provided or on a separate sheet of paper.

_____ ×

INTEGRATE

CONCEPT CONNECTION

The skin is an organ that is composed of both epithelial and connective tissues. Recall that epithelial tissues cover surfaces and are avascular, whereas connective tissues provide support and vary in degree of vascularity. The epidermis of the skin, the outermost layer, is composed of keratinized stratified squamous epithelial tissue. The epidermis aids in temperature regulation, helps prevent excessive water loss, and protects underlying tissues from abrasion and invasion from foreign pathogens. The dermis of the skin is composed of two types of connective tissue. The papillary layer contains areolar connective tissue, whereas the reticular layer contains dense irregular connective tissue. It is critical to relate structure to function. That is, the epidermis contains *stratified* epithelium; therefore, it provides *protection* for underlying structures. The dermis contains largely dense irregular connective tissue, which is composed of large quantities of collagen. Therefore, the dermis provides support and resistance to stress in multiple directions.

THE SCALP—HAIR FOLLICLES AND SEBACEOUS GLANDS

1. Obtain a slide of the **scalp** and place it on the microscope stage. Begin by observing the epidermal and dermal layers of the scalp. The scalp epithelium is thin, but the dermis is thick and contains numerous **hair follicles** (**figure 6.7** and **table 6.4**).

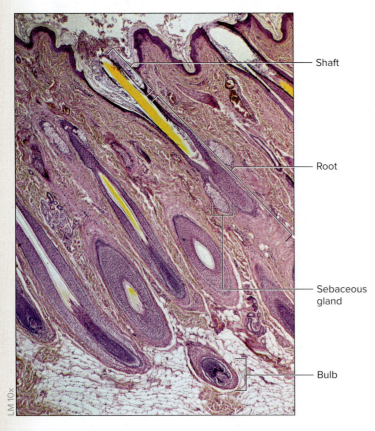

LM 10x

Figure 6.7 Scalp. Skin of the scalp, demonstrating several hair follicles. Arrector pili muscles are not visible in this photomicrograph.
©Ed Reschke/Getty Images

2. Scan the slide until a hair follicle that is sliced longitudinally is visible in the field of view. The entire hair, from the base of the hair follicle to where it exits the skin, should be visible. Notice that the color of the cells lining the hair follicle is similar to the color of the cells within the epidermis. This is because hair follicles are derivatives of the epidermis and they develop as downgrowths of the stratum basale. There are three distinct regions to a hair: (1) the **shaft**, which is the portion of the hair that exits the skin surface; (2) the **root**, which is the portion of the hair within the skin itself; and (3) the **bulb**, which is the swelled base of the hair.

3. Observe the bulb of the hair at higher magnification (**figure 6.8**). Notice the **papilla**, a cone-shaped structure in the middle of the base of the follicle. The papilla is part of the dermis, and is separated from the hair follicle by the basement membrane of the hair follicle epithelium (this basement membrane continues external to the hair follicle as the "glassy membrane"). The papilla contains sensory nerve endings and numerous blood vessels, which are important in supplying nutrients to the developing hair.

4. Return to a lower magnification and look for the oil-secreting **sebaceous glands** (figure 6.7) that connect to the hair follicles in the region of the hair roots. Sebaceous glands secrete an oily substance, **sebum**, into the hair follicle.

5. Carefully observe several hair follicles to locate the small **arrector pili** muscles (*arrector*, that which raises, + *pili*, hair) that attach to the base of the hair follicle (not visible in figure 6.7; see figure 6.1). When these smooth muscles contract, they pull at the base of the hair follicle. This causes the hair to stand up straight rather than lie flat against the surface of the skin. Muscle contraction pulls down on the epidermis of the skin, while the area where the hair shaft exits the epidermis remains elevated. Thus, in humans it gives the appearance of "goose bumps" or "goose pimples."

Table 6.4	Parts of a Hair Follicle
Structure	**Description and Function**
Connective Tissue (Dermal) Root Sheath	The connective tissue of the dermis (mainly dense collagen fibers) that surrounds the entire hair follicle
Cortex	Constitutes the bulk of the hair; composed predominantly of keratin
Cuticle Layer	The outer portion of the hair itself; composed of several layers of hard plates of keratin that surround the cortex of the hair
External Root Sheath	The outer layers of the hair follicle, which are continuous with the stratum basale and stratum spinosum of the epidermis
Glassy Membrane	A specialized basement membrane located external to the external root sheath and internal to the connective tissue that surrounds the hair follicle (the connective tissue root sheath)
Internal Root Sheath	A sheath derived from epithelial tissue that lies between the external root sheath and the hair itself

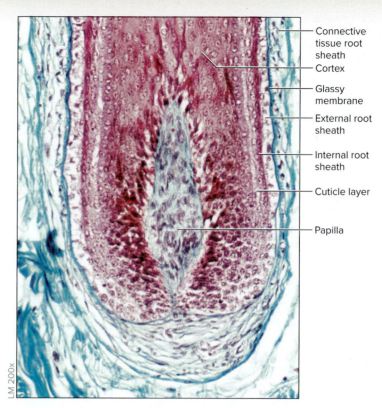

Figure 6.8 **Close-Up View of a Hair Bulb.** ©Biophoto Associates/ Science Source

Labels (top to bottom):
- Connective tissue root sheath
- Cortex
- Glassy membrane
- External root sheath
- Internal root sheath
- Cuticle layer
- Papilla

LM 200x

6. Identify the following structures related to the hair follicle, using figures 6.7 and 6.8 and tables 6.2 and 6.3 as guides:

- ☐ **Arrector pili muscle**
- ☐ **Bulb of hair follicle**
- ☐ **Hair follicle**
- ☐ **Papilla of hair follicle**
- ☐ **Root of hair follicle**
- ☐ **Sebaceous gland**
- ☐ **Shaft of hair follicle**

7. Sketch a hair follicle and associated skin appendages as viewed through the microscope in the space provided or on a separate sheet of paper. Be sure to label all the structures listed in step 6 in the drawing.

8. Next, scan the slide until a hair follicle in cross section is visible within the field of view (**figure 6.9**).

9. Identify the following structures, using figure 6.9 and table 6.4 as guides:

- ☐ **Connective tissue (dermal) root sheath**
- ☐ **Cortex**
- ☐ **Cuticle layer**
- ☐ **External root sheath**
- ☐ **Glassy membrane**
- ☐ **Internal root sheath**

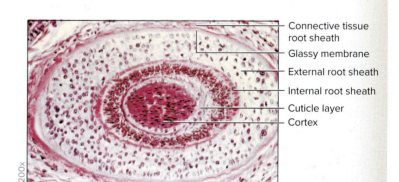

Figure 6.9 **Cross Section of a Hair Follicle.** ©Biophoto Associates/ Science Source

Labels (top to bottom):
- Connective tissue root sheath
- Glassy membrane
- External root sheath
- Internal root sheath
- Cuticle layer
- Cortex

LM 200x

———— ✕

AXILLARY SKIN—APOCRINE SWEAT GLANDS

1. Obtain a slide of **axillary skin** and place it on the microscope stage. Locate a hair follicle on the slide. Notice that the hair follicles in this skin are oriented at a fairly steep angle with respect to the apical surface of the epithelium. This, in part, is what makes the hairs in the axillary region curly.

2. Look for glands that open into the hair follicle. These are **apocrine sweat glands** (*apo-,* away from, + *krino,* to separate or secrete) **(figure 6.10)**. Apocrine sweat glands are predominantly located in the axillary, pubic,

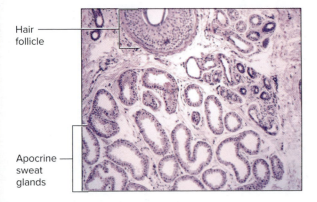

Figure 6.10 Apocrine Sweat Glands. Axillary skin demonstrating apocrine sweat glands. Jose Luis Calvo/Science Source

and anal regions of the body, though they are also located in the areola of the nipple and in men's facial hair. These glands release their secretions via exocytosis, the same mechanism used by merocrine glands. However, apocrine glands produce a secretion that is thicker and oilier than that of the merocrine glands. Although the term *apocrine* historically referred to a different mode of secretion as compared to merocrine glands, these glands are still referred to as apocrine glands.*

3. Sketch a hair follicle from axillary skin and its associated apocrine sweat gland in the space provided or on a separate sheet of paper.

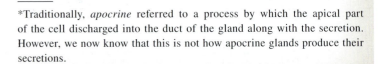

_____ ×

*Traditionally, *apocrine* referred to a process by which the apical part of the cell discharged into the duct of the gland along with the secretion. However, we now know that this is not how apocrine glands produce their secretions.

CONCEPT CONNECTION

The sweat produced by apocrine sweat glands is unique. Similar to merocrine sweat glands, apocrine sweat glands produce sweat that contains water, salt, and nitrogenous wastes. In contrast to merocrine sweat glands, the products of apocrine sweat glands also include lipids and pheromones. A **pheromone** is a type of chemical messenger similar to a hormone. However, whereas hormones act as chemical messengers *within* an individual's own body, pheromones act as chemical messengers *between* two individuals of the same species.

Interestingly, researchers have found that the pheromones produced by apocrine sweat glands are responsible for phenomena such as the coordination of menstrual cycles that occur when two or more women live in the same space for a period of time (such as in college dorm rooms). Pheromones produced

by apocrine sweat glands have also been found to influence mating behavior. This discovery came from a study in which researchers had several male subjects sleep in the same T-shirt two nights in a row. Researchers then had female subjects smell the T-shirts and rate them using an "attraction" scale. Finally, the researchers analyzed the DNA that codes for part of the individual's immune system for both the male and female subjects. The results showed that females consistently rated T-shirts worn by males whose DNA was the most *different* from their own as most attractive. What does this mean? In essence, the better a potential mate smells, the higher the likelihood that mating will produce offspring with a more diverse, hence robust, immune system. Thus, if a potential date smells a little "off," it just might mean that he or she isn't a good genetic match.

EXERCISE 6.9

STRUCTURE OF A NAIL

1. Obtain a slide of a **nail**. A nail is a very specialized skin appendage that arises from epithelial tissue. **Table 6.5** lists the parts of a nail, and **figure 6.11** shows a longitudinal section of a nail.

2. Identify the following structures on the slide, using figure 6.11 as a guide. Then complete the Description column about these structures in table 6.5:

☐ **Body**   ☐ **Hyponychium**

☐ **Eponychium**   ☐ **Nail root**

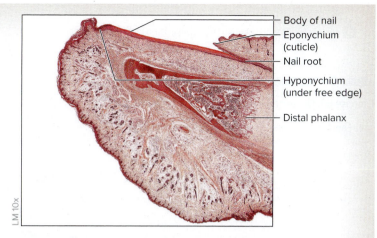

LM 10x

Figure 6.11 **Longitudinal Section of a Nail.** ©McGraw-Hill Education/ Christine Eckel, photographer

3. Sketch a nail as seen through the microscope in the space provided or on a separate sheet of paper. Be sure to label all of the structures listed in step 2 in the drawing.

Table 6.5	Parts of a Nail	
Structure	**Description**	**Word Origins**
Body		
Eponychium		*epi*, upon, + *onyx*, nail
Hyponychium		*hypo*, under, + *onyx*, nail
Lunula		*luna*, moon
Nail Root		

The ❶ corresponds to the Learning Objective(s) listed in the chapter opener outline.

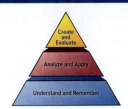

Do You Know the Basics?

Exercise 6.1: Observing Classroom Models of Integument

1. Label the following diagram of the integument. ❶

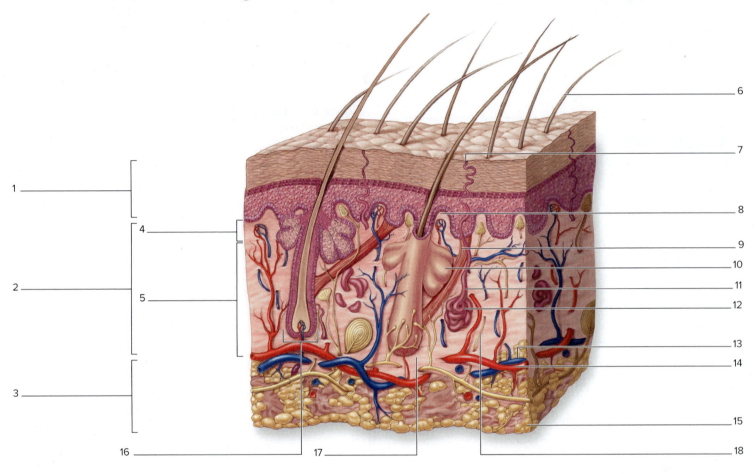

Exercise 6.2: Layers of the Epidermis

2. Which of the following type of tissue is found in the most superficial layer of thick skin? (Circle one.) ❷

   a. adipose connective tissue

   b. areolar connective tissue

   c. dense irregular connective tissue

   d. simple squamous epithelial tissue

   e. stratified squamous epithelial tissue

3. Identify the location(s) on the body where skin contains five layers in the epidermis. (Check all that apply.) ❸

   _____ a. back

   _____ b. face

   _____ c. palms of the hands

   _____ d. scalp

   _____ e. soles of the feet

Exercise 6.3: Fingerprinting

4. Identify structures of the epidermis that are responsible for the formation of fingerprints. (Check all that apply.) **4**

_____ a. dermal papillae

_____ b. epidermal ridges

_____ c. keratinocytes

_____ d. melanocytes

_____ e. tactile corpuscles

Exercise 6.4: Pigmented Skin

5. Although melanocytes are the cells that produce melanin, the cells that actually concentrate the melanin in the apical region of the cell to protect the cell nucleus are

_____ (keratinocytes/tactile cells). **5**

Exercise 6.5: Layers of the Dermis

6. Match the definition listed in column A with the corresponding term listed in column B. **6**

Column A	*Column B*
_____ 1. a cell located in the epidermis that produces melanin	a. dermal papilla
_____ 2. an epithelial cell composing the majority of the epidermis	b. epidermal ridge
_____ 3. a fold of the epidermis that interdigitates with dermal papillae	c. keratinocyte
_____ 4. a "nipplelike" extension of the dermis into the epidermis	d. melanocyte

7. The _____ (papillary/reticular) layer of the dermis is the thickest layer of the dermis. **6**

Exercise 6.6: Merocrine (Eccrine) Sweat Glands and Sensory Receptors

8. Match the description listed in column A with the type of gland listed in column B. **7**

Column A	*Column B*
_____ 1. merocrine glands that empty into hair follicles	a. apocrine sweat glands
_____ 2. holocrine glands that empty into hair follicles	b. eccrine sweat glands
_____ 3. merocrine glands that open to the surface of the skin	c. sebaceous glands

9. Lamellated (Pacinian) corpuscles are located in the_____ (papillary/reticular) layer of the dermis, while tactile (Meissner) corpuscles are located in the

_____ (papillary/reticular) layer of the dermis. **7**

10. Match the function listed in column A with the corresponding sensory receptor listed in column B. **8**

Column A	*Column B*
_____ 1. sensation of fine touch, textures, and shapes	a. free nerve ending
_____ 2. sensation of light touch, temperature, pain, and pressure	b. lamellated (Pacinian) corpuscle
_____ 3. sensation of deep pressure and high-frequency vibration	c. tactile (Meissner) corpuscle
_____ 4. sensation of fine, light touch and texture	d. tactile (Merkel) cell

Exercise 6.7: The Scalp—Hair Follicles and Sebaceous Glands

11. Match the description and function listed in column A with the corresponding structure listed in column B. **9**

Column A	*Column B*
_____ 1. the swelled base of the hair	a. arrector pilli
_____ 2. the portion of the hair that exits the skin surface	b. bulb
_____ 3. the portion of the hair contained within the skin	c. root
_____ 4. small muscles attached to the base of the hair follicle	d. sebaceous gland
_____ 5. oil-secreting glands that connect to the hair follicle	e. shaft

22. A hypodermic needle is used to give certain types of injections.

    a. Based on the name of the needle, what space is the tip of the needle usually directed into?

    _____

    b. What layers of the skin must the hypodermic needle pass through in order to get to this space? (In your answer, include all sublayers of the dermis or epidermis that may apply. Assume the needle is passing through thin skin.)

    _____

    _____

    _____

    _____

    c. What structures do you think are the likely targets of these needles?

    _____

    _____

23. a. What layer of the epidermis represents the transition from living to dead epithelial cells?

    _____

    b. Why do keratinocytes begin to die within this layer?

    _____

    _____

    _____

24. Contrast the epidermal/dermal junction in thick skin with that of thin skin. Specifically note the structure of the epidermal/dermal junction in thick vs. thin skin. Is there a difference in the number of dermal papillae? What function do you think such differences serve?

    _____

    _____

    _____

    _____

    _____

    _____

25. Why is it important that melanin is present in its highest concentration in the keratinocytes at or near the basal layer of cells? (*Hint:* Were cells that were undergoing cell division observed in this layer?)

    _____

    _____

    _____

    _____

26. a. What is the advantage of having tactile (Meissner) corpuscles located near the surface of the skin?

    _____

    _____

    b. Do you think there would be more of these sensory receptors per unit area in the skin on the palm of the hand or in the skin on the back (or would they be the same)?

    _____

    _____

The Skeletal System: Bone Structure and Function

chapter **7**

OUTLINE AND LEARNING OBJECTIVES

Anatomy & Physiology Revealed® 4.0

Module 5: Skeletal System

GROSS ANATOMY

Classification of Bones

Bones of the human skeleton appear in various shapes and sizes, depending on their function(s). The four categories of bone classified by shape are flat, irregular, long, and short. The characteristics of this classification scheme are summarized in **table 7.1** and an example of each is shown in **figure 7.1**.

Table 7.1	Classification of Bones Based on Shape	
Class of Bone	**Description**	**Examples**
Flat	Have thin, flat surfaces; may be slightly curved	Many skull bones (e.g., frontal, parietal)
Irregular	Complex shape that does not fit into other classifications	Vertebrae, some skull bones (e.g., ethmoid, sphenoid)
Long	Greater in length than width	Most limb bones (e.g., humerus, femur, metacarpals)
Short	Nearly equal in length and width	Wrist and ankle bones (e.g., carpals, tarsals)

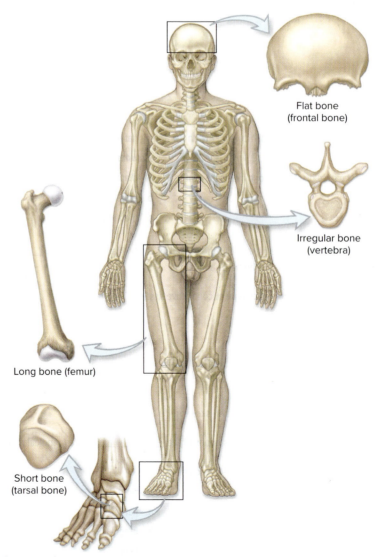

Flat bone
(frontal bone)

Irregular bone
(vertebra)

Long bone (femur)

Short bone
(tarsal bone)

Figure 7.1 **Structural Classifications of Bones.**

IDENTIFYING CLASSES OF BONES BASED ON SHAPE

1. Obtain a box containing disarticulated human bones. Pull each bone out of the box and classify the bone according to its shape. Note that there is a fair amount of variability in the shape of bones within each classification.

2. **Figure 7.2** contains photographs of several disarticulated human bones. In the space next to each bone, name the category to which each bone belongs: flat, irregular, long, or short.

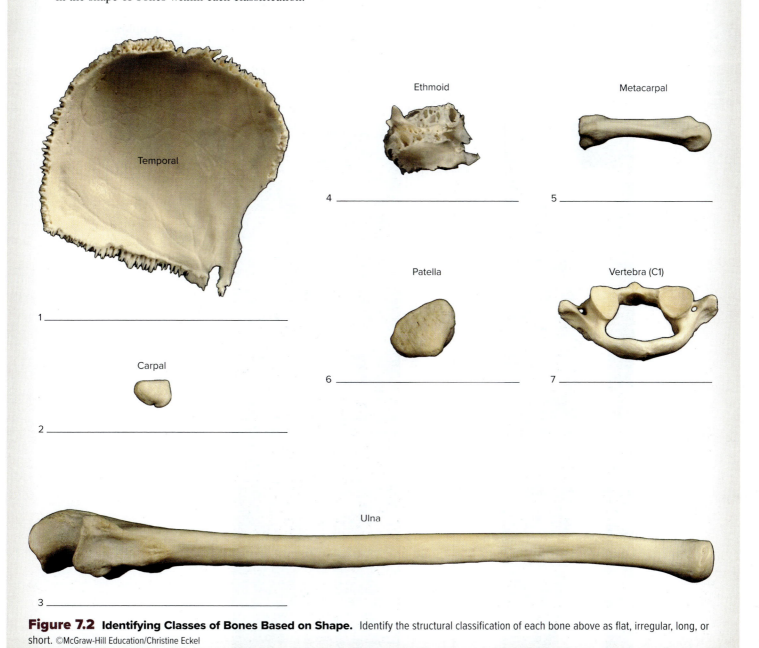

Figure 7.2 Identifying Classes of Bones Based on Shape. Identify the structural classification of each bone above as flat, irregular, long, or short. ©McGraw-Hill Education/Christine Eckel

Structure of a Typical Long Bone

A typical long bone such as the femur (**figure 7.3**) is composed of a long shaft, called the **diaphysis** (*dia,* through, + *physis,* growth); rounded ends, called **epiphyses** (*epi,* upon, + *physis,* growth); and articulation points between the two, called **metaphyses** (*meta,* between, + *physis,* growth). Within the shaft is a large cavity called the **medullary cavity,** which is filled with **yellow bone marrow** (adipose tissue) in the adult. The walls of the diaphysis are composed of a thick layer of **compact bone** tissue. The epiphyses of the bone are surrounded by a thin layer of compact bone and have **articular cartilages** on the ends. **Spongy bone** tissue is found within the epiphyses of the bone. In the fetus, the marrow spaces between the trabeculae of spongy bone are composed of red bone marrow. However, in the adult they are composed mainly of yellow bone marrow because of the conversion of red marrow to yellow marrow that occurs as the skeleton matures. In the adult, red bone marrow is primarily limited to the proximal epiphyses of the humerus and femur, the sternum, and the iliac crest. Observing a fresh bone specimen allows observation of many of the tissues that normally associate with bones, such as periosteum, articular cartilages, muscles, tendons, ligaments, marrow, and blood vessels. The following exercises involve observing a fresh specimen of a long bone from a cow, and an articulated human skeleton. The goal of these exercises is to familiarize students with the major bones of the human body.

COMPONENTS OF A LONG BONE

1. Obtain both an intact femur (not cut) and a femur that has been cut along its longitudinal axis.

2. Identify the following components of a long bone, using figure 7.3 as a guide:

☐ **Compact bone**   ☐ **Epiphysis**

☐ **Diaphysis**   ☐ **Medullary cavity**

☐ **Epiphyseal line**   ☐ **Spongy bone**

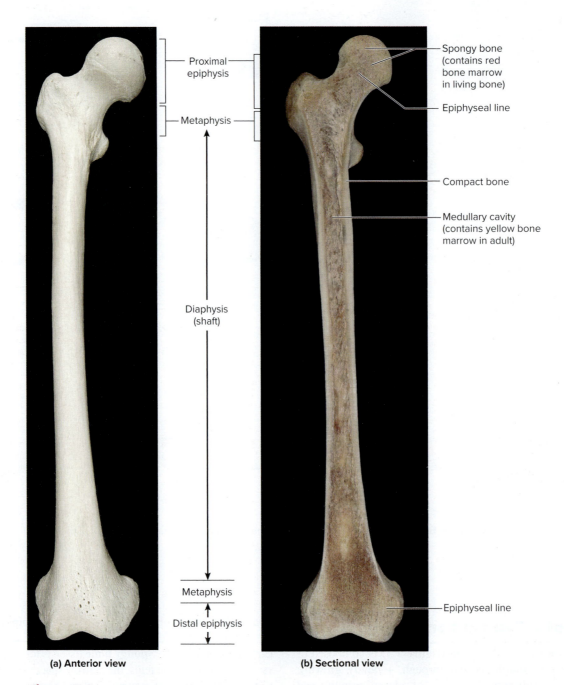

Proximal epiphysis

Metaphysis

Diaphysis (shaft)

Metaphysis

Distal epiphysis

Spongy bone (contains red bone marrow in living bone)

Epiphyseal line

Compact bone

Medullary cavity (contains yellow bone marrow in adult)

Epiphyseal line

(a) Anterior view   (b) Sectional view

Figure 7.3 Gross Anatomy of a Typical Long Bone. The right femur. (*a*) ©McGraw-Hill Education/Christine Eckel, photographer; (*b*) ©Elizabeth Pennefather-O'Brien

COW BONE DISSECTION

1. Obtain a dissecting pan, a blunt probe, forceps, and a fresh cow bone cut in a longitudinal section. Place the bone in the dissecting pan and begin your observations. Find the large medullary cavity that is filled with yellow bone marrow (**figure 7.4**). Notice the thick layer of compact bone that surrounds the medullary cavity.

2. If using a dissecting microscope or magnifying glass, focus in on the compact bone tissue. Notice how "bloody" it appears. Are there any tiny dots of blood within the compact bone? These result from the rupture of tiny blood vessels in the tissue when the bone was sectioned. This observation should reaffirm that living bone is a highly vascular, metabolically active tissue—quite different from the appearance of preserved bones.

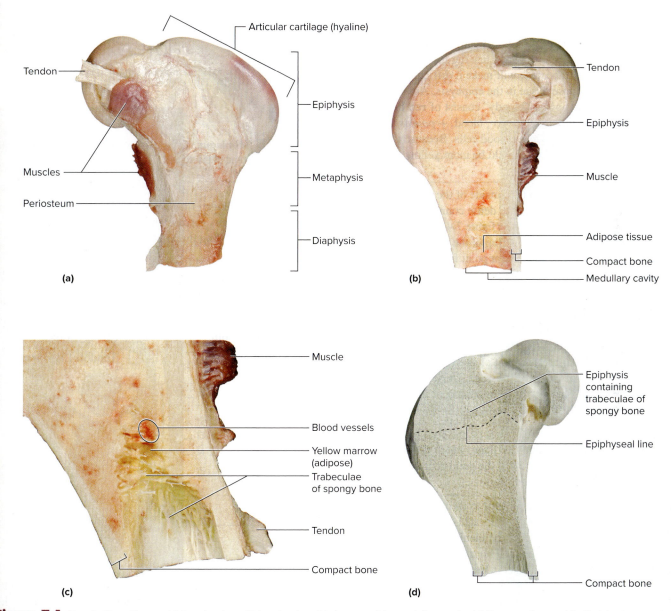

Figure 7.4 **Fresh Cow Bone.** (*a*) Exterior view, (*b*) interior view, (*c*) close-up of the medullary cavity. (*d*) The same bone, with all of the fat, muscle, tendons, cartilage, and blood vessels removed. (*a-d*) ©Christine Eckel

(continued on next page)

COMPACT BONE

1. Obtain a slide of **ground compact bone** (**figure 7.7**) and place it on the microscope stage. Bring the tissue sample into focus on low power and locate an **osteon**. Remember, living cells are not visible in this tissue sample.

2. With the osteon at the center of the field of view, increase the magnification. The **central canals** will be the largest holes visible in the sample. **Concentric lamellae** extend outward from the central canals in concentric rings. **Lacunae**, which house osteocytes and appear very dark, are found at the border between two adjacent lamellae. **Interstitial lamellae** can be seen between osteons. **Circumferential lamellae** can be viewed by scanning the outer border of the bone sample, because circumferential lamellae surround the entire diaphysis of the bone.

   Observe this slide at the highest magnification possible (without using the oil immersion lens). Note the tiny **canaliculi**, which contain cytoplasmic extensions of osteocytes in living bone tissue and appear as tiny "cracks" that run perpendicular to the central canals. Within the canaliculi, osteocytes connect to each other via gap junctions, which allow them to exchange nutrients and other substances with each other.

3. Scan the slide to look for large canals that run perpendicular to the central canals. These canals are **perforating canals**, which convey blood vessels from the outer periosteum into the central canals.

4. Obtain a slide of **decalcified compact bone** (**figure 7.8**) and place it on the microscope stage. Bring the tissue sample into focus on low power and locate an **osteon**. This slide will demonstrate the remnants of living structures (cells, blood vessels, and nerves) within the bone.

5. With an osteon at the center of the field of view, increase the magnification. The **central canals** will be the largest holes visible in the sample. The center of the central canals contains blood vessels and nerves. **Lacunae** will appear to be much smaller holes, and there will only be one cell (an osteocyte) inside each lacuna. Lacunae are located at the border between two adjacent **lamellae**. Scan toward the outer border of the bone sample to identify **circumferential lamellae**, which surround the entire diaphysis of the bone.

6. Scan the outer edge of the bone tissue on the slide and identify the **periosteum** of the bone. The periosteum contains an outer layer of dense irregular connective tissue and an inner layer of **osteoprogenitor cells**, which may or may not be visible.

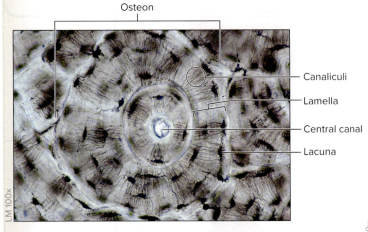

Osteon — Canaliculi — Lamella — Central canal — Lacuna

LM 100×

APR **Figure 7.7** **Ground Compact Bone.**
No perforating canals are visible in this section. ©Ed Reschke/Getty Images

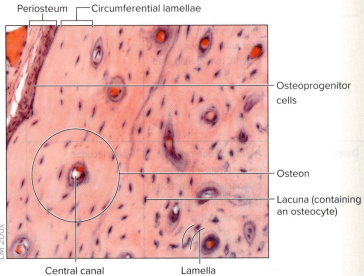

Periosteum — Circumferential lamellae — Osteoprogenitor cells — Osteon — Lacuna (containing an osteocyte)

Central canal — Lamella

LM 200×

Figure 7.8 **Decalcified Compact Bone.** ©McGraw-Hill Education/Al Telser

7. Make sketches of ground compact bone and decalcified compact bone as viewed through the microscope in the spaces provided or on a separate sheet of paper. Label the following on the sketches, using table 7.2 and figures 7.7 and 7.8 as guides:

☐ **Central canal** ☐ **Osteocyte**
☐ **Circumferential lamellae** ☐ **Osteon**
☐ **Concentric lamellae** ☐ **Osteoprogenitor cell**
☐ **Interstitial lamellae** ☐ **Perforating canal**
☐ **Lacuna** ☐ **Periosteum**
☐ **Osteoblast**

Ground compact bone

_____ ×

Decalcified compact bone

_____ ×

EXERCISE 7.6

SPONGY BONE

1. Obtain a slide of decalcified **spongy bone** (**figure 7.9**) and place it on the microscope stage. The sample of spongy bone will appear most similar to the slide of decalcified compact bone. The main difference between the two is that the spongy bone does *not* contain osteons, so neither central canals nor perforating canals are visible.

2. Observe the slide at high magnification and look for areas where several small cells are lined up next to each other on the edge of a trabecula. These are **osteoblasts** (figure 7.9*a*) that actively secrete new bony matrix. Are any very large, multinucleate cells visible on the slide? If so, these are **osteoclasts** involved in bone resorption (figure 7.9*b*). Both osteoclasts and osteoblasts are actively involved in the process of bone remodeling.

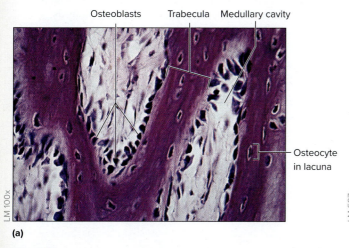

(a)

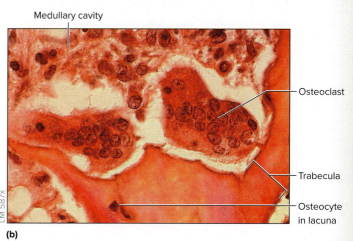

(b)

APR Figure 7.9 **Spongy Bone.** Decalcified sections of spongy bone showing (*a*) osteoblasts and (*b*) osteoclasts. Note how the osteoblasts appear to be lined up in rows along the lacuna in (*a*). The osteoclast in (*b*) is very large compared to adjacent cells and is multinucleate. (*a*) ©Ed Reschke/Getty Images; (*b*) ©Michael Klein/Getty Images

<it>navigation</it>*(continued on next page)*

(continued from previous page)

3. Sketch spongy bone as viewed through the microscope in the space provided or on a separate sheet of paper. Label the following on the sketch, using table 7.2 and figure 7.9 as guides:

☐ **Lacuna** ☐ **Osteoclast**

☐ **Medullary cavity** ☐ **Osteocyte**

☐ **Osteoblast** ☐ **Trabecula**

_____ ×

INTEGRATE

CONCEPT CONNECTION

Recall that bone is a supporting connective tissue that protects organs and allows body movement by providing attachment sites for muscles. Bone matrix is composed of both organic and inorganic components. *Osteoid,* the organic portion of bone, contains a dense collection of collagen fibers and proteoglycans that provide strength and flexibility. This is the "soft" portion of bone that allows for the penetration of blood vessels during development. The "hard," inorganic portion of bone contains crystals of the mineral compound hydroxyapatite, which is composed of calcium and phosphate. This mineralized matrix also provides rigidity and incompressiblility to bone. As bone tissue is loaded by mechanical stresses, osteoblasts are stimulated to secrete osteoid, which leads to calcification of the bone matrix. When mechanical stresses are reduced (e.g., when a limb is immobilized in a cast for some time), osteoclasts are stimulated to secrete substances to break down the osteoid and hydroxyapatite. The result of the breakdown of the bony matrix is the release of calcium and phosphorus into the blood.

EXERCISE 7.7

ENDOCHONDRAL BONE DEVELOPMENT

1. Obtain a slide of **developing long bone (figure 7.10)** and place it on the microscope stage. This slide will typically contain the developing femur of a young mammal. The femur develops using **endochondral ossification**, a process by which a hyaline cartilage model of the bone is gradually replaced by bone tissue.

2. Scan the slide at the lowest magnification and identify the parts of the bone—specifically, the epiphysis, diaphysis, and metaphysis. In the metaphysis of a developing long bone is an **epiphyseal plate**. Name the type of tissue found at this location: _____
_____
_____

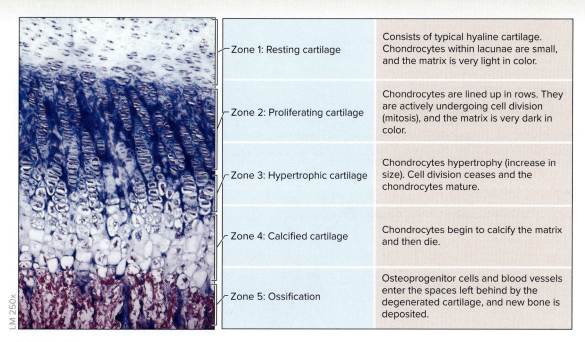

Zone 1: Resting cartilage	Consists of typical hyaline cartilage. Chondrocytes within lacunae are small, and the matrix is very light in color.
Zone 2: Proliferating cartilage	Chondrocytes are lined up in rows. They are actively undergoing cell division (mitosis), and the matrix is very dark in color.
Zone 3: Hypertrophic cartilage	Chondrocytes hypertrophy (increase in size). Cell division ceases and the chondrocytes mature.
Zone 4: Calcified cartilage	Chondrocytes begin to calcify the matrix and then die.
Zone 5: Ossification	Osteoprogenitor cells and blood vessels enter the spaces left behind by the degenerated cartilage, and new bone is deposited.

Figure 7.10 **Epiphyseal Plate.** The five functional layers of the epiphyseal plate. In this photomicrograph, the epiphyseal side of the plate is located at the top of the photo, whereas the diaphyseal side of the plate is located at the bottom of the photo. ©Biophoto Associates/Science Source

3. Position the metaphysis at the center of the field of view and increase the magnification. Identify the five functional layers within the epiphyseal plate, using figure 7.10 as a guide.

4. Sketch the epiphyseal plate as viewed through the microscope in the space provided, or on a separate sheet of paper. Label the following on the sketch, using figure 7.10 and the textbook as guides:

☐ **Zone of calcified cartilage**

☐ **Zone of hypertrophic cartilage**

☐ **Zone of ossification**

☐ **Zone of proliferating cartilage**

☐ **Zone of resting cartilage**

_____ ×

INTEGRATE

CONCEPT CONNECTION

Many hormones influence normal growth of bone tissue. Some hormones stimulate osteoblast activity and promote bone growth, whereas other hormones stimulate an increase in bone resorption by osteoclasts and inhibit bone growth. Thus, disorders involving hormone changes may often manifest themselves as disorders of the skeletal system.

INTEGRATE

LEARNING STRATEGY

A developing long bone has five functional layers in the epiphyseal plate. As the bone develops, cartilage is gradually replaced by bone tissue. There are five functional layers of the epiphyseal plate. A mnemonic to remember the layers, from outermost to innermost, is: **R**esilient **P**eople **H**ave **C**areer **O**ptions.

(1) **R**esting cartilage; (2) **P**roliferating cartilage; (3) **H**ypertrophic cartilage; (4) **C**alcified cartilage; and (5) **O**ssification

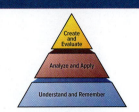

The ❶ corresponds to the Learning Objective(s) listed in the chapter opener outline.

 Do You Know the Basics?

Exercise 7.1: Identifying Classes of Bones Based on Shape

1. Which of the following is the correct classification for a vertebra based on shape? (Circle one.) ❶

   a. flat   b. irregular   c. long   d. short

Exercise 7.2: Components of a Long Bone

2. Label the parts of a typical long bone on the following diagram. ❷

1 _____

2 _____

3 _____

4 _____

5 _____

6 _____

7 _____

8 _____

9 _____

10 _____

Exercise 7.3: Cow Bone Dissection

3. Which of the following is the outer covering of bone that serves as an attachment point for tendons and ligaments? (Circle one.) ❸

   a. articular cartilage   b. medullary cavity   c. periosteum   d. trabeculae

4. The thick layer of bone surrounding the medullary cavity is composed of _____ (compact/spongy) bone tissue. ❸

Exercise 7.4: The Human Skeleton

5. Which of the following bone(s) is/are part of the axial skeleton? (Check all that apply.) ❹

   _____ a. pubis   _____ b. rib   _____ c. scapula   _____ d. skull   _____ e. sternum

6. Which of the following bones compose the os coxae? (Check all that apply.) ❹

   _____ a. coccyx   _____ b. ilium   _____ c. ischium   _____ d. pubis   _____ e. sacrum

7. The pectoral girdle consists of which of the following bones? (Check all that apply.)

   _____ a. clavicle   _____ b. ilium   _____ c. ischium   _____ d. sacrum   _____ e. scapula ❹

Exercise 7.5: Compact Bone

8. Label the components of compact bone on the following diagram. **5** **6**

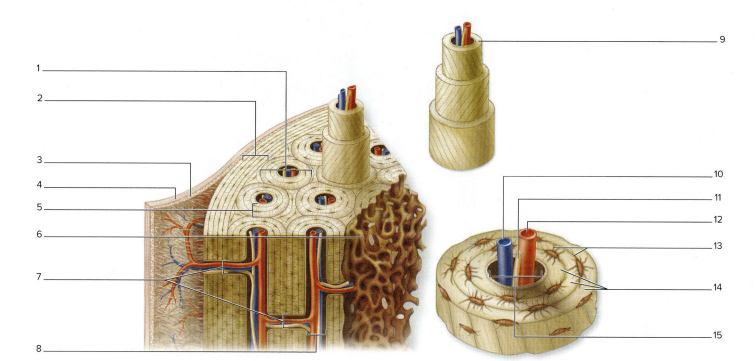

Exercise 7.6: Spongy Bone

9. Which of the following is/are visible on a slide of decalcified spongy bone? (Check all that apply.) **7**

    _____ a. central canal

    _____ b. medullary cavity

    _____ c. osteocyte

    _____ d. osteon

    _____ e. perforating canal

    _____ f. trabecula

Exercise 7.7: Endochondral Bone Development

10. Which of the following bones undergo endochondral bone development (ossification)? (Check all that apply.) **8**

    _____ a. femur

    _____ b. frontal bone

    _____ c. radius

    _____ d. temporal bone

    _____ e. tibia

11. The figure below is a light micrograph of developing endochondral bone. The dotted line divides the epiphyseal plate into two major regions, a and b. One region is **new bone tissue** and the other is **hyaline cartilage**. Label these two components of the developing bone on the following diagram. **9**

    a. _____

    b. _____

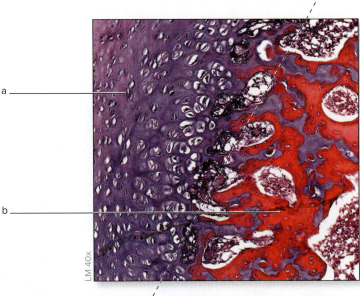

a ————

b ————

LM 40x

©McGraw-Hill Education/Christine Eckel

 Can You Apply What You've Learned?

12. You are a biomedical researcher who is interested in designing a drug to increase bone density. You have decided you can take one of two approaches: (1) design a drug that stimulates bone to be built faster, or (2) design a drug that prevents bone from breaking down. To begin this process, you need to study the cells that would likely respond to either drug 1 or drug 2. What cells are these?

   a. _____ (build bone)   b. _____ (break down bone)

13. What process must stop in order for the epiphyseal plate to close? (That is, which layer of the plate must stop its development first?)

   _____

   _____

14. A forensic anthropologist identifies a bone as a left femur with epiphyses that are detached from the diaphysis. Does this bone belong to a juvenile or an adult? Justify your answer.

   _____

   _____

Can You Synthesize What You've Learned?

15. One effect of growth hormone (GH) is the stimulation of cartilage growth. Discuss how decreased GH production during development might impact a person's height.

   _____

   _____

   _____

16. Why might astronauts, who spend only a few days in outer space under microgravity (reduced gravity), experience a significant loss in bone density?

   _____

   _____

   _____

17. Vitamin D promotes increased dietary calcium and phosphorus absorption in the small intestine. Describe why doctors may recommend vitamin D supplements to patients diagnosed with osteoporosis (weak/brittle bones).

   _____

   _____

   _____

chapter 8

The Skeletal System: Axial Skeleton

OUTLINE AND LEARNING OBJECTIVES

Anatomy & Physiology Revealed® 4.0

Module 5: Skeletal System

INTRODUCTION

The skeletal system is typically composed of 206 bones in an adult and is organized into two divisions: the axial skeleton and the appendicular skeleton. The axial skeleton consists of bones that form the main axis of the body—the skull, vertebrae, ribs, and sternum, which collectively form the body's core structural foundation. Although it may appear that the skull is one solid bone with a movable mandible (jawbone), the skull is actually comprised of 22 bones joined together by sutures. The form ensures that the brain is well protected. When the middle of the back is palpated, the vertebral column also feels solid. In fact, the vertebral column is comprised of 26 bones that allow for movement of the trunk, neck, and head, while still protecting the spinal cord. Similarly, the rib cage consists of 12 pairs of ribs, most connecting to the sternum, that protect the heart and lungs while allowing for the movement associated with breathing and other motions of the trunk.

The exercises in this chapter involve observing the bones that compose the axial skeleton on articulated human skeletons, on disarticulated bones of the human skeleton, or on bone models. While completing the exercises, be sure to view both individual disarticulated bones (e.g., a cervical vertebra) and the same bones on an articulated skeleton. This will increase your understanding of how individual bones fit in with the rest of the axial skeleton (e.g., how 26 bones make up the vertebral column). Each of the bones has identifying features and markings. When learning the processes, projections, foramina (holes), and other markings of the bones, view each structure, study it closely, and contemplate its function. The process may be an attachment point for a ligament, a tendon, or a muscle. The opening or hole may serve as a passageway for a nerve, artery, or vein. The smooth surface may be where the bone articulates with another bone. For each structure, relate form to function.

Finally, learning the names of bones and the bone markings is like learning a new language. In fact, most of the names are derived from Latin or Greek words. It is helpful to learn the meaning of each word as a way to remember each bone. For each bone and bone marking, the word origins listed in the tables in this chapter serve as a guide to learning their names.

List of Reference Tables

These Pre-Laboratory Worksheet questions may be assigned by instructors through their ▇ connect course.

1. Match the description of the bone feature listed in column A with the appropriate name listed in column B.

   Column A

   _____ 1. flattened or shallow depression

   _____ 2. small, flat, shallow, articulating surface

   _____ 3. large, smooth, and round projection

   _____ 4. marked bony prominence

   Column B

   a. condyle

   b. facet

   c. process

   d. fossa

2. Which of the following is/are bony features that function as attachment points for tendons and ligaments? (Check all that apply.)

   _____ a. fossa

   _____ b. process

   _____ c. sinus

   _____ d. tuberosity

3. A round hole that passes through a bone is called a _____ (fissure/foramen), whereas a narrow, slit-like opening through a bone is called a _____ (fissure/foramen).

4. In a typical human, how many vertebrae (individual or fused) does each section of the vertebral column contain? Write the number in the space provided.

   _____ a. cervical

   _____ b. thoracic

   _____ c. lumbar

   _____ d. sacral

   _____ e. coccygeal

5. The sella turcica is a feature of which bone? (Circle one.)

   a. ethmoid bone   b. frontal bone   c. occipital bone   d. sphenoid bone

6. All ribs articulate with which of the following vertebrae? (Circle one.)

   a. cervical   b. coccygeal   c. lumbar   d. sacral   e. thoracic

7. Arrange the parts of the sternum from superior (1) to inferior (3) by placing the numbers 1–3 next to the parts.

   _____ a. body

   _____ b. manubrium

   _____ c. xiphoid process

8. Which is the only skull bone that is mobile (movable)? _____

9. Which of the following bone(s) contain(s) a paranasal sinus? (Check all that apply.)

   _____ a. ethmoid bone

   _____ b. frontal bone

   _____ c. mandible

   _____ d. maxilla

   _____ e. temporal bone

10. The suture that forms between the frontal and parietal bones is called the _____ (coronal/lambdoid/sagittal) suture.

GROSS ANATOMY

Bone Markings

Prior to learning the specific bones of the axial and appendicular skeleton, it is important to become familiar with general terminology that describes the features of each of the bones. Recall from chapter 7 that all bones undergo a process of ossification, whereby osteoblasts lay down new bony matrix (osteoid) during bone development. At the same time, muscles, tendons, and ligaments attach to these bones as they begin to develop, and blood vessels and nerves pass through or between bones as the vessels grow into their target organs. Continual movement during development, particularly when bones are soft and pliable, leads to the formation of distinguishing features on each bone. Once a bone completely ossifies, the markings become solid, recognizable features of the bone. In general, smooth surfaces are found on articulating surfaces (surfaces that form joints), projections represent points of muscle, ligament, and tendon attachment, and foramina are passageways for blood vessels and nerves. **Table 8.1** lists each of the major bone markings and provides a description and general definition for the marking. This information will assist in understanding what kind of structure to look for when identifying individual bones and their respective features in subsequent exercises.

Table 8.1	Bone Markings		
General Structure	**Anatomic Term**	**Description**	**Word Origin**
Articulating Surfaces	Condyle	Large, smooth, round articulating structure	*kondylos*, a knuckle
	Facet	Small, flat, shallow articulating surface	*facet*, a small face
	Head	Prominent, rounded epiphysis	NA
	Trochlea	Smooth, grooved, pulley-like articular process	*trochlea*, a pulley block
Depressions	Alveolus (pl., *alveoli*)	Deep pit or socket in the maxillae or mandible	*alveus*, a hollow cavity
	Fossa (pl., *fossae*)	Flattened or shallow depression	*fossa*, a pit/cavity
	Sulcus	Narrow groove	*sulcus*, a furrow/groove
Projections	Crest	Narrow, prominent, ridge-like projection	*crista*, a crest
	Epicondyle	Projection adjacent to a condyle	*epi*, above + *kondylos*, a knuckle
	Line	Low ridge	*linea*, line
	Process	Any marked bony prominence	*processus*, a projection
	Ramus (pl., *rami*)	Angular extension of a bone relative to the rest of the structure	*ramus*, a branch
	Spine	Pointed, slender process	*spinosus*, a spine
	Trochanter	Massive, rough projection found only on the femur	*trokhanter*, to run
	Tubercle	Small, round projection	*tuberculum*, a small bump
	Tuberosity	Large, rough projection	*tuberosus*, knobby
Openings and Spaces	Canal	Passageway through a bone	*canalis*, channel
	Fissure	Narrow, slit-like opening through a bone	*fissura*, cleft
	Foramen (pl., *foramina*)	Rounded passageway through a bone	*foramen*, a hole
	Meatus	Passageway through a bone	*meatus*, a channel
	Sinus	Cavity or hollow space in a bone	*sinus*, a curve or bay

The Skull

The bones that make up the skull are separated into two functional categories: **cranial bones** (frontal, parietal, temporal, occipital, sphenoid, and ethmoid) and **facial bones** (maxilla, mandible, zygomatic, nasal, lacrimal, palatine, inferior nasal conchae, and vomer). The roof of the cranium—the **calvaria**, or skullcap—is the dome-shaped part of the skull that protects the brain. In an adult, all of the skull bones, with the exception of the mandible, are fused to each other via synarthrotic (immovable) joints called **sutures**.

Table 8.2 describes each individual bone of the skull. An organized approach to learning the bones of the skull begins with viewing the skull from six points of reference: anterior view, lateral view, posterior view, superior view, inferior view, and superior view of the cranial floor. For each view of the skull, first identify the individual bones that are visible. Second, identify all processes, foramina, and major features (often formed from multiple bones) that are visible in each view of the skull. Always relate the bony processes, fossae, and foramina to the individual bone(s) from which they are formed.

Table 8.2	The Axial Skeleton: Skull Bones and Important Bony Markings		
Major Bone	**Bone Features**	**Description and Related Structures of Importance**	**Word Origins**
Ethmoid	Cribriform plate	Forms roof of nasal cavity and part of cranial floor	*cribrum*, sieve + *forma*, form + *platus*, flat
	Cribriform foramina	The olfactory nerve (CN I) passes through to the brain	*cribrum*, sieve + *forma*, form + *foramen*, a hole
	Crista galli	Projection that serves as attachment point for meninges	*crista galli*, cockscomb
	Superior nasal concha	Forms superior lateral wall of nasal cavity; causes turbulent airflow	*superus*, upper + *nasus*, nose + *concha*, shell
	Middle nasal concha	Forms middle lateral wall of nasal cavity; causes turbulent airflow	*middle*, middle + *nasus*, nose + *concha*, shell
	Perpendicular plate	Forms superior part of nasal septum	*perpendiculum*, plumb line + *platus*, flat
Frontal	Frontal sinus	A cavity within frontal bone	*frontellum*, forehead + *sinus*, a hollow
	Supraorbital foramen (notch)	A hole or notch on the superior ridge of orbit	*supra-*, above + *orbit*, eye socket + *foramen*, a hole
	Supraorbital margin	Bony support and protection of superior border of orbit	*supra-*, above + *orbit*, eye socket + *margo*, a border
Inferior Nasal Conchae	NA	Forms inferior part of lateral wall of nasal cavity; causes turbulent airflow	*inferus*, lower + *nasus*, nose + *concha*, shell
Lacrimal	Lacrimal groove	Forms medial, inferior aspect of orbit of eye. Groove connects orbital and nasal cavities.	*lacrima*, tear + *groove*, a pit
Mandible	Alveolar processes	Cavities that form tooth "sockets"	*alveolus*, a trough + *processus*, a projection
	Angle	Portion of mandible connecting the body to the ramus, forming a right angle	*angulus*, a corner
	Body	Anterolateral portion of mandible	NA
	Coronoid process	Attachment point for muscle of the temporomandibular joint	*corona*, crown + *eidos*, resembling + *processus*, a projection
	Head	Forms a joint with the mandibular fossa of temporal bone (temporomandibular joint)	NA
	Mandibular foramen	Passageway for cranial nerve	*mandere*, to chew + *bula*, a means + *foramen*, a hole
	Mental foramen	Passageway for cranial nerve and an artery	*mental*, chin + *foramen*, a hole
	Mental protuberance	Anterior projection of mandible that forms the chin	*mental*, chin + *protuberare*, to swell
	Ramus	Part of bone that forms an angle with body of mandible	*ramus*, branch
Maxilla	Infraorbital foramen	Passageway for cranial nerve and an artery	*infra-*, below *orbit*, eye socket + *foramen*, a hole
	Incisive foramen (fossa)	Contains arteries and nerves passing from nasal cavity into oral cavity	*incidere*, to cut into + *foramen*, a hole
	Palatine process	Forms most of the hard palate (separation between nasal and oral cavities)	*palatin*, the palate + *processus*, a projection

(continued on next page)

Table 8.2		The Axial Skeleton: Skull Bones and Important Bony Markings *(continued)*	
Major Bone	**Bone Features**	**Description and Related Structures of Importance**	**Word Origins**
Nasal	NA	Forms most of the bridge of nose	NA
Occipital	External occipital protuberance	Large projection palpated on the posterior aspect of the head; muscle attachment point	*externus*, outside + *occipital*, occipital bone + *protuberare*, to swell
	Foramen magnum	Large hole for passage of spinal cord	*foramen*, a hole + *magnus*, great
	Hypoglossal canal	Passageway for cranial nerve	*hypo*, under + *glossus*, tongue + *canalis*, channel
	Jugular foramen	Passageway for internal jugular vein and nerves	*jugal*, throat + *foramen*, a hole
	Occipital condyle	Smooth surface for articulation with atlas (first cervical vertebra)	*occipital*, occipital bone + *kondylos*, knuckle
Palatine	NA	L-shaped bone that forms the posterior floor of nasal cavity; part of orbit and hard palate	*palatin*, the palate
Parietal	NA	Forms the lateral, superior wall of cranial cavity	*paries*, wall
Sphenoid	Foramen ovale	Passageway for cranial nerve	*foramen*, a hole + *ovalis*, oval
	Foramen rotundum	Passageway for cranial nerve	*foramen*, a hole + *rotundum*, round
	Foramen spinosum	Passageway for cranial nerve and an artery	*foramen*, a hole + *spinosus*, spine-like
	Greater wing	Forms parts of posterior orbit and middle cranial fossa	NA
	Inferior orbital fissure*	Passageway for cranial nerve and an artery	*inferus*, lower + *orbit*, eye socket + *fissura*, cleft
	Lesser wing	Forms part of anterior cranial fossa	NA
	Optic foramen	Passageway for optic nerve (CN II)	*optikos*, eye + *foramen*, a hole
	Sella turcica	"Turkish saddle"-shaped depression housing the pituitary gland; also called hypophyseal fossa	*sella*, saddle + *turcica*, Turkish
	Superior orbital fissure	Passageway for several cranial nerves	*superus*, upper + *orbit*, eye socket + *fissura*, cleft
Temporal	Carotid canal	Passageway for internal carotid artery and associated nerves	*karotides*, arteries of the neck + *canalis*, channel
	External acoustic (auditory) meatus	Opening into external auditory canal	*externus*, outside + *auditorius*, related to hearing + *meatus*, channel
	Foramen lacerum	Largely covered by cartilage in living human; no structures pass entirely through it	*foramen*, a hole + *lacer*, mangled
	Internal acoustic (auditory) meatus	Passageway for two cranial nerves	*internus*, inside + *auditorius*, related to hearing + *meatus*, channel
	Mandibular fossa	Point of articulation with head of mandible, forming the temporomandibular joint	*mandere*, to chew + *bula*, a means + *fossa*, a pit/cavity
	Mastoid process	Attachment point for muscles of neck	*mastos*, breast + *oideos*, resembling + *processus*, a projection
	Petrous part	Houses structures for hearing and equilibrium; separates middle and posterior cranial cavities	*petrosus*, like a rock
	Squamous part	Forms inferior, posterior part of temporal fossa	*squamosus*, scale-like
	Styloid process	Serves as attachment point for muscles controlling tongue	*stylus*, stylus + *oideos*, resembling + *processus*, a projection
	Zygomatic process	Projection that articulates with temporal process of the zygomatic bone	*zygoma*, a yoke + *processus*, a projection
Vomer	NA	Forms inferior and posterior part of nasal septum	
Zygomatic	Maxillary process	Articulates with zygomatic process of maxillary bone	*maxilla*, jawbone + *processus*, a projection
	Temporal process	Articulates with zygomatic process of temporal bone	*temporalis*, temple/time + *processus*, a projection

* Slit-like opening between maxillary, sphenoid, and zygomatic bones

ANTERIOR VIEW OF THE SKULL
EXERCISE 8.1A Anterior View of the Skull

1. Obtain a skull and observe it from an anterior view
   (**figure 8.1**). An anterior view of the skull reveals much
   of the detail of the facial bones. Facial bones play a
   role in mastication (chewing) and in the protection and
   support of special sensory organs such as the eye.

2. Identify the structures that are listed in figure 8.1 on
   a skull or model, using table 8.2 and the textbook as
   guides. Then label figure 8.1.

3. Many features of the face are made by bones or
   bony markings. What bone makes the forehead?
   _____ What bones make up the bridge
   of the nose? _____

4. *Optional Activity:* **APR** **5: Skeletal System**—Watch the
   "Skull" animation, which demonstrates how the bones
   of the skull fit together. This animation also facilitates
   understanding of difficult concepts such as the location
   of the sphenoid and ethmoid bones with respect to the
   rest of the skull.

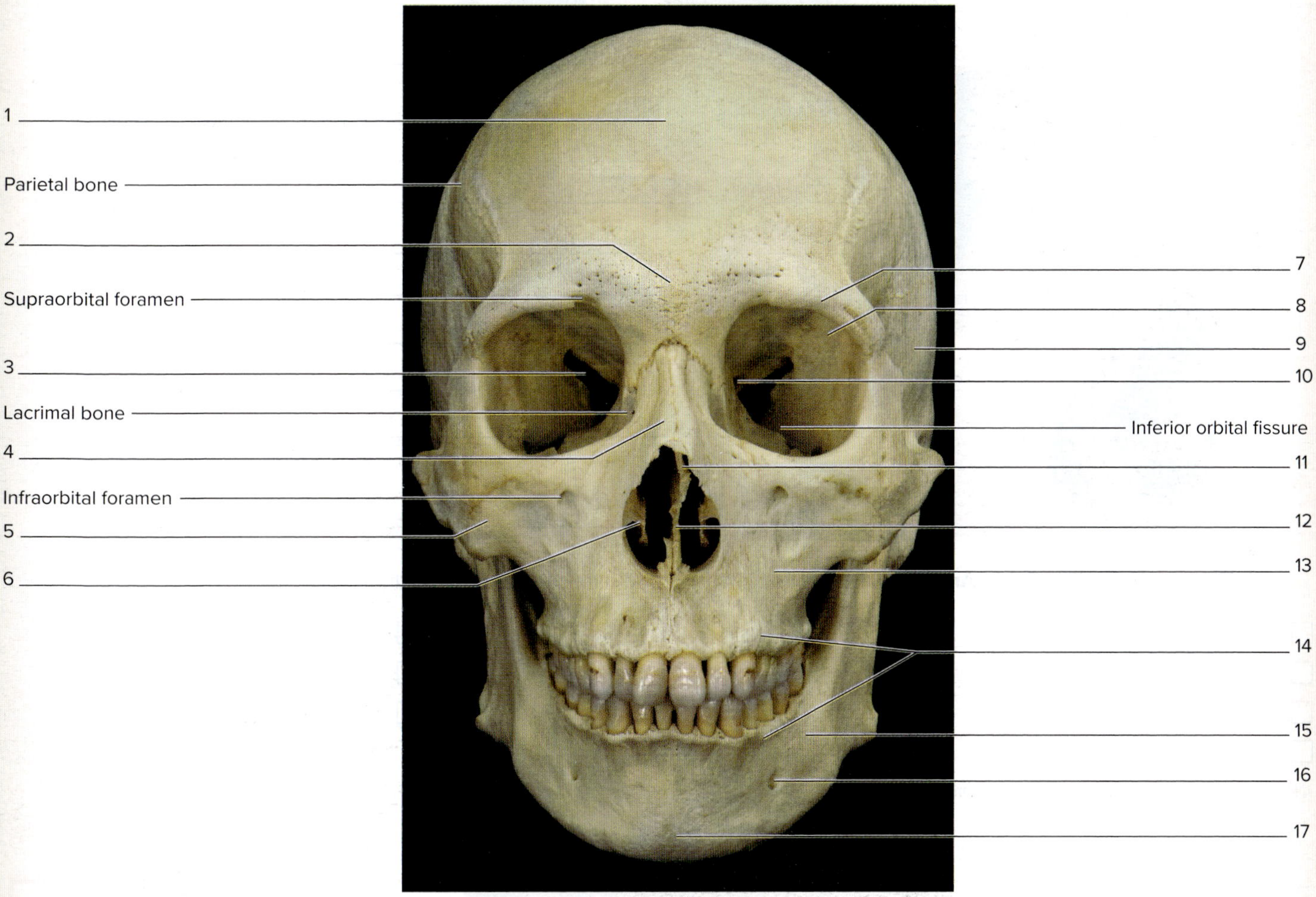

1 _____
Parietal bone _____
2 _____
Supraorbital foramen _____
3 _____
Lacrimal bone _____
4 _____
Infraorbital foramen _____
5 _____
6 _____

7 _____
8 _____
9 _____
10 _____
Inferior orbital fissure
11 _____
12 _____
13 _____
14 _____
15 _____
16 _____
17 _____

Figure 8.1 **Anterior View of the Skull.** Use the terms listed to fill in the numbered labels in the figure.

☐ Alveolar processes
☐ Frontal bone
☐ Glabella
☐ Inferior nasal concha
☐ Mandible
☐ Maxilla

☐ Mental foramen
☐ Mental protuberance
☐ Nasal bone
☐ Optic canal
☐ Perpendicular plate of ethmoid
☐ Sphenoid bone

☐ Superior orbital fissure
☐ Supraorbital margin
☐ Temporal bone
☐ Vomer
☐ Zygomatic bone

(continued on next page)

The Vertebral Column

The vertebral column lies at the core of the human skeleton. It quite literally is the "backbone" that anchors nearly every major component of the skeletal support system. The vertebral column is divided into five major regions: **cervical**, **thoracic**, **lumbar**, **sacral**, and **coccygeal**. The vertebrae themselves change size and shape rather drastically from the cervical region to the coccygeal region. These changes reflect the different weight-bearing demands placed on the vertebrae in each region. **Cervical vertebrae** are small and light because they are not supporting a lot of weight (relatively speaking), and they are specialized to allow a lot of movement of the neck, particularly rotation. **Thoracic vertebrae** are specialized to provide articulation points for the ribs. **Lumbar vertebrae** are very large, bulky vertebrae that are specialized for supporting the weight of the entire vertebral column and body structures above them. They do not allow much movement, but instead are designed to keep the vertebral column stable. The **sacrum**, which consists of 5 fused vertebrae, is specialized to provide a stable anchoring point for the bones of the pelvic girdle. Finally, the **coccyx** consists of 3 to 5 small vertebrae, which have fused together during development. It serves as an attachment point for several ligaments and for muscles of the pelvic floor. Besides vertebrae, the vertebral column also contains fibrocartilaginous pads called **intervertebral discs**. These discs are found between adjacent vertebral bodies from the second cervical vertebra to the sacrum. The discs make up about one-quarter of the vertebral column height and permit movement of the vertebral column. **Table 8.4** summarizes the characteristics of each type of vertebra.

Table 8.4		The Axial Skeleton: Vertebral Column		
Vertebrae	**Number of Vertebrae**	**Bone Features**	**Description and Related Structures of Importance**	**Word Origins**
Representative Vertebra		Lamina	Connects transverse process to spinous process on either side of each vertebra	*lamina*, a saw (or the flap of the ear)
		Pedicle	Connects body to transverse process	*ped*, foot
		Transverse processes	Processes directed laterally (one on each side)	*trans*, across + *versus*, a line + *processus*, a projection
		Spinous process	Process directed posteriorly	*spina*, spine + *processus*, a projection
		Inferior articular process	Contains a facet that forms a joint with the superior articular process of inferior vertebra	*inferus*, lower + *articulo*, a joint + processus, a projection
		Superior articular process	Contains a facet that forms a joint with the inferior articular process of superior vertebra	*superus*, above + *articulo*, a joint + *processus*, a projection
		Vertebral foramen	Large hole within each vertebra; spinal cord extends through "stacked" vertebral foramina	*vertebra*, vertebra of the spine + *foramen*, a hole
		Body	Largest part of vertebra. Intervertebral discs are found between bodies of adjacent vertebrae.	NA
		Intervertebral foramen	Lateral hole formed when two vertebrae come together; passageway for spinal nerves	*inter*, between + *vertebra*, of the spine + *foramen*, a hole
Cervical (C)	7	Body	Small body, oval/kidney bean shape	NA
		Spinous process	Horizontal, bifid (forked) spine of cervical vertebrae 3–6	*spina*, spine + *processus*, a projection
		Vertebral foramen	Large (especially with respect to size of the body); slight oval shape	*vertebra*, vertebra of the spine + *foramen*, a hole
		Transverse processes	Each contains a transverse foramen	*trans*, across + *versus*, a line + *processus*, a projection
		Transverse foramen	Passageway for vertebral artery	*trans*, across + *versus*, a line + *foramen*, a hole

Table 8.4		The Axial Skeleton: Vertebral Column *(continued)*		
Vertebrae	**Number of Vertebrae**	**Bone Features**	**Description and Related Structures of Importance**	**Word Origins**
Atlas (C₁)		Body	Has no body	NA
		Arch	Contains articular surface for dens of axis and posterior tubercle (no spinous process)	*arus, a bow*
Axis (C₂)		Body	Has odontoid process (dens)	NA
Vertebra Prominens (C₇)		Spinous process	Very large and blunt, not bifid; first spinous process easily felt under skin	*spina*, spine + *processus*, a projection
Thoracic (T)	12	Body	Heart-shaped, contains demifacets for articulation of head of rib	NA
		Spinous process	Most point inferiorly	*spina*, spine + *processus*, a projection
		Vertebral foramen	Relatively small; circular in shape; houses spinal cord	*vertebra*, vertebra of the spine + *foramen*, a hole
		Transverse processes	Contain facets for articulation with tubercle of rib	*trans*, across + *versus*, a line + *processus*, a projection
		Costal facets	Located on lateral surface of body and transverse processes; form joints with ribs	*costa*, a rib + *facet*, a small face
Lumbar (L)	5	Body	Very large, heavy	NA
		Spinous process	Short and blunt, square shaped	*spina*, spine + *processus*, a projection
		Vertebral foramen	Small (especially with respect to size of body), round; houses spinal cord	*vertebra*, vertebra of the spine + *foramen*, a hole
		Transverse processes	Short and tapered at the ends	*trans*, across + *versus*, a line + *processus*, a projection
Sacrum (S)	5 (fused)	Anterior sacral foramina	Passageway for exit of anterior (ventral) rami of sacral spinal nerves	*anterior*, foremost + *sacro*, to hold sacred + *foramen*, a hole
		Posterior sacral foramina	Passageway for exit of posterior (dorsal) rami of sacral spinal nerves	*posterus*, that which comes later + *sacro*, to hold sacred + *foramen*, a hole
		Median sacral crest	Represents fused spinous processes of sacral vertebrae (S₁–S₄)	*medianus*, in the middle + *sacro*, to hold sacred + *cristatus*, tufted
		Auricular processes	Ear-like processes that articulate with iliac bones	*auris*, ear, + *processus*, an advance
		Superior articular processes	Contain facets to form joints with inferior articular processes of L₅	*superus*, above, + *auris*, ear, + *processus*, an advance
		Sacral hiatus	Opening at inferior end of sacral canal; formed by unfused laminae of S₅	*sacro*, to hold sacred + *hiatus*, an opening
		Sacral promontory	Anterosuperior border of the body of S₁	*sacro*, to hold sacred + *promunturium*, a mountain ridge
Coccyx (Co)	3–5 (fused)	Cornu (horns)	Small projections that point superiorly (part of Co₁)	*cornu*, horn

VERTEBRAL COLUMN REGIONS AND CURVATURES

1. Observe the vertebral column of an articulated skeleton (**figure 8.14**).

2. Using colored pencils, color and label the regions of the vertebral column in figure 8.14. Use a different color for each region. Count the number of vertebrae that make up the region and write that number in the appropriate space in figure 8.14.

3. As the vertebral column develops, it forms several curvatures because of the stresses placed on it. The first curvatures to develop during the fetal period are **primary curvatures**. These form in the thoracic and sacral regions due to growth of the viscera. The second curvatures, which develop after birth, are **secondary curvatures**. These form in the cervical and lumbar regions. The cervical curvature forms when an infant begins to lift its head and the lumbar curvature forms when an infant begins to stand on its feet.

4. Locate all of the curvatures of the vertebral column on an articulated skeleton, and label the curvatures on figure 8.14.

- ☐ Cervical curvature
- ☐ Cervical vertebrae
- ☐ Coccygeal vertebrae
- ☐ Lumbar curvature
- ☐ Lumbar vertebrae
- ☐ Sacral curvature
- ☐ Sacrum
- ☐ Thoracic curvature
- ☐ Thoracic vertebrae

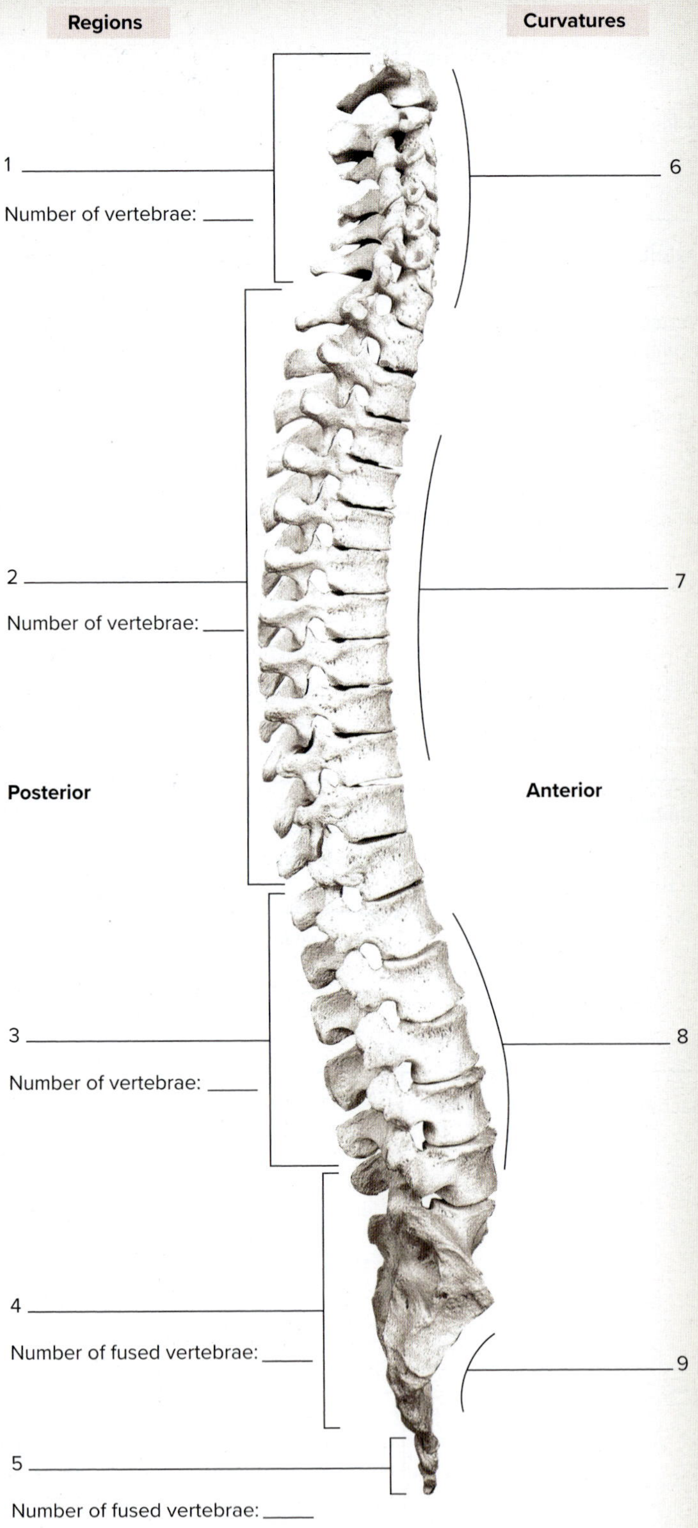

INTEGRATE

LEARNING STRATEGY

Remember the number of vertebrae in each region using mealtimes as an aid. Breakfast is at *seven,* lunch is at *twelve,* dinner is at *five,* and a bedtime snack is at *nine.* That is, there are seven cervical vertebrae (C_1–C_7), twelve thoracic vertebrae (T_1–T_{12}), five lumbar vertebrae (L_1–L_5), five fused primitive vertebrae in the sacrum, and three to five small bones in the coccyx, with an average of four (5 + 4 = 9).

Regions   **Curvatures**

1 _____   6

Number of vertebrae: _____

2 _____   7

Number of vertebrae: _____

Posterior   **Anterior**

3 _____   8

Number of vertebrae: _____

4 _____

Number of fused vertebrae: _____   9

5 _____

Number of fused vertebrae: _____

Figure 8.14 **Lateral View of the Vertebral Column.** Use the terms listed to fill in the numbered labels in the figure. Write the number of vertebrae for each region in the space provided in the figure.

©McGraw-Hill Education/Christine Eckel

STRUCTURE OF A VERTEBRA

1. Obtain a **thoracic vertebra (figure 8.15)**—this will be an example of a representative vertebra. It is helpful to begin studying the vertebral column by first identifying components found in most vertebrae. This will make the study of specific features of vertebrae in the different regions of the vertebral column easier.

2. Looking at the vertebra from a superior view, notice the large **body**. The body is generally the largest part of a vertebra, and connections between adjacent vertebral bodies (with intervertebral discs in between) provide the main support of the vertebral column. Just posterior to the body is the large foramen called the **vertebral (spinal) foramen**. The spinal cord extends through the "stacked" vertebral foramina, which collectively form the vertebral canal. Each vertebral foramen is formed by the **vertebral arch**. The vertebral arch is composed of two sets of processes and the structures that connect them.

3. Observe the vertebral process that projects posteriorly and the processes that project laterally from the vertebral arch. The largest vertebral process is the **spinous process**, which is directed posteriorly, and the **transverse processes**, which are directed laterally.

4. Observe the vertebral arch. Notice the bony connections between the vertebral body and the transverse processes. These structures are called **pedicles** (L. *pediculus*, dim. of *pes*, foot). The word *pedicle* comes from a word meaning "foot." Imagine how the vertebral arch stands upon the body on its "feet." Now notice the bony connections between the transverse processes and the spinous process. These structures are called **laminae** (*lamina*, layer).

5. Next, turn the vertebra to observe it from an anterior view. Notice that there are two prominent structures that project superiorly from the vertebral arch and two that project inferiorly. The projections are respectively called the **superior articular processes** and **inferior articular processes**. Note that on each process there is a smooth, flat surface. These surfaces are called **facets** (*facette*, face). The term *facet* literally means "a little face." (This is the same term used to describe the surfaces on a diamond.) Each vertebra contains a pair of **superior articular facets** and **inferior articular facets** on its superior and inferior processes. These facets are the surfaces that form the joints between vertebrae, as described in the next step 6.

6. Pick up a second vertebra that articulates (forms a joint) with the first vertebra. Put the two together to observe how the superior facets and inferior facets articulate with each other to form a joint. These joints are much more mobile than the intervertebral joints (the joints between the vertebral bodies). They are also the sites where most of the movement is allowed in the vertebral column.

7. Once two vertebrae are articulated with each other, look at them from a lateral view. Notice the foramen that forms between the pedicles of adjacent vertebrae. This is the **intervertebral foramen**. This foramen is the location where spinal nerves (nerves that come off of the spinal cord) exit the vertebral canal to travel to their destinations throughout the body.

8. Identify the structures listed in figure 8.15 on a typical vertebra, using table 8.4 and the textbook as guides. Then label them in figure 8.15.

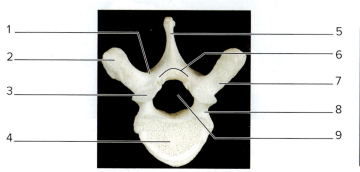

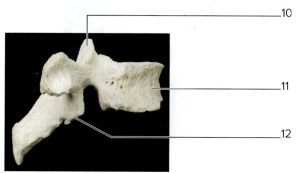

Figure 8.15 **A Vertebra.** Use the terms listed to fill in the numbered labels in the figure. Some answers may be used more than once.

☐ Body

☐ Inferior articular process

☐ Lamina

☐ Pedicle

☐ Spinous process

☐ Superior articular process

☐ Transverse process

☐ Vertebral arch

☐ Vertebral foramen

The Thoracic Cage

The **thoracic cage** consists of the **sternum**, **ribs**, and **thoracic vertebrae**. Its main function is to protect vital organs such as the heart and lungs. However, the bones of the thoracic cage also serve as important attachment sites for muscles involved with respiratory movements and muscles involved with movements of the back, neck, and shoulder. **Table 8.5** summarizes the key features of the sternum and ribs. Refer to table 8.4 for descriptions of the features of the thoracic vertebrae, which articulate with the ribs.

Table 8.5	The Axial Skeleton: Sternum and Ribs		
Bone	**Bone Features**	**Description and Related Structures of Importance**	**Word Origins**
Sternum			
Manubrium	Clavicular notch	Point of articulation with clavicle	*manus*, hand *clavicula*, a small key
	Suprasternal (jugular) notch	Depression at superior border	*supra*, above + *sternon*, the chest
	Notch for rib 1	Location of articulation with costal cartilage of rib 1	NA
	Sternal angle	Joint between manubrium and body; point of articulation with costal cartilage of rib 2	*sternon*, the chest + *angulus*, a corner
Body	Notches for ribs 2–7	Point of articulation for costal cartilages of ribs 2–7; partial notch for rib 2	NA
	Xiphisternal joint	Joint between body and xiphoid process; point of articulation with superior part of costal cartilage of rib 7	*xiphion*, sword-shaped + *sternon*, the chest
Xiphoid process	Partial notch for rib 7	Point of articulation for inferior part of the costal cartilage of rib 7	*xiphion*, sword-shaped
Ribs			
Typical Rib	Head	Part of rib that articulates with bodies of thoracic vertebrae	NA
	Superior articular facet	Facet on head of rib that articulates with inferior costal facet on body of vertebra that lies one level above it (i.e., superior articular facet of rib 6 with T_5)	*superus*, above + *articularis*, pertaining to joints + *facet*, a small face
	Inferior articular facet	Facet on head of rib that articulates with superior costal facet on body of numerically equivalent thoracic vertebra (i.e., inferior articular facet of rib 6 to T_6)	*inferus*, below + *articularis*, pertaining to joints + *facet*, a small face
	Shaft	Main part (body) of rib; begins at angle of rib; projects anteriorly	NA
	Neck	Narrow region where head meets tubercle of rib	NA
	Tubercle	Projection at the junction between shaft and neck; contains facet for articulation with transverse process of thoracic vertebra	*tuberculum*, a small bump
	Angle	Location where rib curves anteriorly	*angulus*, a corner
	Costal groove	Groove on inferior, deep border of shaft	*costa*, a rib + *sulcus*, groove

THE STERNUM

1. Observe the thoracic cage on an articulated skeleton and locate the sternum **(figure 8.22)**. The sternum has three portions: the **manubrium**, the **body**, and the **xiphoid process** (*xiphos*, sword). The depression on the superior part of the manubrium is the **suprasternal notch**.

2. Palpate the suprasternal notch on yourself. Keeping your fingers on the manubrium, move your fingers inferiorly until you feel a rough ridge. This is the **sternal angle**. The sternal angle is located where the manubrium meets the body of the sternum. It is an important clinical landmark because this is where the second rib articulates with the sternum.

3. Identify the structures listed in figure 8.22 on a sternum, using table 8.5 and the textbook as guides. Then label them in figure 8.22.

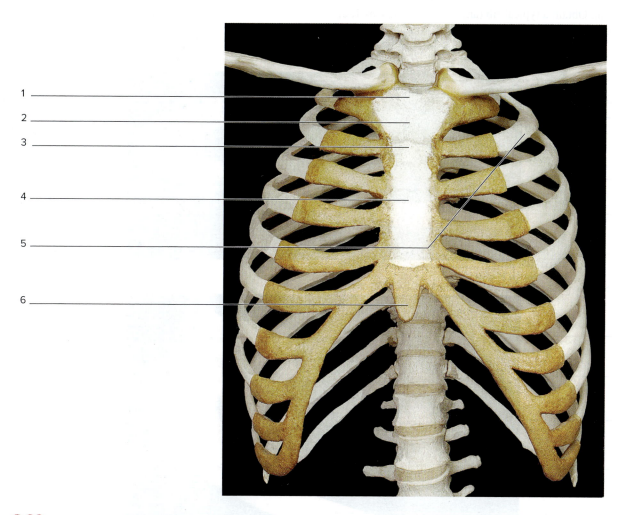

Figure 8.22 **The Sternum.** Anterior view. Use the terms listed to fill in the numbered labels in the figure.

☐ Body

☐ Manubrium

☐ Second rib

☐ Sternal angle

☐ Suprasternal notch

☐ Xiphoid process

©McGraw-Hill Education/Christine Eckel

The corresponds to the Learning Objective(s) listed in the chapter opener outline.

Create and Evaluate

Analyze and Apply

Understand and Remember

Do You Know the Basics?

Exercise 8.1: Anterior View of the Skull

1. Which of the following bone(s) is/are visible from the anterior view of the skull? (Check all that apply.) **1**

   _____ a. ethmoid bone   _____ b. frontal bone   _____ c. occipital bone   _____ d. parietal bone   _____ e. sphenoid bone

2. Which of the following skull bones form the orbit of the eye? (Check all that apply.) **2**

   _____ a. ethmoid bone   _____ b. frontal bone   _____ c. lacrimal bone   _____ d. maxilla   _____ e. nasal bone   _____ f. zygomatic bone

3. The inferior portion in the nasal septum is composed of the _____ (ethmoid bone/vomer). **3**

4. What is the projection of the mandible that forms the anterior part of the chin called? (Circle one.) **4**

   a. alveolar process   b. angle   c. body   d. mental foramen   e. mental protuberance

5. Label the following diagram of an anterior view of the skull. **1 2 3 4**

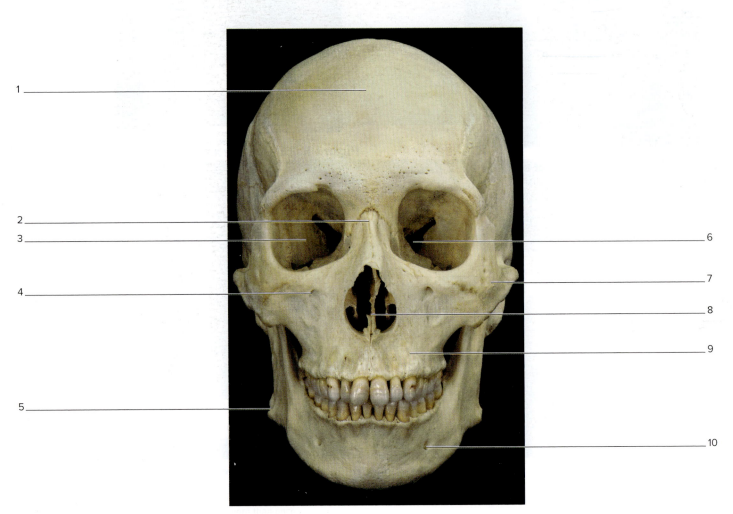

1 _____

2 _____

3 _____

4 _____

5 _____

6 _____

7 _____

8 _____

9 _____

10 _____

©McGraw-Hill Education/Christine Eckel

Exercise 8.2: Additional Views of the Skull

6. Match the bone marking listed in column A with the bone it is associated with, listed in column B. Some answers in column B may be used more than once. **5**

Column A

____ 1. foramen ovale

____ 2. mastoid process

____ 3. jugular foramen

____ 4. styloid process

____ 5. mental foramen

____ 6. foramen spinosum

Column B

a. mandible

b. occipital bone

c. sphenoid bone

d. temporal bone

7. The suture between the two parietal bones is the _____ (coronal/sagittal/lambdoid) suture. **5**

8. The zygomatic arch is composed of which of the following? (Check all that apply.) **6**

____ a. frontal process of the zygomatic bone

____ b. mental foramen of the mandible

____ c. palatine process of the maxilla

____ d. temporal process of the zygomatic bone

____ e. zygomatic process of the temporal bone

Exercise 8.3: Superior View of the Cranial Floor

9. a. What does the term *petrous* mean? **7** _____

b. Why is this term used to describe the petrous part of the temporal bone?

_____

_____

10. Match the bony feature listed in column A with the cranial fossa in which the feature can be found, listed in column B. Some answers in column B may be used more than once. **8**

Column A

____ 1. internal acoustic meatus

____ 2. sella turcica

____ 3. foramen ovale

____ 4. crista galli

Column B

a. anterior cranial fossa

b. middle cranial fossa

c. posterior cranial fossa

Exercise 8.4: Bones Associated with the Skull

11. What is the only bone of the axial skeleton that does not articulate with any other bone? **9** _____.

Exercise 8.5: The Fetal Skull

12. a. What is a fontanel? **10** _____

b. By what age do most of the fontanels completely close? _____

Exercise 8.6: Vertebral Column Regions and Curvatures

13. The _____ (primary/secondary) curvatures of the vertebral column are the *cervical* and *lumbar* curvatures, whereas the _____ (primary/secondary) curvatures are the *thoracic* and *sacral* curvatures. **11**

Exercise 8.7: Structure of a Typical Vertebra

14. Which of the following compose the vertebral arch? (Check all that apply.) **12**

_____ a. inferior articular process _____ b. lamina _____ c. pedicle _____ d. spinous process _____ e. transverse process

INTRODUCTION

The appendicular skeleton is composed of the bones that attach to the axial skeleton. The term **appendicular** comes from the word *appendage*. A dictionary definition of an appendage is *something that is added or attached to an item that is larger or more important*. While the appendages of the human body (the upper and lower limbs) *could* be described as simply "added" structures, most would consider the upper and lower limbs to be essential. Without the limbs, most of the movement essential to the activities of daily living would be absent. Think about all the activities you do within a given day (walking, texting, reaching for an object, etc). These activities are made possible by the bones of the pectoral girdle, the upper limbs, the pelvic girdle, and the lower limbs, and the muscles that attach to them. Each "girdle" serves to attach a limb to the axial skeleton, to provide stability while allowing for the mobility of the limb. The bones of the **pectoral girdle** (the clavicles and scapulae) attach the **upper limbs**, whereas the bones of the **pelvic girdle** (ossa coxae or hip bones) attach the **lower limbs**.

The exercises in this chapter guide you in studying the bones that compose the appendicular skeleton on an articulated human skeleton, on disarticulated bones of the human skeleton, and on parts of bones that can be felt through the skin. The goal of these exercises is to identify the major bones and the markings on each bone. Use the textbook as a reference while completing the labeling exercises. Many of the markings on the bones were created because of the pulling action of muscles or ligaments that attach to them and stress them as they develop. Learning the bony markings now will better prepare you to learn attachment points and actions of muscles later. Bones and their bony features are summarized in separate tables, one table for each major area of the appendicular skeleton. These tables include the name of the bone, the bony feature, a description, and the word origin. Word origins often provide information that makes it easier to remember the structures.

List of Reference Tables

INTEGRATE

LEARNING STRATEGY

When attempting to identify individual bones, determine the different aspects of the bone (e.g., anterior vs. posterior), and determine if a bone is from the right side or left side of the body. Think about what features *you* find most distinctive for each bone. For example, you may look for a large head to distinguish the femur from other bones. You may look for the linea aspera to distinguish the posterior surface from the anterior surface (the linea aspera is on the posterior surface). Both of these features may be considered together to determine if the femur is a right or a left femur. Follow this procedure for each bone of the appendicular skeleton with an understanding that the features that help *you* identify a bone may not be the same features somone else uses to identify that same bone.

These Pre-Laboratory Worksheet questions may be assigned by instructors through their ▥ connect course.

1. The appendicular skeleton is composed of which of the following? (Check all that apply.)

_____ a. lower limb bones

_____ b. pectoral girdle

_____ c. pelvic girdle

_____ d. thoracic cage

_____ e. upper limb bones

2. Which of the following bones compose the pectoral girdle? (Check all that apply.)

_____ a. clavicle

_____ b. humerus

_____ c. ribs

_____ d. scapula

_____ e. sternum

3. Which of the following bones compose part of the os coxae of the pelvic girdle? (Check all that apply.)

_____ a. femur

_____ b. ilium

_____ c. ischium

_____ d. pubis

_____ e. sacrum

4. In the anatomic position, the radius lies _____ (medial/lateral) to the ulna.

5. In the anatomic position, the tibia lies _____ (medial/lateral) to the fibula.

6. The bones in the wrist are the _____ (carpal/tarsal) bones, whereas the bones in the ankle are the _____ (carpal/tarsal) bones.

7. Match the description listed in column A with the appropriate bone listed in column B.

Column A

_____ 1. a bone that has two large tubercles on its proximal end

_____ 2. a bone that has two large trochanters on its proximal end

_____ 3. a bone that contains the olecranon

_____ 4. a bone in the heel

_____ 5. a sesamoid bone found in the knee

_____ 6. the largest bone in the leg

Column B

a. calcaneus

b. femur

c. humerus

d. patella

e. tibia

f. ulna

8. The lateral malleolus is a feature of which bone? (Circle one.)

a. calcaneus

b. femur

c. fibula

d. talus

e. tibia

9. Identify the bone that has both an acromion and a coracoid process. (Circle one.)

a. clavicle

b. humerus

c. radius

d. scapula

e. ulna

10. The bone in the thigh is the _____ (humerus/femur), whereas the bone in the arm is the _____ (humerus/femur).

Table 9.2	The Appendicular Skeleton: Upper Limb *(continued)*		
Bone	**Bony Marking**	**Description**	**Word Origin**
Radius radius, *spoke of a wheel*	Head	The disc-shaped proximal end of the radius; articulates with capitulum of the humerus	
	Neck	Narrow region where the head of the radius meets the shaft of the bone	
	Radial tuberosity	A large projection on the medial surface distal to the neck	*tuber,* a knob
	Styloid process of the radius	A small, pointed projection on the distal radius; forms the lateral aspect of the wrist	*stylos,* pillar, + *eidos,* resemblance
Carpals carpus, *wrist* **Proximal row**	Scaphoid	A large, "boat-shaped" bone; articulates with the radius	*skaphe,* boat, + *eidos,* resemblance
	Lunate	A "moon-shaped" bone; articulates with the radius	*luna,* moon
	Triquetrum	A pyramid-shaped bone located between the pisiform, lunate, and hamate bones	*triquetrus,* three-cornered
	Pisiform	A "pea-shaped" bone; positioned on the medial, palmar surface of the wrist	*pisum,* pea, + *forma,* appearance
Distal row	Trapezium	A "table-shaped" bone located at the base of the first metacarpal (base of the thumb)	*trapezion,* a table
	Trapezoid	A "table-shaped" bone located at the base of the second metacarpal	*trapezion,* a table, + *eidos,* resemblance
	Capitate	A "head-shaped" bone located in the center of the wrist, at the base of the third metacarpal	*caput,* head
	Hamate	A "hook-shaped" bone located at the base of the fifth metacarpal	*hamus,* a hook
Metacarpals		Each bone consists of a head, body, and base	meta-, *after,* + carpus, *wrist*
Phalanges		Bones of the fingers and thumb	*phalanx,* line of soldiers
Digits II through V	Proximal	The phalanx closest to the palm of the hand	*proximus,* nearest
	Middle	The middle phalanx	
	Distal	The small, cone-shaped phalanx forming the tips of the fingers	*distalis,* away
Pollex (I) *pollex, thumb*	Proximal	The phalanx closest to the palm of the hand	*proximus,* nearest
	Distal	The small, cone-shaped phalanx forming the tip of the thumb	*distalis,* away

EXERCISE 9.2

BONES OF THE UPPER LIMB

EXERCISE 9.2A The Humerus

1. Observe the humerus **(figure 9.4)** on an articulated skeleton (or see figure 7.6, which shows a full skeleton). Locate the point of articulation with the scapula to form the glenohumeral joint. Locate the points of articulation with the ulna and radius to form the elbow joint.

2. Palpate your own humerus both at the shoulder joint and at the medial and lateral epicondyle, just proximal from the elbow joint.

3. Obtain a disarticulated humerus. Identify the bony markings of the humerus that are listed in table 9.2.

   a. What distinguishes the proximal end of the humerus from the distal end?

   _____

   _____

   _____

b. What distinguishes the anterior surface of the humerus from the posterior surface?

_____

_____

_____

4. Obtain both a right and a left disarticulated humerus. What distinguishes a right humerus from a left humerus?

_____

_____

_____

5. Label the features of the humerus shown in figure 9.4, using table 9.2 and the textbook as guides.

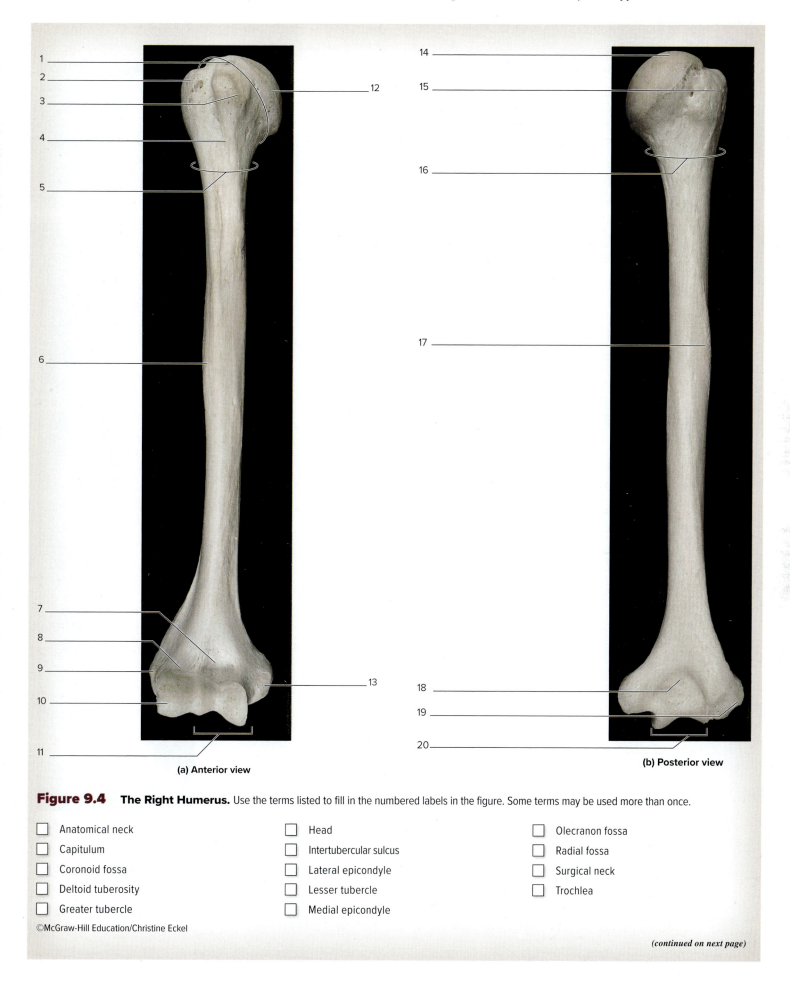

(a) Anterior view

(b) Posterior view

Figure 9.4 **The Right Humerus.** Use the terms listed to fill in the numbered labels in the figure. Some terms may be used more than once.

☐ Anatomical neck

☐ Capitulum

☐ Coronoid fossa

☐ Deltoid tuberosity

☐ Greater tubercle

☐ Head

☐ Intertubercular sulcus

☐ Lateral epicondyle

☐ Lesser tubercle

☐ Medial epicondyle

☐ Olecranon fossa

☐ Radial fossa

☐ Surgical neck

☐ Trochlea

©McGraw-Hill Education/Christine Eckel

(continued on next page)

(continued from previous page)

EXERCISE 9.2B The Radius

1. Observe the radius **(figure 9.5)** on an articulated skeleton (or see figure 7.6, which shows a full skeleton). Locate its articulation with the humerus at the elbow. Also locate its articulation with the carpals at the wrist.

2. Palpate your own radius at the styloid process, which is on the lateral side of the wrist and is aligned with the thumb.

3. Obtain a disarticulated radius. Identify the bony markings of the radius that are listed in table 9.2.

   a. What distinguishes the proximal end of the radius from the distal end?

   _____

   _____

   _____

   b. What distinguishes the anterior surface of the radius from the posterior surface?

   _____

   _____

   _____

4. Obtain both a right and a left disarticulated radius. What distinguishes a right radius from a left radius?

   _____

   _____

   _____

   _____

5. Label the structures on the radius shown in figure 9.5 using table 9.2 and the textbook as guides.

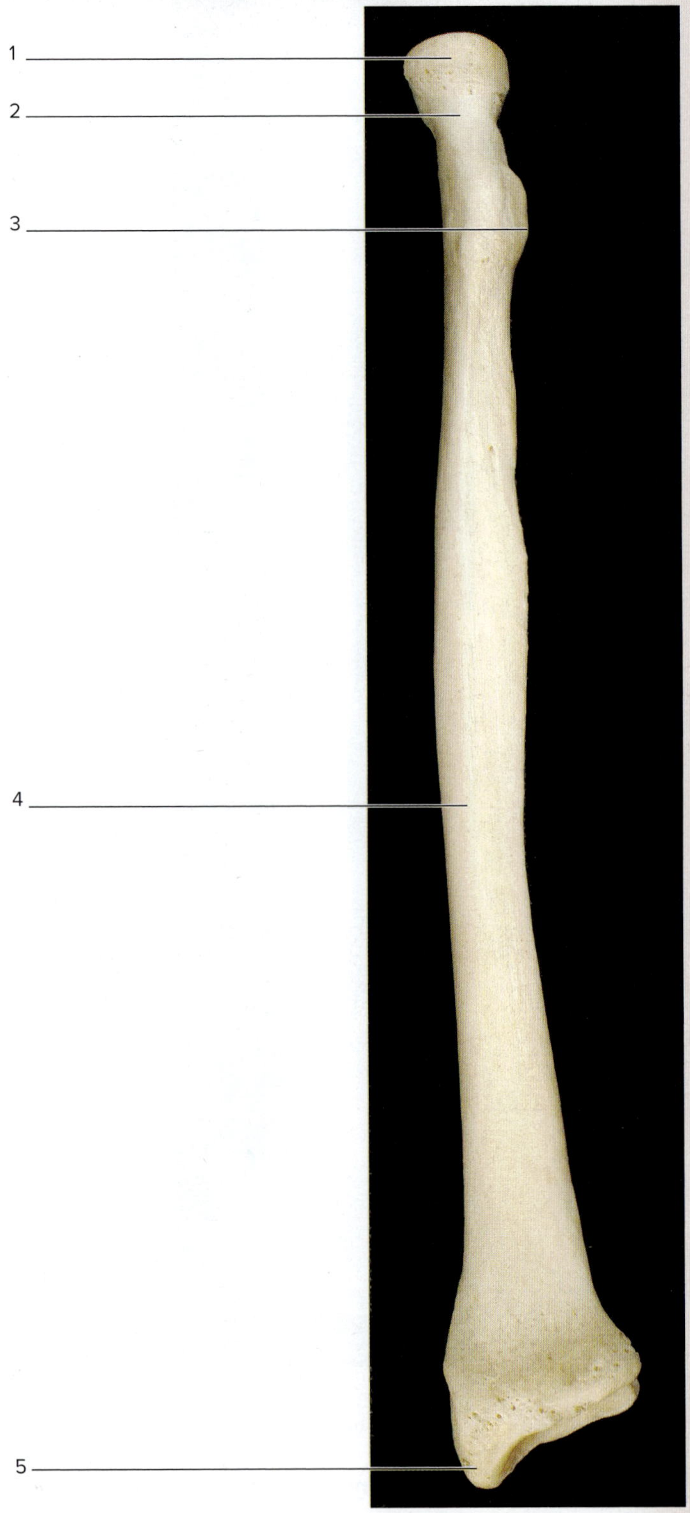

Figure 9.5 **The Radius.** Use the terms listed to fill in the numbered labels in the figure.

☐ Head   ☐ Shaft
☐ Neck   ☐ Styloid process of radius
☐ Radial tuberosity

©McGraw-Hill Education/Christine Eckel

EXERCISE 9.2C The Ulna

1. Observe the ulna **(figure 9.6)** on an articulated skeleton (or see figure 7.6, which shows a full skeleton). Locate its articulation with the humerus at the elbow and its location at the wrist.

2. Palpate your own ulna both at the olecranon at the elbow and at the styloid process on the medial side of the wrist, which is aligned with the "little finger."

3. Obtain a disarticulated ulna. Identify the bony markings of the ulna that are listed in table 9.2.

   a. What distinguishes the proximal end of the ulna from the distal end?

   _____

   _____

   _____

   b. What distinguishes the anterior surface of the ulna from the posterior surface?

   _____

   _____

   _____

4. Obtain both a right and a left disarticulated ulna. What distinguishes a right ulna from a left ulna?

   _____

   _____

   _____

5. Label the structures on the ulna shown in figure 9.6 using table 9.2 and the textbook as guides.

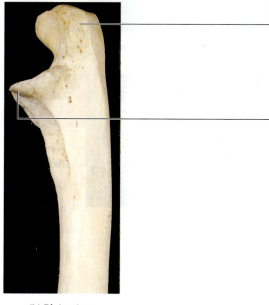

1

2

3

4

5

6

7

8

(b) Right ulna, medial view

☐ Coronoid process   ☐ Styloid process of ulna

☐ Olecranon   ☐ Trochlear notch

☐ Radial notch   ☐ Tuberosity of ulna

☐ Shaft of ulna

(a) Right ulna, anterior view

Figure 9.6   The Ulna. Use the terms listed to fill in the numbered labels in the figure. Some terms may be used more than once.

©McGraw-Hill Education/Photo and Dissection by Christine Eckel

(continued on next page)

(continued from previous page)

EXERCISE 9.2F Articulations in the Upper Extremity

1. Observe the bones around the elbow joint (**figure 9.9**) on an articulated skeleton (or see figure 7.6, which shows a full skeleton). Locate the elbow joint. What bones and bone markings articulate with each other to make the elbow joint?

2. Label the structures shown in figure 9.9, using table 9.2 and the textbook as guides.

3. Observe the bones around the wrist joint on an articulated skeleton (or see figure 7.6, which shows a full skeleton).

4. An x-ray machine is used in the medical field as a diagnostic tool for a variety of conditions. A patient is exposed to the x-ray beam and the resulting image is called a radiograph. Radiographs can be useful in detecting bone fractures or joint dislocations. Identify the radius, ulna, individual carpals, metacarpals, and phalanges on an articulated skeleton. Then label them in the radiograph of the wrist (**figure 9.10**).

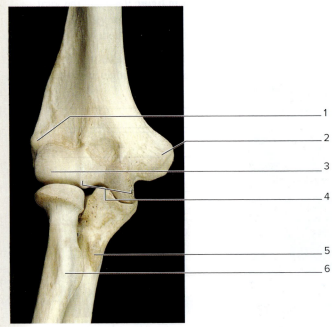

(a) Bones of the elbow, anterior view

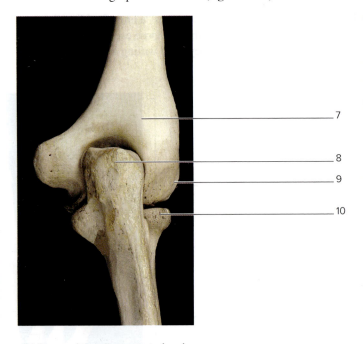

(b) Bones of the elbow, posterior view

Figure 9.9 **The Elbow.** Use the terms listed to fill in the numbered labels in the figure. Some terms may be used more than once.

☐ Capitulum   ☐ Lateral epicondyle   ☐ Radius
☐ Head of radius   ☐ Medial epicondyle   ☐ Trochlea
☐ Humerus   ☐ Olecranon   ☐ Ulna

©McGraw-Hill Education/Christine Eckel

Figure 9.10 **X-Ray of the Wrist.** Use the terms listed to fill in the numbered labels in the figure.

☐ Hamate   ☐ Radius
☐ Metacarpal   ☐ Scaphoid
☐ Middle phalanx   ☐ Ulna
☐ Proximal phalanx

©kravka/Shutterstock

EXERCISE 9.3

SURFACE ANATOMY REVIEW—PECTORAL GIRDLE AND UPPER LIMB

In previous exercises, certain landmarks on the bones of the pectoral gridle and upper limbs were palpated. This exercise serves as an opportunity to review these surface landmarks.

1. Palpate the manubrium of the sternum and suprasternal (jugular) notch on yourself. Move your fingers just lateral from the sternal notch to palpate the joint between the manubrium and the proximal end of the clavicle: the **sternoclavicular joint**. Recall that the sternoclavicular joint is the only bony attachment between the pectoral girdle and the axial skeleton.

2. Palpate along the **clavicle** and make note of the curvatures of the clavicle while moving your fingers from medial to lateral. At the tip of the shoulder, palpate the joint between the lateral aspect of the clavicle and the **acromion** of the scapula: the **acromioclavicular joint**.

3. Continue to palpate along the acromion as it curves posteriorly and becomes the **spine of the scapula**.

4. Palpate the inferior, lateral border of the deltoid muscle, which covers much of the shoulder. You should be able to feel part of the diaphysis of the humerus where the deltoid attaches to the humerus at the **deltoid tuberosity** because there is very little muscle between the bone and the skin at that point.

5. Moving your fingers distally to the elbow, palpate the large **olecranon** of the ulna. This is the bony process that rests on a table when leaning on the elbows.

6. Just proximal from the olecranon on the medial aspect of the elbow, palpate the **medial epicondyle of the humerus**. Place your thumb in the hollow on the posterior part of the elbow between the olecranon of the ulna and the medial epicondyle of the humerus to feel the cable-like **ulnar nerve**. This nerve is what causes the pain or tingly sensations that are felt when hitting the "funny bone."

7. Palpate the olecranon once again. Continue to palpate along the ulna distally until reaching the wrist joint. The bump on the medial aspect of the wrist is the **styloid process of the ulna**. Palpate the corresponding location on the lateral aspect of the wrist to feel the **styloid process of the radius**.

8. Finally, palpate the small metacarpal and phalangeal bones of the hand (see figure 9.8) while reviewing the names of the bones.

9. Label the surface anatomy structures in **figure 9.11** using the textbook as a guide.

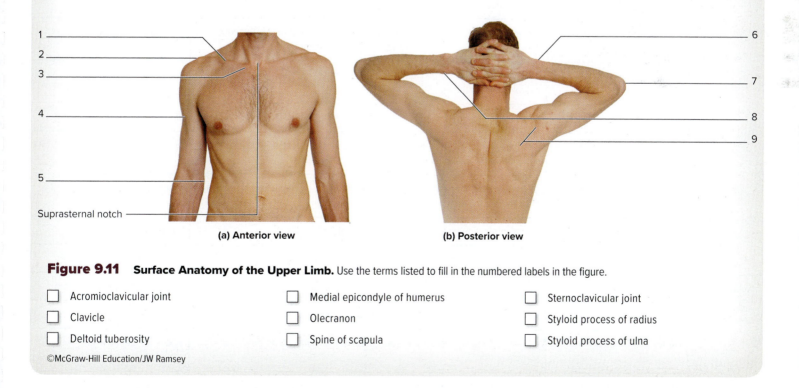

1 _____
2 _____
3 _____
4 _____
5 _____
Suprasternal notch _____

6 _____
7 _____
8 _____
9 _____

(a) Anterior view   **(b) Posterior view**

Figure 9.11   **Surface Anatomy of the Upper Limb.** Use the terms listed to fill in the numbered labels in the figure.

☐ Acromioclavicular joint   ☐ Medial epicondyle of humerus   ☐ Sternoclavicular joint

☐ Clavicle   ☐ Olecranon   ☐ Styloid process of radius

☐ Deltoid tuberosity   ☐ Spine of scapula   ☐ Styloid process of ulna

©McGraw-Hill Education/JW Ramsey

The Pelvic Girdle

The **pelvic girdle** consists of the paired ossa coxae (*os,* bone, + *coxa,* hip) bones. Each os coxae (singular) is composed of three bones: the **ilium**, the **ischium**, and the **pubis** bones. The three bones fuse together at different periods during development. Unlike the bones of the pectoral girdle, the bones of the pelvic girdle fuse together during development to become one solid structure that is joined anteriorly at the pubic symphysis. **Table 9.3** lists the bones composing the os coxae and describes their key features. A complete pelvis is formed from four bones: the right os coxae the left os coxae the sacrum, and the coccyx.

In the following exercises, the features of the os coxae as a whole are described first. The features of the individual bones that compose the os coxae are described next. Finally, the structural differences between the male pelvis and the female pelvis are described.

Table 9.3	The Appendicular Skeleton: Pelvic Girdle		
Bone	**Bony Marking**	**Description**	**Word Origin**
Os Coxae os, *bone,* + coxa, *hip*	Acetabulum	A deep bony socket; articulates with the head of the femur.	*acetabula,* a shallow cup
	Lunate surface	The half-moon-shaped (curved) smooth surface on the superior border of the acetabulum; articulates with the head of the femur.	*luna,* moon
	Obturator foramen	A large, oval hole in the inferior part of the os coxae.	*obturo,* to occlude
Ilium ilium, *flank*	Ala	Concave, curved anterior region of the ilium.	*ala,* wing
	Anterior gluteal line	A rough line running obliquely on the lateral surface of the ilium from the iliac crest to the greater sciatic notch.	*gloutos,* buttock
	Anterior inferior iliac spine	A process inferior to the anterior superior iliac spine.	*spina,* a spine
	Anterior superior iliac spine	A projection at the anteriormost part of the iliac crest.	*spina,* a spine
	Arcuate line	An oblique line between the ilium and ischium that composes the iliac part of the linea terminalis (pelvic brim) of the bony pelvis.	*arcuatus,* bowed
	Auricular surface	An "ear-like" rough surface on the medial aspect of the ilium; articulates with the sacrum.	*auris,* ear
	Greater sciatic notch	A deep notch on the posterior surface of the ilium inferior to the posterior inferior iliac spine; the sciatic nerve is located adjacent to this notch.	*sciaticus,* the hip joint
	Iliac crest	The superior border of the ilium, beginning at the sacrum and ending on the lateral aspect of the hip.	*crista,* a ridge
	Iliac fossa	A large fossa on the anteromedial surface, inferior to the iliac crest.	*ilium,* flank, + *fossa,* a trench
	Inferior gluteal line	A rough line running transversely on the lateral surface of the ilium just superior to the acetabulum.	*gloutos,* buttock
	Posterior gluteal line	A rough line running vertically on the lateral surface of the ilium from the iliac crest to the posterior rim of the greater sciatic notch.	*gloutos,* buttock
	Posterior inferior iliac spine	A small projection on the posterior inferior point of the ilium.	*spina,* a spine
	Posterior superior iliac spine	A projection on the posterior superior point of the ilium.	*spina,* a spine
Ischium ischion, *hip*	Body of ischium	Bulky part of bone superior to the ramus of ischium.	*ischion,* hip, + *spina,* a spine
	Ischial spine	A small, sharp spine on the posterior aspect of the ischium.	*ischion,* hip
	Lesser sciatic notch	A notch located immediately inferior to the ischial spine on the posterior surface of the ilium.	*sciaticus,* the hip joint
	Ramus of ischium	The inferior part of the ischium that connects to the pubis anteriorly; forms the inferior part of the obturator foramen.	*ramus,* branch
	Ischial tuberosity	A large, rough projection on the posterior, inferior surface of the ischium; attachment site for hamstring muscles.	*ischion,* hip, + *tuber,* a knob

Bone	Bony Marking	Description	Word Origin
Pubis pubis, *pubic bone*	Inferior pubic ramus	The inferior part of the pubis that joins with the ischium.	*ramus,* branch
	Pectineal line	Rough ridge on the medial surface of the superior ramus of the pubis.	*pectineal,* relating to the pubis
	Pubic crest	A ridge on the lateral part of the superior ramus of the pubis.	*crista,* crest
	Pubic tubercle	A projection composing the anteriormost point of the bone.	*tuber,* a knob
	Superior pubic ramus	The superior part of the pubis that joins with the ilium.	*ramus,* branch
	Symphysial surface	Site of articulation for pubic bones at the pubic symphysis.	*symphysis,* a growing together

Table 9.3 The Appendicular Skeleton: Pelvic Girdle *(continued)*

EXERCISE 9.4

BONES OF THE PELVIC GIRDLE

EXERCISE 9.4A The Os Coxae

1. Observe the os coxae **(figure 9.12)** on an articulated skeleton (or see figure 7.6, which shows a full skeleton). Locate the point of articulation of the os coxae with the axial skeleton. With what major bone does the os coxae articulate?

_____

2. Palpate your own os coxae at the iliac crest by placing your hands on your hips.

3. Obtain a disarticulated pelvis. Identify the bony markings that are listed in table 9.3.

   a. What distinguishes the superior portion of the os coxae from the inferior portion?

   _____
   _____
   _____

   b. What distinguishes the anterior surface of the os coxae from the posterior surface?

   _____
   _____
   _____

4. Obtain both right and left disarticulated ossa coxae. What distinguishes the right os coxae from the left os coxae?

   _____
   _____
   _____

5. Label the structures on the os coxae shown in figure 9.12, using table 9.3 and the textbook as guides.

EXERCISE 9.4B Male and Female Pelves

1. Obtain a male pelvis and a female pelvis and lay them next to each other on the workspace with the anterior surfaces facing toward you. **Figure 9.13** demonstrates the features of male and female pelves.

2. There are numerous features that help distinguish a **male pelvis** from a **female pelvis**. For example, a female pelvis is generally wider and more flared; it has a broader subpubic angle; a smaller, triangular obturator foramen; a wide, shallow greater sciatic notch; and ischial spines that rarely project into the pelvic outlet. All of these are adaptations that facilitate childbirth. Although not every feature may be true, at the very least, use the guidelines to determine if a pelvis is more male-like than female-like, or vice versa. Observe both male and female pelves to develop a method of differentiating the male pelvis from the female pelvis.

INTEGRATE

LEARNING STRATEGY

There are several methods for determining if a pelvis belonged to a male or a female. One involves estimating the subpubic angles by comparing the angles on the pelves with the angles formed between the digits on the hand with the fingers spread apart (abducted). For example, the angle between the **thumb and index finger** when spread apart approximates the wide subpubic angle of a female pelvis, whereas the angle formed between the **index and middle fingers** when spread apart approximates the narrower subpubic angle of a male pelvis.

The width of the greater sciatic notch may also provide clues. Females typically have a larger greater sciatic notch. Three fingers can be placed in the greater sciatic notch of a female pelvis, whereas only one finger may fit in the greater sciatic notch of a male.

(continued on next page)

The Lower Limb

The bones of the lower limb are the femur, patella, tibia, fibula, tarsals, metatarsals, and phalanges. Nearly all of the projections on these bones are attachment points for the muscles that move the limb. **Table 9.4** lists the bones of the lower limb and describes their key features.

Table 9.4	The Appendicular Skeleton: Lower Limb		
Bone	**Bony Marking**	**Description**	**Word Origin**
Femur femur, *thigh*	Adductor tubercle	A small projection proximal or superior to a medial condyle	*tuber*, a knob
	Fovea	A circular depression within the head of the femur	*fovea*, a dimple
	Gluteal tuberosity	A projection on the proximal aspect of the linea aspera	*gloutos*, buttock, + *tuber*, *a knob*
	Greater trochanter	A very large projection on the lateral surface of the proximal epiphysis; attachment site for both gluteal and thigh muscles	*trochanter*, a runner
	Head	Large, spherical structure on the proximal end of the femur; articulates with the acetabulum of os coxae	
	Intercondylar fossa	A depression on the distal end of the femur between the two condyles	*inter-*, between, + *kondylos*, knuckle
	Intertrochanteric crest	A large ridge that runs between the greater and lesser trochanters on the posterior surface; marking separating shaft and neck of femur	*inter-*, between, + *trochanter*, a runner
	Intertrochanteric line	A shallow ridge that runs between the greater and lesser trochanters on the anterior surface; marking separating shaft and neck of femur	*inter-*, between, + *trochanter*, a runner
	Lateral condyle	A large, rounded surface; articulates with the lateral condyle of the tibia	*kondylos*, knuckle
	Lateral epicondyle	A small projection superior to the lateral condyle; attachment for a ligament	*epi*, above, + *kondylos*, knuckle
	Lesser trochanter	A large projection on the medial surface of the proximal epiphysis; attachment site of iliopsoas muscle	*trochanter*, a runner
	Linea aspera	A "rough line" that runs along the posterior surface of the diaphysis	*linea*, line, + *aspera*, rough
	Medial condyle	A rounded surface; articulates with the medial condyle of the tibia	*kondylos*, knuckle
	Medial epicondyle	A small projection superior to the medial condyle; attachment for a ligament	*epi*, above, + *kondylos*, knuckle
	Neck	The narrow portion where the head meets the shaft of the femur; this is the part of the femur that is fractured in a "broken hip"	
	Patellar surface	A smooth depression on the anterior surface of the distal epiphysis; articulates with the patella	(patella) *patina*, a shallow disk
	Pectineal line	A line on the posterior, superior aspect of the femur	*pectineal*, ridged or comb-like
	Popliteal surface	A triangular region on the posterior aspect of the distal femur	*popliteal*, the back of the knee
Patella		Triangular-shaped sesamoid bone	*patina, a* shallow disk

Table 9.4	The Appendicular Skeleton: Lower Limb *(continued)*		
Bone	**Bony Marking**	**Description**	**Word Origin**
Tibia tibia, *the large shin bone*	Anterior border	A ridge on the anterior surface extending distally from the tibial tuberosity; commonly referred to as the "shin"	
	Fibular articular surface	Smooth region on the proximal, posterolateral surface of the tibia; articulates with the head of the fibula	
Tibia (continued)	Intercondylar eminence	A prominent projection between the two condyles on the proximal epiphysis	*eminentia*, a raised area on a bone
	Lateral condyle	A large, flat surface on the lateral aspect of the proximal epiphysis; articulates with the lateral condyle of the femur	*kondylos*, knuckle
	Medial condyle	A large, flat surface on the medial aspect of the proximal epiphysis; articulates with the medial condyle of the femur	*kondylos*, knuckle
	Medial malleolus	A projection on the medial surface of the distal epiphysis	*malleus*, hammer
	Tibial tuberosity	A projection on the anterior surface of the proximal epiphysis; attachment site for the patellar ligament (quadriceps femoris muscle attachment)	*tibia*, shin bone
Fibula fibula, *a clasp or buckle*	Head	The rounded proximal end of the bone; articulates with lateral tibial condyle	
	Lateral malleolus	A projection on the lateral surface of the distal epiphysis	*malleus*, hammer
	Neck	The narrow portion where the head meets the diaphysis	
Tarsals tarsus, *a flat surface*	Calcaneus	Bone that forms the heel of the foot; attachment site for calcaneal tendon	*calcaneus*, the heel
	Cuboid	A cube-shaped bone; located at the base of the fourth metatarsal	*kybos*, cube, + *eidos*, resemblance
	Intermediate cuneiform	A wedge-shaped bone; located at the base of the second metatarsal	*cuneus*, wedge, + *forma*, shape
	Lateral cuneiform	A wedge-shaped bone; located at the base of the third metatarsal	*cuneus*, wedge, + *forma*, shape
	Medial cuneiform	A wedge-shaped bone; located at the base of the first metatarsal	*cuneus*, wedge, + *forma*, shape
	Navicular	A bone shaped like a boat ("ship"); located just anterior to the talus	*navis*, ship
	Talus	The major weight-bearing bone of the ankle; articulates with the tibia and fibula	*talus*, ankle
Metatarsals		Each bone consists of a head, body, and base	*meta-*, after *tarsus*, a flat surface
Phalanges phalanx, *line of soldiers*			
Digits II through V	Proximal	The phalanx closest to the metatarsal bones	*proximus*, nearest
	Middle	The middle phalanx	
	Distal	The small, cone-shaped phalanx forming the ends of the toes	*distalis*, away
Hallux hallux, *the big toe*	Proximal	The phalanx closest to the sole of the foot	*proximus*, nearest
	Distal	The phalanx forming the end of the big toe	*distalis*, away

6. Match the description listed in column A with the bone marking listed in column B. **4**

   Column A
   _____ 1. proximal end of radius
   _____ 2. narrow part of humerus between the head and tubercles
   _____ 3. narrow part of humerus just distal to the tubercles
   _____ 4. pointed projection on anterior surface of ulna
   _____ 5. small, pointed projection on distal end of radius
   _____ 6. part of the humerus that articulates with the ulna

   Column B
   a. anatomical neck
   b. coronoid process
   c. head
   d. styloid process
   e. surgical neck
   f. trochlea

7. Match the description listed in column A with the carpal bone listed in column B. **5**

   Column A
   _____ 1. shaped like a half moon
   _____ 2. shaped like a "table" and located at the base of the *first* metacarpal (thumb)
   _____ 3. shaped like a "table" and located at the base of the *second* metacarpal (index finger)
   _____ 4. smallest carpal bone; shaped like a "pea"
   _____ 5. shaped like a hook and located at the base of the *fifth* metacarpal (pinky)

   Column B
   a. hamate
   b. lunate
   c. pisiform
   d. trapezium
   e. trapezoid

Exercise 9.3: Surface Anatomy Review—Pectoral Girdle and Upper Limb

8. Which bony feature of the scapula can be palpated at the tip of the shoulder? (Circle one.) **6**

   a. acromion   b. coracoid   c. infraglenoid tubercle   d. spine   e. supraglenoid tubercle

9. Match the description listed in column A with the bony landmark listed in column B. **6**

   Column A
   _____ 1. forms the tip of the elbow
   _____ 2. process of the radius palpated at the wrist
   _____ 3. process of the humerus palpated on the medial aspect of the elbow

   Column B
   a. medial epicondyle
   b. olecranon
   c. styloid process

Exercise 9.4: Bones of the Pelvic Girdle

10. Match the description listed in column A with the bone or marking listed in column B. **7**

    Column A
    _____ 1. three bones fuse to form this bone of the pelvis
    _____ 2. bone that the pelvis rests on when sitting
    _____ 3. most anterior bone of the os coxae
    _____ 4. superior most bone
    _____ 5. large hole in the os coxae
    _____ 6. structure that articulates with the femur at the hip

    Column B
    a. acetabulum
    b. ilium
    c. ischium
    d. obturator foramen
    e. os coxae
    f. pubic bone

11. Identify the bones that form the os coxae. (Check all that apply). **8**
    _____ a. coccyx
    _____ b. ilium
    _____ c. ischium
    _____ d. pubis
    _____ e. sacrum

12. List three of the features that help differentiate a male pelvis from a female pelvis. **7**

    Male Pelvis
    a. _____
    b. _____
    c. _____

    Female Pelvis
    a. _____
    b. _____
    c. _____

Exercise 9.5: Bones of the Lower Limb

13. Match the bones listed in column A with the corresponding joint listed in column B. **9**

    Column A
    _____ 1. os coxae and femur
    _____ 2. femur and tibia
    _____ 3. tibia, fibula, and talus

    Column B
    a. ankle
    b. hip
    c. knee

14. Match the description listed in column A with the bone or bone marking listed in column B. **9**

Column A

_____ 1. rounded, proximal end of the fibula

_____ 2. roughened line on posterior aspect of the femur

_____ 3. large projection on lateral, proximal end of the femur

_____ 4. bone in leg that contains the medial malleolus

_____ 5. bone in leg that contains the lateral malleolus

_____ 6. flat surface on proximal end of the tibia

Column B

a. fibula

b. greater trochanter

c. head

d. linea aspera

e. medial condyle

f. tibia

15. Match the description listed in column A with the tarsal bone listed in column B. **10**

Column A

_____ 1. shaped like a "ship" and located anterior to the talus

_____ 2. major weight-bearing bone that articulates with the tibia and fibula to form the ankle joint

_____ 3. shaped like a "wedge" and located at the base of the first, second, and third metatarsals

_____ 4. forms the heel

Column B

a. calcaneus

b. cuneiforms

c. navicular

d. talus

16. There are _____ (seven/eight) carpal bones and _____ (seven/eight) tarsal bones. **11**

Exercise 9.6: Surface Anatomy Review—Pelvic Girdle and Lower Limb

17. Match the description listed in column A with the bone or bony landmark listed in column B. **12**

Column A

_____ 1. ridge of bone palpated when placing the hands on the hips

_____ 2. palpated on the anterior part of the knee joint

_____ 3. palpated on the lateral aspect of the knee joint

_____ 4. palpated on the anterior leg, distal to the knee joint

_____ 5. palpated on the medial aspect of the ankle joint

_____ 6. palpated on the lateral aspect of the ankle joint

_____ 7. palpated on the medial aspect of the knee joint

Column B

a. iliac crest

b. lateral epicondyle of the femur

c. lateral malleolus of the fibula

d. medial epicondyle of the femur

e. medial malleolus of the tibia

f. patella

g. tibial tuberosity

Can You Apply What You've Learned?

18. What are the structural and functional differences between the anatomical and surgical necks of the humerus?

_____

_____

_____

19. Compare and contrast both the appearance and the location of the carpal and tarsal bones.

_____

_____

_____

20. Explain why one end of the clavicle is called the sternal end and the other is called the acromial end.

_____

_____

_____

Can You Synthesize What You've Learned?

21. What bony landmark of the upper limb is used as a reference point for locating the ulnar nerve at the elbow?

_____

22. Which of the two necks of the humerus do you think is more likely to fracture in an accident? Why?

_____

_____

INTEGRATE

CLINICAL VIEW
Bursitis

Bursae are small round sacs lined with synovial membrane and filled with synovial fluid. They are found both within synovial joints and in other areas of the body that experience a great deal of friction. Although they are structures designed to reduce friction, they themselves may become inflamed when frictional forces are severe and long-lasting. *Bursitis* refers to inflammation of a bursa, and it can be a common source of joint pain. Repetitive movement of the joint and subsequent inflammation of the synovial membrane within the bursa cause bursitis. The inflammation causes the production of excessive synovial fluid. The swelling of the bursa can put pressure on surrounding structures and induce pain. Bursitis can also alter the mobility of the joint, significantly reducing the range of motion. Individuals suffering from bursitis may reduce their joint movements to avoid this pain. The muscles surrounding the joint may also stiffen as a result of bursitis. Common sites of bursitis are the shoulder, elbow, knee, and ankle joints. Treatment typically involves rest, ice, and elevation along with a course of anti-inflammatory drugs to reduce localized swelling and pain. In severe cases, inflamed bursae may be surgically removed.

EXERCISE 10.5

PRACTICING SYNOVIAL JOINT MOVEMENTS

1. All synovial joints are classified as **diarthrotic** (*di-*, two, + *arthron*, a joint) because they are freely movable. There are different types of movement possible with the various types of synovial joints. **Table 10.6** lists the types of movements possible at synovial joints. Practice all of the movements listed in table 10.6 with a laboratory partner to ensure that you can demonstrate or describe them all. It may be helpful to view figures of these movements in the textbook. This knowledge is critically important for success in subsequent chapters, which cover muscle actions. Muscle actions are described in terms of joint movements (e.g., flexor digitorum = muscle that flexes the digits).

2. For each movement completed in step 1, identify the opposite movement. Write the opposite movement in table 10.6 and practice the movement with a laboratory partner.

3. *Optional Activity:* **APR** 5: Skeletal System—Review the series of joint movement animations to see examples of multiple movements at each joint.

Table 10.6	Movements of Synovial Joints		
Movement	**Description**	**Opposite Movement**	**Word Origin**
Abduction	Movement of a body part away from the midline		*ab*, from, + *duco*, to lead
Adduction	Movement of a body part toward the midline		*ad-*, toward, + *duco*, to lead
Circumduction	Movement of the distal part of an extremity in a circle	NA	*circum-*, around, + *duco*, to lead
Depression	Movement of a body part inferiorly (especially mandible and scapula)		*depressio*, to press down
Dorsiflexion	Movement of the ankle such that the foot moves toward the dorsum (back)		*dorsum*, the back, + *flexus*, to bend
Elevation	Movement of a body part superiorly (especially mandible and scapula)		*e-levo-atus*, to lift up
Eversion	Movement of the ankle such that the plantar surface of the foot faces laterally		*e-*, out, + *versus*, to turn
Extension	An increase in joint angle		*extensio*, a stretching out
Flexion	A decrease in joint angle		*flexus*, to bend
Inversion	Movement of the ankle such that the plantar surface of the foot faces medially		*in-*, inside, + *versus*, to turn
Opposition	Placement of the thumb (pollex) such that it crosses the palm of the hand and can touch all of the remaining digits		*op -*, against, + *positio*, a placing
Plantar Flexion	Movement of the ankle such that the foot moves toward the plantar surface		*plantaris*, the sole of the foot, + *flexus*, to bend
Pronation	Movement of the palm from anterior to posterior		*pronatus*, to bend forward
Protraction	Movement of a body part anteriorly (especially mandible and scapula)		*pro-*, before + *tractio*, to draw
Reposition	Movement of the thumb to anatomic position after opposition		*re-*, backward, + *positio*, a placing
Retraction	Movement of a body part posteriorly (especially mandible and scapula)		*re-*, backward, + *tractio*, to draw
Rotation	Movement of a body part around its axis	NA	*rotatio*, to rotate
Supination	Movement of the palm from posterior to anterior		*supinatus*, to bend backward

CLINICAL VIEW
Hip Replacement Surgery

What comes to mind when thinking of someone's "hip"? Many people think this is the part of the body that lies superficial to the iliac crest. In fact, the term *ilium* literally means "hip." When you "put your hands on your hips," you are putting them on your iliac crests. However, when someone falls and suffers a fractured hip, it has little to do with the iliac bones at all. Instead, it concerns a fracture of another bone that composes part of the hip joint: the femur. Specifically, a fractured hip refers to a fracture of the neck of the femur.

The hip joint is the joint formed between the acetabulum of the os coxae and the head of the femur. This joint suffers wear and tear with age and is a common site of osteoarthritis. In the elderly, particularly elderly women, the bones that compose the hip joint may become less dense over time. These individuals are at increased risk of developing a femoral or "hip" fracture if they fall. Unfortunately, hip fractures do not heal particularly well, because the fracture often disrupts the blood supply to the head of the femur. Therefore, instead of repairing the fractured bones, a surgeon might opt to completely replace the hip joint.

Hip replacement surgery involves cleaning out the acetabulum. Previous techniques involved reconstructing the acetabulum with bone grafts. More modern approaches include bone augmentation with platelet-rich plasma (PRP), polymethyl methacrylate (PMMA, bone cement), or Cortoss, a composite that consists of resins and reinforcing glass particles. The purpose of reconstruction or augmentation is to create a more efficient and complete socket that will be a better fit for the two parts of the prosthetic hip. Augmentation techniques allow a "forced fit," which does away with the need for fixation screws. Within the acetabulum, the surgeon places an artificial cup to compose the "socket" of this "ball-and-socket" joint. Next, the fractured head and neck of the femur are removed, and a prosthetic is placed on the proximal end of the femur, thus composing the new "ball" of the "ball-and-socket" joint. **Figure 10.5** shows a few different types of prosthetics. Figure 10.5*a* is an example of a very old hip prosthetic. Notice how large the ball is. Figure 10.5*b* is an example of a prosthetic with a much smaller head, and figure 10.5*c* shows a similar prosthetic that is still within the bone. Notice in figure 10.5*c* that the greater and lesser trochanters of the femur are still intact. This is necessary to maintain the connections between the bone and the gluteal muscles and other muscles that act about the hip. Such muscles are separated during the surgery so the surgeon can access the hip joint. However, they must be reconnected to the bone after the surgery.

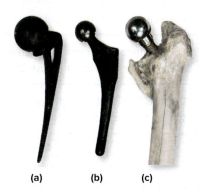

(a) (b) (c)

Figure 10.5 **Three Different Examples of Prostheses Used for Hip Replacements.**

©Christine Eckel

THE KNEE JOINT

This exercise involves observation of the structure of the knee joint, a representative synovial joint. The knee joint is complex, as are most synovial joints. It contains several modifications, such as bursae, tendon sheaths, and menisci. The knee joint also contains several strong ligaments, which help to stabilize the joint. **Table 10.7** summarizes the structures that compose the knee joint.

Most people have some peripheral knowledge of the knee joint, having known others who have suffered a ruptured ACL, torn meniscus, or other knee injury, even without knowledge of the anatomic structure of the knee. The ACL (anterior cruciate ligament) is one of two **cruciate ligaments** (*cruciatus*, resembling a cross) found in the knee joint, and a ruptured ACL is common in football players, downhill skiers, and others involved in contact sports.

1. Observe a model of the knee joint or the knee joint of a cadaver. Notice that the knee joint is actually two joints: the **tibiofemoral joint**, which is classified as a synovial hinge joint that acts in flexion and extension (though it allows for some rotational movement as well), and the **patellofemoral joint**, which is classified as a planar joint. The bony structure of the tibiofemoral joint includes the medial and lateral femoral condyles, which sit on top of the medial and lateral tibial condyles. The bony structure of the patellofemoral joint includes the patellar surface of the femur, which consists of the smooth anterior surface between the femoral condyles, and the medial and lateral facets of the patella.

2. Identify the structures listed in **figure 10.6** on a model of the knee joint or on a cadaver, using table 10.4 and the textbook as guides. Then, label them in figure 10.6.

(continued on next page)

GROSS ANATOMY

Gross Anatomy of Skeletal Muscles

The names of skeletal muscles often seem overly complex. However, understanding the basis of their names makes the task of identifying skeletal muscles of the body seem far less daunting. The name of a muscle often gives a clue to its location, size, shape, action, or attachment points. The exercises in this section include an introduction to the logic behind the naming of skeletal muscles (Exercise 11.1) and an overview of the typical architectures found in skeletal muscles (Exercise 11.2).

EXERCISE 11.1

NAMING SKELETAL MUSCLES

Table 11.1 summarizes some of the common ways skeletal muscles are named and gives word origins for the muscle names. Plan to spend some time mastering these word origins to be better prepared to handle the material to come in chapters 12 and 13. The efforts to learn the Latin and Greek word roots of anatomical terms will become even more valuable when working through the next three chapters.

Table 11.1	Common Methods for Naming Skeletal Muscles		
Name	**Meaning**	**Word Origin**	**Example**
Naming Skeletal Muscles Based on Shape			
Deltoid	Triangular	*delta*, the Greek letter delta (a triangle), + *eidos*, resemblance	Deltoid
Gracilis	Slender	*gracilis*, slender	Gracilis
Lumbrical	Wormlike	*lumbricus*, earthworm	Lumbricals
Rectus	Straight	*rectus*, straight	Rectus abdominis
Rhomboid	Diamond-shaped	*rhombo-*, an oblique parallelogram with unequal sides, + *eidos*, resemblance	Rhomboid major
Teres	Round	*teres*, round	Teres major
Trapezius	A four-sided geometrical figure having no two sides parallel	*trapezion*, a table	Trapezius
Naming Skeletal Muscles Based on Size			
Brevis	Short	*brevis*, short	Adductor brevis
Latissimus	Broadest	*latissimus*, widest	Latissimus dorsi
Longissimus	Longest	*longissimus*, longest	Longissimus capitis
Longus	Long	*longus*, long	Adductor longus
Major	Bigger	*magnus*, great	Teres major
Minor	Smaller	*minor*, smaller	Teres minor

Table 11.1	Common Methods for Naming Skeletal Muscles *(continued)*		
Name	**Meaning**	**Word Origin**	**Example**
Naming Skeletal Muscles Based on the Number of Heads and/or Bellies			
Biceps	2 heads	*bi*, two, + *caput*, head	Biceps brachii
Digastric	2 bellies	*bi*, two, + *gastro*, belly	Digastric
Quadriceps	4 heads	*quad*, four, + *caput*, head	Quadriceps femoris
Triceps	3 heads	*tri*, three, + *caput*, head	Triceps brachii
Naming Skeletal Muscles Based on Position			
Abdominis	Abdomen	*abdomen*, the greater part of the abdominal cavity	Rectus abdominis
Anterior	On the front surface of the body	*ante-*, before, in front of	Serratus anterior
Brachii	Arm	*brachium*, arm	Biceps brachii
Dorsi	Back	*dorsum*, back	Latissimus dorsi
Femoris	Thigh	*femur*, thigh	Rectus femoris
Infraspinatus	Below the scapular spine	*infra-*, below, + *spina*, spine	Infraspinatus
Interosseous	In between bones	*inter*, between + *osseus*, bone	Interossei
Oris	Mouth	*oris*, mouth	Orbicularis oris
Pectoralis	Chest	*pectus*, chest	Pectoralis major
Posterior	On the back surface of the body	*posterus*, following	Serratus posterior
Supraspinatus	Above the scapular spine	*supra-*, on the upper side, + *spina*, spine	Supraspinatus
Naming Skeletal Muscles Based on Depth			
Externus	External	*external*, on the outside	Obturator externus
Internus	Internal	*internal*, away from the surface	Obturator internus
Profundus	Deep	*pro*, before, + *fundus*, bottom	Flexor digitorum profundus
Superficialis	Superficial	*super*, above, + *facies*, face	Flexor digitorum superficialis
Naming Skeletal Muscles Based on Action			
Abductor	Moves a body part away from the midline	*ab*, from, + *ductus*, to bring toward	Abductor pollicis brevis
Adductor	Moves a body part toward the midline	*ad*, toward, + *ductus*, to bring toward	Adductor pollicis
Constrictor	Acts as a sphincter and closes an orifice	*cum*, together, + *stringo*, to draw tight	Superior pharyngeal constrictor
Depressor	Flattens or lowers a body part	*de-*, away, + *pressus*, to press	Depressor anguli oris

(continued on next page)

(continued from previous page)

Table 11.1	Common Methods for Naming Skeletal Muscles *(continued)*		
Name	**Meaning**	**Word Origin**	**Example**
Naming Skeletal Muscles Based on Action *(continued)*			
Dilator	Causes an orifice to open, or dilate	*dilato*, to spread out	Dilator pupillae
Extensor	Causes an increase in joint angle	*ex-*, out of, + *-tensus*, to stretch	Extensor carpi ulnaris
Flexor	Causes a decrease in joint angle	*flectus*, to bend	Flexor carpi ulnaris
Levator	Raises a body part superiorly	*levo* + *atus*, a lifter	Levator scapulae
Pronator	Turns the palm of the hand from anterior to posterior	*pronatus*, to bend forward	Pronator teres
Supinator	Turns the palm of the hand from posterior to anterior	*supino* + *atus*, to bend backward	Supinator

INTEGRATE

LEARNING STRATEGY

Recall from chapter 10 that synovial joints are highly movable joints. Activation of the muscles surrounding a joint creates movement of the joint. For example, flexion of the forearm requires contraction of the brachialis muscle and concurrent relaxation of the triceps brachii muscle. When learning the naming conventions for muscles, it is often helpful to observe the location of the muscle, including the attachment points (i.e., origin and insertion), and use those to predict the motion that is allowed at a particular joint. For example, when the flexor digitorum longus shortens, the digits "flex." It is no surprise, then, that "flex" appears in the name of this muscle based on its action. Be sure to review the movements of synovial joints (table 10.6) in preparation for learning names of axial and appendicular muscles in chapters 12 and 13.

EXERCISE 11.2

ARCHITECTURE OF SKELETAL MUSCLES

The overall architecture of a skeletal muscle affects how the muscle functions. When a whole muscle is observed, the individual fibers and fascicles are visible, making it relatively easy to see how the fascicles are arranged within the muscle. Recall that when skeletal muscle contracts, it generally gets shorter and brings its attachment points closer to each other. Thus, the orientation of the muscle fascicles compared to the attachment points of the muscle will directly affect the force produced by the muscle and the complexity of the muscle's actions. For example, **pennate** architecture (*penna*, feather) allows a muscle to produce greater force per distance shortened than **parallel** architecture. In addition, muscles with more than two attachments (for example, biceps and triceps) produce more complex movements than muscles with only two attachments (one proximal attachment and one distal attachment). **Table 11.2** summarizes the common patterns of fascicle arrangement that contribute to skeletal muscle architecture.

Table 11.2	Common Architectures of Skeletal Muscles					
Diagram						
Name	Unipennate	Bipennate	Multipennate	Circular	Convergent	Parallel
Word Origin	*uni*, one, + *penna*, feather	*bi*, two, + *penna*, feather	*multi*, many, + *penna*, feather	*circum*, around	*cum*, together, + *vergo*, to incline	*para*, alongside

1. Using classroom models of skeletal muscles or a prosected human cadaver, observe the arrangement of fascicles in several different skeletal muscles of the body, using table 11.2 as a guide.

2. Locate the muscles that are listed in **figure 11.1** on classroom models or on a human cadaver, using the textbook as a guide.

3. Complete the table provided by listing the architecture of each of the listed muscles.

	Muscle Name	Architecture
1	Deltoid	
2	Extensor digitorum	
3	Gastrocnemius	
4	Orbicularis oculi	
5	Pectoralis major	
6	Rectus femoris	
7	Sartorius	
8	Trapezius	
9	Triceps brachii	

INTEGRATE

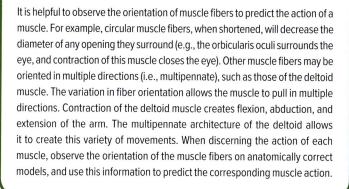

LEARNING STRATEGY

It is helpful to observe the orientation of muscle fibers to predict the action of a muscle. For example, circular muscle fibers, when shortened, will decrease the diameter of any opening they surround (e.g., the orbicularis oculi surrounds the eye, and contraction of this muscle closes the eye). Other muscle fibers may be oriented in multiple directions (i.e., multipennate), such as those of the deltoid muscle. The variation in fiber orientation allows the muscle to pull in multiple directions. Contraction of the deltoid muscle creates flexion, abduction, and extension of the arm. The multipennate architecture of the deltoid allows it to create this variety of movements. When discerning the action of each muscle, observe the orientation of the muscle fibers on anatomically correct models, and use this information to predict the corresponding muscle action.

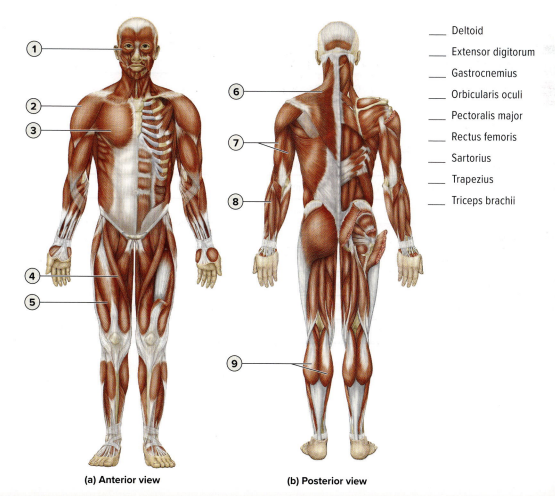

_____ Deltoid
_____ Extensor digitorum
_____ Gastrocnemius
_____ Orbicularis oculi
_____ Pectoralis major
_____ Rectus femoris
_____ Sartorius
_____ Trapezius
_____ Triceps brachii

(a) Anterior view   (b) Posterior view

Figure 11.1 **Muscles of the Human Body.** Match the number of each muscle to the corresponding muscle name, using the labels provided.

(continued from previous page)

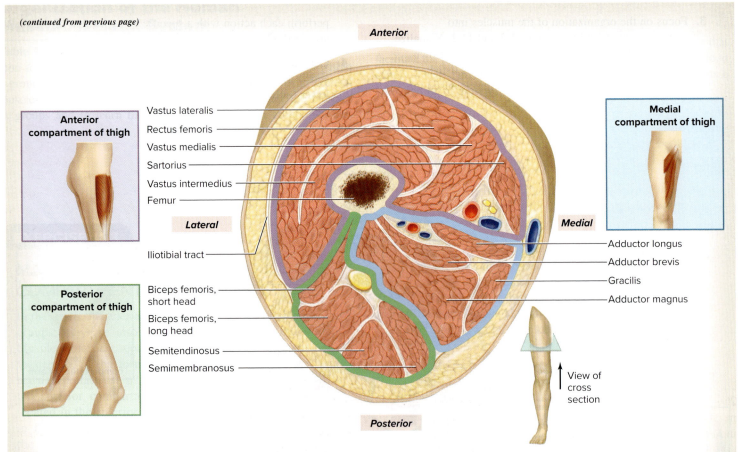

Figure 11.3   **Fascial Compartments of the Right Thigh.**

Table 11.4	Fascial Compartments of the Limbs and Their General Muscle Actions
Compartment	**General Description of Muscle Actions**
Compartments of the Arm	
Anterior	
Posterior	
Compartments of the Forearm	
Anterior	
Posterior	
Compartments of the Thigh	
Anterior	
Posterior	
Medial	
Compartments of the Leg	
Anterior	
Posterior	
Lateral	

HISTOLOGY

Muscle tissue is one of the most metabolically active tissues in the body and is the only tissue capable of creating movement of the body or body organs. All types of muscle tissue are characterized by the following properties: *excitability, contractility, elasticity,* and *extensibility.* **Excitable** tissues are able to respond to a stimulus and generate an electrical signal called an action potential. **Contractile** tissues actively shorten and produce force. **Elastic** tissues return to their original shape following either contraction or stretching. **Extensible** tissues are able to be lengthened by the pull of an external force (such as an external weight or the action of an opposing muscle). **Conductive** tissues allow electrical signals such as action potentials to travel along the plasma membrane of the cell.

The exercises in this section guide the user through observations of the histology of skeletal, cardiac, and smooth muscle tissues. **Table 11.5** lists the characteristics of the three types of muscle tissue and serves as a reference for these exercises.

Skeletal Muscle Tissue

Skeletal muscle cells are some of the largest cells in the body. Their enormous size results from the fusion of hundreds of *myoblasts* (embryonic muscle cells) during development into a single muscle cell, or fiber. A mature muscle fiber is multinucleated, containing 200 to 300 nuclei per millimeter of fiber length. The nuclei of skeletal muscle fibers are located directly beneath the *sarcolemma* (plasma membrane of the muscle fiber).

A muscle, such as the biceps brachii muscle of the arm, consists of several bundles, or **fascicles** (*fascis,* bundle) **(figure 11.4*a*)**.

Each fascicle consists of hundreds of long, cylindrical **muscle fibers** (figure 11.4*b*). Each muscle fiber contains within it a number of cylindrical bundles of contractile proteins called **myofibrils** (figure 11.4*c*). The myofibrils contain the **myofilaments** actin and myosin (the main contractile proteins of muscle), which are arranged into **sarcomeres** (the structural and functional unit of skeletal muscle, figure 11.4*d*). The regular arrangement of actin and myosin into sarcomeres gives each myofibril a striated or banded appearance, with visible **A bands** (dark bands), **I bands** (light bands), and **Z discs** (a dark line in the middle of an I band) (figure 11.4*d,e*). An adult skeletal muscle fiber typically contains about 2000 myofibrils per cell. Because intermediate filaments within the muscle fiber anchor and align the Z discs of adjacent myofibrils, the entire muscle fiber takes on the same regular striated appearance of A bands and I bands found in the myofibrils when viewed through the light microscope. Both thick filaments, composed of myosin, and Z discs appear dark, whereas thin filaments, composed primarily of actin, appear light when skeletal muscle tissue is viewed through a light microscope.

Figure 11.4 demonstrates the relationships between the gross structure of a skeletal muscle and the microstructure of a skeletal muscle fiber. Though myofibrils and myofilaments are not visible when viewing muscle tissue through the microscope, A bands, I bands, and Z discs are usually visible. While viewing the banding pattern of skeletal muscle cells, try to relate the bands to the corresponding arrangement of myofilaments into sarcomeres as shown in the drawing of a sarcomere in figure 11.4.

Table 11.5	Muscle Tissues		
Type of Muscle	**Description**	**Functions**	**Identifying Characteristics**
Skeletal	Elongated, cylindrical cells with multiple nuclei that are peripherally located. Tissue appears striated (alternating light and dark bands along the length of the cell).	Produces voluntary movement of the skin and the skeleton	Length of cells (extremely long); striations; multiple, peripheral nuclei
Cardiac	Short, branched cells with one or two nuclei that are centrally located. Dark lines (intercalated discs) are seen where two cells come together. Tissue appears striated (light and dark bands along the length of each cell).	Performs the contractile work of the heart. Responsible for the pumping action of the heart.	Branched cells connected by intercalated discs; striations; one or two central nuclei
Smooth	Elongated, spindle-shaped cells (fatter in the center, narrowing at the tips) with single, "cigar-shaped" nuclei that are centrally located. No striations.	Produces movement within the walls of visceral organs such as intestines, bladder, uterus, and stomach; produces other specialized involuntary muscle movement (e.g., focusing the lens in the eye)	Spindle shape of the cells; lack of striations; single, central, cigar-shaped or spiral nuclei

INTEGRATE

CONCEPT CONNECTION

Recall from chapter 5 that there are four tissue types in the body: epithelial, connective, muscle, and nervous tissue. Muscle tissue is further divided into three types: skeletal, cardiac, and smooth. The mode of activation for these muscle types differs. Skeletal muscle is voluntary, whereas cardiac and smooth

muscle are involuntary. Involuntary contraction of cardiac and smooth muscle is advantageous, because it would be inconvenient to have to "think" about contracting the heart every time blood circulation is required, or to "think" about propelling food through the gastrointestinal tract when digestion is required. Instead, cardiac muscle contracts ("beats") *intrinsically* (on its own), and smooth muscle in the gut contracts when food is present.

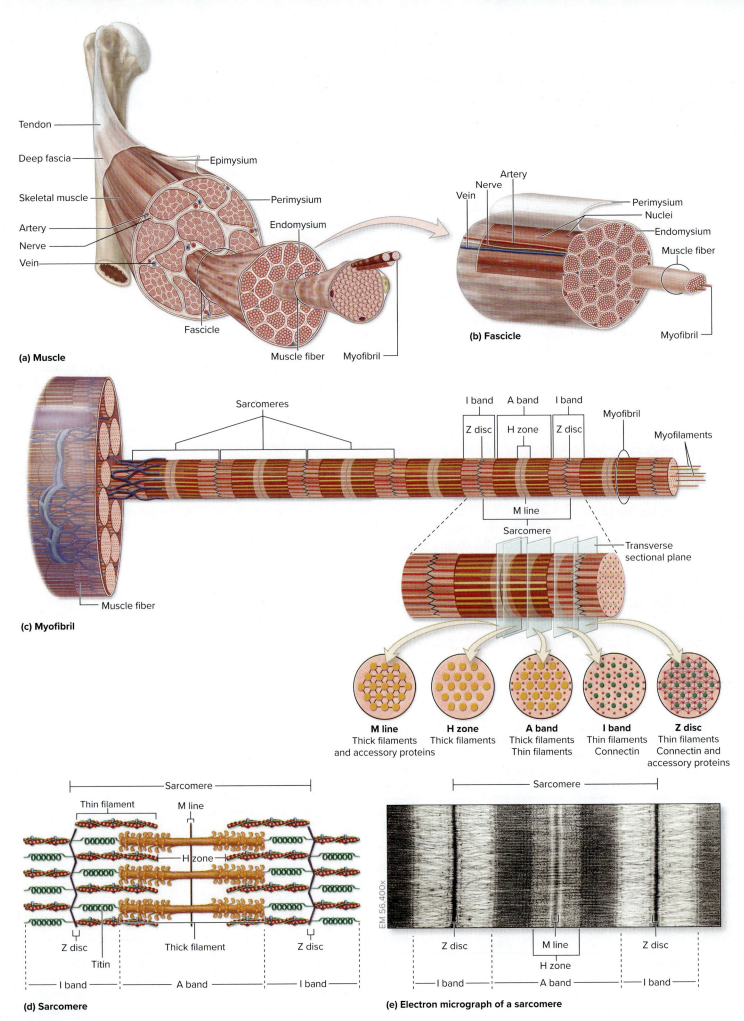

(a) Muscle

Tendon
Deep fascia
Epimysium
Skeletal muscle
Perimysium
Artery
Endomysium
Nerve
Vein
Fascicle
Muscle fiber
Myofibril

(b) Fascicle

Artery
Nerve
Vein
Perimysium
Nuclei
Endomysium
Muscle fiber
Myofibril

(c) Myofibril

Sarcomeres
I band
A band
I band
Z disc
H zone
Z disc
Myofibril
Myofilaments
M line
Sarcomere
Transverse sectional plane
Muscle fiber

M line
Thick filaments and accessory proteins

H zone
Thick filaments

A band
Thick filaments
Thin filaments

I band
Thin filaments
Connectin

Z disc
Thin filaments
Connectin and accessory proteins

(d) Sarcomere

Sarcomere
Thin filament
M line
H zone
Z disc
Thick filament
Z disc
Titin
I band
A band
I band

(e) Electron micrograph of a sarcomere

EM 56,400x

Sarcomere
Z disc
M line
H zone
Z disc
I band
A band
I band

Figure 11.4   Levels of Structural Organization of Skeletal Muscle. (*a*) A whole muscle, (*b*) a muscle fascicle, (*c*) a myofibril, (*d*) a sarcomere, (*e*) an electron micrograph of a sarcomere. **APR**

(e) ©Jim Dennis/Medical Images

HISTOLOGY OF SKELETAL MUSCLE FIBERS

1. Obtain a slide of skeletal muscle tissue and place it on the microscope stage.

2. Bring the tissue into focus using the scanning objective. Switch to low power, bring the tissue sample into focus once again, and then switch to high power. The slide of skeletal muscle contains muscle fibers usually shown in both longitudinal section and cross section. Scan the slide and identify muscle fibers shown in both longitudinal section and cross section (**figures 11.5** and **11.6**).

3. Focus on muscle fibers shown in longitudinal section (figure 11.5) and observe them at high power. Identify an individual muscle fiber. Notice the numerous peripherally located nuclei. Most of the nuclei that are visible on the slide belong to the muscle fibers. However, about 5%

to 15% of the visible nuclei are those of **satellite cells**, myoblast-like cells located between the muscle fibers. Satellite cells give skeletal muscle a limited ability to repair itself after injury. It is not possible to tell which nuclei belong to satellite cells when viewing the tissue through the light microscope.

4. Identify the following structures on the slide of skeletal muscle tissue, using table 11.5 and figure 11.5 as guides:

   ☐ **A Band**   ☐ **Sarcomere (if visible)**

   ☐ **I Band**   ☐ **Z Disc (if visible)**

   ☐ **Nucleus**

5. Sketch skeletal muscle fibers as viewed through the microscope in the space provided, or on a separate sheet of paper. Be sure to label the structures listed in step 4 in the drawing.

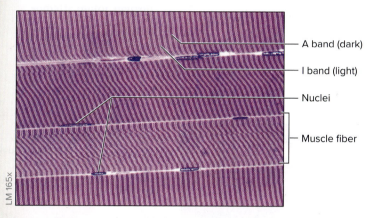

LM 165x

A band (dark)

I band (light)

Nuclei

Muscle fiber

Figure 11.5 **Skeletal Muscle in Longitudinal View.** Z discs and sarcomeres are not visible in this micrograph.

©Victor P. Eroschenko

×

CARDIAC MUSCLE TISSUE

1. Obtain a slide of cardiac muscle tissue and place it on the microscope stage.

2. Bring the tissue sample into focus using the scanning objective. Switch to low power and bring the tissue sample into focus once again. Then switch to high power. Notice that the cells contain only one or two nuclei, and that the cells are short and branched. Where two cells come together, there should be a darkly stained line. This is an **intercalated disc**. Intercalated discs contain numerous **desmosomes**, which function to hold the fibers together, and **gap junctions**, which allow electrical signals to be transmitted very rapidly from one fiber to the next.

3. Identify the following structures on the slide of cardiac muscle tissue, using table 11.5 and figure 11.8 as guides.

   ☐ **A Band**   ☐ **Intercalated disc**

   ☐ **Branching fibers**   ☐ **Nucleus**

   ☐ **I Band**

4. Sketch cardiac muscle fibers as viewed through the microscope in the space provided, or on a separate sheet of paper. Be sure to label the structures listed in step 3.

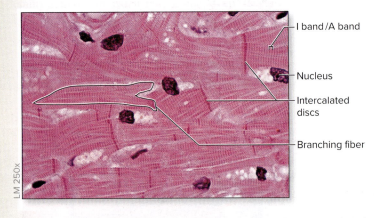

I band/A band

Nucleus

Intercalated discs

Branching fiber

LM 250x

Figure 11.8   **Cardiac Muscle Tissue.**
©Ed Reschke/Getty Images

_____ ✕

Smooth Muscle Tissue

Spindle-shaped cells with cigar-shaped or spiral nuclei and an absence of striations are characteristic of smooth muscle tissue. Individual cells have tapered ends, and there is only one centrally located nucleus per cell. Most of the nuclei will appear to be somewhat cigar shaped. However, in cells that have contracted, the nuclei take on a corkscrew or spiral appearance (**figure 11.9a**), which can be a key identifying feature.

Smooth muscle is generally found in two layers around tubular organs such as the small intestine. The most common arrangement is an *inner circular layer* and an *outer longitudinal layer* of smooth muscle cells (figure 11.9b).

EXERCISE 11.8

SMOOTH MUSCLE TISSUE

1. Obtain a slide containing smooth muscle tissue (figure 11.9) and place it on the microscope stage.

2. Bring the tissue into focus using the scanning objective. Switch to low power and bring the tissue sample into focus once again. Then switch to high power. If viewing a slide of the small intestine, identify smooth muscle cells that have been cut in both longitudinal section and cross section. Note that the cells do not appear to be of uniform diameter when viewed in cross section. This is because some cells are sectioned through the tapered ends, while others are sectioned through the thickest part of the fiber, which contains the nucleus.

3. Identify the following structures on the slide of smooth muscle tissue, using table 11.5 and figure 11.9 as guides.

☐ **Inner circular layer**

☐ **Muscle cell in cross section**

☐ **Muscle cell in longitudinal section**

☐ **Nucleus**

☐ **Outer longitudinal layer**

4. Sketch smooth muscle cells as viewed through the microscope in the space provided, or on a separate sheet of paper. Be sure to label the structures listed in step 3 in the drawing.

_____ ×

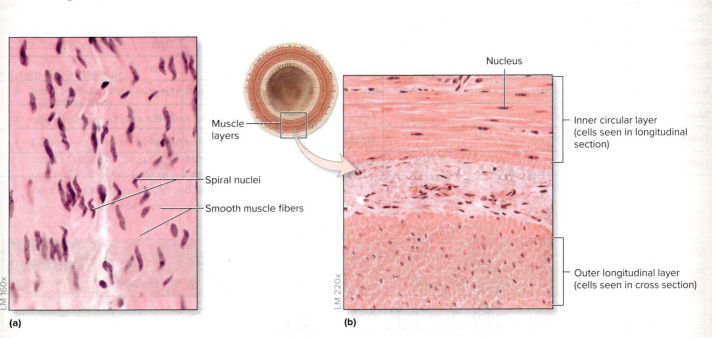

(a)   (b)

Figure 11.9 **Smooth Muscle Tissue.** (*a*) Close-up view of smooth muscle demonstrating spiral nuclei, which are visible when the muscle fibers are contracted; (*b*) circular and longitudinal layers of smooth muscle tissue.

©McGraw-Hill Education/Al Telser

EXERCISE 11.13   **PhILS**

Ph.I.L.S. LESSON 8: PRINCIPLES OF SUMMATION AND TETANUS

Action potential propagation along the sarcolemma of skeletal muscle fibers (cells) results in the release of calcium from the sarcoplasmic reticulum into the surrounding sarcoplasm (cytoplasm within the muscle cell). The duration of the electrical event is rapid (1–2 ms) when compared to the duration of the mechanical contraction (100–200 ms), principally because of the time that it takes to actively pump the calcium back into the sarcoplasmic reticulum. The presence of calcium in the sarcoplasm prolongs the period of crossbridge formation. As action potentials increase in frequency, more calcium is available in the sarcoplasm; therefore, muscle tension increases.

The purpose of this laboratory exercise is to observe the change in muscle tension as the frequency of action potentials increases. As the frequency increases, muscle twitches occur more rapidly, ultimately resulting in summation of muscle tension.

Before beginning, become familiar with the following concepts. Use the textbook as a reference:

- Voltage-gated calcium channels in skeletal muscle
- Crossbridge cycling
- A single muscle twitch
- Summation of muscle twitches
- Incomplete and complete tetanus

State a hypothesis regarding the effect that action potential frequency has on calcium concentration in the sarcoplasm and resulting muscle tension.

_____

_____

_____

_____

_____

_____

_____

1. To begin the experiment, open Ph.I.L.S. Lesson 8: Principles of Summation and Tetanus (see **figure 11.16**).

2. Read the objectives, introduction, and wet lab. Take the pre-lab quiz. The lab exercise will open when the pre-lab quiz has been completed.

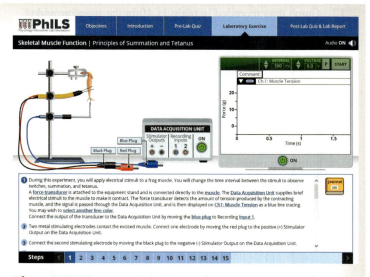

Figure 11.16   Opening Screen for Ph.I.L.S. Lesson 8: Principles of Summation and Tetanus.
©McGraw-Hill Education

3. Click the power button on the virtual computer screen and the data acquisition unit (DAQ).

4. Connect the force transducer to the data acquisition unit by clicking and dragging the blue plug to recording input 1.

5. Connect the stimulating electrodes to the stimulator outputs on the DAQ by clicking and dragging the black plug to the negative terminal and clicking and dragging the red plug to the positive terminal.

6. On the control panel, set the voltage to 1.6 volts by clicking the up arrow.

7. Click the Start button to begin delivering stimuli to the muscle. The default time interval is 500.

8. At a time interval of 500, there are single muscle twitches. To determine the interval that will elicit summation, decrease the interval between stimuli by clicking the down arrow on the control panel. Observe the changes seen in the twitches after decreasing the time interval.

9. Continue the above procedure until summation is achieved, where the blue trace does not always return to the red baseline. Click the Stop button once summation is achieved.

10. Click the Journal in the lower right corner of the screen. Record the time interval where summation is first observed in **table 11.11**. Click the X in the upper right corner of the Journal window to continue the experiment.

11. Click the Start button to begin stimulating the muscle again.

Table 11.11	Results from Ph.I.L.S. Lesson 8: Principles of Summation and Tetanus	
Type of Contraction	**Time Interval of Stimulation**	
Summation		
Incomplete Tetanus		
Complete Tetanus		

12. Continue to decrease the time interval by clicking the down arrow until incomplete tetanus is observed, where the blue muscle tension trace begins to extend beyond the gray line. Click the Journal to record the time interval where the muscle reached incomplete tetanus. Click the X to close the Journal window. Record this time interval in table 11.11.

13. Click the Start button to begin stimulating the muscle again.

14. Continue to decrease the time interval by clicking the down arrow until complete tetanus is observed, where individual twitches are no longer visible. Click the Journal to record the time interval where the muscle reaches incomplete tetanus. Click the X to close the Journal window. Record this time interval in table 11.11.

15. Click the Post-lab Quiz & Lab Report to complete the post-lab quiz.

16. Make note of any pertinent observations here:

_____

_____

_____

EXERCISE 11.14 — PhILS

Ph.I.L.S. LESSON 9: EMG AND TWITCH AMPLITUDE

The purpose of this laboratory exercise is to record the electrical activity from skeletal muscles using virtual surface electrodes to generate an electromyogram (EMG). Pressure generated from a contracting muscle and EMG readings will be obtained in a virtual subject, and this data will be used to estimate the number of muscle fibers recruited to generate muscle tension.

Before beginning, become familiar with the following concepts. Use the textbook as a reference:

- Physiological principles of the neuromuscular junction
- Anatomical and physiological properties of a motor neuron
- Depolarization and repolarization of a muscle cell
- Contractile properties of skeletal muscle
- The "all-or-none" principle of skeletal muscle contraction

State a hypothesis regarding the relationship between muscle force and EMG.

_____

_____

_____

_____

_____

_____

_____

1. Open Ph.I.L.S Lesson 9: EMG and Twitch Amplitude (see **figure 11.17**).

2. Read the objectives, introduction, and wet lab. Take the pre-lab quiz. The lab exercise will open when the pre-lab quiz has been completed.

3. Click the power switch on the virtual computer screen and data acquisition (DAQ) unit to begin the lab.

Figure 11.17 **Opening Screen for Ph.I.L.S. Lesson 9: EMG and Twitch Amplitude.**
©McGraw-Hill Education

4. Connect the pressure transducer by clicking and dragging the orange plug to the recording input number 2.

5. Click and drag the hand dynamometer to the subject's right hand.

6. To set up the EMG, connect the electrodes to the DAQ by clicking and dragging the blue plug to recording input number 1.

7. Click and drag the green electrode to a position just above the subject's right wrist. Click and drag the red electrode to the upper right forearm of the subject (just inside the elbow joint). Click and drag the black electrode to a position between the red and green electrodes on the forearm.

8. Just below the subject's right hand, a slider bar with arrows will appear. This is used to control the amount of forearm contraction. To increase the contraction strength, click the up arrow. To decrease the contraction strength, click the down arrow.

9. To begin the experiment, adjust the grip strength by clicking the down arrow until the setting on the slider bar is set at 1.

(continued on next page)

(continued from previous page)

10. Click Start on the control panel. The EMG is the upper blue tracing and is recorded in millivolts (mV). The pressure is the lower orange tracing. Position the crosshairs on the highest point on the orange tracing. Click on Journal in the lower right-hand corner of the screen. Enter the pressure data in **table 11.12**. Click the X in the upper right corner of the Journal window to continue the experiment.

11. To record the EMG, move the mouse pointer to the highest peak in the EMG recording (blue tracing) and click the Journal. Enter this EMG data in table 11.12.

12. Repeat steps 10–11 by increasing the forearm contraction from 2 to 6. Record data for all points in table 11.12.

Table 11.12	Results from Ph.I.L.S. Lesson 9: EMG and Twitch Amplitude	
Recording (Value)	Pressure Amplitude	EMG Amplitude
1		
2		
3		
4		
5		
6		

13. Graph the data generated in the space provided.

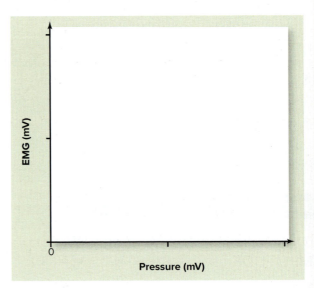

14. Click the Post-lab Quiz & Lab Report to complete the post-lab quiz.

15. Make note of any pertinent observations here:

_____

_____

_____

_____

EXERCISE 11.15

BIOPAC ELECTROMYOGRAPHY (EMG)

This exercise involves observing the electrical activity of muscle at rest and at maximum exertion. A subject will perform the activity (clenching a fist) and observe the corresponding electromyography (EMG) tracing to determine the relationship between motor unit recruitment and whole muscle force production in the subject's dominant versus nondominant forearm.

Obtain the Following:

- **BIOPAC electrode lead set (SS2L)**
- **BIOPAC general purpose electrodes (EL503)**
- **BIOPAC headphones (OUT1, OUT1A, or 40HP for MP45 unit)**
- electrode gel
- **BIOPAC MP36/35/45 recording unit**
- **Laptop Computer with BSL 4 software installed**

1. Prepare the BIOPAC equipment by turning the computer ON and the MP36/35/45 unit OFF. Plug in the following: electrode lead set (SS2L) (CH 1), and headphones (OUT 1; back of unit). Turn the MP36/35/45 unit ON.

2. Prepare a subject for placement of electrodes by rubbing the skin vigorously with an alcohol swab at each of the designated locations on the subject's *dominant* forearm and allowing the skin to dry completely **(figure 11.18)**. Apply a small quantity of electrode gel to the center of the electrodes. Next, place the electrode firmly on the skin of the subject's *dominant* forearm using the color code described in figure 11.18.

3. Have the subject sit in a comfortable position facing the computer monitor. Start the BIOPAC Student Lab Program. Select L01-Electromyography (EMG) I from the Choose a Lesson menu, enter the filename, and click OK.

4. Click Calibrate in the upper left corner. Wait two seconds; then have the subject clench his/her fist as hard as he/she can, and then relax. Once the calibration has stopped, check the graph to ensure that an EMG was recorded. If so, click Continue to proceed to data recording. If no recording was made, repeat the calibration process by clicking on Redo Calibration. See **figure 11.19** for sample calibration data.

5. OBTAIN DATA: During the recording session, the subject will repeat the clench and release task four times. Each time the subject clenches and releases his/her fist,

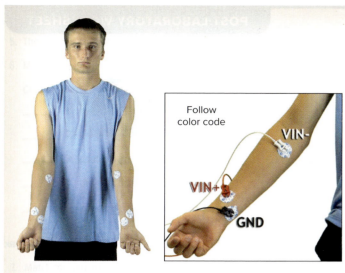

Figure 11.18 **BIOPAC EMG Electrode Placement and Lead Sttachment.**

Courtesy of and ©BIOPAC Systems, Inc. www.biopac.com

Table 11.13	EMG Measurements	
Between Clenches #	**Dominant Arm**	**Nondominant Arm**
	40 Mean	**40 Mean**
1–2		
2–3		
3–4		

he/she should clench it harder than the previous time. Maximum intensity should be obtained on the fourth repetition. To begin recording EMG data for each trial, click Record. After the subject has completed a trial, click Suspend. Review the data recording after each trial to ensure that EMG activity increased in intensity with each trial. If not, click Redo.

6. Repeat the experiment using the subject's *nondominant* forearm by following the same procedure described in steps 2 and 5. Place electrodes on the subject's opposite, or *nondominant* forearm (see step 2 and figure 11.18). Click Continue. Repeat step 5, having the subject complete the task with the opposite fist. Review the data to ensure the EMG increased in intensity with each trial. If so, click Stop. If not, click Redo.

7. To listen to the EMG, place the headphones on and click Listen. Repeat the clench and release task. Adjust the sound as needed to hear the EMG signal through the headphones. Note that data recorded while listening through the headphones is not saved. Click Stop and Redo to listen again. Click Done to conclude the experiment.

8. If analyzing data that was just collected, click on Analyze Current Data File. If opening data that was collected previously, click Review Saved Data. Note that CH 1 displays "EMG," and CH40 displays "Integrated EMG." View the first recorded segment.

9. Measurement boxes appear above the marker region in the data window. Note that CH 40 displays "Mean" (average value).

10. Locate the I-beam cursor. Select the point on the recorded segment that corresponds to the plateau of the first EMG data cluster. Record the values in **table 11.13**.

11. Repeat step 10 for each EMG data cluster for both the subject's dominant forearm and the subject's nondominant forearm.

12. The resting state between the EMG data clusters represents Tonus. Use the I-beam cursor to highlight each tonus for both the dominant and the nondominant forearms. Record the values in **table 11.14**.

13. Make note of any pertinent observations here:

_____

_____

_____

14. When data analysis is complete, click Save or Print; then click Quit to close the program.

Figure 11.19 **Sample Calibration Data for BIOPAC EMG Exercise.**

Courtesy of and ©BIOPAC Systems, Inc. www.biopac.com

Table 11.14	Tonus Measurements	
Clench #	**Dominant Arm**	**Nondominant Arm**
	40 Mean	**40 Mean**
1		
2		
3		
4		

GROSS ANATOMY

Muscles of the Head and Neck

Muscles of the head and neck include the muscles of the face (muscles of facial expression and muscles of mastication), extrinsic eye muscles, muscles that move the tongue, and muscles that move the neck. (Note that extrinsic eye muscles are covered in chapter 18 with the exercises on the eye.)

INTEGRATE

LEARNING STRATEGY

Use the following procedure when studying each major muscle group to help build knowledge and understanding in logical steps. Once the information from step 1 is mastered, move on to subsequent steps. Each successive step requires a more detailed level of understanding. Following these steps will prepare you to deal with the increased level of detail required at each knowledge level.

Stepwise Approach to Learning Muscles of the Body

1. Describe the general location of the muscle group.

2. Describe the general actions that all muscles of the group have in common.

3. List the names of all muscles belonging to that group (or just the ones required by the laboratory instructor).

4. Identify the muscles on a model or a cadaver (this can be done concurrently with step 3).

5. Learn the outliers—the muscles that DO NOT share the common actions of the group.

6. Learn specific points of attachment and actions of the muscles, as required by the laboratory instructor.

Follow these suggestions: Study one muscle group at a time, take a lot of breaks, and study using frequent, short time intervals. The results may be pleasantly surprising.

EXERCISE 12.1

MUSCLES OF FACIAL EXPRESSION

Muscles of the face are separated into two groups based on function and innervation: muscles of facial expression and muscles of mastication (chewing). **Muscles of facial expression (table 12.1)** allow us to express emotions such as fright, delight, confusion, and surprise. These muscles are unique in that they have distal attachments on skin instead of bone. Thus, when the muscles contract, they pull on the skin. This movement is easily seen on the surface of the face as a facial expression. Over time, the pulling of these muscles on the face causes a characteristic wrinkling of the skin. The muscles of facial expression are innervated by the facial nerve (CN VII).

1. Observe muscles of the face on a human cadaver or a classroom model.

2. Identify the muscles of facial expression listed in **figure 12.1** on the cadaver or model of the face, using table 12.1 and the textbook as guides.

3. Label the muscles of facial expression in figure 12.1.

4. Each of the seemingly endless variety of facial expressions is performed by contracting specific muscles of the face. Observe different facial expressions either on yourself (in a mirror) or on the face of a classmate. Determine the muscles involved in the following facial expressions, using table 12.1 as a guide.

☐ **Anger or doubt**

☐ **Happiness (smiling or laughter)**

☐ **Kissing (close mouth, purse cheeks, close eyes)**

☐ **Sadness (frowning)**

☐ **Surprise or delight**

5. *Optional Activity:* **APR** 6: **Muscular System**—Watch the muscle action animations to review the actions of many of the muscles mentioned in chapters 12 and 13. Also try the action and attachment questions found in the quiz area for challenging drill and practice.

Table 12.1	Muscles of Facial Expression*			
Muscle	**Bony Attachment**	**Soft Tissue Attachment**	**Action(s)**	**Word Origin**
Buccinator	Alveolar processes of mandible and maxilla	Orbicularis oris (corners of the lips)	Presses cheek against molar teeth, as in chewing	*bucca*, cheek
Corrugator Supercilii	Superciliary arch	Skin of eyebrow	Pulls eyebrows medially, creating wrinkles above nose, as in worry	*corrugo*, to wrinkle, + *superus*, above, + *cilium*, eyelid

Table 12.1	Muscles of Facial Expression* *(continued)*			
Muscle	**Bony Attachment(s)**	**Soft Tissue Attachment(s)**	**Action(s)**	**Word Origin**
Depressor Anguli Oris	Mandible (antero-lateral surface of the body)	Skin at inferior corner of mouth	Pulls corners of the mouth inferior, as in frowning	*depressus,* to press down, + *angulus,* angle, + *oris,* mouth
Depressor Labii Inferioris	Mandible (between the midline and the mental foramen)	Oribicularis oris and skin of the lower lip	Depresses the lower lip, as in expressions of sadness	*depressus,* to press down, + *labia,* lip, + *inferior,* lower
Epicranius (Occipitofrontalis)**	Epicranial aponeurosis	Skin of the forehead (frontalis); superior nuchal line (occipitalis)	Elevates the eyebrows and causes horizontal wrinkles in the forehead, as in expressions of surprise or delight	*occiput,* the back of the head, + *frontalis,* in front
Levator Anguli Oris	Maxilla (lateral portion)	Skin at the superior corner of the mouth	Elevates the corners of the mouth and pulls them laterally, as in smiling	*levatus,* to lift, + *labia,* lip, + *superus,* above
Levator Labii Superioris	Maxilla (inferior to infraorbital foramen)	Orbicularis oris and skin of the upper lip	Elevates the upper lip, as in snarling or showing teeth	*levatus,* to lift, + *anguli,* angle, + *oris,* mouth
Mentalis	Mandible (inferior to incisors)	Skin of the chin	Wrinkles the skin of the chin; protrudes lower lip as in pouting	*mentum,* the chin
Nasalis	Maxilla and nasal cartilages	Dorsum of nose	Flares the nostrils, widens the anterior nasal aperture	*nasus,* nose
Orbicularis Oculi	Margin of the orbit of the eye	Skin surrounding the eyelids	Closes the eyelids as in blinking	*orbiculus,* a small disk, + *oculus,* eye
Orbicularis Oris	Deep surface of skin of maxilla and mandible	Skin and muscles at margins of lips	Purses and protrudes the lips, closes the mouth	*orbiculus,* a small disk, + *oris,* mouth
Platysma***	Fascia superficial to the deltoid and pectoralis major muscles at ribs 1 and 2	Mandible (lower border) and skin of the cheek	Stretches the skin of the anterior neck, depresses the lower lip, as in expressions of fright	*platys,* flat, broad
Procerus	Nasal bones and nasal cartilages	Aponeurosis at the bridge of the nose and the skin of the forehead	Depresses the eyebrows and wrinkles the skin of the nose, as in frowning and squinting the eyes	*procerus,* long or stretched out
Risorius	Fascia overlying the masseter muscles	Skin of the corner of the mouth	Pulls the corners of the mouth laterally, as in expressions of laughter or smiling	*risus,* to laugh
Zygomaticus (Major and Minor)	Zygomatic bone	Skin and muscle at the corner of the mouth	Pulls the corners of the mouth posteriorly and superiorly, as in smiling	*zygon,* yoke

*All muscles in this table are innervated by the facial nerve (CN VII).

**The epicranius consists of the epicranial aponeurosis and the occipitofrontalis muscle, which has two bellies: frontal belly of occipitofrontalis and occipital belly of occipitofrontalis.

***Platysma is an exception to the attachment rule because its "origin" is inferior on muscles.

(continued on next page)

Muscles of Respiration

The diaphragm, muscles of the thoracic cage (external intercostals, internal intercostals, and transversus thoracis), and the scalenes are the **muscles of respiration (table 12.8)**. Respiration includes inspiration (breathing air in) and expiration (breathing air out). The movement of air occurs due to the changes in the volume of the lungs and thoracic cavity. In order to take air into the lungs during inspiration, the thoracic cavity must expand. The **diaphragm** is the primary muscle of respiration because it causes the largest changes in volume of the thoracic cavity. Contraction of the diaphragm and external intercostals causes the volume of the thoracic cavity to increase, thereby causing inspiration. If further or more forceful inspiration is needed, the scalene muscles in the neck can assist. The natural recoil of the lungs and subsequent relaxation of these inspiratory muscles causes expiration. The internal intercostals and transversus thoracis muscles are used for forced expiration (e.g., sneezing or coughing). These muscles depress the ribs, which decreases the volume of the thoracic cavity and forces air out of the lungs. Other muscles, such as the muscles of the abdominal wall, can assist with forced expiration.

Table 12.8	Muscles of Respiration			
Muscle	**Superior Attachment(s)**	**Inferior Attachment(s)**	**Action(s)**	**Word Origin**
Diaphragm	Central tendon	Lumbar vertebrae; internal surfaces and costal cartilages of ribs 7–12; xiphoid process	Prime mover for inspiration; flattens when contracted, and increases intra-abdominal pressure and the size of the thoracic cavity	*diaphragma*, a partition wall
External Intercostals	Inferior border of superior rib	Superior border of the inferior rib	Elevates the ribs	*externus,* on the outside, + *inter*, between, + *costal*, rib
Internal Intercostals	Superior border of inferior rib	Inferior border of the superior rib	Depresses the ribs	*internus*, away from the surface, + *inter*, between, + *costal*, rib
Transversus Thoracis	Costal cartilages of ribs 2–6 (posterior surface)	Posterior surface of the xiphoid process and inferior sternal body	Depresses the ribs	*transversus*, crosswise, + *thoracis*, thorax

The anterior, middle, and posterior scalene muscles are also considered muscles of respiration. These muscles were discussed in table 12.6.

EXERCISE 12.7

MUSCLES OF RESPIRATION

1. Observe a prosected cadaver or a classroom model of the thorax/abdomen that demonstrates muscles of the thoracic cage.

2. Identify the muscles of respiration listed in **figure 12.7**, using table 12.8 and the textbook as guides.

   a. *Diaphragm*—The primary muscle involved in breathing is the diaphragm. This dome-shaped muscle forms a partition between the thoracic and abdominal cavities. The diaphragm contracts and moves inferiorly during inspiration. The diaphragm relaxes and moves superiorly during expiration. Remove the breastplate from the cadaver or model and observe the diaphragm. The structures that pass through the diaphragm are best seen from an inferior view. If possible, try to observe the diaphragm from both superior and inferior points of view. Observe the broad attachment along the bones that constitutes the lower border of the thoracic cage. The muscle has a unique central attachment, the central tendon of the diaphragm.

   b. *External Intercostals*—The majority of the muscle mass of the **external intercostals** is located on the posterior and lateral thorax, extending from the vertebral column to the **midclavicular line**, a vertical line that passes through the middle of the clavicle (see **figure 12.8a**). The **external intercostal membrane** lies in place of the external intercostal muscles in the space between the midclavicular line and the sternum. Notice that the muscle fibers of the external intercostals are arranged obliquely, pointing in an inferomedial direction. The external intercostals elevate the ribs during inspiration.

   The external intercostal membrane can be identified on a cadaver because the connective tissue fibers parallel the direction of the muscle fibers of the external intercostals. If observing classroom models, this membrane is not identifiable.

   c. *Internal Intercostals*—In contrast to the external intercostals, the majority of the muscle mass of the **internal intercostals** is located on the anterior surface of the thorax. These muscles extend from the sternum to the **scapular line** (a vertical line that passes through the inferior angle of the scapula; **figure 12.8b**) on the posterior thorax. The **internal intercostal membrane** lies in place of the internal intercostal muscles on the posterior

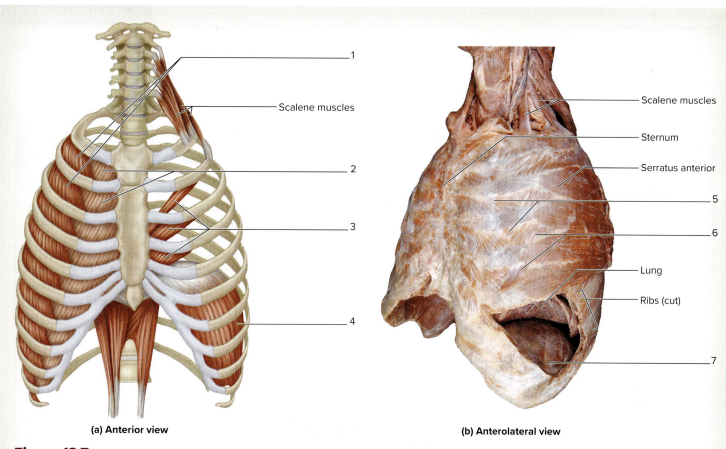

(a) Anterior view   **(b) Anterolateral view**

Figure 12.7 **Muscles of Respiration.** Use the terms listed to fill in the numbered labels in the figure. Some answers may be used more than once.
(b) ©McGraw-Hill Education/Photo and Dissection by Christine Eckel

☐ Diaphragm ☐ External intercostals ☐ Internal intercostals ☐ Transversus thoracis

thorax between the scapular line and the vertebral column. The muscle fibers of the internal intercostals are arranged obliquely, pointing in an inferolateral direction, at right angles to the fibers of the external intercostals The internal intercostal muscles depress the ribs during forced expiration.

d. *Transversus thoracis*—If possible, remove the breastplate on the cadaver (or on a model of the thorax) and

observe its interior surface. Lying adjacent to the inferior part of the sternum is the **transversus thoracis** muscle, which consists of several muscle bellies running obliquely. The transversus thoracis assists in depression of the ribs during forced expiration.

3. Label the muscles of respiration in figure 12.7.

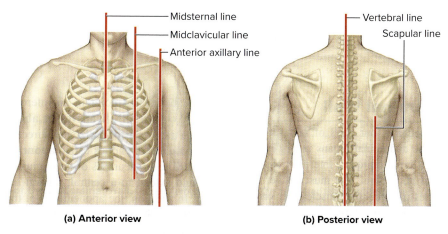

(a) Anterior view   **(b) Posterior view**

Figure 12.8 **Points of Reference on the Anterior and Posterior Thorax.**

(continued on next page)

INTRODUCTION

The **appendicular muscles** are familiar muscles to most people. Anyone who has gone to a gym and lifted weights, or who has admired the muscular body of a basketball player or gymnast, has at least some familiarity with these muscles. They are the muscles used to perform active tasks such as typing, walking, running, and lifting. They are the bulging muscles seen in the arms and legs of elite athletes. When lifting a heavy weight, these muscles are visible in our upper limbs, so it is fairly easy to determine which muscles are used to perform the particular action. If this is not immediately clear from the fatigue or pain experienced immediately in the working muscle, it is definitely clear in the next two days as delayed-onset muscle soreness (DOMS) sets in.

All of these activities give us a fascination and curiosity about the muscles responsible for causing movement. When working through the task of learning names, attachments, and actions for the appendicular muscles, try to identify the muscles on your own body. Practice using them. An excellent way to do this is to go to a gym that has weight machines. Observe the illustrations on the weight machines that demonstrate what muscle or muscles the exercise is meant to work on. Then practice the movement *without using any weights*. When performing the stated action of the muscle(s), feel the muscle(s) produce tension under your skin. Once the connection is made between the muscles in your own body and the muscles seen on the cadaver, on models, or in photographs, you will begin to truly appreciate their functions.

The exercises in this chapter involve identifying, naming, and exploring the structure and function of muscles that move the appendicular skeleton on a model, photo, or cadaver. Prior to beginning, find out from the laboratory instructor which muscles will be assessed on practical exams. Then, place check marks in the summary tables of this chapter so as to focus on those muscles that are required for the course. In addition, prior to beginning the more detailed study of the appendicular musculature in this chapter, be sure to complete the exercises related to appendicular musculature in chapter 11. Those exercises introduced the major muscle groups and the common actions of those groups. Refer to tables 11.3 and 11.4 to review the muscle compartments of the upper and lower limbs, and the major actions of the muscles in each compartment.

List of Reference Tables

Name: _____

Date: _____   Section: _____

These Pre-Laboratory Worksheet questions may be assigned by instructors through their ⬛ connect course.

1. The upper limb includes the _____ (arm/forearm) located between the shoulder and elbow and the _____ (arm/forearm) located between the elbow and wrist.

2. Match the action(s) described in column A with the compartment listed in column B. Actions are those performed by the majority of muscles within the listed compartment.

   Column A
   _____ 1. extension of arm and forearm
   _____ 2. extension of the wrist and digits
   _____ 3. flexion of arm and forearm
   _____ 4. flexion of the wrist and digits

   Column B
   a. anterior compartment of the arm
   b. anterior compartment of the forearm
   c. posterior compartment of the arm
   d. posterior compartment of the forearm

3. Which of the following terms describes a muscle that is primarily responsible for causing a given action about a joint? (Circle one.)

   a. agonist   b. antagonist   c. protagonist   d. synergist

4. Which of the following muscles compose the rotator cuff? (Check all that apply.)

   _____ a. infraspinatus
   _____ b. subscapularis
   _____ c. supraspinatus
   _____ d. teres major
   _____ e. teres minor

5. The portion of the lower limb from the knee to the ankle is the _____ (leg/thigh).

6. The portion of the lower limb from the hip to the knee is the _____ (leg/thigh).

7. Match the action(s) described in column A with the compartment listed in column B. Actions are those performed by the majority of muscles within the listed compartment.

   Column A
   _____ 1. eversion of the ankle
   _____ 2. extension of the leg, flexion of the thigh
   _____ 3. extension of the thigh, flexion of the leg
   _____ 4. dorsiflexion and inversion of the ankle, extension of the toes
   _____ 5. adduction of the thigh
   _____ 6. plantar flexion and inversion of the ankle, flexion of the toes

   Column B
   a. anterior compartment of the leg
   b. anterior compartment of the thigh
   c. lateral compartment of the leg
   d. medial compartment of the thigh
   e. posterior compartment of the leg
   f. posterior compartment of the thigh

8. Which of the following muscles compose the hamstring muscles? (Check all that apply.)

   _____ a. biceps femoris
   _____ b. gracilis
   _____ c. rectus femoris
   _____ d. semimembranosus
   _____ e. semitendinosus

9. Identify the most superficial muscle of the chest. (Circle one.)

   a. deltoid   b. pectoralis major   c. pectoralis minor   d. rhomboid major   e. trapezius

10. Which of the following muscles is found in the posterior compartment of the leg? (Circle one.)

    a. tibialis anterior   b. gastrocnemius   c. fibularis longus   d. rectus femoris

GROSS ANATOMY

Muscles That Act About the Pectoral Girdle and Glenohumeral Joint

Recall from chapter 9 that the pectoral girdle consists of the clavicle and scapula, and that the sternoclavicular joint is the only bony connection holding the upper limb onto the torso. This anatomical arrangement requires very strong muscular attachments between the pectoral girdle and the axial skeleton. The muscles that act about the pectoral girdle both stabilize the scapula and move the pectoral girdle, and they are classified as either anterior or posterior thoracic muscles (**table 13.1**).

The glenohumeral joint (or shoulder joint) consists of the head of the humerus articulating with the relatively shallow glenoid cavity of the scapula. This anatomical arrangement allows for a great deal of flexibility but requires numerous muscular attachments. The muscles acting about the glenohumeral joint both stabilize and create movement of this joint and are classified as muscles originating on the axial skeleton and muscles originating on the scapula (**table 13.2**).

Table 13.1	Axio-Appendicular Muscles			
Muscle	**Proximal Attachment(s)**	**Distal Attachment(s)**	**Action(s)**	**Word Origin**
Anterior Muscles				
Pectoralis Minor	Ribs 3–5 (anterior surface)	Coracoid process of scapula	Protracts and depresses the scapula	*pectus*, chest, + *minor*, smaller
Serratus Anterior	Ribs 1–8 (outer surface)	Medial border of scapula	Protracts the scapula and rotates it superiorly; holds the scapula flat against the rib cage	*serratus*, a saw, + *anterior*, the front surface
Pectoralis Major	Sternum, medial clavicle, and costal cartilages 2–6	Lateral part of intertubercular sulcus of humerus	Flexes and adducts arm	*pectus*, chest, + *major*, larger
Posterior Muscles				
Trapezius	Occipital bone, ligamentum nuchae, spinous processes of C_7–T_{12}	Clavicle (lateral third), acromion, and spine of scapula	Superior fibers: Elevates and superiorly rotates scapula Middle fibers: Retracts scapula Inferior fibers: Depresses scapula	*trapezion*, irregular four-sided figure
Levator Scapulae	Transverse processes of C_1–C_4	Superior angle of scapula	Elevates and inferiorly rotates scapula	*levatus*, to lift, + *scapula*, the shoulder blade
Rhomboid Minor	Spinous processes of C_7–T_1	Medial border of scapula	Retracts and inferiorly rotates scapula	*rhomboid*, resembling an oblique parallelogram, + *minor*, smaller
Rhomboid Major	Spinous processes of T_2–T_5	Medial border of scapula	Retracts and inferiorly rotates scapula	*rhomboid*, resembling an oblique parallelogram, + *major*, larger
Latissimus Dorsi	Spinous process of T_7–L_5, iliac crest, and ribs 10–12	Intertubercular sulcus of humerus	Extends, adducts, and medially rotates arm	*latissimus*, widest, + *dorsum*, the back
Teres Major	Inferior angle of scapula	Medial part of intertubercular sulcus of humerus	Extends, adducts, and medially rotates arm	*teres*, round, + *major*, larger

MUSCLES THAT ACT ABOUT THE PECTORAL GIRDLE AND GLENOHUMERAL JOINT

EXERCISE 13.1A Muscles That Act About the Pectoral Girdle

1. Observe a prosected human cadaver or a classroom model demonstrating muscles of the thorax and upper limb.

2. Identify the **muscles that act about the pectoral girdle** listed in **figure 13.1** on the cadaver or on a classroom model, using table 13.1 and the textbook as guides. Then label them in figure 13.1.

   a. *Anterior Muscles*—The **pectoralis minor** is a small muscle that can have multiple actions, depending upon what other muscles of the thorax and shoulder are contracting at the same time. If the scapula is fixed, the pectoralis minor assists in elevation of the rib cage (as in inspiration). If the scapula is not fixed, the pectoralis minor acts to pull the scapula anteriorly. The **serratus anterior** is a large, flat, fan-shaped muscle positioned between the ribs and the scapula; it is named based on the saw-toothed (serrated) appearance of its origin on the ribs. This muscle helps to both stabilize and move the scapula.

   b. *Posterior Muscles*—The largest posterior muscle that moves the pectoral girdle is the superficial **trapezius** muscle, which acts to anchor the scapula to the entire superior two-thirds of the vertebral column and to the back of the head. As noted in table 13.1, this muscle performs multiple actions on the scapula, depending upon which part of the muscle contracts at a given time. Deep to the trapezius are smaller, more numerous muscles that act as **synergists** of the trapezius muscle. These include the **rhomboids** (major and minor) and the **levator scapulae**. When identifying these muscles on the cadaver or models, think about the actions they share with the trapezius. Practice performing the actions listed in table 13.1 to correlate the actions with the location and fiber orientation of the muscles.

EXERCISE 13.1B Muscles That Act About the Glenohumeral Joint

1. Observe a prosected human cadaver or a classroom model demonstrating muscles of the thorax and upper limb.

2. Identify muscles that act about the glenohumeral joint listed in figure 13.1 on the cadaver or on a classroom model, using tables 13.1 and 13.2 and the textbook as guides. Then label them in figure 13.1.

   a. *Muscles Originating on Axial Skeleton*—The posteriorly located **latissimus dorsi** is a broad, triangular muscle located on the inferior part of the back originating on the axial skeleton and inserting on the humerus. It is often called the "swimmer's muscle"

because its actions are required for many swimming strokes. Contraction of this muscle extends, adducts, and medially rotates the arm. The anteriorly located **pectoralis major** is the large, thick, fan-shaped muscle that covers the superior part of the chest and acts to flex and adduct the arm.

   b. *Muscles Originating on the Scapula*—Numerous muscles originate on the scapula and insert on the humerus (table 13.2). One important group of muscles is the **rotator cuff muscles**. The rotator cuff is a musculotendinous cuff formed by four muscles that impart strength and stability to the glenohumeral joint. These muscles include the **subscapularis**, **supraspinatus**, **infraspinatus**, and **teres minor**. More than any other factor, the strength of these muscles determines the stability of the glenohumeral joint. Weaknesses in these muscles contribute to many musculoskeletal problems in the glenohumeral joint. Rotator cuff injuries are common in baseball players and in older adults. The term *rotator cuff* comes from the fact that these muscles act to medially or laterally rotate the humerus, in addition to stabilizing the glenohumeral joint.

3. Make a representative drawing of the rotator cuff muscles that are seen in a posterior view in the space provided or on a separate sheet of paper.

INTEGRATE

LEARNING STRATEGY

The following mnemonic is helpful for remembering the names of the rotator cuff muscles. "A baseball pitcher who tears his rotator cuff **SITS** out the season." (**SITS: S** = supraspinatus, **I** = infraspinatus, **T** = teres minor, **S** = subscapularis.) To remember it is the teres **minor** (not the teres major) that forms part of the rotator cuff, know that this injured pitcher will then be relegated to the **minor** leagues.

(continued on next page)

COMPARTMENTS OF THE FOREARM

EXERCISE 13.3A Anterior Compartment of the Forearm

The muscles of the **anterior compartment of the forearm** cause flexion of the wrist and the digits (fingers). They are arranged into two layers—a superficial group and a deep group. In most cases, the name of the muscle tells exactly what the muscle does or where the muscle is located. Thus, although the names of the muscles seem long, the meanings of the names are very useful.

1. Observe a prosected human cadaver or a classroom model demonstrating muscles of the upper limb.

2. Identify the muscles of the anterior compartment of the forearm listed in **figure 13.4** on the cadaver or on a classroom model, using **table 13.5** and the textbook as guides. Then label them in figure 13.4.

a. *Palmaris Longus*—Approximately 15% to 20% of humans do not have a palmaris longus muscle. The palmaris longus passes *superficial* to the **flexor retinaculum** (a band of connective tissue that wraps around the anterior part of the wrist) and inserts onto the **palmar aponeurosis**. Perform the following exercise to determine whether you have a palmaris longus muscle. With your forearm supinated, touch the tips of your first and fifth digits together and flex the wrist *slightly*. If the palmaris longus is present, you will see its thin tendon passing longitudinally in the middle of your wrist. If no tendon is visible, the muscle is probably absent.

b. *Flexor Digitorum*—There are two muscles that flex the digits, the **flexor digitorum superficialis** (superficial) and the **flexor digitorum profundus**

Table 13.5	Anterior (Flexor) Compartment of the Forearm			
Muscle	**Proximal Attachment(s)**	**Distal Attachment(s)**	**Action(s)**	**Word Origin**
Superficial Group				
Brachioradialis	Lateral supracondylar ridge of humerus	Styloid process of radius	Flexes forearm	*brachium*, arm, + *radialis*, radius
Flexor Carpi Radialis	Medial epicondyle of humerus	Metacarpals II and III	Flexes wrist; abducts hand	*flex*, to bend, + *carpus*, wrist, + *radialis*, radius
Flexor Carpi Ulnaris	Medial epicondyle of humerus and posterior surface of ulna	Pisiform, hamate, and metacarpal V	Flexes wrist; adducts hand	*flex*, to bend, + *carpus*, wrist, + *ulnaris*, ulna
Flexor Digitorum Superficialis	Medial epicondyle of humerus and coronoid process of ulna	Middle phalanx of digits 2–5	Flexes digits 2–5	*flex*, to bend, + *digit*, finger, + *superficialis*, surface
Palmaris Longus	Medial epicondyle of humerus	Palmar aponeurosis	Flexes wrist	*palmaris*, the palm, + *longus*, long
Pronator Teres	Medial epicondyle of humerus	Radius (lateral shaft)	Pronates and flexes forearm	*pronatus*, to bend forward, + *teres*, round
Deep Group				
Flexor Digitorum Profundus	Ulna (anteromedial surface) and inter-osseous membrane	Distal phalanx of digits 2–5	Flexes digits 2–5	*flexus*, to bend, + *digitorum*, finger, + *profundus*, deep
Flexor Pollicis Longus	Radius (anterior surface) and interosseous membrane	Distal phalanx of the thumb	Flexes thumb	*flexus*, to bend, + *pollex*, the thumb, + *longus*, long
Pronator Quadratus	Ulna (distal anterior shaft)	Radius (distal, anterior shaft)	Pronates forearm	*pronatus*, to bend forward, + *quadratus*, square

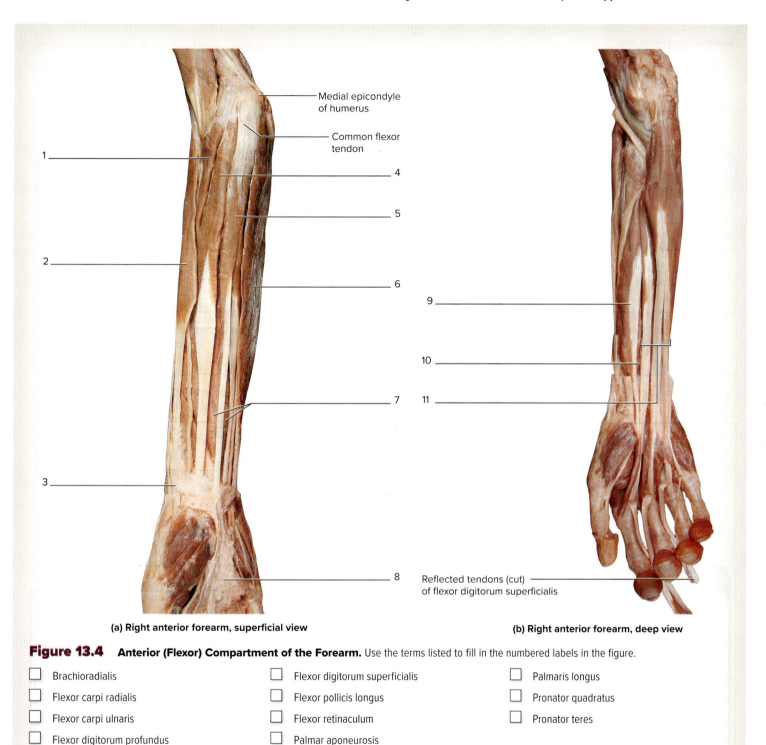

Medial epicondyle
of humerus

Common flexor
tendon

1

4

5

2

6

9

10

7   11

3

8   Reflected tendons (cut)
of flexor digitorum superficialis

(a) Right anterior forearm, superficial view   **(b) Right anterior forearm, deep view**

Figure 13.4   **Anterior (Flexor) Compartment of the Forearm.** Use the terms listed to fill in the numbered labels in the figure.

☐ Brachioradialis   ☐ Flexor digitorum superficialis   ☐ Palmaris longus

☐ Flexor carpi radialis   ☐ Flexor pollicis longus   ☐ Pronator quadratus

☐ Flexor carpi ulnaris   ☐ Flexor retinaculum   ☐ Pronator teres

☐ Flexor digitorum profundus ☐ Palmar aponeurosis

(a, b) ©McGraw-Hill Education/Photo and Dissection by Christine Eckel

(deep) **(figure 13.5)**. These muscles, as their names suggest, flex the fingers. They do so by attaching to the phalanges. These muscles take somewhat unique routes to get to their respective attachments on the phalanges. The **flexor digitorum profundus** tendon attaches to the **distal phalanx**. Thus, it must pass through the tendon of the more superficial muscle (the tendon of the flexor digitorum superficialis) as it

travels to the distal phalanx. Observe a cadaver or a classroom model of the hand, and locate the middle and distal phalanges of one digit. The tendon of the **flexor digitorum superficialis** inserts onto the **middle phalanx** of the digit. Before it reaches its insertion point, it has a slit through which the tendon of the flexor digitorum profundus passes. Thus, though both muscles flex the digits, they do so at different joints.

(continued on next page)

(continued from previous page)

> ### INTEGRATE
>
> ## LEARNING STRATEGY
>
> The **D**orsal interossei **AB**duct the digits (mnemonic is **DAB**), whereas the **P**almar interossei **AD**duct the digits (mnemonic is **PAD**).

1. Observe a prosected human cadaver or a model demonstrating muscles of the hand.

2. Identify the intrinsic muscles of the hand listed in **figure 13.7** on a cadaver or on a classroom model, using table 13.7 and the textbook as guides. Then label them in figure 13.7.

   a. *Interossei*—The two groups of interossei muscles act to abduct and adduct the digits. There are two groups of interossei: dorsal and palmar. The *dorsal interossei* abduct the digits, and the *palmar interossei* adduct the digits. To feel the belly of the first dorsal interosseous muscle, forcibly press your thumb and index finger together, like when holding a key. Then using your other hand, palpate between metacarpals I and II to feel the first dorsal interosseous muscle.

   b. *Lumbricals*—The lumbricals are relatively easy to identify because they appear slim and wormlike, as their name implies. They are also among the few muscles that originate on the tendon of another muscle, the flexor digitorum profundus.

3. Make a representative drawing of the intrinsic muscles of the hand in the space provided or on a separate sheet of paper.

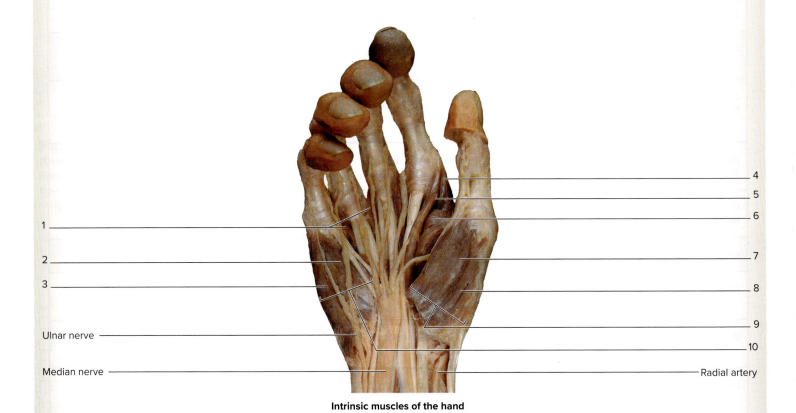

Intrinsic muscles of the hand

Figure 13.7 **Intrinsic Muscles of the Hand.** Use the terms listed to fill in the numbered labels in the figure.

☐ Abductor digiti minimi   ☐ Flexor digiti minimi brevis   ☐ Medial lumbrical

☐ Abductor pollicis brevis   ☐ Flexor pollicis brevis   ☐ Thenar group

☐ Adductor pollicis   ☐ Hypothenar group

☐ First dorsal interosseous   ☐ Lateral lumbricals

Muscles That Act About the Hip Joint/Thigh

The pelvic girdle is composed of the paired os coxae, which articulate with the axial skeleton at the sacrum. Each lower limb is attached to an os coxae at the hip joint, which consists of the head of the femur articulating with the relatively deep acetabulum of the os coxae. Numerous muscles cross this joint to both stabilize and allow for movement—including standing, walking, running, and cycling—at this joint.

The muscles in the gluteal region are broken up into a superficial and a deep group. The large muscles of the superficial group derive their names from the region (e.g., gluteus maximus). The deeper muscles are short rotators of the thigh that also provide stabilization of the hip joint. Several muscles in the thigh have an attachment on the os coxae and cross the hip joint, and thus provide movement at the hip joint. Because some of the thigh muscles also cross the knee joint, the muscles will be discussed in several exercises.

Table 13.8	Muscles of the Gluteal Region			
Muscle	**Proximal Attachment(s)**	**Distal Attachment(s)**	**Action(s)**	**Word Origin**
Superficial Muscles				
Gluteus Maximus	Posterior ilium, sacrum, and coccyx	Gluteal tuberosity of femur; iliotibial tract	Extends and laterally rotates thigh	*gloutos*, buttock, + *maximus*, greatest
Gluteus Medius	Ilium (between the anterior and posterior gluteal lines)	Greater trochanter of femur	Abducts and medially rotates thigh	*gloutos*, buttock, + *medius*, middle
Gluteus Minimus	Ilium (between the anterior and inferior gluteal lines)	Greater trochanter of femur	Abducts and medially rotates thigh	*gloutos*, buttock, + *minimus*, smallest
Tensor Fasciae Latae	Iliac crest (anterior aspect), anterior superior iliac spine (ASIS)	Iliotibial tract (IT band)	Abducts thigh	*tensus*, to stretch, + *fascia*, a band, + *latus*, side
Deep Muscles				
Piriformis	Anterior sacrum	Greater trochanter of femur	Laterally rotates thigh	*pirum*, pear, + *forma*, form
Superior Gemellus	Ischial spine	Greater trochanter of femur	Laterally rotates thigh	*geminus*, twin, + *superus*, above
Obturator Internus	Obturator membrane (posterior surface)	Greater trochanter of femur	Laterally rotates thigh	*obturo*, to occlude, + *internus*, away from the surface
Inferior Gemellus	Ischial tuberosity	Greater trochanter of femur	Laterally rotates thigh	*geminus*, twin, + *inferior*, lower
Quadratus Femoris	Ischial tuberosity	Intertrochanteric crest of femur	Laterally rotates thigh	*quad*, four (sided), + *femur*, thigh

MUSCLES THAT ACT ABOUT THE HIP JOINT/THIGH

1. Observe a prosected human cadaver or a model demonstrating muscles of the hip.

2. Identify the muscles that move the hip joint/thigh listed in **figures 13.9** and **13.10** on the cadaver or on a classroom model, using tables 13.8 through 13.11 and the textbook as guides. Then label them in figures 13.9 and 13.10.

Anterior Muscles

a. *Anterior Compartment of the Thigh*—Two important flexors of the thigh are the iliacus and psoas major. They have separate origins, but have a common insertion on the lesser trochanter of the femur, so together they are called the iliopsoas ("hip flexors"). These muscles are difficult to identify on the cadaver and models because only a small portion of the insertion is visible in the anterior thigh. The bellies of these muscles are located deep within the pelvis on the posterior abdominopelvic wall. Thus, when identifying them on the cadaver or on models, be sure to view the inside of the abdominopelvic cavity where the muscles originate. Both the sartorius and the rectus femoris can also flex the thigh.

b. *Medial Compartment of the Thigh*—The muscles of the medial thigh compartment are responsible for adduction of the thigh. These muscles have a proximal attachment on the pubic bone and, except for the gracilis, all attach to the linea aspera of the femur. Sudden movements at the hip, such as kicking, jumping, or twisting to change directions, can result in injury to these muscles, commonly called a groin strain. These muscles are described in more detail in exercise 13.6.

Posterior Muscles

a. *Gluteal Group*—The gluteus maximus and tensor fasciae latae muscles have an attachment on the iliotibial tract. When these two muscles become stiff from overuse, they put excessive tension on the iliotibial tract. Most commonly this causes pain in the knee joint because the tight iliotibial tract compresses structures in the lateral knee and the underlying vastus lateralis muscle. The gluteus maximus extends and *laterally* rotates the thigh. It plays a large role in helping you to get up from a seated position or to climb stairs.

The gluteus medius and minimus muscles abduct and *medially* rotate the thigh. Consider the action of **thigh abduction**. It may not seem that we perform this action often, except perhaps in a kickboxing class. In fact, the gluteus medius and minimus are extremely important in preventing the opposite pelvis from dropping during walking or when standing on one leg **(figure 13.8)**. To test the function of the gluteus medius and minimus muscles, first take a standing position. Next, place the palm of your right hand flat against your right lateral hip (superficial to the location of the gluteus medius and minimus muscles). Now bring your left foot off the ground so you are balancing on your right limb only. Do you feel tension in the muscles deep to your palm? If you are holding your hip vertical, you will. If you then relax the gluteus medius and minimus muscles, you will find that your left hip drops and the body tilts to the unsupported side. The gluteus medius and minimus hold the hip in the vertical position during single leg stance. By holding the hip in such a position, they allow the foot on the swing limb to move forward without scraping against the ground. It may seem like a minor action, but it can be quite problematic for an individual whose muscles are not functioning properly due to nerve injury or other disease.

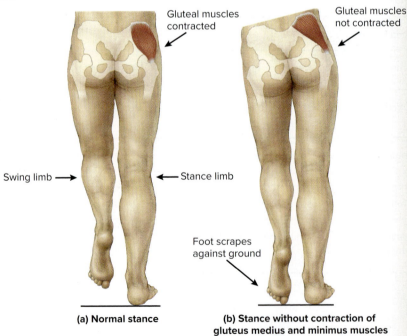

(a) Normal stance

(b) Stance without contraction of gluteus medius and minimus muscles

Figure 13.8   **Actions of the Gluteal Muscles During Locomotion.** The gluteus medius and minimus support the hip on the right side of the body when the right leg is the stance (supporting) leg. This allows the foot on the swing leg to clear the ground.

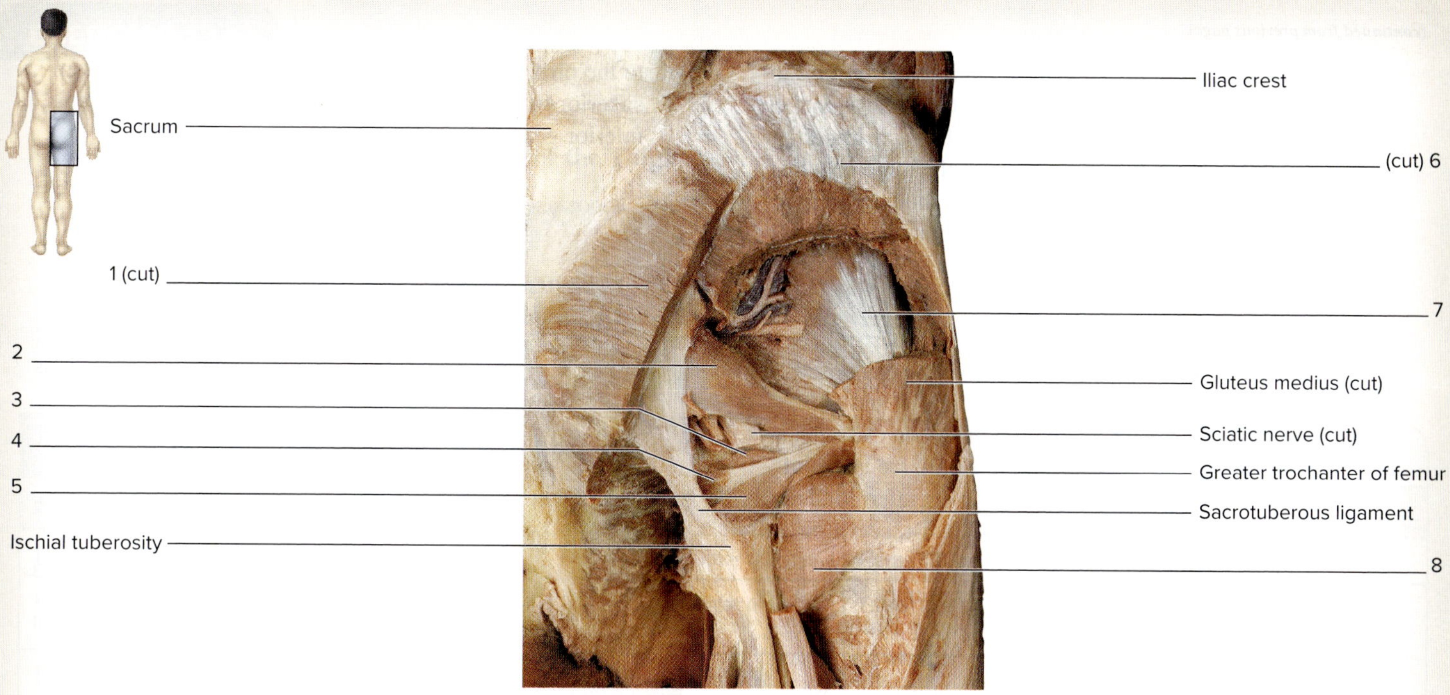

Iliac crest

(cut) 6

Sacrum

1 (cut)

2

3

4

5

Ischial tuberosity

7

Gluteus medius (cut)

Sciatic nerve (cut)

Greater trochanter of femur

Sacrotuberous ligament

8

(a) Right thigh and hip, deep posterior view

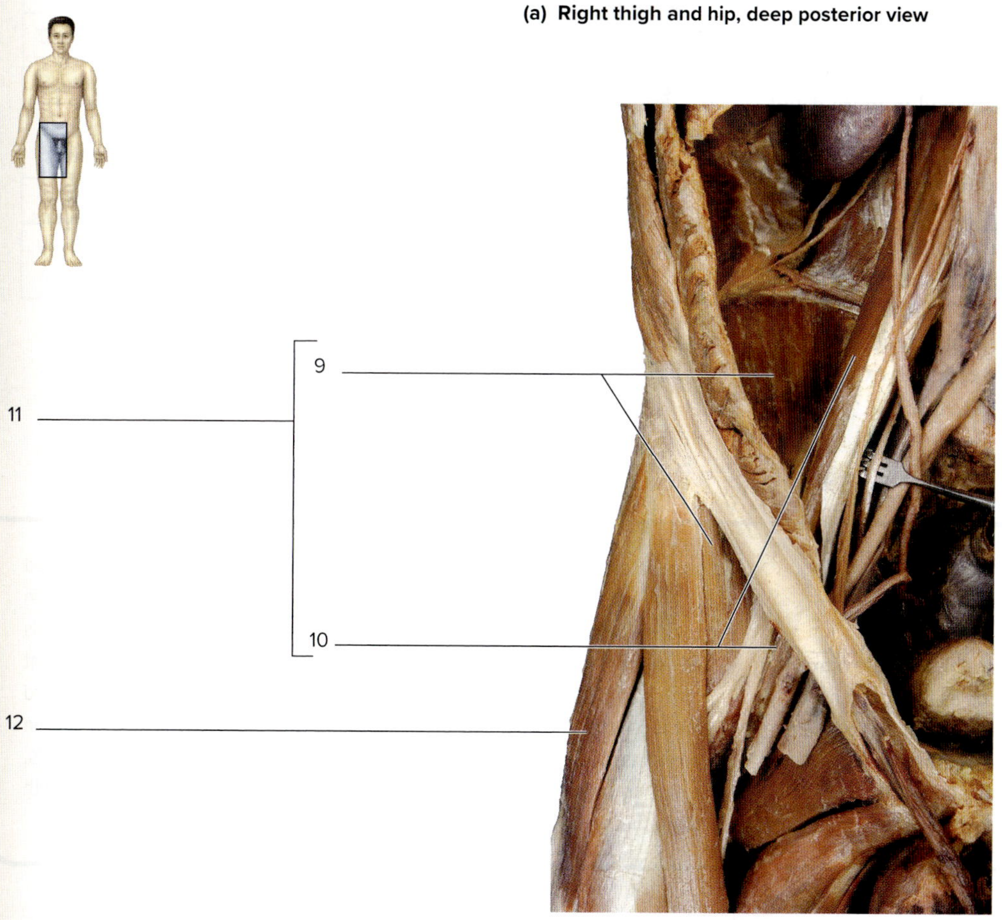

9

11

10

12

(b) Right thigh and hip, deep anterior view

Figure 13.9   **Muscles That Act About the Hip Joint/Thigh.** Use the terms listed to fill in the numbered labels in the figure. Some answers may be used more than once.

☐ Gluteus maximus	☐ Iliacus	☐ Obturator internus	☐ Quadratus femoris
☐ Gluteus medius	☐ Iliopsoas	☐ Piriformis	☐ Superior gemellus
☐ Gluteus minimus	☐ Inferior gemellus	☐ Psoas major	☐ Tensor fasciae latae

(a) ©McGraw-Hill Education/Photo and Dissection by Christine Eckel; (b) ©Christine Eckel

(continued on next page)

(continued from previous page)

Table 13.14	Posterior Compartment of the Leg			
Muscle	**Proximal Attachment(s)**	**Distal Attachment(s)**	**Action(s)**	**Word Origin**
Superficial Muscles				
Gastrocnemius	Femoral condyles	Calcaneus	Plantar flexes foot and flexes leg	*gaster*, belly, + *kneme*, leg
Soleus	Proximal tibia and fibula	Calcaneus	Plantar flexes foot	*solea*, a sandal, sole of foot
Plantaris	Femur (supracondylar ridge)	Calcaneus	Plantar flexes foot	*plantar*, relating to sole of foot
Deep Muscles				
Flexor Digitorum Longus	Posterior tibia	Distal phalanx of digits 2–5	Flexes digits 2–5	*flexus*, to bend, + *digit*, a toe, + *longus*, long
Flexor Hallucis Longus	Fibula (inferior two-thirds)	Distal phalanx of the great toe	Flexes great toe	*flexus*, to bend, + *hallux*, great toe, + *longus*, long
Popliteus	Lateral condyle of femur	Posterior, proximal surface of tibia	Flexes leg, unlocks knee joint	*poplit*, back of knee
Tibialis Posterior	Proximal tibia and fibula	Medial cuneiform and navicular	Inverts and plantar flexes foot	*tibia*, shinbone, + *posterior*, back surface

INTEGRATE

CONCEPT CONNECTION

Sensory receptors embedded in muscles and tendons sense precise information regarding muscle length and tension. These specialized receptors, called *proprioceptors* (*proprio-*, one's own), relay this information to the central nervous system, enabling perception of body position. For example, when closing your eyes, you generally still have some awareness of the position of your limbs. Likewise, when walking into a dark room, you are still able to navigate through the space, in part due to proprioception. Two important types of proprioceptors are muscle spindles and Golgi tendon organs, which detect muscle length and tension, respectively. Muscle spindles lie parallel to muscle fibers so that they experience the same changes in length as the muscle fibers. Golgi tendon organs are embedded within the tendons

of muscles. As muscle force builds during a muscle contraction, the sensory receptor becomes compressed. Both muscle spindles and Golgi tendon organs also play a role in reflexes, or preprogrammed responses to stimuli. For example, the stretch reflex involves the muscle spindle, and it is often tested clinically with a hammer tap to the patellar ligament (see chapter 16). The tap causes a rapid stretch of the quadriceps muscle group, which also stretches the muscle spindles. The result is a contraction of the entire quadriceps group, thereby causing knee extension, which counteracts the stretch that was applied. It is a protective mechanism to prevent too much force on a muscle from causing injury. Proprioceptors are distributed in all skeletal muscles. However, the number of muscle spindles and Golgi tendon organs varies from one muscle to another.

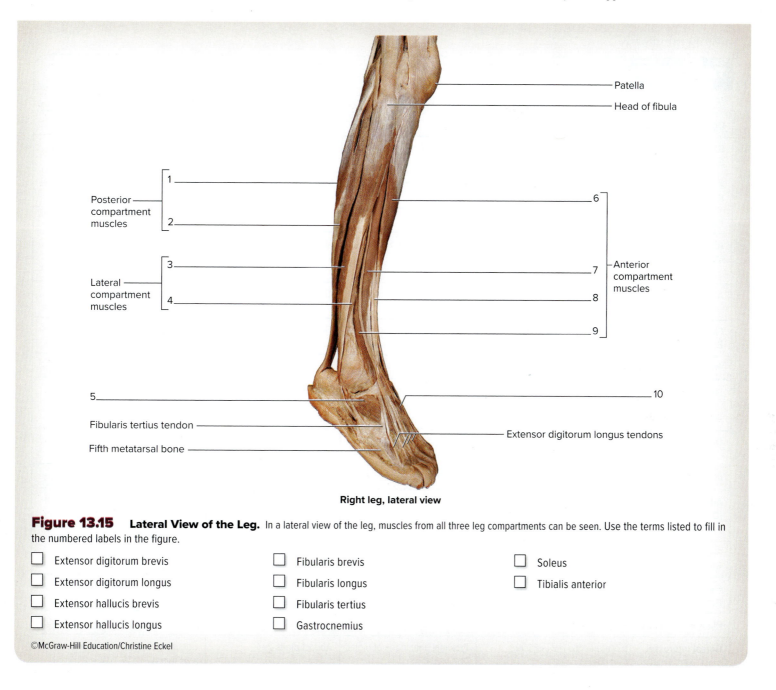

Figure 13.15 **Lateral View of the Leg.** In a lateral view of the leg, muscles from all three leg compartments can be seen. Use the terms listed to fill in the numbered labels in the figure.

☐ Extensor digitorum brevis   ☐ Fibularis brevis   ☐ Soleus

☐ Extensor digitorum longus   ☐ Fibularis longus   ☐ Tibialis anterior

☐ Extensor hallucis brevis   ☐ Fibularis tertius

☐ Extensor hallucis longus   ☐ Gastrocnemius

©McGraw-Hill Education/Christine Eckel

EXERCISE 13.8

INTRINSIC MUSCLES OF THE FOOT

The intrinsic muscles of the plantar surface of the foot are arranged in four layers, named layer 1 to layer 4 from superficial to deep. The foot also contains two neurovascular planes (a neurovascular plane is a region where the nerves and blood vessels that serve the region are located). The first is located between muscle layers 1 and 2, and the second is located between layers 3 and 4.

1. Observe a prosected human cadaver or a classroom model demonstrating muscles of the foot. Then label them in **figure 13.16**.

2. Identify the **intrinsic muscles of the foot** listed in figure 13.16 on the cadaver or on a classroom model, using table 13.15 and the textbook as guides. Then label them in figure 13.16.

(continued on next page)

(continued from previous page)

Table 13.15	Intrinsic Muscles of the Foot			
Muscle	**Proximal Attachment(s)**	**Distal Attachment(s)**	**Action(s)**	**Word Origin**
Dorsal Musculature				
Extensor Digitorum Brevis	Calcaneus and extensor retinaculum	Middle phalanx of digits 2–4	Extends proximal phalanx of digits 2–4	*dorsal*, the back, + *inter-*, between, + *os*, bone
Extensor Hallucis Brevis	Calcaneus and extensor retinaculum	Proximal phalanx of great toe	Extends the proximal phalanx of great toe	*palmar*, the palm of the hand (foot), + *inter-*, between, + *os*, bone
Plantar Musculature—Layer 1				
Abductor Digiti Minimi	Calcaneus	Proximal phalanx of digit 5	Abducts digit 5	*abduct*, to move away from the median plane, + *digitus*, toe, + *minimi*, smallest
Abductor Hallucis	Calcaneus	Proximal phalanx of digit 1	Abducts great toe	*abduct*, to move away from the median plane, + *hallux*, the great toe
Flexor Digitorum Brevis	Calcaneus	Middle phalanx of digits 2–5	Flexes digits 2–5	*flex*, to bend, + *digitus*, toe, + *brevis*, short
Plantar Musculature—Layer 2				
Lumbricals	Tendons of the flexor digitorum longus	Tendons of the extensor digitorum longus	Flex the proximal phalanges and extend distal phalanges of digits 2–5	*lumbricus*, earthworm
Quadratus Plantae	Calcaneus	Tendon of the flexor digitorum longus	Flexes digits 2–5	*quadratus*, having four sides, + *plantae*, sole of the foot
Plantar Musculature—Layer 3				
Adductor Hallucis (Oblique Head)	Oblique head: Metatarsals II-V Transverse head: Capsules of metatarsophalangeal joints III-V	Base of the proximal phalanx of great toe	Adducts great toe	*adduct*, to move toward the median plane, + *hallux*, the great toe
Flexor Digiti Minimi Brevis	Metatarsal V	Proximal phalanx of digit 5	Flexes the proximal phalanx of digit 5	*flex*, to bend, + *digitus*, a toe, + *minimi*, smallest, + *brevis*, short
Flexor Hallucis Brevis	Cuboid and lateral cuneiform	Proximal phalanx of the great toe	Flexes the proximal phalanx of great toe	*flex*, to bend, + *hallux*, the great toe, + *brevis*, short
Plantar Musculature—Layer 4				
Dorsal Interossei	Metatarsals I–V (adjacent surfaces)	Proximal phalanges of digits 2–4 (sides)	Abduct digits 2–4	*dorsal*, the back, + *inter-*, between, + *os*, bone
Plantar Interossei	Bases of metatarsals III–V	Bases of proximal phalanges of digits 3–5 (medial side)	Adduct digits 2–4	*plantae*, the sole of the foot, + *inter-*, between, + *os*, bone

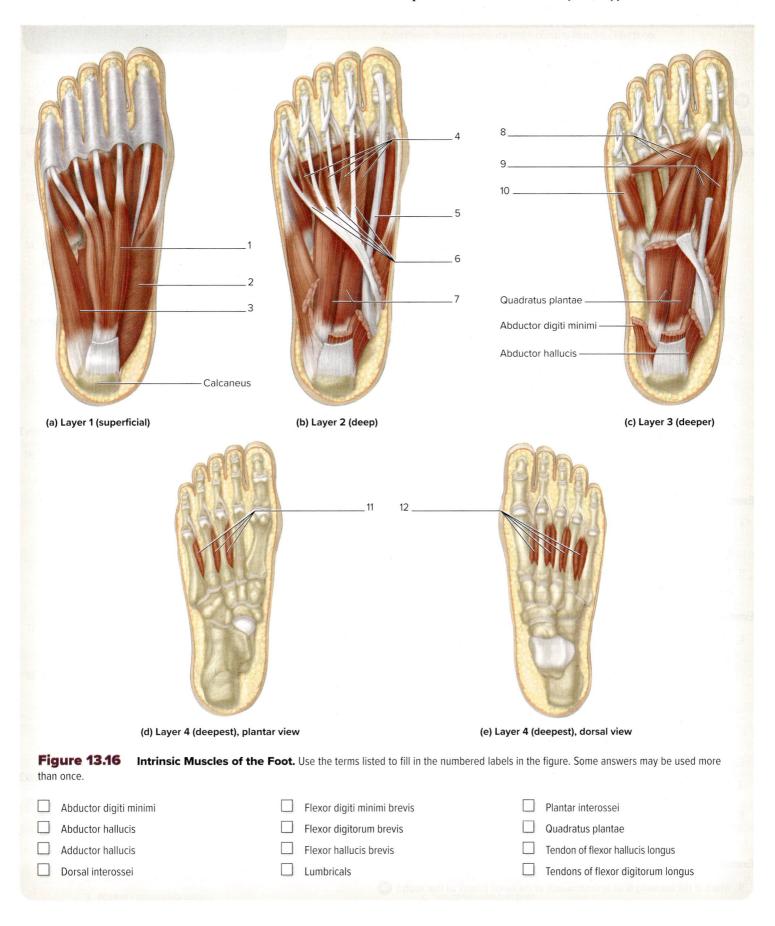

(a) Layer 1 (superficial)

(b) Layer 2 (deep)

(c) Layer 3 (deeper)

Calcaneus

Quadratus plantae

Abductor digiti minimi

Abductor hallucis

(d) Layer 4 (deepest), plantar view

(e) Layer 4 (deepest), dorsal view

Figure 13.16   **Intrinsic Muscles of the Foot.** Use the terms listed to fill in the numbered labels in the figure. Some answers may be used more than once.

- ☐ Abductor digiti minimi
- ☐ Abductor hallucis
- ☐ Adductor hallucis
- ☐ Dorsal interossei

- ☐ Flexor digiti minimi brevis
- ☐ Flexor digitorum brevis
- ☐ Flexor hallucis brevis
- ☐ Lumbricals

- ☐ Plantar interossei
- ☐ Quadratus plantae
- ☐ Tendon of flexor hallucis longus
- ☐ Tendons of flexor digitorum longus

Can You Apply What You've Learned?

25. Name three muscles that attach to the coracoid process of the scapula.

    a. _____ b. _____ c. _____

26. List the three muscles that abduct the hip.

    _____

    _____

27. Which muscle of the medial compartment of the thigh does *not* insert onto the linea aspera of the femur? _____

28. Which muscle of the quadriceps muscle group is the only muscle to cross the hip joint? _____ What is the *origin* (proximal

    attachment) of this muscle? _____. What is the *insertion* (distal attachment)? _____

29. Describe how to distinguish between the semitendinosus muscle and the semimembranosus muscle.

    _____

    _____

    _____

    _____

30. The tendons of which three of the rotator cuff muscles hold the humerus into the glenoid cavity on the posterior surface? _____

    _____The tendon of the rotator cuff muscle that holds the humerus into the

    glenoid cavity on the anterior surface is the tendon of the _____ muscle.

31. List all the muscles that have an attachment on the ischial tuberosity.

    _____

    _____

Can You Synthesize What You've Learned?

32. A patient is told he has a "torn hamstring." List the possible muscles that could have been torn in this case. Then, describe the actions that would demonstrate weakness in the muscle as a result of the tear.

    _____

    _____

    _____

33. Compartment syndrome is a disorder in which inflammation within the compartment of a limb causes pressure such that blood flow is reduced, nerves become damaged, and the muscles start to become nonfunctional (and eventually necrotic, if the situation is not treated). A patient is experiencing compartment syndrome in the anterior compartment of the leg, which has weakened the muscles in this compartment. As a result, what actions would this patient have difficulty performing?

    _____

    _____

    _____

34. A common exercise performed in the gym to increase muscle strength and size is a pull-up. List the muscles that might be involved in a pull-up and their corresponding actions.

    _____

    _____

    _____

chapter 14

Nervous Tissues

OUTLINE AND LEARNING OBJECTIVES

Anatomy & Physiology Revealed® 4.0

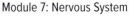

Module 7: Nervous System

GENERAL MULTIPOLAR NEURONS— ANTERIOR HORN CELLS

1. Obtain a slide of the *spinal cord in cross section.*

2. Place the slide on the microscope stage and bring the tissue sample into focus on low power (**figure 14.3***a*). Again, note the inner core of gray matter surrounded by an outer region of white matter. Move the microscope stage so the inner gray matter is in the center of the field of view and then switch to high power (figure 14.3*b*).

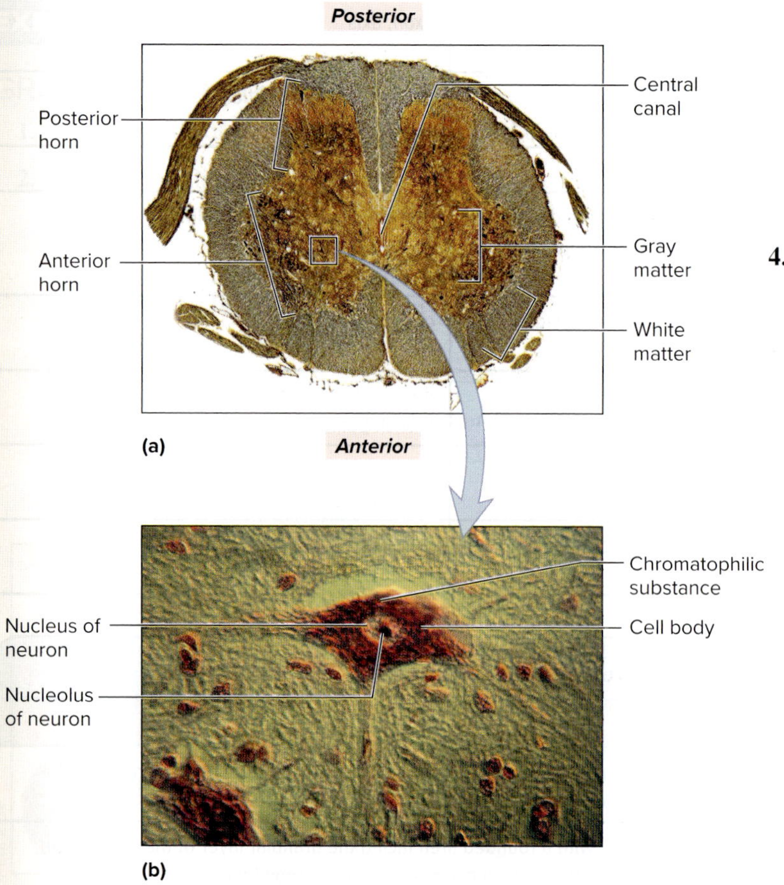

Look for very large, multipolar neurons found in the anterior (ventral) horn of the gray matter. These cells are called **anterior horn cells**, based on their location. They are large somatic motor neurons whose axons exit the spinal cord and travel through peripheral nerves to skeletal muscles in the body. Note the dark granules in these neurons. This chromatophilic substance represents the extensive rough ER network of the cells.

3. Identify the following structures on the spinal cord slide, using tables 14.1 and 14.2 and figure 14.3 as guides:

   ☐ **Cell body of anterior horn cell**

   ☐ **Chromatophilic substance**

   ☐ **Gray matter**

   ☐ **Nucleolus of anterior horn cell**

   ☐ **Nucleus of anterior horn cell**

   ☐ **White matter**

4. Sketch anterior horn cells as seen through the microscope in the space provided or on a separate sheet of paper. Be sure to label all of the structures listed in step 3 in the drawing.

Figure 14.3 **Gray Matter of the Spinal Cord.** (*a*) Cross section of spinal cord. (*b*) Close-up showing large motor neurons (anterior horn cells).

(*a*) ©Jose Luis Calvo/Shutterstock; (*b*) ©Steve Gschmeissner/Science Photo Library/Alamy Stock Photo

_____ ✕

CEREBRUM—PYRAMIDAL CELLS

1. Obtain a slide of the cerebrum that has been stained with Nissl stain. Nissl stain colors the **rough endoplasmic reticulum** and free ribosomes (the chromatophilic substance) of the neurons an intense blue color.

2. Place the slide on the microscope stage and bring the tissue sample into focus on low power. Identify areas of gray matter and white matter on the slide (**figure 14.4**).

3. Bring an area of gray matter to the center of the field of view and switch to high power. Note the very large cells located within the gray matter. These cells are **neurons**. Note the very large nuclei of the neurons. Neurons are surrounded by much smaller cells called *glial cells*. Locate large neurons whose cell bodies appear triangular in shape. These neurons of the cerebrum are called **pyramidal cells** because the cell body's three-dimensional shape is pyramidal. (figure 14.4, table 14.2).

Figure 14.4 Pyramidal Cells in the Cerebral Cortex. The pyramidal cells (neurons) are the large triangular cells. The smaller nuclei visible in the slide are mainly those of glial cells.

©Rick Ash

4. Sketch pyramidal cells as seen through the microscope in the space provided or on a separate sheet of paper. Label the following in the drawing:

☐ **Glial cells**   ☐ **Pyramidal cells**

_____ ×

CEREBELLUM—PURKINJE CELLS

1. Obtain a slide of the **cerebellum**. Place it on the microscope stage and bring the tissue sample into focus on low power **(figure 14.5a)**.

2. Note that the cerebellum has folds of tissue that are divided into histologically distinct regions (figure 14.5a). The outermost region contains mainly axons, dendrites, and projections of neuroglial cells. The innermost region contains mostly small circular neurons closely packed together. Move the microscope stage to the junction between these two layers and center this in the field of view. Switch to high power to identify the large multipolar neurons located in this middle region (figure 14.5b). These are **Purkinje cells** (table 14.2). Deep to these layers is the white matter of the cerebellum.

3. Sketch Purkinje cells as seen through the microscope at high power in the space provided or on a separate sheet of paper. Label the following in the drawing:

☐ **Purkinje cells**   ☐ **White matter**

_____ ×

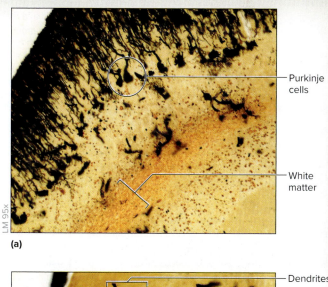

(a)

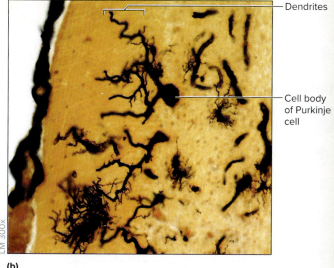

(b)

Figure 14.5 Cerebellum—Purkinje Cells. The cerebellum, demonstrating Purkinje cells between the outer and inner regions. Note that axons are not visible in this image. (*a*) Low power. (*b*) High power.

©Rick Ash

Glial Cells

Glial cells, or **neuroglia** (literally, "nerve glue"), are much more abundant than neurons in the nervous system. Unlike neurons, these cells can undergo cellular division; thus, most brain tumors arise from glial cells (*gliomas*). The central nervous system (CNS) contains four distinct glial cell types, the most abundant of which are **astrocytes** and **oligodendrocytes**. Because of their large size, astrocytes and oligodendrocytes are collectively referred to as *macroglia*. The central nervous system also contains resident macrophages called **microglia**. As their name implies, microglia are very small cells. Their name is a bit of a misnomer, however, because these cells become very large, phagocytic cells when tissue injury or infection occurs. Finally, special epithelial cells called **ependymal cells** are found lining the fluid-filled ventricles of the brain and central canal of the spinal cord.

The peripheral nervous system contains only one type of glial cell, but it is named differently—and functions differently—depending upon its location. These glial cells have the same embryonic origin, but not the same function, which is why they are given different names. Thus, we typically refer to the peripheral nervous system (PNS) as having two glial cell types: **neurolemmocytes (Schwann cells)** and **satellite cells**. Table 14.3 summarizes the characteristics of each type of glial cell.

Table 14.3	Glial Cells		
Cell Name	**Description**	**Function(s)**	**Word Origin**
Central Nervous System			
Astrocytes	Star-shaped cells; most abundant cell of the central nervous system	General supporting cells in the CNS. Transfer nutrients to neurons from the blood. Reinforce the blood-brain barrier. Maintain the extracellular environment around neurons.	*astron*, star, + *kytos*, cell
Ependymal Cells	Cuboidal to columnar-shaped epithelial cells with microvilli and cilia on their apical surfaces	Ependymal cells line the ventricles (fluid-filled spaces) of the brain and the central canal of the spinal cord. They play a role in the production and circulation of CSF.	*ependyma*, an upper garment
Microglia	Small cells with oval nuclei and multiple branching processes	Microglia are derived from blood monocytes, and they are the resident macrophages in the CNS. These cells transform into large, phagocytic cells when tissues are injured or infected.	*mikros*, small, + *glia*, glue
Oligodendrocytes	Cells with several long processes that wrap around axons in the CNS	Each oligodendrocyte myelinates multiple axons in the CNS. Myelination allows for faster nerve signal propagation.	*oligos*, few, + *dendro-*, like a tree, + *kytos*, cell
Peripheral Nervous System			
Neurolemmocytes (Schwann Cells)	Large cells that wrap around axons in the PNS	Each neurolemmocyte can myelinate only part of one axon; typically many are needed to myelinate an entire axon. Myelination allows for faster nerve signal propagation.	*neuron*, nerve, + *lemma*, husk, + *kytos*, cell; *Theodor Schwann*, German histologist and physiologist
Satellite Cells	Small glial cells that surround the cell bodies within a ganglion (e.g., posterior root ganglion)	Satellite cells surround the cell bodies of somatic sensory neurons, hence the appearance of "satellites" around those neurons. They provide general support for the neurons and are analogous in function to astrocytes in the CNS.	*satelles*, attendant

Glial Cells of the Central Nervous System

EXERCISE 14.5

ASTROCYTES

1. Obtain a slide of the cerebrum or cerebellum that has been stained with silver stain. Silver stain makes the general supporting cells of the central nervous system, the **astrocytes**, stain very dark so they are visible through the microscope (**figure 14.6**).

2. Place the slide on the microscope stage and bring the tissue sample into focus on low power. Then switch to high power and bring the tissue sample into focus once again.

3. The two most prominent cell types visible are the large neurons (e.g., pyramidal cells in the cerebrum or Purkinje cells in the cerebellum) and the smaller astrocytes. Astrocytes, as their name implies, are shaped like stars (*astron-*, star). They have multiple long cellular processes that wrap themselves around neurons and around blood vessels in the central nervous system. These processes allow astrocytes to perform one of their main functions, which is to transport nutrients from the blood to the neurons.

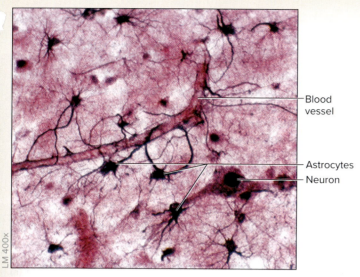

Figure 14.6 **Astrocytes**. Two astrocytes in the center of the slide have processes that can be seen touching a blood vessel.

©Rick Ash

Blood vessel

Astrocytes

Neuron

LM 400x

4. Sketch an astrocyte as seen through the microscope in the space provided or on a separate sheet of paper.

_____ ×

EXERCISE 14.6

EPENDYMAL CELLS

1. Obtain a slide of the brain and place it on the microscope stage. Bring the tissue sample into focus on low power. Then switch to high power and bring the tissue sample into focus once again. Look for the "empty" spaces on the slide, because those spaces will likely be the ventricular spaces that are lined with ependymal cells.

2. **Ependymal cells** (**figure 14.7**) are cuboidal to columnar-shaped epithelial cells with *cilia* and *microvilli* on their apical surfaces. These cells line the ventricles of the brain and the central canal of the spinal cord. They also form the outer layer of the choroid plexus. The cilia present on the apical surfaces of these cells help to move cerebrospinal fluid along.

3. Locate ependymal cells on the slide, using figure 14.7 as a guide.

4. Sketch ependymal cells as seen through the microscope in the space provided or on a separate sheet of paper. Label the following in the drawing:

☐ **Ependymal cells**   ☐ **Ventricular space**

_____ ×

Ventricular space

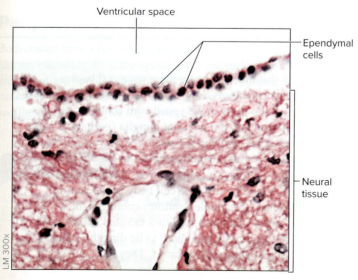

Ependymal cells

Neural tissue

LM 300x

Figure 14.7 **Ependymal Cells.** Cuboidal ependymal cells lining a ventricular surface of the brain.

©Rick Ash

EXERCISE 14.12 ₪₪PhILS

Ph.I.L.S. LESSON 12:
THE COMPOUND ACTION POTENTIAL

The structure of a nerve is such that there are multiple axons contained within a bundle of connective tissue. An external stimulus (voltage) that allows the membrane of an axon to reach threshold will initiate an action potential. Increasing the stimulus intensity will recruit more neurons, thereby producing compound action potentials. Compound action potentials can be recorded at the postsynaptic membrane. The purpose of this exercise is to vary the intensity of external stimuli on an excised frog sciatic nerve to demonstrate both threshold and recruitment.

Before beginning the exercise, become familiar with the following concepts (use the main textbook as a reference):

- The role of sodium (Na^+) and potassium (K^+) in generating and maintaining membrane potential
- How the concept of threshold applies to the generation of an action potential
- The functionality of voltage-gated channels
- The definition of depolarization and hyperpolarization in terms of membrane potential

Before beginning the experiment, state a hypothesis regarding the effect of stimulus intensity on threshold and recruitment.

_____

_____

_____

1. Open Ph.I.L.S. Lesson 12: Action Potentials: The Compound Action Potential (**figure 14.14**).

2. Read the objectives and introduction. Then take the pre-lab quiz. The laboratory exercise will open when the pre-lab quiz has been completed.

3. Drag the blue plug to recording input 1 on the data acquisition unit. Click and drag the recording cables to the matching colored posts on the nerve chamber.

4. Drag the red plug to the positive stimulator output on the data acquisition unit, and drag the red recording cable to the matching post on the nerve chamber.

5. Drag the black plug to the negative stimulator output on the data acquisition unit, and drag the black recording cable to the matching post on the nerve chamber.

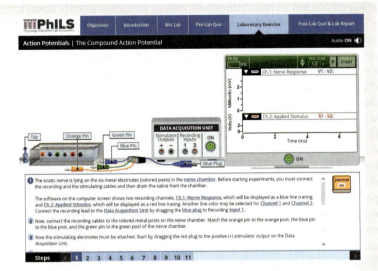

Figure 14.14 Setup of the Wet Lab for Ph.I.L.S. Lesson 12: The Compound Action Potential. An isolated nerve has been placed in the apparatus and is hooked up to both stimulating and recording electrodes.

©McGraw-Hill Education

6. Click and hold the "tap" until all saline is drained from the chamber.

7. The default setting for the voltage is 1.0 volt. Click the Start button in the upper right corner of the virtual computer screen to apply a stimulus to the nerve.

8. You will now measure the nerve response by measuring the amplitude (height) of the compound action potential (CAP). One cursor is centered on the flat part of the line tracing, just prior to the action potential. Locate the second cursor and center it on the peak of the compound action potential. The difference between these values, represented by the V1-V2 value, represents the strength of the action potential.

9. Click the Journal panel to enter the value into the Journal. A table with volts and amplitude (amp) and a graph will appear. Note the range in values for volts from 0 to 1.6 volts. Close the Journal window by clicking on the X in the upper right corner.

10. To complete the table and produce the graph, repeat steps 9 and 10 for the entire range of voltages (0–1.6 mV). Record the data in **table 14.7**.

Table 14.7	Membrane Potentials with Applied External Stimuli	
Volts	**Amplitude**	
0.0		
0.1		
0.2		
0.3		
0.4		
0.5		
0.6		
0.7		
0.8		
0.9		
1.0		
1.1		
1.2		
1.3		
1.4		
1.5		
1.6		

11. After finishing the laboratory exercise, print the line tracing by clicking on the P in the upper left corner of the virtual computer screen.

12. Construct a graph of a compound action potential in the space provided or on a separate sheet of paper. Be sure to indicate where voltage-gated sodium channels and voltage-gated potassium channels are opened and closed.

13. Open the post-lab quiz and lab report by clicking on it. Answer the post-lab questions that appear on the computer screen. Click Print Lab to print a hard copy of the report or click Save PDF to save an electronic copy of the report.

14. Make note of any pertinent observations here.

_____

_____

_____

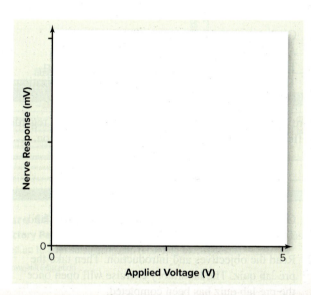

INTEGRATE

CONCEPT CONNECTION

The membrane potential of an excitable cell is regulated by active and passive **membrane transport** mechanisms. Active pumps move Na⁺ out of the cell and K⁺ into the cell. The movements of Na⁺ and K⁺ in this case are caused by active transport because the pump is moving the ions *against* their respective concentration gradients. This leads to a separation of charge across the plasma membrane due to the membrane's selective permeability. The movement of Na⁺ or K⁺ ions into or out of the cell is driven by passive transport mechanisms. These ions pass freely through their respective leak channels along their individual electrochemical gradients (e.g., K⁺ moves out of the cell, from an area of high concentration to an area of low concentration). In addition to leak channels, ions may pass through either chemically gated or voltage-gated channels. Opening many gates simultaneously creates a large permeability change. Again, passive transport mechanisms allow for these ions to move along their respective electrochemical gradients through these channels. Rather than memorizing the movement of ions into or out of the plasma membrane, simply refer to the governing principles of membrane transport for guidance.

GROSS ANATOMY

The Human Brain

These exercises involve identifying structures that are visible in four views of the human brain: superior, lateral, inferior, and midsagittal. **Tables 15.1, 15.2, 15.3, 15.4,** and **15.5** list the main brain structures visible in each view and include functional descriptions of each structure.

When identifying parts of the brain, keep in mind that many are visible in the multiple views of the brain that follow. Follow these steps to make things easier:

• Name the structures in a logical order, such as the order in which the structures appear from anterior to posterior.

Learning the structures in an orderly fashion will allow for improved information recall later on.

• Do not think of brain regions as isolated structures. The structures are easier to identify within the context of their surroundings. Think about this the next time you are traveling to your anatomy & physiology class—would you know how to get to the classroom if all of the buildings on campus (with the exception of the one to which you are traveling) were suddenly moved around?

• Always associate a function with each structure identified. Use tables 15.1 through 15.5 as a reference.

Table 15.1	Brain Structures Visible from the Superior View		
Brain Structure	**Description**	**Function(s)**	**Word Origin**
Central Sulcus	A deep groove that extends along the coronal plane	Separates the frontal lobe from the parietal lobe in the cerebrum.	*central*, in the center, + *sulcus*, a furro
Frontal Lobe	Lobe of cerebrum deep to the frontal bone	Controls conscious movement of skeletal muscle; contains Broca's area, which controls motor speech. Higher-level functions include judgment and foresight (the ability to think before acting).	*frontal*, in the front, + *lobos*, lobe
Precentral Gyrus	Fold of brain tissue located directly anterior to the central sulcus	Primary somatic motor area of the brain. Somatic motor neurons here initiate motor signals to control voluntary muscle activity.	*pre*, before, + *central*, relating to the central sulcus, + *gyros*, circle
Longitudinal Fissure	A deep fissure between the two cerebral hemispheres	Separates the two cerebral hemispheres; the falx cerebri occupies this fissure during life.	*Longus*, long + *fissure*, a deep furrow
Occipital Lobe	Lobe of cerebrum deep to the occipital bone	Primary visual area (area of the cerebral cortex where visual information synapses, after the thalamus); visual association area	*occiput*, the back of the head, + *lobos*, lobe
Parietal Lobe	Lobe of cerebrum deep to parietal bone	Receives sensory input from the skin and proprioceptors. Higher-level functions include logical reasoning (math, problem solving).	*parietal*, a wall, + *lobos*, lobe
Postcentral Gyrus	Fold of brain tissue located directly posterior to the central sulcus	Primary somatic sensory area of the brain. Sensory information from the body travels to this area of the cerebral cortex.	*post*, after, + *central*, relating to the central sulcus, + *gyros*, circle

Table 15.2	Brain Structures Visible from the Lateral View		
Brain Structure	**Description**	**Function(s)**	**Word Origin**
Cerebellum	The second largest part of the brain; characteristic white matter branching pattern (*arbor vitae*)	Regulates muscle tone, coordinates muscle activity, and maintains balance and equilibrium.	*cerebellum*, little brain
Medulla Oblongata	The most inferior aspect of the brainstem; between pons and spinal cord	Contains major respiratory and cardiac centers; contains nuclei of the *reticular activating system* that regulate wakefulness and attention.	*medius*, middle, + *oblongus*, rather long
Lateral Sulcus	A horizontal groove between the frontal/parietal lobes and the temporal lobe	Separates the frontal and parietal lobes from the temporal lobe.	*latus*, the side, + *sulcus*, a furrow
Pons	Large mass of the brainstem superior to the medulla oblongata	"Bridge" of fiber tracts connecting the cerebral hemispheres to the cerebellar hemispheres; contains centers for control of respiration	*pons*, bridge
Temporal Lobe	Lobe of cerebrum deep to the temporal bone	Primary auditory and auditory association area of the brain; conscious perception of smell	*tempus*, time
Transverse Fissure	A deep fissure between the cerebrum and the cerebellum	Separates the cerebral hemispheres from the cerebellar hemispheres. The tentorium cerebelli lies in this fissure.	*transversus*, across, + *fissure*, a deep furrow

INTEGRATE

CONCEPT CONNECTION

Recall from chapter 11 that each skeletal muscle fiber in a muscle is innervated by a single motor neuron. Impulses travel down this motor neuron and cause all the muscle fibers innervated by this neuron to contract. These motor neurons that directly innervate the muscle fibers and excite the skeletal muscle are called lower motor neurons.

However, the initial impulse actually arises higher within the central nervous system (CNS), specifically in the motor cortex of the frontal lobe. Here, impulses are first generated in upper motor neurons before being passed down through the brainstem and spinal cord. In the spinal cord, they synapse with the lower motor neurons that carry the impulse to the muscle fibers.

EXERCISE 15.1

SUPERIOR VIEW OF THE HUMAN BRAIN

1. Obtain a human brain or models of a human brain and observe the superior surface.

   - Numerous grooves are associated with the brain surface. A large groove is a fissure, whereas a small groove is a sulcus. The most prominent feature in this view is the **longitudinal fissure**, which is a deep groove that separates the two cerebral hemispheres from each other. Observe the many sulci on the brain surface. One major sulcus to identify is the **central sulcus**. Identification of the central sulcus is difficult, though not impossible, on a real human brain. The following are two features to look for:

   - The **precentral gyrus** and the **postcentral gyrus** (two raised areas approximately in the middle of the brain's

superior surface) should become continuous with each other on the lateral aspect of the central sulcus just above the lateral sulcus (a groove that separates the temporal lobe from the frontal and parietal lobes). This means the central sulcus will not enter the lateral sulcus.

   - The central sulcus will dip down into the longitudinal fissure.

   - Three of the five lobes of the cerebrum are visible on the superior surface view.

2. Identify the structures listed in **figure 15.1** on the superior view of the brain, using table 15.1 and the textbook as guides. Then label them in figure 15.1.

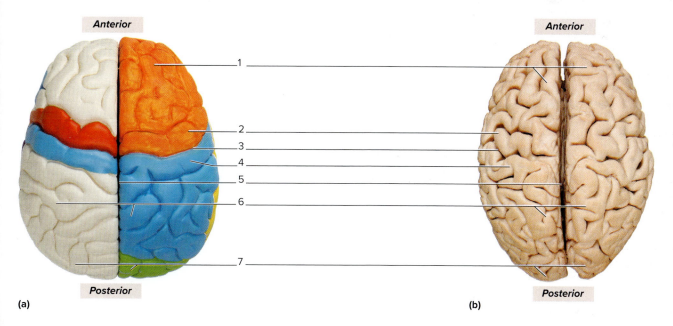

Figure 15.1   Superior View of the Brain. (*a*) Classroom model of the brain. (*b*) Preserved human brain. Use the terms listed to fill in the numbered labels in the figure.

☐ Central sulcus   ☐ Longitudinal fissure   ☐ Parietal lobe   ☐ Precentral gyrus

☐ Frontal lobe   ☐ Occipital lobe   ☐ Postcentral gyrus

(*a*) Model # C22 [1000228] ©3B Scientific GmbH, Germany, 2013 www.3bscientific.com; (*b*) ©McGraw-Hill Education/Photo and Dissection by Christine Eckel

EXERCISE 15.2

LATERAL VIEW OF THE HUMAN BRAIN

1. Obtain a human brain or models of a human brain and observe the lateral surface.

   • As with the superior view, the **central sulcus** should be visible by identifying the location where the pre- and postcentral gyri become continuous with each other just above the **lateral sulcus**.

2. Identify the structures listed in **figure 15.2** on the lateral view of the brain, using tables 15.1 and 15.2 and the textbook as guides. Then label them in figure 15.2.

3. *Optional Activity:* **APR** 7: **Nervous System**—Watch the "Divisions of Brain" animation for an overview of the regions of the brain and their general functions.

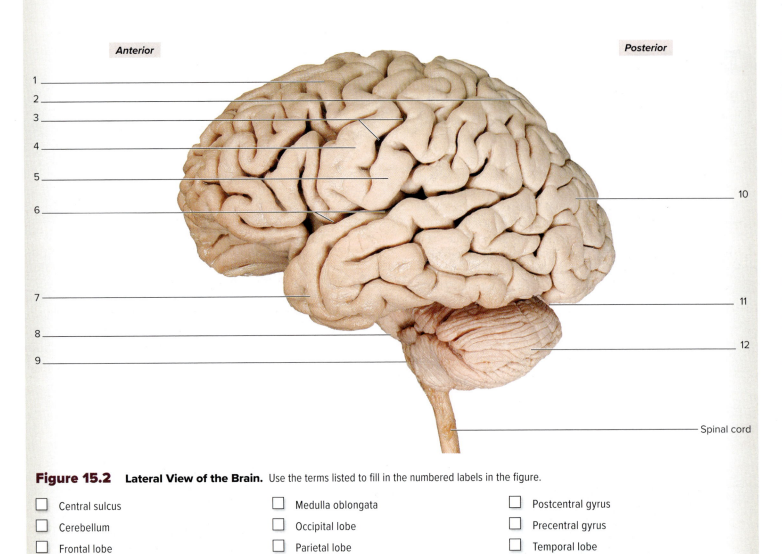

Figure 15.2   **Lateral View of the Brain.** Use the terms listed to fill in the numbered labels in the figure.

☐ Central sulcus	☐ Medulla oblongata	☐ Postcentral gyrus
☐ Cerebellum	☐ Occipital lobe	☐ Precentral gyrus
☐ Frontal lobe	☐ Parietal lobe	☐ Temporal lobe
☐ Lateral sulcus	☐ Pons	☐ Transverse fissure

©McGraw-Hill Education/Photo and Dissection by Christine Eckel

Table 15.3	Brain Structures Visible from the Inferior View		
Brain Structure	**Description**	**Function(s)**	**Word Origin**
Infundibulum	A funnel-shaped inferior extension of the brain located immediately posterior to the optic chiasm	Consists of tracts that connect the hypothalamus to the posterior pituitary	*infundibulum,* a *funnel*
Mammillary Bodies	Two small bump-like structures of the hypothalamus located immediately posterior to the infundibulum	Involved in short-term memory processing; part of the limbic system (the emotional brain). Also involved with suckling and chewing reflexes.	*mammillary,* shaped like a breast
Olfactory Bulbs	Swellings at the anterior end of the _____ surface of the frontal lobe lateral to the longitudinal fissure	Location of initial synapse of the olfactory nerve (CN I) after passing through the cribriform plate of the ethmoid bone; part of the limbic system	_____ _____ _____ular structure
Olfactory Tracts	Nerve fibers extending from the olfactory bulbs posteriorly to the junction where the frontal lobes meet the optic chiasm	Carry the axons of neurons from the olfactory bulbs toward structures in other areas of the brain involved with olfaction; part of the limbic system	*olfactus,* to smell, + *tractus,* a drawing out
Optic Chiasm	X-shaped structure formed where the two optic nerves meet; located just anterior to the infundibulum	Location where some fibers from both optic nerves cross over and travel in the optic tract on the opposite side.	*optikos,* relating to the eye or vision, + *chiasma,* two crossing lines
Optic Nerves	Anterior to the optic chiasm	Sensory neurons carrying visual information from the retina to the optic chiasm.	*optikos,* relating to the eye or vision, + *nervus,* a white, cordlike structure
Optic Tracts	Posterior to the optic chiasm	Sensory neurons carrying visual information from the optic chiasm to the thalamus.	*optikos,* relating to the eye or vision, + *tractus,* a drawing out

EXERCISE 15.3

INFERIOR VIEW OF THE HUMAN BRAIN

1. Obtain a human brain or models of a human brain and observe the inferior surface. This view is considerably more complicated than the superior or lateral views because of the cranial nerves that arise from the brain. Cranial nerve identification is covered later in this chapter.

   - Both the brainstem and the cerebellum are visible from this view.

   - Prominent features associated with the cerebrum are the **optic chiasm** and the **optic tracts**, which extend from it into the brain.

   - The **mammillary bodies** are two small projections posterior to the optic chiasm.

   - One of the more problematic structures to identify in this view on a real brain is the infundibulum. When a brain is removed from the cranium, the pituitary gland almost always gets removed from the brain. The only structure left connected to the brain is the stalk of tissue that connects the pituitary gland to the hypothalamus, which is the **infundibulum**, or **pituitary stalk**. Observation of a model of the brain shows the pituitary gland to be intact. The infundibulum can be identified as a small strand of tissue that is located directly posterior to the optic chiasm and directly anterior to the mammillary bodies.

2. Identify the structures listed in **figure 15.3** on the inferior view of the brain, using tables 15.1 through 15.3 and the textbook as guides. Then label them in figure 15.3.

(continued on next page)

(continued from previous page)

3. Label the meningeal structures in **figure 15.6**.

4. *Optional Activity:* **APR** 7: **Nervous System**—Watch the "Meninges" and "Dural Sinus Blood Flow" animations to reinforce your understanding of these structures and their relationships.

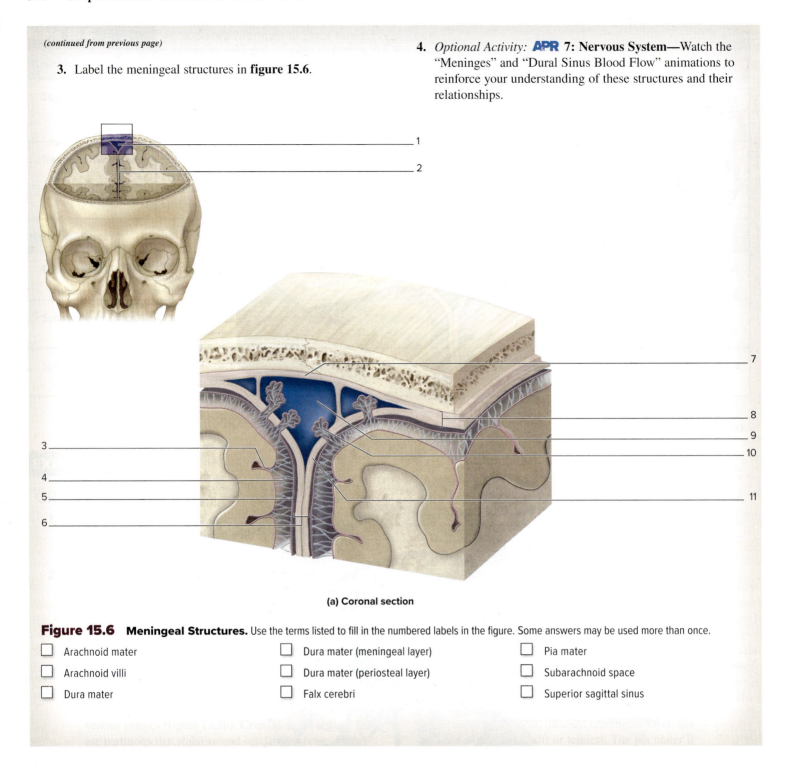

(a) Coronal section

Figure 15.6   **Meningeal Structures.** Use the terms listed to fill in the numbered labels in the figure. Some answers may be used more than once.

☐ Arachnoid mater   ☐ Dura mater (meningeal layer)   ☐ Pia mater

☐ Arachnoid villi   ☐ Dura mater (periosteal layer)   ☐ Subarachnoid space

☐ Dura mater   ☐ Falx cerebri   ☐ Superior sagittal sinus

INTEGRATE

CLINICAL VIEW
Meningiomas

Meningiomas are brain tumors that originate in the meninges of the CNS. These slow-growing tumors develop from cells within the arachnoid villi and project into the venous sinuses within the brain. Incidence is higher in women than in men, although the reason is unknown. Most commonly, meningiomas develop either superior to the frontal and parietal lobes near the falx cerebri (parasagittal meningioma) or between the cerebral hemispheres (falcine meningiomas). Some may grow directly under the skull (convexity meningiomas), and some develop within the ventricles (ventricular meningiomas). Other tumors are considered skull base meningiomas due to their location. Subtypes of skull base meningiomas include

those located near the sphenoid bone (sphenoid wing meningioma), near the posterior fossa (petrous meningioma), or near the ethmoid bone (paranasal/olfactory meningioma). Meningiomas are dome-shaped, encapsulated tumors that typically attach to the dura mater. In some cases, the tumor may project into and infiltrate surrounding bone tissue. Many meningiomas are asymptomatic and benign. As a tumor grows in size, it may cause symptoms such as headaches, seizures, muscle weakness, sensory and visual disturbances, and increased intracranial pressure. Surgery is required to remove tumors that become symptomatic or malignant (this means that the tumor has spread to other tissues). Some meningiomas are associated with genetic disorders, some are linked to radiation exposure, and some have no known cause at all. Because many meningiomas are asymptomatic, they may not be discovered unless an autopsy is performed.

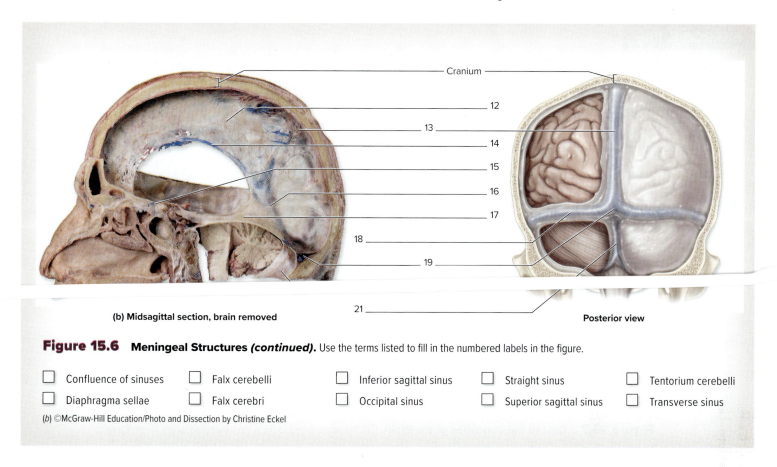

(b) Midsagittal section, brain removed

Posterior view

Figure 15.6 **Meningeal Structures *(continued).*** Use the terms listed to fill in the numbered labels in the figure.

☐ Confluence of sinuses ☐ Falx cerebelli ☐ Inferior sagittal sinus ☐ Straight sinus ☐ Tentorium cerebelli

☐ Diaphragma sellae ☐ Falx cerebri ☐ Occipital sinus ☐ Superior sagittal sinus ☐ Transverse sinus

(b) ©McGraw-Hill Education/Photo and Dissection by Christine Eckel

Ventricles of the Brain

The **ventricles** are fluid-filled spaces within the CNS that are complex in shape. Exercise 15.6 involves viewing a *cast* of the ventricles, which is produced by filling the ventricular spaces with plastic, allowing the plastic to harden, and then removing the brain tissue so only the cast is left. A cast allows one to visualize the three-dimensional structure of the ventricles without the brain literally "getting in the way." If casts of the brain ventricles are not available, this exercise can be performed using **table 15.7** and figures in the textbook.

The CNS initially develops as a neural tube. As it grows, the neural tube begins to change size and shape. The cephalic end develops into the brain, while the rest develops into the spinal cord. Both the brain and the spinal cord contain fluid-filled spaces inside. The pattern of growth of the neural tissue surrounding the neural tube changes the size and shape of the fluid-filled spaces within. Thus, because the spinal cord remains mostly a tubular structure as it grows, the fluid-filled space inside, the **central canal,** remains tubular. On the other hand, because the cephalic (brain) end of the neural tube undergoes extensive folding as it grows, the fluid-filled spaces within develop into irregular shapes. These shapes tell a story about how the parts of the brain developed.

The cephalic end of the neural tube first develops into three **primary vesicles** (prosencephalon, mesencephalon, and rhomben-cephalon) and then into five **secondary vesicles** (telencephalon, diencephalon, mesencephalon, metencephalon, and myelencepha-lon). Table 15.7 lists the secondary vesicles of the brain and the

parts of the ventricular system that develop from each of them. For instance, the telencephalon undergoes extensive growth as it develops into the cerebral hemispheres. This growth results in the horseshoe-shaped structure of the **lateral ventricles** in the adult brain. On the other hand, the mesencephalon does not undergo such extensive growth as it develops into the midbrain. Hence the fluid-filled space inside, the **cerebral aqueduct**, remains tubular in shape within the adult brain.

INTEGRATE

LEARNING STRATEGY

When observing whole brains or brain models in the laboratory, always begin by locating the ventricular spaces and associating each ventricular space with a secondary brain vesicle (table 15.7). Next, identify the adult brain structures that surround the ventricular spaces. Finally, correlate each adult brain structure to the secondary brain vesicle from which it formed. For example, the lateral ventricles (ventricular space), which are part of the telencephalon (secondary brain vesicle), are surrounded by the cerebral hemispheres (adult brain structures). Thus, the cerebral hemispheres (brain structure) are derived from the telencephalon (secondary brain vesicle).

BRAIN VENTRICLES

1. Obtain a cast of **the ventricles of the brain (figure 15.7).**

2. Identify the ventricles of the brain listed in figure 15.7 on the cast of the ventricles, using table 15.7 and the textbook as guides. Then label them in figure 15.7. When identifying each of the ventricles, relate each ventricle to the **secondary brain vesicle** from which it developed (table 15.7).

Lateral view

Figure 15.7   Cast of the Ventricles of the Brain. Use the terms listed to fill in the numbered labels in the figure.

☐ Cerebral aqueduct   ☐ Interventricular foramen   ☐ Third ventricle

☐ Fourth ventricle   ☐ Lateral ventricles

Model # VH410 [1001262] ©3B Scientific GmbH, Germany, 2013 www.3bscientific.com/Photo by Christine Eckel, Ph.D.

Table 15.7	Secondary Brain Vesicles and Associated Structures of the Brain			
Secondary Brain Vesicles	**Ventricles of the Brain**	**Location and Description of Ventricles**	**Brain Structures Associated with Secondary Vesicles**	
Telencephalon	Lateral ventricles	Horseshoe-shaped ventricles within the cerebral hemispheres; contain anterior and posterior horns whose shape follows the development of the cerebral hemispheres	Cerebrum	
Diencephalon	Third ventricle	Narrow, quadrilateral-shaped ventricle located in the midsagittal plane inferior to the corpus callosum and medial to the thalamic nuclei; surrounded by the diencephalon	Epithalamus Thalamus Hypothalamus	
Mesencephalon (midbrain)	Cerebral aqueduct	Narrow channel that lies in the midbrain between the cerebral peduncles and the tectal plate (corpora quadrigemina)	Midbrain	
Metencephalon	Fourth ventricle (superior part)	Diamond-shaped ventricle located anterior to the cerebellum and posterior to the pons	Pons, cerebellum	
Myelencephalon	Fourth ventricle (inferior part); part of central canal	Diamond-shaped ventricle located anterior to the cerebellum and posterior to the pons	Medulla oblongata	
Neural canal				

EXERCISE 15.7

CIRCULATION OF CEREBROSPINAL FLUID (CSF)

Cerebrospinal fluid (CSF) is a clear, colorless fluid produced from blood plasma, which is filtered by the choroid plexus within each brain ventricle. Approximately 500 mL of CSF is produced each day. Following its production within the lateral ventricles, CSF moves through the following structures: the interventricular foramen, the third ventricle, the cerebral aqueduct, and the fourth ventricle. CSF exits the fourth ventricle through the median and lateral apertures to enter the subarachnoid space, which surrounds both the brain and the

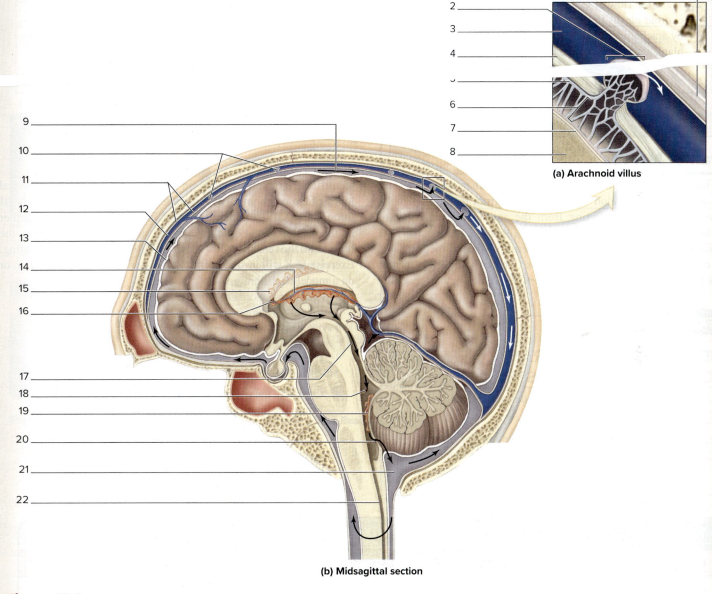

(a) Arachnoid villus

(b) Midsagittal section

Figure 15.8 **Cerebrospinal Fluid (CSF) Production and Circulation.** Note that arrows show the directional flow of CSF. Use the terms listed to fill in the numbered labels in the figure. Some terms may be used more than once.

☐ Arachnoid mater	☐ Cerebral cortex	☐ Dura mater	☐ Meningeal dura
☐ Arachnoid villi	☐ Choroid plexus of fourth ventricle	☐ Fourth ventricle	☐ Periosteal dura
☐ Arachnoid villus	☐ Choroid plexus of lateral ventricle	☐ Interventricular foramen	☐ Pia mater
☐ Central canal	☐ Choroid plexus of third ventricle	☐ Lateral aperture	☐ Subarachnoid space
☐ Cerebral aqueduct	☐ CSF flow	☐ Median aperture	☐ Superior sagittal sinus

(continued on next page)

Table 15.
Nerve Numb
I
II
III
IV
V
V₁
V₂
V₃
VI
VII
VIII
IX
X
XI
XII

(continued from previous page)

EXERCISE 15.10A Olfactory (CN I)

The olfactory nerves **(figure 15.18)** are unique as cranial nerves in that there are more than two of them (the rest of the cranial nerves are paired—a right and left for each), and they are constantly being replaced. The nerves lie within the nasal epithelium, and their axons project through the **olfactory foramina** within the cribriform plate of the ethmoid bone (table 15.8). The olfactory neurons then synapse with neurons within the **olfactory bulbs**, and the signals are sent to the brain via the **olfactory tracts**.

1. Obtain vials of peppermint, lemon, and vanilla oils.

2. While the subject's eyes are closed, pass an open vial of peppermint oil just under a laboratory partner's nose. Was the subject able to identify the smell?

3. Repeat this process with the vials of lemon and vanilla oils. Allow some time between applications of the different oils. Damage to the olfactory nerves results in an inability to identify odors. Excessive smoking or inflammation of the nasal mucosa as a result of a viral infection can inhibit the sense of smell, and the sense of smell also declines with age.

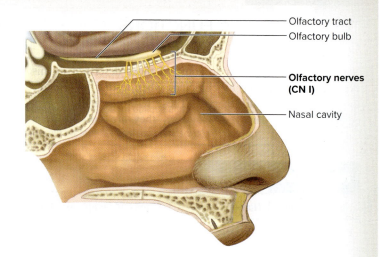

Figure 15.18   **Location of the Olfactory Nerves (CN I).**

EXERCISE 15.10B Optic (CN II)

Each of the optic nerves **(figure 15.19)** begins as axons of ganglion cells located within the **retina** of the eye. Different parts of the retina correspond to different portions of the visual field. The axons of these nerves exit the eye at the optic disc, or "blind spot," of the eye, at which point they become the **optic nerve**, which then travels posteriorly toward the diencephalon. Anterior to the infundibulum, most of the fibers cross over at the prominent **optic chiasm** (*chiasma*, a crossing of two lines) to the other side of the brain. The fibers then continue to travel posteriorly to reach the visual cortex in the occipital lobe of the brain. Figure 15.19 demonstrates the pattern of flow of visual information from the retina to the brain. When damage to the retina or optic nerve is suspected, visual field tests are performed to discover the location of the damage. These tests of visual function are beyond the scope of this course and will not be performed in this laboratory session.

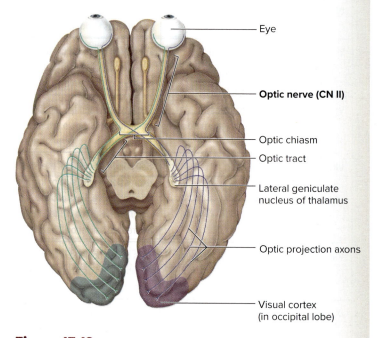

Figure 15.19   **The Optic Nerve (CN II).**

EXERCISE 15.10C Oculomotor (CN III)

The oculomotor nerves **(figure 15.20)** send motor fibers to the majority of the extrinsic eye muscles as well as to the muscles that control pupil diameter and ciliary muscles involved in focusing (tables 15.10 and 15.11).

1. Obtain a small flashlight. Look into one of a laboratory partner's eyes and observe the size of the pupil. While looking into the subject's eye, gently shine the light into the eye (if it is a bright light, just bring it near the eye so that more light enters the eye—the goal here is not to blind the individual with the light).

Was there a change in pupil diameter? _____

If so, what happened? _____

What other cranial nerve is involved as the sensory component of this reflex? _____

2. Repeat the preceding activity, but this time observe the pupil of the other eye.

Was there a change in pupil diameter? _____

If so, what happened? _____

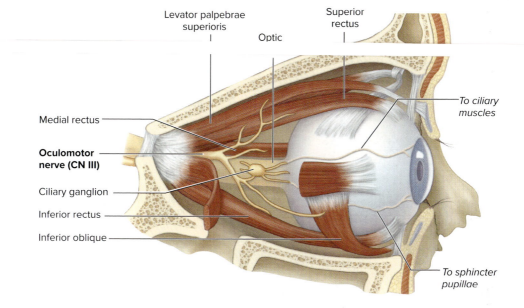

Figure 15.20 **The Oculomotor Nerve (CN III).**

EXERCISE 15.10D Trochlear (CN IV)

The trochlear nerve **(figure 15.21)** controls only one extraocular eye muscle—the superior oblique (tables 15.10 and 15.11). Ask a laboratory partner to look down and out (inferior and lateral). Weakness or an inability to perform this action indicates a weak superior oblique muscle or damage to the trochlear nerve.

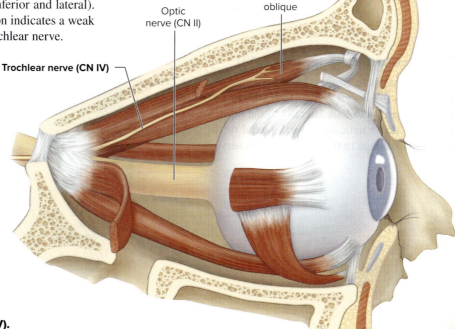

Figure 15.21 **The Trochlear Nerve (CN IV).**

(continued on next page)

(continued from previous page)

EXERCISE 15.10K Hypoglossal (CN XII)

The hypoglossal nerve **(figure 15.30)** innervates intrinsic and extrinsic muscles of the tongue (tables 15.10 and 15.11). Have the subject stick out the tongue. The tongue should protrude straight out. If the hypoglossal nerve is damaged, the tongue will deviate to the side of the damaged nerve.

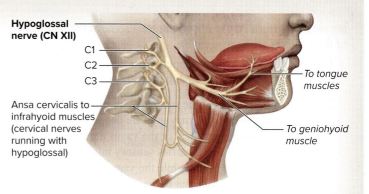

Figure 15.30 **The Hypoglossal Nerve (CN XII).**

Testing Brain Function

Previous exercises in this chapter involve identifying brain structures on models and/or cadavers, and relating structure to function. An **electroencephalogram (EEG)** is used to record electrical activity of the brain using surface electrodes placed on the skull. Clinically, an EEG is used to assess brain function. The technique is similar to recording electrical activity of skeletal muscles with an electromyogram (EMG) or recording electrical activity of the heart with an ECG. Changes in the electrical activity of the brain observed on an EEG may reveal abnormalities in the frontal lobe or cerebellum of a patient who is experiencing motor dysfunction (e.g., gait disturbances). Similarly, an EEG may reveal abnormalities in the parietal lobe of a patient experiencing sensory dysfunction (e.g., slow reactions to painful stimuli).

INTEGRATE

CLINICAL VIEW
Epilepsy

Epilepsy is a neurological condition characterized by excessive patterns of synchronous electrical activity of the brain. The resulting **seizure** may vary in severity depending upon the strength, duration, and location of synchronous electrical activity. Seizures may impact a local area or a large portion of the brain. An EEG is used clinically to diagnose epilepsy. Possible symptoms that occur during a seizure include confusion, sensitivity to light and sounds, slurred speech, and a glazed stare. Seizures that impact large portions of the brain may be classified as *tonic-clonic* (formerly *grand mal*) seizures. In a tonic-clonic seizure, abnormal electrical activity in the brain causes a patient's muscles to contract, leading to whole-body stiffening (tonic phase) and loss of consciousness. The muscles then begin to rapidly relax and contract, causing whole-body convulsions (clonic phase). Once patients regain consciousness, they often have no recollection of the event. Treatment depends upon the severity of the seizures. Drugs that influence the excitability of neurons may be prescribed. In patients where the electrical activity is localized to one portion of the brain and there is no response to drug therapy, doctors may implant stimulators that can apply a "resetting" current to nerves or brain tissue. This is similar to a pacemaker that resets the electrical activity in the heart. In the most severe cases, doctors may surgically remove affected brain tissue.

EXERCISE 15.11

BIOPAC Electroencephalography (EEG)

This exercise involves observing the electrical activity of the brain with a subject's eyes open and closed. Alpha, beta, delta, and theta waves will be compared in the subject's EEG for two states.

Obtain the Following:
- **BIOPAC electrode lead set (SS2L)**
- **cot and pillow**
- **electrode gel**
- **surface electrodes**
- **swim cap or wrap**

Before beginning the experiment, become familiar with the following concepts (use the textbook as a reference):
- Brain structure and function
- **Depolarization and repolarization of an excitable cell**

State a hypothesis regarding how the electrical activity of the brain will change with a subject's eyes open and closed.

_____

_____

_____

_____

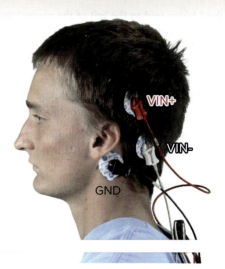

Figure 15.31 BIOPAC Equipment and Electrode Placement for EEG. Place electrodes and leads on the subject's scalp in the designated locations using the color code illustrated.

Courtesy of and ©BIOPAC Systems, Inc.

1. Prepare the BIOPAC equipment by turning the computer ON and the MP3X Data Acquisition Unit (DAQ) OFF (in addition to MP3x hardware, this lesson is also compatible with the 2-channel MP45 DAQ unit). Plug in the following: electrode lead set (SS2L) (CH 1). Turn the MP3X DAQ ON.

2. Allow the subject to assume a relaxed position. It is recommended that the subject lie supine on a cot with the subject's head resting comfortably on a pillow (or rolled-up jacket or sweater) with eyes closed.

3. Apply a small quantity of electrode gel to the center of the electrodes, and place the electrode firmly on the subject's scalp in the designated locations using the color code, as illustrated in **figure 15.31**. Be sure that hair has been kept away from the electrodes as much as possible. A swim cap or wrap may be placed around the subject's head to hold the electrodes firmly in place.

4. Start the BIOPAC Student Lab Program. Select L03-Electroencephalography (EEG) I from the Choose a Lesson menu, enter the file name, and click OK.

5. Click Calibrate in the upper left corner of the Setup window. Ensure proper placement of the subject's electrodes and click OK. Once the calibration has stopped after 8 seconds of recording, check to ensure that the baseline remains unchanged at approximately 0 μV. If spikes appear in the trace, Redo Calibration. If calibration data are good, click Continue.

6. During the 60-second recording session, the subject will remain in a relaxed position. The subject will begin with eyes closed (0–20 seconds), then eyes open (21–40 seconds), and then eyes closed (41–60 seconds). During

the period when the eyes are open, the subject should try not to blink. One lab member should serve as the **director**, informing the subject when to open and close the eyes, and one lab member will serve as the **recorder**, marking the events on the EEG trace.

7. The subject will close his/her eyes and remain in a relaxed position. To begin recording EEG data, click Record. After 20 seconds, the director will inform the subject to open his/her eyes. The recorder will press F4 to add an Eyes Open event marker to the data trace. After 20 seconds with the eyes open, the director will inform the subject to close his/her eyes once again. The recorder will press F5 to add an Eyes Closed event marker to the data trace. After the subject has completed all three conditions, click Stop. Review the data to ensure that the EEG resembles **figure 15.32**. If not, click Redo.

8. To observe the component of the EEG trace that corresponds to the various frequency components, click on each of the frequency buttons in the following order: alpha, beta, delta, and theta. The amplitude should decrease during the "eyes open" condition (see **figure 15.33**). If not, be sure that the electrodes make good contact with the scalp and click Redo. If the data resemble those in figure 15.33, click Done to conclude the experiment. Remove the subject's electrodes, and clean the skin using soap and water.

9. To analyze saved data, click Review Saved Data. Note that CH 1 displays EEG and CH 2–5 display alpha, beta, delta, and theta waves, respectively. Measurement boxes appear above the marker region in the data window. To set up, indicate the channel number and measurement type for each of four boxes: CH 2–5, Stddev.

(continued on next page)

Exercise 15.4: Midsagittal View of the Human Brain

5. Match the description listed in column A with the corresponding structure listed in column B. **4**

Column A

_____ 1. controls hormone secretion from the pituitary gland

_____ 2. fiber tract that connects the left cerebral hemisphere to the right cerebral hemisphere

_____ 3. involved in suckling reflex and chewing

_____ 4. primary relay center for sensory information coming into the brain

_____ 5. secretes the hormone melatonin from its precursor molecule, serotonin

Column B

a. corpus callosum

b. hypothalamus

c. mammillary bodies

d. pineal body (gland)

e. thalamus

Exercise 15.5: Cranial Meninges

6. Match the cranial dural septum listed in column A with the corresponding description listed in column B. **5**

Column A

_____ 1. diaphragma sellae

_____ 2. falx cerebelli

_____ 3. falx cerebri

_____ 4. tentorium cerebelli

Column B

a. drapes across the cerebellar hemispheres horizontally within the transverse fissure

b. located between the two cerebellar hemispheres

c. located between the two cerebral hemispheres

d. superior to the sella turcica of the sphenoid bone

7. Order the layers of the meninges from most superficial 1 to deep 3. **5**

_____ arachnoid mater

_____ dura mater

_____ pia mater

8. Place the following structures through which CSF and venous blood flow in the correct order as they drain from the subarachnoid space 1 into the internal jugular vein 4. **6**

_____ a. confluence of sinuses

_____ b. sigmoid sinuses

_____ c. superior sagittal sinus

_____ d. transverse sinuses

Exercise 15.6: Brain Ventricles

9. Match the description listed in column A with the appropriate structure listed in column B. **7**

Column A

_____ 1. capillaries that aid in production of CSF

_____ 2. a channel located between the third and fourth ventricles

_____ 3. a hole between the lateral ventricle and the third ventricle

_____ 4. a membrane that separates the two lateral ventricles

_____ 5. the two most superior ventricles

_____ 6. the ventricle located between the brainstem and cerebellum

_____ 7. the ventricle surrounded by the diencephalon

Column B

a. cerebral aqueduct

b. choroid plexus

c. fourth ventricle

d. interventricular foramen

e. lateral ventricle

f. septum pellucidum

g. third ventricle

10. Match the ventricular space listed in column A with the secondary brain vesicle from which it developed listed in column B. **8**

Column A

_____ 1. cerebral aqueduct

_____ 2. fourth ventricle

_____ 3. lateral ventricles

_____ 4. third ventricle

Column B

a. diencephalon

b. mesencephalon

c. myelencephalon

d. telencephalon

Exercise 15.7: Circulation of Cerebrospinal Fluid (CSF)

11. Place the following structures through which CSF flows in the correct order as CSF flows through the ventricular system of the brain. **9**

_____ a. cerebral aqueduct

_____ b. fourth ventricle

_____ c. interventricular foramen

_____ d. lateral ventricles

_____ e. third ventricle

Exercise 15.8: Identification of Cranial Nerves on a Brain or Brainstem Model

12. Complete the table by listing the cranial nerves that extend from the three regions of the brainstem. **10**

Region of Brainstem	Cranial Nerves Extending from This Region of Brainstem
Medulla Oblongata	
Pons	
Midbrain	

23. You should have noticed that the tectal plate (corpora quadrigemina) is much larger in the sheep brain, relative to total brain size, than in the human brain. Using this information, answer the following questions:

    a. What is the function of the superior colliculus?

    _____

    _____

    b. What is the function of the inferior colliculus?

    _____

    _____

    c. What does the difference in size between the superior and inferior colliculi tell you about the influence this region of the brain has on the overall functioning of a sheep versus a human? (That is, compare how much influence this area of the brain has on control over body functions.)

    _____

    _____

    _____

24. Compare and contrast structures of the human brain and those of the sheep brain. Complete the table with information about the relative size of the structure compared to the size of the entire brain. Then, based on function, explain why the structure might be more important for survival of the human or the sheep.

Brain Structure	Human Brain	Sheep Brain
Frontal Lobe		
Inferior Colliculi		
Mammillary Bodies		
Medulla Oblongata		
Olfactory Bulbs		
Pineal Body (Gland)		
Superior Colliculi		

25. What would be the effect of severing the corpus callosum?

    _____

    _____

    _____

Can You Synthesize What You've Learned?

26. If the passage of fluid is blocked at the confluence of sinuses, into which sinuses will fluid back up?

_____

27. An *acoustic neuroma* is a tumor that arises from neurolemmocytes (Schwann cells) surrounding the vestibular portion of the vestibulocochlear nerve (CN VIII). The tumor is benign, but generally grows within the confined space of the petrous part of the temporal bone, thus compressing the nerve and creating problems with balance and hearing loss. What nerve other than CN VIII would you expect to be affected by this tumor (due to its close proximity)?

_____

_____

_____

28. When a light is shined into a patient's right eye, an examiner expects to see a change in pupil diameter in both eyes. The response, called the *consensual light reflex,* is used to test the function of two cranial nerves. The reflex involves one cranial nerve sending the afferent (sensory) signal toward the brain and another cranial nerve sending the efferent (motor) signal out to the pupil.

   a.   Which cranial nerve carries the afferent (sensory) signal to the brain?

   _____

   b.   Which cranial nerve carries the efferent (motor) signal from the brain?

   _____

29. An 8-year-old male suffers from epilepsy. EEG results reveal epileptiform spikes recorded from the left temporal lobe. Doctors performed a left temporal lobe resection, and symptoms resolved. Discuss the consequences of a left temporal lobe resection, and state how these consequences may differ for an adult.

_____

_____

_____

INTRODUCTION

When learning that someone has fractured a vertebra in an accident, people probably think the worst-case scenario: "They are going to be paralyzed!" Though paralysis is common when the spinal cord is damaged, the amount of paralysis and subsequent loss of function are highly dependent upon what part of the spinal cord is injured and the extent of the injury.

The spinal cord transmits nerve signals between the body and the brain. The spinal cord contains bundles of axons called fasciculi that take sensory information to the brain and motor information from the brain. The fasciculi then have connections to the spinal nerves that convey information to and from the body. An understanding of where nerve signals enter and exit the spinal cord is important for understanding the degree of paralysis that might result from trauma. For instance, damage to the beginning of the spinal cord near C_1 or C_2 levels can be fatal. This is not because of paralysis or loss of sensation from all the limbs, but because the nerve that controls the diaphragm (the phrenic nerve) no longer can receive the signal from the brain, and breathing ceases. In contrast, damage to the lower thoracic spinal cord may result in loss of function or sensation in the lower limbs, while the upper limbs and diaphragm remain intact.

Injury to a spinal nerve or the peripheral nerves that branch from them are most often the result of a nerve compression or irritation. Some familiar examples are *carpal tunnel syndrome*, which causes pain, weakness, or numbness in the hand, and *sciatica*, a condition in which pain or numbness occurs in the lower back and may radiate to the buttock, posterior thigh, leg, and foot. Understanding these clinical conditions requires an appreciation for the organization of the spinal cord and spinal nerves.

The exercises in this chapter explore the structure and function of the spinal cord and spinal nerves, with a focus on major nerves of the body and the structures they serve.

It may be necessary to look at more than one model to be able to see all of the anatomic structures of the spinal cord and spinal nerves. For instance, seeing all the parts of the brachial plexus may require both observing a model of the head, neck, and thorax and then viewing a model of the upper limb. As far as the spinal cord is concerned, understanding the cross-sectional anatomy of the spinal cord is critical for understanding the links between the central and peripheral nervous systems.

Following a detailed study of the spinal cord and spinal nerves, further exercises explore somatic reflexes. Reflexes are preprogrammed responses that the nervous system uses to detect and respond to sensory input. A classic example is the reflex that causes a person to immediately pull his or her hand away from a hot stove. The pain is only noticeable after pulling the hand away. That is because the movement of the hand away from the stimulus is part of a reflex.

The exercises in this chapter involve not only learning more about the anatomy of these reflexes, but also performing physiological tests to elicit these reflexes in test subjects. Such tests are important in a clinical situation because they allow a clinician to test for signs of dysfunction in the region of the nervous system involved with the reflex. When performing the tests, be sure to relate the components of each reflex to anatomical structures covered in the spinal cord histology and gross anatomy sections of this chapter.

List of Reference Tables

These Pre-Laboratory Worksheet questions may be assigned by instructors through their 🔴 connect course.

1. Which of the following meningeal layers are found in both the brain and the spinal cord? (Check all that apply.)

   _____ a. arachnoid mater   _____ b. dura mater   _____ c. pia mater

2. In the spinal cord, the _____ (anterior/posterior) root contains axons of motor neurons, and the _____ (anterior/posterior) root contains axons of sensory neurons.

3. Myelinated axons are found in the _____ (gray/white) matter of the spinal cord.

4. Which of the following are part of the peripheral nervous system? (Check all that apply.)

   _____ a. spinal nerves

   _____ c. cranial nerves

   _____ d. brain

5. Which of the following sections of the spinal cord does not form a nerve plexus? (Circle one.)

   a. cervical   b. lumbar   c. sacral   d. thoracic

6. Match the description listed in column A with the corresponding part of a spinal cord listed in column B.

Column A	Column B
_____ 1. contains cell bodies of somatic motor neurons	a. anterior horn
_____ 2. contains cell bodies of somatic sensory neurons	b. gray matter
_____ 3. contains cell bodies of visceral motor neurons	c. lateral horn
_____ 4. inner butterfly-shaped portion of spinal cord	d. posterior root ganglion
_____ 5. outer portion of spinal cord	e. white matter

7. Which of the following plexuses is composed of rami, trunks, divisions, and cords? (Circle one.)

   a. brachial   b. cervical   c. lumbar   d. sacral

8. Match the representative nerve listed in column A with the corresponding plexus listed in column B.

Column A	Column B
_____ 1. femoral nerve	a. brachial plexus
_____ 2. median nerve	b. cervical plexus
_____ 3. phrenic nerve	c. lumbar plexus
_____ 4. tibial nerve	d sacral plexus

9. Cell bodies of somatic motor neurons are found in the _____ (anterior/posterior) horn of the spinal cord.

10. Match the description listed in column A with the corresponding component of a reflex arc listed in column B.

Column A	Column B
_____ 1. carries out the response	a. effector
_____ 2. detects the stimulus	b. interneuron
_____ 3. neuron that sends impulses away from CNS	c. motor neuron
_____ 4. neuron that sends impulses to the CNS	d. receptor
_____ 5. neuron within the CNS	e. sensory neuron

GROSS ANATOMY

The Spinal Cord

The spinal cord, which is part of the central nervous system, is located within the vertebral canal and extends from the medulla oblongata of the brainstem to about L1 vertebral level in the adult. The distal end of the spinal cord is cone-shaped and thus is called the **conus medullaris**. Like the brain, the spinal cord is protected by three meninges: the dura mater, arachnoid mater, and pia mater. However, the dura mater surrounding the spinal cord consists of only a single tissue layer, whereas the dura mater surrounding the brain consists of two layers (periosteal layer and meningeal layer). The dura mater and arachnoid mater extend beyond the end of the spinal cord, resulting in a large subarachnoid space filled with cerebrospinal fluid inferior to the spinal cord. In addition, there are extensions of pia mater that help stabilize the spinal cord within the vertebral canal: the denticulate ligaments and the filum terminale. **Table 16.1** lists some of the key features of structures associated with the spinal cord identified upon gross observation and their functions.

Table 16.1	Gross Anatomy of the Spinal Cord	
Structure	**Description**	**Word Origin**
Anterior Rami	Branch off the spinal nerve that innervates skin and muscles of the anterolateral trunk, upper limb, and lower limb	*ramus*, branch
Anterior Root*	A nerve root extending from the anterolateral surface of the spinal cord that contains axons of somatic motor neurons and visceral motor neurons	*anterior*, the front, + *root*, the beginning part
Cauda Equina	A collection of anterior and posterior roots that extend inferiorly from the lumbar and sacral parts of the spinal cord and lie within the vertebral canal. It is named for the fact that the bundle of nerve roots resembles a horse's tail.	*cauda*, tail, + *equinus*, horse
Conus Medullaris	The cone-shaped distal tip of the spinal cord	*konos*, cone, + *medius*, middle
Denticulate Ligaments	Lateral extensions of the pia mater located between anterior and posterior roots. These "ligaments" anchor the spinal cord to the arachnoid and dura mater at intervals.	*denticulus*, a small tooth, + *ligamentum*, a bandage
Filum Terminale	An inferior extension of pia mater beyond the distal end of the spinal cord. It begins at the conus medullaris and attaches to the distal end of the dura mater at the coccyx.	*filum*, thread, + *terminatio*, ending
Posterior Rami	Branch off the spinal nerve that innervates skin of the back and deep muscles of the back	*ramus*, branch
Posterior Root*	A nerve root extending from the posterolateral surface of the spinal cord that contains axons of sensory neurons	*posterior*, the back, + *root*, the beginning part
Posterior Root Ganglion*	A swelling on the posterior root that contains cell bodies of somatic sensory neurons	*posterior*, the back, + *root*, the beginning part, + *ganglion*, a swelling
Rootlets (Radicular Fila)	Small branches of nerve fibers coming off the spinal cord that come together to form the anterior and posterior roots	*radicula*, a spinal nerve root, + *filum*, thread
Spinal Nerves	Nerves that form when anterior and posterior roots converge. A spinal nerve exits the vertebral canal through an intervertebral foramen.	*spinal*, relating to the spinal cord, + *nevus*, a white, cord-like structure

*The anterior root is also known as the ventral root. The posterior root is also known as the dorsal root.

INTEGRATE

CLINICAL VIEW
Lumbar Puncture versus Epidural Anesthesia

A **lumbar puncture** (spinal tap) and an epidural both involve inserting a needle between two vertebrae (L₃–L₄) in the region of the cauda equina of the spinal cord (below where the spinal cord typically ends in an adult at L₁ vertebra). In a lumbar puncture, the needle pierces the dura mater and arachnoid mater within the vertebral canal to enter the subarachnoid space, allowing the withdrawal of cerebrospinal fluid (CSF). In contrast, an **epidural** (a procedure to administer an anesthetic to a woman during childbirth) does not penetrate any meningeal layers within the vertebral canal. Medications are injected into the epidural space, which contains mostly blood vessels and fat. Of note, when a lumbar puncture is performed on a young child or infant, the needle must be placed lower than vertebral level L₃–L₄ because the spinal cord ends at a lower level in these individuals.

GROSS ANATOMY OF THE SPINAL CORD

1. Observe a whole spinal cord from a cadaver (meninges intact) or a model of the spinal cord.

2. Identify the structures listed in **figure 16.1** on a human spinal cord or a model of the spinal cord, using table 16.1 and the textbook as guides. Then label them in figure 16.1.

3. *Optional Activity:* **APR** 7: **Nervous System**—Watch the "Typical Spinal Nerve" animation to help visualize how spinal nerves relate to the spinal cord.

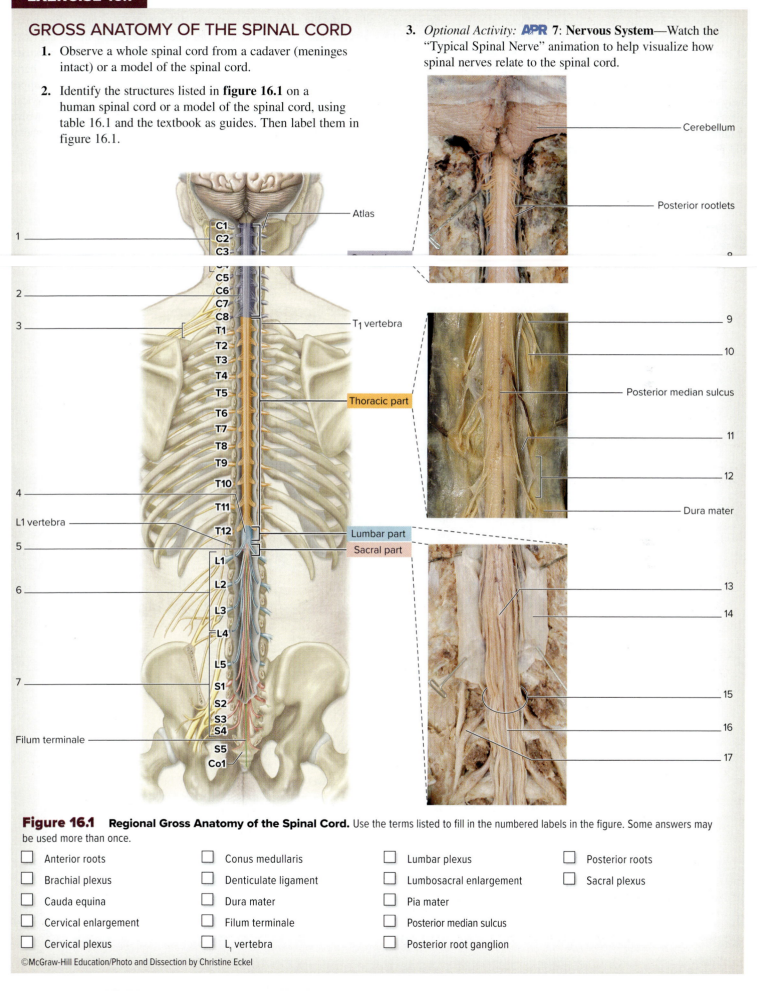

Figure 16.1 **Regional Gross Anatomy of the Spinal Cord.** Use the terms listed to fill in the numbered labels in the figure. Some answers may be used more than once.

- [ ] Anterior roots
- [ ] Brachial plexus
- [ ] Cauda equina
- [ ] Cervical enlargement
- [ ] Cervical plexus
- [ ] Conus medullaris
- [ ] Denticulate ligament
- [ ] Dura mater
- [ ] Filum terminale
- [ ] L₁ vertebra
- [ ] Lumbar plexus
- [ ] Lumbosacral enlargement
- [ ] Pia mater
- [ ] Posterior median sulcus
- [ ] Posterior root ganglion
- [ ] Posterior roots
- [ ] Sacral plexus

Peripheral Nerves

The peripheral nervous system consists of cranial nerves and spinal nerves. The structure and function of the cranial nerves was covered in chapter 15. The next series of laboratory exercises investigate the structure and function of the **spinal nerves**. After a spinal nerve exits the vertebral canal through an intervertebral foramen, it immediately branches into posterior and anterior **rami**. The **posterior rami** innervate both skin and muscles of the back that move the vertebral column (this excludes muscles located on the back that move the pectoral girdle and upper limb). In comparison, the structures innervated by the anterior rami depend upon the location at which these rami extend from the vertebral column. In the thoracic region, the **anterior rami** become **intercostal nerves** that supply skin, bone, and muscle of the thoracic cage. The anterior rami in the cervical, lumbar, and sacral regions form complex networks called **plexuses** (*plexus*, a braid). These plexuses then branch to form peripheral nerves that innervate skin, muscle, and bones of the neck, upper limbs, and lower limbs. There are four major nerve plexuses: cervical, brachial, lumbar, and sacral. These laboratory exercises involve investigating each of these plexuses and exploring the major peripheral nerves that arise from each plexus.

EXERCISE 16.2

THE CERVICAL PLEXUS

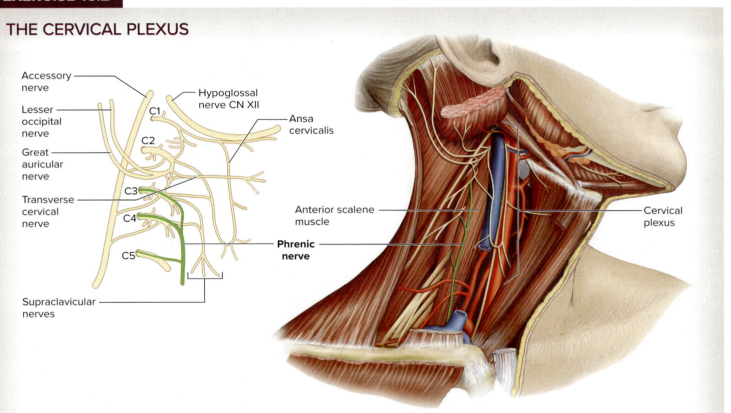

Figure 16.2   The Cervical Plexus and Phrenic Nerve in the Posterior Triangle of the Neck.

1. Observe a prosected human cadaver or a model of the head, neck, and thorax demonstrating nerves of the cervical plexus.

    The cervical plexus arises from the anterior rami of spinal nerves C1–C4. Most of the nerves extending from the cervical plexus transmit nerve signals from skin of the neck and portions of the head and shoulders, and motor information to anterior neck muscles. Perhaps the single most important nerve branching from the cervical plexus is the **phrenic nerve**, which innervates the diaphragm. The phrenic nerves can be seen on the inferior part of the anterior scalene muscle within the neck region, but they are difficult to identify in this location (**figure 16.2**). They are most easily identified within the thoracic cavity (**figure 16.3**). Here they travel within the mediastinum, between the pleural and pericardial cavities. If the heart or lungs have been removed from the cadaver, the phrenic nerves will be easy to identify (if they are still intact). If a model of the thorax is used, look at the location where the pleural and pericardial cavities meet to locate the phrenic nerves.

2. Identify the structures listed in figure 16.3 on a human cadaver or a human torso model, using the textbook as a guide. Then label them in figure 16.3.

INTEGRATE

LEARNING STRATEGY

The phrenic nerve arises from the anterior rami of spinal nerves C3–C5, with the majority of fibers coming from C4. A handy way to remember this is "C3, C4, and C5 keep the body alive." If the phrenic nerves are unable to stimulate the diaphragm, breathing will cease and death will follow.

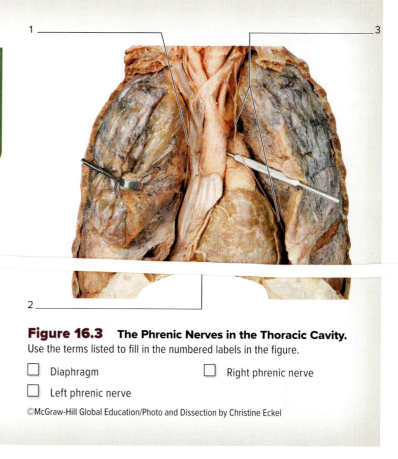

Figure 16.3 **The Phrenic Nerves in the Thoracic Cavity.**
Use the terms listed to fill in the numbered labels in the figure.

☐ Diaphragm ☐ Right phrenic nerve

☐ Left phrenic nerve

©McGraw-Hill Global Education/Photo and Dissection by Christine Eckel

---

EXERCISE 16.3

THE BRACHIAL PLEXUS

1. Observe a prosected human cadaver or a model of the axilla and upper limb demonstrating nerves of the brachial plexus. The organization of the brachial plexus is summarized in **table 16.2**.

   The **brachial plexus** arises from the anterior rami of spinal nerves C5–T1. The overall organization of the brachial plexus follows an organized pattern and will be explored in detail in this laboratory exercise. However, note that the branching pattern of the brachial plexus is unique to the brachial plexus. Other plexuses (cervical, lumbar, and sacral) have their own branching patterns.

   The focus of this exercise is on identification of the **trunks**, **cords**, and **branches** of the brachial plexus. There are three **trunks**, which pass between the anterior and middle scalene muscles of the neck. The **superior trunk** forms from the anterior rami of C5 and C6, the **middle trunk** is a continuation of the anterior ramus of C7, and the **inferior trunk** forms from the anterior rami of C8 and T1. Each trunk divides into two **divisions**, anterior and posterior. All posterior divisions come together to form the **posterior cord**, and the anterior divisions come together to form either the **medial cord** or the **lateral cord**. The cords are named for their location relative to the axillary artery.

INTEGRATE

LEARNING STRATEGY

The segments of the brachial plexus, from proximal to distal, are **R**ami, **T**runks, **D**ivisions, **C**ords, and **B**ranches. A mnemonic for remembering this is, "**R**eally **T**ired, **D**rink **C**offee—**B**lack."

2. First locate the medial and lateral cords around the axillary artery and then spread apart the terminal branches that arise from the cords. Notice that the connections between the cords and the branches appear to form a letter 'M' (**figure 16.4**).

• *The Posterior Cord:* There are two major nerves that arise from the posterior cord: the axillary nerve and radial nerve (**table 16.3**). The **axillary nerve** remains in the axillary region to innervate the deltoid and teres minor muscles. The **radial nerve** is a large nerve that continues to travel posteriorly and innervates muscles and skin in the posterior arm, including the triceps brachii, and posterior forearm. Thus, a nerve from the *posterior* cord innervates all structures in the *posterior* compartments of both the arm and the forearm.

(continued on next page)

(continued from previous page)

Table 16.2	Organization of the Brachial Plexus		
Structure	**Description**	**Number**	**Names**
Rami	Anterior rami of cervical spinal nerves combine to form trunks as they pass between the anterior and medial scalene muscles of the neck.	5	C5, C6, C7, C8, T1
Trunks	Located in the neck and named for their relationship to each other.	3	Superior, middle, inferior
Divisions	Each trunk divides into an anterior and a posterior division behind the clavicle.	6	Anterior and posterior division (for each trunk)
Cords	Located in the axilla and named for their location relative to the axillary artery	3	Medial, lateral, and posterior
Branches (Terminal Nerves)	Each terminal nerve innervates a specific portion of the arm, forearm, and/or hand.	5	Axillary, radial, musculocutaneous, median, ulnar

- *The Medial and Lateral Cords:* The medial and lateral cords form three main terminal nerves: the musculocutaneous, median, and ulnar nerves. The **musculocutaneous nerve** forms from the lateral cord and innervates muscles of the anterior compartment of the arm (biceps brachii, coracobrachialis, and brachialis). It is most easily identified where it pierces through the coracobrachialis muscle. The **ulnar nerve** forms from the medial cord and innervates muscles on the ulnar surface of the forearm along with most intrinsic muscles of the hand. The ulnar nerve is most easily identified where it passes superficially behind the medial epicondyle of the humerus. In this location it is vulnerable to injury; this region of the elbow supplied by the ulnar nerve is commonly referred to as the "funny bone." The **median nerve** forms from

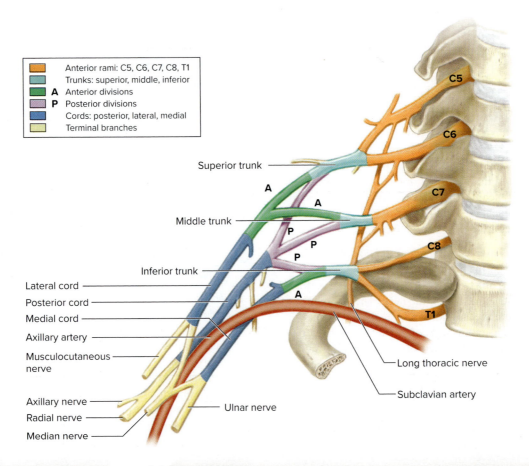

Figure 16.4   Organizational Scheme of the Brachial Plexus.

Table 16.3	Major Nerves of the Brachial Plexus			
Nerve	**Description**	**Motor Innervation**	**Cutaneous Innervation**	**Word Origin**
Posterior Cord				
Axillary (C5–C6)	Arises from the posterior cord and runs around the surgical neck of the humerus en route to the shoulder	Deltoid and teres minor	Skin on the superolateral arm	*axilla*, the armpit
Radial (C5–T1)	Arises from the posterior cord and extends deep within the posterior compartment of the arm adjacent to the humerus	Posterior (extensor) compartments of the arm and forearm	Skin overlying the posterior compartments of the arm and forearm and the lateral dorsum of the hand (except the tips of the fingers and the entire fifth digit)	*radialis*, relating to the radius
Medial and Lateral Cords				
Musculocutaneous (C5–C7)	Arises from the lateral cord and pierces through the coracobrachialis muscle. Courses between the brachialis and biceps brachii muscles.	Anterior compartment of the arm; biceps brachii, brachialis, and coracobrachialis	Skin overlying the lateral surface of the forearm	*muscus*, a mouse (muscle), + *cutaneous*, relating to the skin
Median (C5–T1)	Arises from both the medial and lateral cords. Extends along the anteromedial region of the arm, passes through the cubital fossa, and then enters the forearm. After passing through the forearm, it courses deep to the flexor retinaculum through the carpal tunnel and into the hand.	Anterior compartment of the forearm (except for the flexor carpi ulnaris and ulnar half of the flexor digitorum profundus) and thenar muscles in the hand	Skin on the lateral palm and palmar aspect of the lateral 3.5 digits, including the dorsal tips	*median*, in the middle
Ulnar (C8–T1)	Arises from the medial cord and continues along the medial aspect of the arm. It travels posterior to the medial epicondyle of the humerus. After passing through the forearm, it enters the palm on the anteromedial side.	Anterior forearm muscles (flexor carpi ulnaris and ulnar half of flexor digitorum profundus); most intrinsic hand muscles	Skin of the dorsal and palmar aspects of the medial hand and the medial 1.5 digits	*ulnar*, relating to the ulna

INTEGRATE

CLINICAL VIEW
Additional Nerves of the Brachial Plexus and Spinal Cord Injury

There are several other small nerves that arise from the rami, trunks, and cords of the brachial plexus. These smaller nerves innervate muscles that are responsible for moving the pectoral girdle or glenohumeral joint. These muscles include the rotator cuff muscles, pectoralis major, rhomboids, and latissimus dorsi. One example is the **long thoracic nerve** that arises from the anterior rami of C5–C7 and innervates the **serratus anterior muscle**. Review the discussion of the serratus anterior muscle in chapter 13, which describes how to diagnose a paralyzed or nonfunctional serratus anterior muscle. Through the small and terminal branches, the brachial plexus innervates all the muscles that move the upper limb, with the exception of the trapezius muscle, which is innervated by the spinal accessory nerve (CN XI). Thus, damage to the spinal cord above C5 can result in loss of sensation and function of the upper limb. Damage to the spinal cord below T1 will have no effect on the upper limb's function or sensation.

(continued on next page)

(continued from previous page)

branches of both medial and lateral cords. The median nerve is most easily identified anterior to the elbow as it passes through the **cubital fossa** and in the anterior wrist, where it passes through the **carpal tunnel** into the hand. The carpal tunnel is a narrow passageway that contains the median nerve and tendons of flexor muscles. As the flexor muscles of the forearm contract, their tendons can rub against the median nerve, causing inflammation that results in **carpal tunnel syndrome**. If you are observing the median nerve on a cadaver, try to pass a blunt probe through the carpal tunnel alongside the median nerve.

INTEGRATE

LEARNING STRATEGY

The *cord* of the brachial plexus is *medial*, as in located medial to the axillary artery; the *nerve* is *median*, as in located in the middle (of the branches that form from medial and lateral cords).

3. Identify the structures of the brachial plexus listed in **figure 16.5** on a human cadaver or on models of the upper limb, using tables 16.2 and 16.3 and the textbook as guides.

4. Make a simple line drawing of the brachial plexus in the space provided or on a separate sheet of paper. Then label the structures using the terms listed in figure 16.5.

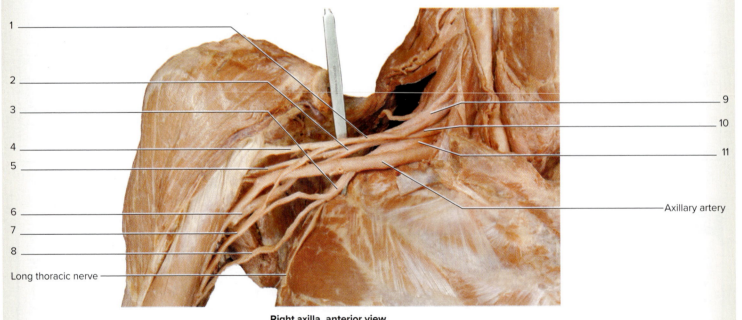

Right axilla, anterior view

Figure 16.5 **Major Nerves of the Brachial Plexus.** Use the terms listed to fill in the numbered labels in the figure.

☐ Axillary nerve ☐ Median nerve ☐ Radial nerve
☐ Inferior trunk ☐ Middle trunk ☐ Superior trunk
☐ Lateral cord ☐ Musculocutaneous nerve ☐ Ulnar nerve
☐ Medial cord ☐ Posterior cord

THE LUMBAR AND SACRAL PLEXUSES

1. Observe a prosected human cadaver or a model of the abdomen and lower limb demonstrating nerves of the lumbar plexus. The lumbar plexus forms from the anterior rami of spinal nerves L1–L4 (**table 16.4**). The sacral plexus forms from the anterior rami of spinal nerves L4–S4 (**table 16.5**). The major nerves arising from the lumbar and sacral plexuses innervate the muscles in the gluteal region and lower limb.

It will be easiest to learn the nerves of the lower limb by relating each nerve to a single compartment. Recall the limb compartments from chapter 13; there are three compartments of the thigh (anterior, posterior, and medial) and three compartments of the leg (anterior, posterior, and lateral). Each compartment, for the most part, receives innervation from a single nerve from either the lumbar or the sacral plexus. First associate one nerve with one compartment. Then, learn the "exceptions" to the rules.

Table 16.4	Major Nerves of the Lumbar Plexus			
Nerve	**Description**	**Motor Innervation**	**Cutaneous Innervation**	**Word Origin**
Femoral (L2–L4)	Runs along the lateral border of the psoas major muscle, travels under the inguinal ligament into the femoral triangle and the anterior compartment of the thigh	Anterior compartment of the thigh	Skin of the anterior thigh and medial leg	*femoral*, relating to the femur or thigh
Obturator (L2–L4)	Runs along the medial border of the psoas major muscle, travels through the obturator foramen into the medial compartment of the thigh	Medial compartment of the thigh	Skin on the medial surface of the thigh	*obturatus*, to occlude or stop up

Table 16.5	Major Nerves of the Sacral Plexus			
Nerve	**Description**	**Motor Innervation**	**Cutaneous Innervation**	**Word Origin**
Common Fibular (L4–S2)	Arises from the sciatic nerve proximal to the popliteal fossa, passes superficially near the head of the fibula, wraps around the neck of the fibula, and then divides into superfcial and deep branches	Short head of biceps femoris	Skin on the proximal posterolateral surface of the leg	*fibular*, relating to the fibula
Deep Fibular (L4–S1)	Arises from the common fibular nerve at the neck of the fibula, passes through the extensor digitorum longus muscle into the anterior compartment of the leg	Anterior compartment of the leg, dorsal musculature of the foot	Skin of the webbing between first and second toe	*deep*, situated at a deeper level than a corresponding structure, + *fibular*, relating to the fibula
Inferior Gluteal (L5–S2)	Exits the pelvis through the greater sciatic foramen inferior to the piriformis	Gluteus maximus	NA	*inferior*, lower, + *gloutos*, the buttock
Posterior Femoral Cutaneous (S1–S3)	Exits pelvis through the greater sciatic foramen inferior to the piriformis, just medial to the sciatic nerve	NA	Skin on the posterior thigh	*posterior*, behind, + *femur*, the thigh, + *cutis*, the skin
Pudendal (S2–S4)	Exits the pelvis through the greater sciatic foramen inferior to the piriformis muscle, then travels through the lesser sciatic foramen to enter the perineum	Perineal muscles, external anal sphincter, and external urethral sphincter	External genitalia in both males and females	*pudendal*, that which is shameful
Sciatic (L4–S3)	Exits the pelvis through the greater sciatic foramen inferior to the piriformis muscle, then enters the posterior compartment of the thigh through a groove between the ischial tuberosity and the greater trochanter of the femur	Posterior compartment of the thigh (except for the short head of the biceps femoris)	NA	*sciaticus*, the hip joint

(continued on next page)

(continued from previous page)

Table 16.5	Major Nerves of the Sacral Plexus *(continued)*			
Nerve	**Description**	**Motor Innervation**	**Cutaneous Innervation**	**Word Origin**
Superficial Fibular (L5–S2)	Arises from the common fibular nerve at the neck of the fibula and descends within the lateral compartment of the leg	Lateral compartment of the leg	Skin on the distal, lateral surface of the leg and the dorsal surface of the foot	*superficialis*, the surface, + *fibular*, relating to the fibula
Superior Gluteal (L4–S1)	Exits the pelvis through the greater sciatic foramen superior to the piriformis	Gluteus medius, gluteus minimus, and tensor fascia lata	NA	*superus*, above, + *gloutos*, the buttock
Tibial (L4–S3)	Begins at the bifurcation of the sciatic nerve proximal to the popliteal fossa, runs along the tibialis posterior muscle, and then branches into two plantar nerves at the ankle	Posterior compartment of the leg, plantar musculature of the foot	Skine of plantar surface of the foot	*tibial*, relating to the tibia

2. *Nerves of the Lumbar Plexus:* The **femoral nerve** travels under the inguinal ligament and through the femoral triangle. The femoral nerve innervates the muscles of the anterior thigh, including the quadricep femoris group. The **obturator nerve** runs through the obturator foramen and innervates skin and muscles of the medial thigh, including the adductor group. The general rules for the lumbar plexus are as follows:

Compartment	Nerve
Anterior thigh	Femoral nerve
Medial thigh	Obturator nerve

3. Identify the nerves of the lumbar plexus and nearby associated muscles listed in **figure 16.6** on a human cadaver or a model of the lower limb, using table 16.4 and the textbook as guides. Then label them in figure 16.6.

4. *Nerves of the Sacral Plexus:* The **gluteal nerves** (superior and inferior) are named for their exit location relative to the piriformis muscle of the deep buttock **(figure 16.7)**. The **superior gluteal nerve** arises superior to the piriformis, while the **inferior gluteal nerve** arises inferior to the piriformis. A small yet important nerve that also arises in this region is the **pudendal nerve** (*pudendal*, that which is shameful). The **sciatic nerve** arises inferior to the piriformis muscle and travels into the posterior compartment of the thigh, passing just lateral to the ischial tuberosity. The sciatic nerve is the largest nerve in the body because it is actually two nerves, the **common fibular nerve** and the **tibial nerve**, bundled together in a common connective tissue sheath. Most commonly the two nerves separate from each other just proximal to the popliteal fossa of the knee. It is the **tibial division** of the sciatic nerve that is responsible for innervating the posterior compartment of the thigh. The general rules for the sacral plexus are as follows:

Compartment	Nerve
Posterior thigh	Sciatic nerve (tibial division)
Anterior leg	Deep fibular nerve
Lateral leg	Superficial fibular nerve
Posterior leg	Tibial nerve

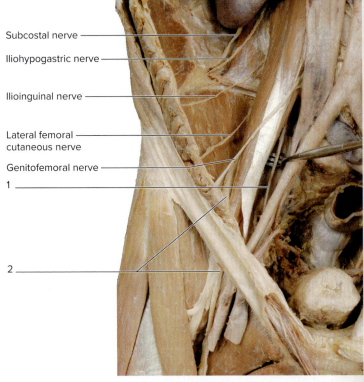

Subcostal nerve

Iliohypogastric nerve

Ilioinguinal nerve

Lateral femoral cutaneous nerve

Genitofemoral nerve

1 _____

2 _____

Right pelvic region, anterior view

Figure 16.6 **Nerves of the Lumbar Plexus.** Deep dissection of the pelvis and anterior thigh. Use the terms to fill in the numbered labels in the figure.

☐ Femoral nerve   ☐ Obturator nerve

©McGraw-Hill Education/Photo and Dissection by Christine Eckel

5. Identify the nerves of the sacral plexus and nearby associated muscles listed in figure 16.7 on a human cadaver or a model of the lower limb, using table 16.5 and the textbook as guides. Then label them in figure 16.7.

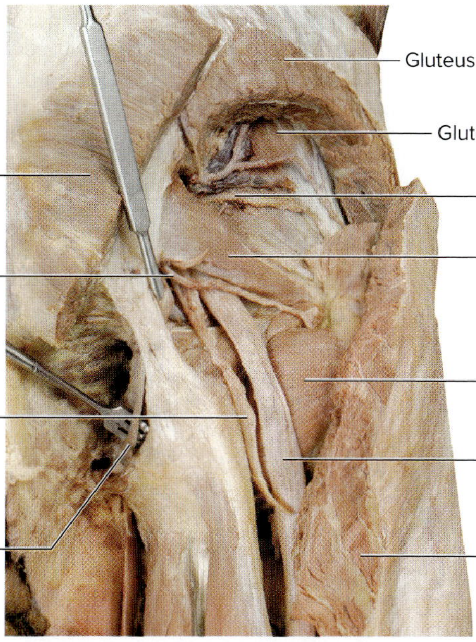

Gluteus medius muscle (cut)

Gluteus minimus muscle

Gluteus maximus muscle (cut)

4

Piriformis muscle

1

Quadratus femoris muscle

2

5

3

Gluteus maximus muscle (cut)

(a) Right lower limb, posterior view (gluteal region)

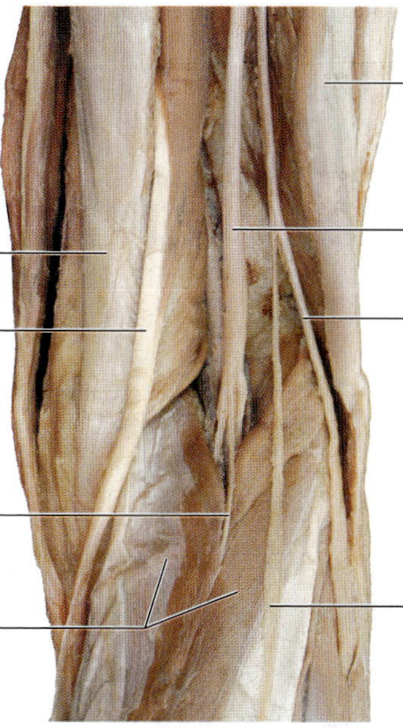

Biceps femoris muscle

Semimembranosus muscle

6

Semitendinosus muscle

7

Medial

Lateral

Medial sural cutaneous nerve

Gastrocnemius muscle

Lateral sural cutaneous nerve

(b) Right lower limb, posterior view (popliteal region)

Figure 16.7 **Nerves of the Sacral Plexus Within the Gluteal and Popliteal Regions.** Use the terms listed to fill in the numbered labels in the figure.

☐ Common fibular nerve   ☐ Pudendal nerve   ☐ Tibial nerve

☐ Inferior gluteal nerve   ☐ Sciatic nerve

☐ Posterior femoral cutaneous nerve   ☐ Superior gluteal nerve

©McGraw-Hill Education/Photo and Dissection by Christine Eckel

INTEGRATE

LEARNING STRATEGY

The word *pudendal* comes from Latin, meaning "that which is shame." This is an archaic reference to the external genitalia, which were to be hidden from view by clothing or a fig leaf. The term is now used for structures associated with the external genitalia. The pudendal nerve innervates the skin of the external genitalia and external sphincters surrounding the anus and urethra. A very minimal anesthetic given during childbirth is a **pudendal nerve block**. It blocks sensation from the birth canal (vagina), but the mother still receives sensations from the contracting uterus.

HISTOLOGY

Spinal Cord Organization

The spinal cord is organized with an outer cortex of white matter surrounding an inner core of gray matter. The gray matter of the spinal cord is organized into horns, so named because of their appearance. Sensory neurons extend from a receptor to the posterior horn. Posterior horns contain the dendrites and cell bodies of interneurons (association neurons). **Anterior horns** contain dendrites and cell bodies of somatic motor neurons, whose axons exit the spinal cord through the anterior roots and extend to skeletal muscle. **Lateral horns**, located only in the thoracic region and first two lumbar segments of the spinal cord, contain the dendrites and cell bodies of visceral motor neurons, with axons that exit the spinal cord through the anterior roots to ultimately innervate visceral effectors (smooth muscle, cardiac muscle, and glands). The **gray commissure** is a horizontal bar of gray matter that surrounds the narrow central canal.

The white matter on each side of the spinal cord is partitioned into three **funiculi** (*funis,* cord), which are identified based on their anatomic position: posterior funiculus, lateral funiculus, and anterior funiculus. Each funiculus is composed of bundles of myelinated axons, called fasciculi, that transmit nerve signals between the brain and the body. The anterior funiculi are interconnected by the **white commissure**.

The spinal cord varies in diameter in different regions. It is the largest in the cervical region (called the **cervical enlargement**) and lumbar region (called the **lumbar enlargement**). These enlargements exist because of the increased number of neuron cell bodies in these regions that make the additional connections to the upper and lower limbs, respectively. Make note of similarities and differences in organization between gray matter and white matter in the spinal cord as compared to the brain.

Table 16.6	Histology of the Spinal Cord in Cross Section	
Structure	**Description**	**Word Origin**
White Matter	Outer portion of the spinal cord consisting of bundles of myelinated axons organized into tracts	literally, a substance that appears white in color
Anterior Funiculus	White matter that occupy the anterior region of the spinal cord between the anterior gray horn and the anterior median fissure	*anterior,* the front, + *funis,* cord
Lateral Funiculus	White matter that occupy lateral sides of the spinal cord	*latus,* to the side, + *funis,* cord
Posterior Funiculus	White matter that occupy the posterior region of the spinal cord between the posterior gray horn and the posterior median sulcus	*posterior,* the back, + *funis,* cord
White Commissure	White matter that interconnects the anterior funiculi	*commissura,* a seam
Gray Matter	Butterfly-shaped inner region of the spinal cord. It contains cell bodies of motor neurons or interneurons, depending upon the location.	literally, a substance that appears gray in color
Anterior Horns	Gray matter on the anterolateral part of the spinal cord that primarily houses the cell bodies of somatic motor neurons (neurons that extend to skeletal muscle)	*anterior,* the front, + *horn,* resembling a horn in shape
Lateral Horns	Small lateral extensions of gray matter found in the T_1–L_2 region of the spinal cord, and containing cell bodies of visceral motor neurons (neurons that extend to smooth muscle, cardiac muscle, and glands)	*latus,* to the side, + *horn,* resembling a horn in shape
Posterior Horns	Gray matter on the posterolateral part of the spinal cord that primarily houses the axons of sensory neurons and the cell bodies of interneurons (neurons contained completely within the central nervous system)	*posterior,* the back, + *horn,* resembling a horn in shape
Gray Commissure	A horizontal bar of gray matter that surrounds the central canal	*commissura,* a seam
Other		
Anterior Median Fissure	Deep groove on the anterior surface of the spinal cord	*anterior,* the front, + *median,* the middle, + *fissure,* a deep furrow
Central Canal	Small channel within the gray commissure of the spinal cord that is continuous with the ventricles of the brain; appears as a hole in a microscopic slide of the spinal cord	*central,* in the center, + *canalis,* a duct or channel
Posterior Median Sulcus	Shallow groove located on the posterior surface of the spinal cord	*posterior,* the back, + *median,* the middle, + *sulcus,* a furrow

HISTOLOGICAL CROSS SECTIONS OF THE SPINAL CORD

1. Observe a model of a cross section of the spinal cord.

2. Identify the structures listed in **figure 16.8** on the spinal cord model, using tables 16.1 and **16.6** and the textbook as guides. Then label them in figure 16.8.

3. Obtain slides containing cross sections of the spinal cord.

4. First observe each slide with the naked eye and try to identify the different parts of the spinal cord (cervical, thoracic, lumbar, and sacral) based on cross-sectional area and relative amounts of gray versus white matter.

5. Place the slide on the microscope stage and bring the tissue sample into focus on low power.

6. Identify the structures listed in table 16.6 on the histology slides of the spinal cord using **figures 16.8** and **16.9** as guides.

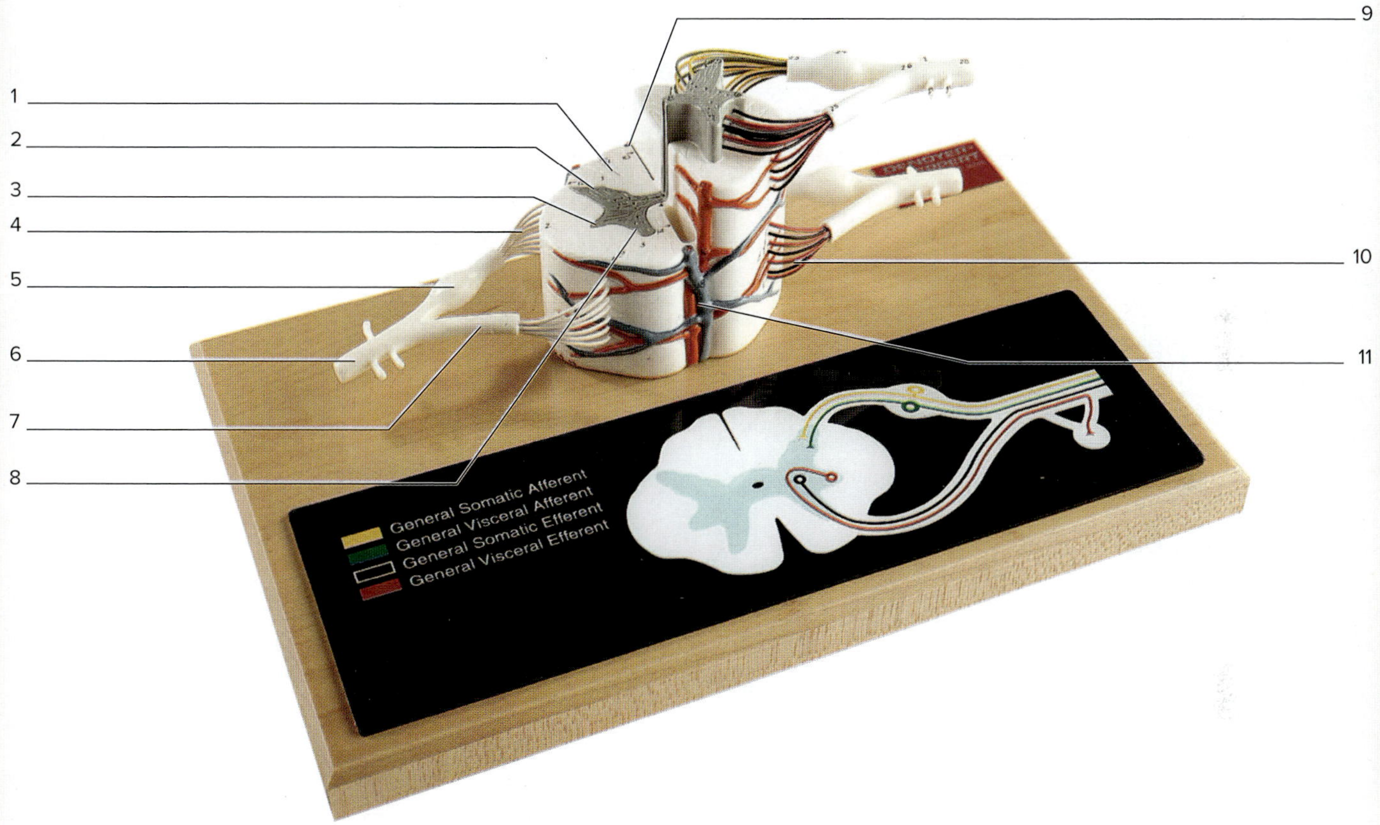

Figure 16.8 **Model of a Spinal Cord in Cross Section.** Use the terms listed to fill in the numbered labels in the figure.

☐ Anterior horn
☐ Anterior rootlets
☐ Posterior horn
☐ Posterior rootlets
☐ Anterior median fissure
☐ Lateral horn
☐ Posterior median sulcus
☐ Spinal nerve
☐ Anterior root
☐ Posterior funiculus
☐ Posterior root ganglion

Copyright by Denoyer-Geppert

(continued on next page)

(continued from previous page)

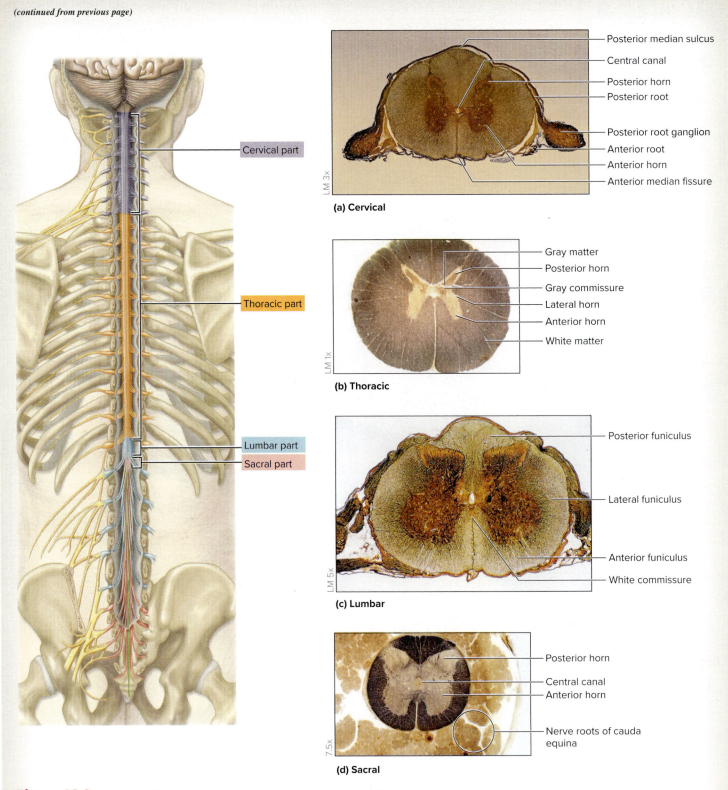

Figure 16.9 **Regional Histology of the Spinal Cord in Cross Section.**

(*a, c*) ©Ed Reschke; (*b*) ©Getty Images USA, Inc.; (*d*) ©Al Telser/McGraw-Hill Education

7. Observe the cross-sections of different regions of the spinal cord in figure 16.9. Look at the differences in the amount of white matter versus gray matter in the different sections and the differences in shapes between the different regions. Then complete the following table with the regional characteristics of the spinal cord.

Region of the Spinal Cord	Relative Size and Shape of the Spinal Cord	Predominantly White or Gray Matter?	Other Distinguishing Characteristics
Cervical			
Thoracic			
Lumbar			
Sacral			

PHYSIOLOGY

Reflexes

Reflexes involve a rapid response to a stimulus that is automatic and involuntary. All reflexes involve the same key components: receptor, sensory neuron, control center, motor neuron, and effector. In somatic reflexes, the sensory receptors can be cutaneous receptors within the skin (e.g., lamellated corpuscles, Meissner corpuscles, or free nerve endings) or proprioreceptors found within muscles (e.g., Golgi tendon organs and muscle spindles). The motor output is always to skeletal muscle (remember: somatic is associated with *voluntary* movement). This is precisely the type of reflex that allows the rapid withdrawal of a hand from a painful stimulus or the contraction of a muscle to maintain balance. The reflex is so quick that you might not be aware it is occurring until after it is completed. The following exercise explores the components of a reflex. Exercises within the physiology section of this chapter allow students to elicit various somatic reflexes using standard reflex tests. Abnormal reflex responses may indicate certain diseases (e.g., syphilis, late-stage alcoholism, diabetic neuropathy), damage to peripheral nerves, or damage to the spinal cord.

INTEGRATE

CONCEPT CONNECTION

Spinal nerves, like cranial nerves, are part of the peripheral nervous system. All sensory information coming into the central nervous system "travels" within peripheral nerves. Although processing by the brain is necessary to provide this conscious awareness of sensory input, the brain is not required for "processing" *all* sensory and motor information. For example, spinal reflexes such as the patellar reflex are neural circuits that require only the functioning of spinal nerves and the spinal cord. There are also reflexes that involve the cranial nerves and the brainstem, such as the pupillary light reflex and the gag reflex.

WITHDRAWAL AND CROSSED-EXTENSOR REFLEX

The withdrawal reflex is difficult to reproduce because it involves withdrawing a limb away from a noxious or painful stimulus, such as a hot, cold, or sharp object. When nociceptors (*noci-*, pain) are stimulated, the subject withdraws quickly and unconsciously from the painful stimulus, exciting flexor muscles ipsilaterally (on the same side of the body). Simultaneously, extensor muscles are excited on the contralateral side (the opposite side of the body). To elicit a true withdrawal reflex, the subject must be unaware that the stimulus is coming, which makes it particularly difficult to reproduce in a laboratory setting.

Rather than eliciting a withdrawal and crossed-extensor reflex for this laboratory exercise, recall one of the following scenarios: touching a hot surface or stepping on a sharp object. Answer the following questions:

1. What happened to the limb (upper or lower) that experienced the painful stimulus?

   _____

   _____

2. What happened to the opposite limb?

   _____

   _____

3. Which happened first, withdrawal of the limb away from the stimulus or perception of the stimulus (circle one)? Based on an understanding of the anatomy of spinal reflexes, justify your answer.

   _____

   _____

   _____

PLANTAR REFLEX

1. Obtain a rubber percussion hammer and choose a student to be the subject.

2. Have the subject remove his or her shoe and sock from one foot and lie supine on the lab table.

3. Move the metal end of the rubber hammer firmly over the lateral aspect of the sole of the foot from the heel to the base of the great toe (**figure 16.12**) and note observations:

   _____

   _____

   _____

4. This exercise tests the function of cutaneous receptors. Stimulation of cutaneous receptors on the plantar surface of the foot evokes a spinal reflex that, in turn, stimulates the flexor muscles of the toes (e.g., flexor digitorum longus). What somatic nerve innervates this group of muscles (Hint: it extends down the posterior leg)?

   _____

   _____

   From what somatic nerve plexus does this nerve extend?

   _____

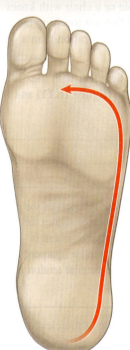

Figure 16.12   Testing the Plantar Reflex. Have the subject lie down on the lab table. Use the metal end of the percussion hammer so that you can press firmly on the foot. Press the bottom of the foot, following the arc denoted by the arrow in this figure. Note the reaction of the subject.

What spinal nerves contribute to this plexus? _____

_____

CLINICAL VIEW
Babinski Reflex

In infants, the normal response to the plantar reflex test (figure 16.12) is opposite the normal response of adults because the axons within this nerve are not yet completely myelinated. That is, the infant response to the plantar reflex test involves the toes spreading apart (abduction) and the foot dorsiflexing.

This reflex is called the *Babinski reflex* (toe abduction and foot dorsiflexion denotes a "positive" Babinski sign). If an adult responds to the plantar reflex test with a positive Babinski sign, it indicates an abnormal response. Specifically, this abnormality suggests damage to the corticospinal pathway (or pyramidal neurons) within the central nervous system (CNS), because this motor pathway is responsible for modulating the spinal reflex. The cause of such damage must be explored further by the physician.

The ➊ corresponds to the Learning Objective(s) listed in the chapter opener outline.

Create and Evaluate

Analyze and Apply

Understand and Remember

▲ Do You Know the Basics?

Exercise 16.1: Gross Anatomy of the Spinal Cord

1. To anchor the spinal cord, the pia mater extends laterally as the _____ (denticulate ligaments/filum terminale) and inferiorly as the _____ (denticulate ligaments/filum terminale). ➊

2. Which of the following is a structure that consists of posterior and anterior roots extending from the inferior end of the spinal cord into the lower vertebral canal and the sacral canal? (Circle one.) ➋

   a. cauda equina

   b. conus medullaris

   c. filum terminale

   d. posterior root ganglion

3. The _____ (cervical/lumbar) enlargement of the spinal cord is associated with the upper limbs, whereas the _____ (cervical/lumbar) enlargement is associated with the lower limbs. ➋

Exercise 16.2: The Cervical Plexus

4. The cervical plexus is formed from the anterior rami of these spinal nerves. (Circle one.) ➌

   a. C1–C3

   b. C1–C4

   c. C1–C5

   d. C1–C6

   e. C1–C7

Exercise 16.3: The Brachial Plexus

5. Rank the levels of organization of the brachial plexus in order, beginning with anterior rami. ➍

   _____ a. branches

   _____ b. cords

   _____ c. divisions

   _____ d. roots (anterior rami)

   _____ e. trunks

6. Match the region or compartment of the upper limb listed in column A with the nerve that innervates the majority of muscles in that region/compartment listed in column B. (Answer choices may be used more than once.) ➏

   Column A

   _____ 1. anterior arm (e.g., biceps brachii)

   _____ 2. anterior forearm (medial aspect and intrinsic muscles of the hand)

   _____ 3. most of the anterior forearm (lateral aspect)

   _____ 4. posterior arm (e.g., triceps brachii muscle)

   _____ 5. posterior forearm

   _____ 6. shoulder (e.g., deltoid muscle)

   Column B

   a. axillary n.

   b. median n.

   c. musculocutaneous n.

   d. radial n.

   e. ulnar n.

7. Using colored pencils, color in the segments of the brachial plexus in the illustration as per the colors in the key. Then label the nerves indicated by the leader lines. **5**

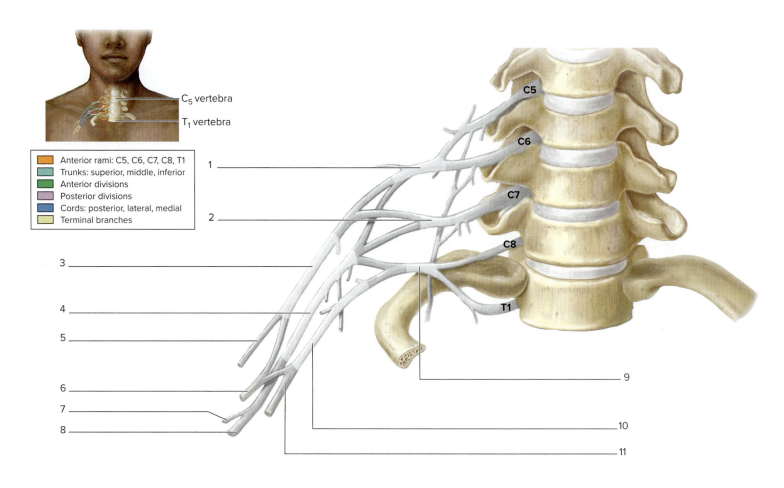

Exercise 16.4: The Lumbar and Sacral Plexuses

8. Identify a nerve that arises from the lumbar plexus. _____ **7**

9. The femoral nerve innervates the _____ (anterior/medial/posterior) thigh, whereas the obturator nerve innervates the _____ (anterior/medial/posterior) thigh. **8**

10. Identify a nerve that arises from the sacral plexus. _____ **9**

11. Match the region or compartment of the lower limb listed in column A with the nerve that innervates the majority of muscles in that region/compartment listed in column B. **10**

Column A

_____ 1. anterior leg

_____ 2. lateral leg

_____ 3. posterior leg

_____ 4. posterior thigh

Column B

a. deep fibular n.

b. sciatic n.

c. superficial fibular n.

d. tibial n.

Exercise 16.5: Histological Cross Sections of the Spinal Cord

12. The outer region of the spinal cord consists of _____ (gray/white) matter, whereas the inner region of the spinal cord consists of

_____ (gray/white) matter. **11**

13. Label the parts of a cross section of a spinal cord indicated by the leader lines. **12**

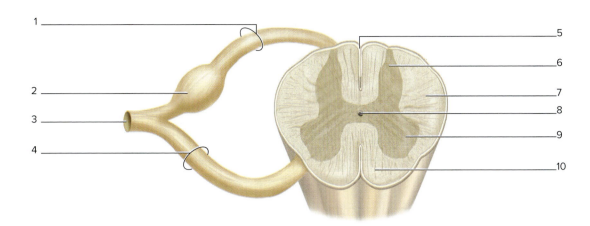

1 _____

2 _____

3 _____

4 _____

5

6

7

8

9

10

Exercise 16.6: Identifying Components of a Reflex

14. Rank the components of a reflex arc in the correct order from stimulus to response. **13**

_____ a. control center

_____ b. effector organ

_____ c. motor neuron

_____ d. receptor

_____ e. sensory neuron

Exercise 16.7: Patellar Reflex

15. The effect of the patellar reflex is the _____ (stretch/contraction) of the quadriceps femoris muscle group. **14**

16. The patellar reflex tests spinal nerves in which of the following regions of the spinal cord? (Circle one.) **15**

a. cervical

b. lumbar

c. sacral

d. thoracic

Exercise 16.8: Withdrawal and Crossed-Extensor Reflex

17. Identify the type of sensory receptor that is stimulated when eliciting the withdrawal and crossed-extensor reflex. (Circle one.) **16**

a. chemoreceptors

b. mechanoreceptors

c. nociceptors

d. photoreceptors

18. When eliciting the withdrawal and crossed-extensor reflexes, flexion withdrawal occurs in the _____ (contralateral/ipsilateral) limb, whereas

extension occurs in the _____ (contralateral/ipsilateral) limb. **17**

Exercise 16.9: Plantar Reflex

19. A normal adult response in the plantar reflex is _____ (flexion/extension) of the toes. **18**

20. The plantar reflex tests spinal nerves in which of the following regions of the spinal cord? (Check all that apply.) **19**

_____ a. cervical

_____ b. lumbar

_____ c. sacral

_____ d. thoracic

Can You Apply What You've Learned?

21. Fill in the following table with features of the main nerve plexuses.

Plexus	Formed from Anterior Rami of These Spinal Nerves	Major Nerves Formed from the Plexus
Cervical		
Brachial		
Lumbar		
Sacral		

22. Fill in the following table with the nerve that innervates each compartment, what plexus the nerve originates from, and one possible action that would be weak or absent if the nerve were damaged.

Compartment	Nerve	Plexus	Potential Motor Deficit
Anterior Compartment of the Arm			
Lateral Compartment of the Leg			
Posterior Compartment of the Thigh			

Can You Synthesize What You've Learned?

23. If a posterior *root* is severed, what loss of function will result?

_____

24. Where in the spinal cord could an injury occur that would result in loss of sensation and function of the hands and lower limbs but preserve the ability to breathe and move the shoulders?

_____

_____

25. Susan, a 50-year-old female, is experiencing motor and sensory deficits in her limbs. Specifically, she exhibits hyperactive stretch reflexes and bilateral positive Babinski signs. Explain each of these results. Include in your explanation a description of how each can be tested in a clinical setting. Finally, discuss possible locations (i.e., spinal cord levels) of the injury.

_____

_____

_____

_____

_____

_____

_____

The Autonomic Nervous System

OUTLINE AND LEARNING OBJECTIVES

Anatomy & Physiology Revealed 4.0

Module 7: Nervous System

INTRODUCTION

Why is it that heart rate increases and sweat glands increase secretions during stressful or frightening conditions? How is it that we begin salivating at the sight, smell, or even thought of something tasty? The answer is the **autonomic nervous system (ANS)**. The human nervous system is divided into two major functional components: the somatic **nervous system** and the autonomic nervous system. The **somatic nervous system (SNS)** can be thought of as the "voluntary" division—it involves sensations that are usually conscious. This includes somatic sensory input from the five special senses (i.e., touch, taste, hearing, vision, and smell) and proprioceptors within joints, tendons, and muscles. Motor output to skeletal (voluntary) muscle tissue is also considered part of the somatic nervous system. In contrast, the ANS can be thought of as the "involuntary" division—it involves sensory stimuli that we are usually unaware of (e.g., blood pressure or oxygen levels in the blood) and motor output to cardiac muscle, smooth muscle, and glands. The ANS is composed of two divisions: the parasympathetic division, which controls "rest-and-digest" activities, and the sympathetic division, which controls the "fight-or-flight" response.

The ANS and SNS differ structurally and functionally. You previously learned that somatic motor neurons innervate skeletal muscle, and that skeletal muscles are controlled consciously. In the SNS, there is only a single somatic motor neuron extending from the central nervous system (CNS). In addition, excitation of the somatic motor neuron results in an excitatory response of skeletal muscle.

Cardiac muscles, smooth muscles, and glands are controlled unconsciously. Autonomic motor neurons are in a two-neuron ANS chain, whereby two neurons form a chemical synapse in an autonomic ganglion **(figure 17.1)**. Neurotransmitters are released by the preganglionic neurons, which modulate the excitability of the ganglionic neuron. The postganglionic axon of the ganglionic neuron innervates the effector organ (i.e., cardiac muscle, smooth muscle, or gland). Autonomic ganglia serve as integration centers regulating output to the effector. The response of the effector may be excitatory or inhibitory. While the overall structure of two neurons in this cascade is conserved in both divisions of the ANS, there are structural and functional differences between the parasympathic and sympathetic divisions, which is the focus of this chapter.

This laboratory session begins by reviewing the histological and gross anatomical organization of the parasympathetic and sympathetic divisions of the ANS. Exercises that are designed to test the function of specific autonomic reflexes will follow. Use knowledge from the previous chapters on the nervous system when completing these relatively condensed but very important laboratory exercises.

Reference Table

Table 17.1: Comparison of Parasympathetic
and Sympathetic Divisions   p. 441

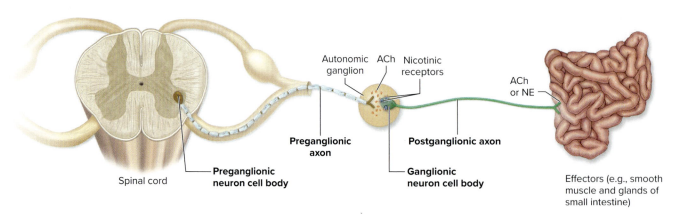

Figure 17.1 **Lower Motor Neurons of the Autonomic Nervous System.**

INTEGRATE

LEARNING STRATEGY

When learning the two divisions of the ANS, remember "rest-and-digest" for the parasympathetic division and "fight-or-flight" for the sympathetic division. Based on these descriptions, many assume that the parasympathetic division is always inhibitory and that the sympathetic division is always excitatory. However, that is not the case. For example, while the parasympathetic division is inhibitory to pacemaker cells of the heart, causing heart rate to decrease, it is stimulatory to smooth muscle in the digestive tract, promoting digestion and motility of the GI tract. Likewise, while the sympathetic division is stimulatory to pacemaker cells of the heart, causing heart rate to increase, it is inhibitory to smooth muscle surrounding the bronchioles (small airways), which causes dilation of the airways. Rather than simply associating "inhibition" or "excitation" with each division, think about the main effects you would expect of each division to predict the effector response, and remember that the actions of autonomic nerves can be inhibitory or excitatory for either division of the ANS.

These Pre-Laboratory Worksheet questions may be assigned by instructors through their ▦ connect course.

1. The _____ (parasympathetic/sympathetic) division of the autonomic nervous system (ANS) controls "rest-and-digest" activities, whereas the _____ (parasympathetic/sympathetic) division of the ANS controls the "fight-or-flight" response.

2. The _____ (preganglionic/postganglionic) neuron innervates the effector organ.

3. The division of the ANS also known as the *craniosacral* division is the _____ (parasympathetic/sympathetic) division. The division of the ANS also known as the *thoracolumbar* division is the _____ (parasympathetic/sympathetic) division.

4. Which of the following statements describes an action performed by the parasympathetic division of the ANS? (Check all that apply.)

   _____ a. increased gastrointestinal (GI) motility

   _____ b. increased heart rate

   _____ c. increased salivation

   _____ d. pupillary constriction

5. Match the feature listed in column A with the corresponding division listed in column B. (Answers will be used more than once.)

   Column A

   _____ 1. effector response is excitatory only

   _____ 2. effector response may be excitatory or inhibitory

   _____ 3. involuntary control of effector

   _____ 4. two-neuron chain

   _____ 5. voluntary control of effector

   Column B

   a. autonomic

   b. somatic

6. Activation of the sympathetic division of the ANS will cause the pupils of the eye to _____ (constrict/dilate).

7. Identify the structure(s) that is/are innervated by autonomic motor neurons. (Check all that apply.)

   _____ a. cardiac muscle

   _____ b. glands

   _____ c. skeletal muscle

   _____ d. smooth muscle

GROSS ANATOMY

Autonomic Nervous System

The autonomic nervous system (ANS) regulates cardiac muscle, smooth muscle, and glands. Because of the involuntary nature of the ANS, most visceral effectors are dually innervated by the ANS. This means that effector organs (e.g., heart, gastrointestinal tract) receive innervation from both divisions of the ANS. These two divisions are the **parasympathetic** (craniosacral) **division**, which functions to control "rest-and-digest" functions, and the **sympathetic** (thoracolumbar) **division**, which is activated in emergency situations and during the "fight-or-flight" response **(figure 17.2)**.

The term "parasympathetic" refers to the fact that the nerve fibers of this division originate from the CNS adjacent to where the nerve fibers of the sympathetic division originate (*para*, alongside). Recall from chapter 16 that cell bodies of autonomic (sympathetic) motor neurons are located in the lateral horns of the thoracic and lumbar regions of the spinal cord. The cell bodies of the parasympathetic division arise from portions of the CNS that are superior and inferior

to the regions that give rise to the sympathetic division. Specifically, cell bodies of neurons that compose the parasympathetic division arise from the brain (with axons that travel within cranial nerves III, VII, IX, and X) and also from lateral horns within the sacral region of the spinal cord. This is the origin of the alternate terms *craniosacral* for the parasympathetic division and *thoracolumbar* for the sympathetic division. In addition to the difference in the locations of preganglionic neuron cell bodies, there is a difference in the location of the ganglia for each of the divisions. Parasympathetic ganglia are located either close to the effector (these are called **terminal ganglia**), or within the wall of the target organs (these are called **intramural ganglia**) **(table 17.1)**. Sympathetic ganglia are located in either the **sympathetic trunk ganglia** (also called paravertebral ganglia) or the **prevertebral ganglia**. The following exercise examines the gross anatomy of each division of the ANS, and involves noting locations of preganglionic and postganglionic neurons and tracing the pathway of motor output to target organs of the ANS throughout the body.

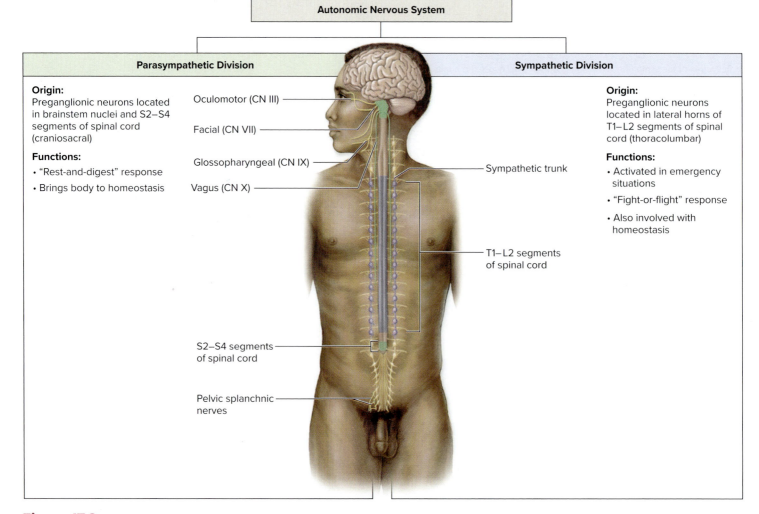

Figure 17.2 Comparison of the Parasympathetic and Sympathetic Divisions of the ANS.

Table 17.1	Comparison of Parasympathetic and Sympathetic Divisions	
Feature	Parasympathetic "Rest and Digest"	Sympathetic "Fight or Flight"
Structure		
Length of Axons	Long preganglionic Short postganglionic	Short preganglionic Long postganglionic
Divergence of Axons	Few (1 axon innervates < 4 ganglionic cell bodies)	Extensive (1 axon innervates > 20 ganglionic cell bodies)
Location of Ganglia	Close to an effector (terminal ganglia); within the effector (intramural ganglia)	Close to the spinal cord (sympathetic trunk ganglia); on either side of the spinal cord or anterior to the spinal cord (prevertebral ganglia)
Function	Conserving energy and replenishing nutrient stores	Maintaining homeostasis during exercise or times of stress or emergency

EXERCISE 17.1

PARASYMPATHETIC DIVISION

The pathways for the parasympathetic division are simpler than those for the sympathetic. There are more targeted responses for "rest-and-digest." Preganglionic neurons originate from either the brainstem or the pelvic splanchnic nerves, are smaller in diameter, and have slower nerve signal propagation than the sympathetic division. This exercise involves tracing these pathways and considering the response of each organ that is innervated by each pathway.

Complete **table 17.2** and identify the structures labeled in **figure 17.3**, using the textbook as a guide. The figure has been grayed out, and structures to identify appear in color.

Table 17.2	Parasympathetic Division Outflow			
Nerve(s)	Origin of Preganglionic Neurons	Autonomic Ganglia	Effectors Innervated	Example Effector Response
Oculomotor (CN III)	Midbrain	Ciliary		
Facial (CN VII)	Pons	Pterygopalatine; submandibular		
Glossopharyngeal (CN IX)	Medulla oblongata	Otic		
Vagus (CN X)	Medulla oblongata	Terminal and intramural		
Pelvic Splanchnic Nerves	S2 – S4 (spinal cord)	Terminal and intramural		

(continued on next page)

(continued from previous page)

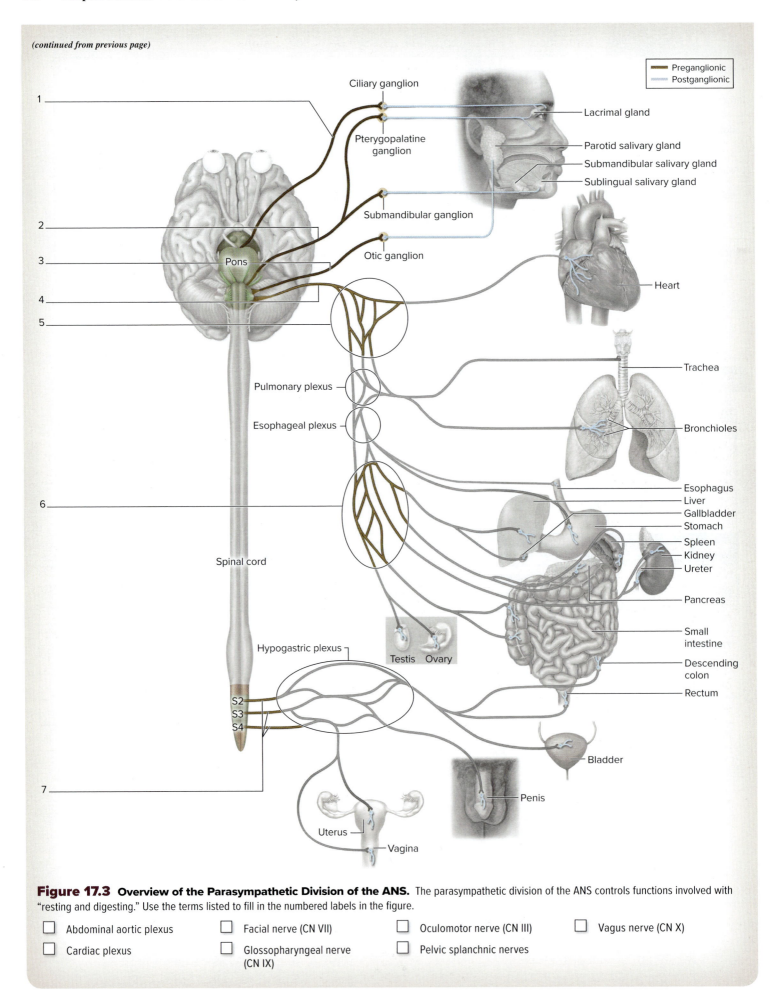

Figure 17.3 Overview of the Parasympathetic Division of the ANS. The parasympathetic division of the ANS controls functions involved with "resting and digesting." Use the terms listed to fill in the numbered labels in the figure.

☐ Abdominal aortic plexus

☐ Cardiac plexus

☐ Facial nerve (CN VII)

☐ Glossopharyngeal nerve (CN IX)

☐ Oculomotor nerve (CN III)

☐ Pelvic splanchnic nerves

☐ Vagus nerve (CN X)

LEARNING STRATEGY

The adrenal gland receives input from preganglionic neurons of the sympathetic division of the ANS, which innervate chromaffin cells in the medulla. These cells are responsible for synthesizing and releasing epinephrine and norephinephrine into the systemic circulation. Once released, epinephrine and norepinephrine bind to adrenergic receptors on effectors to induce sympathetic effector responses.

When attempting to commit to memory the role that the adrenal gland plays in the sympathetic division of the ANS, think of the gland as the "C and C Hormone Factory."

"C" stands for chromaffin cells and catecholamines (epinephrine and norepinephrine)

EXERCISE 17.2

SYMPATHETIC DIVISION

The pathways for sympathetic fibers are more complex than those for parasympathetic fibers. One reason for this is that in order for the sympathetic response to have widespread and immediate effects, it must activate organs throughout the entire body in unison to initiate the "fight-or-flight" response. There are three separate pathways for preganglionic fibers to reach these ganglia from the spinal cord, with additional nerve fibers that extend to the **adrenal medulla**. This exercise involves tracing these pathways and considering the response of each organ that is innervated by each pathway.

Complete **table 17.3** and identify the structures labeled in **figure 17.4**, using the textbook as a guide. The figure has been grayed out, and structures to identify appear in color.

Table 17.3	Sympathetic Division Pathways			
Pathway	**Origin (Spinal Segment)**	**Destination**	**Effectors Innervated**	**Effector Response**
Spinal Nerve	T1 – L2	Integumentary structures		
Postganglionic Sympathetic Nerve	T1 – T5	Head and neck viscera; thoracic organs		
Splanchnic Nerve	T5 – L2	Abdominal and pelvic organs		
Adrenal Medulla	T8 – T12	Adrenal gland		

(continued on next page)

(continued from previous page)

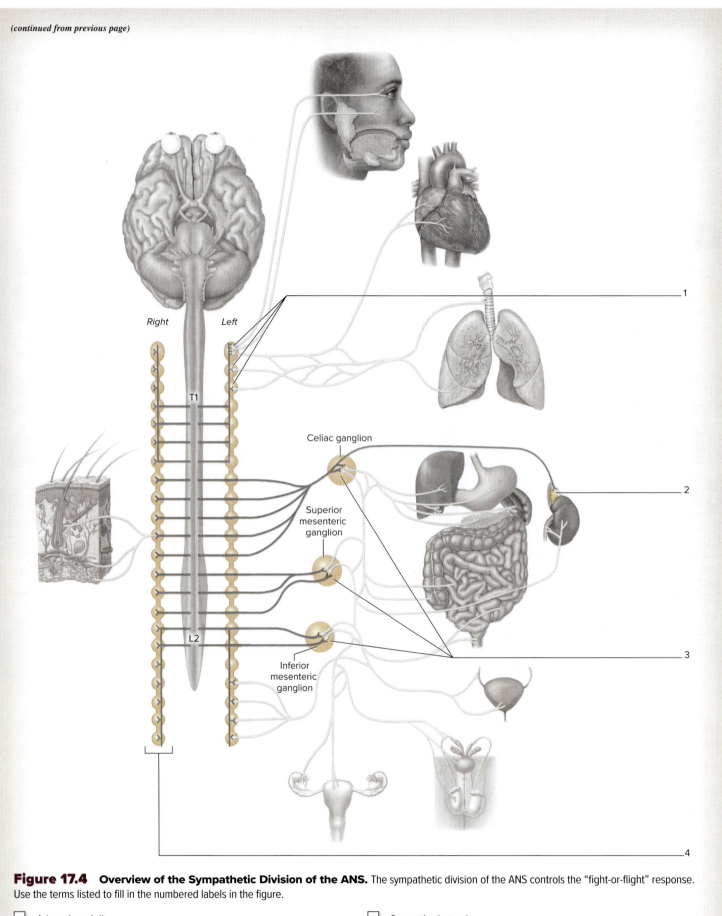

Right

Left

T1

Celiac ganglion

Superior
mesenteric
ganglion

L2

Inferior
mesenteric
ganglion

1

2

3

4

Figure 17.4   Overview of the Sympathetic Division of the ANS. The sympathetic division of the ANS controls the "fight-or-flight" response. Use the terms listed to fill in the numbered labels in the figure.

☐ Adrenal medulla

☐ Prevertebral ganglia

☐ Sympathetic trunk

☐ Sympathetic trunk ganglia (paravertebral)

PHYSIOLOGY

Autonomic Reflexes

Recall from chapter 16 that a typical reflex contains five main components: receptor, sensory neuron, control center, motor neuron, and effector organ. Autonomic reflexes, like somatic reflexes, require both sensory input and motor output. Somatic reflexes involve somatic motor neurons innervating skeletal muscle as the effector organ, whereas autonomic reflexes involve autonomic motor neurons innervating smooth muscle, cardiac muscle, or glands. Thus, the reflex occurs below the level of conscious awareness. Autonomic reflexes can involve either the spinal cord (spinal reflexes) or the brain (cranial reflexes) as the control center. The reflexes observed in this exercise are examples of cranial reflexes. Cranial reflexes involve transmission of nerve signals along cranial nerves. They are important in clinical tests to determine the function of specific cranial nerves or regions of the brain where the appropriate cranial nerves arise.

EXERCISE 17.3

PUPILLARY REFLEXES

The pupillary reflex is an example of a protective reflex. The **pupillary reflex** involves photoreceptors located within the retina, which detect light entering the eye. Sensory input is relayed along the optic nerve (CN II) to nuclei within the brain. Motor output is then relayed from the brain along the oculomotor nerve (CN III) to the pupillary sphincter muscle within the iris of the eye. When a bright light is shone into the eye, the normal pupillary response is constriction of the pupil. This is a protective function because too much light entering the eye can damage the retina. Testing for the pupillary reflex is important when brain trauma or other brain damage (e.g., a stroke) is suspected. For example, increased intracranial pressure can crush the cranial nerves that transmit either afferent or efferent signals. If this happens, the pupils will remain dilated when a light is shone into the eye. In addition, the size of the pupil is controlled by both divisions of the autonomic nervous system. Activation of the sympathetic division of the ANS results in dilation of the pupil to increase visual acuity. Conversely, activation of the parasympathetic division of the ANS results in the constriction of the pupil.

(continued on next page)

(continued from previous page)

1. In the spaces provided, list the structures that comprise each component of the pupillary reflex.

   Sensory receptor _____

   Sensory neuron _____

   Control center _____

   Motor neuron _____

   Effector organ _____

2. Obtain a flashlight and a small ruler, and choose a study subject.

3. Measure the size of the subject's pupil using the ruler. DO NOT touch the ruler to the subject's eyes. Simply hold it up in front of each eye and get as close as possible to a precise measurement. Record the initial pupil diameters:

   Right pupil diameter: Left pupil diameter:

   _____ mm   _____ mm

4. Shielding one eye (by holding a hand vertically between the subject's eyes), quickly shine a flashlight into the other eye (**figure 17.5**).

5. Measure the resulting pupil diameters:

   Right pupil diameter: Left pupil diameter:

   _____ mm   _____ mm

Figure 17.5 **Testing the Pupillary Reflex.** Have the subject shield one eye by holding his or her hand in front of the nose as shown above. Shine a light in one eye and note the change in pupil diameter in *both* eyes.
©Jill Braaten

EXERCISE 17.4

BIOPAC GALVANIC SKIN RESPONSE

The electrodermal response (EDA), also known as galvanic skin response (GSR), is a response of the skin to arousing stimuli. That is, the skin is known to conduct electricity better when a person is having a "sympathetic response." This exercise involves observing the changes in the GSR during varying emotional states. The GSR is altered due to changes in sympathetic nervous system stimulation, particularly to sweat glands and cutaneous blood vessels. That is, increasing sympathetic output increases sweat production and dilates cutaneous blood vessels, resulting in increased blood flow to the skin. This is the underlying physiology behind a polygraph test, which is a test that can be used to detect whether or not a subject is lying or telling the truth. Lying can increase sympathetic output, which alters the GSR. A true polygraph test also assesses respiratory and heart rates to determine whether or not a subject is telling the truth. The following BIOPAC exercise measures GSR, heart rate, and respiratory rate (breaths per minute, BMP) during a series of tasks.

Obtain the Following:
- **BIOPAC electrode lead set (SS2L)**
- **BIOPAC EDA Setup–EDA Lead (SS57LA) and 2 gelled electrodes (EL507)**
- **BIOPAC respiratory transducer (SS5LB)**
- **Colored paper (9 colors: white, black, green, red, blue, yellow, orange, brown, pink).**
- **BIOPAC general purpose electrodes (EL503)**

Before beginning the experiment, become familiar with the following concept (use the textbook as a reference):
- **Ohm's law**

1. Prepare the BIOPAC equipment by turning the computer ON and the MP36/35 unit OFF. Plug in the following: respiration transducer (SS5LB) (CH 1), electrode lead set (SS2L) (CH 2), and EDA (SS57LA) (CH 3). Turn the MP36/35 unit ON.

2. Place the respiratory transducer around the subject's chest. The band should fit firmly around the chest, below the subject's armpits and above the subject's nipples. The band can be placed directly over the subject's skin or over thin clothing (e.g., a thin T-shirt) (**figure 17.6a**).

3. For best recording results, it is helpful if the subject's hands are slightly sweaty before beginning calibration or data collection. Fill the two electrode (EL507) cavities with electrode gel to enhance conduction between the skin and the electrodes. Connect the gelled electrodes to the subject's index and middle fingers. Attach the leads (SS57L) to the electrodes and wrap the Velcro tape around each finger. The contact should be

Figure 17.6 **Setup for BIOPAC Galvanic Skin Response Exercise.** (*a*) Respiratory transducer placement. (*b*) Finger electrode placement. (*c*) ECG electrode placement.
Courtesy of and © BIOPAC Systems, Inc.

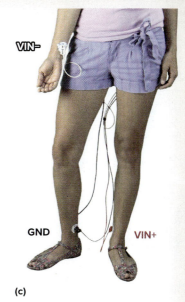

(a)   (b)   (c)

firm, but not so tight that it cuts off circulation. Wait at least five minutes prior to recording. Be sure that the transducer is placed on the skin beneath the fingertip rather than under the fingernail (figure 17.6*b*).

4. Prepare the subject's skin for electrode placement on the electrode sites (figure 17.6*c*). Apply a small quantity of electrode gel to the center of the surface ECG electrodes. Place the electrodes firmly on the subject's skin in the designated locations. Attach the leads (SS2L) to each electrode according to the color code: right anterior forearm above the wrist (white), medial surface of the right leg just above the ankle (black), and medial surface of the left leg just above the ankle (red).

5. Start the BIOPAC Student Lab Program. Select L09 EDA & Polygraph from the Choose a Lesson menu, enter the file name, and click OK.

6. When beginning data collection, have the subject assume a relaxed position in a chair and make sure the subject cannot see the computer screen. This is important, because visual feedback may skew results. The room must also be quiet, and the subject should remain as quiet as possible during data collection. One lab member will serve as the **director**, asking questions and performing specific tasks, and one lab member will serve as the **recorder**.

7. Click Calibrate in the upper left corner of the Setup window. Three seconds after calibration begins, a beep will sound. When heard, the subject will inhale once quickly and deeply, and then return to normal breathing. The calibration will stop automatically after 10 seconds of recording. Once the calibration has stopped, check the data (**figure 17.7**). The recording should show some fluctuations, which correspond to the deep breathing. If no fluctuations are present, click Redo Calibration. If calibration data are good, click Continue and proceed to Data Recording.

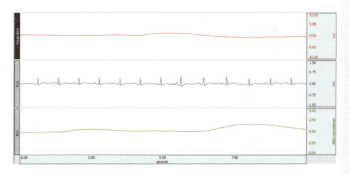

Figure 17.7 **Sample Calibration Data for BIOPAC Galvanic Skin Response Exercise.**
Courtesy of and © BIOPAC Systems, Inc.

8. OBTAIN DATA: The director will instruct the subject to perform a series of tasks. Prior to each task, the recorder will mark the event on the trace. The director will ensure that the data trace returns to baseline conditions before having the subject initiate another task. Click Record and begin the following tasks 5 seconds after recording begins:

   a. Director: Ask the subject to say his/her name quietly. Recorder: Press F2 and wait 5 seconds.

   b. Director: Ask the subject to count backward from 10. Recorder: Press F3 and wait 5 seconds.

   c. Director: Ask the subject to count backward from 30, subtracting increasing odd numbers. Recorder: Press F4 and wait 5 seconds.

   d. Director: Touch the side of the subject's face. Recorder: Press F5 and wait 5 seconds.

   Once data collection is complete, the recorder will click on Suspend. At this time, review the data to ensure that the traces resemble **figure 17.8a**. If not, click Redo. If so, click, Continue.

(continued on next page)

(continued from previous page)

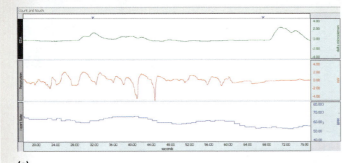

(a)

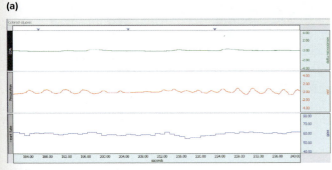

(b)

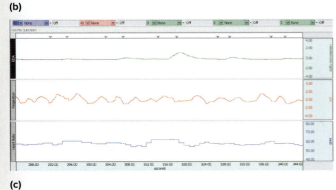

(c)

Figure 17.8   Sample Data for Galvanic Skin Response Exercise. (*a*) Count and touch. (*b*) Colored squares. (*c*) Yes-No.
Courtesy of and © BIOPAC Systems, Inc.

9. The director will then resume data collection for another task: showing the subject different colored papers. The recorder will mark the trace (F9) with event markers each time the paper colors are changed. Nine colors are shown to the subject, in this order: white, black, red, blue, green, yellow, orange, brown, and pink. Once data collection is complete, the recorder will click on Suspend. Review the data to ensure that the traces resemble figure 17.8*b*. If not, click Redo. If so, click, Continue.

10. The director will ask the subject a series of 10 yes or no questions. The recorder will insert markers (F6 when question is asked; F7 if answered "Yes"; F8 if answered "No") at the end of each of the following questions:

    a. Are you currently a student?

    b. Are your eyes blue?

c. Do you have any brothers?

d. Do you drive a motorcycle?

e. Are you less than 25 years of age?

f. Have you ever traveled to another planet?

g. Have aliens from another planet visited you?

h. Do you like dogs?

i. Have you answered all of the preceding questions truthfully?

Once data collection is complete, the recorder will click on Suspend. Review the data to ensure that traces resemble figure 17.8*c*. If not, click Redo. If so, click Done to conclude the experiment. Be sure to save the data to the computer's hard drive before closing the program.

11. ANALYZE THE DATA: If analyzing data that was just collected, click on Analyze Current Data. If opening data that was collected previously, click on Review Saved Data. To prepare to analyze the data, do the following:

    a. Use the zoom tool to select the first five seconds of data.

    b. Click Display menu at the top of the screen and select Autoscale Waveforms.

    c. Set up the measurement boxes at the top of the screen using the drop-down menus. Set CH 41 to Heart Rate. Set CH 40 to BPM. Set CH 3 to EDA.

12. Locate the I-beam tool. Use the activated I-beam tool to select the baseline EDA value and corresponding heart rate (at 2 seconds). Record this value in **table 17.4**. Then select the portion of the respiration trace that corresponds to one respiratory cycle from one inhalation to the next inhalation. Record this baseline respiratory rate (BPM) in table 17.4.

13. Scroll through all of the segment 1 data, repeating step 12 for each subsequent task. For each task, there should be a corresponding heart rate, respiratory rate, and EDA. Record these values in table 17.4.

14. Repeat steps 12 and 13 for the second and third segments of data, which correspond to the colored paper task and the series of yes or no questions, respectively. Record values for heart rate, respiratory rate, and EDA for each condition in **tables 17.5** and **17.6**. Save or print the data and exit the program.

15. Make note of any pertinent observations here: _____

    _____

    _____

    _____

    _____

Table 17.4 Segment 1 Data	Heart Rate (CH 41 Value)	Respiratory Rate (CH 40 BPM)	EDA (CH 3 Value)
Procedure			
Resting (baseline)			
Quietly say name			
Count from 10			
Count from 30			
Face touched			

Table 17.5 Segment 2 Data	Heart Rate (CH 41 Value)	Respiratory Rate (CH 40 BPM)	EDA (CH 3 Value)
Color			
White			
Black			
Red			
Blue			
Green			
Yellow			
Orange			
Brown			
Pink			

Table 17.6 Segment 3 Data					
Question	**Answer**	**Truth**	**Heart Rate (CH 41 Value)**	**Respiratory Rate (CH 40 BPM)**	**EDA (CH 3 Value)**
Student?	Y N	Y N			
Blue eyes?	Y N	Y N			
Brothers?	Y N	Y N			
Earn "A"?	Y N	Y N			
Motorcycle?	Y N	Y N			
Less than 25?	Y N	Y N			
Another planet?	Y N	Y N			
Aliens visit?	Y N	Y N			
Like dogs?	Y N	Y N			
Truthful?	Y N	Y N			

Chapter 17: The Autonomic Nervous System

POST-LABORATORY WORKSHEET

The corresponds to the Learning Objective(s) listed in the chapter opener outline.

Do You Know the Basics?

Exercise 17.1: Parasympathetic Division

1. The parasympathetic division of the ANS is also called the _____ (craniosacral/thoracolumbar) division because of its anatomical location. **1**

2. Identify the cranial nerves that carry parasympathetic information. (Check all that apply.) **1**

   _____ a. facial nerve (CN VII)

   _____ b. glossopharyngeal nerve (CN IX)

   _____ c. oculomotor nerve (CN III)

   _____ d. optic nerve (CN II)

   _____ e. vagus nerve (CN X)

Exercise 17.2: Sympathetic Division

3. The sympathetic division of the ANS is also called the _____ (craniosacral/thoracolumbar) division because of its anatomical location. **2**

4. The major sympathetic ganglia that lie on top of the unpaired abdominal blood vessels (celiac trunk, superior mesenteric artery, inferior mesenteric artery) are also known as _____ (paravertebral/prevertebral) ganglia because of their anatomical location relative to the vertebral column. **2**

Exercise 17.3: Pupillary Reflexes

5. The optic nerve (CN II) carries _____ (afferent/efferent) input of the pupillary reflex, whereas the oculomotor nerve (CN III) carries _____ (afferent/efferent) output of the pupillary reflex. **3**

6. Shining a bright light into the eye causes the pupil to _____(constrict/dilate) in response. **3**

Exercise 17.4: BIOPAC Galvanic Skin Response

7. The division of the ANS responsible for an increase in GSR, heart rate, and BPM is the _____ (parasympathetic/sympathetic) division. **4** **5**

Can You Apply What You've Learned?

8. Match each of the effectors listed in column A with the cranial nerve that innervates the effector listed in column B.

   Column A

   _____ 1. bronchioles

   _____ 2. ciliary muscles and iris of eye

   _____ 3. heart

   _____ 4. kidneys

   _____ 5. lacrimal gland

   _____ 6. salivary gland (parotid)

   _____ 7. salivary glands (under tongue)

   _____ 8. stomach and other digestive organs

   Column B

   a. facial nerve (CN VII)

   b. glossopharyngeal nerve (CN IX)

   c. oculomotor nerve (CN III)

   d. vagus nerve (CN X)

Can You Synthesize What You've Learned?

9. A patient has suffered a spinal cord injury that completely severed the spinal cord at level T2. Identify which of the following will be negatively affected by the paralysis. (Check all that apply.)

_____ a. involuntary control of muscles that reduce the amount of light entering the eye

_____ b. involuntary control of sphincters that relax to drain urine from the urinary bladder

_____ c. release of epinephrine from the adrenal medulla

_____ d. voluntary control of both lower limbs

10. Match each of the descriptions in column A with the division of the autonomic nervous system listed in column B. (Answers will be used more than once.)

Column A

_____ 1. conserves energy and replenishes nutrient stores

_____ 2. contains ganglia that are located next to the spinal cord

_____ 3. contains ganglia that are close to an effector

_____ 4. maintains homeostasis during times of stress or emergency

_____ 5. outflow pathways include cranial nerves and pelvic splanchnic nerves

_____ 6. preganglionic neurons innervate the adrenal medulla

Column B

a. parasympathetic division

b. sympathetic division

11. Patients experiencing hypertension (high blood pressure) may be prescribed drugs called beta-blockers. These drugs bind to receptors on cardiac muscle cells and block the action of the sympathetic nerves to the heart. Describe how the administration of a beta-blocker will influence heart rate and strength of contraction.

_____

_____

_____

_____

_____

12. A visit to the eye doctor often involves "dilation" of the pupils. Doctors will place drops of a drug called *tropicamide* in each eye and wait a period of time for the drops to take effect. The drug binds to receptors on the pupillary sphincter muscles to block action of the parasympathetic nerves to these muscles. Describe why application of these eye drops leads to increased sensitivity to light.

_____

_____

_____

_____

_____

13. John, a 28-year-old male, enters the emergency room after experiencing a severe blow to the head while playing football. Doctors test pupillary reflexes in both eyes and note asymmetrical pupillary response. Describe these results, including how the pupillary reflex can be tested in a clinical setting. Finally, discuss possible locations of the injury.

_____

_____

_____

_____

_____

General and Special Senses

OUTLINE AND LEARNING OBJECTIVES

31 Describe how to test for the blind spot and colorblindness, and explain how each test works

32 Explain the use of the Rinne and Weber tests, and what abnormal findings for each of the tests may indicate

33 Explain the purpose of the Romberg and Barany tests, and what abnormal findings for each of the tests may indicate

Module 7: Nervous System

INTRODUCTION

Why is it that the first sip of coffee in the morning seems so invigorating? Is it the sound of the coffee being poured into a cup, the aroma as it is being brought toward the face, or the taste when taking that first sip? Whatever it may be, we seem to be instantly energized. These characteristics that we appreciate so much in our morning cup of coffee are processed and interpreted by the nervous system. These senses (and others) are considered **special senses** with complex sensory receptors located in the head (**Table 18.1**). But be careful with that cup, it might be hot! Senses that allow us to perceive temperature, pressure, and pain are considered **somatic senses** because the sensory receptors are located within somatic tissues. In reality, our enjoyment and perception of the morning coffee are due to *all* of our senses and, importantly, the brain's ability to integrate and interpret this input.

These are just a sampling of the myriad of sensory input the brain processes daily. The laboratory exercises within this chapter explore the intricate structure and function of these important sensory receptors and organs throughout the body. The exercises within this chapter explore the gross anatomy of two special sensory organs (the eye and ear) though classroom models and by dissection of a cow eye. Exercises explore the histology of

a select few somatic sensory receptors and special sensory organs. Finally, the physiology exercises explore specific tests that are used to examine the function of both general and special senses.

Table 18.1	Criteria for Classifying Sensory Receptors	
Classification	**Description**	**Examples**
SENSORY RECEPTOR DISTRIBUTION (LOCATION OF RECEPTOR)		
General senses	Distributed thoughout the body; structurally simple	
Somatic sensory receptors	Located in skin and mucous membranes Located in joints, muscles, and tendons	Tactile (touch) receptors Joint receptors, muscle spindles, Golgi tendon organs
Visceral sensory receptor	Located within walls of viscera and blood vessels	Stretch receptors for in stomach wall, chemoreceptors in blood vessels
Special sensess	Located only in the head; structurally complex sense organs	Sensory receptors for smell, taste, vision, hearing, and equilibrium

These Pre-Laboratory Worksheet questions may be assigned by instructors through their ▓ connect course.

1. Which of the following sensory receptors is/are involved in general sensation? (Check all that apply.)

   _____ a. cochlear hair cells _____ b. free nerve endings _____ c. olfactory receptors _____ d. photoreceptors _____ e. tactile corpuscles

2. Which of the five special senses have receptor organs located within the petrous part of the temporal bone? (Check all that apply.)

   _____ a. equilibrium _____ b. gustation _____ c. hearing _____ d. olfaction _____ e. vision

3. Tactile corpuscles are located within the _____ (papillary/reticular) layer of the dermis.

4. Lamellated corpuscles are located within the _____ (papillary/reticular) layer of the dermis.

5. Match the special sense listed in column A with the corresponding cranial nerve that transmits the sensation to the brain, listed in column B. (Some answers may be used more than once because some special senses are transmitted by more than one cranial nerve.)

   Column A

   _____ 1. equilibrium

   _____ 2. gustation

   _____ 3. hearing

   _____ 4. olfaction

   _____ 5. vision

   Column B

   a. facial nerve (CN VII)

   b. glossopharyngeal nerve (CN IX)

   c. olfactory nerve (CN I)

   d. optic nerve (CN II)

   e. vestibulocochlear nerve (CN VIII)

6. Olfactory nerves extend through which part of the ethmoid bone? (Circle one.)

   a. crista galli

   a. cribriform plate

   b. middle nasal concha

   c. perpendicular plate

   d. superior nasal concha

7. Which of the following are internal structures of the eye? (Check all that apply.)

   _____ a. cornea

   _____ b. fovea centralis

   _____ c. lens

   _____ d. optic nerve

   _____ e. retina

   _____ f. sclera

8. Taste buds are lemon-shaped structures housed within _____ (papillae/crypts) on the tongue.

9. The _____ (Rinne/Weber) test is a test for conduction deafness.

10. The sense of gustation is influenced by the sense of _____ (olfaction/hearing).

GROSS ANATOMY

General Senses

Sensory receptors are classified in multiple ways, one being receptor distribution in the body. The two distribution classifications are the general senses and the special senses. Receptors of the **general senses** are simple structures composed of dendritic endings of sensory neurons located throughout the body. General sensory receptors are organized into somatic sensory receptors and visceral sensory receptors. **Somatic sensory receptors** are housed within somatic tissue—for example, the skin, where they monitor tactile sensations (e.g., pressure, vibration, and pain). These receptors are also within joints, muscles, and tendons for the detection of stretch and pressure relative to position and movement of the skeleton and muscles. **Visceral sensory receptors** are located within the walls of the viscera (internal organs); they respond to temperature, chemicals, stretch, and pain. In contrast, receptors of the **special senses** are complex structures located only in the head, and they are associated with taste, smell, vision, hearing, and equilibrium.

Most sensory receptors responsible for general sensation (things such as touch, pain, pressure, temperature) are located in the skin. Exercise 18.1 involves observing a classroom model of the skin, with special emphasis on sensory receptors located within the skin.

EXERCISE 18.1

SENSORY RECEPTORS IN THE SKIN

1. Observe a classroom model of thick skin **(figure 18.1)**. Some somatic sensory receptors in the skin, such as tactile menisci and free nerve endings, are too small to view under the microscope. Therefore, these structures will be observed on classroom models.

2. Locate a tactile disc. **Tactile discs** (Merkel discs) are located at the dermal/epidermal junction of thick skin, although their location is not restricted to dermal papillae, unlike tactile corpuscles. The function of tactile discs is the sensation of light touch.

3. Find some free nerve endings on the skin model. **Free nerve endings** are just that—nerve endings with no specialized cells surrounding them. The ends of these neurons are located near the epidermal/dermal junction, with many extending into the epidermis. These endings are also located in hair follicles and glands. Free nerve endings function in the sensation of sustained touch, temperature, itching, and pain. Recall a time when you had a blister that ripped open. A blister forms when the dermis separates from the epidermis, and fluid accumulates between the layers. When the epidermis is removed from a blister, it is very painful! Why? The free nerve endings in the dermis are now exposed to the environment, and this causes them to generate action potentials. This is why large superficial wounds ("scrapes") on the skin are much more painful than deep wounds. A deep wound is perceived as a more severe wound (as it typically is) but is often confusing because it causes less pain than a more superficial wound. The greater the surface area exposed, the greater the number of nerve endings stimulated. Hence, the common exclamation, "It's only a scrape, but it hurts like crazy!"

4. Identify the following on a model of the skin, using figure 18.1, table 18.6, and the textbook as guides:

   ☐ **Dermal papillae**   ☐ **Lamellated corpuscle**
   ☐ **Free nerve endings**   ☐ **Tactile corpuscle**

5. Sketch thick skin and label the locations of the sensory receptors listed in table 18.6 in the space provided or on a separate sheet of paper.

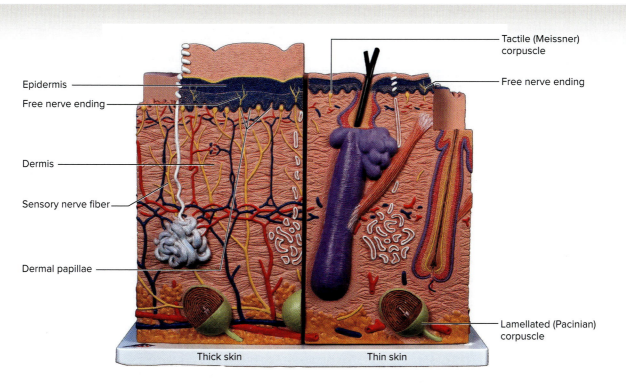

Epidermis

Free nerve ending

Dermis

Sensory nerve fiber

Dermal papillae

Tactile (Meissner) corpuscle

Free nerve ending

Lamellated (Pacinian) corpuscle

Thick skin   Thin skin

Figure 18.1   Skin. Model of the skin demonstrating sensory receptors, such as free nerve endings and lamellated corpuscles.

Model # J16 [1000294] ©3B Scientific GmbH, Germany, 2013 www.3bscientific.com/Photo by Christine Eckel, Ph.D.

Special Senses

Remember that the special senses are those with receptors in the head and that they are rather complex. These exercises focus on two special sensory organs: the eye and the ear.

EXERCISE 18.2

GROSS ANATOMY OF THE EYE

This laboratory exercise explores the accessory structures of the eye. Identify all structures listed in **table 18.2** on a classroom model or on yourself. Subsequent exercises will explore internal structures of the eye and the structure and function of both external and internal structures of a cow eye.

Table 18.2	External and Accessory Structures of the Eye		
Structure	**Description**	**Function**	**World Origins**
Fibrous Tunic	Tough outer connective tissue covering of the eye	Protects and maintains the shape of the eye, serves as an attachment site for extrinsic eye muscles, and refracts light	*fibro-*, fiber + *tunic*, a coat
Cornea	Transparent tissue on the anterior surface of the eye consisting of an external layer of stratified squamous epithelium, a middle layer of regularly arranged collagen fibers, and an inner layer of endothelium	Refracts (bend) light waves in the eye	*corneus*, horny
Sclera	Dense irregular connective tissue that surrounds the entire eye except for the cornea	Protects the eye, serves as an attachment point for extrinsic eye muscles, and helps maintain the round shape of the eye	*skleros*, hard

(continued on next page)

(continued on next page)

Table 18.2	External and Accessory Structures of the Eye *(continued)*		
Structure	**Description**	**Function**	**World Origins**
Lacrimal Apparatus	Structures producing and draining lacrimal fluid	Produces and drains lacrimal fluid	*lacrima*, a tear
Lacrimal Gland	Almond-shaped serous gland located in the superior and lateral aspect of the orbit	Secretes tears	*lacrima*, a tear
Lacrimal Puncta	Two small openings in the lacrimal caruncle; the tiny holes on the "bump" at the inferomedial aspect of the eye	Drains lacrimal fluid into the lacrimal sac	*lacrima*, a tear, + *punctum*, a prick or point
Lacrimal Sac	A swelling at the superior part of the nasolacrimal duct, medial to the lacrimal bone, lateral to the nasal bone, and deep to the maxilla	Receives lacrimal fluid from the lacrimal canals and transports it to the nasolacrimal duct	*lacrima*, a tear
Nasolacrimal Duct	A duct that runs from the lacrimal sac into the nasal cavity	Conducts lacrimal fluid from the lacrimal sac into the nasal cavity	*nasal*, relating to the nose, + *lacrima*, a tear, + *ductus*, to lead
Optic Nerve (CN II)	A large nerve exiting the posteromedial region of the eye; consists of myelinated axons of ganglion cells that exit the orbit through the optic foramen	Transmits nerve signals from the eye to the brain	*optikos*, relating to the eye or vision

EXERCISE 18.2A Accessory Structures of the Eye

1. Observe a classroom model of the eye (**figure 18.2a**).

2. Identify the following structures on a model of the eye, using figure 18.2a, table 18.2, and the textbook as guides:

   ☐ **Cornea**   ☐ **Optic disc**
   ☐ **Iris**   ☐ **Optic nerve**
   ☐ **Lacrimal gland**   ☐ **Sclera**

3. Obtain a mirror and observe the externally visible structures of your eye. If you have no mirror, perform this observation on a lab partner. Using table 18.2 and the textbook as guides, identify all of the structures listed in figure 18.2 on your eye (or your lab partner's eye), and then label them in figure 18.2b.

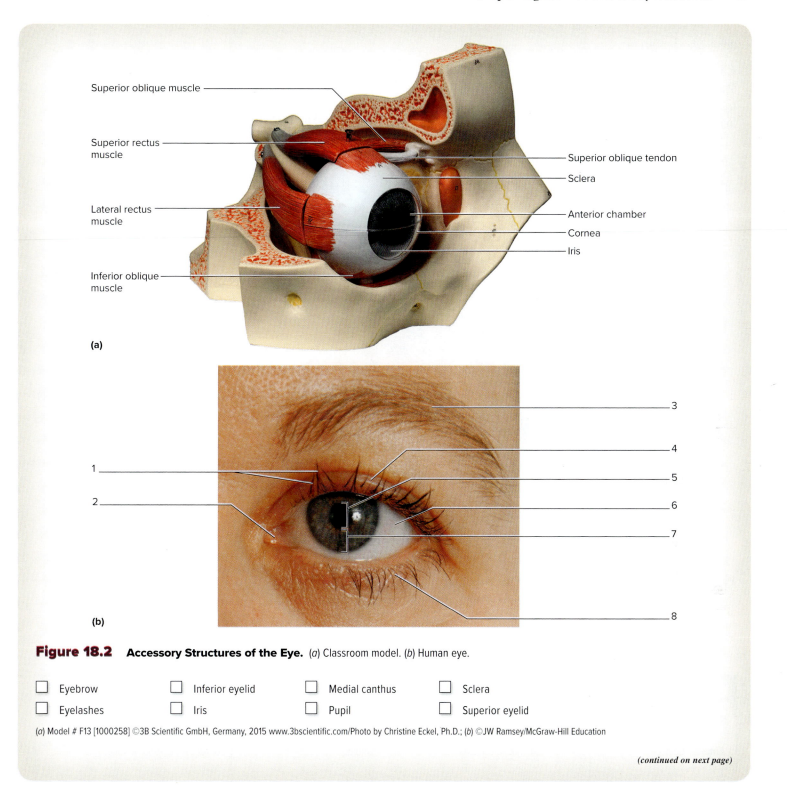

Figure 18.2 **Accessory Structures of the Eye.** (*a*) Classroom model. (*b*) Human eye.

☐ Eyebrow   ☐ Inferior eyelid   ☐ Medial canthus   ☐ Sclera

☐ Eyelashes   ☐ Iris   ☐ Pupil   ☐ Superior eyelid

(*a*) Model # F13 [1000258] ©3B Scientific GmbH, Germany, 2015 www.3bscientific.com/Photo by Christine Eckel, Ph.D.; (*b*) ©JW Ramsey/McGraw-Hill Education

(*continued on next page*)

(continued from previous page)

EXERCISE 18.2B Internal Structures of the Eye

The internal structures of the eye **(table 18.3)** function in the transmission of light, the nourishment of the eye, and the processing of visual information.

1. Observe a classroom model of the eye where internal structures are visible **(figure 18.3)**. Many classroom models of the eye contain both external and internal eye structures. Viewing internal eye structures may require disassembly of the eye model to access the structures.

2. Identify the following structures on the model of the eye, using figure 18.3, table 18.3, and the textbook as guides:

☐ **Anterior cavity**   ☐ **Fovea centralis**

☐ **Anterior chamber**   ☐ **Iris**

☐ **Choroid**   ☐ **Lens**

☐ **Ciliary body**

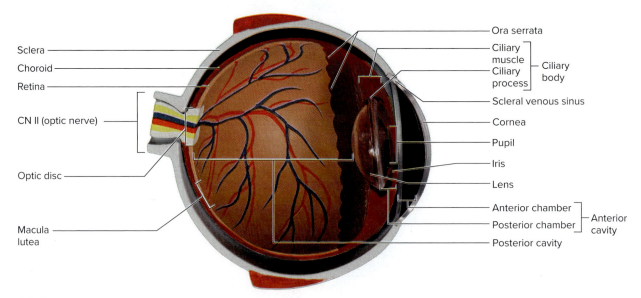

Figure 18.3   **Classroom Model of the Internal Eye.**

Model # F15 [1000259] ©3B Scientific GmbH, Germany, 2015 www.3bscientific.com/Photo by Christine Eckel, Ph.D.

Table 18.3	Internal Structures of the Eye		
Structure	**Description**	**Function**	**Word Origin**
Anterior Cavity	A space anterior to the lens	Filled with aqueous humor	*anterior*, the front surface, + *cavus*, hollow
Anterior Chamber	The space between the cornea and the iris	Filled with aqueous humor, a watery fluid that circulates within the chambers of the eye	*anterior*, the front surface, + *camera*, an enclosed space
Posterior Chamber	The space between the lens and the iris	Filled with aqueous humor	*posterior*, the back surface, + *camera*, an enclosed space
Choroid Layer	The pigmented, vascular layer located between the retina and the sclera	Blood vessels of the choroid supply nutrients to the tissues of the retina and sclera; pigment in the choroid absorbs light after it passes through the retina	*choroideus*, like a membrane

Table 18.3	Internal Structures of the Eye *(continued)*		
Structure	**Description**	**Function**	**Word Origin**
Ciliary Body	The thickened extension of the vascular tunic, located between the choroid and the iris	Produces aqueous humor; contraction of the ciliary muscle within the ciliary body alters the shape of the lens	*cilium*, eyelid, + *bodig*, a thing or substance
Ciliary Muscle	Smooth muscle found within the ciliary body that is composed of both circular and radial fibers	Contraction of this muscle relaxes the suspensory ligaments that attach it to the lens, which increases the lens curvature to accommodate for near vision	*cilium*, eyelid
Ciliary Process	Ridges extending off the ciliary body	Secretes aqueous humor	*cilium*, eyelid
Fovea Centralis	The depression ("central pit") in the macula lutea that contains only cones and lacks blood vessels	The area of highest visual acuity in the eye	*fovea*, a pit, + *centralis*, in the center
Iris	The colored portion of the eye, which makes up the anterior portion of the vascular tunic; the dilator pupillae and sphincter pupillae muscles are located within the iris	Controls the amount of light entering the eye through the pupil; contraction of the radially arranged dilator pupillae muscle increases pupil diameter, whereas contraction of the circularly arranged sphincter pupillae muscle decreases pupil diameter	*iris*, rainbow
Lens	A transparent, biconvex structure composed of a highly specialized, modified epithelium	Refracts (bends) light waves so they hit the retina optimally for clear vision	*lens*, a lentil
Optic Disc ("Blind Spot")	An area of the retina where there is an absence of photoreceptors	The location where axons of ganglion cells exit the eye to become the optic nerve	*optikos*, the eye, + *discus*, disc
Posterior Cavity (Vitreous Chamber)	A space posterior to the lens and anterior to the retina	Occupied by the vitreous humor, a gelatinous mass that maintains shape of eye and keeps the retina against the wall of the eye	*vitreus*, glassy, + *camera*, an enclosed space
Pupil	The space (opening) in the center of the iris	The size of the pupil determines the amount of light entering the eye	*pupilla*, pupil
Retina	Also called the neural tunic of the eye; the inner layer of the eye composed of a pigmented layer, rods, cones, bipolar cells, and ganglion cells	Transduces light that enters the eye as light waves into nerve signals (action potentials) that are interpreted by the brain	*rete*, a net
Suspensory Ligaments	Ligaments that extend between the ciliary muscles and the lens	Attach the lens to the ciliary muscles so that contraction and/or relaxation of ciliary muscles can alter the shape of the lens	*suspensio*, to hang up, + *ligamentum*, a bandage

EXTRINSIC EYE MUSCLES

1. Observe a model of the eye with extrinsic eye muscles.

2. The **extrinsic**, or extraocular (*extra-*, outside of, + *oculus,* eye), muscles of the eye **(table 18.4)** allow us to move our eyes up, down, side to side, and at an angle. These muscles originate on bone and insert onto the sclera of the eye. They are named based on location and shape, so they are relatively easy to identify and remember. Recall from chapter 15 that cranial nerves innervate the extrinsic muscles of the eye. Therefore,

observations of impairment in eye movements are helpful in assessing cranial nerve disorders.

3. Ask a laboratory partner to look in different directions and observe his or her eye movements. As his or her eyes move, name the muscles (in *both eyes* because they will be different!) used to cause the movement (use table 18.4 as a guide).

4. Identify the **extrinsic eye muscles** listed in **figure 18.4** and on the model of the eye, using table 18.4 and the textbook as guides. Then label them in figure 18.4.

Table 18.4	Extrinsic Eye Muscles				
Muscle	**Orbital attachment (O)**	**Eyeball attachment (E)**	**Action**	**Innervation**	**Word Origin**
Inferior Oblique	Maxilla (anterior portion of orbit)	Sclera on the anterior, lateral surface of the eyeball, deep to the lateral rectus muscle	Elevates, abducts, and laterally rotates the eyeball	Oculomotor (CN III)	*inferior,* lower, + *obliquus,* slanting
Inferior Rectus	Sphenoid (tendinous ring around optic canal)	Sclera on the anterior, inferior surface of the eyeball	Depresses, adducts, and medially rotates the eyeball	Oculomotor (CN III)	*inferior,* lower, + *rectus,* straight
Lateral Rectus	Sphenoid (tendinous ring around optic canal)	Sclera on the anterior, lateral surface of the eyeball	Abducts the eyeball	Abducens (CN VI)	*lateralis,* lateral, + *rectus,* straight
Medial Rectus	Sphenoid (tendinous ring around optic canal)	Sclera on the anterior, medial surface of the eyeball	Adducts the eyeball	Oculomotor (CN III)	*medialis,* middle, + *rectus,* straight
Superior Oblique	Sphenoid (tendinous ring around optic canal)	Sclera on the posterior, superiolateral surface of the eyeball just deep to the belly of the superior rectus muscle	Depresses, abducts, and laterally rotates the eyeball	Trochlear (CN IV)	*superus,* above, + *obliquus,* slanting
Superior Rectus	Sphenoid (tendinous ring around optic canal)	Sclera on the anterior, superior surface of the eyeball	Elevates, adducts, and medially rotates the eyeball	Oculomotor (CN III)	*superus,* above, + *rectus,* straight

LEARNING STRATEGY

The following "chemical formula" can help you remember the eye muscle innervation:

$$[(SO_4)(LR_6)]_3$$

In words, the superior oblique **(SO)** is innervated by cranial nerve IV **(4)**, the lateral rectus **(LR)** is innervated by cranial nerve VI **(6)**, and the rest of the eye muscles are innervated by cranial nerve III **(3).**

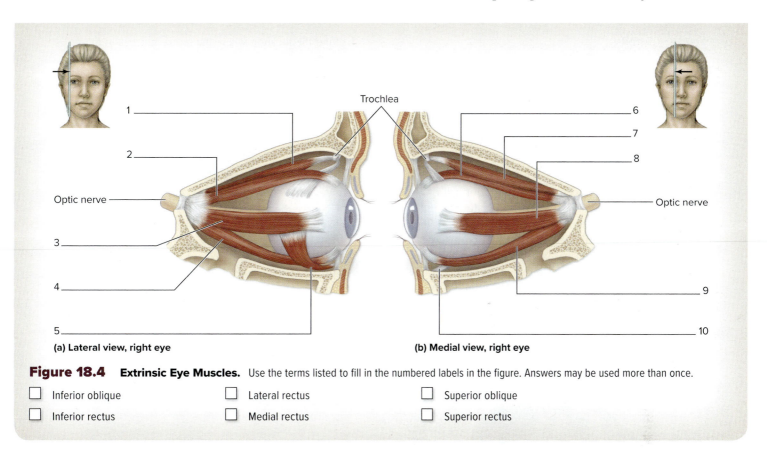

Trochlea

1
2
Optic nerve
3
4
5

(a) Lateral view, right eye

6
7
8
Optic nerve
9
10

(b) Medial view, right eye

Figure 18.4 **Extrinsic Eye Muscles.** Use the terms listed to fill in the numbered labels in the figure. Answers may be used more than once.

☐ Inferior oblique   ☐ Lateral rectus   ☐ Superior oblique

☐ Inferior rectus   ☐ Medial rectus   ☐ Superior rectus

EXERCISE 18.4

COW EYE DISSECTION

In exercise 18.4 we will explore the anatomy of the eye, using a cow eye as a model organ. Although there is some variance in structure between a cow eye and a human eye, cow eyes are much easier to obtain and they are larger, which greatly facilitates making internal observations of the eye. This exercise is designed to be done using fresh cow eyes, although preserved cow or pig eyes may be substituted if necessary. If dissecting a preserved cow or pig eye, the tissues will be tougher and the cornea will be opaque instead of transparent.

1. Obtain a dissecting pan, dissecting tools, and a fresh cow eye. Observe the gross structure of the eye before making any cuts. Identify the following structures, using tables 18.3 and 18.4 and **figure 18.5** as guides:

    ☐ **Cornea**   ☐ **Orbital fat**

    ☐ **Extrinsic eye muscles**   ☐ **Sclera**

    ☐ **Optic nerve**

2. ⚠️ Using scissors and forceps (the tissue will be slippery!), remove the orbital fat and extraocular muscles, leaving the optic nerve intact (figure 18.5b). Once these structures have been removed, the entire eye should be visible. Notice the toughness of the outer covering or **sclera** of the eye. This is the layer of tissue to cut through in order to see structures within the eye.

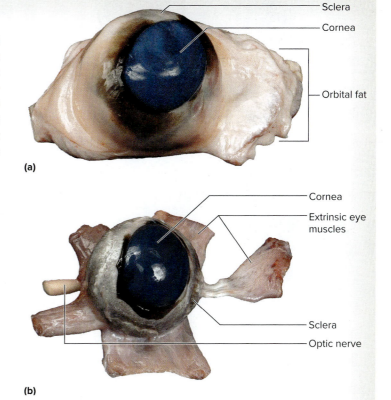

Sclera
Cornea

Orbital fat

(a)

Cornea

Extrinsic eye muscles

Sclera
Optic nerve

(b)

Figure 18.5 **Fresh Cow Eye.** (*a*) Before dissection with orbital fat pad intact. (*b*) After dissecting away the orbital fat pad to expose the extraocular muscles and optic nerve.

©McGraw-Hill Education/Photo and Dissection by Christine Eckel

(continued on next page)

(continued from previous page)

3. Using scissors and forceps, cut the eye open by making a coronal incision through the sclera that completely encircles the eye approximately 1/4 of an inch posterior to the cornea. Once this has been done, notice that a jelly-like fluid oozes out of the posterior cavity of the eye. This fluid is the **vitreous humor**, which fills the posterior cavity **(figure 18.6b)**. This fluid's functions include holding the **retina** against the posterolateral walls of the eye. Identify structures in the posterior half of the dissected eye. Look for the retina, a yellowish, thin membrane that is connected to the posterior wall of the eye at only one spot (figure 18.6*a*). It contains neurons responsible for detecting visual stimuli and initiating nerve signals to the brain (histology of the retina will be covered in detail in exercise 18.10). The retina is very delicate and easily falls away from the posterior wall of the eye when the vitreous humor is not present to hold it in place.

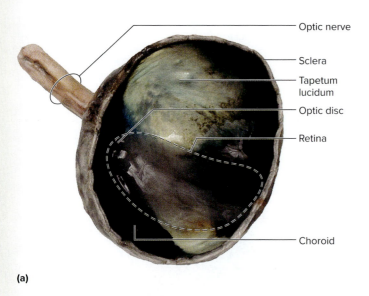

Optic nerve

Sclera

Tapetum lucidum

Optic disc

Retina

Choroid

(a)

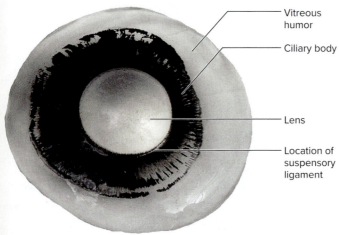

Vitreous humor

Ciliary body

Lens

Location of suspensory ligament

(b)

Figure 18.6 **Coronal Sections of Cow Eye.** (*a*) Posterior part demonstrating the retina, optic disc, and tapetum lucidum. (*b*) Anterior part demonstrating the choroid, lens, and iris.

4. Find the location where the retina attaches to the posterior wall of the eye (it will "pucker" in this area). To help you to find this area, first find the optic nerve on the outside of the eye and then look inside the eye for the location where the optic nerve leaves the eye. This spot within the eye is the **optic disc** (blind spot). It is called a "blind" spot because it is devoid of photoreceptors. This is where axons from the ganglion cell layer of the retina (see exercise 18.10) leave the eye and extend to the brain as the optic nerve (CN II).

5. Observe the inner walls of the posterior half of the eye. Notice the very colorful, iridescent **tapetum lucidum** (figure 18.6*a*). This structure is not present in humans, but it is present in animals that must be able see well in dim light, such as cows. The tapetum lucidum reflects light. Thus, when it is dark outside and very little light enters the eye, the tapetum lucidum increases the frequency with which light rays stimulate the retina. This makes things more visible, but the image is not sharp. In humans, the inside of the eye is completely coated with a black choroid, which absorbs excess light. This makes it more difficult for us to see things in the dark. On the plus side, the images seen by humans are sharper.

6. Now focus on the anterior portion of the eye (figure 18.6*a*). It may be difficult to see some structures because the **choroid**, containing black pigment, makes the structures appear dark. Notice the semitransparent **lens** is suspended in place by the black-colored **ciliary body**. Anterior to the lens and posterior to the cornea is the **anterior cavity** of the eye. This cavity is further subdivided by the iris into an **anterior chamber** (between the cornea and the iris) and a **posterior chamber** (between the iris and the lens). In a living organism, the anterior cavity is filled with a clear, watery fluid called **aqueous humor**. Try to find the fine, delicate structures composing the **suspensory ligament**, which extends between the ciliary body and the lens.

7. Carefully remove the lens from the eye. Notice that it is somewhat, but not completely, transparent. Place it on a piece of paper containing text. As shown in **figure 18.7**, place the lens over a letter or two of text and make note of the change in appearance of the text, if any, as seen through the lens:

_____

_____

_____

_____

_____

_____

_____

8. Identify the following structures on the interior of the dissected cow eye (use tables 18.2 and 18.3, and figures 18.5 through 18.7 as guides):

☐ **Anterior cavity**	☐ **Posterior cavity**
☐ **Choroid**	☐ **Retina**
☐ **Ciliary body**	☐ **Suspensory ligament**
☐ **Lens**	☐ **Tapetum lucidum**
☐ **Optic disc**	☐ **Vitreous humor**

9. When finished with the dissection, clean up the workspace: Dispose of the cow eye debris in the organic waste receptacle. Dispose of used scalpel blades in the sharps container. Dispose of used paper towels and other paper waste in the wastebasket. Rinse off the dissecting tray and dissection instruments, and lay them out to dry. Finally, wipe down the laboratory workstation with disinfectant so it is clean for the next person who comes into the laboratory.

Lens

Figure 18.7 **Lens.** After removing the lens from the cow eye, place it over some text to see how it changes the image of the text.
©Christine Eckel

EXERCISE 18.5

GROSS ANATOMY OF THE EAR

The ear is responsible for two important special sensory modalities: equilibrium (balance) and hearing. It houses the organs for both within the *petrous part of the temporal bone.* These tiny sensory organs are nearly impossible to identify on a cadaver; thus, exploration of the gross anatomy of the ear will be accomplished using models of the ear.

1. Obtain a model of the ear **(figure 18.8)**. On the model, first distinguish between the external-ear, middle-ear, and inner-ear cavities, and the structures that link the cavities to each other **(table 18.5)**. The **tympanic membrane** is the link between the external-ear and the middle-ear cavities, while the **oval window** is the link between the middle- and inner-ear cavities.

2. Identify the following structures on the model, and label the structures in figure 18.8. Use table 18.5 and the textbook as a guide.

☐ **Auditory tube**	☐ **Perilymph**
☐ **Auricle**	☐ **Round window**
☐ **Cochlea**	☐ **Saccule**
☐ **Endolymph**	☐ **Semicircular canals**
☐ **External acoustic meatus**	☐ **Stapedius muscle**
	☐ **Stapes**
☐ **Incus**	☐ **Tensor tympani muscle**
☐ **Malleus**	
☐ **Ossicles**	☐ **Tympanic membrane**
☐ **Oval window**	☐ **Utricle**

3. Sketch the ear and label the locations of the structures listed in step 2 in the space provided or on a separate sheet of paper.

(continued on next page)

(continued from previous page)

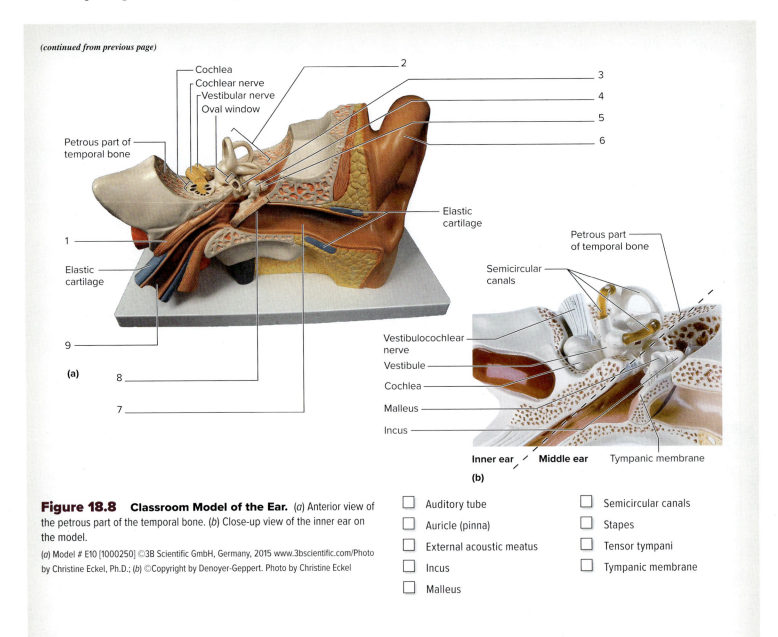

Figure 18.8   Classroom Model of the Ear. (*a*) Anterior view of the petrous part of the temporal bone. (*b*) Close-up view of the inner ear on the model.

(*a*) Model # E10 [1000250] ©3B Scientific GmbH, Germany, 2015 www.3bscientific.com/Photo by Christine Eckel, Ph.D.; (*b*) ©Copyright by Denoyer-Geppert. Photo by Christine Eckel

☐ Auditory tube

☐ Auricle (pinna)

☐ External acoustic meatus

☐ Incus

☐ Malleus

☐ Semicircular canals

☐ Stapes

☐ Tensor tympani

☐ Tympanic membrane

Table 18.5	Structures of the External, Middle, and Internal Ear		
Structure	**Description**	**Function**	**Word Origin**
External Ear			
Auricle (Pinna)	External ear, composed of an elastic cartilage skeleton covered by skin	Funnels sound waves from the environment into the external acoustic meatus	*pinna*, a wing
External Acoustic (Auditory) Meatus	Canal leading from the auricle to the tympanic membrane	Transmits sound waves from the auricle to the tympanic membrane, causing vibration	*externa*, outside, + *acoustic*, relating to sound, + *meatus*, a passage
Tympanic Membrane	Drumlike, tight, thin membrane that separates the external ear cavity from the middle-ear cavity	Vibrates in response to sound waves that strike it as they reach the end of the external acoustic meatus; vibrations are transferred to the ossicles of the middle-ear cavity (malleus, incus, and stapes)	*tympanon*, drum, + *membrana*, a membrane

Table 18.5	Structures of the External, Middle, and Internal Ear　　*(continued)*		
Structure	**Description**	**Function**	**Word Origin**
Middle Ear	Air-filled cavity between the external ear and the inner ear		
Auditory Ossicles	Three tiny bones (malleus, incus, and stapes) found within the middle ear	Transmit movements caused by pressure vibrations from the tympanic membrane to the oval window of the cochlea, causing fluid pressure waves in the perilymph of the scala vestibuli	*ossiculum*, a bone
Auditory Tube (Pharyngotympanic or Eustachian Tube)	Tube connecting the middle-ear cavity to the nasopharynx	Opening of this channel allows air to enter or leave the middle-ear cavity such that the pressure in the middle ear equilibrates with the environmental pressure; this allows the tympanic membrane to vibrate freely.	*audio*, to hear, + *tubus*, a canal
Oval Window	Membrane-covered opening into the scala vestibuli that is covered by the foot of the stapes	Vibrations of the stapes at the oval window cause fluid pressure waves in the perilymph of the scala vestibule.	*oval*, egg-shaped, + *window*, an opening
Stapedius Muscle	Small muscle connecting the neck of the stapes to the temporal bone	Contraction of this muscle acts to dampen vibrations of the stapes as a protective measure against excessive movement at the oval window from very loud noises.	*stapedius*, relating to the stapes
Tensor Tympani Muscle	Small muscle connecting the handle of the malleus to the auditory tube	Contraction of this muscle pulls the malleus medially and tenses the tympanic membrane as a protective measure against excessive vibration from very loud noises.	*tensus*, to stretch, + *tympani*, relating to the tympanic membrane
Inner Ear	Fluid-filled space located within the petrous part of the temporal bone that contains the cochlea, vestibule, and semicircular canals		
Cochlea	Spiral-shaped organ found within the inner ear	Contains the spiral organ and associated structures that are involved in the special sense of hearing	*cochlea*, a snail shell
Semicircular Canals	Three ring-like canals that are oriented at right angles to each other	Detect angular acceleration and equilibrium	*semicircular*, shaped like a half circle, + *canalis*, a duct or channel
Spiral Organ (Organ of Corti)	Organ composed of specialized epithelium found within the scala media (cochlear duct) of the cochlea	Special sensory organ for hearing	*spiralis*, a coil, + *organon*, a tool or instrument
Vestibule	Portion of the inner ear located between the cochlea and the semicircular canals; contains the saccule and utricle		
Saccule	Smallest membranous sac in the vestibule; connects with the cochlear duct	Contains receptors that sense linear vertical acceleration	*saccus*, a sac
Utricle	The largest membranous sac in the vestibule	Contains receptors that sense linear horizontal acceleration	*uter*, leather bag
Vestibulocochlear Nerve (CN VIII)	Travels through the internal acoustic meatus	Cranial nerve transmitting nerve signals associated with balance, equilibrium, and hearing to the brain	*vestibulo-*, referring to the vestibule, + *cochlea*, referring to the cochlea

(continued on next page)

(continued from previous page)

4. After identifying all of the gross structures of the ear, review the sequence of events required for the transmission of sound waves from the environment to the cochlea **(figure 18.9)**. Name all of the structures involved in the sequence. The sequence is as follows:

INTEGRATE

CLINICAL VIEW

Pressure Changes in the Middle Ear

The auditory tube is contains elastic cartilage and remains collapsed unless there is a large difference in pressure between the environment and the middle-ear cavity. When a difference in pressure exists, the auditory tube opens briefly and air moves to equalize the pressure in the middle ear with the pressure in the environment. The opening and closing of the auditory tube is what accounts for the "popping" sound made when pressure is equalized. Without correction, these differences in pressure between the external environment and the middle-ear cavity can result in injury, such as a ruptured tympanic membrane.

---

(1) Sound waves are "funneled" into the **external acoustic (auditory) meatus** by the contours of the outer ear **(auricle)** and cause vibrations of the **tympanic membrane**. The **auditory tube** ensures that air pressure in the middle ear is the same as air pressure in the environment so the tympanic membrane can vibrate freely.

(2) Vibrations of the tympanic membrane cause the **auditory ossicles** (malleus, incus, and stapes) to vibrate. Excessive vibrations (e.g., from a loud noise) cause a reflexive contraction of the **tensor tympani** and **stapedius** muscles to dampen the vibrations of the ear ossicles and help protect the delicate cells of the inner ear.

(3) Vibration of the foot of the stapes against the **oval window** causes pressure waves of the **perilymph** within the scala vestibuli.

(4) Vestibular membrane movements cause pressure waves in the endolymph within the scala media (cochlear duct). This displaces the basilar membrane (in different regions depending on the frequency). Hair cells of the spiral organ bend, initiating nerve signals that are transmitted along the cochlear division of the vestibulocochlear nerve (CN VIII) to the brain.

(5) Pressure waves are absorbed by the round window.

5. *Optional Activity:* **APR 7: Nervous System**—Watch the "Hearing" animation to review the sequence of events involved in hearing.

Figure 18.9 **Sound-Wave Pathways Through the Ear.** Sound waves enter the external ear, are conducted through the ossicles of the middle ear, and then are detected by a specific region of the spinal organ in the inner ear.

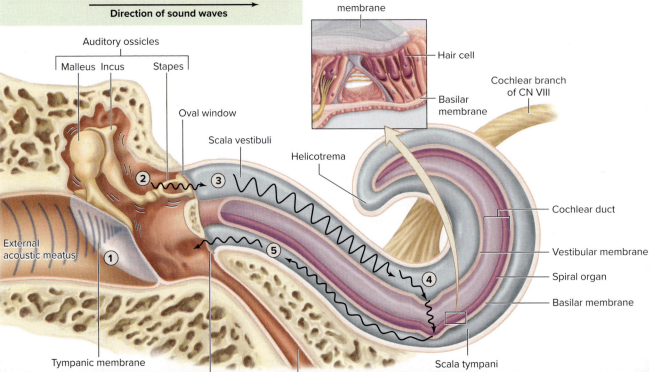

HISTOLOGY

General Senses

The definitions of general and special senses are given at the beginning of the gross anatomy section of this chapter. If you are starting laboratory observations with the histology exercises, read the introduction to senses in the gross anatomy section before proceeding.

This section focuses on the histology of somatic sensory receptors in the skin (**table 18.6**).

Table 18.6	Sensory Receptors in Thick Skin			
Receptor	**Location**	**Structure**	**Senses**	**Word Origin**
Unencapsulated Tactile Receptors				
Free Nerve Ending	Primarily the papillary layer of the dermis with some extending into the epidermis; associated with glands and hair follicles	An unmodified nerve ending	Temperature, pain, and pressure	*free*, referring to the fact that there are no connective tissue coverings
Tactile Disc (Merkel Disc)	At the junction between the dermis and the epidermis	An association between a modified keratinocyte in the epidermis, called a tactile cell, and a specialized nerve ending in the dermis, called a tactile disc	Fine, light touch	*tactus*, to touch
Encapsulated Tactile Receptors				
Lamellated (Pacinian) Corpuscle	Deep within the reticular layer of the dermis	Concentric layers of inner neurolemmocytes and outer connective tissue surrounding a nerve ending	Deep pressure and high-frequency vibration	*lamina*, plate, + *corpus*, body
Tactile (Meissner) Corpuscle	Within dermal papillae, especially in areas lacking hair follicles.	An oval structure consisting of modified neurolemmocytes and connective tissue encapsulating a nerve ending	Fine, discriminative touch to determine textures and shapes; light touch	*tactus*, to touch, + *corpus*, body

EXERCISE 18.6

TACTILE (MEISSNER) CORPUSCLES

1. Obtain a histology slide of thick skin and place it on the microscope stage. Bring the tissue sample into focus using the scanning objective.

2. Observe the slide on low power and review the layers of the skin (**figures** 6.1 and **18.10**).

   Epidermis: stratum basale, stratum spinosum, stratum granulosum, stratum lucidum, and stratum corneum

   Dermis: papillary layer and reticular layer

3. Move the microscope stage so the junction between the dermis and epidermis is in the center of the field of view, and locate the **dermal papillae** (figure 18.10).

Figure 18.10 **Thick Skin and Tactile (Meissner) Corpuscles.**
Tactile corpuscles are found within the dermal papillae of the skin and are sensory receptors for fine touch.

©McGraw-Hill Education/Al Telser

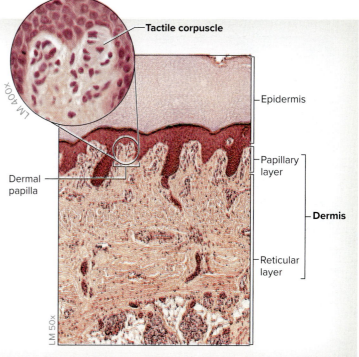

(continued on next page)

(continued on next page)

Next, move the stage so a single papilla is in the center of the field of view, then change to high power.

4. Sensory receptors called **tactile corpuscles** are located within the dermal papillae of thick skin. These receptors are oval in shape with a surrounding capsule of connective tissue (figure 18.10 and table 18.6), and they function in sensing light touch. Look for tactile corpuscles within the dermal papilla. If a tactile corpuscle is not visible, scan the slide for other papillae that may have tactile corpuscles within.

5. Sketch a tactile corpuscle within a dermal papilla as seen through the microscope in the space provided or on a separate sheet of paper.

_____ ×

EXERCISE 18.7

LAMELLATED (PACINIAN) CORPUSCLES

1. Obtain a histology slide of thick skin and place it on the microscope stage.

2. Observe the slide on low power and identify the dermis and epidermis (figure 18.10).

3. Move the microscope stage so the deepest part of the reticular layer of the dermis is in the center of the field of view. Within this portion of the dermis there are cross sections through numerous sweat glands and blood vessels. Cross sections of **lamellated corpuscles** should also be visible. Lamellated corpuscles resemble onions in cross section because they are composed of concentric layers of connective tissue. The inner core of the corpuscle is composed of neurolemmocytes that enclose the dendritic endings of sensory neurons (**figure 18.11** and table 18.6).

4. Locate a lamellated corpuscle, and then move the microscope stage so this structure is in the center of the field of view. Change to a higher power to view the structure in more detail. Lamellated corpuscles function in the sensation of deep pressure. When enough pressure is applied to the surface of the skin, the layers of connective tissue surrounding the central sensory receptor are compressed and initiate nerve signals to the brain.

5. Sketch a lamellated corpuscle as seen through the microscope, making note of its location in the skin, in the space provided or on a separate sheet of paper.

_____ ×

LM 30x

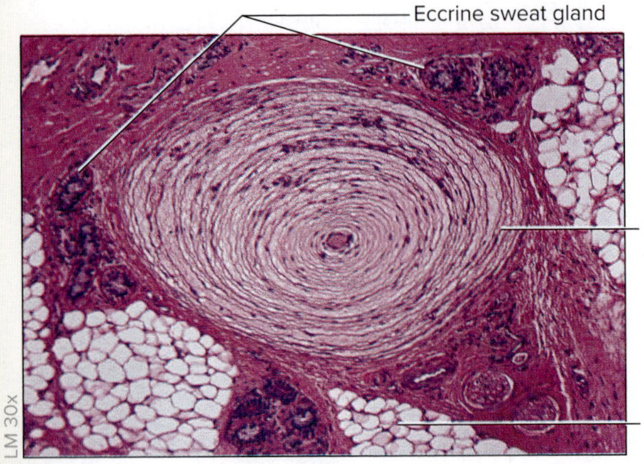

Figure 18.11 **Lamellated Corpuscles.** Lamellated corpuscles are located deep within the reticular layer of the dermis and are sensory receptors for deep pressure.

©Carolina Biological Supply Company/Phototake

Special Senses

Special senses are specialized organs within the head that respond specifically to the modalities of olfaction, taste, vision, hearing, and equilibrium. The next set of laboratory exercises explores the structure and function of these special sensory organs.

EXERCISE 18.8

GUSTATION (TASTE)

One reason gustation (taste) is so pleasurable is that this pathway relays sensory input to the limbic system, the emotional brain. Yet to have a complete sense of gustation, our olfactory sense must also be functioning. Like gustation, the olfactory pathway relays signals to the limbic system, and without the ability to smell, taste suffers tremendously.

There are four types of papillae located on the tongue: filiform, fungiform, vallate (circumvallate), and foliate. The detailed structure, function, and locations of the tongue papillae are shown in **figure 18.12**. Although all papillae (with the exception of filiform) contain taste buds, more than half of these taste buds are located in the large vallate papillae on the posterior region of the tongue. **Taste buds** (gustatory buds) are sensory receptors specialized for gustatory sensation (**table 18.7** and **figure 18.13**). Taste buds are particularly concentrated on the papillae of the tongue, but they are also located throughout the oral cavity and pharynx. This exercise explores the types of tongue papillae and the location and function of taste buds associated with the papillae.

1. Obtain a histology slide of the tongue or a histology slide demonstrating mammalian *vallate papillae* (figure 18.12*d*). Scan the slide at low power and identify the papillae on the surface of the tongue. Move the microscope stage so one or two papillae are in the center of the field of view. Then increase the magnification, first to medium and then to high power. Locate a taste bud at the edge of the papilla (figure 18.13).

2. Identify the following structures, using figures 18.12 and 18.13, and table 18.7 as guides. (Note that all of the types of papillae may not be visible on a single slide.)

- ☐ **Basal cells**
- ☐ **Filiform papillae**
- ☐ **Foliate papillae**
- ☐ **Fungiform papillae**
- ☐ **Gustatory cells**
- ☐ **Supporting cells**
- ☐ **Taste pores**
- ☐ **Vallate (circumvallate) papillae**

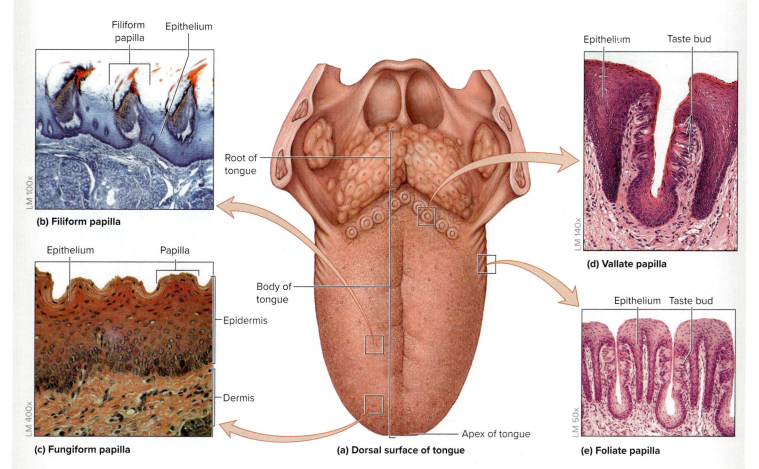

Figure 18.12 **Taste Buds.** Gustation (taste) requires taste buds, which are associated with tongue papillae. (*a*) Dorsal surface of tongue, (*b*) Filiform papilla, (*c*) Fungiform papilla, (*d*) Vallate papilla, (*e*) Foliate papilla.

(*b*) ©CNRI/Science Photo Library/Corbis; (*c*) ©McGraw-Hill Education/Christine Eckel, photographer; (*d*) ©McGraw-Hill Education/Alvin Telser; (*e*) ©Jose Luis Calvo/Shutterstock.com

(continued on next page)

(continued from previous page)

3. Locate a taste bud at the edge of the papilla in the crevice (crypt) between two papillae (figure 18.13).

4. Sketch a vallate papilla as seen through the microscope in the space provided or on a separate sheet of paper. Be sure to label the taste buds.

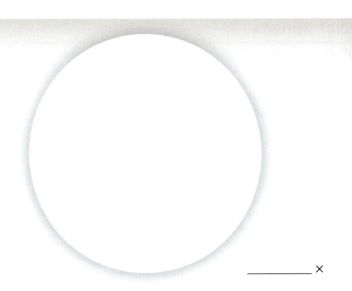

_____ ×

Table 18.7	Cells Associated with Taste Buds		
Structure	**Description**	**Function**	**Word Origin**
Basal Cells	Small stem cells found at the base of the taste bud	Precursor cells to the supporting cells and gustatory cells	*basalis*, situated near the base, + *cella*, a chamber
Gustatory Cells	Light-staining cells with round nuclei; contain modified microvilli (called *taste hairs*) on the apical surface that extend through taste pores	Detect chemicals dissolved in solution; transmit nerve signals for taste sensation to the CNS	*gustus*, a tasting, + *oriusi*, having to do with, + *cella*, a chamber
Supporting Cells	Dark-staining cells with oval-shaped nuclei; located between gustatory cells	Support the gustatory cells by producing a glycoprotein	*supporto*, to carry, + *cella*, a chamber

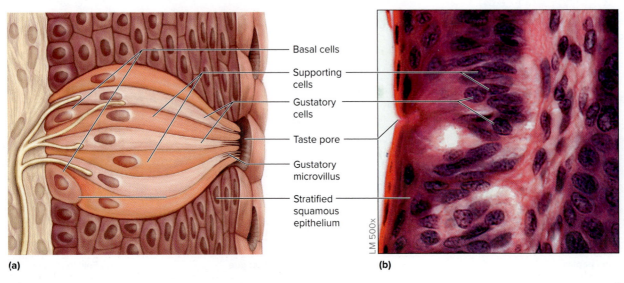

(a)   (b)

Figure 18.13   **Detailed Structure of a Taste Bud.** (*a*) Illustration. (*b*) Photomicrograph.

(*b*) ©McGraw-Hill Education/Alvin Telser

OLFACTION (SMELL)

Olfaction is the sense of smell. It is an important sensory modality, not just for smell, but also for gustation (taste). Olfactory sensation is detected by special sensory cells found within the epithelium lining the roof of the nasal cavity—**olfactory epithelium** (**figure 18.14a** and **table 18.8**). The **olfactory receptor cells** are neurons, but they are unique because they are continuously replaced. The olfactory receptor cells compose the **olfactory nerves (CN I)**.

1. Obtain a histology slide of olfactory epithelium (figure 18.14b) and place it on the microscope stage.

2. Bring the tissue sample into focus on low power. Increase magnification to medium power, and then to high power so the cells composing the olfactory epithelium are clearly visible. It may be difficult to distinguish between the three major cell types of the olfactory epithelium: basal cells, olfactory receptor cells, and supporting cells. In general, the nuclei closest to the basement membrane of the epithelium are the nuclei of *basal cells* (the cells appear triangular in shape), the nuclei in the middle of the epithelium are the nuclei of *olfactory receptor cells*, and the nuclei closest to the apical surface of the epithelium are the nuclei of *supporting cells*.

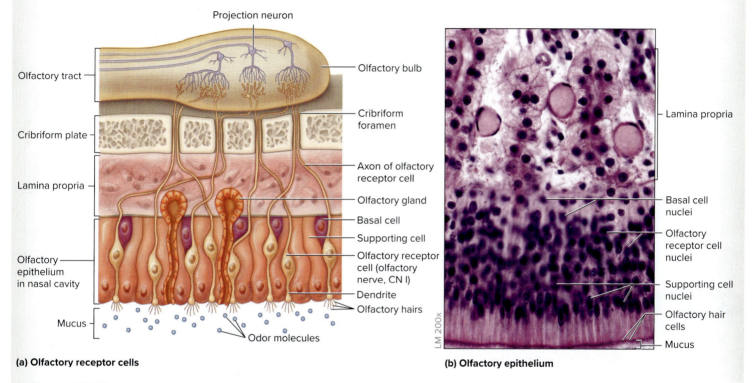

(a) **Olfactory receptor cells**

(b) **Olfactory epithelium**

Figure 18.14 **Olfactory Epithelium.** (a) Cell types within the olfactory epithelium. (b) Histology of the olfactory epithelium.

(b) ©Alvin Telser/Media Bakery

Table 18.8	Olfactory Epithelium		
Structure	**Description**	**Function**	**Word Origin**
Basal Cells	Triangular or cone-shaped cells; nuclei located near the basal surface of the olfactory epithelium	Neural stem cells; continually replace both olfactory receptor cells and supporting cells	*basalis*, situated near the base, + *cella*, a chamber
Olfactory Receptor Cells	Bipolar neurons containing large, centrally located nuclei; have a single dendrite with olfactory hairs (modified cilia) and an unmyelinated axon	Sensory receptors for olfaction (smell); olfactory hairs (cilia) detect dissolved odoriferous substances.	*olfactus*, to smell, + *recipio*, to receive, + *cella*, a chamber
Supporting Cells	Columnar epithelial cells with nuclei located near the apical surface of the olfactory epithelium	Surround and support the specialized olfactory receptor cells	*supporto*, to carry, + *cella*, a chamber

(continued on next page)

(continued from previous page)

3. Identify the following structures on the histology slide of olfactory epithelium, using figure 18.14 and table 18.18 as guides:

☐ **Basal cells**   ☐ **Olfactory receptor cells**
☐ **Ofactory hairs (cilia)**   ☐ **Supporting cells**

4. Sketch the olfactory epithelium as seen through the microscope in the space provided or on a separate sheet of paper. Label all of the structures listed in step 3.

_____ ×

INTEGRATE

CONCEPT CONNECTION

An anatomical relationship exists between the structures associated with olfaction and the ethmoid bone, which forms the roof and lateral walls of the nasal cavity. The superior and middle nasal conchae cause incoming air to swirl within the nasal cavity, which provides more time for odors to stimulate olfactory receptor cells within the olfactory epithelium. In addition, the superior portion of the ethmoid bone contains the **cribriform foramina**, tiny holes within the cribriform plate. Axons of the olfactory nerves (CN I) extend through the cribriform foramina to synapse with neurons within the **olfactory bulbs**, which rest on the superior aspect of the cribriform plate. Neurons within the olfactory bulbs subsequently transmit nerve signals to the brain via the **olfactory tract**. Damage to any of these structures, including damage to the ethmoid bone itself (as might occur with a fracture to the bone), may cause a reduced ability or an inability to smell (anosmia).

EXERCISE 18.10

VISION (THE RETINA)

The **retina (figure 18.15)** of the eye is called the **neural tunic** because this layer is composed of neural tissue. The retina develops as a direct outgrowth of the brain. Thus, the retina is the only part of the brain visible without surgical intervention (though an ophthalmoscope is required). Axons from neurons within the retina travel to the brain through the **optic nerve (CN II)**. The retina is responsible for transducing light rays into electrical signals (action potentials) to the brain. This information is relayed from photoreceptor cells (rods and cones) to bipolar cells to the ganglion cells within the retina. Axons of ganglion cells extend from the back of the eye as the optic nerve to synapse with neurons within the **thalamus**. These neurons extend to the **occipital lobe** of the brain. Here, the visual information is processed and interpreted. The retina is a very complex yet beautifully organized structure. Rods function in dim light and cones function in high-intensity light and in color vision. This laboratory exercise explores the structure and function of the retina by observing the cells that are visible histologically. Laboratory exercises within the gross anatomy section of this chapter will place the retina in the context of other structures of the eye.

1. Obtain a histology slide of the retina and place it on the microscope stage.

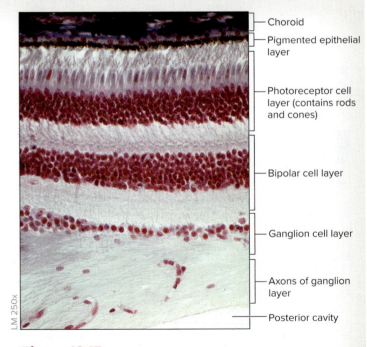

LM 250x

Choroid
Pigmented epithelial layer
Photoreceptor cell layer (contains rods and cones)
Bipolar cell layer
Ganglion cell layer
Axons of ganglion layer
Posterior cavity

Figure 18.15 **Histology of the Retina.**

©McGraw-Hill Education/Al Telser

2. Bring the tissue sample into focus on low power, then move the microscope stage so the retina (figure 18.15) is in the center of the field of view. Switch to medium power and bring the tissue sample into focus once again.

3. Identify the following structures on the slide of the retina, using **table 18.9** and figure 18.15 as guides. (High power may be required to view all of the structures or to see them in greater detail.)

- ☐ **Bipolar cell layer**
- ☐ **Choroid**
- ☐ **Ganglion cell layer**
- ☐ **Photoreceptor cell layer**
- ☐ **Pigmented layer**
- ☐ **Sclera**

4. With the medium-power objective in place, scan the slide and locate the **fovea centralis (figure 18.16a)**. The fovea centralis is a thinner-than-normal area of the retina. The fovea centralis contains photoreceptor cells and is devoid of bipolar and ganglion cell layers. This area has the highest concentration of cones of the entire retina, providing the highest visual acuity. We turn our head so that we are using the fovea centralis to generate the sharpest image of the object of interest.

5. Scan the slide and locate the **optic disc**, the location where the **optic nerve** leaves the eye (figure 18.16b). The **optic disc** (blind spot) is easily identifiable because all retinal layers are absent at this location. Notice that the optic disc and optic nerve are approximately the same color as the cells in the ganglion cell layer of the retina. This is helpful in understanding that the ganglion cell layer, the optic disc, and the optic nerve all are composed of the axons of ganglion cells. The axons of the ganglion cell layer leave the eye at the optic disc and extend from the eye to the brain as the optic nerve. Axons of the optic nerve are the only portion of the ganglionic axons that are myelinated.

6. Sketch the retina as seen through the microscope in the space provided or on a separate sheet of paper. Label the following layers: rod and cone layer, bipolar cell layer, and ganglion cell layer.

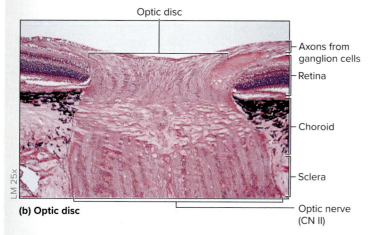

(a) Fovea centralis

Fovea centralis

- Ganglion cell layer
- Bipolar cell layer
- Photoreceptor cell layer
- Pigmented layer
- Choroid

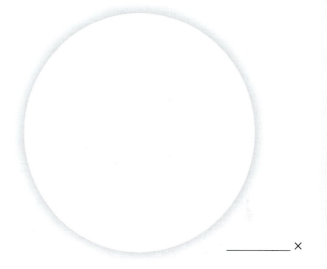

(b) Optic disc

Optic disc

- Axons from ganglion cells
- Retina
- Choroid
- Sclera
- Optic nerve (CN II)

_____ ✕

7. *Optional Activity:* **APR** 7: **Nervous System**—Watch the "Vision" animation to learn the sequence of events involved in vision and the functions of cells in the retina.

Figure 18.16 **Specialized Areas of the Neural Tunic of the Eye.** (*a*) The fovea centralis is the area of the retina where visual acuity is the highest. Ganglion and bipolar cell layers are absent, and there is an abundance of cones in the photoreceptor layer. (*b*) The optic disc is the area of the retina where the axons of ganglion cells exit the retina to become the optic nerve. There are no rods or cones in this area, which is why the optic disc is also referred to as the "blind spot" of the retina.

(*a*) ©Gene Cox/Science Source; (*b*) ©Victor P. Eroschenko

(continued on next page)

(continued from previous page)

Table 18.9	The Retina		
Structure	**Description**	**Function**	**Word Origin**
Retina	The "neural tunic" of the eye; consists of numerous layers of neurons involved in phototransduction	Phototransduction: transduction of light waves that enter the eye into nerve signals (action potentials)	*rete*, a net
Bipolar Cell Layer	Middle layer composed of bipolar neurons with intermediate-sized nuclei	Receives signals from rods and cones and transmits electrical signals to ganglion cells	*bipolar*, relating to bipolar neurons
Ganglion Cell Layer	Innermost layer of the retina, composed of neurons with very large nuclei	Receives information from bipolar cells and sends that information to the brain	*ganglion*, a swelling or knot
Pigmented Layer	Outermost portion of retina attached to choroid	Absorbs extraneous light; provides vitamin A for photoreceptor cells	NA
Photoreceptor Cell Layer	Outermost layer of the retina (closest to the choroid and sclera), containing the light-transducing portions of photoreceptor cells (rods and cones); rods and cones are the smallest and most numerous nuclei of the retina	Layer of the retina where light waves are initially transduced into neuronal action potentials; the cells in this layer synapse with neurons in the bipolar cell layer	NA
Cones	Photoreceptor cells with a light-transducing portion located in the outermost layer of the retina; not distinguishable from rods using a light microscope	Photoreceptor cell specializing in color vision; **fovea centralis** is an area composed exclusively of cones and is the area of highest visual acuity	*conus*, shaped like a cone
Rods	Photoreceptor cells with a light-transducing portion located in the outermost layer of the retina; not distinguishable from cones using a light microscope	Photoreceptor cells specializing in black-and-white vision; very sensitive, most useful when light is dim	*rod*, shaped like a rod

EXERCISE 18.11

HEARING

Hearing is a function of the **cochlea**. This special sensory organ is located within the petrous part of the temporal bone. The gross anatomy section of this chapter focuses on the location, gross structure, and function of this organ. This section focuses on the histological features of the highly specialized epithelium that lines the cochlea, the **spiral organ** (organ of Corti), which will provide an insight into how this organ performs its function: transformation of sound waves into nerve signals that can be interpreted by the brain.

The cochlea **(figure 18.17)** is the organ responsible for transducing fluid vibrations received at the oval window into electrical signals that are sent to the thalamus and then on to the temporal lobe of the brain, where they are interpreted. Within the cochlea, the spiral organ rests upon the **basilar membrane**, within the scala media (cochlear duct).

1. Obtain a slide of the cochlea **(figure 18.18)** and place it on the microscope stage. Bring the tissue sample into focus on low power and then increase the magnification. Move the microscope stage until a single cross section through the cochlea is in the center of the field of view.

2. Identify the three chambers within the cochlea and the membranes that separate the chambers from each other, using figures 18.17 and 18.18 and **table 18.10** as guides:

□ **Basilar membrane**

□ **Scala media (cochlear duct)**

□ **Scala tympani**

□ **Scala vestibuli**

□ **Vestibular membrane**

3. Once the *scala media* (cochlear duct) has been identified, move the microscope stage so the scala media is in the center of the field of view. Increase the magnification to high power and focus in on the spiral organ (figure 18.18).

4. Identify the following structures on the slide of the cochlea, using **tables** 18.10 and **18.11** and figure 18.18 as guides:

□ **Basilar membrane**

□ **Cochlear nerve**

□ **Endolymph**

□ **Perilymph**

□ **Scala media (cochlear duct)**

□ **Scala tympani**

□ **Scala vestibuli**

□ **Spiral ganglion**

□ **Supporting cells**

□ **Tectorial membrane**

□ **Vestibular membrane**

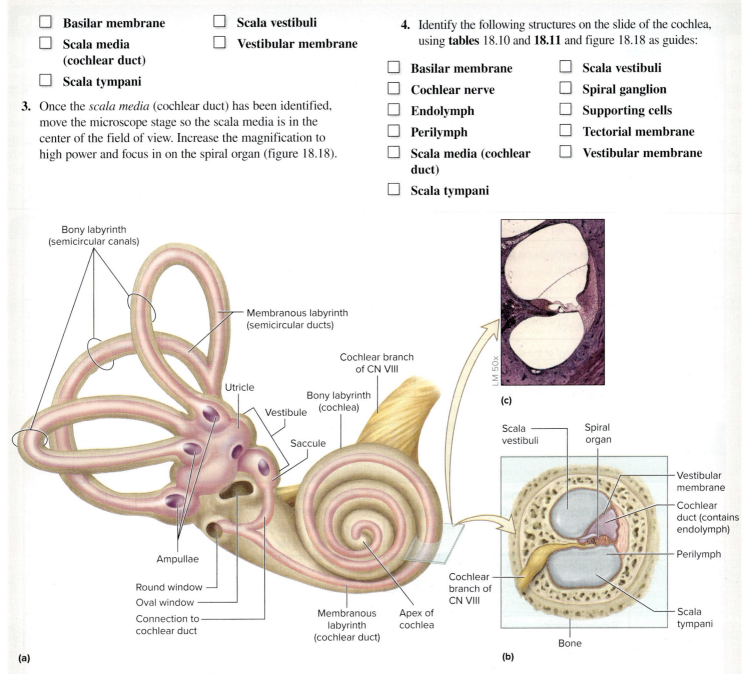

(a)

Bony labyrinth (semicircular canals)

Membranous labyrinth (semicircular ducts)

Utricle

Vestibule

Saccule

Ampullae

Round window

Oval window

Connection to cochlear duct

Membranous labyrinth (cochlear duct)

Apex of cochlea

Cochlear branch of CN VIII

Bony labyrinth (cochlea)

LM 50x

(c)

(b)

Scala vestibuli

Spiral organ

Vestibular membrane

Cochlear duct (contains endolymph)

Perilymph

Scala tympani

Cochlear branch of CN VIII

Bone

Figure 18.17 **The Cochlea.** (*a*) The semicircular canals and cochlea are part of the inner ear. (*b*) The cochlea houses the spiral organ, which contains specialized cells that translate sound waves into sensory impulses. (*c*) Light micrograph demonstrating a cross section through the cochlea.

(c) ©Biophoto Associates/Science Source

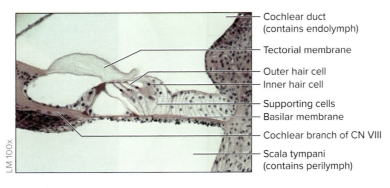

Cochlear duct (contains endolymph)

Tectorial membrane

Outer hair cell

Inner hair cell

Supporting cells

Basilar membrane

Cochlear branch of CN VIII

Scala tympani (contains perilymph)

LM 100x

Figure 18.18 **Histology of the Spiral Organ.**

©Biophoto Associates/Science Source

(continued on next page)

TACTILE LOCALIZATION

This exercise tests the ability of the brain to detect the precise location on the body that has been touched by a stimulus. This is called **tactile localization**. Perception of the locale of a stimulus differs for areas of the body. This variation in stimulus locale is related to the size of the receptive field. When touching the subject during the tactile localization task, the subject will not necessarily know exactly what point was touched. Instead, the subject will perceive that a larger area surrounding the touch point was stimulated. This larger area is a *receptive field*.

Before beginning, state a hypothesis regarding the difference (if any) in the receptive field of sensory receptors within the skin on the back of the hand as compared to the skin on the palm of the hand:

_____

_____

_____

_____

1. Obtain two or three washable markers of different colors.

2. Choose a person to be the subject. Have the subject close his or her eyes.

3. Touch the palm of the subject's hand with a marker (**figure 18.20**).

4. Hand the subject a marker of a different color than the one previously used. Then ask the subject (still keeping his or her eyes closed) to try to touch the exact point previously touched, using his or her marker.

5. Measure the distance between the two points in millimeters (mm) and record it in **table 18.13**.

6. Repeat the process two more times. As the examiner, be sure to always touch the same point. However, the subject will not necessarily touch that same point. Calculate the average distance for the three trials and record it in table 18.13.

7. Complete steps 2–6 for each of the regions listed in table 18.13. Record the results in the table.

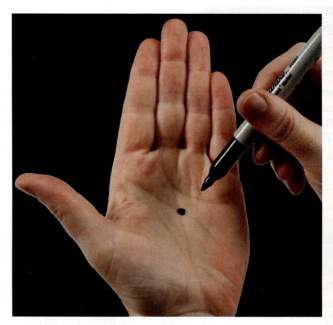

Figure 18.20   **Tactile Localization Test.** Testing tactile localization on the palm of the hand.

©Christine Eckel

Table 18.13	Results of the Tactile Localization Test			
Body Region	**Test 1 Distance (mm)**	**Test 2 Distance (mm)**	**Test 3 Distance (mm)**	**Average Distance (mm)**
Palm of Hand				
Back of Hand				
Fingertip				
Anterior Forearm				
Back of Neck				
Anterior Leg				

EXERCISE 18.14

GENERAL SENSORY RECEPTOR TESTS: ADAPTATION

This exercise tests the ability of the brain to detect a constantly applied stimulus over time. A general property of the nervous system is that the brain tends to place emphasis and focus on incoming stimuli that are *changing* as opposed to stimuli that remain constant. Thus, when initially putting on clothing, glasses, jewelry, or other items in the morning, you feel the clothing slide across your skin, the glasses rest upon your ears and nose, and your watch rest on your wrist. Soon afterwards, if these items don't move considerably, then you stop noticing them. Have you ever found yourself looking for your glasses only to discover they are right there on your head? This phenomenon is called **sensory adaptation**. Here, sensory receptors adapt so that the initial sensation (e.g., pressure) becomes a new "set point" for the brain. The brain subsequently ignores the constant signals coming in from that stimulus. This allows the brain to respond only when there is a *change* in the stimulus.

1. Obtain five pennies.

2. Choose a person to be the subject. Have the subject close his or her eyes.

3. Place a coin on the anterior surface of the subject's forearm approximately 2 cm proximal from the wrist **(figure 18.21a)**.

4. Ask the subject to state when he or she can no longer feel the coin on his or her arm. Record the time elapsed in **table 18.14**.

5. Repeat the test, placing coins at the locations on the forearm listed in table 18.14. Record the time elapsed for each test in table 18.14.

6. Finally, perform the test one more time on the spot used in step 3, only this time perform the test using five pennies stacked on top of each other (figure 18.21b).

(a)

(b)

Figure 18.21 **Sensory Adaptation Test.** Testing sensory adaptation on the anterior surface of the forearm with (*a*) one object and (*b*) multiple objects.
©Christine Eckel

Table 18.14	Results of the Adaptation Test	
Coin Placement on Anterior Forearm	**Number of Coins**	**Time Elapsed When Subject No Longer Perceives the Coins (Seconds)**
2 cm Proximal from Wrist	1	
Midway Between Wrist and Elbow	1	
1 cm Distal from Elbow	1	
2 cm Proximal from Wrist	5	
Midway Between Wrist and Elbow	5	
1 cm Distal from Elbow	5	

Special Senses

Gustation (Taste)

There are approximately 4000 gustatory receptors ("taste buds") in the body, many of which are associated with the papillae of the tongue observed in the histology section of this chapter. In addition to this location of receptors on the tongue, there are also taste receptors located on the soft palate, pharynx, and epiglottis. Each receptor is specialized to respond to one of five primary taste sensations: salty, sweet, sour, bitter, and umami ("meaty"). **Table 18.15** summarizes the types of substances that elicit each of the five basic taste sensations. This laboratory exercise tests the specificity and location of the different types of taste sensed by these taste receptors.

Table 18.15	The Five Basic Taste Sensations	
Taste Sensation	**Substances Stimulating the Sensation**	**Examples**
Bitter	Alkaloids, which are plant-derived, nitrogen-containing basic compounds	Quinine, nicotine, caffeine, unsweetened cocoa
Salt	Metal ions	Table salt (NaCl)
Sour	Acids, which are substances in foods that release hydrogen ions (H^+)	Lemon juice, vinegar
Sweet	Natural sugars or artificial sweeteners	Corn syrup, Splenda®, granulated sugar
Umami	Amino acids such as glutamate and aspartate	Meats, cheeses, tomatoes, anchovies

EXERCISE 18.15

GUSTATORY TESTS

Before beginning, state a hypothesis regarding the region(s) of the oral cavity and pharynx that may have the largest number of receptors for each of the taste sensations.

_____

_____

_____

1. Obtain seven cotton swabs and a small sample of each of the following:

   ☐ **Monosodium glutamate solution (umami)**

   ☐ **Dalt solution (salt)**

   ☐ **Sugar solution (sweet)**

   ☐ **Tonic water (bitter)**

   ☐ **Vinegar (sour)**

   ☐ **Water (for rinsing mouth)**

2. One student will be the tester and one student will be the subject. As the tester, do not tell the subject which solutions are being tested. Dip a cotton swab in one of the solutions and apply it to the roof of the subject's mouth (the palate), the inside of his or her cheeks, and the surface of his or her tongue.

3. Place a check mark in **table 18.16** next to all of the regions where the subject is able to taste the solution.

4. Discard the swab in a biohazard waste container.

5. Have the subject drink some plain water to rinse out his or her mouth.

6. Dip a clean swab into a different solution and repeat step 2. Continue to do this until the subject has tasted all of the solutions.

7. Use two cotton swabs. Dip one in the salt solution and one in the monosodium glutamate solution. Swab the subject's tongue with both solutions simultaneously and ask the subject to describe the taste. Record the subject's description of the taste here:

   _____

   _____

   _____

Table 18.16	Results of the Gustatory Tests		
Place a check in the boxes where the subject was able to detect each taste.			
Taste	**Palate**	**Cheeks**	**Tongue**
Bitter			
Salty			
Sour			
Sweet			
Umami			

Olfaction (Smell)

As observed in exercise 18.9, the sense of olfaction involves sensory receptors that are embedded in a special epithelium that lines the roof of the nasal cavity, the **olfactory epithelium**. This epithelium also contains goblet cells, which produce mucin that when hydrated forms mucus. As odorants enter the nasal cavity, they dissolve in the mucus. The olfactory receptor cells bind the odorant molecules and are stimulated. Nerve signals are initiated in the brain that are interpreted as the sense of olfaction (smell). As likely experienced before, the sense of smell is highly associated with the sense of gustation (taste). Recall a time when you had a "stuffy nose." The blockage of the sense of smell may have greatly affected your appetite because of its effect on gustation. This exercise will test the effect of olfaction on the perception of gustatory (taste) sensation.

EXERCISE 18.16

OLFACTORY TESTS

EXERCISE 18.16A Effect of Olfaction on Taste I

1. Obtain six cotton swabs, a blindfold, and vials containing samples of each of the following:

   ☐ **Almond extract**   ☐ **Peppermint oil**

   ☐ **Clove oil**   ☐ **Vanilla**

   ☐ **Lemon extract**   ☐ **Wintergreen oil**

2. One student will be the tester and one student will be the subject. As the tester, do not tell the subject which substances are being tested. Have the subject close his or her eyes (or use the blindfold if that is easier) and pinch his or her nose closed.

3. Dip a cotton swab in one of the solutions and apply it to the subject's tongue. Ask the subject if he or she can taste the substance. Place a check in the appropriate box in **table 18.17** if the subject was able to taste the odorant with the nose closed.

4. After a few seconds, have the subject release his or her nose and breathe in. Ask him or her to identify the odorant again. Place a check in the appropriate box in table 18.17 if the subject was able to taste the odorant with the nose open.

5. Briefly note the conclusions in the space below:

   _____

   _____

   _____

Table 18.17	Results of the Tests for the Effect of Olfaction on Taste I	
Place a check in the boxes where the subject was able to detect each taste.		
Odorant	**Nose Closed**	**Nose Open**
Almond		
Clove		
Lemon		
Peppermint		
Vanilla		
Wintergreen		

(continued on next page)

(continued from previous page)

EXERCISE 18.16B Effect of Olfaction on Taste II

1. Obtain the following:

   ☐ **Apple slices**   ☐ **Potato slices**

   ☐ **Blindfold**

2. One student will be the tester and one student will be the subject. As the tester, do not tell the subject which substances are being tested. Have the subject close his or her eyes (or use the blindfold if that is easier). Then have the subject pinch his or her nose closed and stick out his or her tongue.

3. Apply either the apple or the potato slice to the subject's tongue. Ask the subject if the substance can be tasted. Place a check in the appropriate box in **table 18.18** if the subject was able to correctly identify the taste of the substance.

4. After a few seconds, have the subject once again close his or her eyes and stick out his or her tongue. Place the apple on the tongue and the potato under the nose (or vice versa). Place a check in the appropriate box in table 18.18 if the subject was able to correctly identify the substance.

5. Briefly note the conclusion in the space below:

   _____

   _____

   _____

EXERCISE 18.16C Olfactory Adaptation

1. Obtain six cotton swabs, a blindfold, a timer or other mechanism for keeping time, and vials containing samples of each of the following:

   ☐ **Almond extract**   ☐ **Peppermint oil**

   ☐ **Clove oil**   ☐ **Vanilla**

   ☐ **Lemon extract**   ☐ **Wintergreen oil**

2. One student will be the tester and one student will be the subject. As the tester, do not tell the subject which odors are being tested. Have the subject close his or her eyes (or use the blindfold if that is easier).

3. Dip a cotton swab in one of the solutions and hold the swab 1–2 inches away from the subject's nose. Start the timer as soon as the swab is placed next to the subject's nose. Ask the subject to identify the odor and state when he or she can no longer smell the odor. The odor will "go away" with time because of adaptation of the olfactory receptor cells to the odor. Record the time taken for adaptation to occur in **table 18.19**.

4. Repeat step 3 for all of the remaining odorants.

5. Briefly note the conclusion in the space below:

   _____

   _____

   _____

Table 18.18	Results of Tests for the Effect of Olfaction on Taste Sensation	
Subject pinches nose closed and tester places item on subject's tongue.		
Correct Identification of Taste?	**Yes**	**No**
Apple		
Potato		

Tester places one item on subject's tongue and the other under subject's nose (subject's nose is open).		
Item on Tongue	**Item Under Nose**	**Correct Taste Perceived?**
Apple	Potato	
Potato	Apple	

Table 18.19	Results of Olfactory Adaptation Tests
Record the time for adaptation for each of the following odorants.	
Odorant	**Time for Adaptation (Seconds)**
Almond	
Clove	
Lemon	
Peppermint	
Vanilla	
Wintergreen	

Vision

Recall from exercise 18.10 that the special sense of vision involves stimulation of photoreceptors housed in the neural tunic of the eye, called the **retina**. Specifically, light changes the conformation of light-sensitive proteins in the photoreceptors of the retina: the rods and cones. **Rods** are photoreceptors that specialize in black-and-white vision, whereas **cones** are photoreceptors that specialize in color vision. When light stimulates these photoreceptors, nerve signals are transmitted along the optic nerve to the thalamus and then to the occipital lobe of the cerebrum, where the information is perceived as a visual image. Incidentally, there are no rods or cones where the optic nerve exits the eye, making that location a visual *blind spot*. Refer to the textbook as a guide when reviewing the specific pathways involved in processing visual information.

EXERCISE 18.17

VISION TESTS

The vast majority of us have first-hand knowledge of the special sense of vision. A trip to the eye doctor involves testing **visual acuity**. An ophthalmologist assesses **vision** with a Snellen eye chart (**figure 18.22**) and from there determines if corrective lenses are necessary. Also, sometimes age forces us to hold things we are reading farther away from the eyes in order to focus on an image or the words on a page. This **near-point accommodation** decreases dramatically with age due to the decreased elasticity in the lens of the eye. The result is an inability to change the shape of the lens, as is required when focusing on near objects. The following exercises test such aspects of vision as acuity, near-point accommodation, and the visual blind spot.

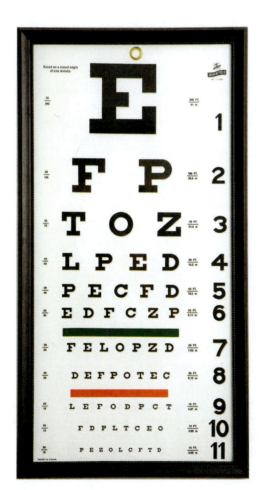

Figure 18.22 **Snellen Eye Chart.** Testing visual acuity using the Snellen eye chart.

©McGraw-Hill Education/Rick Brady, photographer

EXERCISE 18.17A Visual Acuity

1. Obtain a Snellen eye chart to test visual acuity. Hang the eye chart at eye level on a wall. Be sure that the wall is well illuminated.

2. Have the subject stand 20 feet from the eye chart, while a partner remains near the eye chart to validate the subject's responses.

3. Have the subject cover his or her left eye and read the lowest possible line of letters with his or her right eye. Record the visual acuity in **table 18.20**.

4. Repeat step 3 with the subject's left eye. Note that if the subject wears corrective lenses, this test can be conducted with and/or without glasses. A ratio of 1 (e.g., 20/20) indicates normal vision. A ratio of greater than 1 (20/15) indicates greater visual acuity; a ratio of less than 1 (20/30) indicates lesser visual acuity.

Table 18.20	Results of Visual Acuity Tests
Record visual acuity for the right and left eye, with and without corrective lenses.	
Eye	**Visual Acuity**
Right Eye	
Right Eye (with corrective lenses)	
Left Eye	
Left Eye (with corrective lenses)	

EXERCISE 18.17B Near-Point Accommodation

1. Obtain a pen or pencil and hold the object at arm's length in front of you. Cover the left eye and slowly move the object toward the right eye until the image of the object is no longer clear (i.e., is blurry or appears as two objects). Have a lab partner measure the distance

(continued on next page)

(continued from previous page)

from the right eye to the object. Record the near-point accommodation for the right eye in **table 18.21**.

2. Repeat step 1 by covering the right eye and moving the object toward the left eye. Record near-point accommodation for the left eye in table 18.21. Note that if the subject wears corrective lenses, this test can be conducted with and/or without glasses.

Table 18.21	Results of Near-Point Accommodation Tests

Record near-point accommodation for the right and left eye, with and/or without corrective lenses.

Eye	Near-Point Accommodation (cm)
Right Eye	
Right Eye (with corrective lenses)	
Left Eye	
Left Eye (with corrective lenses)	

EXERCISE 18.17C Blind Spot Determination

1. Hold **figure 18.23** an arm's length away from your face, approximately 46 cm (18 inches) away. Close the left eye and focus on the X with your right eye.

2. Move the figure toward your face while continuing to focus on the X. Stop when the black dot no longer appears in your field of view. Have a lab partner measure the distance between the eye and the image. Record the distance between the image and the right eye in **table 18.22**.

3. Flip the lab book upside-down so the dot appears on the left of the image in figure 18.23. Repeat steps 1–2 by closing the right eye and focusing on the X with the left eye. Record the distance between the image and the left eye in table 18.22.

Figure 18.23   **Blind Spot Determination.**

Table 18.22	Results for Blind Spot Determination Test

Record the blind spot distance for the right and for the left eye.

Eye	Distance (cm)
Right Eye	
Left Eye	

EXERCISE 18.17D Color Blindness

1. Have a lab partner hold **figure 18.24** (Ishihara color plate number 7) approximately 30 inches away from your face directly in your line of vision.

2. Examine the Ishihara color plate number 7 (figure 18.24) for 3 seconds. Is a number visible on the plate? _____ (yes/no). If yes, what number is visible? _____

3. Have a lab partner repeat steps 1–2. Are the results the same? _____

4. Compare your results to those of other classmates. Note observations here: _____

_____

5. If a series of Ishihara color plates is available in the laboratory, repeat steps 1–4 with all available color plates. For each color plate, be sure to record if numbers are visible or not, and if so, what number is visible. Report observations here: _____

_____

_____

Figure 18.24   **Ishihara Color Test Plate.** Subjects with normal color vision should see the number 74. Subjects with some degree of color blindness may see the number 21 or no number at all.

©Steve Allen/Getty Images

Hearing and Equilibrium

The inner ear is composed of the cochlea, vestibule, and semicircular canals. The **cochlea** is responsible for the sense of hearing.

INTEGRATE

CLINICAL VIEW
Tinnitus

If you have ever heard low ringing or buzzing sounds that seem to come out of nowhere, then you have experienced **tinnitus**, the perception of sound despite no outside stimulus or source. For many people, tinnitus is chronic and is a symptom of an underlying condition. These conditions are not usually life threatening but can permanently affect hearing. Damage to hair cells in the cochlea due to continuous exposure to loud noises, infections of the ear, or benign tumors that arise on vestibulocochlear nerve (CN VIII) can all result in tinnitus. Although there are no cures for tinnitus, there are treatments that make this symptom more manageable.

The vestibule and semicircular canals are also contained within the inner ear. The **vestibule** monitors *static equilibrium,* when the body is not in motion, and *linear dynamic equilibrium,* when the body is accelerating or decelerating in a linear plane. The **semicircular canals** detect *angular dynamic equilibrium.* The semicircular canals are three fluid-filled canals that are arranged orthogonally (at right angles) to detect motion in three dimensions. For example, the vestibule detects acceleration or deceleration (i.e., traveling along a straight path), whereas semicircular canals detect angular motion (i.e., sharply turning a corner). Interestingly, hair cells, the same type of receptors responsible for detecting hearing in the cochlea, also detect static and dynamic equilibrium in the vestibule and semicircular canals.

Stimulation of the receptors within the vestibule and semicircular canals initiates nerve signals that are transmitted along the vestibular branch of the vestibulocochlear nerve (CN VIII) to the thalamus and cerebral cortex, as well as to the nuclei for the three cranial nerves that control the movement of the eye: oculomotor (CN III), trochlear (CN IV), and abducens (CN VI). Refer to your textbook as a guide for reviewing the specific neural pathways involved in hearing and equilibrium.

EXERCISE 18.18

HEARING TESTS

Health care providers may perform tests for proper hearing function if they suspect a problem. The following laboratory exercises involve performing tests of hearing. The sensory receptors for hearing are housed within the inner ear, and sensory information is transmitted along the vestibulocochlear nerve. There are tests for the different types of **deafness** (absence of sound perception) and for vestibular dysfunction.

Distinguishing between **nerve deafness** and **conduction deafness** requires important clinical tests, namely, the Rinne and Weber tests. The **Rinne test** evaluates a person's ability to perceive sound from vibrations in the air as compared to vibration applied directly to the temporal bone. The **Weber test** evaluates a person's ability to perceive sound in both ears equally, which is a specific test for nerve conduction deafness.

EXERCISE 18.18A Hearing Test: Rinne

1. Obtain a tuning fork and choose an individual to be the subject. When the tuning fork is tapped on a hard surface, it will begin to vibrate at a certain wavelength and produce sound.

2. Tap the tuning fork on a hard surface and hold it near the subject's ear **(figure 18.25a)**. Ask the subject to tell you about the sound, and make a note of the subject's response here: _____

   _____

3. Gently tap the tuning fork on a hard surface and place the vibrating instrument on the subject's mastoid process (figure 18.25b). Again, ask the subject about the sound and about any differences in perceived sound. Make a note of the subject's response here:

   _____

   _____

EXERCISE 18.18B Hearing Test: Weber

1. Obtain a tuning fork and choose an individual to be the subject.

2. Tap the tuning fork on a hard surface. Place the vibrating instrument on the center of the subject's forehead **(figure 18.26)**. Ask the subject to compare his or her perception of sound in the right versus left ear, and make a note of the subject's response here:

   _____

   _____

(a)   **(b)**

Figure 18.25 **Rinne Hearing Test.** (*a*) Strike the tuning fork and place the device near the ear. (*b*) Place the vibrating tuning fork on the mastoid process.

©McGraw-Hill Education/Jill Braaten

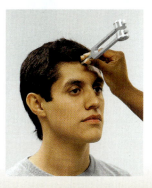

Figure 18.26 **Weber Hearing Test.** Place the vibrating tuning fork on the center of the forehead.

©McGraw-Hill Education/Jill Braaten

EQUILIBRIUM TESTS

Health care providers may perform tests for proper vestibular function if they suspect a problem. Like those for hearing, sensory receptors for equilibrium are housed within the inner ear, and sensory information from these receptors is transmitted along the vestibulocochlear nerve. Equilibrium tests may be complicated to interpret because many anatomical structures, such as the vestibular apparatus, eyes, proprioceptors, cerebellum, and cranial nerves are involved. The following laboratory exercises explore the Romberg test and the Barany test. A **Romberg test** is administered during a neurological exam when a patient exhibits motor or sensory deficits. It is also performed when a person is suspected of drunk driving. There is an exercise that demonstrates the influence of vestibular input on eye movement. The **Barany test** demonstrates the reflex between the movement of the fluid in the semicircular canals and the extrinsic eye muscles.

EXERCISE 18.19A Equilibrium Test: Romberg

1. Choose an individual to be the subject. Ask the subject to stand for one minute with both feet together and both hands by his side **(figure 18.27a)**. For ease, have the subject stand in front of a surface where the subject's shadow is visible or mark the subject's location (i.e., on a whiteboard). Note any exaggerated swaying movements to the left or right here:

   _____

   _____

   _____

2. Have the subject repeat step 1 with his eyes closed. Compare the magnitude of sway in the eyes-open versus eyes-closed condition. Make a note of it here:

   _____

   _____

   _____

3. Have the subject rotate 90 degrees, such that his shoulder is adjacent to the whiteboard (figure 18.27b). Be sure that the position is easily visible with a shadow or outline of the initial position. Look for any forward-backward swaying motion when the subject's eyes are open. Make a note of it here:

   _____

   _____

   _____

4. Have the subject close his eyes and repeat step 3. Look for any differences in forward-backward sway in the eyes-open versus eyes-closed condition. Make a note of any differences in sway here:

   _____

   _____

   _____

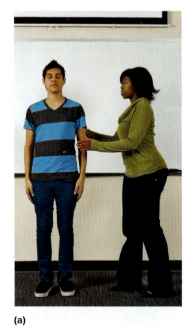 

(a)   (b)

Figure 18.27   **Romberg Test.** The subject stands with feet together and hands by his side, with his (*a*) back toward the whiteboard and (*b*) shoulders adjacent to the whiteboard. Stand close to the subject so as to catch him if he loses his balance.
©McGraw-Hill Education/Jill Braaten

INTEGRATE

CONCEPT CONNECTION

A Romberg test does more than simply test the signals of the vestibular system. It evaluates the integration of vestibular information, visual information, and proprioception. All three signals are used to maintain an upright posture. When one or more of these senses is compromised, our ability to balance (particularly when impaired) is compromised. A Romberg test involves intentionally removing one signal, the visual input, and then observing any disruption in balance. Disruptions in balance are observed when there is an increase in sway. If a person exhibits a positive Romberg sign (increased sway with eyes closed), it indicates that there may be deficits with proprioception. Specifically, there is likely impairment of the dorsal column motor pathway, which transmits proprioceptive information from the sensory receptors to the central nervous system. The Romberg test demonstrates that vestibular input is not sufficient by itself to allow a person to maintain an upright posture.

EXERCISE 18.19B Equilibrium Test: Barany

CAUTION:

 Any person subject to dizziness or nausea should not perform this test. Any rotation should cease if the subject reports dizziness or nausea. Testers should be prepared to hold or catch a subject if the test causes the subject to lose balance.

1. Choose an individual to be the subject. Have the subject sit in a swivel chair, holding onto the arms of the chair firmly. Be sure to have several students standing nearby to catch the subject in the event that the subject loses her balance.

2. Have the subject sit with head tilted forward approximately 30 degrees. This position of the head is optimal for stimulation of the lateral semicircular canals **(figure 18.28*a*)**. Spin the chair 10 revolutions to the right while the subject keeps her eyes open.

3. Stop the chair and have the subject sit up and look forward. Note the movement, if any, of the subject's eyes here: _____

Have the subject describe her perception of movement while sitting still. Make a note of it here:

_____

_____

4. Wait several minutes. Repeat steps 2 and 3 with the head tilted forward at 90 degrees (figure 18.28*b*). Note eye movements and the subject's perception of movement here: _____

_____

_____

(a)   (b)

Figure 18.28 **Barany Test.** The subject sits in a swivel chair with the head tilted forward (a) 30 degrees or (b) 90 degrees.

©McGraw-Hill Education/Jill Braaten

INTEGRATE

CLINICAL VIEW
Vertigo

Vertigo is a condition in which a person experiences sensations of dizziness or unsteadiness even when standing still. That is, the individual is not experiencing any outward rotational movement. In essence, vertigo occurs when one's vestibular pathways are stimulated in the absence of outward movement. Some sensations of vertigo may occur immediately following rotational movement. For example, one may experience vertigo immediately after riding a roller coaster by experiencing the sensation of still moving even after the ride is over.

People may also experience vertigo when they have a common cold or the flu. Ménierè's disease, which is characterized by excessive fluid in the inner ear, causes a severe sensation of vertigo. In Ménierè's disease, the excess fluid in the inner ear causes excessive depolarization of hair cells within the vestibule and semicircular canals, which leads to the increased perception of spinning. Often vertigo will also impact hearing, because both hearing and equilibrium are transmitted to the central nervous system via the vestibulocochlear nerve (CN VIII). Sensations of vertigo may also be associated with extreme nausea and vomiting as well as *nystagmus*, a rapid, involuntary, horizontal movement of the eyes.

Chapter 18: General and Special Senses

The corresponds to the Learning Objective(s) listed in the chapter opener outline.

Do You Know the Basics?

Exercise 18.1: Sensory Receptors in the Skin

1. Match the location listed in column A with its appropriate sensory receptor listed in column B. **1**

Column A

_____ 1. located at the dermal/epidermal junction

_____ 2. located deep in the reticular layer of the dermis

_____ 3. located within the dermal papillae

_____ 4. located throughout the dermis

Column B

a. free nerve ending

b. lamellated corpuscle

c. tactile corpuscle

d. tactile disc

Exercise 18.2: Gross Anatomy of the Eye

2. Match the description listed in column A with the part of the eye listed in column B. **2**

Column A

_____ 1. anterior-most part of the retina, which appears serrated

_____ 2. colored part of the eye

_____ 3. ligament extending between ciliary muscles and the lens

_____ 4. metallic-appearing, opalescent inner layer of the sclera; it is present in many animals (e.g., the cow eye), but not the human eye

_____ 5. neural tunic of the eye; composed of several layers of neurons involved with transducing light energy into nerve signals

_____ 6. smooth muscle within the ciliary body composed of both circular and radial muscle fibers

_____ 7. transparent, biconvex structure composed of highly specialized, modified epithelium

_____ 8. watery fluid that circulates within the anterior and posterior chambers of the eye

Column B

a. aqueous humor

b. ciliary muscle

c. iris

d. lens

e. ora serrata

f. retina

g. suspensory ligament

h. tapetum lucidum

Exercise 18.3: Extrinsic Eye Muscles

3. Which of the following extrinsic eye muscles is/are innervated by CN III? (Check all that apply.) **3**

_____ a. inferior oblique

_____ b. inferior rectus

_____ c. lateral rectus

_____ d. medial rectus

_____ e. superior oblique

_____ f. superior rectus

Exercise 18.4: Cow Eye Dissection

4. A cow eye has a colorful structure on the posterior wall called the _____ (tapedum lucidum/choroid body), which functions to _____ (increase/decrease) the amount of light stimulating the retina. **4** **5**

Exercise 18.5: Gross Anatomy of the Ear

5. Match the description in column A with the part of the ear listed in column B. **6** **7**

Column A

_____ 1. cavity between the external ear and inner ear; contains ossicles

_____ 2. drumlike, tight, thin membrane that separates the external-ear cavity to the middle ear

_____ 3. external ear, composed of an elastic cartilage skeleton that is covered with skin

_____ 4. largest membranous sac in the vestibule; contains receptors for sensing horizontal acceleration

_____ 5. portion of ear located within the petrous part of the temporal bone that includes the cochlea, vestibule, and semicircular canals

_____ 6. three ringlike canals that are oriented at right angles to each other and communicate with the vestibule

Column B

a. auricle

b. inner ear

c. middle ear

d. utricle

e. semicircular canals

f. tympanic membrane

Exercise 18.6: Tactile (Meissner) Corpuscles

6. Tactile (Meissner) corpuscles are found in the _____ (dermal papillae/reticular dermis) of skin and are responsible for detecting _____ (fine touch/deep pressure) stimuli. **8** **9**

Exercise 18.7: Lamellated (Pacinian) Corpuscles

7. Which of the following corresponds to the mode of sensation detected by lamellated (Pacinian) corpuscles? (Check all that apply.) **10** **11**

_____ a. deep pressure

_____ b. fine touch

_____ c. light touch

_____ d. vibration

Exercise 18.8: Gustation (Taste)

8. The large papillae that contain more than half of the taste buds in the body are the _____ (foliate/vallate) papillae. **12**

9. Cells that detect chemicals dissolved in solution and transmit nerve signals for taste sensation to the CNS are known as _____ (basal/gustatory) cells. **13**

Exercise 18.9: Olfaction (Smell)

10. The following table lists the cell types associated with olfactory epithelium. Next to each cell type, give a brief description of the location of the cell within the olfactory epithelium and the function of the cell. **14** **15**

Cell Type	Location	Function
Basal Cells		
Olfactory Receptor Cells		
Supporting Cells		

Exercise 18.10: Vision (The Retina)

11. Trace the path of light as it stimulates the cells in the retina. Use numbers to indicate the order in which these cells are activated (1 = first; 3 = last). **16** **17**

_____ a. bipolar cell layer

_____ b. ganglion cell layer

_____ c. photoreceptor cell layer

12. The area of highest visual acuity and largest density of cones in the retina is the _____ (fovea centralis/optic disc). **18**

Exercise 18.11: Hearing

13. The cochlea is innervated by _____ (CN II/CN VIII). **19**

14. The scala tympani and vestibuli both contain _____ (endolymph/perilymph), whereas the scala media contains _____ (endolymph/perilymph). **20**

15. Which of the following structures in the cochlea corresponds to the location of the spiral organ? (Circle one.) **21**

    a. scala media

    b. scala tympani

    c. scala vestibuli

16. Vibration of which membrane allows for sound transmission in the cochlea? (Circle one.) **22**

    a. basilar membrane

    b. tectorial membrane

    c. vestibular membrane

Exercise 18.12: Two-Point Discrimination

17. There is a(n) _____ (direct/inverse) relationship between the density of sensory receptors in the skin and the size of the receptive field in that area of the body. **23**

18. The density of sensory receptors in the fingertip is _____ (larger/smaller) compared to the density of sensory receptors in the anterior forearm. **24**

Exercise 18.13: Tactile Localization

19. When stimulating the skin, it is _____ (more/less) difficult to locate the exact source of the sensory input if multiple receptive fields are stimulated. **25**

Exercise 18.14: General Sensory Receptor Tests: Adaptation

20. The nervous system adapts to a constant stimulus by becoming _____ (less/more) sensitive to that stimulus over time. **26**

Exercise 18.15: Gustatory Tests

21. Match each of the substances listed in column A with an example of a taste sensation listed in column B. **27**

Column A	Column B
_____ 1. dark chocolate bar	a. bitter
_____ 2. dill pickle	b. salty
_____ 3. maple syrup	c. sour
_____ 4. seasoning salt	d. sweet
_____ 5. steak	e. umami

Exercise 18.16: Olfactory Tests

22. When tasting two substances that have similar consistencies (such as an apple and a potato), closing one's nose _____ (decreases/increases) the ability to distinguish one from the other. **28**

Exercise 18.17: Vision Tests

23. An individual has the following results on a visual acuity test: 20/10. This individual's vision is _____ (better/worse) than "normal" vision. **29**

24. As we age, the flexibility of the lens of the eye decreases. This causes the near point of accommodation to move _____ (farther from/closer to) the eye. **30**

25. The area of the retina that completely lacks photoreceptors and therefore cannot detect the source of light is the _____ (optic disc/choroid). **31**

26. Color blindness can have multiple causes, including a reduced distribution or defective function of the _____ (rods/cones). **31**

Exercise 18.18: Hearing Tests

27. The _____ (Rinne/Weber) test is a test for neural deafness. **32**

Exercise 18.19: Equilibrium Tests

28. The _____ (Barany/Romberg) test is used to determine if vertigo is caused by a disorder in the patient's inner ear or somewhere

in the patient's brain. **33**

Can You Apply What You've Learned?

29. A fracture of the cribriform plate of the ethmoid bone can result in a loss of the sense of smell. Given your knowledge of the location of olfactory receptor cells and the pathway taken by their axons to reach the brain, explain why this can happen.

_____

_____

_____

_____

30. The optic disc is referred to as the "blind spot" of the eye. Based on your histological observation of the optic disc, explain why this is the case.

_____

_____

_____

_____

31. Ossicle chain dislocation, usually caused by trauma to the ear, can significantly reduce a person's ability to perceive sound. Based on your knowledge of hearing, describe how a fracture to the incus can lead to hearing loss.

_____

_____

_____

_____

Can You Synthesize What You've Learned?

32. Why do you think tactile (Meissner) corpuscles are located relatively close to the surface of the skin rather than deep within the dermis?

_____

_____

_____

_____

33. An individual who suffers a strong blow to the head may end up with a detached retina that causes visual problems. Why do you think the retina easily detaches from the posterior wall of the eye? What structure normally holds the retina in place?

_____

_____

_____

_____

34. Susan, a 67-year-old female, is experiencing ataxia, or poor muscle coordination. Doctors perform a neurological exam, and they find that Susan is not positive for the Romberg test. Describe what these findings suggest about the source of Susan's ataxia.

_____

_____

_____

_____

35. Anosmia is a condition characterized by an inability to smell. While some may be unable to perceive a particular odorant, most suffering from this condition are unable to perceive multiple odorants. Based on your knowledge of the olfactory sense, propose several scenarios that would lead to a patient's diagnosis of anosmia.

   _____

   _____

   _____

   _____

36. Patients suffering from hearing loss or impairment may be eligible for a cochlear implant, a device that mimics the action of the cochlea. Specifically, the cochlear implant contains external components that detect, process, and transmit sounds to an internal receiver. The internal components stimulate the vestibulocochlear nerve directly. Based on your knowledge of hearing, describe how the cochlear implant allows for the perception of sound.

   _____

   _____

   _____

   _____

The Endocrine System

OUTLINE AND LEARNING OBJECTIVES

Anatomy & Physiology Revealed® 4.0

Module 8: Endocrine System

INTRODUCTION

How long has it been since your last meal? Even if it has been several hours, blood glucose and blood calcium levels remain remarkably stable, fluctuating only minor amounts around the body's normal physiological level (unless, of course, there is an underlying condition such as diabetes). Maintenance of blood glucose and blood calcium levels are physiological imperatives, for if their levels are too high or too low, severe impairment of nervous and muscular activity will occur. However, it is not often that blood glucose or calcium levels move drastically out of the normal range. This is because such variables are tightly regulated by the **endocrine system**. For example, the hormones insulin and glucagon, produced by the pancreas, regulate blood glucose levels, and the hormones calcitonin and parathyroid hormone, produced by the thyroid and parathyroid glands, respectively, regulate blood calcium levels.

This brief description of these hormones and the variables they regulate provides only a glimpse into the functioning of the endocrine system, a system of chemical messengers (called hormones) that are transported in the blood and act on distant target cells. The endocrine system consists of a number of "classical" endocrine organs, such as the pituitary gland, adrenal glands, and thyroid gland. However, many organs in the body also contain cells or tissues that produce and secrete hormones. For instance, cells in the walls of the stomach secrete hormones that regulate appetite, gastric motility, and acid secretion. Cells in the testes (males) and ovaries (females) are responsible for the secretion of hormones that regulate the maturation of sperm and eggs, respectively. While the thymus is considered an endocrine organ due to its secretion of the hormone *thymosin*, it plays a larger role in the immune system, as described in chapter 23.

The exercises in this chapter explore the structure and function of the classical endocrine organs. Many of the organs explored contain cells that secrete hormones (thus, the organs have an endocrine role). However, an entire organ may not necessarily be called an endocrine *gland* because it has other functions as well. This chapter begins with a study of the gross anatomy of the endocrine system, with a focus on the major (large) endocrine organs. Following the study of gross anatomy is an exploration of the endocrine system at the microscopic level. When examining histology slides, make associations between the tissues viewed under the microscope and the gross anatomic location where the tissue is found. In addition, consider the names and function(s) of the hormone(s) secreted by each organ. Finally, this chapter contains an activity that explores the physiology of metabolism, and a clinical case study that involves applying the principles of the endocrine system to a problem involving an imbalance in hormones. Subsequent chapters covering the cardiovascular, lymphatic, digestive, urinary, and reproductive systems contain exercises that further explore the structure and function of endocrine cells and tissues (and associated hormones) related to each system.

List of Reference Tables

These Pre-Laboratory Worksheet questions may be assigned by instructors through their connect course.

1. Glands that produce and release chemical messengers (hormones) to be transported in the blood are _____ (endocrine/exocrine) glands.

2. Glands that produce a product that is released into a duct, which transports the product directly to its target organ or tissue,

   are _____ (endocrine/exocrine) glands.

3. Identify the gland where parafollicular cells are located. (Circle one.)

   a. adrenal gland

   b. pineal gland

   c. pituitary gland

   d. thymus

   e. thyroid gland

4. Match the description listed in column A with the appropriate endocrine gland listed in column B.

   Column A

   _____ 1. consists of a cortex and medulla, with each part having different embryological origins

   _____ 2. consists of follicles lined with a simple cuboidal epithelium

   _____ 3. four small endocrine glands that secrete a hormone that regulates blood calcium levels

   _____ 4. secretes hormones to regulate hormone release by the anterior pituitary gland

   _____ 5. secretes the hormone melatonin

   Column B

   a. adrenal gland

   b. hypothalamus

   c. parathyroid glands

   d. pineal gland

   e. thyroid gland

5. Which of the following is the major endocrine gland that secretes sex steroid hormones in the female? (Circle one.)

   a. adrenal gland

   b. ovaries

   c. pineal gland

   d. pituitary gland

   e. testes

6. Which of the following hormone(s) is/are secreted by the posterior pituitary gland? (Check all that apply.)

   _____ a. antidiuretic hormone

   _____ b. growth hormone

   _____ c. oxytocin

   _____ d. prolactin

   _____ e. thyroid-stimulating hormone

7. The hormone released by the anterior pituitary gland that induces ovulation in females is _____ (follicle-stimulating hormone/luteinizing hormone).

8. A drop in blood glucose levels is corrected by the secretion of _____ (insulin/glucagon) by _____ (alpha/beta) cells in the pancreas.

9. Which of the following occurs when thyroid hormone levels increase? (Check all that apply.)

   _____ a. decreased metabolic rate

   _____ b. increased glycogenesis

   _____ c. increased body temperature

   _____ d. increased oxygen consumption

   _____ e. increased lipolysis

10. A decrease in thyroid hormone _____(inhibits/stimulates) the release of thyrotropin-releasing hormone by the hypothalamus.

GROSS ANATOMY

This section will focus on the gross anatomy of the endocrine system. Some endocrine glands are readily seen on the cadaver, while others are more difficult to identify. Using a cadaver or classroom models, you will observe the overall structure and location of the major endocrine glands of the body.

Endocrine Organs

INTEGRATE

CLINICAL VIEW
Anabolic Steroids

Anabolic steroids are lipid-soluble hormones that are structurally and functionally similar to the endogenous hormone testosterone. When introduced exogenously (from outside the body) into the systemic circulation, either by injection, in pill form, or as a cream applied transdermally, anabolic steroids cross the plasma membrane of target cells and bind to androgen receptors. Overall, anabolic steroids promote protein synthesis, particularly in muscle cells. Other effects of anabolic steroids include increased appetite, increased bone growth, and increased erythropoiesis (red blood cell synthesis). Anabolic steroids also promote the development of secondary sex characteristics, including increased hair growth, increased sweat production by sweat glands, and elongation of vocal cords, which causes a deepening of the voice. Dangerous side effects of anabolic steroids include cardiovascular changes such as increased blood cholesterol levels and increased blood pressure.

Endogenous testosterone levels are regulated by negative feedback mechanisms. Therefore, exogenous administration of anabolic steroids acts as a stimulus to suppress the release of luteinizing hormone (LH) by the anterior pituitary gland, which in turn reduces testosterone production by the testes. In males, whose testosterone levels are typically much higher than in females, administration of anabolic steroids can reduce libido (sex drive), testicle size, and sperm production. It can also promote gynecomastia (enlargement of breast tissue). Anabolic steroids are sometimes used medically; however, they can also be abused, as when individuals take anabolic steroids to stimulate increases in body mass and strength (e.g., by bodybuilders). Individuals who use anabolic steroids recreationally may become dependent and may experience some moderate to severe psychological effects (e.g., "roid rage") during and following (i.e., withdrawal) use of the drugs.

EXERCISE 19.1

GROSS ANATOMY
OF ENDOCRINE ORGANS

1. Observe a human cadaver or classroom models of the brain, thorax, abdomen, and skull.

2. Identify the structures listed in **figure 19.1** on a human cadaver or on classroom models, using the textbook as a guide. Then label them in figure 19.1.

3. *Optional Activity:* **APR 8: Endocrine System**—Watch the endocrine system animations to review the structure and function of the hypothalamus and pituitary gland, pancreas, thyroid and parathyroid glands, and adrenal (suprarenal) glands.

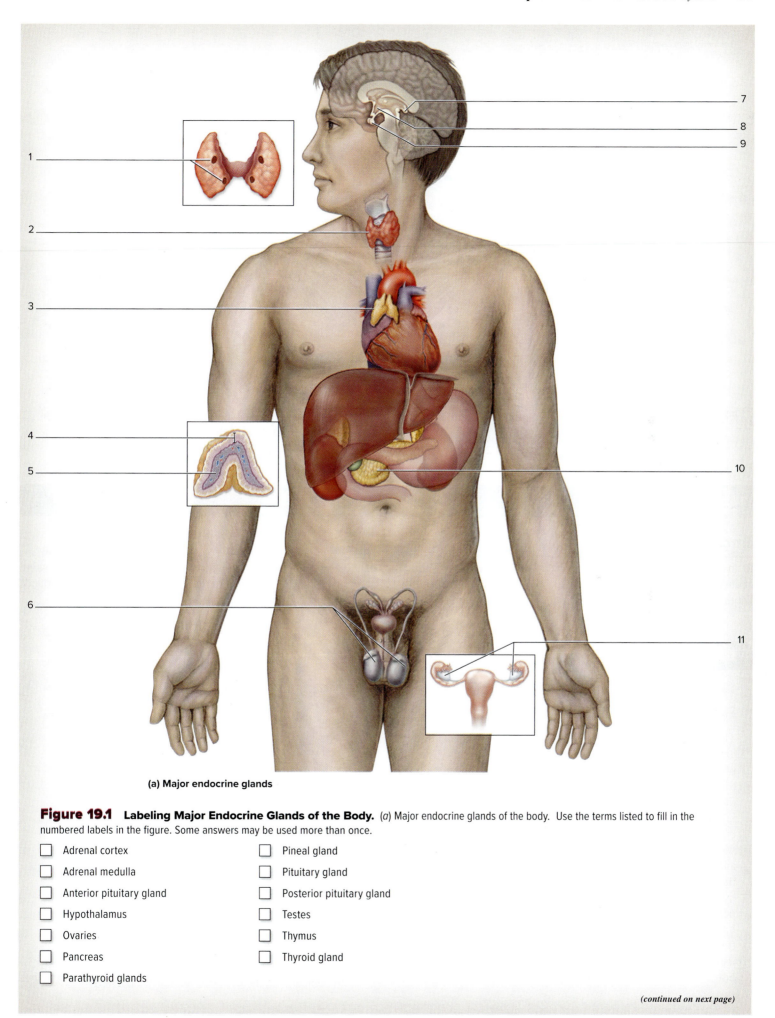

(a) Major endocrine glands

Figure 19.1 **Labeling Major Endocrine Glands of the Body.** (*a*) Major endocrine glands of the body. Use the terms listed to fill in the numbered labels in the figure. Some answers may be used more than once.

☐ Adrenal cortex

☐ Adrenal medulla

☐ Anterior pituitary gland

☐ Hypothalamus

☐ Ovaries

☐ Pancreas

☐ Parathyroid glands

☐ Pineal gland

☐ Pituitary gland

☐ Posterior pituitary gland

☐ Testes

☐ Thymus

☐ Thyroid gland

(continued on next page)

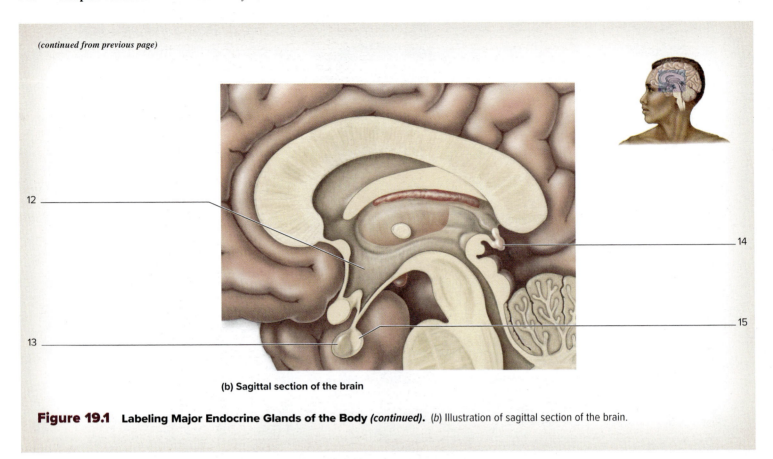

(continued from previous page)

12
13
14
15

(b) Sagittal section of the brain

Figure 19.1   **Labeling Major Endocrine Glands of the Body** *(continued).*   (*b*) Illustration of sagittal section of the brain.

HISTOLOGY

Endocrine Glands

In the following exercises, the microscopic anatomy of the various endocrine glands is described in detail. The tables in these exercises list the glands, the hormones they produce, and the functions of each of the listed hormones. A major goal in completing the following exercises will be to learn to differentiate the various endocrine glands from each other when viewing them under the microscope.

EXERCISE 19.2

THE HYPOTHALAMUS AND PITUITARY GLAND

The **pituitary** gland, or hypophysis (*hypophysis*, an undergrowth), is a remarkable organ. About the size and shape of a pea, this small organ secretes hormones that regulate the growth and development of nearly every other organ in the body. Because of the significant role this organ plays in endocrine regulation, its function must also be tightly controlled. Control of the pituitary gland is mediated by cells within the **hypothalamus**, a part of the brain whose structure and function were discussed in chapter 15. The textbook covers the structure of, function of, and relationships between the hypothalamus and pituitary gland in detail. Because it is not possible to visualize the details of these relationships in the laboratory, the focus here is on the structure and function of the pituitary gland alone. **Table 19.1** summarizes the cells and structures that compose the pituitary gland and lists the hormones secreted by each cell type.

1. Obtain a histology slide of the pituitary gland **(figure 19.2)**. Before placing it on the microscope stage, observe the slide with the naked eye. Notice there is a distinctive difference in color between the two parts, or lobes, of the pituitary gland. The darker area is the **anterior pituitary** (also called the anterior lobe, or **adenohypophysis** [*adeno-*, a gland, + *hypophysis*, pituitary]), whereas the lighter area is the **posterior pituitary** (also called the posterior lobe, or **neurohypophysis** [*neuro-*, relating to nervous tissue, + *hypophysis*, pituitary]).

2. Place the slide on the microscope stage and bring the tissue sample into focus on low power. Once again, identify the anterior and posterior lobes (figure 19.2*a*). The anterior pituitary is derived embryologically from an outpocketing of the roof of the mouth and consists of epithelial tissue. Thus, the cells have an appearance that is characteristic of glandular epithelial tissue (figure 19.2*b*). In comparison, the posterior pituitary is derived

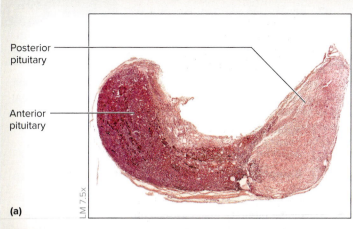

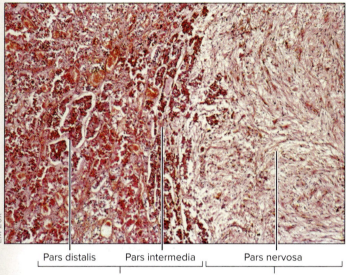

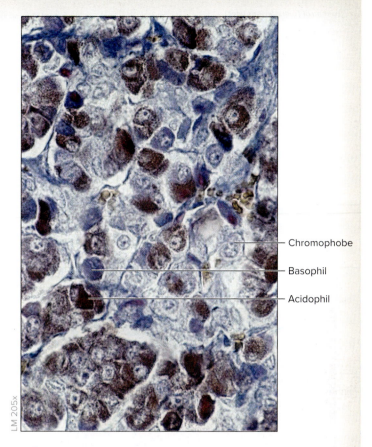

Figure 19.3 **Anterior Pituitary.** The three general cell types in the anterior pituitary are basophils, which stain blue, acidophils, which stain red, and chromophobes, which don't take up biological stains.
©Victor P. Eroschenko

Figure 19.2 **Pituitary Gland.** (*a*) Low-magnification view of both lobes of the pituitary. (*b*) Medium-magnification view of the anterior and posterior pituitary.

(*a*) ©McGraw-Hill Education/Al Telser; (*b*) ©Astrid & Hanns-Frieder Michler/Science Source

embryologically from a downgrowth of the diencephalon of the brain and consists of nervous tissue. Thus, the cells have an appearance that is characteristic of nervous tissue (figure 19.2*b*).

3. *Anterior Pituitary*—Identify the anterior pituitary gland using figures 19.2 and **19.3** as guides. Then move the microscope stage so the anterior pituitary is in the center of the field of view.

4. Increase the magnification to observe the glandular nature of the cells (figure 19.3). There are three cell types within the anterior pituitary: acidophils, basophils, and chromophobes. The first two kinds of cells are named for their "love" (-*phil*, to love) of acidic or basic dyes. *Acidophils* attract acidic dyes and appear red. *Basophils* attract basic dyes and appear blue. *Chromophobes* (*chroma*, color, + phobos, fear) are cells that attract neither acidic nor basic dyes, and they are thought to be cells that have released their hormone(s). Identification of the various cell types within the anterior pituitary is

INTEGRATE

LEARNING STRATEGY

Simple mnemonics for remembering which hormones are produced by which cells of the anterior pituitary are as follows:

Mnemonic for acidophils: **GPA** (as in Grade Point Average: If a student does not remember this information, it could be harmful to his or her GPA.)

G = growth hormone (GH)

P = prolactin (PRL)

A = acidophil

Mnemonic for basophils: **B-FLAT** (as in the musical note: If a student remembers this information, it will be beneficial to his or her GPA and he or she will be happily singing to the tune of B-flat!)

B = basophil

F = follicle-stimulating hormone (FSH)

L = luteinizing hormone (LH)

A = adrenocorticotropic hormone (ACTH)

T = thyroid-stimulating hormone (TSH)

(*continued on next page*)

THE PINEAL GLAND

The **pineal gland** (*pineus*, relating to a pine, shaped like a pinecone), also called the **pineal body**, is a small region of the epithalamus of the brain. Its primary cells, called **pinealocytes (figure 19.5)**, secrete the hormone **melatonin**. Melatonin's effects in humans are unsubstantiated, but in other organisms melatonin is responsible for regulation of circadian rhythms. In humans, it may have a role in determining the onset of puberty. Pinealocytes are innervated by neurons from the sympathetic nervous system, and their secretion of hormone is affected by the amount of light received by the individual, which is relayed to the pineal gland through these neurons. Melatonin secretion increases when light levels are low (at night) and decreases when light levels are high (during the day). Pinealocytes appear in groups of cells within the pineal gland. They are surrounded by glial cells,

whose function is similar to that of astrocytes in other parts of the brain. Clinically, one of the most important features of the pineal gland is the presence of calcium concretions, termed **"pineal sand"** (corpora arenacea). These concretions are easily visible in radiographs of the head and provide radiologists with a landmark that is consistent and easy to identify. The number of concretions in the pineal gland increases with age.

1. Obtain a histology slide of the pineal gland. Place the slide on the microscope stage and bring the tissue sample into focus on low power.

2. Identify the following structures on the slide, using figure 19.5 as a guide:

   ☐ **Pinealocytes**   ☐ **Pineal sand**

3. Sketch the pineal gland as seen through the microscope at medium or high magnification in the space provided or on a separate sheet of paper. Be sure to label the structures in step 2.

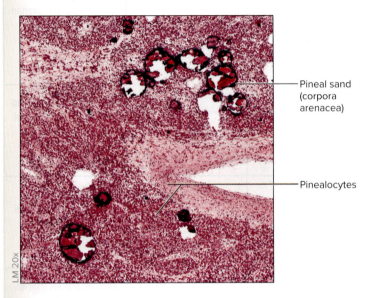

Pineal sand (corpora arenacea)

Pinealocytes

LM 20x

Figure 19.5   The Pineal Gland. Histology of the pineal gland.
©McGraw-Hill Education/Al Telser

THE THYROID AND PARATHYROID GLANDS

The **thyroid gland** (**figure 19.6; table 19.2**) is a butterfly-shaped gland located anterior to the trachea and inferior to the thyroid cartilage of the larynx. It consists of two main lobes connected to each other anteriorly by a narrow **isthmus** (*isthmus*, neck). The functional units of the thyroid gland are **thyroid follicles**, which are lined with a simple epithelium that may range from simple squamous to simple columnar. Inside each follicle is a mass of **colloid**, consisting largely of thyroglobulins. **Thyroglobulins** are the precursor molecules for the formation of the **thyroid hormones** (T3 and T4). The areas between the follicles contain

another cell type, called **parafollicular cells**. These cells secrete the hormone **calcitonin**.

Embedded within the posterior aspect of the thyroid gland are a series of small glands (usually four) called **parathyroid glands**. These glands consist of two cell types: **chief (principal) cells**, which are smaller, more abundant cells with relatively clear cytoplasm that produce **parathyroid hormone** (PTH, parathormone), and **oxyphil cells**, which are larger, less abundant cells with granular pink cytoplasm, and whose function is unknown.

1. Obtain a histology slide of the thyroid and parathyroid glands and place it on the microscope stage. Bring the tissue sample into focus on low power and then change to high power.

2. Identify the following structures on the slide, using figure 19.6 and table 19.2 as guides:

Thyroid gland
☐ **Colloid**
☐ **Follicular cells**
☐ **Parafollicular cells**
☐ **Thyroid follicles**

Parathyroid gland
☐ **Chief (principal) cells**
☐ **Oxyphil cells**

3. Sketch the thyroid and parathyroid as seen through the microscope in the space provided or on a separate sheet of paper. Be sure to label all of the structures listed in step 2.

_____ ✕

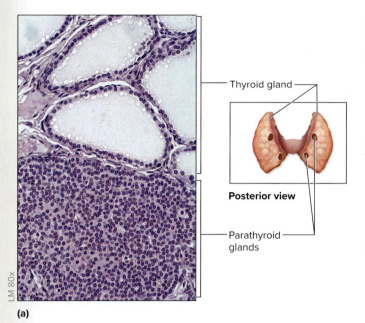

Follicular cell
(cuboidal epithelial
cell)
Parafollicular
cell
Thyroid follicle

Colloid within a
thyroid follicle

LM 250x

(b) Thyroid gland

Thyroid gland

Posterior view

Parathyroid
glands

Chief cells

Oxyphil cells

LM 135x

(c) Parathyroid gland

LM 80x

(a)

Figure 19.6 **Thyroid and Parathyroid Glands.** (*a*) Micrograph showing histological relationship between the large follicles of the thyroid and the tightly packed cells of the parathyroid gland. (*b*) Micrograph showing thyroid follicles, which produce the thyroid hormones, and parafollicular cells, which produce calcitonin. (*c*) High-magnification view of a parathyroid gland demonstrating chief cells and oxyphil cells.

(*a*) & (*c*) ©Victor P. Eroschenko; (*b*) ©McGraw-Hill Education/Alvin Telser

Table 19.2	Histology and Function of the Thyroid and Parathyroid Glands				
Cell Types	**Description**	**Hormones Produced**	**Action of Hormone**	**Mechanism of Action**	**Word Origin**
Thyroid Gland	*A shield- or butterfly-shaped gland located anterior to the trachea and inferior to the thyroid cartilage. Consists of two lobes connected by a narrow isthmus anteriorly.*				*thyroid, shaped like an oblong shield*
Follicular Cells	Cells may be simple squamous, simple cuboidal, or simple columnar; nuclei stain very dark.	Thyroid hormones (T3 and T4/ Thyroxine)	Increases basal metabolic rate (BMR); important in early development of the central nervous system	Stimulates or inhibits transcription of certain genes in target cells	*folliculus, a small sac*

(continued on next page)

(continued from previous page)

Table 19.2	Histology and Function of the Thyroid and Parathyroid Glands *(continued)*				
Cell Types	**Description**	**Hormones Produced**	**Action of Hormone**	**Mechanism of Action**	**Word Origin**
Parafollicular Cells	Light-staining cells found in areas surrounding follicles; larger than follicular cells; may be difficult to identify with routine stains	Calcitonin	Decreases blood calcium levels	Inhibits the action of osteoclasts, increases urinary excretion of calcium	*para*, next to, + *folliculus*, a small sac
Parathyroid Glands	*4–6 small glands located on the posterior surface of the thyroid gland*				*para*, next to, + *thyroid*, shaped like an oblong shield
Chief (Principal) Cells	Small cells that contain a centrally located, round nucleus with one or more nucleoli	Parathyroid hormone	Increases blood calcium levels	Indirectly increases the action of osteoclasts, decreases urinary excretion of calcium, and increases dietary absorption of calcium	*principal*, the predominant cell type of a gland
Oxyphil Cells	Larger than chief cells and more reddish in color	Unknown	NA	NA	*oxys*, sour acid, + *-phil*, to love

EXERCISE 19.5

THE ADRENAL GLANDS

The adrenal (suprarenal) glands are located directly superior to each kidney (*ad*, to, + *ren*, kidney). They are similar to the pituitary gland in that they are composed of two regions (**figure 19.7**), each with a separate embryological origin. The outer region, the **adrenal cortex**, is derived from mesoderm and has the appearance of typical glandular epithelium. The inner region, the **adrenal medulla**, is derived from modified postganglionic sympathetic neurons and has the appearance of nervous tissue. The cells of the adrenal cortex synthesize steroid hormones (specifically, **corticosteroids**, *cortico*, relating to the adrenal cortex, + *steroid*, steroid hormone such as cortisol), whereas the cells of the adrenal medulla synthesize

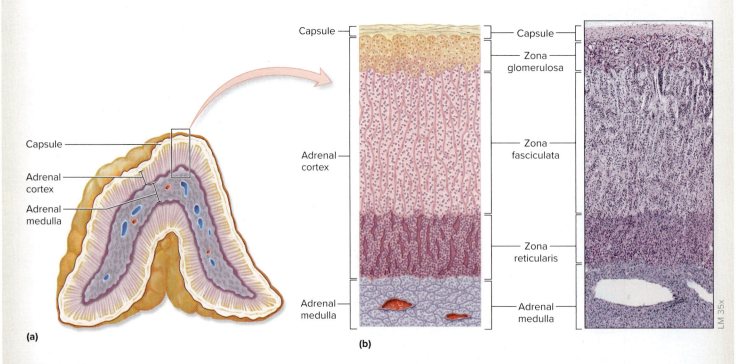

Figure 19.7 **Adrenal Glands.** (*a*) Cross section of the adrenal gland. (*b*) Micrograph demonstrating the three layers of the adrenal cortex and part of the adrenal medulla.

©McGraw-Hill Education/Electronic Publishing Services, Inc., NY

catecholamine hormones (that is, hormones derived from the amino acid tyrosine and that contain a catechol ring; includes epinephrine and norepinephrine). The entire gland is surrounded by a dense irregular connective tissue **capsule**, which protects the gland and helps anchor it to the superior border of the kidney. The adrenal cortex has three recognizable zones, and each zone has cells that predominantly secrete one category of corticosteroid hormones. Characteristics of the zones, and descriptions of the hormones secreted by cells within each zone, are summarized in **table 19.3**.

1. Obtain a histology slide of the adrenal gland (figure 19.7) and place it on the microscope stage. Bring the tissue sample into focus on low power and identify the two major regions, the cortex and the medulla (figure 19.7*b*).

2. Move the stage so the *adrenal cortex* (figure 19.7*b*) is in the center of the field of view. Then change to high power.

3. Identify the following zones of the adrenal cortex from outermost to innermost: zona glomerulosa, zona fasciculata, and zona reticularis, using figure 19.7 and table 19.3 as guides.

Table 19.3	Histology and Function of the Adrenal Glands					
Adrenal Gland Region	**Zone and/or Cells**	**Description**	**Hormones Produced**	**Action of Hormone(s)**	**Word Origin**	
Adrenal Cortex	Zona glomerulosa	Outermost region of cortex deep to the capsule of the adrenal gland; cells arranged into circular clusters, or "balls"	Mineralocorticoids: aldosterone	Increases sodium and water retention by the kidneys (thus increasing blood volume and blood pressure)	*zona*, zone, + *glomus*, a ball of yarn	
	Zona fasciculata	Middle and largest region of adrenal cortex; cells arranged in long cords, or "bundles"	Glucocorticoids: cortisol	Increases glucose synthesis through gluconeogenesis (production of new glucose from amino acids), and lipolysis	*zona*, zone, + *fasciculus*, a bundle	
	Zona reticularis	Innermost region of cortex located between the zona fasciculata and medulla; cells arranged as a branched "network"	Glucocorticoids and gonadocorticoids: androgens	Androgens are similar in structure and function to the male sex steroid hormone testosterone; secreted in very low amounts compared to testosterone secretion by the testes (males)	*zona*, zone, + *rete*, a net	
Adrenal Medulla	Chromaffin cells	Large, spherical cells with yellowish-brown tint when stained due to their reaction with chrome salts	Catecholamines: epinephrine and norepinephrine	Epinephrine (80–90% of total hormone secretion by the medulla) increases heart rate and contraction of heart muscle (force); norepinephrine stimulates vasoconstriction.	*chroma*, color, + *affinis*, affinity, attraction for	

4. Change back to the low-power objective, move the microscope stage so the *adrenal medulla* is in the center of the field of view, and then change back to high power. The nuclei visible within the adrenal medulla are nuclei of *chromaffin cells*, which are modified postganglionic sympathetic neurons. What hormone(s) do these cells secrete? _____

5. Identify the following structures on the slide of the adrenal glands, using figure 19.7 and table 19.3 as guides:

   ☐ **Adrenal cortex**   ☐ **Zona fasciculata**
   ☐ **Adrenal medulla**   ☐ **Zona glomerulosa**
   ☐ **Capsule**   ☐ **Zona reticularis**
   ☐ **Chromaffin cells**

6. Sketch the adrenal gland as seen through the microscope in the space provided or on a separate sheet of paper. Be sure to label all of the structures listed in step 5.

_____ ✕

7. *Optional Activity:* **APR** 8: **Endocrine System**—Review the histology slides of the adrenal (suprarenal) gland, as well as the pituitary gland, thyroid gland, and endocrine pancreas.

(continued on next page)

PHYSIOLOGY

Metabolism

Metabolism is regulated by **thyroid hormone (TH)**. The release of TH is controlled indirectly through the hypothalamus. The hypothalamus detects a decrease in blood levels of TH (and other stimuli), which triggers the release of **thyrotropin-releasing hormone (TRH)** into the hypothalamic-hypophyseal portal vein. TRH triggers cells in the anterior pituitary to release **thyroid-stimulating hormone (TSH)**. TSH is transported in the general circulation to target cells within the **thyroid gland**. TSH stimulates the **follicular cells** of the thyroid gland to release thyroid hormone in its two forms: **triiodothyronine (T_3)** and **tetraiodothyronine (T_4, thyroxine)**. T_3 is the more active hormone. Most T_4 is converted to T_3 when it reaches its target cells. For the purposes of this exercise, T_3 and T_4 will be collectively referred to as thyroid hormone (TH).

Thyroid hormone (a lipid-soluble hormone) exerts its effects by binding to intracellular receptors and stimulating the transcription of genes that code for certain proteins. For example, TH stimulates the transcription of genes that code for the Na^+/K^+ pump in excitable cells, which causes these cells to increase production of these pumps. To meet the demands of increased numbers of these pumps, the cell increases its rate of cellular respiration to synthesize the additional ATP that is required for this active transport process. TH acting elsewhere in the body stimulates the increase of heart and respiratory rates. This increases the transport of oxygenated blood to active cells throughout the body. The overall effect of TH is to **increase metabolic rate**, **increase oxygen consumption**, and **increase body temperature**. TH also decreases the level of circulating amino acids in the blood by promoting their uptake into cells for protein synthesis. In addition, thyroid hormone stimulates **lipolysis** and **glycogenolysis**, thereby increasing both fatty acid and blood glucose levels in the blood.

EXERCISE 19.7   PhILS

Ph.I.L.S. LESSON 19: THYROID GLAND AND METABOLIC RATE

The purpose of this laboratory exercise is to observe the effects of TH on body temperature and metabolic rate. Metabolic rates of two female white mice will be measured based on rates of oxygen consumption. This simulation will calculate the least square linear regression, which determines the *dependent* (*y*) and *independent* (*x*) variables, plots the data points, and determines the line of best fit for the data points. The equation that generates the line ($y = mx + b$) can then be used to predict the value of the dependent variable if the value of the independent variable is known.

The two mice in the simulation have eaten either (1) normal mouse chow, or (2) mouse chow containing 0.15% propylthiouracil (PTU). The production of TH is significantly reduced by the compound PTU. Finally, the rate at which each mouse consumes oxygen will be measured at different temperatures in order to determine the effect of cooling on metabolic rate.

Prior to beginning the experiment, become familiar with the following concepts (use the textbook as a reference):

- negative feedback regulation
- the hypothalamus and its anatomical and physiological relationship with the anterior pituitary gland
- follicular tissue of the thyroid gland
- the cellular effects of TH
- the relationship between oxygen consumption and metabolic rate

Prior to beginning the experiment, state a hypothesis regarding the influence of intake of mouse chow with TPU on the levels of TH, oxygen consumption, and metabolic rate in the mouse:

_____

_____

_____

_____

1. Open the Ph.I.L.S Lesson 19: Endocrine Function: Thyroid Gland and Metabolic Rate.

2. Read the objectives, introduction, and wet lab sections. Then take the pre-lab quiz. The laboratory exercise will open when the pre-lab quiz has been completed **(figure 19.9)**.

3. Click Tare to set the scale to zero. To weigh a mouse, click on one of the mice and drag it to the scale. After weighing the mouse, place it in the cage by clicking and dragging it to the cage in the lower left corner of the screen.

4. Remove the pipette from the beaker, which contains bubbles, and place the tip of the pipette to the calibration tube (at the 10).

5. Measure the initial position of the soap bubbles (if at the end of the tube it will be 10 mm) and record the position at 0:00 time in the table in the program.

Figure 19.9   Opening Screen for the Laboratory Exercise on Endocrine Function: Thyroid Gland and Metabolic Rate.
©McGraw-Hill Education.

6. After clicking the Start button, measure the position of the soap bubble within the calibration tube at 15-second intervals (a beep will be heard every five seconds) and record it in the data table. (The timer will begin when you hit Start.) It may be more manageable to click the Pause button after each 15-second interval, measure, record, and then click Start to resume the experiment. When finished, click Pause.

7. Click Calc to see the graph (linear regression) in the program. The Journal will open and show a graph of the data collected up to this point in the laboratory exercise. The Journal also includes a table. Notice that the table (with temperatures ranging from 8°C to 24°C) will have the first data point entered for the average oxygen consumed per minute (total amount of oxygen divided by 2 minutes). Enter these data in **table 19.5**. No points will appear on the graph until all data points are entered. Close the table and graph windows by clicking the X in the upper right corner.

8. To complete the process again with the same animal, repeat steps 4–7 but decrease the temperature 2 degrees (by clicking on the down arrow on the thermostat) prior to beginning the steps until data are recorded for all temperatures. Be sure to record the data in table 19.5 every time one trial is completed.

9. After returning the first animal to the cage, repeat steps 3–8 for the second animal.

10. Construct a graph that plots the linear regression of oxygen consumption versus temperature in the space provided or on a separate sheet of paper.

11. Open the post-lab quiz and lab report by clicking on it. Answer the post-lab questions that appear on the computer screen. Click Print Lab to print a hard copy of the report or click Save PDF to save an electronic copy of the report.

12. Make note of any pertinent observations here:

_____

_____

_____

Table 19.5	Data for Ph.I.L.S. Lesson 19: Thyroid Gland and Metabolic Rate	
Temp (°C)	**Normal (24.2 g)**	**+PTU (g)**
24		
22		
20		
18		
16		
14		
12		
10		
8		

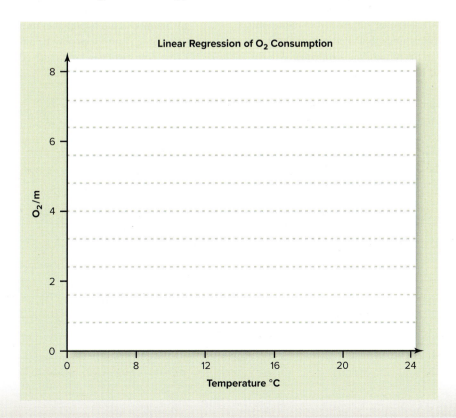

Linear Regression of O_2 Consumption

EXERCISE 19.8

A CLINICAL CASE IN ENDOCRINE PHYSIOLOGY

The purpose of this exercise is to apply concepts related to the endocrine system to a real scenario, or clinical case. This exercise requires information presented in this chapter, in the histology, gross anatomy, and physiology sections, as well as in previous chapters. Integrating multiple systems in a case study reinforces understanding of homeostatic processes.

Prior to beginning the case, ask a laboratory instructor if the exercise will be completed individually, in pairs, or in groups. Review the topics of the nervous system, as well as the endocrine system, including histology and gross anatomy of the pituitary gland, hormone production, and physiological effects of each hormone, as needed. Use the textbook and this lab manual as guides.

Case Study

Christopher, an 80-year-old male, was diagnosed with stage III non–small cell carcinoma of the lung. There was no evidence of metastasis on computed tomography (CT) scans at the time of diagnosis. Even so, doctors treated his cancer aggressively with surgical removal of the tumor followed by radiation therapy. One year following his cancer diagnosis and treatment, Christopher complained of extreme fatigue, cold intolerance, headache, mental confusion, and visual disturbances. He experienced a severe fall, and was admitted to the hospital. Upon admission, blood tests revealed that Christopher's TSH and TH levels were low. In addition, he had elevated blood calcium levels as well as elevated PTH levels. He was prescribed *levothyroxine* (synthetic TH) and *prednisone* (a glucocorticoid). A magnetic resonance imaging (MRI) examination of his brain and spinal cord revealed slight abnormalities.

Doctors concluded that Christopher suffered from metastatic small cell carcinoma of the lung that had spread to his pituitary gland. The lesions in his pituitary protruded superiorly, which compressed nerves in his optic chiasm, thus causing his visual disturbances. It is likely that his altered consciousness and sudden onset of headache were caused by bleeding into the subarachnoid space surrounding his brain. To confirm the diagnosis, doctors performed an MRI of his head before and after administration of a contrast material. The MRI showed evidence of thickening of the infundibulum, enlargement of the pituitary gland, and enlargement of the third ventricle. There also was evidence of hypothalamic involvement. A biopsy was deemed too risky, and Christopher died just 3 weeks after his initial hospital admission.

Answer the following questions regarding Christopher's case:

1. Describe how a tumor in the hypothalamus and pituitary gland could lead to lowered levels of TSH and TH. Be sure to identify which cells are responsible for secreting each hormone, and describe feedback mechanisms that regulate their release.

   _____

   _____

   _____

   _____

   _____

2. Justify Christopher's cold intolerance and fatigue given lowered levels of TSH and TH. Be sure to state specifically how TH influences metabolism and temperature regulation.

   _____

   _____

   _____

   _____

   _____

3. Discuss the pros and cons of *prednisone* administration. Be sure to address specific targets of glucocorticoids, and positive and negative effects of the drug.

   _____

   _____

   _____

   _____

   _____

4. Are there any other hormones whose effects may be altered by a tumor in the hypothalamus and pituitary gland? If so, state which hormones may be altered, and discuss possible symptoms associated with altered levels of each hormone.

   _____

   _____

   _____

   _____

   _____

This case is adapted from the following: Case 35–2001. *The New England Journal of Medicine,* 1483–1488, 2001 November 15, Vol. 345, Issue 20.

The ❶ corresponds to the Learning Objective(s) listed in the chapter opener outline.

Do You Know the Basics?

Exercise 19.1: Gross Anatomy of Endocrine Organs

1. The adrenal glands are located _____ (superior/inferior) to each kidney. ❶

Exercise 19.2: The Hypothalamus and Pituitary Gland

2. The anterior pituitary gland is composed of glandular tissue, which consists of acidophils, basophils, and chromophobes; thus it generally stains _____
   (darker/lighter) than the posterior pituitary gland. ❷

3. Which of the following hormone(s) is/are produced by acidophils of the anterior pituitary gland? (Check all that apply.) ❸

   _____ a. adrenocorticotropic hormone (ACTH)

   _____ b. follicle-stimulating hormone (FSH)

   _____ c. growth hormone (GH)

   _____ d. luteinizing hormone (LH)

   _____ e. prolactin

4. Axon terminals of the supraoptic nuclei primarily store and release _____ (ADH/oxytocin), while the axon terminals of the paraventricular nuclei
   primarily store and release (ADH/oxytocin). ❹

Exercise 19.3: The Pineal Gland

5. One defining histological feature of the pineal gland, which also serves as an important radiological landmark, is the small islands of "crystal"-looking structures that
   compose _____ (corpora arenacea/pinealocytes). ❺ ❼

6. Melatonin, the hormone secreted by pinealocytes of the pineal gland, _____(decreases/increases) when light levels are low and _____
   (decreases/increases) when light levels are high. ❻

Exercise 19.4: The Thyroid and Parathyroid Glands

7. The thyroid gland consists of follicles that may be lined with _____ (simple cuboidal/stratified columnar) epithelial tissue. ❽

8. Follicular cells of the thyroid gland produce _____ (calcitonin/thyroid hormone), whereas parafollicular cells produce _____
   (calcitonin/thyroid hormone). ❾

9. The major action of parathyroid hormone, produced by the parathyroid gland, is to _____ (decrease/increase) blood calcium levels by
   _____ (decreasing/increasing) the action of osteoclasts and _____ (decreasing/increasing) urinary excretion of calcium. ❿

Exercise 19.5: The Adrenal Glands

10. Place the zones of the adrenal cortex in the correct order, from superficial 1 to deep 3. ⓫

    _____ a. zona fasciculata

    _____ b. zona glomerulosa

    _____ c. zona reticularis

11. Match the region of the adrenal gland listed in column A with the corresponding hormones produced by that region, listed in column B. **12**

Column A

_____ 1. medulla

_____ 2. zona fasciculata

_____ 3. zona glomerulosa

_____ 4. zona reticularis

Column B

a. androgens and glucocorticoids

b. catecholamines

c. glucocorticoids

d. mineralcorticoids

12. Match the hormone description and/or action(s) listed in column A with the appropriate hormone, listed in column B. **13**

Column A

_____ 1. increased heart rate, contractility, and vasoconstriction

_____ 2. increased glucose synthesis through protein breakdown, gluconeogenesis, and lipolysis

_____ 3. increased sodium and water retention by the kidneys

_____ 4. similar in structure and function to testosterone

Column B

a. aldosterone

b. androgens

c. cortisol

d. epinephrine and norepinephrine

Exercise 19.6: The Endocrine Pancreas—Pancreatic Islets (Islets of Langerhans)

13. The small islands of lighter-staining cells of the pancreas are known as pancreatic _____ (acini/islets). **14**

14. The pancreatic islets compose the portion of the pancreas dedicated to _____(endocrine/exocrine) function, whereas the pancreatic acini compose the

portion of the pancreas dedicated to _____ (endocrine/exocrine) function. **15**

15. Insulin is released in response to a(n) _____ (increase/decrease) in blood glucose levels, whereas glucagon is released in response to a(n)

_____ (increase/decrease) in blood glucose levels. **16**

16. Match the cell type of the pancreatic islets, listed in column A, with the hormones secreted by each cell, listed in column B. **17**

Column A

_____ 1. alpha cells

_____ 2. beta cells

_____ 3. delta cells

_____ 4. F cells

Column B

a. glucagon

b. insulin

c. pancreatic polypeptide

d. somatostatin

Exercise 19.7: Ph.I.L.S. Lesson 19: Thyroid Gland and Metabolic Rate

17. Cooling the body _____ (decreases/increases) thyroid hormone release, which in turn _____(decreases/increases) metabolic and oxygen

consumption rate. **18** **19**

Exercise 19.8: A Clinical Case in Endocrine Physiology

18. Decreased levels of thyroid-stimulating hormone and thyroid hormone could lead to which of the following symptoms? (Check all that apply.) **20**

_____ a. cold intolerance

_____ b. fatigue

_____ c. weight gain

_____ d. weight loss

Can You Apply What You've Learned?

19. Identify each of the following glands by writing in the name of the gland under the photo.

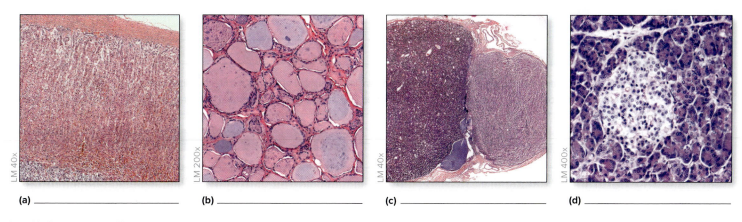

(a) _____   (b) _____   (c) _____   (d) _____

(*a–c*) ©McGraw-Hill Education/Christine Eckel; (*d*) ©Victor P. Eroschenko

20. Based on your understanding of the location of the pituitary gland with respect to other brain structures, explain why individuals with pituitary gland tumors often also experience visual disorders.

_____

_____

21. Predict the oxygen consumption in an individual with hypothyroidism. Then compare the rate of oxygen consumption in an individual with hypothyroidism with that of an individual with hyperthyroidism.

_____

_____

22. Bodybuilders have been known to inject insulin to increase muscle mass. Explain this reasoning, and discuss any risks that may be posed by this practice.

_____

_____

23. Identify some health risks that may come from long-term use of corticosteroids.

_____

_____

24. A pheochromocytoma is a tumor located in the medulla of the adrenal gland. This tumor results in excessive release of hormones from the adrenal medulla. Based on your understanding of the adrenal glands, what symptoms might patients with this kind of tumor have?

_____

_____

_____

INTRODUCTION

Blood is a bodily fluid that has fascinated humans for centuries. Its rich red color adds to both its beauty and its mystery. Early Greek physicians believed that blood was one of the four bodily "humors" (*humor*, a bodily fluid). The other bodily fluids were phlegm, black bile, and yellow bile (**table 20.1**). Greek physicians believed that a balance of the four humors was required for optimal health of both the body and the mind, and that disease was the result of an imbalance between the bodily fluids. It was this concept of disease that made bloodletting a popular treatment for disease. By draining blood from a patient, the physician supposedly was putting the patient's humors back into balance. While removing blood from the body does not cure disease, early Greek physicians were not entirely wrong. Their insight that the humors may reveal information about the patient's state is similar to today's use of blood for the purpose of disease diagnosis. Thinking about the advances in medical treatments since ancient times may be enough to transform one's mood from melancholy to "sanguine," or lively and optimistic. Are you feeling lively and optimistic yet? Then it is time to tackle the current "sanguine" topic: blood!

The exercises in this chapter explore the form and function of normal constituents of human blood. Knowledge of the normal appearance and abundance of the various types of blood cells provides a basis for recognizing when cells or amounts have become abnormal and for understanding what that means. Exercises within this chapter involve identifying the formed elements of blood, exploring the functional relevance of each of the components, and learning how the different cell types play a critical role in both the cardiovascular system and the immune system. Exercises also guide students in performing common blood tests such as determination of the hemoglobin content of blood, determination of blood type, and measurement of blood cholesterol levels. More detail on the structure and function of blood is covered in the main textbook. Thus, it is important to read the section in the textbook that covers blood before performing these laboratory exercises.

List of Reference Tables

Table 20.1	The Four Greek Humors				
Bodily Fluid (Humor)	**Element**	**Characteristics**	**Source**	**Mood**	**Word Origin**
Black Bile	Earth	Dry and cold	Spleen	Melancholy	*melas*, black, + *chole*, bile
Blood	Air	Hot and wet	Heart	Sanguine	*sanguis*, blood
Phlegm	Water	Cold and wet	Brain	Phlegmatic	*phlegma*, inflammation
Yellow Bile	Fire	Hot and dry	Liver	Choleric	*chole*, bile

These Pre-Laboratory Worksheet questions may be assigned by instructors through their ▦ connect course.

1. Number the following components of whole blood in order of abundance, from most abundant (1) to least abundant (3).

   _____ a. buffy coat

   _____ b. erythrocytes

   _____ c. plasma

2. Basophils, eosinophils, and neutrophils belong to the category of leukocytes known as _____ (agranuloctyes/granulocytes).

3. Match the description of the leukocyte listed in column A with the type of leukocyte listed in column B.

   Column A

   _____ 1. bilobed nucleus; blue cytoplasmic granules

   _____ 2. bilobed nucleus; red cytoplasmic granules

   _____ 3. large, horseshoe-shaped nucleus; much larger than an erythrocyte

   _____ 4. multi-lobed nucleus; lavender cytoplasmic granules

   _____ 5. round nucleus surrounded by pale blue cytoplasm; similar in size to an erythrocyte

   Column B

   a. basophil

   b. eosinophil

   c. lymphocyte

   d. monocyte

   e. neutrophil

4. Which of the following is the cell responsible for producing platelets? (Circle one.)

   a. basophils
   b. erythrocytes
   c. lymphocytes
   d. megakaryocytes
   e. neutrophils

5. Which of the following is the formed element responsible for transporting oxygen and carbon dioxide in the blood? (Circle one.)

   a. basophils
   b. erythrocytes
   c. lymphocytes
   d. megakaryocytes
   e. neutrophils

6. Which of the formed elements is responsible for initiating coagulation? (Circle one.)

   a. basophils
   b. erythrocytes
   c. megakaryocytes
   d. neutrophils
   e. platelets

7. Number the following leukocytes in order of abundance, from most abundant (1) to least abundant (5).

   _____ a. basophil

   _____ b. eosinophil

   _____ c. lymphocyte

   _____ d. monocyte

   _____ e. neutrophil

HISTOLOGY

EXERCISE 20.1

IDENTIFICATION OF FORMED ELEMENTS ON A PREPARED BLOOD SMEAR

EXERCISE 20.1A Making a Human Blood Smear

Exercise 20.1 may involve use of a prepared slide of human blood, or a blood smear made from your own blood. To make a blood smear using your own blood, use the following procedure.

Obtain the Following:
- **four glass slides**
- **lancet**
- **cotton balls**
- **alcohol prep pads**
- **vial of Wright's stain**

1. Clean the slides with soap and water and let them dry. It takes two slides to make a blood smear, and it may require several attempts to obtain a good preparation. For this reason, clean at least four slides.

2. Using the alcohol prep pad, clean the tip of the middle or ring finger on the hand you use the least (e.g., if you are right-handed, use the tip of the finger on your left hand).

3. Place the tip of the lancet on the clean finger, and pull the trigger so it pierces the skin. If using a lancet without a trigger **(figure 20.1)**, poke the skin with a quick jab. Dispose of the lancet in a sharps container.

4. Squeeze the finger until a small drop of blood appears on the surface. Blot away the first drop of blood using an alcohol prep pad and then squeeze the finger again until another drop appears. Gently touch the drop of blood to the top surface of the microscope slide about 1.5–2 cm from the edge of the slide **(figure 20.2, step 1)**. Preparing the blood smear does not require a lot of blood, so try not to squeeze the finger too hard.

5. Orient the slide so the end with the drop of blood on it is closest to you. Then place the edge of a second microscope slide on the drop of blood and hold it at a 45-degree angle to the slide with the drop of blood on it (figure 20.2, step 2).

6. Quickly push the second slide over the first slide (push away from you) to smear the blood (figure 20.2, step 3). The goal is to obtain a very thin, transparent layer of blood on the slide. If the layer is too thick, it will be difficult to see individual cells because they will be too

Figure 20.1   **Lancing Finger for Blood Sample.**
©Christine Eckel

closely packed together. If the smear is not even or well spread out, start over with two new slides. It might take a couple of tries to create a good, even smear. Discard any slides that made contact with blood in a bowl containing a bleach solution or sharps container. Allow the slide with the blood smear on it to air dry.

7. Place a couple of drops of Wright's stain on the blood smear slide so there is enough stain on the slide to cover the smear, but not so much that it overflows the slide. Keep track of the number of drops used (number of drops = _____).

① Place a small drop of blood approximately 2 cm from the edge of the slide.   ② Place the edge of a second slide on the drop of blood and hold it at a 45° angle to the first slide.   ③ Push the second slide away from you to smear the blood on the first slide.

Figure 20.2   **Preparation of a Blood Smear.**

8. Let the slide stand for 3 to 4 minutes, until the smear begins to take on a blue/green color.

9. Add to the slide a volume of distilled water equal to the volume of Wright's stain used. Generally this means using the same number of drops of distilled water as Wright's stain. Gently move the slide back and forth and blow on it a little bit to mix the distilled water and stain on the slide.

10. Let the slide stand for 5 to 10 minutes. During this time, blow on the slide occasionally to keep the distilled water and stain mixed.

11. Rinse the slide with distilled water for 2 to 3 minutes, and then tilt the slide to allow the excess water to drip off the slide. Allow the slide to air dry. Once the slide is dry, it will be ready to examine.

EXERCISE 20.1B Viewing Formed Elements on a Prepared Blood Smear

1. Obtain a prepared slide of human blood or use the slide prepared in exercise 20.1A. Place the slide on the microscope stage and observe at low power. Not much is visible at this magnification, but, as always, it is important to progress from low to high power to keep the sample in focus before progressing to the oil immersion objective.

2. Change from low to medium, then to high power, making sure to bring the slide into focus at each power. After the slide is in focus on high power, obtain a vial of immersion oil. Rotate the nosepiece (the part of the microscope that holds the objective lenses) so the high-power and

oil immersion objectives lie on either side of the slide (**figure 20.3**). Place a drop of immersion oil over the center of the slide (where the light can be seen coming through the slide), and then carefully rotate the nosepiece to bring the immersion objective into the oil over the slide. Bring the slide into focus.

3. Identify the following formed elements on the prepared blood slide, using **table 20.2** and **figure 20.4** as guides. Note that some elements will be much easier to locate than others due to their relative abundance in whole blood.

☐ **Basophils**   ☐ **Monocytes**

☐ **Eosinophils**   ☐ **Neutrophils**

☐ **Erythrocytes**   ☐ **Platelets**

☐ **Lymphocytes**

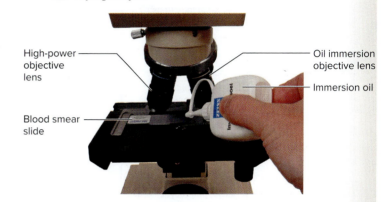

Figure 20.3 **Orienting the Objective Lenses for Placement of a Drop of Immersion Oil on the Prepared Blood Slide.**

©Christine Eckel

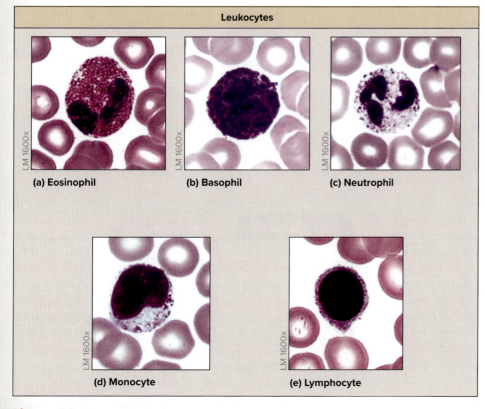

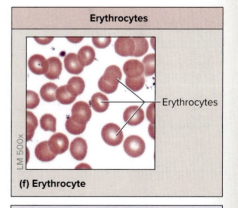

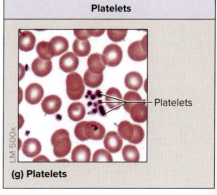

Figure 20.4 **Formed Elements in the Blood.**

(a–g) ©McGraw-Hill Education/Alvin Telser

(continued on next page)

(continued from previous page)

Table 20.2	Characteristics of the Formed Elements of Blood				
Formed Element	**Function**	**Histological Features**	**Size**	**Abundance**	**Word Origin**
Erythrocytes (Red Blood Cells)	Transport oxygen and carbon dioxide; aid in blood pH regulation	Biconcave discs; lack a nucleus; orange and red in color due to eosinophilic stains	~7.5 μm	99% of formed elements	*erythros,* red, + *kytos,* a hollow (cell)
Leukocytes (white blood cells)	Protect the body from pathogens, fight infections, and remove dead or damaged cells from the body	Blue to purple in color due to basophilic staining properties of nuclei	7–21 μm	<1% of formed elements	*leukos,* white + *kytos,* a hollow (cell)
Neutrophils	Important in fighting bacterial infections; phagocytize bacteria and damaged cells	Very light, lavender-colored cytoplasmic granules; multi-lobed nucleus (2–5 lobes)	~1.3x erythrocyte	50–70% of leukocytes	*neutro-,* neutral, + *philos,* to love
Lymphocytes	Responsible for the adaptive immune response to infection	Large, blue/purple nuclei surrounded by a halo of pale blue cytoplasm	1–2x erythrocyte	20–40% of leukocytes	*lympho-,* referring to the lymphatic system, + *kytos,* a hollow (cell)
Monocytes	Circulating cells that migrate out of the bloodstream to become large, phagocytic cells called macrophages	Large, blue/purple nucleus that is horseshoe-shaped	2–3x erythrocyte	2–8% of leukocytes	*monos,* single, + *kytos,* a hollow (cell)
Eosinophils	Fight parasitic infections and mediate (neutralize) the effects of histamines; phagocytic cells	Orange to reddish cytoplasmic granules; bluish colored, bi-lobed nucleus	~1.3x erythrocyte	1–4% of leukocytes	*acidus,* sour, referring to acidic dyes such as eosin, + *philos,* to love
Basophils	Release histamine and heparin; involved in the inflammatory response	Cytoplasmic granules stain very dark blue; bi-lobed nucleus may not be visible	~1.3x erythrocyte	0.5–1% of leukocytes	*baso-,* referring to basic dyes, + *philos,* to love
Platelets	Play a role in hemostasis (blood clotting)	Small, purple-colored structures that are fragments of megakaryocytes	~2 μm	1% of formed elements	*platys,* flat

4. After completing this activity, clean the microscope and slide thoroughly to remove all traces of oil from the instruments. To do this, rotate the nosepiece so the high-power and oil immersion lenses are on either side of the slide once again (figure 20.3). Use *lens paper* (do not use anything else or it may damage the lenses!) and lens cleaning solution to carefully and thoroughly clean the oil from both the oil immersion objective and the blood smear slide. Blood smear slides prepared using your own blood should be disposed of by placing them in a bleach solution for disinfection.

INTEGRATE

LEARNING STRATEGY

Although lymphocytes usually have circular nuclei and monocytes usually have indented, or "horseshoe-shaped," nuclei, sometimes monocytes can have nuclei that appear circular. In these cases, monocytes can appear very similar to lymphocytes. To keep from misidentifying a monocyte as a lymphocyte, always consider the size of the cell in addition to the shape of the nucleus. A monocyte is two to three times larger than an erythrocyte, whereas a lymphocyte is much closer to the same size as an erythrocyte.

EXERCISE 20.2

IDENTIFICATION OF MEGAKARYOCYTES ON A BONE MARROW SLIDE

All the formed elements of the blood are produced in the red bone marrow through the process of **hemopoiesis** (*hemo-*, blood, + *poiesis*, a making). Identifying the precursor cells of circulating blood cells is a complicated endeavor, which is generally undertaken only in upper-level histology courses. The goal of this laboratory exercise is to locate the precursor cells of platelets (*megakaryocytes*) and investigate their structure and function. However, while engaging in the process of identifying megakaryocytes, try to also obtain an appreciation for the general appearance of the precursor cells of erythrocytes and leukocytes that are also visible on the slide.

1. Obtain a prepared slide of red bone marrow and place it on the microscope stage. Bring the slide into focus on high power. Identify the following on the slide, using **figure 20.5** as a guide:

   ☐ **Bone tissue**   ☐ **Platelets**

   ☐ **Megakaryocytes**   ☐ **Red bone marrow**

2. **Megakaryocytes** (*megas*, big, + *karyo*, kernel [referring to the nucleus] + *kytos*, a hollow, cell) are extremely large cells with enormous nuclei. They are easily identifiable on prepared slides of bone marrow. Megakaryocytes remain within the bone marrow. However, their products, **platelets**, are continuously

delivered to the bloodstream. Megakaryocytes produce long extensions called proplatelets. While still attached to the megakaryocyte, these proplatelets extend through the blood vessel wall (between the endothelial cells) in the bone marrow. The force from the blood flow "slices" these proplatelets into the fragments known as platelets.

3. Sketch a megakaryocyte as seen through the microscope in the space provided or on a separate sheet of paper. Be sure to indicate the relative size of the megakaryocyte as compared to the size of the developing blood cells that are also visible on the bone marrow slide.

_____ ×

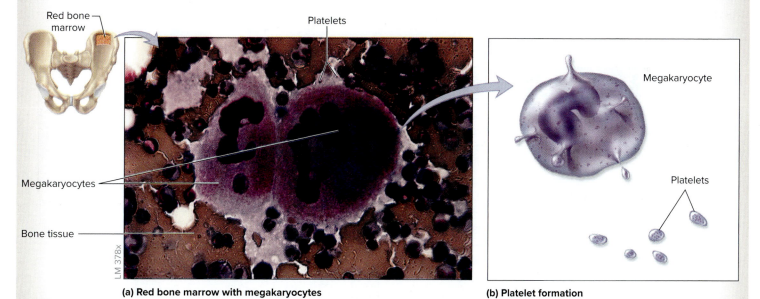

Red bone marrow

Platelets

Megakaryocytes

Bone tissue

LM 378x

(a) Red bone marrow with megakaryocytes

Megakaryocyte

Platelets

(b) Platelet formation

Figure 20.5 **Bone Marrow Slide.** The megakaryocytes in red bone marrow give rise to platelets.
©Dr. Dorothea Zucker-Franklin/Phototake

CLINICAL VIEW
Bone Marrow Transplants

Leukemia is a disease characterized by excessive proliferation of abnormal leukocytes. Patients who have experienced a progression of the disease may opt to undergo a bone marrow transplant. Leukocytes are produced in a process called leukopoiesis, which occurs in the bone marrow. A bone marrow transplant procedure involves first destroying the leukemia patient's existing bone marrow through radiation or chemotherapy, and then harvesting hemopoietic stem cells from a donor. The harvested cells are injected into the leukemia patient's blood. Recall that adult bone marrow still undergoing

hemopoiesis is located in the proximal epiphyses of the humerus and femur, in the sternum, and in the iliac crest. Typically, bone marrow is harvested from the donor's iliac crest because it is relatively easy to access and causes the fewest complications for the donor.

After the donor's cells are transplanted into the recipient, they will normally migrate to the recipient's spleen, and the process of leukopoiesis will begin anew. Care must be taken to ensure that the donor's bone marrow is a good match and will not elicit an immune response in the patient. While patients undergo this procedure, their leukocytes become depleted, making them very susceptible to infection. For this reason, patients must limit their exposure to pathogens until their leukocyte count has risen to safe levels once again.

PHYSIOLOGY

Blood Diagnostic Tests

Blood is an amazing fluid with a beautiful red color due to the presence of hemoglobin within erythrocytes. **Hemoglobin** is a protein in erythrocytes that is essential for transporting oxygen throughout the body. However, blood transports much more than just oxygen. Blood also transports nutrients, waste products, carbon dioxide, hormones, and heat throughout the body. Proper circulation of blood to the tissues is essential for their survival. In medicine, a sample of a patient's blood can yield important clues as to what diseases may be affecting an individual (**table 20.3**). A blood test involves first collecting a blood sample, typically performed by a *phlebotomist* (*phleps,* vein, + *tome-,* to cut), and placing the sample in a test tube (**figure 20.6**, step 1). The test tube is then placed in a centrifuge, which spins the

sample at high speed to separate the component parts (figure 20.6, step 2). Because the formed elements like leukocytes and erythrocytes are heavy, they fall to the bottom of the test tube, while the blood plasma remains floating on the top (figure 20.6, step 3). Leukocytes and platelets form a thin layer called a *buffy* coat, which sits on top of the layer of erythrocytes. The iron present in the hemoglobin in erythrocytes makes erythrocytes heavier than leukocytes, which explains why erythrocytes lie in the layer below leukocytes in the test tube.

The following laboratory exercises explore purpose and procedure for performing several clinically relevant blood tests, including a differential leukocyte count, hematocrit, hemoglobin concentration, coagulation time, blood typing, cholesterol testing, and blood glucose testing.

Table 20.3	Normal Ranges for Laboratory Blood Tests
Blood Glucose (8–12 hours after a meal)	70–100 mg/dl
Coagulation Time	5–8 min
Hematocrit	Males: 42–56%, Females: 38–46%
Hemoglobin	Males: 13.5–17.5 g/dL, Females: 12.0–16.0 g/dL
Platelets	150,000–400,000 cells/mm³
Erythrocytes	Males: ~5.4 million cells/mm³; Females: ~4.8 million cells/mm³
Leukocytes	4,500–11,000 cells/mm³
Total Cholesterol = HDL + LDL (HDL = High density lipoprotein, LDL = Low density lipoprotein)	
LDL Cholesterol	Recommended: <130 mg/dL, Moderate risk: 130–159 mg/dL, High risk: >160 mg/dL
HDL Cholesterol	40–80 mg/dL

Source: Goldman, L. and A.I. Schafer, *Goldman's Cecil Medicine 24e.* New York, NY: Elsevier, 2012.

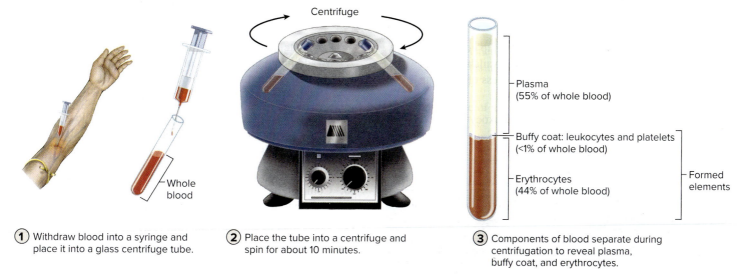

① Withdraw blood into a syringe and place it into a glass centrifuge tube.

② Place the tube into a centrifuge and spin for about 10 minutes.

③ Components of blood separate during centrifugation to reveal plasma, buffy coat, and erythrocytes.

Figure 20.6 **Separation of a Whole Blood Sample by Centrifugation.** Using a centrifuge to separate whole blood into plasma (55%) and formed elements (45%) is the first step in determining the composition of whole blood.

INTEGRATE

CONCEPT CONNECTION

Recall that lymphocytes are the second-most abundant leukocytes and participate in immune responses to infection. More specifically, B- and T-lymphocytes increase in number when they become activated in response to an antigen. These B- and T-lymphocytes allow for an adaptive (or acquired) immune response that is *specific* for a particular pathogen and exhibits lasting *memory*. This adaptive immune response involves proliferation of

B- and T-lymphocytes and the production by B-lymphocytes of antibodies (i.e., immunoglobulins, or plasma proteins) that can assist in the removal of antigens from the body by making them more susceptible to phagocytosis. The formation of antigen-antibody complexes also increases the likelihood of agglutination, which occurs in blood transfusion reactions. To avoid a rejection response with blood transfusion, patients typically receive blood that is typed-matched to the donor's blood.

EXERCISE 20.3

DETERMINATION OF LEUKOCYTE COUNTS

A **leukocyte count** performed on a sample of the patient's blood can determine whether or not the patient has a bacterial infection. An excess of neutrophils will indicate that the patient has such an infection, whereas a normal abundance of neutrophils might be a clue that an infection is viral rather than bacterial. An abnormally low number of leukocytes (leukopenia) may be indicative of a patient's inability to fight infection. In addition, certain pathological conditions may lead to the uncontrolled proliferation of abnormal leukocytes as in leukemia (see Clinical View: Bone Marrow Transplants).

1. Obtain a slide of a human blood smear or prepare one using your own blood. See exercise 20.1A for preparing a human blood smear using your own blood.

2. Place the human blood smear slide on the microscope stage. Using the scanning objective (see exercise 3.1), move the slide on the stage so the highest density of blood cells appears in the field of view.

3. Rotate the coarse adjustment knob until leukocytes are visible and in focus within the field of view. Because of the stain, leukocytes will appear as small *purple* dots in a sea of erythrocytes.

4. To properly view blood cells, it is necessary to use the oil immersion (highest-power) objective of the microscope. See exercise 20.1B for instructions on how to view specimens using an oil immersion objective. Rotate the nosepiece so the highest-power or oil immersion objective is directed toward the blood smear slide.

5. Turn the fine focus adjustment knob until the leukocytes come into focus.

6. Gently move the slide systematically in a back-and-forth motion, making sure to span the entire width of the blood smear on each pass (**figure 20.7**). Count leukocytes as

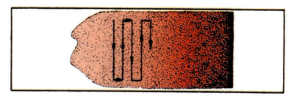

Figure 20.7 **Method for Performing Differential Leukocyte Count.** Scan the slide by moving the microscope stage so the view follows the pattern illustrated here.

(continued on next page)

(continued from previous page)

6. Use the hemolysis applicator to stir the blood such that the erythrocytes rupture (lyse) and the blood appears clear (~45 seconds) (figure 20.9*b*).

7. Place the smaller, flat piece of glass onto the U-shaped area as a coverslip.

8. Slide both pieces of glass into the metal clips of the blood chamber, and replace the blood chamber into the side slot of the hemoglobinometer (figure 20.9*c*). Be sure that the blood chamber is secure within the hemoglobinometer.

9. Locate the light switch on the underside of the hemoglobinometer. Using your left hand, hold the hemoglobinometer while resting your left thumb on the switch. Look into the eyepiece and turn the light on (figure 20.9*d*).

10. Notice there are two green areas in the field of view. Use your right hand to adjust the slide on the right side of the hemoglobinometer until the two halves of the green field match in color.

11. Record the resulting blood hemoglobin level:
    _____ g/dl

12. To clean, remove the blood chamber and place the glass plates and clip into a 10% bleach solution.

13. **Alternate Method:** The reasonably accurate and inexpensive *Tallquist method* may be used to estimate blood hemoglobin level. If performing this activity, use a **Tallquist test kit**, which contains absorbent paper and an associated color scale. Repeat step 5 and place the drop of blood on the absorbent paper. Allow the blood to dry (ensuring that the blood has not dried so much so that it appears brown) and match the blood to the color scale provided in the kit.

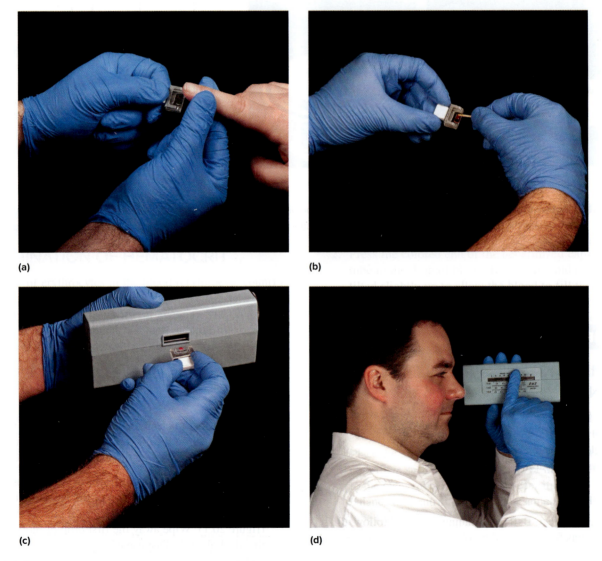

(a) (b) (c) (d)

Figure 20.9 **Determining Hemoglobin Content Using a Hemoglobinometer.** (*a*) Gently place a drop of blood on the U-shaped part of the glass in the blood chamber. (*b*) Use a hemolysis applicator stick to stir the blood. (*c*) Slide the blood chamber into the slot on the side of the hemoglobinometer. (*d*) Use your thumb to turn on the light while using your index finger to move the slide until the colors seen through the window have matching intensity. Read the hemoglobin content from the side of the hemoglobinometer.

(*a–d*) ©Christine Eckel

EXERCISE 20.6

DETERMINATION OF COAGULATION TIME

A test to determine the time required for blood to clot, or **coagulation time**, may be used to assess platelet number and function. When a blood vessel is damaged, platelets are the formed elements involved in **hemostasis** (*hemo-,* blood, + *stasis,* stability), the process that stops blood flow. Hemostasis involves three phases: vascular spasm, platelet plug formation, and coagulation. Both the formation of a platelet plug and the reactions required for coagulation involve platelets. **Platelet plug** formation occurs when a blood vessel is damaged and collagen fibers beneath the endothelium are exposed. Platelets stick to the exposed collagen, undergo morphologic changes, and secrete chemicals that attract additional platelets and promote both coagulation and vessel repair. **Coagulation** involves a cascade of reactions that lead to the formation of a blood clot, called a **thrombus**. In addition to platelets, these reactions require substances such as calcium, clotting factors, and vitamin K. The result is a meshwork of the insoluble protein fibrin, which traps substances contained within the blood. When the inside of a vessel is damaged, the formation of a thrombus typically takes 3 to 6 minutes. Both the formation of the platelet plug and coagulation are regulated by positive feedback mechanisms in order to minimize blood loss.

Obtain the Following:

- **nonheparinized capillary tube**
- **alcohol swab**
- **lancet**
- **triangular file**

1. Choose a subject. Before beginning, read and follow all instructions for safe handling of human blood (see the warning at the start of the Histology section).

2. Use an alcohol swab to cleanse the tip of the subject's middle or ring finger and prick the finger with a lancet. Wipe away the first drop of blood with the alcohol swab. Dispose of the lancet in a sharps container.

3. Load the nonheparinized capillary tube with blood (exercise 20.4, step 2).

4. Fill the capillary tube with blood approximately three-fourths of the length of the tube.

5. Lay the capillary tube flat on a paper towel and record the initial time: _____

6. After 1 minute, use the triangular file to weaken a small area of the capillary tube just adjacent to its end **(figure 20.10a)**. Once the tube is filed slightly, use both hands to hold the tube on either side of the filed section. Break the tube away from you, ensuring that both ends remain close so that you can observe a fibrin clot (if any). If the clot has not yet formed, the tube will break cleanly (figure 20.10b). If the clot has formed, fibrin will extend from each end (figure 20.10c).

7. Repeat step 6 at 1-minute intervals until a fibrin clot is observed. Record the time required for coagulation: _____ minutes

(a)

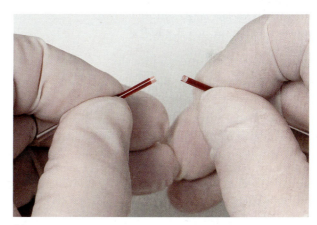

(b)

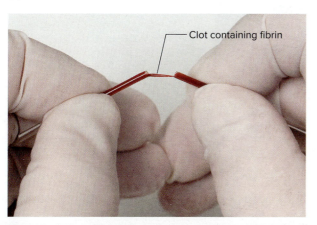

Clot containing fibrin

(c)

Figure 20.10 **Determining Coagulation Time.**
(*a*) Use a triangular file to score the slide. (*b*) Break the slide to see if coagulation has occurred (no coagulation has yet occurred in this photo). (*c*) Appearance of fibrin within a clot after coagulation has occurred.

(*a*) ©Christine Eckel; (*b–c*) ©J and J Photography

DETERMINATION OF BLOOD TYPE

Another clinically relevant blood test that may be performed is determination of a person's blood type (A, B, AB, or O). **Blood type determination** requires some understanding of an immune response; when erythrocytes clump together (agglutination), it is indicative of an immune response called *agglutination*. Agglutination occurs when antibodies circulating in the blood bind to antigens on the surface of erythrocytes. As concerns blood types, people with blood type A have surface antigen A and produce antibodies to type B blood. People with blood type B have surface antigen B and produce antibodies to type A blood. People with blood type AB have both surface antigens A and B and produce no antibodies. People with blood type O have no surface antigens and produce both type A and type B antibodies. Another common surface antigen on erythrocyte membranes determines the Rh blood type. The Rh blood type is determined by the presence or absence of the Rh surface antigen, often called either Rh factor or surface antigen D. When the Rh factor is present, the individual is said to be Rh positive (Rh+). When the Rh factor is absent, the individual is said to be Rh negative (Rh–). Individuals with type O negative (O–) blood are typically referred to as *universal donors* because their blood can be donated to individuals of any blood type without causing a transfusion reaction in the recipient. Individuals with AB positive (AB+) blood are referred to as *universal recipients* because they produce no antibodies that would cause a transfusion reaction.

Obtain the Following:
- **blood sample (human, animal, or artificial)**
- **two clean microscope slides**
- **antibody sera (anti-A, anti-B, anti-Rh)**
- **droppers**
- **toothpicks**
- **wax pencil**
- **alcohol swabs and lancets (if using human blood)**

1. Before beginning, read and follow all instructions for safe handling of human blood (see the warning at the start of the Histology section). Draw a vertical line down the center of one microscope slide with a wax marking pencil, dividing the slide into equal halves. Label the left side "anti-A" and the right side "anti-B" (**figure 20.11a**).

2. Place one drop of blood on the left side of the dividing line and place one drop of blood on the right side of the dividing line of the first slide. If using human blood, use an alcohol swab to cleanse the finger and prick the finger with a lancet to obtain the drop of blood (figure 20.1). Dispose of the lancet in a sharps container.

3. Place one drop of blood in the center of the second slide.

4. Place one drop of anti-A serum on the drop of blood on the left side of the dividing line on the first slide. Place one drop of anti-B serum on the drop of blood on the right side of the first slide. Place one drop of anti-Rh serum (also called "anti-D" serum) on the drop of blood on the second slide.

5. Mix each blood/anti-serum sample with a clean toothpick (figure 20.11*b*). Be sure to use a separate toothpick for each mixture. Transferring the toothpick to multiple mixtures will contaminate the samples. Note that an Rh typing box may be required if using human blood because thorough mixing of the blood and anti-Rh serum at a slightly elevated temperature is required.

6. After 2 minutes, observe the blood samples. Agglutination in any of the samples indicates a positive result (figure 20.11*c*).

7. Place the microscope slides in a bowl containing a 10% bleach solution. Consult with the laboratory instructor for proper disposal of all contaminated materials.

8. If the entire class is reporting blood types to determine relative abundance of blood types in the the class (see Clinical View: Blood Type Abundance Within the U.S. Population), record the blood type in the location indicated by the instructor. After all groups have reported their results, record the numbers and calculate abundances in the table on this page. Make note of any pertinent observations here:

_____

_____

_____

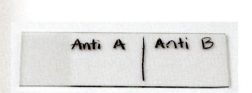

(a)   (b)   (c)

Figure 20.11 **Blood Typing.** (*a*) Prepare the slide as demonstrated here. (*b*) Mix anti-sera and blood sample with toothpick. (*c*) Observe the slides after 2–3 minutes, looking for agglutination reactions.

(*a–c*) ©Christine Eckel

INTEGRATE

CLINICAL VIEW
Blood Type Abundance Within the U.S. Population

In the United States, the most abundant blood type is O⁺. However, there are slightly different relative abundances of blood types when broken down by ethnicity. Other antigens present in the blood may also cause

agglutination reactions in some instances. In cases where the risk of such a reaction in a patient is particularly high, it may be important to transfuse blood from a donor that is not only matched by ABO blood type, but also comes from a donor of the same ethnic group as the patient.

Blood Type Abundance Within the U.S. Population					
Blood Type	**Class Results (%)**	**Caucasian (%)**	**African American (%)**	**Hispanic (%)**	**Asian (%)**
O⁺		37	47	53	39
O⁻		8	4	4	1
A⁺		33	24	29	27
A⁻		7	2	2	0.50
B⁺		9	18	9	25
B⁻		2	1	1	0.40
AB⁺		3	4	2	7
AB⁻		1	0.30	0.20	0.10

Source: http://www.redcrossblood.org/learn-about-blood/blood-types

INTEGRATE

CLINICAL VIEW
Blood Transfusions

Blood transfusions involve the transfer of blood from one person, the **donor**, to another person, the **recipient**. It is critical that the blood from both donor and recipient are compatible, or the consequences of the transfusion may be lethal. For a recipient to safely receive a donor's blood, the donor blood must not be "recognized" as foreign. That is, the donor's erythrocytes must not contain antigens that the recipient's erythrocytes will bind to. If a recipient

with A⁺ blood receives A⁻ blood from a donor, the recipient will not reject the blood because both donor and recipient have surface antigen A on their erythrocytes and produce anti-B antibodies. On the other hand, if a recipient with A⁻ blood receives A⁺ blood from a donor, the recipient will potentially reject the blood if the anti-Rh antibodies are present in the recipient's blood (A⁻). If present, anti-D antibodies will bind to the Rh antigen that is on the surface of the donor's erythrocytes (A⁺), thus identifying the donor blood as "foreign." Rejection involves an agglutination reaction, as observed in exercise 20.7.

EXERCISE 20.8

DETERMINATION OF BLOOD CHOLESTEROL

A test to determine **total blood cholesterol**, including the blood values for low density lipoproteins (LDLs) and high density lipoproteins (HDLs), may be important in evaluating an individual's risk for cardiovascular disease. Excess LDLs not used by peripheral

tissues may be deposited in the walls of blood vessels and are therefore considered to be "bad cholesterol." Conversely, HDLs *remove* lipids from the walls of blood vessels and transport the lipids to the liver for waste removal. Therefore, HDLs are considered to be "good cholesterol." Clinically, elevated levels of total blood cholesterol, increased numbers of LDLs, and decreased numbers of HDLs have all been implicated as risk factors for

(continued on next page)

Exercise 20.5: Determination of Hemoglobin Content

9. When tested with a hemoglobinometer, blood containing a decreased level of hemoglobin as compared to normal blood will appear _____(darker/lighter) in color. **9**

Exercise 20.6: Determination of Coagulation Time

10. Which of the following are conditions that may cause coagulation time to increase? (Check all that apply.) **10**

_____ a. decreased calcium

_____ b. decreased megakaryocytes

_____ c. decreased platelet count

_____ d. decreased vitamin K

_____ e. increased hematocrit

Exercise 20.7: Determination of Blood Type

11. If an agglutination reaction only occurs when anti-B antibody is applied, then the blood type must be type _____ (A/B) _____ (+/−). **11**

Exercise 20.8: Determination of Blood Cholesterol

12. When considering components of total blood cholesterol, _____ (HDLs/LDLs) are considered to be the "good" type. **12**

Exercise 20.9: Determination of Blood Glucose

13. Which of the following is considered an abnormal fasting blood glucose level? (Circle one.) **13**

a. 70 mg/dl

b. 90 mg/dl

c. 100 mg/dl

d. 150 mg/dl

Can You Apply What You've Learned?

14. In determining hematocrit, if the length of erythrocytes in the capillary tube is 45 millimeters and the length of total blood is 95, what is the hematocrit?

_____

15. What are some symptoms you might expect an individual suffering from anemia to exhibit? _____

_____

16. Since leukocytes and erythrocytes are produced from the same precursor cells, predict the change in leukocyte count that might occur in a patient who has an

elevated erythrocyte count (polycythemia). _____

17. Predict the change in leukocyte count in a patient suffering from leukemia (cancer involving leukocytes). _____

18. Erythropoietin (EPO), a hormone produced by the kidneys, is released into the blood when blood oxygen levels are low. EPO stimulates the production of erythrocytes by red bone marrow. Predict the resulting EPO level in the blood and erythrocyte count of a person who has spent considerable time living at high altitudes, where atmospheric oxygen levels are reduced. _____

_____

19. A patient has been injured and is in need of a blood transfusion. When she arrives at the ER, doctors determine that she has type A⁺ blood. What antigens are present

on her erythrocytes? _____ What antibodies are present in her blood?

_____ What blood type(s) can she receive in her blood transfusion?

_____

Can You Synthesize What You've Learned?

20. Bonnie, a 50-year-old female, complains of severe fatigue, weakness, and dizziness. Her complete blood count (CBC) is shown below. Consult table 20.3. Discuss which of her variables are out of the expected range, and propose a diagnosis.

Leukocytes	1000 cells/mm^3
Erythrocytes	2 million cells/mm^3
Hemoglobin (Hb)	7.5 g/dl
Hematocrit	25%
Neutrophils	31%
Eosinophils	6%
Basophils	2%
Lymphocytes	53%
Monocytes	8%
Platelets	50,000 cells/mm^3

_____

_____

_____

21. Elizabeth is 20 weeks pregnant with her second child. Elizabeth's blood type is AB negative (AB$^-$). Her physician is concerned about the possibility of a reaction between Elizabeth's blood and the baby's blood because of the blood type of the baby's father. The physician prescribes RhoGAM for Elizabeth. RhoGAM contains immunoglobulins that prevent Elizabeth from producing Rh antibodies (anti-D antibodies). What does the physician fear may be a complication with this pregnancy in the event that RhoGAM is not administered?

_____

_____

The Cardiovascular System: The Heart

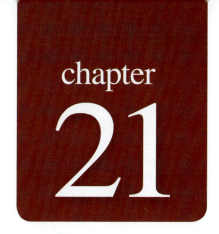

chapter

21

OUTLINE AND LEARNING OBJECTIVES

Module 9: Cardiovascular System

INTRODUCTION

The heart is an amazing organ that has been viewed with awe for ages. For many centuries physicians thought the heart, not the brain, was the control center and spiritual/emotional center of the body. Perhaps this is because the structural and functional relationships in the heart are relatively straightforward and easy to see, whereas such relationships are nearly impossible to discover through gross observation of the brain. Students of anatomy and physiology who have studied the brain know that our emotions come not "from the heart" but from the brain. Even though scientists and laypeople alike recognize that the brain, not the heart, controls the functioning of the rest of the body, they also recognize that the heart *is* essential for the survival of all organs and tissues in the body: If the heart fails to pump blood, and that failure results in a lack of flow of oxygenated blood to the tissues, the tissues will die.

Consider for a moment the incredible fact that the heart continues to beat, day and night, year after year, without stopping. It is an enormous job. Failure of this organ is often fatal. Perhaps it is no surprise that heart disease is the number one cause of death for Americans (it accounts for approximately 23% of all deaths, while cancer, at number two, accounts for 21% of all deaths). A thorough understanding of the structure and function of the heart is critical for everyone, whether an individual intends to go into a health science field or not. Nearly all health-care-related fields require practitioners to deal with individuals suffering from heart disease on a daily basis. Even if an individual's career does not involve health care, how one takes care of his or her heart now may very well determine the length and overall quality of that person's life.

In this laboratory session, you will identify heart structures on a preserved human heart or a model of the heart, and through dissection of a sheep heart. You will then review the structure of cardiac muscle tissue that enables the heart to contract and pump blood throughout the body. While working through the exercises, be aware that most textbook figures of the heart are drawn to make identification of the chambers and vessels very straightforward. When observing a real heart, identification of structures is more challenging because the chambers do not lie directly superior, inferior, or lateral to each other as they are often depicted in textbook drawings. In fact, most of the heart structures labeled "right" (such as the right atrium and right ventricle) lie not only on the right side of the heart, but also on the *anterior* surface of the heart. Likewise, most of the structures labeled "left" (such as the left atrium and left ventricle) lie not only on the left side of the heart, but also on the *posterior* surface of the heart.

Physiology exercises within this chapter will involve measuring electrical activity of the heart using electrocardiography (ECG), listening to heart sounds and relating these sounds to the events of the cardiac cycle, and exploring the length–tension relationship in cardiac muscle.

List of Reference Tables

These Pre-Laboratory Worksheet questions may be assigned by instructors through their ![connect] course.

1. The _____ (pulmonary/systemic) circuit pumps blood to all organs of the body, and the

   _____ (pulmonary/systemic) circuit pumps blood to the lungs.

2. A(n) _____ (artery/vein) is a vessel that always carries blood away from the heart.

3. The _____ (pulmonary/systemic) circuit pumps blood at a higher pressure than the
   _____ (pulmonary/systemic) circuit.

4. What are the muscles that attach to tendinous cords (chordae tendineae) called? (Circle one.)

   a. papillary muscles

   b. pectinate muscles

   c. pectoral muscles

   d. trabeculae carneae

5. The region where the great vessels such as the aorta and pulmonary trunk are attached to the heart is called the _____ (apex/base) of the heart.

6. The wall of the _____ (right/left) ventricle is much thicker than the wall of the _____ (right/left) ventricle.

7. The inner layer of the pericardial sac is the _____ (parietal/visceral) layer of the serous pericardium.

8. Most of the anterior surface of the heart receives oxygenated blood from a branch of the _____ (left/right) coronary artery.

9. The valve between the right atrium and the right ventricle is the _____ (bicuspid/tricuspid) valve.

10. The left ventricle pumps blood into the _____ (aorta/pulmonary trunk).

11. The superior and inferior vena cava drain blood into the _____ (left/right) atrium.

12. What component of an ECG represents the electrical activity that precedes ventricular contraction? (Circle one.)

    a. S-T segment

    b. P wave

    c. QRS complex

    d. T wave

13. Stimulation of the heart's contraction is initiated by what structure, which is sometimes called the "pacemaker"? (Circle one.)

    a. AV node

    b. SA node

    c. AV bundle

    d. Purkinje fibers

14. The first heart sound heard is caused by turbulent blood flow as the _____ (AV/semilunar) valves close.

(continued from previous page)

Keep in mind that the left atrium and left ventricle will be on *your* right, and the right atrium and ventricle will be on *your* left. The right ventricle is anterior and the left ventricle is posterior. Notice that the mass of the left ventricle fills up nearly the entire palm of your hand because it has a much thicker myocardium than the right ventricle. The **right atrium** is superior and lateral to the right ventricle. The left atrium is not visible in this view. You will need to rotate the heart to see the left atrium.

5. *Layers of the Heart Wall*: Both the atrial and the ventricular wall have three layers. The outermost layer is the **epicardium**, also called the visceral layer of the serous pericardium. Next is the **myocardium**, a thick muscular layer responsible for the contraction of the heart. The **endocardium** is the thin, innermost layer of the heart wall, and its epithelial layer is continuous with the lining of the blood vessels. Identify the following components of the heart wall on a heart model or on a preserved human heart, using figure 21.2 as a guide.

☐ **Endocardium**   ☐ **Myocardium**

☐ **Epicardium**

INTEGRATE

LEARNING STRATEGY

One way to remember which of the atrioventricular (AV) valves is on the right side of the heart and which is on the left is to use the saying, "**Try** before you **Buy**." The **TRI**cuspid ("Try") valve, which is on the right side of the heart, comes before the **BI**cuspid ("Buy") valve, which is on the left side of the heart. Note that this mnemonic only works for blood flow returning from the body.

---

6. *The Right Atrium*: Holding the heart in your right hand once again, with the anterior surface directed toward you, identify the **superior vena cava** attached to the right atrium (figure 21.3). Within the thorax, the superior vena cava extends vertically and is attached to the upper portion of the right atrium. Now locate the inferior vena cava, which extends vertically and is attached to the lower portion of the right atrium. Next, look inside the right atrium. Notice the thin strands of **pectinate muscle** in the wall of the right atrium **(figure 21.4)**. Pectinate muscle is found in the wall of the right atrium especially within its *auricles*, which are wrinkled, flat extensions of the atria. Locate the shallow depression covered with a thin membrane in the interatrial septum. This is the **fossa ovalis**, a remnant of a fetal shunt between the right and left atria called the **foramen ovale**. What is the function of the foramen ovale in a fetus? _____

_____

Is the opening completely closed off in the specimen? If not, what might some of the consequences be (this condition is called a *patent* foramen ovale [*pateo*, to lie open])? _____ Just inferior to the fossa ovalis, look for the small opening of the **coronary sinus**, a **vein** that drains nearly all deoxygenated blood from the heart wall. Finally, observe the **right atrioventricular (AV) valve** and count the cusps. How many cusps are there? _____ Based on that information, is the right AV valve a tricuspid or bicuspid valve? _____

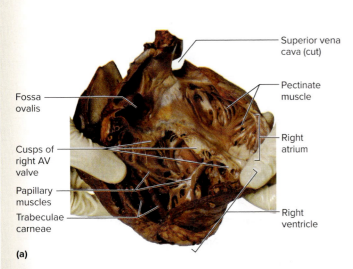

(a)

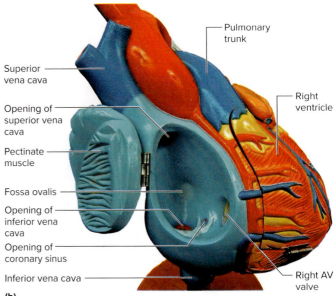

(b)

Figure 21.4 **The Right Atrium of a Human Heart.** (*a*) Gross specimen. (*b*) Heart model.

(*a*) ©Christine Eckel; (*b*) ©Denoyer-Geppert. Photo by Christine Eckel

7. *The Right Ventricle:* Place a blunt probe in the right atrium, and direct the tip into the right ventricle (**figure 21.5**). If the right ventricle is not already cut open, ask the instructor to cut it open to identify the internal structures. A prominent feature in the wall of the right ventricle is the folds of cardiac muscle called **trabeculae carneae** (*trabs*, a beam, + *carneus*, fleshy). Extending into the chamber are nipple-like **papillary muscles** (*papilla*, a nipple) that attach to the **cusps of the right AV valve** by stringlike structures called **tendinous cords** (or *chordae tendineae*). When the ventricles contract, the papillary muscles also contract and pull down on the cusps of the AV valve. Because blood is being pushed up against the inferior portion of the valve cusps as the ventricles contract, the action of the papillary muscles pulling down on the valve cusps keeps the valve closed. This prevents blood from flowing back into the right atrium. Consequently, the blood is forced out through the **pulmonary trunk**.

8. Identify the following structures in the right ventricle of a human heart or heart model, using figure 21.5 as a guide:

   ☐ **Cusps of the right AV valve**   ☐ **Tendinous cords**

   ☐ **Papillary muscles**   ☐ **Trabeculae carneae**

9. Place the tip of a blunt probe in the right ventricle and pass it out through the **pulmonary trunk**. To enter the pulmonary trunk, the probe will have to pass through the **pulmonary semilunar valve**. If the vessel has been cut open, the cusps of the semilunar valve will be visible. Notice they are shaped like "half moons"—hence the name (*semi-*, half + *lunar*, moon). How many cusps are there?_____ Where is the blood in the pulmonary trunk transported?

   _____

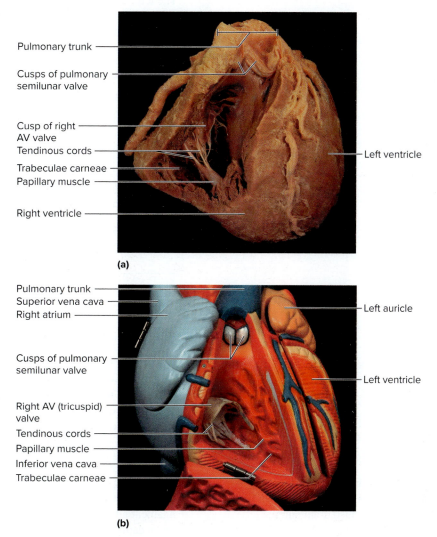

(a)

(b)

Figure 21.5 **The Right Ventricle of a Human Heart.** (*a*) Gross specimen. (*b*) Heart model.

(*a*) ©McGraw-Hill Education/Photo and Dissection by Christine Eckel; (*b*) ©Denoyer-Geppert. Photo by Christine Eckel

(*continued on next page*)

(continued from previous page)

10. Sketch the right atrium and right ventricle in the space provided or on a separate sheet of paper. Use arrows to indicate the flow of blood from the right atrium to the pulmonary trunk. Be sure to include and label all of the structures listed below in your drawing.

☐ **Fossa ovalis** ☐ **Pulmonary trunk**

☐ **Inferior vena cava** ☐ **Right atrium**

☐ **Opening of coronary sinus** ☐ **Right AV valve**

☐ **Right ventricle**

☐ **Papillary muscles** ☐ **Superior vena cava**

☐ **Pectinate muscle** ☐ **Tendinous cords**

☐ **Pulmonary semilunar valve** ☐ **Trabeculae carneae**

11. *The Left Atrium:* Rotate the heart until its posterior surface is visible (**figure 21.6**). Open the left atrium and notice the four **pulmonary veins**, which drain into the left atrium. Look inside the left atrium. Notice that the wall of the left atrium is thin and smooth and does *not* have pectinate muscles (although the wall of the left *auricle* does). Recall that the right atrium has pectinate muscles in both its wall and auricle. This difference provides a means of distinguishing the right atrium from the

left atrium when viewing the internal surface of the heart. The left atrium is little more than an expansion of the tissue where the four pulmonary veins come together. Thus, its walls are not always easy to identify. Now that both the right and left atria have been identified, place your index finger in the right atrium and your thumb in the left atrium to feel for the **fossa ovalis**, which lies in the **interatrial septum**. What is the name of the fetal structure of which the fossa ovalis is a remnant?_____ Next, observe the **left atrioventricular (AV) valve** and count the cusps. How many cusps are there?_____ Based on that information, is the left AV valve a tricuspid or bicuspid valve?_____ What is another name for this valve?_____

12. *The Left Ventricle:* Place a blunt probe in the left atrium and pass it into the **left ventricle (figure 21.7)**. If the left ventricle is not already cut open, ask the instructor to cut it open so you can identify the structures within. The left ventricle contains all the same structures as the right ventricle, and the **left atrioventricular (AV) valve** functions the same way the right AV valve functions.

13. Identify the following structures in the left ventricle, using figure 21.7 as a guide:

☐ **Left AV valve** ☐ **Tendinous cords**

☐ **Papillary muscles** ☐ **Trabeculae carneae**

14. Note the difference in thickness between the myocardium in the wall of the left ventricle and that of the right ventricle. Which is thicker? _____ What is the reason for this difference?

_____

_____

_____

Does the chamber size (volume) of the left ventricle appear to be greater than that of the right ventricle? _____ Do the two chambers pump the same volume of blood? _____

15. Place a probe in the left ventricle and pass it out through the **aorta**. To enter the aorta, the probe will have to pass through the **aortic semilunar valve**. If the vessel has been cut open, the cusps of the semilunar valve will be visible. How many cusps are there?_____ Where is the blood in the aorta transported? _____

_____

16. Observe the outside of the arch of the aorta where it passes just superior to the pulmonary trunk. Look for a small ligament attaching the pulmonary trunk to the aorta in this location (see figure 21.3 and figure 21.7*b*). This is the **ligamentum arteriosum**. The ligamentum arteriosum is a remnant of what fetal structure?_____ This fetal structure shunts blood from the _____ to the _____, thereby allowing blood to bypass the _____.

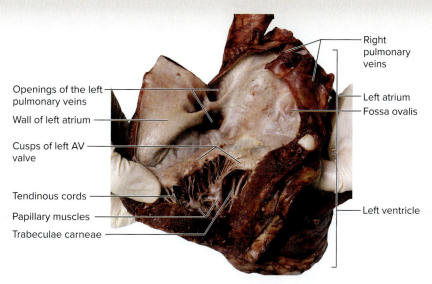

Figure 21.6 **The Left Atrium of a Human Heart.** Posterior view of a human heart with left atrium and left ventricle cut open.

©Christine Eckel

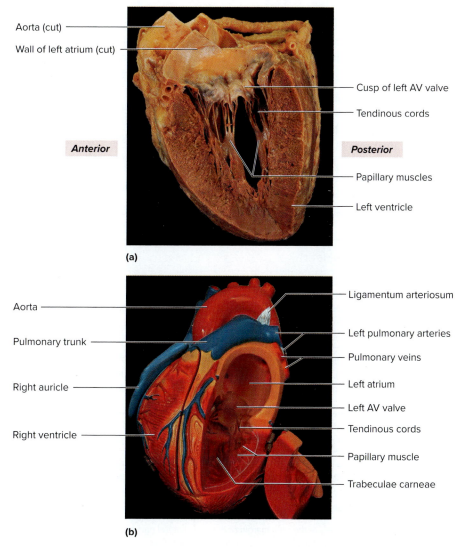

Figure 21.7 **The Left Ventricle of a Human Heart.** (*a*) Gross specimen. (*b*) Classroom model.

(*a*) ©McGraw-Hill Education/Photo and Dissection by Christine Eckel; (*b*) ©Denoyer-Geppert. Photo by Christine Eckel

(continued on next page)

(continued from previous page)

17. Sketch the left atrium and left ventricle in the space provided or on a separate sheet of paper. Use arrows to indicate the flow of blood from the lungs to the aorta. Be sure to include and label all of the structures listed below in your drawing.

☐ **Aorta** ☐ **Left ventricle**

☐ **Aortic semilunar valve** ☐ **Papillary muscles**

☐ **Fossa ovalis** ☐ **Pulmonary veins**

☐ **Left atrium** ☐ **Tendinous cords**

☐ **Left AV valve** ☐ **Trabeculae carneae**

18. *Optional activity:* **APR** **9: Cardiovascular System—** Watch the "Heart" animation to gain a 3D fly-through perspective of the internal heart.

EXERCISE 21.3

THE CORONARY CIRCULATION

1. Obtain a preserved human heart or a classroom model of the heart.

2. The **coronary circulation** is the circulation to the heart wall itself. The blood supply to the heart arises from two main coronary vessels, the **right and left coronary arteries (table 21.1)**. The openings into these arteries arise immediately superior to the cusps of the aortic semilunar valve. **Figure 21.8** shows the relationship between the cusps of the aortic semilunar valve and the openings to the right and left coronary arteries. Observe

the aortic semilunar valve and identify the openings for the right and left coronary arteries (figure 21.8). When the left ventricle contracts and pushes blood into the aorta, the cusps of the aortic semilunar valve cover the openings to the coronary arteries. What consequence does this have in terms of blood flow to the heart wall during ventricular contraction (systole)?

_____

_____

_____

Table 21.1	Arterial Supply to the Heart		
Vessel	**Description**	**Areas Served**	**Word Origin**
Left Coronary Artery	Located posterior to the pulmonary trunk; branches into the anterior interventricular and circumflex arteries just after it emerges from behind the pulmonary trunk	Branches into anterior interventricular and circumflex arteries	*corona,* a crown
Anterior Interventricular Artery	Branch of the left coronary artery; located in the anterior interventricular sulcus (groove). Physicians typically refer to this as the "LAD" (left anterior descending).	Anterior parts of the right and left ventricles and most of the interventricular septum	*inter,* between, + *ventricular,* referring to the ventricles of the heart (from *ventriculus,* belly)
Circumflex Artery	Branch of the left coronary artery; located in the coronary sulcus between the left atrium and left ventricle	Left atrium and left ventricle (lateral part)	*circum,* around, + *flexus,* to bend
Right Coronary Artery	Located in the coronary sulcus between the right atrium and the right ventricle. Branches include the marginal artery, posterior interventricular artery, and SA nodal artery, which supplies blood to the sinoatrial node within the right atrium.	Right atrium; branches into marginal and posterior interventricular arteries	*corona,* a crown
Marginal Artery	Branch of the right coronary artery at the right margin of the heart and located on the lateral part of the right ventricle	Lateral part of the right ventricle	*margo,* border or edge
Posterior Interventricular Artery	Continuation of the right coronary artery located in the posterior interventricular sulcus	Posterior parts of the right and left ventricles	*inter,* between, + *ventricular,* referring to the ventricles of the heart (from *ventriculus,* belly)

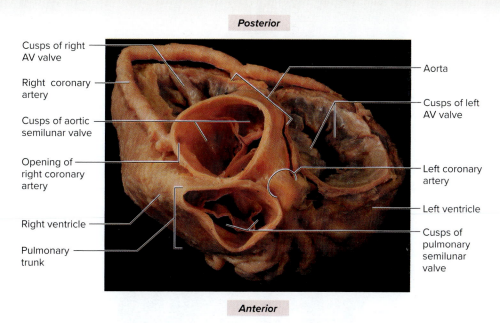

Posterior

Cusps of right AV valve

Right coronary artery

Cusps of aortic semilunar valve

Opening of right coronary artery

Right ventricle

Pulmonary trunk

Aorta

Cusps of left AV valve

Left coronary artery

Left ventricle

Cusps of pulmonary semilunar valve

Anterior

Figure 21.8 **The Coronary Circulation.** Superior view of the human heart with the atria removed. All four valves of the heart can be seen in this view. In this photo you can see where the right and left coronary arteries come off of the aorta, and you can see the cusps of both the pulmonary semilunar valve within the pulmonary trunk and the aortic semilunar valve within the aorta.

©McGraw-Hill Education/Photo and Dissection by Christine Eckel

When the ventricles relax, the cusps of the aortic semilunar valves close as they fill with blood. Thus, they are no longer covering the openings to the coronary arteries. What consequence does this have in terms of blood flow to the heart wall during ventricular relaxation (diastole)?

_____

_____

_____

3. *Cardiac Veins:* All venous blood draining the heart wall (with one exception; see **table 21.2**) eventually drains into one large vessel, the **coronary sinus**. The coronary sinus is located on the posterior surface of the heart

within the coronary sulcus. What chamber does the coronary sinus empty into?

_____

4. Identify the vessels shown in **figure 21.9** on the cadaver heart or on the classroom model of the heart, using tables 21.1 and 21.2 and the textbook as guides. Then label them in figure 21.9.

5. *Optional Activity:* **APR** 9: **Cardiovascular System—** Study the "Heart" dissections to review the vasculature and features of the heart, then use the Quiz feature to test yourself on these structures.

Table 21.2	Venous Drainage of the Heart		
Vessel	**Description**	**Areas Drained**	**Word Origin**
Coronary Sinus	Located within the coronary sulcus on the posterior surface of the heart; it opens into the right atrium just inferior to the fossa ovalis	Entire heart; all veins of the heart drain into the coronary sinus, with the exception of a few small veins of the right ventricle, which drain directly into the right atrium	*corona*, a crown, + *sinus*, cavity
Great Cardiac Vein	Located in the anterior interventricular sulcus next to the anterior interventricular artery	Anterior parts of the right and left ventricles	*cardiacus*, heart, + *vena*, vein
Middle Cardiac Vein	Located in the posterior interventricular sulcus next to the posterior interventricular artery	Posterior parts of the right and left ventricles	*cardiacus*, heart, + *vena*, vein
Small Cardiac Vein	Located on the lateral part of the right ventricle, near the marginal artery	Lateral part of the right ventricle	*cardiacus*, heart, + *vena*, vein

(continued on next page)

(continued from previous page)

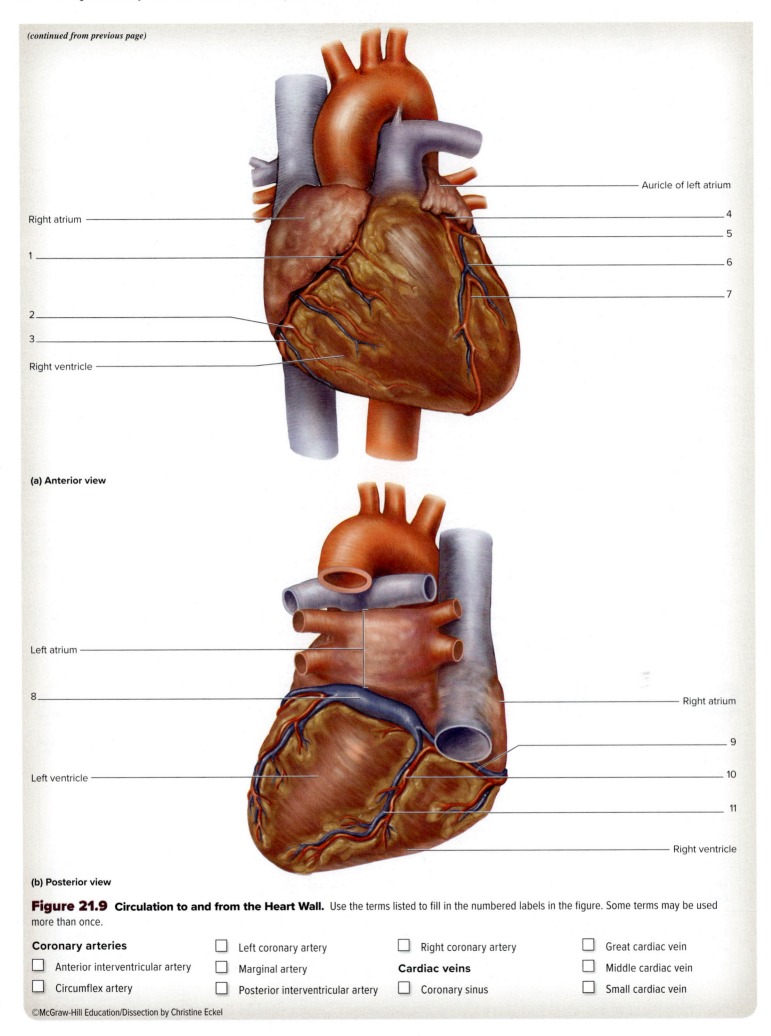

Auricle of left atrium

Right atrium

4

5

1

6

7

2

3

Right ventricle

(a) Anterior view

Left atrium

8

Right atrium

9

Left ventricle

10

11

Right ventricle

(b) Posterior view

Figure 21.9 **Circulation to and from the Heart Wall.** Use the terms listed to fill in the numbered labels in the figure. Some terms may be used more than once.

Coronary arteries

☐ Anterior interventricular artery

☐ Circumflex artery

☐ Left coronary artery

☐ Marginal artery

☐ Posterior interventricular artery

☐ Right coronary artery

Cardiac veins

☐ Coronary sinus

☐ Great cardiac vein

☐ Middle cardiac vein

☐ Small cardiac vein

INTEGRATE

CLINICAL VIEW
Myocardial Infarction

Adequate blood flow through the coronary arteries is essential for the functioning of the heart. If any coronary vessel becomes blocked due to disease or other processes, the region of the heart served by the vessel may become **ischemic**, meaning it lacks blood flow (*ischio*, to keep back, + *chymos*, juice). Prolonged ischemia to heart muscle leads to **hypoxia** (*hypo-*, too little, + *oxia*, oxygen). When cardiac muscle lacks an oxygen supply for more than a few minutes, the tissue dies, or becomes **necrotic** (*nekrosis*, death). The area of dead tissue is referred to as a **myocardial infarction** (*myocardial*, referring to the myocardium, + *in-farcio*, to stuff into), otherwise known as a "heart attack" or "MI." If a myocardial infarction results in vast tissue destruction, the heart may no longer be effective as a pump, which can lead to the death of the individual. If the individual survives the attack, the body will replace the damaged tissue with scar tissue. Scar tissue is mainly composed of a dense collection of collagen fibers. **Figure 21.10** demonstrates a human heart with evidence of a healed myocardial infarction. If observing a human heart, does the specimen show any evidence of (scarring from) past myocardial infarctions (scar tissue is generally clear to whitish in appearance and is much tougher than muscle tissue)? _____

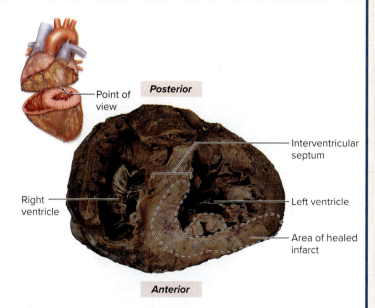

Figure 21.10 **Inferior View of a Transverse Section Through a Human Heart.** Scar tissue, which is evidence of a previous myocardial infarction, can be seen in the interventricular septum and posterior wall of the left ventricle.

©McGraw-Hill Education/Photo and Dissection by Christine Eckel

---

EXERCISE 21.4

SUPERFICIAL STRUCTURES OF THE SHEEP HEART

1. Obtain a dissecting pan, dissecting instruments, gloves, and a preserved sheep heart. Rinse the heart with water to remove any dried blood or other debris, and place it in the dissecting pan to begin observations of superficial structures.

2. **Figure 21.11** demonstrates superficial structures of the sheep heart from both anterior and posterior views. Begin by distinguishing the anterior surface from the posterior surface. One way to identify the anterior surface of the heart is the fairly distinctive, ruffled borders of both the right and left **auricles**, which are extensions of the right and left **atria**. Both auricles are visible in this view.

3. Observe the surface of the heart closely to locate the **visceral pericardium** (epicardium). Then observe the outer surfaces of the great vessels of the heart to identify remnants of the **parietal pericardium** where it was attached to these vessels. Note the large amount of fatty tissue deep to the epicardium, called **epicardial fat**. One

of the functions of this fatty tissue is to help cushion the heart within the pericardial cavity.

4. Identify the following superficial features on the sheep heart, using figure 21.11 as a guide:

   - ☐ **Anterior interventricular sulcus**
   - ☐ **Apex**
   - ☐ **Coronary sulcus**
   - ☐ **Left atrium**
   - ☐ **Left auricle**
   - ☐ **Left ventricle**
   - ☐ **Posterior interventricular sulcus**
   - ☐ **Right atrium**
   - ☐ **Right auricle**
   - ☐ **Right ventricle**

5. *Great vessels:* Identify the great vessels: the aorta, pulmonary trunk, pulmonary veins, superior vena cava, and inferior vena cava. These vessels are often cut very close to their attachments to the heart, which can make identification difficult. To make the task easier, carefully remove as much of the epicardial fat as possible from the superior aspect of the heart (leave the fat in place on the ventricles for now).

(continued on next page)

CONCEPT CONNECTION

As discussed in exercise 21.7, there are similarities between cardiac muscle tissue and skeletal muscle tissue. Cardiac muscle, like skeletal muscle, contains striations. The striations are due to the presence of sarcomeres, the fundamental contractile unit of cardiac and skeletal muscle. Sarcomeres are composed of overlapping thick filaments composed of myosin protein and thin filaments composed of actin,

troponin, and tropomyosin. The mechanism of muscle contraction in cardiac muscle is similar to that of skeletal muscle. Calcium binds to the regulatory protein troponin, causing a conformational change in the troponin–tropomysin complex that exposes the myosin binding sites on actin. This allows myosin heads to bind to actin, forming crossbridges. The formation of crossbridges in both cardiac and skeletal muscle is what allows the muscle to produce force and generate movement.

EXERCISE 21.8

LAYERS OF THE HEART WALL

Both the atrial and ventricular heart wall are composed of three layers: the endocardium, myocardium, and epicardium. The layers of the heart wall are compared in **table 21.4**. Note that when referring to the outer layer of the heart wall as part of the pericardial sac, the appropriate term is *visceral layer of serous pericardium* (instead of epicardium).

1. Obtain a slide demonstrating the atria of the heart. Place it on the microscope stage and observe on low power.

2. Identify the following on the slide of the atrium, using **figure 21.15** and table 21.4 as guides. You may also

see cross sections of the coronary vessels deep to the epicardium.

☐ **Endocardium**   ☐ **Myocardium**

☐ **Epicardium**

3. Sketch the layers of the heart wall as seen through the microscope in the space provided or on a separate sheet of paper. Be sure to include and label all of the structures listed in step 2 in your drawing.

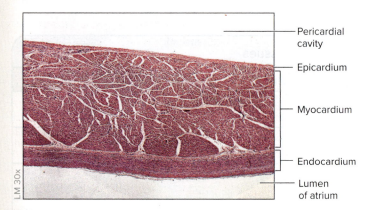

LM 30×

- Pericardial cavity
- Epicardium
- Myocardium
- Endocardium
- Lumen of atrium

Figure 21.15 Histology of the Heart Wall. The myocardium is the thickest layer of the heart wall, and the endocardium is relatively thicker than the epicardium.

©Rick Ash

_____ ✕

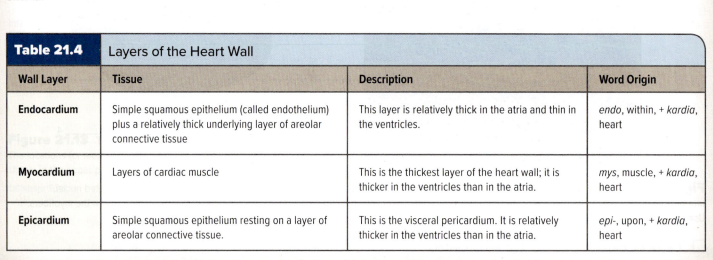

Table 21.4	Layers of the Heart Wall		
Wall Layer	**Tissue**	**Description**	**Word Origin**
Endocardium	Simple squamous epithelium (called endothelium) plus a relatively thick underlying layer of areolar connective tissue	This layer is relatively thick in the atria and thin in the ventricles.	*endo*, within, + *kardia*, heart
Myocardium	Layers of cardiac muscle	This is the thickest layer of the heart wall; it is thicker in the ventricles than in the atria.	*mys*, muscle, + *kardia*, heart
Epicardium	Simple squamous epithelium resting on a layer of areolar connective tissue.	This is the visceral pericardium. It is relatively thicker in the ventricles than in the atria.	*epi-*, upon, + *kardia*, heart

PHYSIOLOGY

Cardiac Cycle and Heart Sounds

The **cardiac cycle** refers to the events that occur in one heartbeat. One complete cardiac cycle occurs when both atria and ventricles contract and then relax. The two main phases of the cardiac cycle are **diastole**, or relaxation, and **systole**, or contraction. These phases may be further subdivided as follows: (1) diastole = isovolumetric ventricular relaxation + ventricular filling, and (2) systole = isovolumetric ventricular contraction + ejection. A typical description of the cardiac cycle includes both atrial and ventricular events that occur on *the left side of the heart* only because the pressures on that side of the heart are much greater than those on the right. The volume of blood in the ventricles at the end of filling is the **end diastolic volume (EDV)**, and the volume of blood remaining in the ventricles following contraction is the **end systolic volume (ESV)**. The volume of blood ejected with each beat is the **stroke volume** (EDV – ESV). **Cardiac output (CO)** is simply the stroke volume (SV) multiplied by the heart rate (HR).

Blood is transported through the heart along pressure gradients, from areas of higher pressure to areas of lower pressure. **Atrioventricular (AV)** and **semilunar valves** ensure one-way flow of blood through the heart. Once pressure in the atria exceeds the pressure in the ventricles, the AV valves open and blood flows into the ventricles. Contraction of the right and left atria pushes any remaining blood into the ventricles. At this time the ventricles are relaxed and fill with blood. During ventricular systole, the pressure rises in the ventricles. When pressure in the ventricles exceeds pressure in the atria, the AV valves are forced closed. Once pressure in the ventricles exceeds pressure in the pulmonary trunk and aorta, blood pushes open the cusps of the pulmonary and aortic semilunar valves and blood is ejected from the heart into the pulmonary and systemic circuits.

The duration of a cardiac cycle varies among individuals and also varies throughout one's lifetime. Typically, a cardiac cycle lasts 0.7 to 0.8 second, which corresponds to an average heart rate of 75–85 beats per minute. Heart sounds (*lub dupp*) can be associated with the cardiac cycle. Turbulent blood flow that occurs as the AV valves close creates the first heart sound (*lub*), whereas turbulent

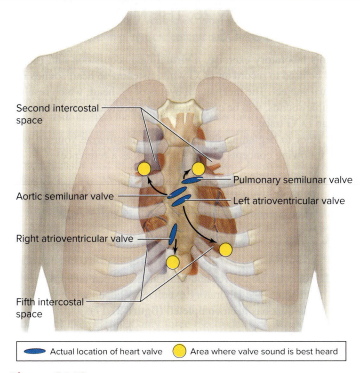

Figure 21.16 Optimal Listening Sites for Heart Sounds.

blood flow that occurs as the semilunar valves close creates the second heart sound (*dupp*). **Figure 21.16** shows the location of each of the heart valves with respect to thoracic wall surface anatomy. The yellow dots indicate the areas where each heart sound is best heard at the surface of the body. The semilunar valves are best heard in the second intercostal spaces on each side of the sternum, whereas the AV valves are best heard in the fifth intercostal spaces. Listen for the right AV valve in the fifth intercostal space, just to the right of the sternum. Listen to the left AV valve in the fifth intercostal space at the midclavicular line (see figure 12.8). **Auscultation** is listening to the internal body sounds, such as the heart sounds.

EXERCISE 21.9

AUSCULTATION OF HEART SOUNDS

1. Obtain alcohol swabs and a stethoscope.

2. Clean the earpieces and diaphragm of the stethoscope with the alcohol swabs and allow them to dry.

3. Place the earpieces of the stethoscope into the ears of the person doing the auscultating, then press the diaphragm of the stethoscope to the subject's skin directly over the left fifth intercostal space at the midclavicular line (see figure 21.16). This space overlies the apex of the heart and is the optimal placement for auscultation of the first heart sound. Listen for a *lub* sound. While auscultating, breathe slowly, deeply, and quietly to minimize extra noise.

4. Next, press the diaphragm to the subject's skin directly over the second intercostal space to the left of the subject's

sternum (see figure 21.16). This is the optimal placement for auscultation of the second heart sound. Listen for a *dupp* sound.

5. Have the subject take deep breaths, and listen carefully for any differences in the heart sounds that occur.

6. Make a note of any pertinent observations here:

_____

_____

_____

recorded on the ECG paper. Lead I corresponds to the voltage difference between the left arm and the right arm; lead II corresponds to the voltage difference between the right arm and the left leg; lead III corresponds to the voltage difference between the left arm and the left leg.

A normal ECG measures an overall change in voltage across the atria and ventricles and consists of three deflection waves. Each deflection wave corresponds to a distinct electrical event within the heart **(figure 21.20)**. The **P wave** corresponds to atrial depolarization, the **QRS complex** corresponds to ventricular depolarization, and the **T wave** corresponds to ventricular repolarization. Atrial repolarization does occur; however, the ECG deflection wave associated with it is masked by the large deflection caused by ventricular depolarization that occurs at approximately the same time. Figure 21.20 also depicts several other measures that reflect heart function: the P-Q segment, S-T segment, P-R interval, and Q-T interval. The **P-Q segment** corresponds to the atrial plateau that occurs on the sarcolemma of cardiac muscle cells when the muscle cells within the atria are contracting. The **S-T segment** corresponds to the ventricular plateau when the cardiac muscle cells within the ventricles are contracting. These segments can be measured directly and are physiologically relevant, as will be explored in the following exercises. The **P-R interval** is the time that it takes for the depolarization to spread from the SA node through the AV node, which typically ranges from 0.12 to 0.20 second. The **Q-T interval** is the time required for the ventricles to depolarize and repolarize, which typically ranges from 0.2 to 0.4 second. An ECG can be used to detect heart rates that are faster (*tachycardia*: >100 beats/min) or slower (*bradycardia*: <60 beats/min) than normal; the influence of drugs on cardiac function; or the existence of ectopic pacemakers (ectopic = outside of the normal conduction system). Thus, the ECG is a valuable diagnostic tool.

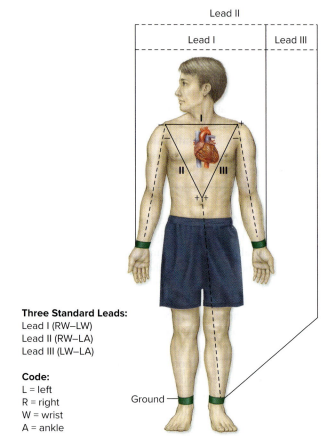

Three Standard Leads:
Lead I (RW–LW)
Lead II (RW–LA)
Lead III (LW–LA)

Code:
L = left
R = right
W = wrist
A = ankle

Figure 21.19 **Standard Limb Leads for ECG.** The standard limb leads for an ECG roughly place the heart at the center of Einthoven's triangle. Each lead (I–III) corresponds to a voltage difference between two measurement points.

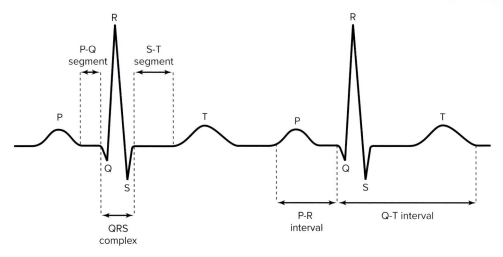

Figure 21.20 **Normal Components of an ECG.**

EXERCISE 21.11

ELECTROCARDIOGRAPHY USING STANDARD ECG APPARATUS

Obtain the Following:
- **alcohol swabs**
- **electrodes**
- **electrode gel**
- **lead selector switch**
- **electrocardiograph (or alternative ECG recording device)**

1. Choose a person to be the subject. Prepare to place the electrodes by cleaning the skin where the electrodes will be placed. To do this, rub the indicated area vigorously with an alcohol swab. First clean the skin on the anterior surface of the subject's forearms approximately 2 to 3 inches proximal to the left and right wrists. Next, clean the skin approximately 2 to 3 inches superior to the medial malleoli on both the left and right legs. Note that skin free of hair will make the best contact with the electrode.

2. If not using electrodes containing gel as a component of the electrode, place a small quantity of electrode gel on the center of each electrode. Place one electrode firmly on the skin in each of the following locations (that were cleaned in step 1): right arm, left arm, right leg, and left leg. The right leg will serve as the ground.

3. Observe the leads of the ECG machine. A lead is a wire that has a clip on the end that attaches to an electrode. The leads have labels that identify which limb each must be attached to. Attach the leads to the electrodes. RA = Right Arm, LA = Left Arm, RL = Right Leg, LL = Left Leg

4. Turn on the electrocardiograph. Set the paper speed to 25 mm/second, the standard for recording an ECG. At this speed, each millimeter of paper corresponds to a time interval of 1 second. Set the gain of the system to 10 mm/mV. Note: Depending on the machine in use, the gain may need to be set lower or higher. Ask the instructor if in doubt about where the gain should be set.

5. Have the subject lie supine or sit in a comfortable position.

6. Record the ECG for 1 minute. To begin recording, press the record button. If the machine is working properly, the ECG paper should start coming out of the machine and three ECG tracings (one for each lead: I–III) should be seen on the paper. When finished recording the ECG, write the subject's name and "Resting ECG" on the indicator paper.

7. Remove the leads from the electrodes (do *not* remove the electrodes at this time). Have the subject exercise vigorously for at least 3 minutes.

8. Have the subject return to the same seated or supine position as in step 5 immediately post-exercise.

Reconnect the leads to the electrodes, taking care to ensure the leads are connected to the proper locations. Record a post-exercise ECG for 3 minutes. When finished recording the ECG, write the subject's name and "Post-Exercise ECG" on the indicator paper.

9. Calculate the subject's heart rate at rest and post-exercise using the following procedure, which assumes the paper speed is set to 25 mm/second:

   a. Start as close as possible to the peak of one QRS complex and count out a 6-second interval on the ECG paper **(figure 21.21)**. With a paper speed of 25 mm/second, one second corresponds to the length of five large squares (each large square = 0.2 second; each small square = 0.04 second).

   b. Measure the number of QRS complexes for the lead II tracing within the 6-second interval that was counted out in step a.

   c. Multiply the number of QRS complexes counted in step b by 10 to get the heart rate ($6 \times 10 = 60$ seconds = 1 minute)

   d. Record the heart rate values in **table 21.6**.

10. Measure other important features of the ECG, including the durations of the QRS complex, Q-T interval, and P-T interval (see figure 21.20). Record these values in table 21.6.

11. Make note of pertinent observations comparing resting versus post-exercise ECG tracings here:

   _____

   _____

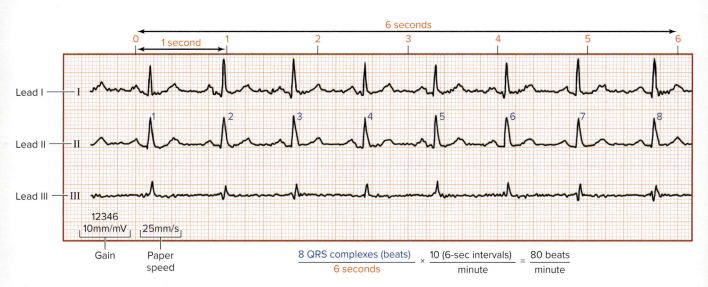

Figure 21.21 Interpreting an ECG Tracing. This ECG tracing was recorded at a paper speed of 25 mm/second. Red numbers indicate 1-second time intervals. Blue numbers indicate number of QRS complexes. The calculation demonstrates how to calculate heart rate in beats per minute using a 6-second time interval.

(continued on next page)

(continued from previous page)

Table 21.6	ECG Component Data	
ECG Component	**Duration at Rest (sec)**	**Duration Post-Exercise (sec)**
QRS Complex		
Q-T Interval		
P-T Interval		
Heart Rate		

INTEGRATE

CLINICAL VIEW
ECG

Clinicians use ECGs to diagnose abnormalities with the conduction system of the heart. While this laboratory exercise involves using a three-lead system, a more comprehensive 12-lead system is typically used in a clinic or hospital setting. Trained specialists evaluate the ECG trace, looking for any abnormalities. **Figure 21.22** presents three examples of abnormal ECG traces that might be detected. Observe the tracing in figure 21.22*a*. Are QRS complexes visible? (yes/no) Is each QRS complex followed by a T wave? (yes/no) Do all of the QRS complexes look the same? (yes/no) Look for regularly spaced P waves between the QRS complexes. Are they visible? (yes/no) Tracing (a) is a case of **atrial fibrillation**, which involves chaotic depolarization of cardiac muscle fibers resulting in uncoordinated contraction of the atria. A patient suffering from atrial fibrillation will exhibit many QRS complexes, but no clear P waves. The continuous and chaotic depolarization of the atria may result in incomplete filling of the ventricles. Given this information, why do you think the QRS complexes also look a little bit irregular in spacing and amplitude?

Next, observe the tracing in figure 21.22*b*. Are QRS complexes visible? (yes/no) Is each QRS complex followed by a T wave? (yes/no) Next, look for P waves. Are any P waves visible? (yes/no)

Tracing (b) is a case of **ventricular fibrillation**. In ventricular fibrillation there are no clear P, QRS, or T waves due to aberrant electrical activity. While atrial fibrillation may lead to incomplete filling of the ventricles, ventricular fibrillation leads to inefficient ejection of blood into the systemic circulation. Without intervention, the latter results in certain death.

Observe the tracing in figure 21.22*c*. Are QRS complexes visible? (yes/no) Is each QRS complex followed by a T wave? (yes/no) Is each QRS complex preceded by a P wave? (yes/no)

Tracing (c) is a case of premature ventricular contraction (PVC). PVCs are characterized by QRS complexes that are not always preceded by a P wave. While not necessarily as detrimental as ventricular fibrillation, PVCs can also lead to a decreased cardiac output.

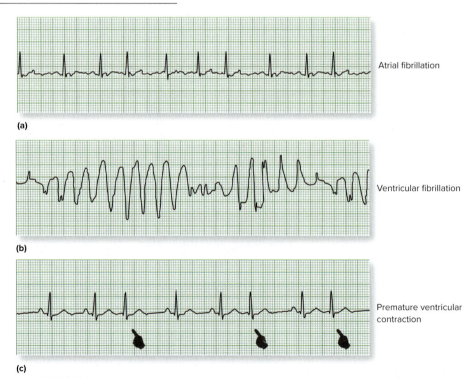

(a)

(b)

(c)

Figure 21.22 **Abnormal ECG Traces.** (*a*) Atrial fibrillation, (*b*) ventricular fibrillation, (*c*) premature ventricular contraction.

©doc-stock GmbH/Phototake

EXERCISE 21.12

BIOPAC LESSON 5: ELECTROCARDIOGRAPHY I

This experiment introduces students to the procedure for measuring the electrical activity of the heart using a three-lead electrocardiogram (ECG). An ECG will be recorded for the subject under three conditions: lying down, sitting up, and after a bout of exercise. ECG parameters will be measured for each of the experimental conditions.

Obtain the Following:

- **BIOPAC electrode lead set (SS2L)**
- **BIOPAC general purpose electrodes (EL503)**
- **BIOPAC MP36/35/45 recording unit**
- **laptop computer with BSL 4 software installed**

1. Before beginning, choose one student to be the experimental subject and one student to be the recorder.

2. Prepare the subject for placement of electrodes by cleaning the skin where the electrodes will be placed. To do this, rub the indicated area vigorously with an alcohol swab. First clean the skin on the medial surface of each leg, just above the ankle. Next, clean the skin on the anterior surface of the right anterior forearm at the wrist. Note that skin that is free of hair will make the best contact with the electrodes.

3. Place one electrode (EL503) firmly on the skin in each of the locations that was cleaned in step 2. Wait approximately 5 minutes before attaching the leads (step 5) to give the electrodes a chance to affix firmly to the skin. Proceed with step 4 while waiting for the electrodes to affix firmly to the skin.

4. Turn the computer on. Plug the electrode lead set (SS2L) into channel 2 of the BIOPAC MP36/35 unit. Turn on the BIOPAC unit.

5. Have the subject lie down in a comfortable position. Attach the electrode leads (SS2L) to the electrodes that were placed in step 2 following the color scheme outlined in **figure 21.23**. Attach the WHITE lead to the electrode

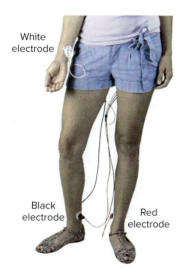

White electrode

Black electrode

Red electrode

Figure 21.23 Lead Placement.
Courtesy of and ©BIOPAC Systems, Inc. www.biopac.com

on the right forearm. Attach the BLACK lead to the electrode on the medial surface of the subject's right leg. Attach the RED lead to the electrode on the medial surface of the subject's left leg.

6. Clip the unit containing the lead cables to the subject's waistband. Be certain that the cables coming from the leads are not pulling on the electrodes and that clothing is not rubbing on the electrodes.

7. Start the Biopac Student Lab Program and choose Lesson 5 (L05-Electrocardiography (ECG) I).

8. Type in a file name using the subject's first name and last initial. Click OK.

9. Click on the Calibrate button in the upper left corner. Wait for the calibration procedure to finish. The calibration procedure stops automatically after about 8 seconds.

10. If the calibration readings look like an ECG **(figure 21.24)** and look stable, proceed to step 11. If the calibration readings do not look right, click Redo Calibration before clicking Continue and proceeding to step 11.

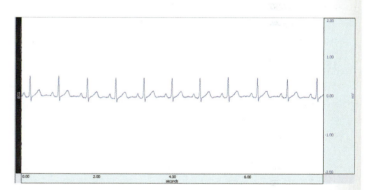

Figure 21.24 **Correct Calibration Reading.** The reading should look like a normal ECG.

Courtesy of and ©BIOPAC Systems, Inc. www.biopac.com

11. **OBTAIN RESTING VALUES:** Ensure the subject is lying down and resting comfortably before starting to record data. Click on Record to start data collection. Record data for 20 seconds to obtain resting values. Next, click Suspend to stop the device from continuing to record the ECG. If the subject made any large movements, talked, or laughed during the recording (activities that interfere with collection of ECG data), click on Redo and record for another 20 seconds to obtain resting values. Otherwise, click Continue.

12. **OBTAIN VALUES FOR SITTING UP WITH NORMAL BREATHING:** Have the subject sit upright and breathe normally. When the subject is ready, click on Record to start data collection. Record data for 20 seconds to obtain values for the "sitting" up condition. Next, click Suspend to stop the device from continuing to record the ECG. If the subject made any large movements, talked, or laughed during the recording (activities that interfere with collection of ECG data), click on Redo and record for another 20 seconds. Otherwise, click Continue.

(continued on next page)

(continued from previous page)

13. **OBTAIN POST-EXERCISE VALUES:** Unhook the leads from the ECG electrodes and unclip the unit containing the lead cables from the subject's waistband to "release" the subject. Have the subject exercise vigorously for at least 5 minutes (e.g., running up and down stairs, doing pushups or jumping jacks). When the subject has finished exercising, have the subject sit down next to the recording apparatus. Immediately reattach the leads to the electrodes, taking care to connect the leads to the proper locations (figure 21.23). Click on Record to start data collection. Record data for 60 seconds to obtain post-exercise values. Next, click Suspend to stop the device from continuing to record the ECG. If the subject made any large movements, talked, or laughed during the recording (activities that interfere with collection of ECG data), click on Redo and record for another 60 seconds If the data are good, click Done.

14. When finished, click Done. Remove the ECG leads and electrodes from the subject. The electrodes can be thrown away in the garbage and should not be reused. To record data for another subject, click on Record Data from Another Subject. When finished collecting data for all subjects, proceed to step 15: ANALYZE THE DATA.

15. **ANALYZE THE DATA:** If analyzing data that was just collected, click on Analyze current data file. If opening data that were collected previously, click on Review Saved Data. To prepare to analyze the data, do the following:

    a. Use the zoom tool to select approximately 4 seconds' worth of ECG waveforms.

    b. Click on the Display menu at the top of the screen and select Autoscale Waveforms.

    c. Set up the measurement boxes at the top of the screen using the drop-down menus. Set CH 40 to Value. Set CH 1 to Delta T. Set CH 1 to P-P. Set CH 1 to BPM.

    d. When using the I-beam tool, the indicator box for Delta T will show the time elapsed during the selected region in seconds; the indicator box for P-P will show the peak to-peak measurement; the indicator box for BPM will show the heart rate in beats per minute (bpm). The computer calculates a heart rate using the time between two R-waves of adjacent ECG waveforms.

16. Scroll along the ECG tracing for the subject until the data from the first segment (20 seconds) of ECG recording is in the middle of the screen. Using the I-beam tool, measure the following parameters of the ECG and record them in **table 21.7**: Heart Rate; P wave; P-R Interval; P-R Segment; QRS Complex: S-T Segment; Q-T Interval; T Wave.

17. Scroll along the ECG tracing for the subject until the data from the second segment (20 seconds) of ECG recording is in the middle of the screen. Using the I-beam tool, measure the following parameters of the ECG and record them in table 21.7: Heart Rate; P wave; P-R Interval; P-R Segment; QRS Complex: S-T Segment; Q-T Interval; T Wave.

18. Scroll along the ECG tracing for the subject until the data from the third segment (60 seconds) of ECG recording is in the middle of the screen. Using the I-beam tool, measure the following parameters of the ECG and record them in table 21.7: Heart Rate; P wave; P-R Interval; P-R Segment; QRS Complex: S-T Segment; Q-T Interval; T Wave.

19. Make a note of any pertinent observations here:

    _____

    _____

    _____

20. When data analysis is complete, click Save or Print, then click Quit to close the program. Before leaving the workstation, throw away any garbage, clean the work area, and return the BIOPAC unit and computer to the condition in which they were found.

Table 21.7	Data for BIOPAC ECG Exercise			
ECG Parameter	**Resting ECG**	**Sitting Up ECG**	**Post-Exercise ECG**	**Expected Values (seconds)**
Heart Rate				60–80 bpm resting; >100 bpm post-exercise
P Wave				0.06–0.11
P-R Interval				0.12–0.20
P-R Segment				0.08
QRS Complex				<0.12
S-T Segment				0.12
Q-T Interval				0.31–0.41
T Wave				0.16

EXERCISE 21.13 **PhILS**

Ph.I.L.S. LESSON 22: REFRACTORY PERIOD OF THE HEART

Cardiac muscle tissue must contract in a coordinated way to effectively pump the blood to the lungs and systemic tissues. Between contractions, cardiac muscle must relax to allow time for the ventricles to fill with blood. Cardiac muscle, like skeletal muscle, has a period when it is insensitive to an electrical stimulus, known as the **refractory period**. The refractory period within cardiac muscle tissue is extended in comparison to skeletal muscle tissue. This extended refractory period results when cardiac muscle cells remain in the depolarized state because both voltage-gated K^+ channels and voltage-gated Ca^{2+} channels are open. This extended depolarized state is called the **plateau**. Because repolarization does not immediately follow depolarization, the mechanical events of cardiac muscle contraction and relaxation have time to occur prior to restimulation of cardiac muscle cells. (This prevents a sustained contraction [tetanus] from occurring.)

The purpose of this laboratory exercise is to measure the duration of the refractory period in cardiac muscle tissue. The exercise involves delivering brief electrical shocks to exposed frog cardiac muscle tissue and observing the contractions that are, or are not, evoked in the muscle as a result of stimulation. Shocks will be delivered at various times throughout the cardiac cycle to observe when evoked contractions are and are not possible.

Prior to beginning the experiment, familiarize yourself with the following concepts (use the textbook as a reference):

- Voltage-gated channels of cardiac cells
- Cardiac action potentials
- Refractory period

1. Open Ph.I.L.S Lesson 22: Refractory Period of the Heart.

2. Read the objectives and introduction. Then, take the pre-lab quiz. The laboratory exercise will open when the pre-lab quiz has been completed (**figure 21.25**).

3. Click and drag the blue plug to recording input 1 on the data acquisition unit (DAQ).

4. Click and drag the red plug to the stimulator output + position on the DAQ. Click and drag the black plug to the stimulator output − position on the DAQ.

5. Click the Start button on the control panel of the virtual monitor. A Blue trace should appear, which represents ventricular contractions. The red trace indicates when a shock has been applied to the tissue.

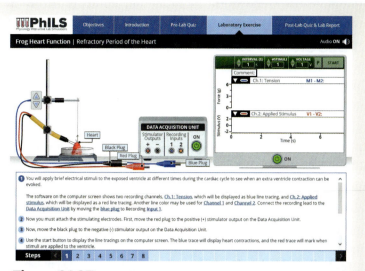

Figure 21.25 Opening Screen for the Laboratory Exercise on Refractory Period of the Heart.

©McGraw-Hill Education

6. To mimic the heart filling with more blood, click the blue up arrow on the clamp (left side of the screen) to stretch the myocardium and increase the heart response on the screen.

7. Click the stimulus button to stimulate the heart every second. Pay special attention after stimulating the heart because changes in the blue tracing will indicate the effect of the stimuli.

8. Continue recording until you have seen two or three extra ventricular contractions, then click the Stop button to halt the recording.

9. Open the post-lab quiz & lab report by clicking on it. Answer the post-lab questions that appear on the computer screen. Once completed, click on Lab Results. Then click Print Lab to print a hard copy of the report or click Save PDF to save an electronic copy of the report.

10. Make a note of any pertinent observations here:

_____

_____

_____

_____

_____

The corresponds to the Learning Objective(s) listed in the chapter opener outline.

Do You Know the Basics?

Exercise 21.1: Location of the Heart and Pericardium

1. Match the following descriptions in column A with the corresponding structure in column B. **1**

   Column A

   _____ 1. inner portion of the pericardial sac

   _____ 2. layer of the pericardium that is tightly adhered to the heart

   _____ 3. outer portion of the pericardial sac

   Column B

   a. fibrous pericardium

   b. parietal layer of the serous pericardium

   c. visceral layer of the serous pericardium

Exercise 21.2: Gross Anatomy of the Human Heart

2. The atria and ventricles of the heart both contain three layers of the heart wall. In order of superficial to deep these layers are the _____, _____, and _____ (endocardium, epicardium, myocardium). **2**

3. The chamber that contains pectinate muscles in its walls is the _____ (atrium/ventricle). **3**

4. The structures that increase the forcefulness of contraction within the right atrium are _____ (papillary/pectinate) muscles, whereas the structures that prevent the eversion of AV valves are _____ (papillary/pectinate) muscles. **4**

5. Papillary muscles attach directly to _____ (atrioventricular/semilunar) valve cusps. **5**

6. The right atrioventricular valve is also known as the _____ (bicuspid/tricuspid) valve, whereas the left atrioventricular valve is also known as the _____ (bicuspid/tricuspid) valve. **6**

7. The right ventricle pumps blood to the _____ (body/lungs), while the left ventricle pumps blood to the _____ (body/lungs). **7**

8. Pulmonary and aortic semilunar valves each contain _____ (two/three) cusps and prevent backflow of blood into the _____ (atria/ventricles). **8**

9. Place the following structures in the order in which a drop of blood would travel through the right side of the heart, beginning with the superior and inferior venae cavae and ending with the lungs. **9**

   _____ a. pulmonary arteries

   _____ b. pulmonary trunk

   _____ c. right atrioventricular valve

   _____ d. right atrium

   _____ e. pulmonary semilunar valve

   _____ f. right ventricle

10. Match the description listed in column A with the structure listed in column B. (Answers will be used more than once.) **10**

    Column A

    _____ 1. located between the pulmonary trunk and aorta

    _____ 2. located between the right and left atria

    _____ 3. remnant of the ductus arteriosis

    _____ 4. remnant of the foramen ovale

    Column B

    a. fossa ovalis

    b. ligamentum arteriosum

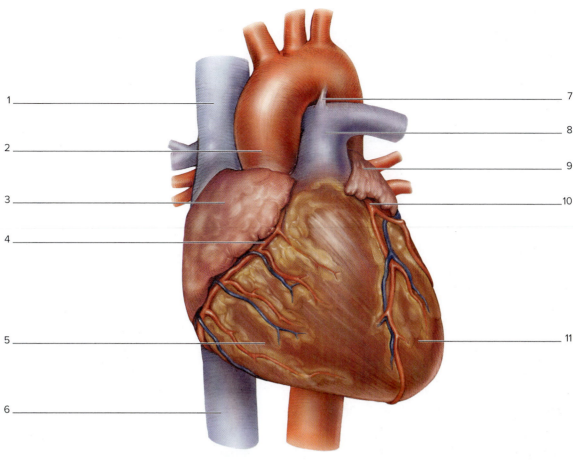

1 _____

2 _____

3 _____

4 _____

5 _____

6 _____

7 _____

8 _____

9 _____

10 _____

11 _____

11. Label the following diagram of the anterior view of the heart. **2** **7** **9** **11**

Exercise 21.3: The Coronary Circulation

12. The anterior interventricular artery branches from the _____ (right/left) coronary artery. **11**

13. The coronary sinus is located on the _____ (anterior/posterior) surface of the heart. **12**

Exercise 21.4: Superficial Structures of the Sheep Heart

14. The right ventricle is visible in a(n) _____ (anterior/posterior) view of the sheep heart. **13**

15. The _____ (parietal/visceral) layer of the serous pericardium is tightly adhered to the heart, whereas the _____ (parietal/visceral) layer of the serous pericardium is tightly adhered to the fibrous pericardium. **14**

16. When viewing the anterior surface of the sheep heart, the _____ (aorta/pulmonary trunk) is located most anteriorly. **15**

17. Oxygenated blood returns to the left atrium through the pulmonary _____ (arteries/veins). **16**

Exercise 21.5: Coronal Section of the Sheep Heart

18. A coronal section of the sheep heart reveals that the left ventricular wall is significantly _____ (thicker/thinner) than the right ventricular wall. **17**

19. The systemic circuit has _____ (more/less) resistance than the pulmonary circuit. Therefore, the left ventricle contracts with a _____ (greater/lesser) force than the right ventricle. **18**

20. Tendinous cords in the sheep heart tether the cusps of _____ (atrioventricular/semilunar) valves to prevent the backflow of blood. **19**

Exercise 21.6: Transverse Section of the Sheep Heart

21. When viewing a transverse section of the sheep heart, the right ventricular wall is _____ (thicker/thinner) than the left ventricular wall. **20**

22. The right and left ventricles pump _____ (the same/a different) volume of blood and have _____ (the same/a different) wall thickness. **21**

Exercise 21.7: Cardiac Muscle

23. Cardiac muscle is _____ (voluntary/involuntary), whereas skeletal muscle is _____ (voluntary/involuntary). **22**

Exercise 21.8: Layers of the Heart Wall

24. The layer of the heart wall that is composed on simple squamous epithelium is the _____ (endocardium/myocardium). **23**

Exercise 21.9: Auscultation of Heart Sounds

25. The first heart sound corresponds to the closing of the _____ (atrioventricular/semilunar) valves, whereas the second heart sound corresponds to the closing of _____ (atrioventricular/semilunar) valves. **24**

26. Optimal placement of a stethoscope for auscultation of the first heart sound is the _____ (second/fifth) intercostal space. **25**

27. An incompetent semilunar valve may lead to a(n) _____ (normal/abnormal) heart sound. **26**

Exercise 21.10: Ph.I.L.S. Lesson 26: The Meaning of Heart Sounds

28. The mechanical event of the heart associated with the P wave is _____ (atrial/ventricular) _____ (contraction/relaxation). **27**

Exercise 21.11: Electrocardiography Using Standard ECG Apparatus

29. The optimal limb lead and surface electrode placement for recording a three-lead ECG involves placing surface electrodes on the right arm, the left arm, and the _____ (right/left) leg. **28**

30. The QRS complex on a normal ECG trace represents _____ (atrial/ventricular) depolarization. **29**

Exercise 21.12: BIOPAC Lesson 5: Electrocardiography I

31. On a normal ECG trace, which segment/interval represents the time the ventricles are contracting? (Circle one.) **30**

    a. P-Q segment

    b. P-R interval

    c. S-T segment

    d. Q-T interval

32. Compared to resting, the post-exercise R-R interval had _____ (increased/decreased), indicating the heart rate had _____ (increased/decreased). **31**

Exercise 21.13: Ph.I.L.S. Lesson 22: Refractory Period of the Heart

33. Which of the following ion movements causes the plateau in the cardiac action potential? (Check all that apply.) **32**

    _____ a. influx of calcium

    _____ b. outflux of sodium

    _____ c. influx of potassium

34. Cardiac muscle has a _____ (shorter/longer) refractory period than skeletal muscle, which prevents force summation in cardiac muscle. **33**

Exercise 21.14: Ph.I.L.S. Lesson 29: ECG and Heart Block

35. Which of the following accurately describes an ECG tracing for a patient who presents with an AV node block? (Circle one.) **34**

    a. absent P waves

    b. absent QRS complexes

    c. disconnected P waves and QRS complexes

    d. inverted T waves

36. A patient with a damaged SA node is likely to experience a(n) _____ (decreased/increased) heart rate. **35**

Exercise 21.15: Ph.I.L.S. Lesson 30: Abnormal ECGs

37. Several P waves occurring before a QRS complex would be indicative of _____ (atrial flutter/ventricular fibrillation). **36**

38. Which of the following accurately describes a patient with ventricular fibrillation? (Circle one.) **37**

    a. P wave followed by a QRS complex

    b. longer P-R interval

    c. no recognizable waveforms

    d. disconnected P waves and QRS complexes

Can You Apply What You've Learned?

39. Will the surgeon need to cut through the visceral layer of serous pericardium to remove the heart from the pericardial sac? Why or why not?

_____

_____

40. Describe the effect that scar tissue would have on the spread of depolarization through the ventricles. _____

_____

41. If a patient suffers from heart block, would you expect the rate of ventricular contraction to be faster or slower than normal? Explain your answer.

_____

_____

42. When listening for heart sounds, only two heart sounds are typically audible. Propose a clinical scenario that might cause additional sounds to be heard.

_____

_____

Can You Synthesize What You've Learned?

43. Pulmonary edema (accumulation of fluid in the lungs) can result if there is a mismatch in the volumes of blood pumped by the right and left ventricles.

Discuss how differences in these blood volumes may cause this condition. _____

_____

_____

44. External pacemakers are often used to correct problems related to electrical conduction from the atria to the ventricles. These pacemakers have two "leads," which are wires that are threaded into the heart. Where do you think the two leads are directed within the heart? Why would

two leads be threaded into the heart rather than one? _____

_____

_____

45. A cardiac stress test is often used as a diagnostic tool to investigate underlying heart conditions. During a stress test, the doctor has the patient exercise on a treadmill at maximum intensity for a short duration while observing the patient's ECG for abnormalities. Describe how the stress test might influence

components of the ECG, and how this may be used as a diagnostic tool. _____

_____

46. A patient complains of chest pains. Upon urging from his friends, he goes to the hospital. Doctors perform an ECG upon his arrival at the hospital. His ECG trace reveals that P waves and QRS complexes are occurring at different rates. More specifically, each QRS complex is not preceded by a P wave. Describe what is happening with the patient's electrical conduction system, and propose a diagnosis.

_____

_____

The Cardiovascular System: Vessels and Circulation

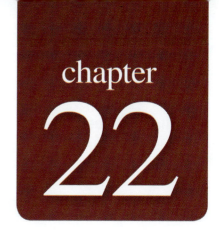

chapter
22

OUTLINE AND LEARNING OBJECTIVES

Anatomy & Physiology Revealed® 4.0

Module 9: Cardiovascular System

INTRODUCTION

There are roughly 100,000 miles of blood vessels within the average adult body. Some of these vessels are large enough to see on the surface of the skin or feel as the blood pulsates through them. Others are so small that their diameter is equivalent to the size of a red blood cell. These vessels form a closed-loop circuit, connected to the heart and lungs, that carries blood, oxygen, nutrients, and waste throughout the body.

There are three main types of blood vessels: arteries, capillaries, and veins. **Arteries** are blood vessels that carry blood *away* from the heart, whereas **veins** are blood vessels that carry blood *toward* the heart. As arteries carry blood toward the tissues, they branch into smaller and smaller vessels, ultimately forming **arterioles** (*arteriole*, a small artery). Arterioles have the special function of controlling the flow of blood into **capillary beds**. Capillaries are the site of exchange for substances (e.g., gases, nutrients, wastes) between the blood and the tissues.

Blood flows out of capillaries into small veins called **venules**, which merge to form the larger veins that return blood to the heart. **Veins** transport blood back to the heart and serve as blood reservoirs (*reservoir,* a receptacle). Generally, a vein is positioned alongside each major artery and has the same name as the artery it accompanies. However, there are typically more veins draining a structure than there are arteries supplying it. In the limbs, most of the veins that do not accompany an artery are superficial veins, located just under the skin. For example, the brachium (arm) is supplied by the brachial artery, which has the brachial vein traveling next to it. The brachial artery and vein are located fairly deep within the arm, where they are protected by the musculature of the arm. In addition to the brachial vein, two superficial veins also drain blood from the arm. These are the cephalic and basilic veins. Note that there is considerably more variation among individuals in the branching patterns and locations of veins than there is with arteries. Such variation often has clinical significance. For example, when blood samples need to be collected from a patient, blood is commonly drawn from the median cubital vein. However, not all individuals have a median

cubital vein. When tracing blood flow through the venous system, remember that veins *drain* blood from an area of the body. This means that descriptions of blood flow through veins start by naming the most distal veins first, and then name the veins blood travels through as it proceeds toward the heart.

The exercises in this chapter begin with identification of the major arteries and veins of the body on a human cadaver or on classroom models of the cardiovascular system. Upon completion of the gross anatomy exercises in this chapter, a student should be able to describe the pathway a drop of blood takes as it is transported from the heart to a target organ and back to the heart once again. Subsequent exercises explore the histological characteristics of the different types of blood vessels (arteries, arterioles, capillaries, and veins). Finally, physiology exercises in this chapter explore the concept of **blood pressure**, which is the pressure that blood exerts against a blood vessel wall. The goal of these exercises is to investigate the relationships among blood pressure, blood flow, and blood vessel resistance.

List of Reference Tables

These Pre-Laboratory Worksheet questions may be assigned by instructors through their ▣ connect course.

1. The tunica _____ (externa/intima/media) is the innermost layer of a blood vessel wall, whereas the tunica _____ _____ (externa/intima/media) is the outermost layer.

2. Which of the following consists of only an endothelium and a basement membrane, allowing for exchange of nutrients? (Cicle one.)

   a. arteriole

   b. capillary

   c. elastic artery

   d. vein

   e. venule

3. The most permeable type of capillary is a _____ (continuous/fenestrated/sinusoidal) capillary, whereas the least permeable type of capillary

   is a _____ (continuous/fenestrated/sinusoidal) capillary.

4. Which of the following is an anatomic feature unique to veins? (Circle one.)

   a. smooth muscle in the tunica media

   b. valves to prevent backflow of blood

   c. vasa vasorum in the tunica externa

   d. tunica media consists only of endothelium and a basement membrane

5. The hepatic portal system is a system of veins that drains blood into the _____ (inferior vena cava/liver).

6. Which of the following is a paired branch off the abdominal aorta? (Circle one.)

   a. celiac trunk

   b. inferior mesenteric artery

   c. renal artery

   d. superior mesenteric artery

7. The ligamentum arteriosum is a remnant of the fetal _____ (ductus arteriosus/foramen ovale), whereas the fossa ovalis is a remnant of

   the fetal _____ (ductus arteriosus/foramen ovale).

8. The major vein that drains blood from the inferior half of the body and empties into the right atrium of the heart is the _____ (inferior/superior) vena cava.

9. Clinically, blood pressure is written with the _____ (diastolic/systolic) pressure on top and the _____ (diastolic/systolic) pressure on the bottom.

10. The first heart sound (lub) is created by turbulent blood flow that occurs when the _____ (atrioventricular/semilunar) valves of the heart close.

(continued from previous page)

3. Deoxygenated blood from the head, neck, and upper extremity return to the heart via the **right** and **left brachiocephalic veins**, which merge to form the **superior vena cava** (figure 22.2). Blood from the abdominal cavity and lower extremity flow through the **inferior vena cava** to return to the heart.

4. Identify the *arteries* listed in figure 22.2a that branch off the aortic arch, using the textbook as a guide. Then label them in figure 22.2a.

5. Identify the *veins* listed in figure 22.2b that return blood to the heart, using the textbook as a guide. Then label them in figure 22.2b.

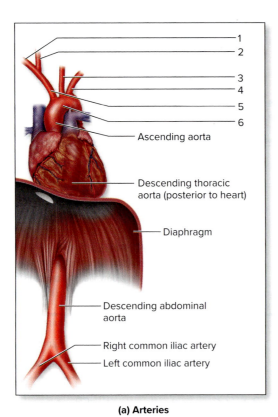

(a) Arteries

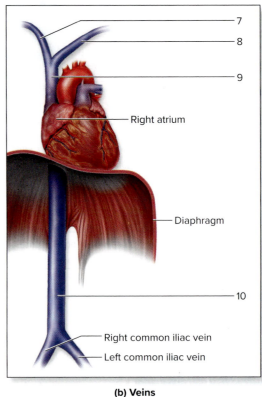

(b) Veins

Figure 22.2   Great Vessels of the Heart. Use the terms listed to fill in the numbered labels in the figure.

Arterial Supply

☐ Aortic arch

☐ Brachiocephalic trunk

☐ Left common carotid artery

☐ Left subclavian artery

☐ Right common carotid artery

☐ Right subclavian artery

Venous Drainage

☐ Inferior vena cava

☐ Left brachiocephalic vein

☐ Right brachiocephalic vein

☐ Superior vena cava

EXERCISE 22.3

CIRCULATION TO THE HEAD AND NECK

1. Observe the head and neck regions of a prosected human cadaver or observe classroom models or charts showing blood vessels of the head and neck.

2. The major arteries carrying blood to structures of the head and neck are the external and internal carotid arteries (**figure 22.3**). The **external carotid arteries** supply most superficial structures of the head and neck. The **external jugular veins** drain most superficial areas of the scalp, face, and neck.

3. Identify the *arteries* listed in figure 22.3a that supply blood to the head and neck, using the textbook as a guide. Then label them in figure 22.3a.

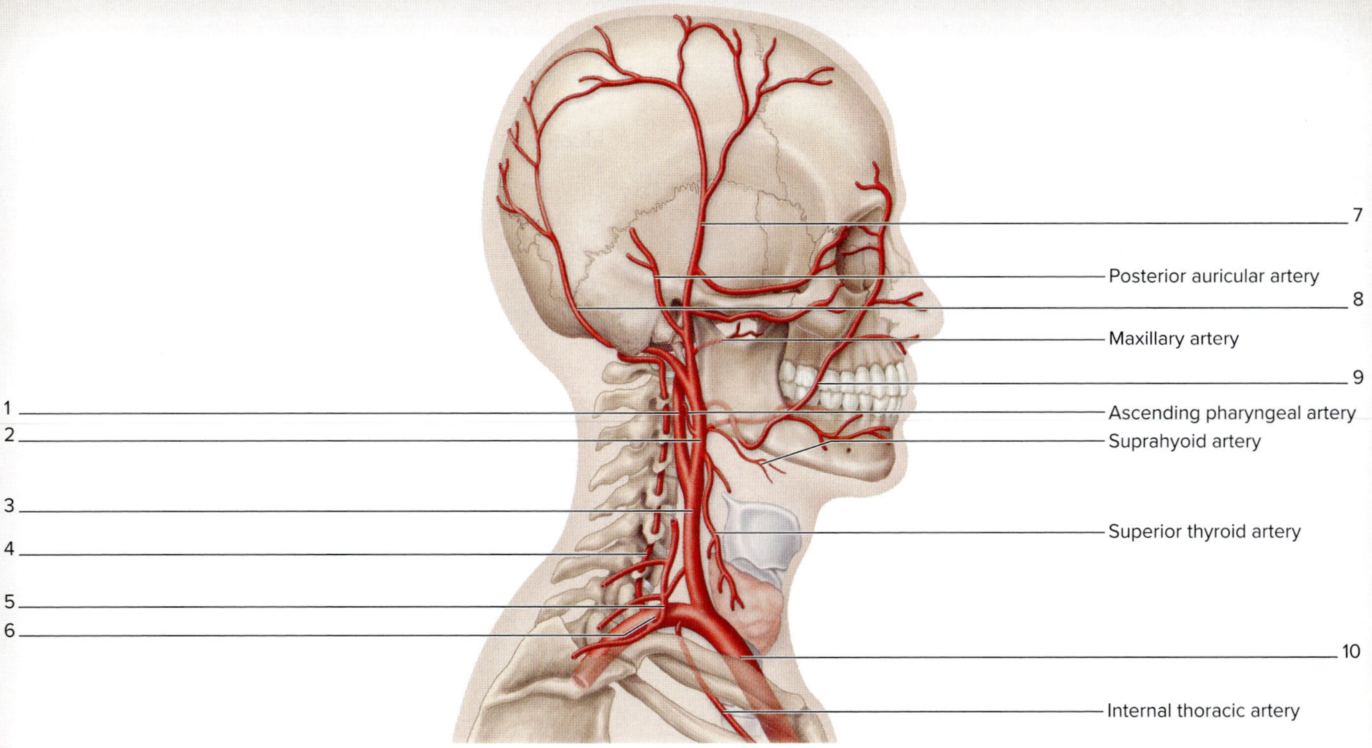

7

Posterior auricular artery

8

Maxillary artery

9

Ascending pharyngeal artery

Suprahyoid artery

Superior thyroid artery

1

2

3

4

5

6

10

Internal thoracic artery

(a) Arteries, right lateral view

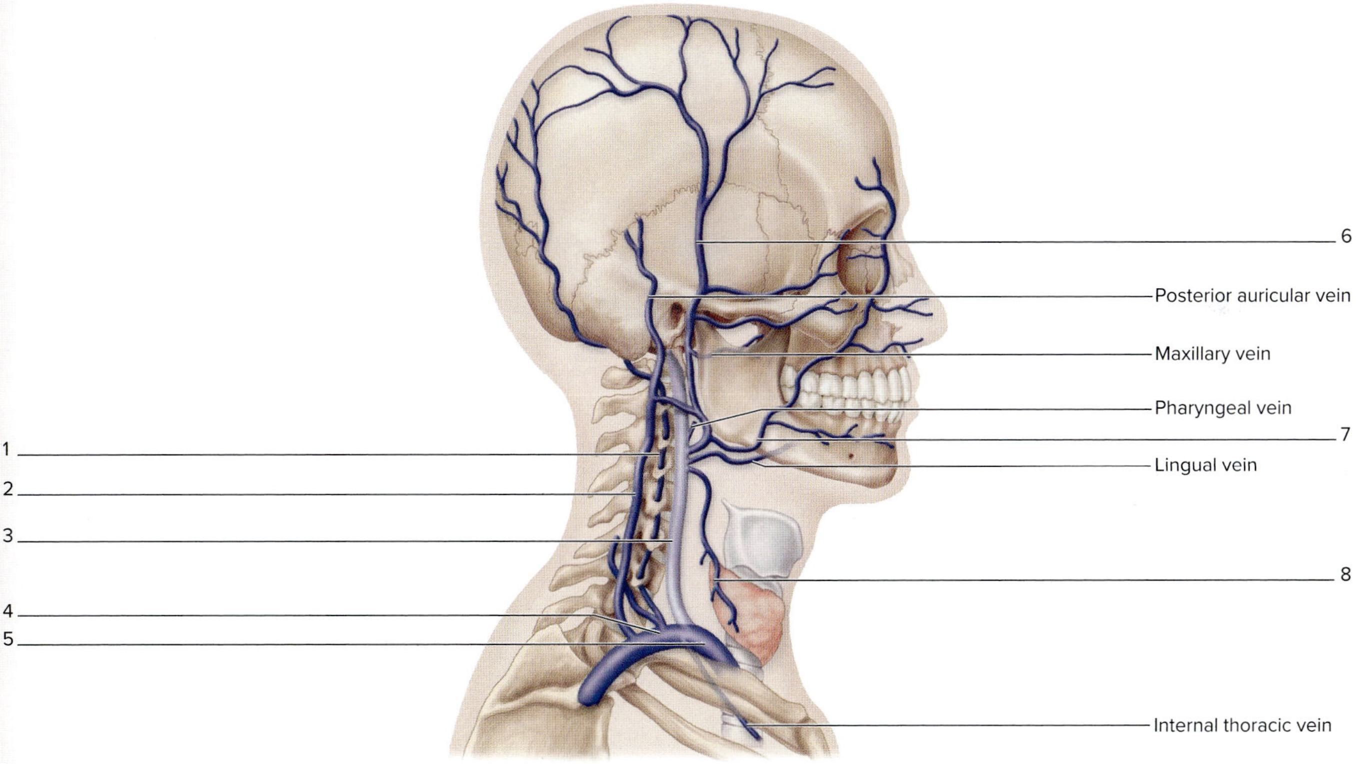

6

Posterior auricular vein

Maxillary vein

Pharyngeal vein

7

Lingual vein

1

2

3

4

5

8

Internal thoracic vein

(b) Veins, right lateral view

Figure 22.3 **Circulation to the Head and Neck.** Use the terms listed to fill in the numbered labels in the figure.

(a) Arterial Supply

☐ Brachiocephalic trunk
☐ Common carotid artery
☐ External carotid artery
☐ Facial artery
☐ Internal carotid artery

☐ Occipital artery
☐ Subclavian artery
☐ Superficial temporal artery
☐ Thyrocervical trunk
☐ Vertebral artery

(b) Venous Drainage

☐ External jugular vein
☐ Facial vein
☐ Internal jugular vein
☐ Right brachiocephalic vein

☐ Subclavian vein
☐ Superficial temporal vein
☐ Superior thyroid vein
☐ Vertebral vein

(continued on next page)

(continued from previous page)

4. Identify the *veins* listed in figure 22.3*b* that drain blood from the head and neck, using the textbook as a guide. Then label them in figure 22.3*b*.

5. The pathway a drop of blood takes to get from the aortic arch to the *skin overlying the anterior part of the right parietal bone of the skull* and back to the superior vena cava is shaded in **figure 22.4**. Trace this flow of blood by writing in the names of the vessels in the figure.

Label the vessels in order, starting at number 1, so as to figuratively trace the pathway by naming the vessels the blood flows through.

6. *Optional Activity*: **APR** **9: Cardiovascular System**— Anatomy & Physiology Revealed includes numerous dissections showing vascular supply to all body regions; review these dissections and use the Quiz feature to review each region.

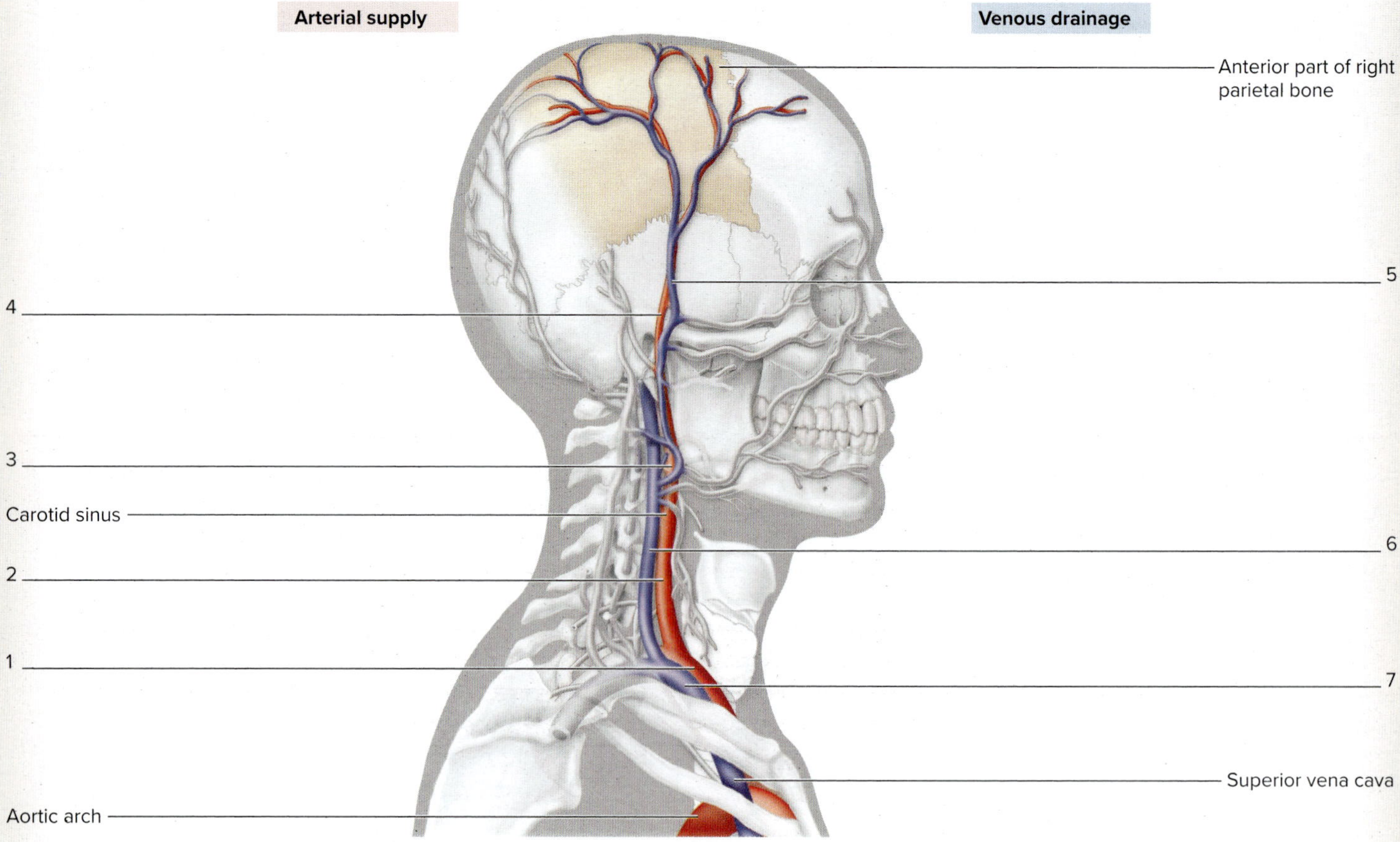

Figure 22.4 **Circulation from the Aortic Arch to the Anterior Part of the Right Parietal Bone and Back to the Superior Vena Cava.** Use the terms listed to fill in the numbered labels in the figure.

- ☐ Brachiocephalic trunk
- ☐ Right brachiocephalic vein
- ☐ Right common carotid artery
- ☐ Right external carotid artery
- ☐ Right internal jugular vein
- ☐ Right superficial temporal artery
- ☐ Right superficial temporal vein

CIRCULATION TO THE BRAIN

1. Observe the cranium and brain of a prosected human cadaver or observe classroom models or charts demonstrating blood vessels of the cranium and brain.

2. The major arteries carrying blood to structures of the brain are the internal carotid arteries and the vertebral arteries (see figure 22.3 and **figure 22.5**). The **internal carotid arteries** supply ~75% of the blood flow to the brain, while the **vertebral arteries** supply ~25% of the blood flow to the brain. Both pairs of vessels supply blood to the **cerebral arterial circle** (figure 22.5a). The major veins draining blood from the head, neck, and brain are the external and internal jugular veins. The **internal jugular veins** drain all blood from inside the cranial cavity plus some superficial areas of the face. Blood draining from brain tissues first enters the **dural venous sinuses**, which collectively drain into the internal jugular vein (figure 22.5b).

3. Identify the *arteries* listed in figure 22.5a that supply blood to the brain, using the textbook as a guide. Then label them in figure 22.5a.

4. Identify the *veins* listed in figure 22.5b that drain blood from the brain, using the textbook as a guide. Then label them in figure 22.5b.

5. The pathway a drop of blood takes to get from the aortic arch to the *right parietal lobe of the brain* and back to the right brachiocephalic vein is shaded in **figure 22.6**. Trace this flow of blood by writing in the names of the vessels in the figure. Label the vessels in order, starting at number 1, so as to figuratively trace the pathway by naming the vessels the blood flows through.

(continued on next page)

INTEGRATE

CLINICAL VIEW
Stroke

Maintaining adequate blood flow to organs ensures that oxygen and nutrients are delivered to the tissues. A stroke may occur when blood flow is restricted to the brain. This, in turn, restricts oxygen delivery to the tissues, a condition referred to as ischemia (*ischaemus*, stopping blood). Prolonged periods of ischemia may result in tissue death, or necrosis. As previously discussed, blood flow is influenced by changes in blood pressure and resistance. Common causes of ischemia include atherosclerotic plaques that dislodge and travel to the brain (embolism). When an embolus is present in a vessel, there is a decrease in blood vessel diameter and an increase in resistance. Due to the inverse relationship between resistance and blood flow, this increased resistance results in a dramatic decrease in blood flow. Deficits observed following a stroke depend on the location of the ischemic event. For instance, a stroke in Broca's area may result in slurred speech, and an ischemic event in the vermis of the cerebellum may lead to motor deficits. Clinically, tests such as magnetic resonance imaging (MRI) or computed tomography (CT) are used to observe the area of infarct (tissue death).

EXERCISE 22.5

CIRCULATION TO THE THORACIC AND ABDOMINAL WALLS

1. Observe the thoracic cavity on a prosected human cadaver or observe classroom models or charts demonstrating blood vessels of the thoracic and abdominal cavities.

2. Identify the *arteries* listed in **figure 22.7a** that supply blood to the thoracic and abdominal walls, using the textbook as a guide. Then label them in figure 22.7a.

3. Identify the *veins* listed in figure 22.7b that drain blood from the thoracic and abdominal walls, using the textbook as a guide. Then label them in figure 22.7b.

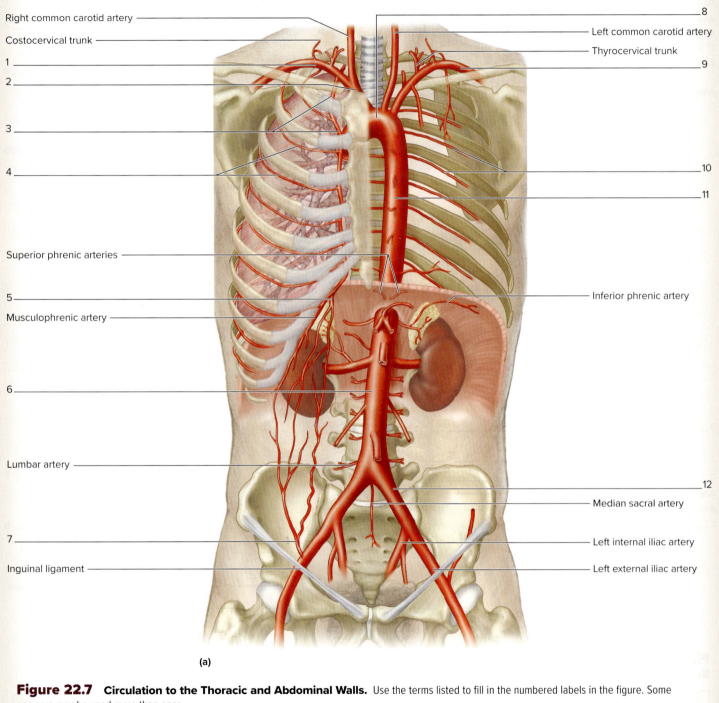

(a)

Figure 22.7 **Circulation to the Thoracic and Abdominal Walls.** Use the terms listed to fill in the numbered labels in the figure. Some answers may be used more than once.

(a) Arterial Supply

☐ Anterior intercostal arteries

☐ Aortic arch

☐ Brachiocephalic trunk

☐ Common iliac artery

☐ Descending abdominal aorta

☐ Descending thoracic aorta

☐ Inferior epigastric artery

☐ Internal thoracic artery

☐ Left subclavian artery

☐ Posterior intercostal arteries

☐ Right subclavian artery

☐ Superior epigastric artery

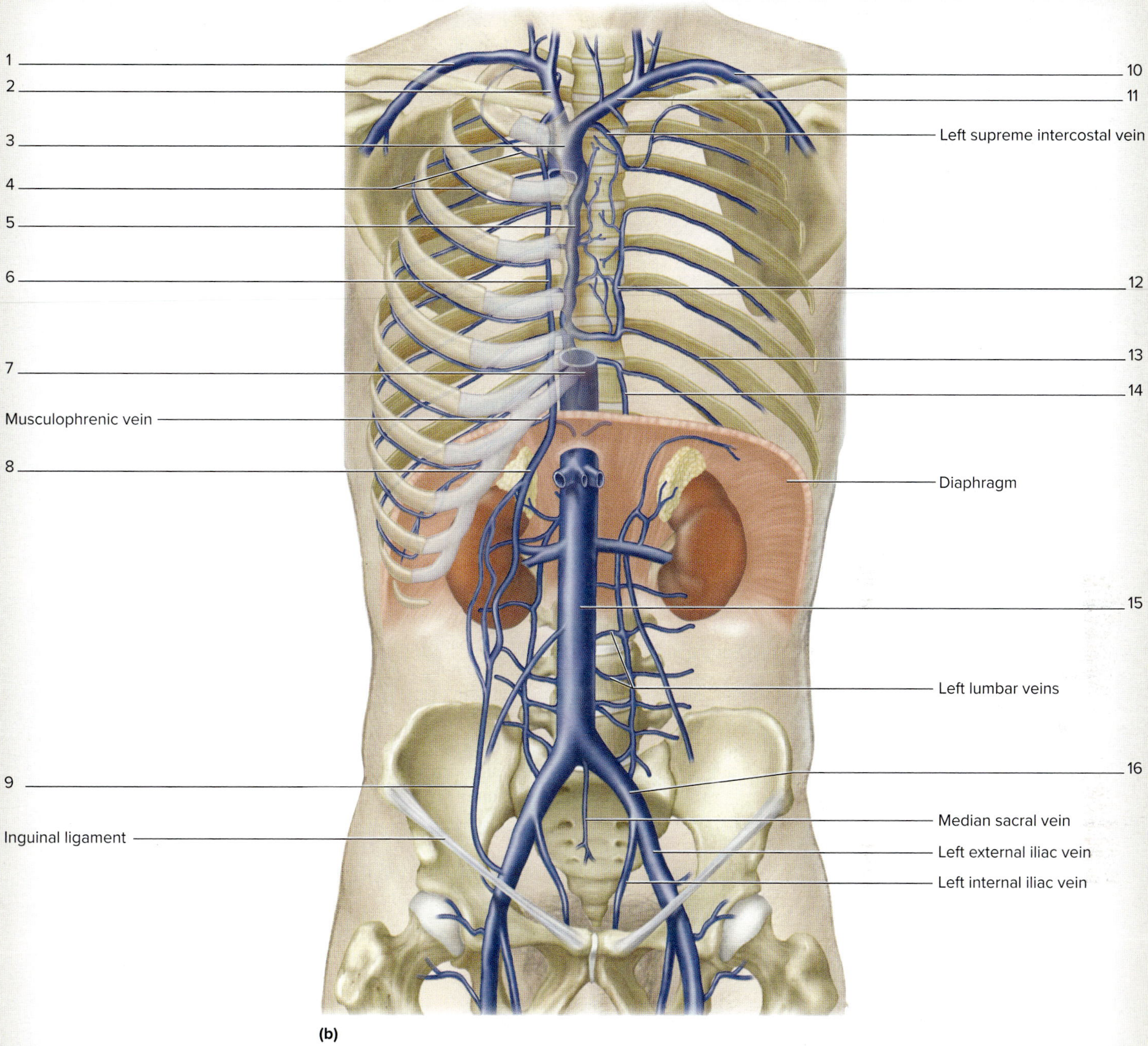

(b)

Figure 22.7 **Circulation to the Thoracic and Abdominal Walls *(continued)*.** Use the terms listed to fill in the numbered labels in the figure. Some answers may be used more than once.

(b) Venous Drainage

- ☐ Accessory hemiazygos vein
- ☐ Anterior intercostal veins
- ☐ Azygos vein
- ☐ Hemiazygos vein
- ☐ Inferior epigastric vein
- ☐ Inferior vena cava
- ☐ Internal thoracic vein
- ☐ Left brachiocephalic vein
- ☐ Left common iliac vein
- ☐ Left subclavian vein
- ☐ Posterior intercostal vein
- ☐ Right brachiocephalic vein
- ☐ Right subclavian vein
- ☐ Superior epigastric vein
- ☐ Superior vena cava

CIRCULATION TO THE ABDOMINAL CAVITY

1. Observe the abdominal cavity on a prosected human cadaver and/or observe classroom models or charts demonstrating blood vessels of the abdominal cavity.

2. Identify the *arteries* listed in **figure 22.8** that supply blood to structures in the abdomen, using the textbook as a guide. Then label them in figure 22.8.

3. Venous drainage of abdominal organs is unique in that it is an example of a portal system, called the **hepatic portal system**. In this system there are two capillary beds—the first in an abdominal organ, and the second in the liver—connected to each other by a **portal vein**. An artery supplies blood to the first capillary bed, which is located in an abdominal organ such as the stomach, intestine, or spleen. Blood drains from abdominal organs into three veins: the **splenic**, **inferior mesenteric**, and **superior mesenteric** veins. These veins then drain into the **hepatic portal vein**, which carries blood to the second capillary bed in the liver. This blood is high in nutrient content, but it also may be transporting toxins, bacteria, and other potentially dangerous substances. Because the blood flows from the abdominal organs directly to the liver, nutrients, drugs, and pathogens may be removed from the blood before the blood enters the general circulation. Thus, the liver is said to have "first pass" at the blood that drains from the abdominal organs. Finally, venous blood drains from liver capillaries into **hepatic veins**, which carry it to the inferior vena cava and back to the heart.

4. Identify the *veins* listed in **figure 22.9** that compose the hepatic portal system, using the textbook as a guide. Then label them in figure 22.9.

5. In the space provided, write out the structures blood would flow through to get from the left ventricle of the heart to the *duodenum* and back to the right atrium of the heart.

_____

_____

_____

_____

6. Some abdominal organs are paired, and thus the blood supply is bilateral. One example is the paired kidneys, which are supplied by the renal arteries and veins. The pathway a drop of blood takes to get from the heart to the *right kidney* and back to the right atrium of the heart is shaded in **figure 22.10**. Trace this flow of blood by writing in the names of vessels in the figure. Label the vessels in order, starting at number 1, so as to figuratively trace the pathway by naming the vessels the blood flows through.

7. The pathway a drop of blood takes to get from the abdominal aorta to the *spleen* and back to the right atrium of the heart is shaded in **figure 22.11**. Trace this flow of blood by writing in the names of the vessels in the figure. Label the vessels in order, starting at number 1, so as to figuratively trace the pathway by naming the vessels the blood flows through.

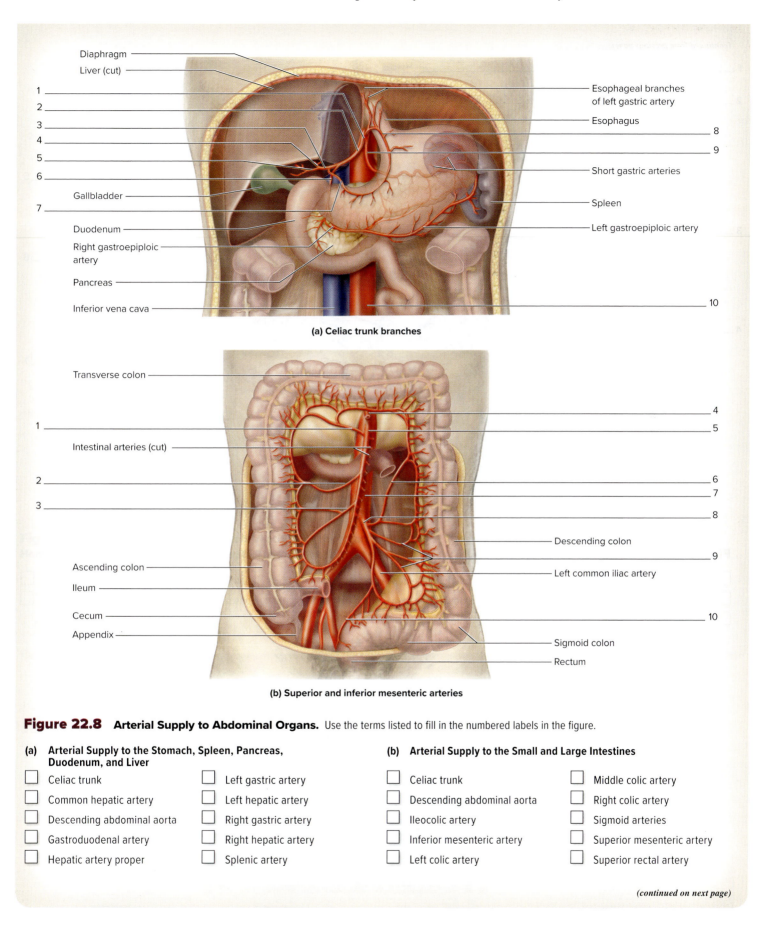

Diaphragm
Liver (cut)
1
2
3
4
5
6
Gallbladder
7
Duodenum
Right gastroepiploic artery
Pancreas
Inferior vena cava

Esophageal branches of left gastric artery
Esophagus
8
9
Short gastric arteries
Spleen
Left gastroepiploic artery
10

(a) Celiac trunk branches

Transverse colon
1
Intestinal arteries (cut)
2
3
Ascending colon
Ileum
Cecum
Appendix

4
5
6
7
8
Descending colon
9
Left common iliac artery
10
Sigmoid colon
Rectum

(b) Superior and inferior mesenteric arteries

Figure 22.8 **Arterial Supply to Abdominal Organs.** Use the terms listed to fill in the numbered labels in the figure.

(a) **Arterial Supply to the Stomach, Spleen, Pancreas, Duodenum, and Liver**

☐ Celiac trunk
☐ Common hepatic artery
☐ Descending abdominal aorta
☐ Gastroduodenal artery
☐ Hepatic artery proper

☐ Left gastric artery
☐ Left hepatic artery
☐ Right gastric artery
☐ Right hepatic artery
☐ Splenic artery

(b) **Arterial Supply to the Small and Large Intestines**

☐ Celiac trunk
☐ Descending abdominal aorta
☐ Ileocolic artery
☐ Inferior mesenteric artery
☐ Left colic artery

☐ Middle colic artery
☐ Right colic artery
☐ Sigmoid arteries
☐ Superior mesenteric artery
☐ Superior rectal artery

(continued on next page)

(continued from previous page)

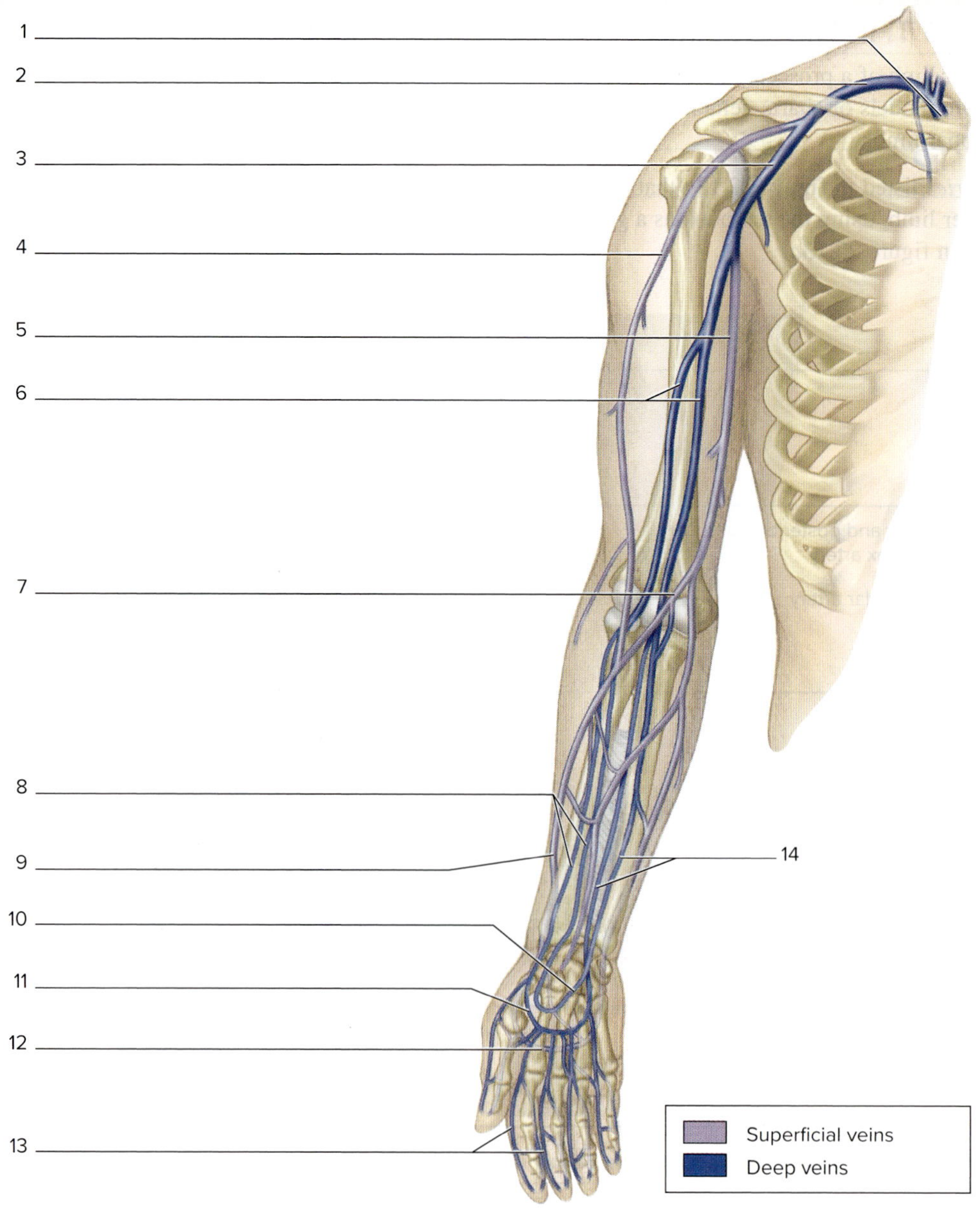

1 _____
2 _____
3 _____
4 _____
5 _____
6 _____
7 _____
8 _____
9 _____
10 _____
11 _____
12 _____
13 _____
14 _____

| Superficial veins |
| Deep veins |

(b) Veins of right upper limb

Figure 22.13 **Circulation to the Upper Limb *(continued).*** Use the terms listed to fill in the numbered labels in the figure. Some answers may be used more than once.

(b) Venous Drainage

☐ Axillary vein
☐ Basilic vein
☐ Brachial veins
☐ Brachiocephalic vein

☐ Cephalic vein
☐ Deep palmar venous arch
☐ Digital veins
☐ Dorsal venous network
☐ Median cubital vein

☐ Radial veins
☐ Subclavian vein
☐ Superficial palmar venous arch
☐ Ulnar veins

4. *Superficial Trace*: The pathway a drop of blood takes to get from the aortic arch to the *anterior surface of the index finger* and back along a superficial route to the superior vena cava is shaded in **figure 22.14**. Trace this flow of blood by writing in the names of the vessels in the figure. Label the vessels in order, starting at number 1, so as to figuratively trace the pathway by naming the vessels the blood flows through.

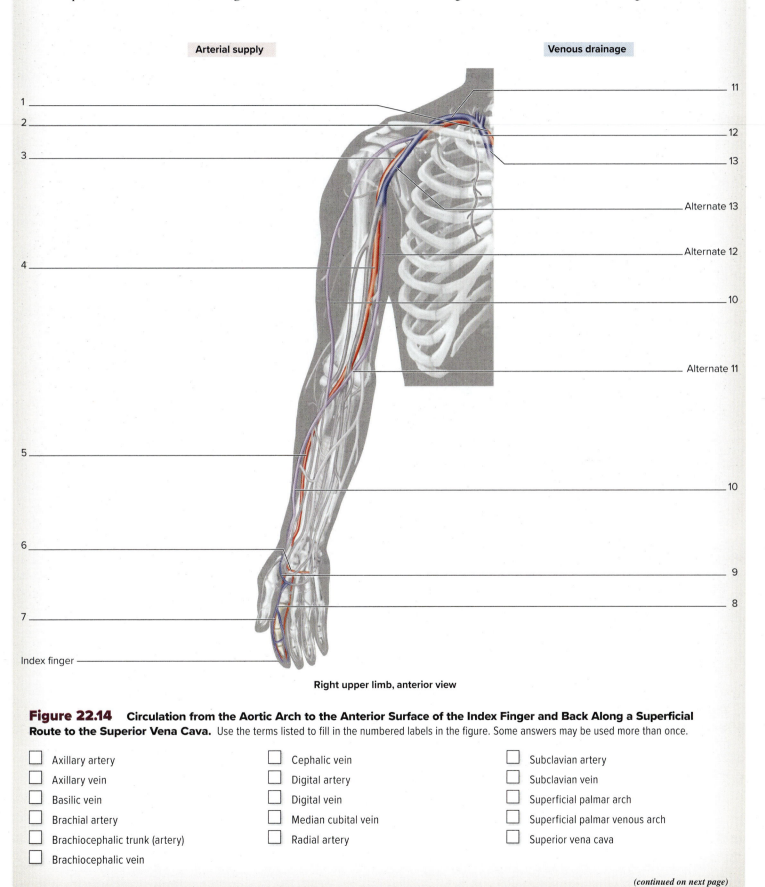

Figure 22.14 **Circulation from the Aortic Arch to the Anterior Surface of the Index Finger and Back Along a Superficial Route to the Superior Vena Cava.** Use the terms listed to fill in the numbered labels in the figure. Some answers may be used more than once.

- ☐ Axillary artery
- ☐ Axillary vein
- ☐ Basilic vein
- ☐ Brachial artery
- ☐ Brachiocephalic trunk (artery)
- ☐ Brachiocephalic vein
- ☐ Cephalic vein
- ☐ Digital artery
- ☐ Digital vein
- ☐ Median cubital vein
- ☐ Radial artery
- ☐ Subclavian artery
- ☐ Subclavian vein
- ☐ Superficial palmar arch
- ☐ Superficial palmar venous arch
- ☐ Superior vena cava

(continued on next page)

(continued from previous page)

5. *Deep Trace:* The pathway a drop of blood takes to get from the aortic arch to the *capitate bone in the wrist* and back along a deep route to the superior vena cava is shaded in **figure 22.15**. Trace this flow of blood by writing in the names of the vessels in the figure. Label the vessels in order, starting at number 1, so as to figuratively trace the pathway by naming the vessels the blood flows through.

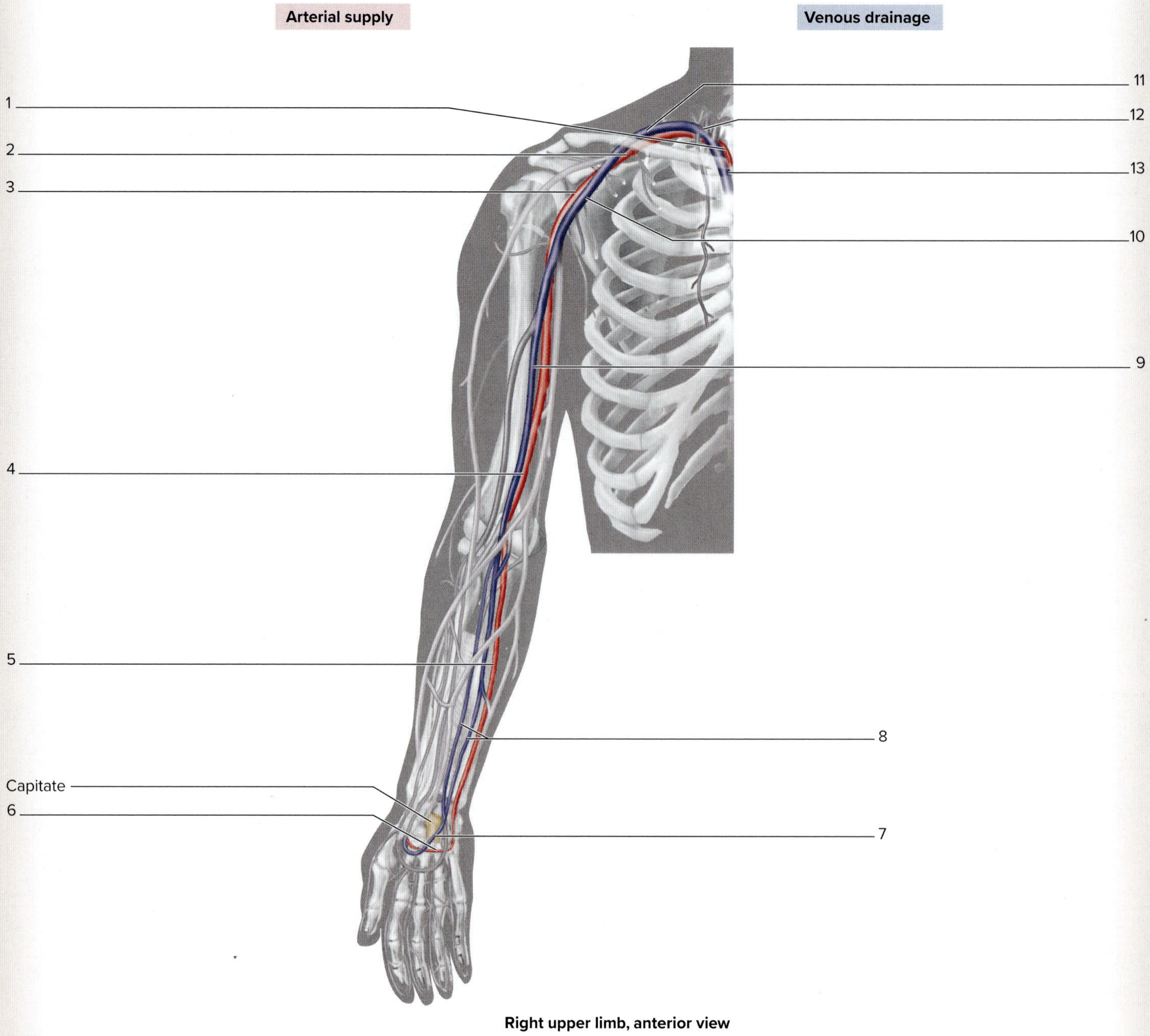

Right upper limb, anterior view

Figure 22.15   **Circulation from the Aortic Arch to the Capitate Bone of the Wrist and Back Along a Deep Route to the Superior Vena Cava.** Use the terms listed to fill in the numbered labels in the figure.

- [ ] Axillary artery
- [ ] Axillary vein
- [ ] Brachial artery
- [ ] Brachial vein
- [ ] Brachiocephalic trunk
- [ ] Brachiocephalic vein
- [ ] Deep palmar arch
- [ ] Deep palmar venous arch
- [ ] Subclavian artery
- [ ] Subclavian vein
- [ ] Superior vena cava
- [ ] Ulnar artery
- [ ] Ulnar veins

CIRCULATION TO THE LOWER LIMB

1. Observe the lower limb of a prosected human cadaver or observe classroom models or charts demonstrating blood vessels of the lower limb.

2. Identify the *arteries* listed in **figure 22.16a** that supply blood to the lower limb, using the textbook as a guide. Then label them in figure 22.16a.

3. Identify the *veins* listed in figure 22.16b that drain blood from the lower limb, using the textbook as a guide. Then label them in figure 22.16b.

4. *Superficial Trace:* The pathway a drop of blood takes to get from the abdominal aorta to the *dorsal surface of the big toe (hallux)* and back along a superficial route to the inferior vena cava is shaded in **figure 22.17**. Trace this flow of blood by writing in the names of the vessels in the figure. Label the vessels in order, starting at number 1, so as to figuratively trace the pathway by naming the vessels the blood flows through.

5. *Deep Trace:* The pathway a drop of blood takes to get from the abdominal aorta to the *cuboid bone in the foot* and back along a deep route to the inferior vena cava is shaded in **figure 22.18**. Trace this flow of blood by writing in the names of the vessels in the figure. Label the vessels in order, starting at number 1, so as to figuratively trace the pathway by naming the vessels the blood flows through.

INTEGRATE

CLINICAL VIEW
Cardiac Catheterization via the Femoral Artery

When a patient with a blockage in one of the coronary arteries requires balloon angioplasty and stenting to open up the blockage, the physician performing the procedure takes the route of least resistance to thread the catheter with the balloon through the circulatory system to get to the blocked artery. Ironically, this is not the shortest path. Instead, the catheter is introduced into the cardiovascular system via the femoral artery within the *femoral triangle* of the thigh (see chapter 13). The femoral artery is relatively easy to access because the only tissues that lie superficial to it are skin, connective tissue, and fat. Once a catheter is introduced into the femoral artery, it can then be threaded "backward" to the heart via the following pathway: femoral artery, external iliac artery, common iliac artery, abdominal aorta, thoracic aorta, aortic arch, right or left coronary artery. Although this seems like a long distance, the path is relatively straightforward, and the catheter travels through these relatively large vessels until it gets to the coronary artery itself. Once the catheter is within the target vessel, a balloon is inflated, which presses the blockage against the wall of the artery, thus opening up the blockage. After the angioplasty is complete, a stent may be placed to prevent the artery from closing off again. A stent is a piece of mesh that holds the artery open.

(continued on next page)

(continued from previous page)

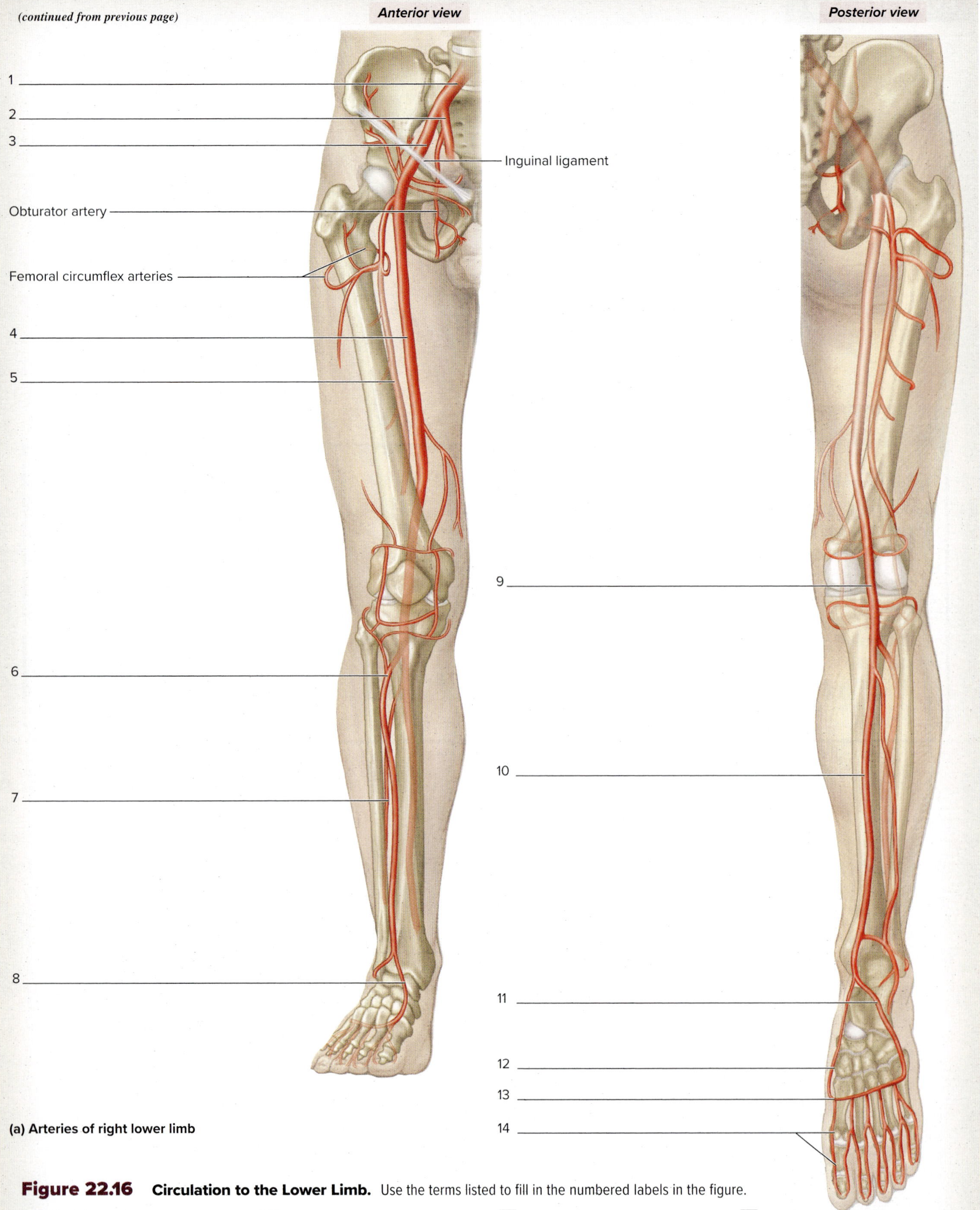

Anterior view

Posterior view

1 _____

2 _____

3 _____

Inguinal ligament

Obturator artery _____

Femoral circumflex arteries _____

4 _____

5 _____

9 _____

6 _____

10 _____

7 _____

8 _____

11 _____

12 _____

13 _____

(a) Arteries of right lower limb

14 _____

Figure 22.16 **Circulation to the Lower Limb.** Use the terms listed to fill in the numbered labels in the figure.

(a) Arterial Supply

☐ Anterior tibial artery
☐ Common iliac artery
☐ Deep femoral artery

☐ Digital arteries
☐ Dorsalis pedis artery
☐ External iliac artery
☐ Femoral artery

☐ Fibular artery
☐ Internal iliac artery
☐ Lateral plantar artery
☐ Medial plantar artery

☐ Plantar arterial arch
☐ Popliteal artery
☐ Posterior tibial artery

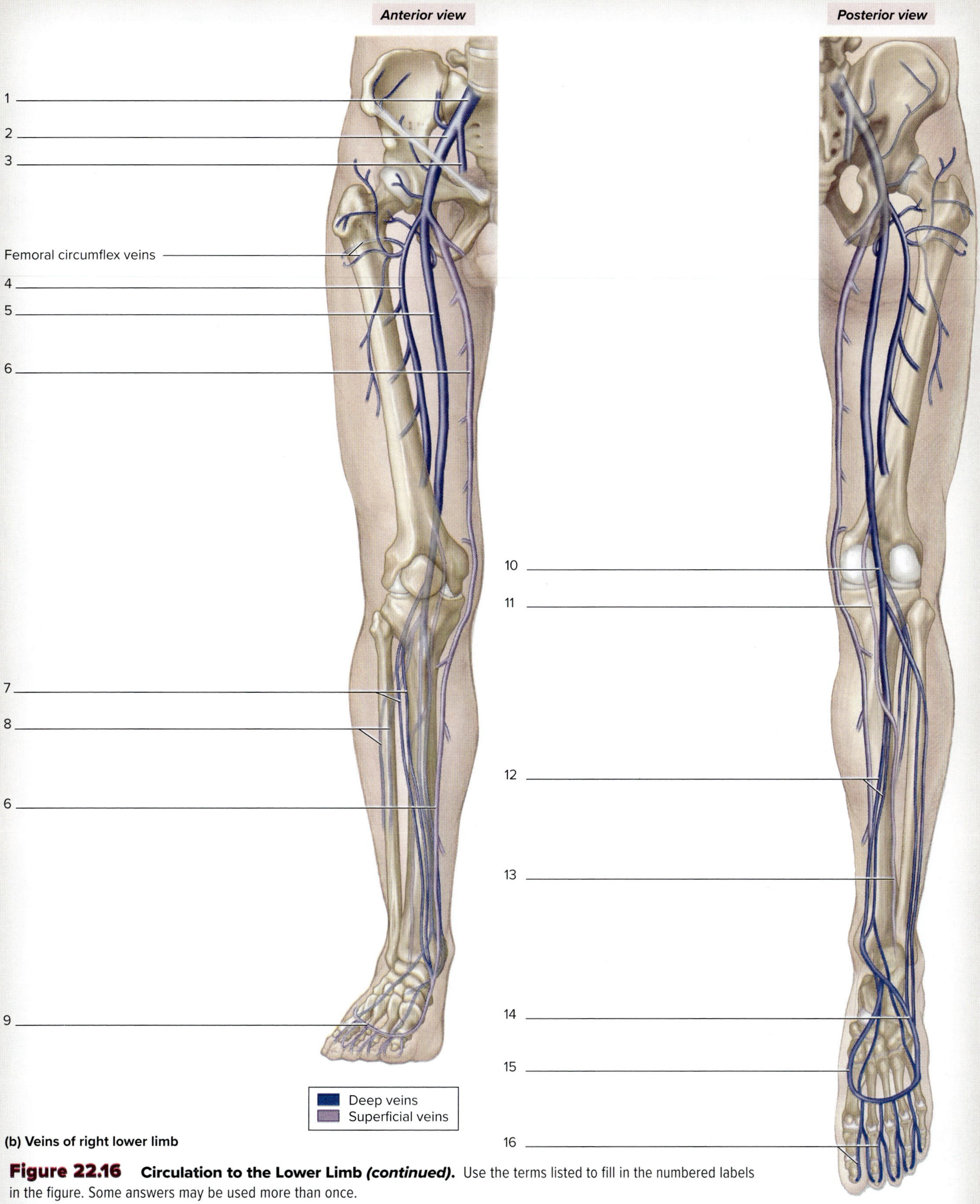

Anterior view

Posterior view

1 _____

2 _____

3 _____

Femoral circumflex veins _____

4 _____

5 _____

6 _____

10 _____

11 _____

7 _____

8 _____

6 _____

12 _____

13 _____

14 _____

15 _____

9 _____

16 _____

Deep veins
Superficial veins

(b) Veins of right lower limb

Figure 22.16 **Circulation to the Lower Limb *(continued).*** Use the terms listed to fill in the numbered labels in the figure. Some answers may be used more than once.

(b) Venous Drainage

☐ Anterior tibial veins

☐ Common iliac vein

☐ Deep femoral vein

☐ Digital veins

☐ Dorsal venous arch

☐ External iliac vein

☐ Femoral vein

☐ Fibular veins

☐ Great saphenous vein

☐ Internal iliac vein

☐ Lateral plantar vein

☐ Medial plantar vein

☐ Popliteal vein

☐ Posterior tibial veins

☐ Small saphenous vein

(continued on next page)

(continued from previous page)

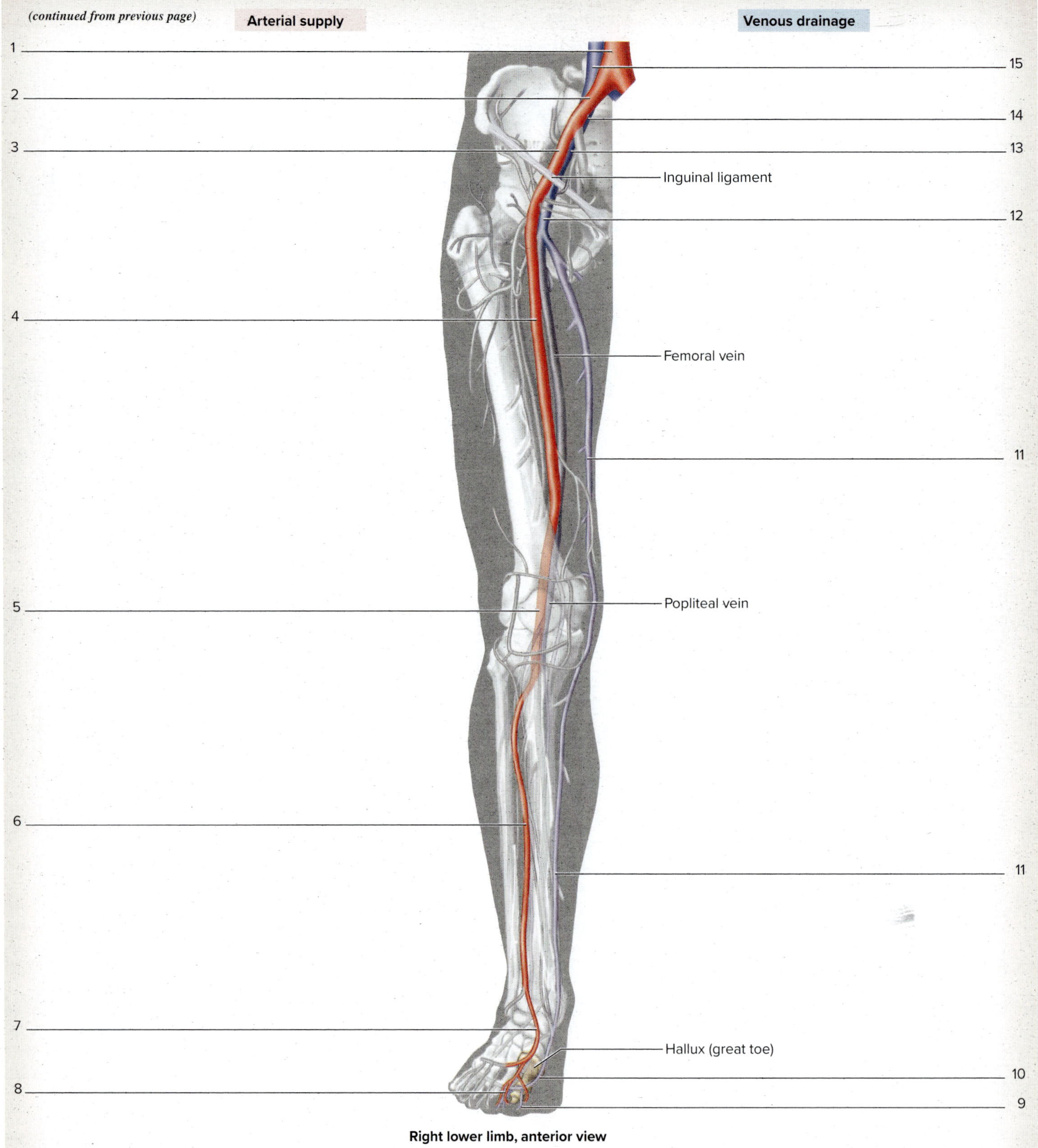

Right lower limb, anterior view

Figure 22.17 Circulation from the Abdominal Aorta to the Dorsal Surface of the Big Toe (Hallux) and Back Along a Superficial Route to the Inferior Vena Cava. Use the terms listed to fill in the numbered labels in the figure.

- ☐ Abdominal aorta
- ☐ Anterior tibial artery
- ☐ Common iliac artery
- ☐ Common iliac vein
- ☐ Digital artery
- ☐ Digital vein
- ☐ Dorsal venous arch
- ☐ Dorsalis pedis artery
- ☐ External iliac artery
- ☐ External iliac vein
- ☐ Femoral artery
- ☐ Femoral vein
- ☐ Great saphenous vein
- ☐ Inferior vena cava
- ☐ Popliteal artery

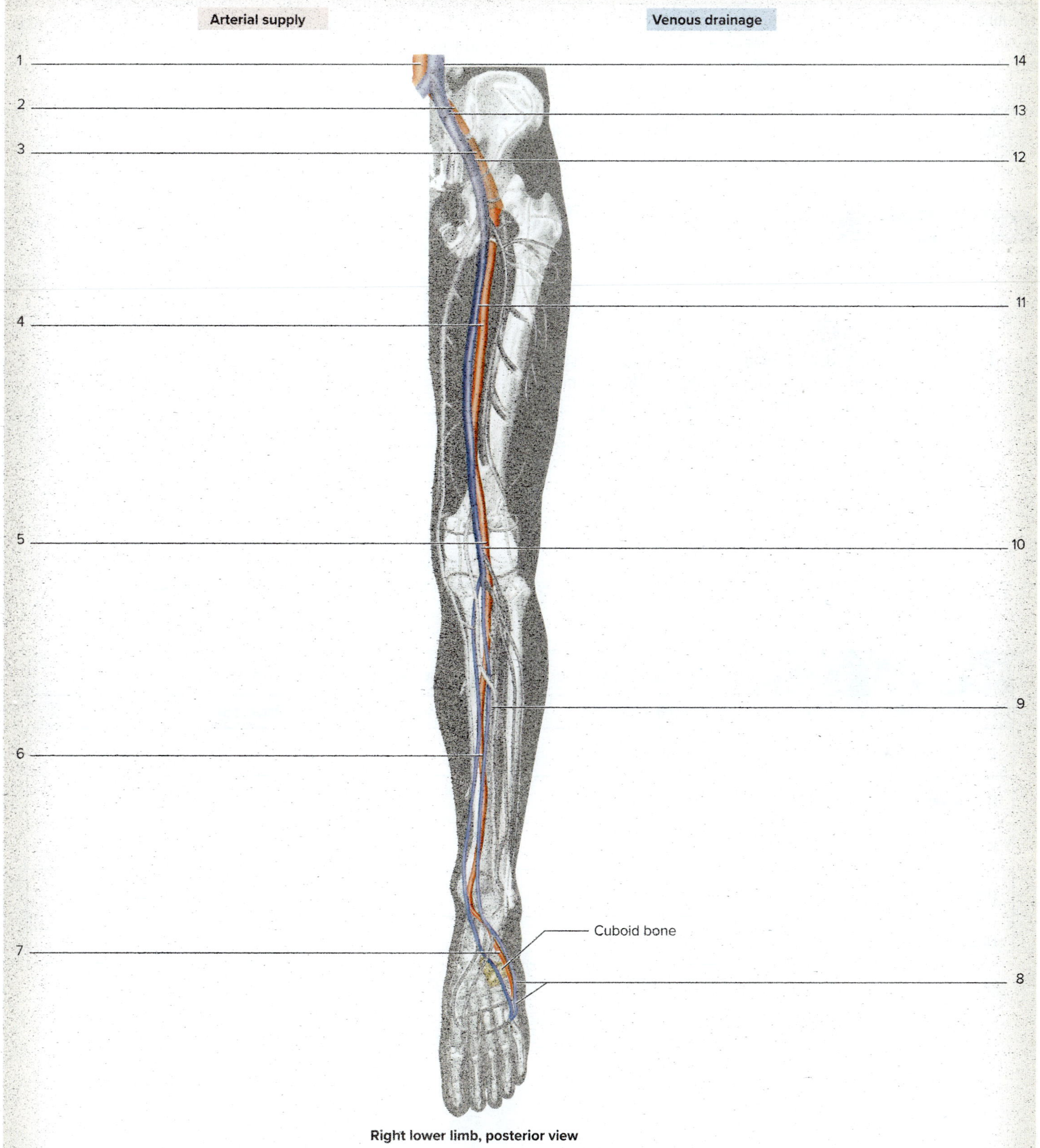

Arterial supply

Venous drainage

1

2

3

4

5

6

7

14

13

12

11

10

9

8

Cuboid bone

Right lower limb, posterior view

Figure 22.18 **Circulation from the Abdominal Aorta to the Cuboid Bone in the Foot and Back Along a Deep Route to the Inferior Vena Cava.** Use the terms listed to fill in the numbered labels in the figure.

☐ Abdominal aorta

☐ Common iliac artery

☐ Common iliac vein

☐ External iliac artery

☐ External iliac vein

☐ Femoral artery

☐ Femoral vein

☐ Inferior vena cava

☐ Lateral plantar artery

☐ Lateral plantar veins

☐ Popliteal artery

☐ Popliteal vein

☐ Posterior tibial artery

☐ Posterior tibial vein

Fetal Circulation

The lungs are nonfunctional in the fetus and need only a small amount of blood to support the developing lung tissue. This blood must be oxygenated blood coming from the fetal respiratory organ: the placenta. Once the fetus is born, the circulation must change as the lungs replace the placenta as the respiratory organs. In addition, the blood returning from the placenta via the umbilical vein bypasses the liver through the ductus venosus. Thus, there are a number of **shunts** present in the fetal circulation that direct blood away from the lungs, to and from the placenta, and away from the liver. These shunts must close at birth to establish the normal postnatal circulatory pathways. The following exercise involves identifying the unique cardiovascular structures of the fetal circulation, tracing the flow of blood through the fetal circulation, and identifying the postnatal structures that are remnants of the fetal circulation.

EXERCISE 22.9

FETAL CIRCULATION

1. Identify the fetal circulatory system structures listed in **figure 22.19**, using the textbook as a guide.

2. List the structures that the blood passes through as it is transported from the left ventricle of the fetal heart to the placenta and back to the right atrium of the fetal heart.

_____

_____

_____

_____

_____

_____

_____

_____

_____

_____

3. Write in the names of the postnatal structures that are remnants of the fetal circulation, and describe each structure's function in the fetus in **table 22.1**.

Table 22.1	Fetal Cardiovascular Structures and Associated Postnatal Structures	
Fetal Cardiovascular Structure	**Postnatal Structure**	**Function of Fetal Cardiovascular Structure**
Ductus Arteriosus		
Ductus Venosus		
Foramen Ovale		
Umbilical Arteries		
Umbilical Vein		

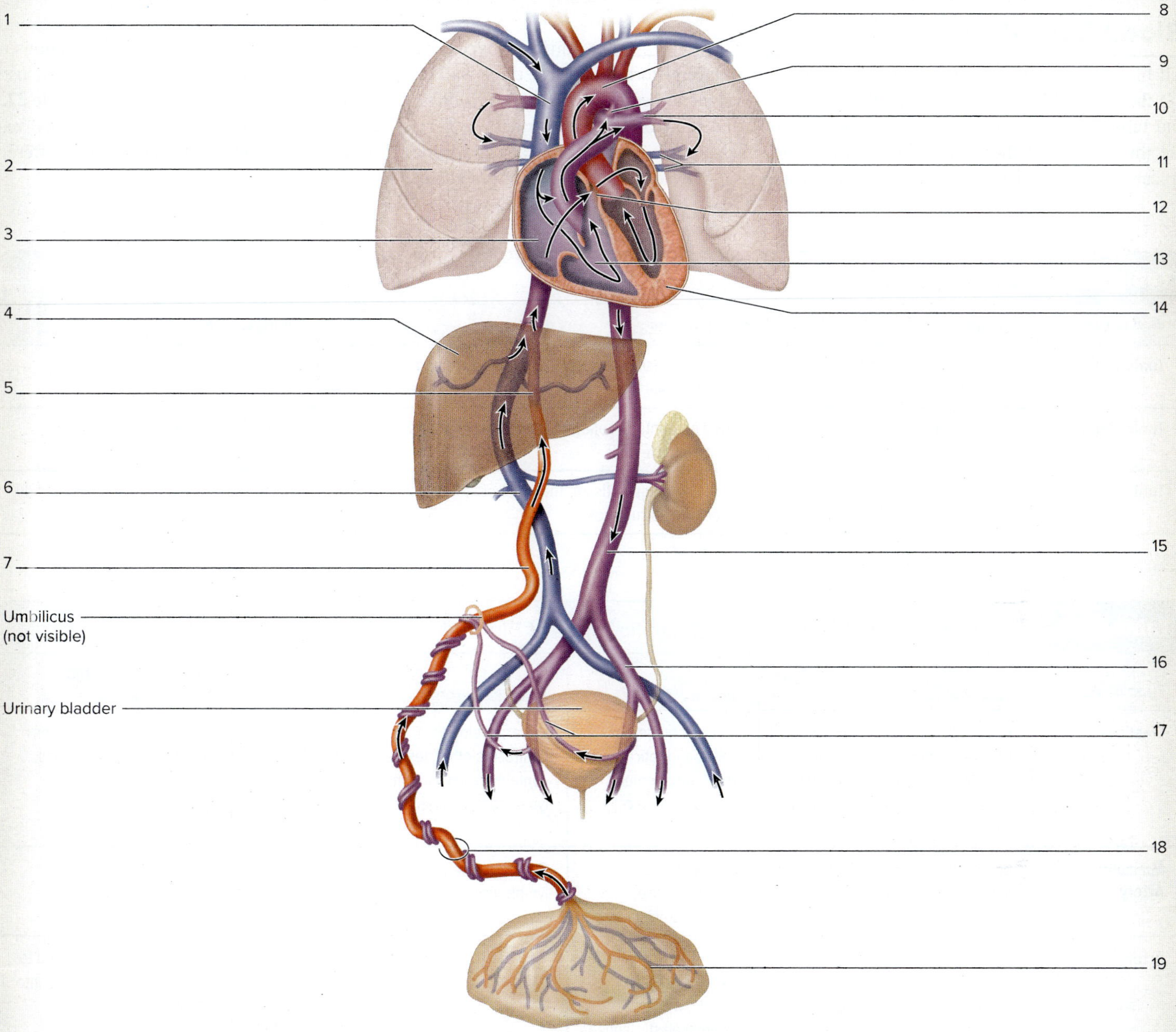

Figure 22.19   Fetal Circulation. Use the terms listed to fill in the numbered labels in the figure.

Aortic arch

Common iliac artery

Descending abdominal aorta

Ductus arteriosus

Ductus venosus

Foramen ovale

Heart

Inferior vena cava

Liver

Lung

Placenta

Pulmonary artery

Pulmonary veins

Right atrium

Right ventricle

Superior vena cava

Umbilical arteries

Umbilical cord

Umbilical vein

HISTOLOGY

Blood Vessel Wall Structure

All blood vessels except capillaries have three tunics (layers) forming their walls. These layers, called the tunica intima, tunica media, and tunica externa, are analogous in both structure and function to the three layers of the heart wall (endocardium, myocardium, and epicardium). The differences in structure and function between the different types of blood vessels come mainly from modifications of these three wall layers. In particular, the type of tissue that composes the tunica media greatly affects the function of the vessel. **Table 22.2** describes the general composition of the three layers of a blood vessel wall, and **table 22.3** summarizes unique features in the different types of blood vessels.

Table 22.2	Layers of a Blood Vessel Wall		
Wall Layer	**Location**	**Components**	**Word Origin**
Tunica Intima	Innermost layer; in contact with the lumen of the vessel	Endothelium (simple squamous epithelium) and a subendothelial layer composed of areolar connective tissue	*tunic*, a coat, + *intimus*, innermost
Tunica Media	Middle layer	Varied amounts of collagen fibers, elastic fibers, and smooth muscle cells	*tunic*, a coat, + *medius*, middle
Tunica Externa	Outermost layer	Areolar connective tissue that anchors the vessel to surrounding structures	*tunic*, a coat, + *externus*, on the outside

Table 22.3	Characteristics of Wall Layers in Specific Types of Blood Vessels				
Type of Vessel	**Tunica Intima**	**Tunica Media**	**Tunica Externa**	**Diameter**	**Characteristics and Special Functions**
Elastic Artery	Endothelium and subendothelial layer; an internal elastic lamina is present but not easily distinguished from the elastic tissue of the tunica media	Contains numerous elastic and reticular fibers; also contains smooth muscle cells	Underdeveloped in contrast to other vessels; contains vasa vasorum, lymphatics, and nerves	2.5 cm–1 cm	Expansion and contraction of elastic tissues smooths out the flow of blood
Muscular Artery	Endothelium and subendothelial layer; contains a very prominent internal elastic lamina	Multiple layers of smooth muscle; numerous elastic fibers	Contains vasa vasorum, lymphatics, and nerves	1 cm–3 mm	Recoil of wall continues to propel blood through the arteries
Arteriole	Endothelium and subendothelial layer; an internal elastic lamina is present only in the largest arterioles	Contains less than six layers of smooth muscle, with no external elastic lamina	Very thin	3 mm–10 µm	Size of lumen is regulated to control the flow of blood into capillary beds
Capillary	Endothelium and a basement membrane only	NA	NA	8–10 µm	Thin wall allows for exchange between the blood and tissues
Venule	Endothelium and a thin subendothelial layer	Very thin with very few smooth muscle cells	Thickest layer of the wall	50–100 µm	Venules are simply small veins, and are the counterpart to arterioles
Vein	Endothelium and subendothelial layer; infoldings form valves, which prevent the backflow of blood. Not all veins have valves.	Very thin with a small amount of smooth muscle	Thickest layer of the wall; contains vasa vasorum	Greater than 100 µm	Low-pressure conduits; valves aid in preventing backflow of blood

EXERCISE 22.10

BLOOD VESSEL WALL STRUCTURE

1. Obtain a slide showing an artery and a vein (they may both be on the same slide, or they may be on two different slides).

2. Place the slide on the microscope stage and bring the tissue sample into focus on low power. Scan the slide and look for the circular or oval cross section of a vessel. If there is more than one vessel on the slide, determine which vessel is an artery and which is a vein. In general, arteries have relatively thick walls and small lumens, whereas veins have relatively thin walls and large lumens **(figure 22.20)**. In addition, the lumens of veins are often collapsed because of the relative thinness of the blood vessel wall.

3. After identifying an artery and a vein, move the microscope stage so the wall of the *artery* is in the center of the field of view. Increase the power on the microscope until all the layers of the artery wall are visible.

4. Identify the structures listed below using figure 22.20 and table 22.2 as guides. Keep in mind that on low power the innermost layer of the vessel (the tunica intima) will be incredibly thin and difficult to identify. Most likely only the flattened nuclei of the endothelial cells, and very little of the rest of the cells, will be visible.

- ☐ **Artery**
- ☐ **Lumen**
- ☐ **Tunica externa**
- ☐ **Tunica intima**
- ☐ **Tunica media**
- ☐ **Vein**

INTEGRATE

CONCEPT CONNECTION

The tunica intima of a blood vessel consists of endothelium and a basement membrane. This smooth, thin layer provides a physical barrier to separate blood from surrounding tissues. In addition, the endothelial cells produce various molecules that help regulate the clotting process, capillary exchange, inflammation, and vessel diameter. Thus, an intact endothelial layer is critical for normal blood flow and health. If this layer is damaged, resulting in exposure of the deeper layer of collagen fibers, a thrombus, or blood clot, will form. Recall from the discussion of cellular transport mechanisms that small molecules may cross a membrane according to concentration gradients, whereas large molecules and cells are transported via vesicles or receptors. A widening of endothelial junctions or an abnormal increase in vesicular transport may erode the barrier effect, leading to an abnormal deposit of molecules, such as lipids, in the blood vessel wall. The result would be local inflammation, a feature in the early stages of atherosclerosis.

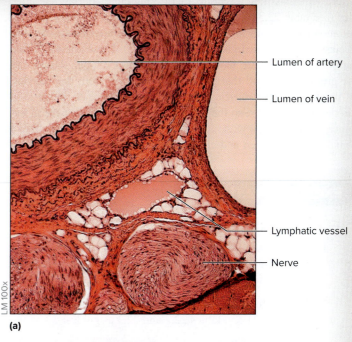

(a)

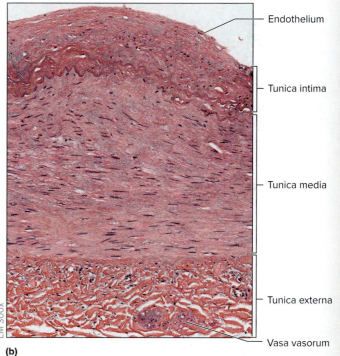

(b)

Figure 22.20 **Blood Vessel Wall Structure.** (*a*) Cross section through the center of a neurovascular bundle containing a nerve, artery, vein, and lymphatic vessel. (*b*) The three layers of the wall of a blood vessel: tunica intima, tunica media, and tunica externa. The tunica externa has its own blood supply, the vasa vasorum (literally, "the vessels of the vessels").

©McGraw-Hill Education/Christine Eckel, photographer

Elastic Arteries

Arteries are classified as elastic arteries, muscular arteries, and arterioles (see table 22.3). The aorta, and the pulmonary, brachiocephalic, common carotid, subclavian, and common iliac arteries are classified as elastic arteries (see figure 22.21, table 22.3). These arteries, located very close to the heart, have walls that are thick enough to withstand the pressure of blood that is pumped into them from the ventricles of the heart. The ventricles generate enough force to move blood through these vessels, so they need very little smooth muscle in their tunica media to assist with blood flow. Instead, elastic arteries have an abundance of collagen and elastic fibers in their tunica media, which makes them both tough (collagen fibers) and expandable (elastic fibers). The ability of the vessel wall to expand as it receives blood from the ventricles, and then recoil, greatly smooths out the flow of blood through the arteries.

Large vessels such as the aorta have tiny blood vessels called *vasa vasorum* (literally, "the vessels of the vessels," see figure 22.20*b*) in the tunica externa. The vasa vasorum are analogous in both structure and function to the coronary arteries in the outer layer of the heart wall (epicardium). That is, the vessels of the vasa vasorum supply blood to the walls of larger vessels just as the coronary arteries supply blood to cardiac muscle tissue.

EXERCISE 22.11

ELASTIC ARTERY—THE AORTA

1. Obtain a slide of the aorta (**figure 22.21**) and place it on the microscope stage. Bring the wall of the aorta into focus on low power.

2. Identify the following on the slide of the aorta, using figure 22.21 and tables 22.2 and 22.3 as guides:

   ☐ **Elastic fibers**

   ☐ **Lumen**

   ☐ **Tunica externa**

   ☐ **Tunica intima**

   ☐ **Tunica media**

3. Sketch the wall of the aorta as seen through the microscope in the space provided or on a separate sheet of paper.

_____ ×

Figure 22.21 **Elastic Artery.** The wall of the aorta, an elastic artery, contains numerous elastic fibers (black) in the tunica media.

©McGraw-Hill Education/Christine Eckel, photographer

Muscular Arteries

As blood moves through the elastic arteries and travels farther away from the heart, the pressure exerted by the heart is no longer great enough to keep the blood moving through the vessels. Thus, the amount of elastic tissue in the tunica media of the arteries starts to decrease and the amount of smooth muscle in the tunica media starts to increase. Contraction of the smooth muscle keeps blood moving through the arteries as the blood gets farther away from the heart.

Most named vessels, including the brachial, anterior tibial, and inferior mesenteric arteries, are muscular arteries. **Muscular arteries** are easily distinguished from elastic arteries by the presence of two prominent bands of elastic fibers: the **internal elastic lamina**, which is the outermost layer of the tunica intima, and the **external elastic lamina** (adjacent to the tunica externa). There are several layers of smooth muscle sandwiched in between the two prominent elastic laminae (see figure 22.22, table 22.3).

EXERCISE 22.12

MUSCULAR ARTERY

1. Obtain a slide of a small, muscular artery, and place it on the microscope stage. Bring the tissue sample into focus on low power, then locate the wall of the vessel **(figure 22.22)**.

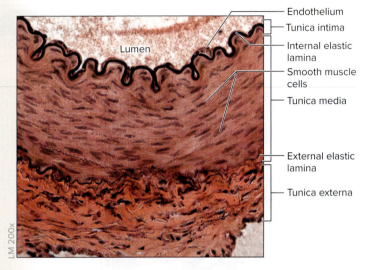

Figure 22.22 **Muscular Artery.** Muscular arteries contain distinct internal and external elastic laminae bordering the tunica media, which is predominantly smooth muscle.

©McGraw-Hill Education/Christine Eckel

2. Identify the following on the slide of the muscular artery, using figure 22.22 and table 22.3 as guides:

- ☐ **External elastic lamina**
- ☐ **Internal elastic lamina**
- ☐ **Lumen**
- ☐ **Smooth muscle cells**
- ☐ **Tunica externa**
- ☐ **Tunica intima**
- ☐ **Tunica media**

3. Sketch the wall of a muscular artery as seen through the microscope in the space provided or on a separate sheet of paper.

_____ ×

Arterioles

Arterioles have a unique function in the cardiovascular system; they control the flow of blood into capillary beds (see table 22.3). To match this function, a prominent feature of an arteriole is layers of circular smooth muscle in the tunica media. The smooth muscle cells of these layers are regulated to contract and relax to change the diameter of the lumen of the vessel, thus regulating the flow of blood into the capillary beds.

EXERCISE 22.13

ARTERIOLE

1. Obtain a slide of an arteriole and place it on the microscope stage. Bring the tissue sample into focus on low power. Scan the slide to locate an arteriole in cross section **(figure 22.23)**.

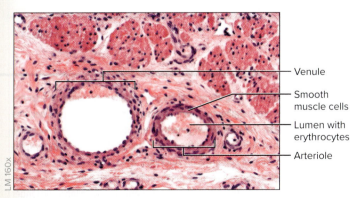

Figure 22.23 **Arteriole.** The tunica media of the arteriole contains circular smooth muscle that functions to alter the diameter of the vessel.

©McGraw-Hill Education/Al Telser

2. Identify the following on the slide of the arteriole, using figure 22.23 and tables 22.2 and 22.3 as guides:

- ☐ **Lumen**
- ☐ **Smooth muscle**

3. Sketch an arteriole as seen through the microscope in the space provided or on a separate sheet of paper.

_____ ×

PHYSIOLOGY

Blood Pressure and Pulse

The blood vessels observed in the histology and gross anatomy exercises in this chapter are the conduits that provide the passageways for blood to flow through the body. As the blood flows through these vessels, it exerts force on the vessel walls. This force (per unit area) is termed **blood pressure**. In both the systemic and the pulmonary circulations, blood flow is driven by blood pressure. Specifically, there must be a pressure gradient to drive the flow from the heart through the vessels. Contraction of the ventricles of the heart generates this pressure gradient. It follows, then, that the greater the blood pressure gradient, the greater the blood flow through the vessels. Significant drops in blood pressure may lead to insufficient oxygen and nutrient delivery to the tissues. Without sufficient oxygen and nutrients, tissue death (necrosis) may occur.

Blood pressure within blood vessels is pulsatile (not smooth); that is, blood pressure waxes and wanes with each heartbeat (cardiac cycle). Pressure is highest as the ventricles contract **(systole)** and lowest when the ventricles relax **(diastole)**. In a clinical setting, we refer to these two pressures (maximum and minimum) as the **systolic** and **diastolic** pressures. A clinically "normal" or "safe" blood pressure is 110/70 mm Hg (110 "over" 70), where 110 corresponds to the systolic pressure and 70 corresponds to the diastolic pressure. A clinically relevant pressure is the **pulse pressure**, or the difference between the systolic blood pressure and diastolic blood pressure. For the blood pressure of 110/70 mm Hg, the pulse pressure would be 40 mm Hg (110 mm Hg – 70 mm Hg = 40 mm Hg). The expansion and recoil of an artery associated with the pulse pressure can be palpated in arteries that are close to the body's surface—locations called **pulse points**. This is referred to simply as "taking a pulse." Counting the number of "pulses" in an artery per minute is an indirect measure of heart rate. Another clinically significant measurement is the **mean arterial blood pressure (MAP)**, which is the average pressure exerted on the arterial blood vessels over time. Because the ventricles spend significantly more time in relaxation than in contraction during a single cardiac cycle, the MAP is closer in magnitude to the diastolic blood pressure. Mathematically, MAP can be calculated as follows:

$$\text{MAP} = 1/3 \text{ pulse pressure} + \text{diastolic pressure}$$

so for a blood pressure of 110/70, the MAP would be roughly 83 (40/3 + 70 = 83).

Blood pressure can also be measured indirectly with the **auscultatory method** (*auscultatare*, to listen), which involves the use of a stethoscope (*stetho-*, chest + *skop*, to look at) and **sphygmomanometer** (*sphygmo-*, pulsation + *manometer*, sparse measure). The sounds heard are referred to as **Korotkoff sounds**. To measure blood pressure using the auscultatory method, an inflatable cuff is placed around the arm and inflated until the pressure in the cuff exceeds systolic blood pressure in the brachial artery (see figure 22.27). This causes arterial blood flow in the artery to cease. Air is then released from the cuff through a valve, causing the pressure in the cuff to drop. When pressure in the the cuff falls below systolic blood pressure, blood momentarily pushes open the artery and flows through. This creates a sharp tapping sound, the first Korotkoff sound. The pressure at which the first Korotkoff sound is heard is an approximate measure of systolic blood pressure. As pressure continues to drop in the cuff, blood pressure is able to keep the brachial artery open for longer periods of time with each subsequent cardiac cycle. Each time the artery closes (during diastole), the blood flow becomes turbulent (*turbulentus*, restless), and creates a noise that can be heard through the stethoscope. Sounds are audible until the pressure in the cuff drops below the diastolic blood pressure, at which time the blood flow is smooth (no detectable turbulence). The pressure at which there is a cessation (stopping) of the sounds is an approximate measure of diastolic blood pressure.

INTEGRATE

CONCEPT CONNECTION

Blood flow is directly proportional to *changes* in pressure and inversely proportional to the resistance of the vessels.* Thus, as blood pressure increases, blood flow increases, and as blood pressure decreases, blood flow decreases. In contrast, as resistance increases, blood flow decreases, and as resistance decreases, blood flow increases. The following equation shows this relationship: $F = \Delta P/R$, where F is flow, ΔP is a change in pressure, and R is the resistance. Resistance is a property of both the blood and the blood vessels and is influenced by three main factors: viscosity of the blood, length of the vessel, and diameter of the vessel. **Viscosity** is a measure of the "thickness" of the blood. Molasses, for instance, has a much higher viscosity than water. In blood, the number of formed elements, particularly red blood cells, is the major factor that affects blood viscosity. For instance, a patient with an elevated hematocrit would have an increase in blood viscosity, thereby increasing the resistance in the vessels. In addition, increasing the length of the vessels, as in the case of obesity, increases resistance.

While viscosity and length influence overall resistance, they are factors that do not change on a minute-to-minute basis. The most dramatic way to alter the resistance of a vessel is to alter the diameter of the lumen of the vessel.

Specifically, decreasing the diameter of lumen of the vessel dramatically increases the resistance. Imagine water flowing through a garden hose (small diameter) versus a fire hose (large diameter). The flow of water out of the fire hose is much greater than that out of the garden hose. In other words, an increase in the diameter of the blood vessel (vasodilation) causes a decrease in resistance and thus an increase in blood flow. Changing blood flow to a specific organ (like a muscle) can be accomplished by changing the diameter of the arteriole that feeds that organ. Recall the smooth muscle in the tunica media of arterioles. When the smooth muscle contracts, it closes down the arteriole, dramatically increasing resistance and decreasing blood flow.

This relationship is also important when looking at the blood vasculature system as a whole. When systemic blood pressure drops (e.g., in an accident victim who loses a great deal of blood), the sympathetic nervous system stimulates widespread constriction of blood vessels all over the body. This causes an increase in total peripheral resistance, which helps to increase blood pressure.

*In a direct relationship, when variable A increases, variable B also increases. In an inverse relationship, when variable A increases, variable B decreases.

PhILS

Ph.I.L.S. LESSON 27: ECG AND FINGER PULSE

In this laboratory exercise, blood flow and pulse will be monitored in a virtual subject, and the subject's heart sounds and ECG will then be related to the events of the cardiac cycle.

Prior to beginning the experiment, familiarize yourself with the following concepts (use the main textbook as a reference):

- **Cardiac cycle**
- **Systemic circuit of blood flow**
- **Pulsatile flow of blood**
- **Arterial distensibility**

1. Open Ph.I.L.S. Lesson 27: ECG and Finger Pulse.

2. Read the objectives and introduction. Then take the pre-lab quiz. The laboratory exercise will open when the pre-lab quiz has been completed (**figure 22.26**).

3. Click to the Sound Test button in the middle of the window to check if the sound is working. Heart sounds should be audible; if not, contact the instructor or campus internet/computer support for assistance. Click End Test.

Figure 22.26 **Opening Screen for Ph.I.L.S. Lesson 27: ECG and Finger Pulse.**

©McGraw-Hill Education

4. Connect the finger pulse unit to the recording input by dragging the orange plug to input 2 on the data acquisition unit (DAQ).

5. Click and drag the finger pulse unit to the fingers of the subject's left hand.

6. Now, connect the ECG electrodes by clicking and dragging the blue plug to recording input 1 on the DAQ.

7. Connect the ECG recording electrodes to the subject by clicking and dragging the black electrode to the subject's left wrist, the red electrode to the subject's right wrist, and the green electrode to the subject's left ankle.

8. To record and measure ECG, pulse, and heart sounds, click the Start button on the control panel to begin recording. On the recording, the ECG is shown at the top, the finger pulse appears in the middle, and the heart sounds are displayed on the bottom. Click the Stop button on the control panel after at least six heartbeat cycles.

9. Now compare the ECG with the finger pulse. Measure the time interval between an R wave and the peak of the pulse recording. Move the cursor to the top of the peak of the R wave. Then, move the other cursor over the peak of the pulse wave. The time interval, represented by the T1–T2 value, is displayed in the data window on the control panel. Enter the data in the Journal by clicking the Journal button in the lower-right corner of the screen. Enter the data below:

   Time Interval: _____

10. Open the post-lab quiz & lab report by clicking on it. Answer the post-lab questions that appear on the computer screen. Click Print Lab to print a hard copy of the report or click Save PDF to save an electronic copy of the report.

11. Make a note of any pertinent observations here:

   _____

   _____

   _____

EXERCISE 22.17

BLOOD PRESSURE AND PULSE USING A STANDARD BLOOD PRESSURE CUFF

CAUTION:

⚠️ Cuff pressure should not exceed 160 mm Hg. Cuff pressures greater than 120 mm Hg should not be maintained for longer than 1 minute. Ensure that each subject is in good health and that he/she has refrained from activities or substances that may elevate heart rate and/or blood pressure (e.g., caffeine, exercise, smoking) at least 1 hour prior to testing.

Obtain the Following:

- **alcohol swabs**
- **sphygmomanometer**
- **stethoscope**

1. To clean the stethoscope, wipe the earpieces and diaphragm with alcohol swabs.

2. Allow the subject's right arm to rest at heart level. The subject should remain as relaxed as possible throughout the exercise.

3. Locate the subject's radial artery (figure 22.13a) on the anterior surface of the right wrist.

4. Count the subject's pulse rate for 1 minute. Record this value in **table 22.5**.

5. Locate the right brachial artery, approximately 1.5 to 2 inches above the antecubital fossa (figure 22.13a). Wrap the cuff of the sphygmomanometer evenly and snugly around the subject's right arm such that the lower edge of the cuff is directly over the brachial artery, and attach Velcro® to hold in place **(figure 22.27)**. Ensure that the tubing and cables are not tangled or pinched.

6. Close the valve and inflate the cuff. Observe the sphygmomanometer while inflating the cuff, and record the pressure at which the subject's pulse disappears. This corresponds to the systolic blood pressure. Deflate the cuff by opening the valve and letting air escape.

7. Firmly press the diaphragm of the stethoscope over the right brachial artery. Reinflate the cuff until reaching a pressure that is approximately 30 mm Hg greater than the systolic pressure observed in step 6.

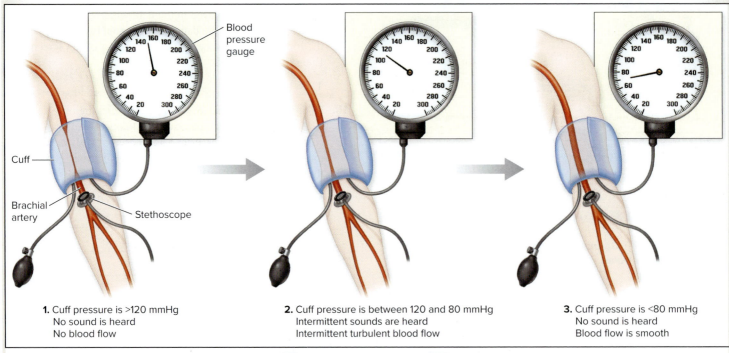

1. Cuff pressure is >120 mmHg
No sound is heard
No blood flow

2. Cuff pressure is between 120 and 80 mmHg
Intermittent sounds are heard
Intermittent turbulent blood flow

3. Cuff pressure is <80 mmHg
No sound is heard
Blood flow is smooth

Systolic pressure
- Top number of a blood pressure reading
- Pressure in the arteries when the ventricles contract
- Recorded when the first pulsating sound is heard (when pressure in the brachial artery is greater than pressure in the cuff, reestablishing blood flow)

Diastolic pressure
- Bottom number of a blood pressure reading
- Pressure in the arteries when the ventricles relax
- Recorded when sounds are no longer heard (because pressure in the cuff is no longer compressing the artery)

Figure 22.27   Blood Pressure. Blood pressure is commonly taken using a sphygmomanometer placed around the brachial artery and a stethoscope placed on the brachial artery in the cubital fossa. Based on when sounds are first heard (systolic pressure) and first disappear (diastolic pressure), a reading is recorded.

8. Deflate the cuff at a rate of 2 to 3 mm Hg per second while listening for the first Korotkoff sound from the brachial artery. Make note of the pressure at which the first Korotkoff sound is heard. This is the systolic pressure. Continue to deflate the cuff while listening for the sounds to disappear. Make note of the pressure at which the sound can no longer be heard. This is the diastolic pressure. Record both the systolic blood pressure and the diastolic blood pressure in **table 22.6**.

9. Deflate the cuff completely.

10. Repeat this process in each arm (pulse rate: steps 3–4, blood pressure: steps 5–9). Allow the subject to rest for 2 to 3 minutes between recordings.

11. Repeat this process for the following three conditions: 1. after allowing the subject to lie down for 3 to 5 minutes; 2. after allowing the subject to stand for 3 to 5 minutes; 3. after having the subject perform vigorous exercise for 3 to 5 minutes. Record one trial using the left arm and the second using the right arm (or vice versa) (pulse rate: steps 3–4; blood pressure: steps 5–9). Record the results in table 22.6.

12. Make a note of any pertinent observations here:

_____

_____

_____

Table 22.5	Pulse and Blood Pressure Readings While Sitting			
	Sitting			
	Trial 1		**Trial 2**	
	Right Arm	**Left Arm**	**Right Arm**	**Left Arm**
Pulse Rate				
Systolic Blood Pressure				
Diastolic Blood Pressure				

Table 22.6	Pulse and Blood Pressure Readings While Lying Down and Standing, and After Vigorous Exercise					
	Lying Down		**Standing**		**Exercise**	
	Right Arm	**Left Arm**	**Right Arm**	**Left Arm**	**Right Arm**	**Left Arm**
Pulse Rate						
Systolic Blood Pressure						
Diastolic Blood Pressure						

EXERCISE 22.18

BIOPAC LESSON 16: BLOOD PRESSURE

CAUTION:

⚠ Cuff pressure should not exceed 160 mm Hg. Cuff pressures greater than 120 mm Hg should not be maintained for longer than 1 minute. Ensure that each subject is in good health and that he/she has refrained from activities or substances that may elevate heart rate and/or blood pressure (e.g., caffeine, exercise, smoking) at least 1 hour prior to testing.

Obtain the Following:

- **BIOPAC blood pressure cuff (SS19L) that fits securely around the subject's arm**
- **BIOPAC MP36/35 recording unit**
- **Laptop computer with BSL 4 software installed**
- **BIOPAC electrode lead set (SS2L)**
- **BIOPAC general purpose electrodes (EL503)**
- **BIOPAC stethoscope (SS30L)**

1. Ensure that all the air has been expelled from the blood pressure cuff by turning the release valve counterclockwise. Turn the valve clockwise to close.

(continued on next page)

(continued from previous page)

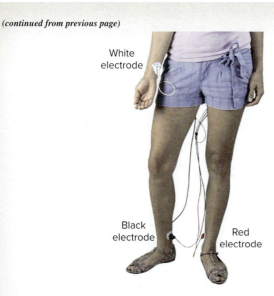

White electrode

Black electrode

Red electrode

Figure 22.28 **Lead Placement for ECG Recording.**
Courtesy of and ©BIOPAC Systems, Inc. www.biopac.com

2. Prepare the BIOPAC equipment by turning the computer ON and the MP36/35 unit OFF. Plug in the following: BP cuff (SS19L/LA/LB)-CH 1, stethoscope (SS30L)-CH 2, and electrode lead set (SS2L)-CH 3. Turn the MP36/35 unit ON.

3. Clean the stethoscope earpieces and diaphragm using alcohol swabs. Prepare the subject for placement of electrodes by cleaning the skin where the electrodes will be placed. To do this, rub the indicated area vigorously with an alcohol swab. First, clean the skin on the medial surface of each leg, just above the ankle. Next, clean the skin on the anterior surface of the right anterior forearm at the wrist. Note that skin that is free of hair will make the best contact with the electrodes.

4. Have the subject sit in a relaxed position. Apply a small quantity of electrode gel to the center of the electrodes, and place the electrode firmly on the skin in the designated locations: right arm, right leg, and left leg. Wait approximately 5 minutes before attaching the leads (step 5) to give the electrodes a chance to affix firmly to the skin.

5. Attach the electrode leads (SS2L) to the electrodes, following the color scheme outlined in **figure 22.28**. Attach the WHITE lead (VIN-) to the electrode on the right forearm. Attach the BLACK (GND) lead to the electrode on the medial surface of the subject's right leg. Attach the RED (VIN+) lead to the electrode on the medial surface of the subject's left leg.

6. Start the BIOPAC Student Lab Program and choose Lesson 16—Blood Pressure and click OK. Enter a file name.

7. To calibrate, ensure that the cuff is fully deflated. Click Calibrate in the upper left corner of the Setup window. Inflate the cuff to 100 mm Hg and click OK. Deflate the cuff to 40 mm Hg. Note that cuff pressure should be released at a rate of 2–3 mm Hg per second. In 20 to 30 seconds, the pressure should drop approximately 60 mm Hg. Click OK. Once calibration data have started recording, tap the stethoscope diaphragm twice and wait for the calibration to stop. Check the calibration data to ensure that two sounds are recorded in the middle box, that the ECG is visible in the bottom box, and that the calibration data resemble **figure 22.29**. If so, click Continue to proceed with Data Recording; otherwise, click Redo Calibration.

8. Allow the subject's left arm to rest at heart level and position the "artery" label over the subject's brachial artery, approximately 1.5–2 inches above the antecubital fossa. Wrap the cuff of the sphygmomanometer evenly and snugly around the subject's left arm such that the lower edge of the cuff is directly over the brachial artery, and attach Velcro® to hold in place. Ensure that the tubing and cables are not tangled or pinched. Position the pressure dial indicator such that it can be visualized easily. Firmly press the stethoscope diaphragm over the brachial artery.

9. OBTAIN VALUES FOR SITTING: Inflate the cuff to 160 mm Hg. Click Record. Release pressure at a rate of 2–3 mm Hg/second and insert an event marker by clicking F4 when the Korotkoff sound for systolic pressure is heard. Continue listening and note the time

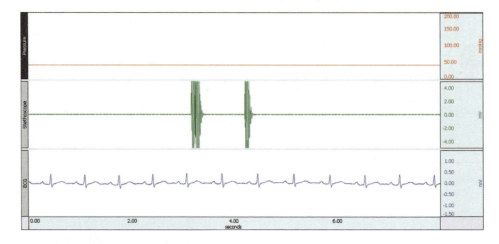

Figure 22.29 **Calibration Data for ECG Recording.**
Courtesy of and ©BIOPAC Systems, Inc. www.biopac.com

(F5) when sounds are no longer heard. Click Suspend. Fully deflate the cuff. If incorrect, click Redo. Otherwise, click Continue.

10. Repeat step 9 for the left arm "seated" condition. Remove the cuff and place on the subject's right arm. Repeat step 9 two times (Trials 1 and 2) for the right arm "seated" condition.

11. OBTAIN VALUES FOR SUPINE: Have the subject lie down and relax. Repeat step 9 two times (Trials 1 and 2) for the right arm "supine" condition.

12. OBTAIN POST-EXERCISE VALUES: Disconnect the electrode lead cables from the subject by releasing the electrodes at the metal clip. Have the subject perform an exercise to elevate her heart rate (e.g., running or jumping jacks). Reattach the electrode lead cables when exercise is complete. Click Continue.

13. ANALYZE THE DATA: If analyzing data that were just collected, click on "Analyze current data file." To perform the data analysis on data that were collected previously, click "Review saved data." Note that CH 1 displays pressure (mm Hg), CH 2 displays the stethoscope readings (mV), and CH 3 displays ECG (mV).

14. To prepare to analyze the data, do the following:

    a. Use the zoom tool for optimal viewing of the first recording.

    b. Set up measurement boxes at the top of the screen using the drop-down menus. Set CH 1 to Value, set CH 2 to BPM, and set CH 3 to Delta T.

15. Locate the I-beam cursor. Select the point on the recording segment that corresponds to the first event marker, or systolic blood pressure. Obtain the amplitude in the Value measurement box, and record this value in **table 22.7**. Repeat for each event marker. Calculate the averages for each condition and record in table 22.7.

16. Use the I-beam tool to select an area from one R wave to the next R wave. View the BPM measurement box and record the value in **table 22.8**. Repeat for two additional R waves. Calculate the average BPM for three cycles within each trial, and calculate the average across trials for each condition. Record the answers in table 22.8. Use these values, and those obtained in step 14, to calculate the mean arterial pressures and pulse pressures. Record these answers in **table 22.9**.

17. Zoom in on one ECG trace between the systolic and diastolic pressure. Use the I-beam tool to select the area from the peak of the R wave to the beginning of the sound detected by the stethoscope. Record the Delta T. Zoom out, locate the next recording segment, and repeat the measurement. Repeat for each recording segment and record the values in **table 22.10**.

18. Make a note of any pertinent observations here:

    _____

    _____

    _____

    _____

Table 22.7	Systolic and Diastolic Pressures				
		Systolic Pressure (mm Hg)		**Diastolic Pressure (mm Hg)**	
Condition	**Trial**	**Sound Detection**	**Sound Average (calculate)**	**Sound Detection**	**Sound Average (calculate)**
Left Arm, Seated	1				
	2				
Right Arm, Seated	1				
	2				
Right Arm, Supine	1				
	2				
Right Arm, After Exercise	1				
	2				

Source: BIOPAC Systems, Inc. Data from BIOPAC Tables 16.2, 16.4.

(continued on next page)

4. Label the diagram with the appropriate vein names. **3** – **14**

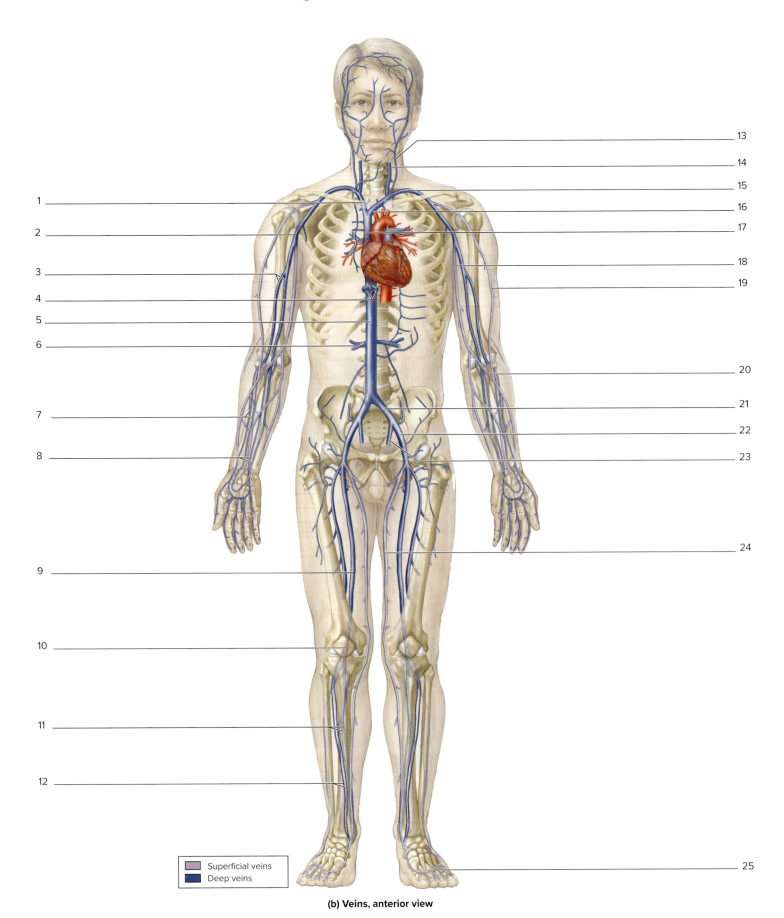

(b) Veins, anterior view

5. Which of the following are veins that compose the hepatic portal circulation? (Check all that apply.) **9**

_____ a. celiac trunk

_____ b. inferior mesenteric vein

_____ c. inferior vena cava

_____ d. splenic vein

_____ e. superior mesenteric vein

Exercise 22.9: Fetal Circulation

6. In the fetus, the umbilical vein carries _____ (oxygenated/deoxygenated) blood, whereas the umbilical arteries carry

_____ (oxygenated/deoxygenated) blood. **15**

7. In fetal circulation, oxygenated blood flows from the placenta, to the umbilical _____ (arteries/vein), to the right atrium of the heart. Blood

returns to the placenta via the umbilical _____ (arteries/vein). **16**

8. When referencing a structure present in the interatrial septum, the prenatal structure is called the _____ (foramen ovale/fossa ovalis),

whereas the postnatal structure is called the _____ (foramen ovale/fossa ovalis). **17**

Exercise 22.10: Blood Vessel Wall Structure

9. The tunic in blood vessels that is composed of simple squamous epithelium (or endothelium) and a subendothelial layer composed of areolar connective tissue is

known as the tunica _____ (interna/externa). **18** **19**

Exercise 22.11: Elastic Artery—The Aorta

10. Which of the following statements correctly describes an elastic artery? (Check all that apply.) **20**

_____ a. contains an internal elastic lamina

_____ b. contains multiple layers of smooth muscle

_____ c. tunica media contains numerous elastic fibers

_____ d. typical diameters are 1–2.5 cm in diameter

_____ e. very thin with no smooth muscle cells

11. The ability of an elastic artery to expand as it receives blood from the ventricles is due to _____ (collagen/elastic) fibers. **21**

12. Large arteries such as the aorta are nourished by tiny blood vessels called _____ (arterioles/vasa vasorum). **22**

Exercise 22.12: Muscular Artery

13. The amount of smooth muscle in the tunica media of a muscular artery is _____ (greater/less) than that of elastic arteries. **23**

14. Elastic arteries are typically located closest to the heart where pressures are _____ (highest / lowest). **24**

Exercise 22.13: Arteriole

15. Constriction of arterioles _____ (decreases/increases) blood flow into capillary beds. **25**

Exercise 22.14: Vein

16. Veins are considered _____ (high/low) pressure vessels. **26**

17. A venous valve is formed by an infolding of the tunica _____ (intima/media/externa). **27**

Exercise 22.15: Capillaries

18. Rank the following types of capillaries from most to least permeable. **28**

_____ a. continuous

_____ b. fenestrated

_____ c. sinusoidal

Exercise 22.16: Ph.I.L.S. Lesson 27: ECG and Finger Pulse

19. Blood is propelled forward through vessels during _____ (diastole/systole). **29**

20. An increase in heart rate should correspond to a(n) _____ (increase/decrease) in an R-R interval and a(n) _____

    (increase/decrease) in pulse rate. **30**

Exercise 22.17: Blood Pressure and Pulse Using a Standard Blood Pressure Cuff

21. When measuring blood pressure using a blood pressure cuff, the first Korotkoff sound corresponds to _____ (diastolic/systolic) pressure,

    whereas the second Korotkoff sound corresponds to _____ (diastolic/systolic) pressure. **31** **33**

22. The blood pressure changes that occur during the cardiac cycle are detected by measuring the pulse in the _____

    (internal jugular vein/radial artery). **32**

Exercise 22.18: BIOPAC Lesson 16: Blood Pressure

23. Mean arterial pressure (MAP) is closer in value to a subject's _____ (diastolic/systolic) blood pressure. **34**

24. Vigorous exercise _____ (decreases/increases) systemic blood pressure. **35**

Can You Apply What You've Learned?

25. Explain why large blood vessels have their own blood vessels (vasa vasorum) in the tunica externa.

    _____

    _____

26. A physician wishes to place a balloon catheter into the left coronary artery of his patient. A catheter is placed into the femoral artery just inferior to the inguinal ligament, and is threaded backward through the arterial system until it reaches the ascending aorta. From there, the catheter will be threaded into the left coronary artery. List all of the arteries, in order, that the catheter will pass through as it travels from the femoral artery to the left coronary artery.

    Femoral artery ⟶ _____ ⟶ _____ ⟶ _____ ⟶ _____ ⟶ _____ ⟶ _____ ⟶ Left
    coronary artery

27. A physician wishes to place a central line (a catheter used to repeatedly administer drugs such as chemotherapy drugs into a patient's circulatory system) into the right atrium of the patient's heart. To do this, the physician will place the catheter into the basilic vein just superior to where it branches from the median cubital vein. The physician will then guide the catheter along the venous system until it reaches the right atrium. List, in order, the veins through which the central line passes as it travels from the basilic vein to the right atrium of the heart.

    Basilic vein ⟶ _____ ⟶ _____ ⟶ _____ ⟶ _____ ⟶ Right atrium

28. After the central line was placed (question 27), it pinched off and failed to work when the patient held a heavy weight in her hand and the inferior movement of her clavicle compressed the central line within the subclavian vein. The physician decided to remove the central line from the basilic vein and place another central line into the internal jugular vein (IJV). The IJV provides a more direct route to the heart and avoids the problem of catheter pinch-off. List the veins, in order, through which the central line passes as it travels from the internal jugular vein to the right atrium of the heart.

    IJV ⟶ _____ ⟶ _____ ⟶ _____ ⟶ Right atrium

29. Discuss why blood pressure might differ in the right and left arms.

    _____

    _____

30. As we age, our arteries become stiff and less flexible. Discuss what might happen to resistance and blood flow through the arteries in aging adults.

    _____

    _____

Can You Synthesize What You've Learned?

31. Local factors are important for regulating blood flow to tissue capillary beds. For example, in skeletal muscle tissue that is actively contracting (as during exercise), levels of oxygen start to decrease as the oxygen is utilized, and levels of carbon dioxide increase as it is produced during cellular respiration. Would increasing levels of carbon dioxide in the tissues cause the smooth muscle forming the sphincters in arterioles to contract or relax? Justify your answer.

_____

_____

_____

32. Specialized capillary beds are located throughout the body and are adapted for their specific physiological function. Recall from chapter 14 that the blood vessels that supply nervous tissues of the brain are surrounded by glial cells called astrocytes, which collectively form the blood-brain barrier. Similarly, blood vessels that form the choroid plexus within the brain ventricles are surrounded by glial cells called ependymal cells, which collectively form the blood-cerebrospinal fluid (CSF) barrier. What type of capillary (continuous, fenestrated, or sinusoids) would you expect to find in each of these areas of the brain, and why?

_____

_____

_____

33. Given what you have just learned about the contents of the blood entering the liver from the hepatic portal system, discuss the type of capillaries located within the liver, and state why these capillaries might be advantageous given their structure and function.

_____

_____

_____

34. Preeclampsia is a condition that can occur in pregnant women that results in elevated blood pressure. While the origin of preeclampsia is unclear, there is evidence that substances that cause vasoconstriction may be released in the blood. Describe how these substances will increase blood pressure. Then discuss why this could be life threatening for both mother and child.

_____

_____

_____

35. A patient enters a clinic for a routine physical examination. The nurse assigned to the patient begins to record the patient's blood pressure using a sphygmomanometer and stethoscope. As the nurse inflates the cuff to 160 mm Hg, there is an emergency down the hall that requires her attention. Rather than releasing the valve, she keeps the cuff inflated. Describe the risks to the patient that are involved when leaving the cuff inflated for longer than 1 minute.

_____

_____

_____

36. Dan is diagnosed with atherosclerosis and coronary artery disease. He undergoes coronary bypass surgery to restore blood flow to his heart. Discuss why doctors might suggest that Dan limit vigorous exercise during his recovery.

_____

_____

_____

The Lymphatic System and Immunity

OUTLINE AND LEARNING OBJECTIVES

GROSS ANATOMY   622

EXERCISE 23.1: GROSS ANATOMY OF LYMPHATIC STRUCTURES   622

1 Identify gross anatomical lymphatic structures on classroom models

HISTOLOGY   625

Lymphatic Vessels   625

EXERCISE 23.2: LYMPHATIC VESSELS   625

2 Identify lymphatic vessels and valves when viewed through a dissecting microscope

3 Identify lymphatic vessels and valves when viewed through a compound microscope

4 Compare and contrast the structure and function of a lymphatic vessel with the structure and function of a vein

Mucosa-Associated Lymphatic Tissue (MALT)   626

EXERCISE 23.3: TONSILS   627

5 Identify the three tonsils and their characteristic structures when viewed through a microscope

6 Distinguish pharyngeal, palatine, and lingual tonsils from each other histologically

7 Describe the structure and function of a tonsil

EXERCISE 23.4: PEYER PATCHES   628

8 Identify Peyer patches when viewed through a microscope

9 Describe the structure and function of a Peyer patch

EXERCISE 23.5: THE VERMIFORM APPENDIX   629

10 Identify the vermiform appendix and its structures when viewed through a microscope

11 Describe the structure and function of the vermiform appendix

Lymphatic Organs   629

EXERCISE 23.6: THE THYMUS   629

12 Identify the thymus gland and its structures when viewed through a microscope

13 Describe the structure and function of the thymus

EXERCISE 23.7: LYMPH NODES   631

14 Identify the parts of a lymph node and its structures when viewed through a microscope

15 Describe the structure and function of a lymph node

16 Trace the flow of lymph through a lymph node

EXERCISE 23.8: THE SPLEEN   634

17 Identify the spleen and its structures when viewed through a microscope

18 Explain the structural and functional differences between red pulp and white pulp of the spleen

PHYSIOLOGY   637

EXERCISE 23.9: A CLINICAL CASE STUDY IN IMMUNOLOGY   637

19 Discuss the role of lymphatic organs, innate defenses, and adaptive defenses in fighting foreign pathogens

20 Apply concepts from the nervous, endocrine, cardiovascular, lymphatic, and immune systems to a clinical case

Anatomy & Physiology
Revealed 4.0

Module 10: Lymphatic System

INTRODUCTION

When a capillary bed is perfused with blood, a small amount of fluid is lost into the interstitial (tissue) space. Over the course of a day, given the many capillary beds in the body, upward of 3 liters of fluid may be lost. Gone uncorrected, this would negatively affect fluid balance, blood volume, and blood pressure.

Fortunately, lymphatic vessels are closely associated with blood capillary beds, and these vessels reabsorb this lost fluid, now called lymph, and return it to the heart. The terms *lymph* and *lymphatic* come from the Latin word *lympha*, which means "pure spring water." Although lymph is not exactly comparable to spring water in its composition, the fluid is relatively clear and normally free of suspended material because it contains no red blood cells and only small amounts of plasma proteins.

Returning the fluid to circulation ensures that blood volume and pressure are unaffected by this type of fluid loss. This might seem like a strange mechanism to maintain fluid balance. However, there is a crucial benefit: On the way back to the heart, lymph is screened by specialized tissues and organs that contain immune cells capable of defending against foreign invaders. Therefore, the lymphatic system functions not only to regulate fluid balance and volume but also to detect and defend against foreign invaders wherever they are located in the body.

Because most lymphatic structures are anatomically small and difficult to see in a human cadaver or on isolated organ preparations, the majority of the exercises in this chapter involve histological observations of lymphatic tissues and organs. Histological observations will be complemented by observations of classroom models that show lymphatic organs such as lymph nodes, the spleen, and the thymus. If human cadavers are used in the laboratory, most lymphatic structures will be studied on the cadaver in conjunction with other organ systems, such as the respiratory and digestive systems, rather than within this laboratory exercise. Exercises within this chapter will also explore the functional relevance of lymphatic tissues and lymph, with a review of the immune system and the cell types involved in eliminating infectious agents.

List of Reference Tables

These Pre-Laboratory Worksheet questions may be assigned by instructors through their ▣ connect course.

1. Which of the following is *not* a function of the lymphatic system? (Circle one.)

   a. absorption of dietary fats

   b. production and proliferation of lymphocytes

   c. production of erythrocytes

   d. transport excess interstitial fluid back into the blood

2. Which of the following is *not* a named aggregation of mucosa-associated lymphatic tissue (MALT)? (Circle one.)

   a. lymph node

   b. palatine tonsil

   c. Peyer patch

   d. pharyngeal tonsil

   e. vermiform appendix

3. The structure of lymphatic vessels most closely resembles _____ (arteries/veins).

4. The right lymphatic duct drains lymph from the (Circle all that apply):

   a. right side of the head.

   b. right side of the thorax.

   c. right upper limb.

   d. right lower limb.

5. Which of the following is a lymphatic organ that filters blood?

   a. lymph node   b. Peyer patch   c. spleen   d. thymus   e. vermiform appendix

6. Lymphatic vessels contain _____ (valves/ducts) that prevent the backflow of lymph.

7. Which of the following is a lymphatic organ located in the thoracic cavity just deep to the sternum?

   a. Peyer patch   b. spleen   c. thymus   d. tonsil   e. vermiform appendix

8. The basic unit of lymphatic tissue is called a _____ (nodule/Peyer patch).

9. Which of the following is a lymphatic organ that has both afferent and efferent lymphatic vessels? (Circle one.)

   a. lymph node   b. spleen   c. thymus   d. tonsil   e. vermiform appendix

10. Which of the following takes place in a primary lymphoid organ? (Circle one.)

    a. development of lymphocytes

    b. activation of lymphocytes

    c. filtration of lymph

    d. trapping of pathogens

(continued from previous page)

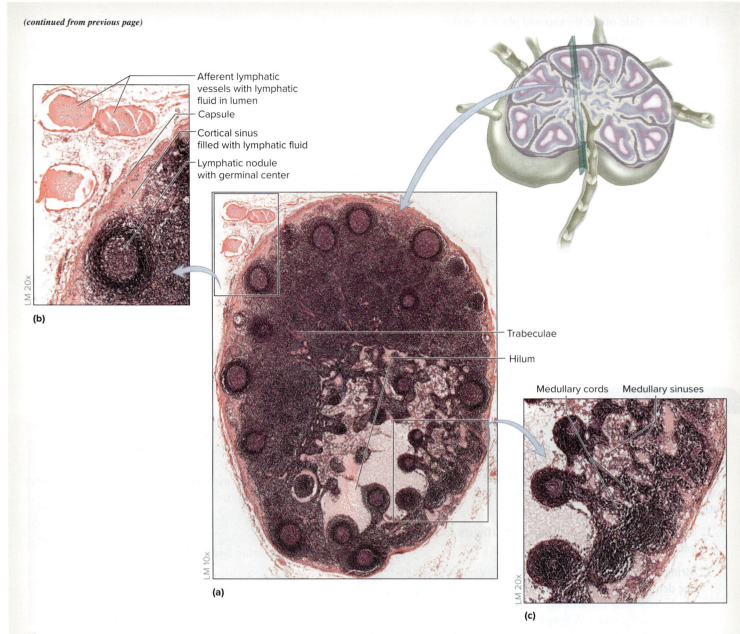

Afferent lymphatic vessels with lymphatic fluid in lumen

Capsule

Cortical sinus filled with lymphatic fluid

Lymphatic nodule with germinal center

LM 20x

(b)

Trabeculae

Hilum

Medullary cords   Medullary sinuses

LM 20x

(c)

LM 10x

(a)

Figure 23.12 **Lymph Node.** (*a*) Photomicrograph of a cross section of a lymph node that has been sectioned through the hilum. On the surface opposite the hilum, cross sections of three afferent lymphatic vessels are visible, as well as the capsule and a cortical sinus filled with lymphatic fluid (*b*). Medullary cords and medullary sinuses are visible in (*c*).

©McGraw-Hill Education/Christine Eckel

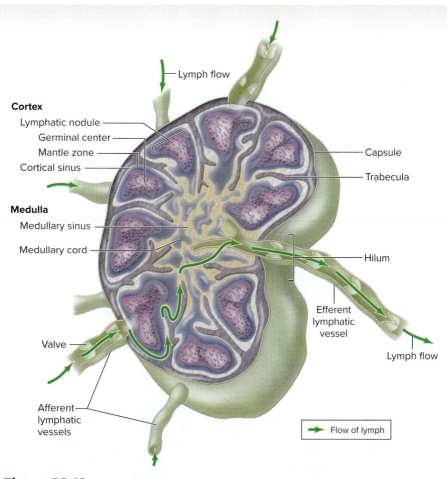

Figure 23.13 Lymph Node. Pathway that lymph takes as it flows through a lymph node.

Table 23.4	Parts of a Lymph Node	
Structure	**Description**	**Word Origin**
Afferent Lymphatic Vessel	Vessel that transports lymph into a lymph node	*affero,* to bring to
Capsule	Dense irregular connective tissue that surrounds the entire lymph node to support and protect it	*capsula,* a box
Cortex	Outer portion of the lymph node that lies deep to the capsule and contains lymphatic nodules	*cortex,* bark
Cortical sinuses	Spaces located between lymphatic nodules and the capsule or the trabecula in a lymph node through which lymphatic fluid flows	*cortex,* bark + *sinus,* hollow curve
Efferent Lymphatic Vessel	Vessel that transports lymph away from the lymph node; located at the hilum	*effero,* to bring out
Hilum	Indented area where efferent lymphatic vessels (and blood vessels, nerves) enter	*hilum,* a small bit
Lymphatic Nodule	Aggregate of B-lymphocytes within a germinal center, surrounded by T-lymphocytes macrophages, and dendritic cells	*lympha,* clear spring water, + *nodulus,* a small knot
Medulla	Inner region of the lymph node containing medullary cords of B-lymphocytes, T-lymphocytes, and macrophages	*medius,* middle
Medullary Cord	Strands of B-lymphocytes, T-lymphocytes, and macrophages supported by connective tissue that are located in the medulla of a lymph node	*medius,* middle, + *chorda,* a string
Medullary Sinuses	Spaces between the medullary cords in a lymph node through which lymphatic fluid flows	*medius,* middle + *sinus,* hollow curve
Trabeculae	Invaginations of the dense irregular connective tissue capsule of a lymph node that partition the cortex into smaller compartments	*trabs,* a beam

(continued on next page)

(continued from previous page)

8. Sketch a lymph node as seen through the microscope in the space provided or on a separate sheet of paper. Label all of the structures listed in steps 5 and 7.

9. Describe the pathway taken by lymph as it flows through a lymph node in the space provided.

_____

_____

_____

_____

_____ ✕

INTEGRATE

CONCEPT CONNECTION

As discussed in chapter 5, reticular tissue is a type of connective tissue found in lymphatic organs, including the spleen and lymph nodes. Recall that reticular tissue is composed of cells and reticular fibers. Reticular fibers are a fine type of collagen that provide a loose supporting framework for the cells and aid in trapping substances. The reticular fibers help to slow the flow of lymph, allowing the contents of the lymph to come in contact with immune cells within the organ. Once pathogens become trapped by the reticular fibers of the lymph nodes, B- and T-cells can launch an immune response. The details of the immune response are discussed in the physiology section of this chapter.

EXERCISE 23.8

THE SPLEEN

The spleen is similar to a lymph node in many ways, with one major exception: The spleen filters *blood*, whereas a lymph node filters *lymph*. However, both organs filter their fluid for the purpose of identifying and eliminating foreign antigens that may be present in the fluid. The spleen receives blood through the **splenic artery**, which is a branch of the celiac trunk. Blood leaves the spleen through the **splenic vein**, which drains into the hepatic portal vein (see figures 22.8 and 22.9).

When a fresh spleen is cut, two different types of tissue are observed: red pulp and white pulp. **Red pulp** consists of splenic sinusoids and is red due to the large amount of blood contained within the sinusoids. Red pulp contains erythrocytes, platelets, macrophages, and B-lymphocytes. It composes the majority of the splenic tissue. **White pulp** consists of aggregations of T-lymphocytes, B-lymphocytes, dendritic cells, and macrophages, and is white due to the white blood cells (leukocytes) within each mass of tissue. Because the spleen tissue sample that will be viewed through the microscope has been stained, red and white pulp will not appear red and white. Instead, the red pulp will usually be reddish-pink, but the white pulp will be darker purple, with a lighter-colored central region.

1. Obtain a slide of the spleen and place it on the microscope stage. Bring the tissue sample into focus on low power and locate the dense irregular connective tissue capsule of the spleen, if present in the section on the slide (**figure 23.14**). Look for several connective tissue trabeculae, which are invaginations of the capsule that separate the spleen into distinct regions.

2. Notice the circular purple aggregations of cells. These aggregations constitute the white pulp of the spleen, which consists of B-lymphocytes, T-lymphocytes, dendritic cells, and some macrophages.

3. Look for a **central artery** within each mass of white pulp. Central arteries receive blood from **trabecular arteries**. Blood coming into the spleen travels from the splenic artery to the trabecular arteries to the central arteries, and finally empties into the sinusoids or surrounding tissues of the spleen before draining into splenic veins. Note that the central artery may not be in the exact center of the white pulp.

4. Move the microscope stage until a mass of white pulp is in the center of the field of view and then change to high power. At this magnification, individual small purple-staining cells, which are lymphocytes, will be visible.

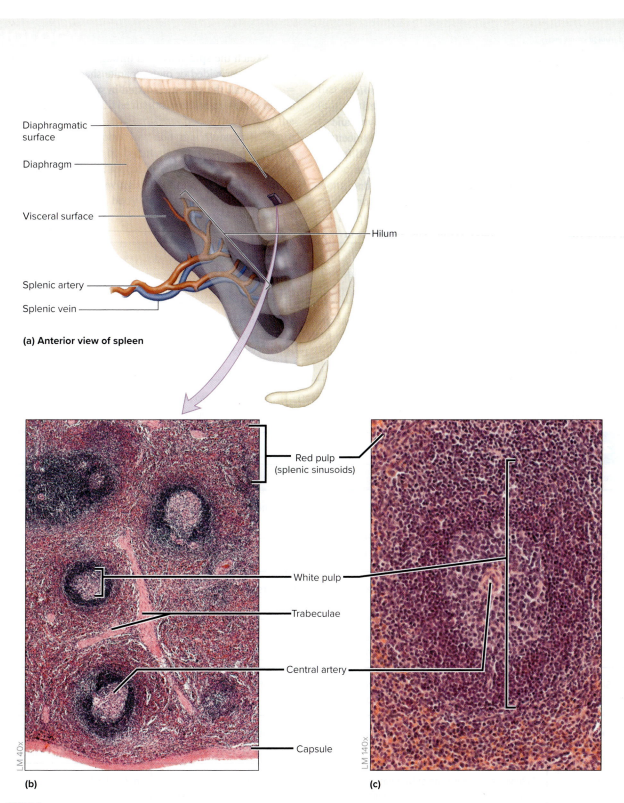

Diaphragmatic surface

Diaphragm

Visceral surface

Hilum

Splenic artery

Splenic vein

(a) Anterior view of spleen

Red pulp (splenic sinusoids)

White pulp

Trabeculae

Central artery

Capsule

LM 40x

(b)

LM 140x

(c)

Figure 23.14 **The Spleen.** The spleen contains two predominant tissues: red pulp, which consists of splenic sinusoids containing red blood cells, and white pulp, which contains lymphocytes. (*a*) Location of spleen. Photomicrograph of the spleen as seen under (*b*) low magnification and (*c*) high magnification.
(*b, c*) ©McGraw-Hill Education/Al Telser

(continued on next page)

(continued from previous page)

exposed to a child who had an upper respiratory tract infection. A buccal swab test was negative for both influenza A and B. Suspecting that Greg was suffering from pneumonia and lymphadenopathy (enlarged lymph nodes), doctors prescribed a broad-spectrum antibiotic.

The antibiotic failed to treat Greg's symptoms. Instead, his symptoms worsened and he began to vomit blood. His pulse remained elevated and his respiratory rate also became elevated. In addition, the amount of fluid in his lungs increased, causing him to experience dyspnea (shortness of breath) and labored breathing. An electrocardiogram revealed sinus tachycardia, and a transthoracic echocardiogram revealed a decreased ejection fraction of his heart. In addition, Greg's lymphadenopathy worsened, and mediastinal and hilar (lung) lymph nodes became enlarged. Greg also suffered from splenomegaly (enlarged spleen). Greg's mean arterial pressure dropped suddenly, and he developed hypoxemia and renal failure. Doctors administered *epinephrine* and high doses of *methylprednisolone* (a glucocorticoid).

Still in search of the pathogen responsible for Greg's rapid decline, doctors placed a bronchoscope into his lungs to obtain a specimen. From the specimen, doctors determined that Greg tested positive for the H1N1 influenza virus (i.e., "swine flu"). Once the diagnosis was confirmed, doctors prescribed an antiviral drug. Unfortunately, Greg's condition continued to decline, and he died 9 days following his onset of symptoms. Greg was one of many victims of the 2009 H1N1 pandemic.

Answer the following questions regarding Greg's case:

1. Describe innate and adaptive defenses that would be employed to prevent entry of a pathogen such as the H1N1 influenza virus into the respiratory tract.

_____

_____

_____

_____

_____

2. Why did Greg suffer from lymphadenopathy, particularly in his mediastinal and hilar lymph nodes, given his diagnosis of H1N1 influenza? Be sure to describe the structure and function of lymph nodes, and describe why some lymph nodes may become enlarged but not others.

_____

_____

_____

_____

_____

3. Why was Greg's spleen enlarged, given his diagnosis of H1N1 influenza? Describe the spleen's role in both cardiovascular and immune function.

_____

_____

_____

_____

_____

4. Justify the administration of *methylprednisolone* given Greg's symptoms. Where are glucocorticoids typically synthesized, and what is their typical function?

_____

_____

_____

_____

_____

_____

Adapted from Case 40-2009, *The New England Journal of Medicine*, 361, no. 26 (2009): 2558–69.

Chapter 23: The Lymphatic System and Immunity

Name: _____

Date: _____   **Section:** _____

The ❶ corresponds to the Learning Objective(s) listed in the chapter opener outline.

  Do You Know the Basics?

Exercise 23.1: Gross Anatomy of Lymphatic Structures

1. Imagine a bacterium has entered a lymph capillary in the right arm. Place the following structures in order as the bacterium travels from a lymph node in the axilla to the right atrium of the heart (1 = first; 5 = last). ❶

   _____ a. efferent lymphatic vessel

   _____ b. right brachiocephalic vein

   _____ c. right lymphatic duct

   _____ d. right subclavian vein

   _____ e. superior vena cava

Exercise 23.2: Lymphatic Vessels

2. The valves located in lymphatic vessels are extensions of the _____ (tunica intima/tunica media). ❷ ❸

3. Lymphatic vessels are most similar in structure to _____ (arteries/veins). ❹

Exercise 23.3: Tonsils

4. Match the description listed in column A with the type of tonsil listed in column B. (Some answers may be used more than once.) ❺ ❻ ❼

   Column A

   _____ 1. covered with pseudostratified ciliated columnar epithelium

   _____ 2. covered with stratified squamous (nonkeratinized) epithelium

   _____ 3. located at the base of the tongue

   _____ 4. located at the junction of the nasopharynx and oropharynx at the posterior aspect of the soft palate

   _____ 5. projects from the roof of the nasopharynx

   Column B

   a. lingual

   b. palatine

   c. pharyngeal

Exercise 23.4: Peyer Patches

5. Peyer patches have a _____ (lighter/darker) staining germinal center where antibody-secreting cells are activated. ❽

6. Peyer patches are a defining characteristic of the: (Circle one.) ❾

   a. tonsil.

   b. lymph node.

   c. ileum.

   d. spleen.

Exercise 23.5: The Vermiform Appendix

7. The vermiform appendix is lined with which type of epithelial tissue? (Circle one.) ❿

   a. pseudostratified ciliated columnar epithelium

   b. simple columnar epithelium

   c. simple cuboidal epithelium

   d. simple squamous epithelium

   e. stratified squamous epithelium

8. An individual suffering from an inflamed appendix will most commonly have pain in the _____ (upper/lower) _____ (left/right) abdominopelvic quadrant. **11**

Exercise 23.6: The Thymus

9. The thymus is divided into lobules by: (Circle one.) **12**

   a. connective tissue septa

   b. medullary sinuses

   c. epithelial derived cells

   d. afferent lymphatic vessels

10. The thymus is the site of _____ (B-lymphocyte/T-lymphocyte) maturation. **13**

Exercise 23.7: Lymph Nodes

11. Match the descriptions listed in column A with the structure listed in column B. **14** **15**

   Column A

   _____ 1. space through which lymph travels within a lymph node

   _____ 2. structure in the medulla of the lymph node that contains T-lymphocytes and macrophages

   _____ 3. vessel bringing lymph into the lymph node

   _____ 4. vessel taking lymph away from the lymph node

   Column B

   a. afferent

   b. efferent

   c. medullary cord

   d. sinus

12. Imagine a virus is circulating in the lymph. Place the following structures in the order in which the virus travels (1 = first; 4 = last). **16**

   _____ a. afferent lymphatic vessel

   _____ b. cortical sinus

   _____ c. efferent lymphatic vessel

   _____ d. medullary sinus

Exercise 23.8: The Spleen

13. Splenic sinusoids filter _____ (blood/lymph). **17**

14. The _____ (white/red) pulp of the spleen contains lymphocytes and serves to fight pathogens, whereas the _____ (white/red) pulp of the spleen contains sinusoids and filters blood. **18**

Exercise 23.9: A Clinical Case Study in Immunology

15. The clinical term for enlarged lymph nodes is _____ (mononucleosis/lymphadenopathy) **19**

16. Which of the following systems would be affected by splenomegaly (enlarged spleen)? (Check all that apply.) **20**

   _____ a. cardiovascular

   _____ b. digestive

   _____ c. lymphatic

   _____ d. respiratory

Can You Apply What You've Learned?

17. Lymphatic vessels are more permeable than blood vessels. How could this increased permeability be advantageous? How could this increased permeability be disadvantageous?

_____

_____

18. Using the information learned in this chapter, complete the empty boxes in the following figure with an appropriate functional description of the structures indicated.

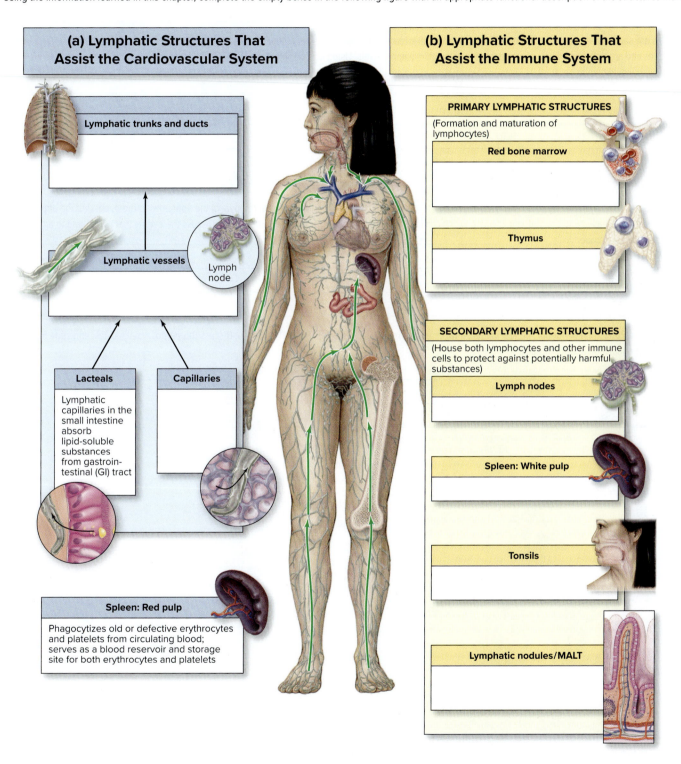

(a) Lymphatic Structures That Assist the Cardiovascular System

Lymphatic trunks and ducts

Lymphatic vessels

Lymph node

Lacteals

Lymphatic capillaries in the small intestine absorb lipid-soluble substances from gastrointestinal (GI) tract

Capillaries

Spleen: Red pulp

Phagocytizes old or defective erythrocytes and platelets from circulating blood; serves as a blood reservoir and storage site for both erythrocytes and platelets

(b) Lymphatic Structures That Assist the Immune System

PRIMARY LYMPHATIC STRUCTURES
(Formation and maturation of lymphocytes)

Red bone marrow

Thymus

SECONDARY LYMPHATIC STRUCTURES
(House both lymphocytes and other immune cells to protect against potentially harmful substances)

Lymph nodes

Spleen: White pulp

Tonsils

Lymphatic nodules/MALT

19. When pharyngeal tonsils become inflamed, they are referred to as adenoids. Based on the location of the adenoids (pharyngeal tonsils), what kinds of symptoms might a patient experience due to the presence of swollen adenoids?

_____

_____

_____

20. When an individual suffers from a ruptured spleen, the spleen is surgically removed so the patient does not die from internal bleeding. Using knowledge of the circulation of the spleen, explain why a ruptured spleen bleeds so profusely. (Hint: What type of capillaries are found in the spleen?)

_____

_____

21. Discuss why lymphatic nodules are so abundant in the wall of the small intestine. _____

_____

_____

22. Lacteals are specialized lymphatic capillaries found in the small intestine that have the special function of absorbing dietary fats. Absorbed fats first enter lymphatic capillaries and are transported as a component of lymph within lymph vessels. Using knowledge of the flow of both lymph and blood in the body, describe the route a dietary lipid would take to get from its location of absorption in the small intestine to its entry into the right atrium of the heart. _____

_____

_____

_____

🔺 Can You Synthesize What You've Learned?

23. Acute appendicitis (inflammation of the vermiform appendix) sometimes requires immediate surgical removal of the appendix. In the event that the appendix bursts, its contents are released into the abdominal cavity, which can lead to widespread infection, sepsis, and death. Discuss the events that might lead to an enlarged and inflamed vermiform appendix. _____

_____

_____

_____

24. Following the 2009 H1N1 pandemic, annual influenza vaccines began to include strains for H1N1 influenza. Discuss the purpose of vaccines, and state how vaccines may protect against disease outbreaks.

_____

_____

_____

Design Elements: Integrate: Clinical View icon (clipboard): ©Laia Design Studio/Shutterstock.com; Integrate Learning Strategy (pencil): ©Slavoljub Pantelic/Shutterstock.com

The Respiratory System

OUTLINE AND LEARNING OBJECTIVES

Anatomy & Physiology Revealed 4.0

Module 11: Respiratory System

INTRODUCTION

Breathe in deeply. What muscles were used to do this? Recall from chapter 11 that the **diaphragm** and the **external intercostal muscles** are the key muscles used for **inspiration**. Contraction of these muscles increases the volume of the thoracic cavity, causing the lungs to expand. As the lungs expand and the volume of the lungs increases, the pressure within the lungs becomes lower than atmospheric pressure and air flows into the lungs. This air brings a fresh supply of oxygen that is picked up by **erythrocytes** in the blood flowing through the pulmonary capillaries. Freshly oxygenated blood will be transported back to the heart and pumped out to the body. Take another deep breath. This time focus on the process of **expiration** (breathing out). Quiet expiration is a passive process, which does not require muscular effort. As soon as the diaphragm and external intercostal muscles relax, the elastic tissues within the lungs and chest cavity wall recoil, causing a decrease in the volume of the thoracic cavity (and the lungs), which increases intrapulmonary pressure. As intrapulmonary pressure becomes greater than atmospheric pressure, air flows out of the lungs. This air has a relatively high concentration of carbon dioxide, a waste product from cellular respiration, which must be removed from the body.

Each breath you take is only a small fraction of the air that could be exchanged by your lungs. Your lungs can hold and expel more air than you do at rest. This reserve allows for a greater intake of oxygen when needed by the body, such as during exercise. Try to exhale all the air out of your lungs. Despite your best efforts, you cannot exhale all the air. This **residual volume** of air ensures that your lungs, and the small **alveoli** within, can expand on the next breath.

The exercises in this chapter explore the gross and histological structure of the respiratory tract and the lungs. They begin with observations of the gross anatomy of the upper airways and the lungs, using cadaveric specimens, fresh sheep specimens, models, or some combination of these. This is followed by an investigation into the general layering pattern of the walls of the respiratory tract to prepare you to identify specific histological structures located within each layer of major respiratory tract structures, including the lungs. Finally, the processes of ventilation, gas exchange, and gas transport will be explored through a series of physiological experiments.

List of Reference Tables

Name: _____

Date: _____   Section: _____

These Pre-Laboratory Worksheet questions may be assigned by instructors through their ▨ connect course.

1. Respiratory epithelium, which lines the nasal cavity and upper respiratory tract, is classified as: (Circle one.)

   a. simple columnar with cilia

   b. simple cuboidal with microvilli

   c. simple columnar with cilia and goblet cells

   d. pseudostratified columnar with cilia and goblet cells

   e. pseudostratified columnar with microvilli

2. The left lung contains _____ (two/three) lobes, whereas the right lung contains _____ (two/three) lobes.

3. Which of the following bones does *not* form a border of the nasal cavity? (Circle one.)

   a. ethmoid   b. mandible   c. maxilla   d. palatine   e. vomer

4. The pulmonary arteries carry _____ (oxygenated/deoxygenated) blood from the _____ (right/left) ventricle of the heart to the lungs, and the pulmonary veins carry _____ (oxygenated/deoxygenated) blood from the lungs to the _____ (right/left) atrium of the heart. This circuit is referred to as the _____ (pulmonary/systemic) circuit.

5. The amount of air inhaled or exhaled during quiet breathing is called _____ (tidal volume/vital capacity).

6. Which of the following is *not* a component of the respiratory membrane? (Circle one.)

   a. capillary endothelial cell

   b. fused basement membranes

   c. type I cell

   d. type II cell

7. Which of the following are cells that produce surfactant? (Circle one.)

   a. alveolar macrophages

   b. capillary endothelial cell

   c. fused basement membranes

   d. type I cell

   e. type II cell

8. A _____ (lobar/main/segmental) bronchus leads into each lung; a _____ (lobar/main/segmental) bronchus leads into a lobe of a lung; a _____ (lobar/main/segmental) bronchus leads into a bronchopulmonary segment.

9. The _____ (parietal/visceral) pleura covers the outer surface of each lung, whereas the _____ (parietal/visceral) pleura lines the inner surfaces of the thoracic cavity.

10. Structures that compose the _____ (conducting/respiratory) division of the respiratory tree contain alveoli.

The Pleural Cavities and the Lungs

Within the thoracic cavity, the lungs are located within separate **pleural cavities** (*pleura*, a rib). The space within the thoracic cavity between the two lungs is the **mediastinum** (*medius*, middle). Tightly adhered to each lung is a serous membrane called the **visceral pleura**. Adhered to the inner wall of the thoracic cavity is a serous membrane called the **parietal pleura**.

On a human cadaver, the parietal pleura can often be seen as a shiny tissue attached to the innermost part of the rib cage or the superior surface of the diaphragm. Between the two serous membranes is a fluid-filled space called the **pleural cavity**. Note that the lungs are contained within the pleurae, which contain serous fluid. The serous fluid reduces friction and increases surface tension of the pleurae. The latter allows the lungs to expand and contract as the thoracic cavity changes volume, because it keeps the visceral and parietal layers "stuck" together. Incidentally, if air is introduced into this pleural space, a condition called pneumothorax (*pneumo-*, air + thorax), the lungs will collapse, thereby making ventilation impossible.

EXERCISE 24.3

THE PLEURAL CAVITIES

1. Observe the thoracic cavity of a human cadaver or a model of the thorax (**figure 24.5**).

2. Identify the structures listed in figure 24.5 on a cadaver or model of the thorax, using the textbook as a guide. Then label figure 24.5.

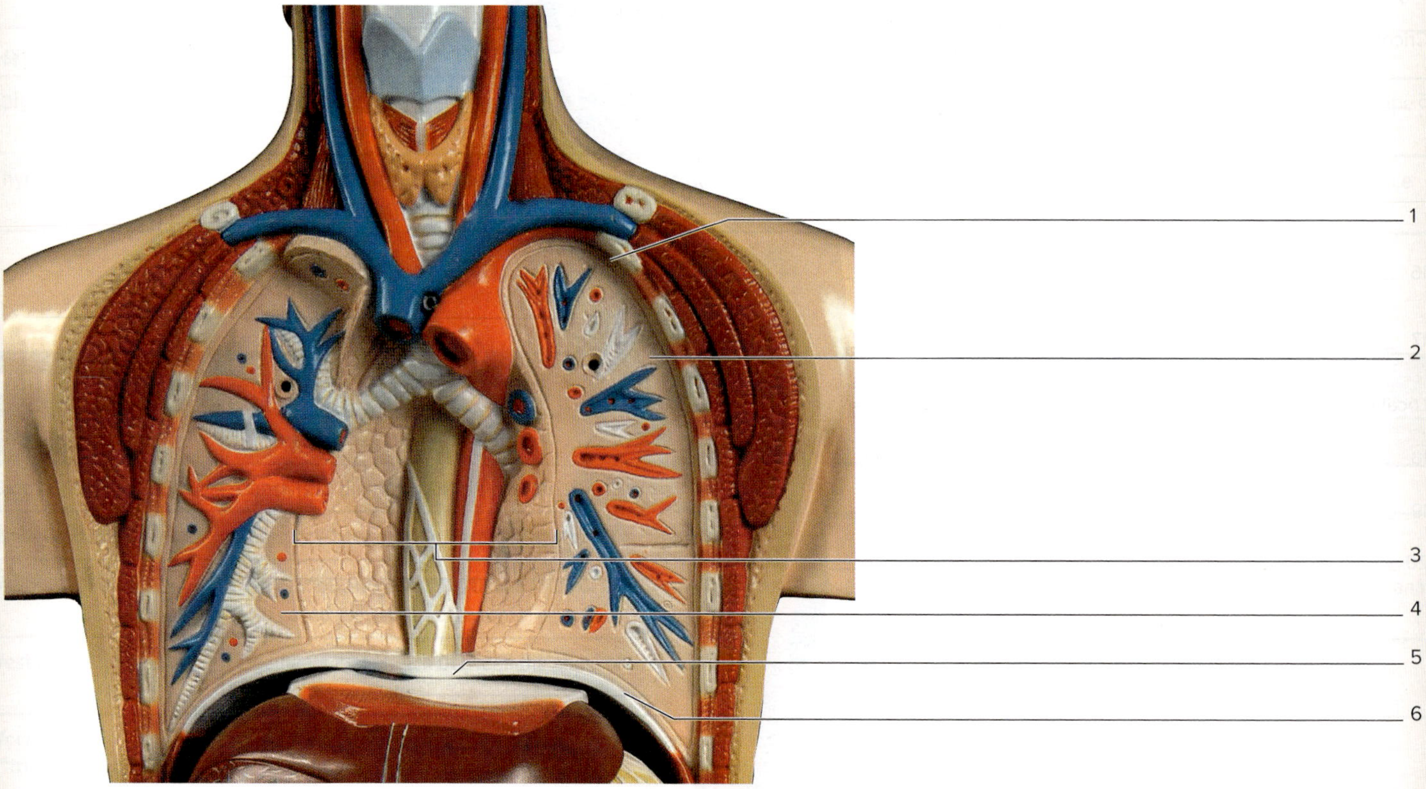

Figure 24.5   The Pleural Cavities.   Use the terms listed to fill in the numbered labels in the figure.

©Denoyer-Geppert. Photo by Christine Eckel

☐ Diaphragm   ☐ Mediastinum   ☐ Right lung

☐ Left lung   ☐ Parietal pleura   ☐ Visceral pleura

THE LUNGS

This exercise involves comparing and contrasting the structures of the right and left lungs, and observing the branching pattern of the respiratory tree. Although it is easy to distinguish the right and left lungs from each other based on the number of lobes (two for the left, three for the right), locating the structures that enter the hilum of the lung (pulmonary arteries, pulmonary veins, and bronchi) is more challenging. There are patterns for recognizing these structures that are described in this laboratory exercise. In addition, observations of several **impressions** made in the lungs by adjacent structures will be made. These impressions are visible in preserved human cadaver lungs and on classroom models of the lungs, but may not be visible in fresh lungs. This is because the process of fixing the lungs with preservative also fixes the impressions of adjacent organs. The impressions are not found in fresh lungs because they have not been fixed with preservative.

EXERCISE 24.4A The Right Lung

1. Observe the lungs of a human cadaver, a fresh or preserved sheep pluck (a *pluck* contains the heart, lungs, and trachea), or a classroom model of the lungs.

2. Begin by observing the right lung (**figure 24.6**). How many lobes does the right lung have? _____

3. Turn the lung so the hilum is visible (medial view; figure 24.6*b*). The hilum of the right lung contains branches of the pulmonary arteries and pulmonary veins, bronchi, and small bronchial arteries and veins (which represent the systemic circulation to the lungs). Which of the vessels (pulmonary arteries or pulmonary veins) should have thicker walls? Explain your answer:

_____

_____

_____

In general, the pulmonary arteries are located on the superior aspect of the hilum of the right lung, the pulmonary veins are located on the inferior and anterior aspect of the hilum of the right lung, and the bronchi are located on the superior and posterior aspect of the hilum of the right lung. If viewing cadaveric lungs, these structures will be more difficult to differentiate from each other because they are not color-coded. Thus, relying on the texture of the vessels and their locations will be necessary for proper identification.

4. Identify the structures listed in figure 24.6 on the right lung, using the textbook as a guide. Then label figure 24.6.

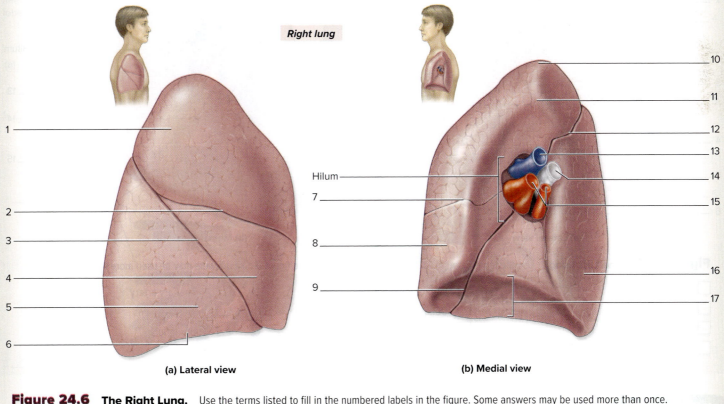

Right lung

(a) Lateral view

(b) Medial view

Figure 24.6 **The Right Lung.** Use the terms listed to fill in the numbered labels in the figure. Some answers may be used more than once.

- [ ] Apex
- [ ] Base
- [ ] Horizontal fissure
- [ ] Inferior lobe
- [ ] Middle lobe
- [ ] Oblique fissure
- [ ] Primary bronchi
- [ ] Pulmonary artery
- [ ] Pulmonary veins
- [ ] Superior lobe

(continued on next page)

(continued from previous page)

4. Identify the following structures on the slide of the trachea, using table 24.4 and figure 24.10 as guides:

☐ **Adventitia**
☐ **Basal cells**
☐ **Ciliated cells**
☐ **Epithelium**
☐ **Goblet cells**
☐ **Lamina propria**
☐ **Mucosa**

☐ **Submucosa**
☐ **Submucous glands**
☐ **Tracheal cartilages**
☐ **Trachealis muscle**

5. Sketch a cross section of the trachea as viewed through the microscope in the space provided or on a separate sheet of paper. Be sure to label all of the structures listed in step 4 in the drawing.

_____ ×

Table 24.4	Histology of the Trachea
Structure	**Description and Function**
Mucosa	Innermost layer that lines the tracheal wall; consists of respiratory epithelium resting on a basement membrane and underlying lamina propria
Epithelium	Ciliated pseudostratified columnar epithelium
Basal Cells	Small triangular cells that lie on the basal lamina and do not reach the apical surface of the epithelium. These cells are the precursor cells to the other cell types and are responsible for regeneration of the epithelium.
Ciliated Cells	The most prominent and abundant cells. These are ciliated pseudostratified columnar cells of the epithelium. They function to propel mucus and particulate matter via the "mucus escalator" along the epithelial sheet toward the pharynx.
Goblet Cells	Large round cells that appear white or clear. They secrete mucin onto the surface of the epithelium. The mucus traps particulate matter that enters the trachea while the cilia move the mucus superiorly (a "mucus escalator") toward the pharynx so that it may be swallowed.
Lamina Propria	A layer of areolar connective tissue that underlies the respiratory epithelium. In the trachea this layer can be seen in the area between the epithelial folds and the trachealis muscle or C-shaped cartilages.
Submucosa	The middle layer of the tracheal wall containing seromucous glands, blood vessels, and nerves
Submucous Glands	These produce a substance that is part watery (serous) and part viscous (mucus)
Tracheal Cartilages	C-shaped sections of cartilage that serve to support the wall of the respiratory tract.
Trachealis Muscle	A layer of smooth muscle found on the posterior aspect of the trachea where the trachea lies against the esophagus. Laterally, the trachealis muscle is anchored to the ends of the cartilage C rings, and its contraction decreases the diameter of the trachea, which is important for coughing and sneezing. The decreased diameter of the trachea causes the air to exit more forcefully, which helps dislodge substances within the airways.
Adventitia	Loose connective tissue on the outermost surface of the tracheal wall

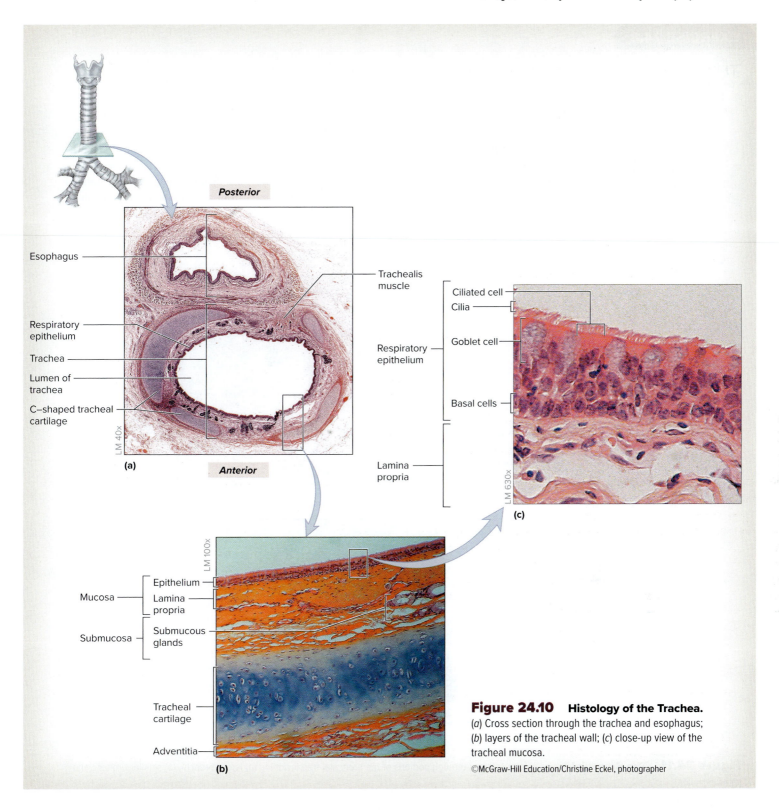

Esophagus

Respiratory
epithelium

Trachea

Lumen of
trachea

C–shaped tracheal
cartilage

Posterior

LM 40x

(a)

Anterior

Trachealis
muscle

Respiratory
epithelium

Lamina
propria

Ciliated cell

Cilia

Goblet cell

Basal cells

LM 630x

(c)

Mucosa

Submucosa

Epithelium

Lamina
propria

Submucous
glands

Tracheal
cartilage

Adventitia

LM 100x

(b)

Figure 24.10 **Histology of the Trachea.**
(*a*) Cross section through the trachea and esophagus;
(*b*) layers of the tracheal wall; (*c*) close-up view of the
tracheal mucosa.

©McGraw-Hill Education/Christine Eckel, photographer

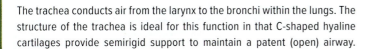

LEARNING STRATEGY

The trachea conducts air from the larynx to the bronchi within the lungs. The structure of the trachea is ideal for this function in that C-shaped hyaline cartilages provide semirigid support to maintain a patent (open) airway.

You can remember that the trachea is composed largely of cartilage by remembering that it is C-shaped cartilage (C is for cartilage). The absence of cartilage on the posterior portion of the trachea provides flexibility to the airway. This flexibility allows the esophagus to expand anteriorly when food is being transported within.

THE BRONCHI AND BRONCHIOLES

1. Obtain a slide of the lungs and place it on the microscope stage. Bring the tissue sample into focus on low power and scan the slide to locate cross sections of **large bronchi, small bronchi,** and **bronchioles (figure 24.11).**

2. The different characteristics of the conducting zone structures are listed in **table 24.5.** In general, as the airways travel deeper into the lung and become progressively smaller, three changes take place:

(1) The epithelium transitions from ciliated pseudostratified columnar to ciliated simple columnar, and the number of cilia decrease; (2) large plates of hyaline cartilage in the walls give way to smaller and smaller pieces of cartilage, with no cartilage present in bronchioles; and (3) the relative amount of smooth muscle in the airways increases. Note: The lack of cartilage in bronchioles distinguishes these portions of the respiratory tract from bronchi, which do contain cartilage.

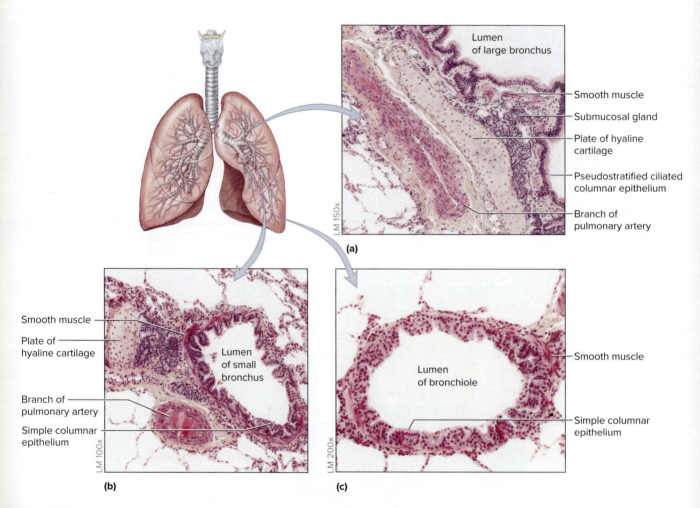

Figure 24.11   The Bronchial Tree.   Cross sections through portions of the bronchial tree including (*a*) larger bronchi, (*b*) smaller bronchi, and (*c*) bronchioles.

©McGraw-Hill Education/Christine Eckel

Table 24.5	Histology of the Bronchial Tree		
Structure	**Epithelium**	**Hyaline Cartilage**	**Smooth Muscle**
Large Bronchi	Ciliated pseudostratified columnar	Large plates, which keep airway open	Encircles the lumen
Small Bronchi	Ciliated simple columnar	Small plates	Encircles the lumen
Bronchioles	Ciliated simple columnar	None	Encircles the lumen

3. Identify the following structures on the slide of the lungs, using figure 24.11 and table 24.5 as guides:

   ☐ **Branch of pulmonary artery**

   ☐ **Bronchiole**

   ☐ **Hyaline cartilage**

   ☐ **Large bronchus**

   ☐ **Small bronchus**

   ☐ **Smooth muscle**

4. Sketch a cross section of a bronchiole as viewed through the microscope in the space provided or on a separate sheet of paper.

Lungs

The lungs consist of functional units called **alveoli** (*alveus*, hollow sac), which are the sites of gas exchange between the air within the alveoli and the blood within the pulmonary capillaries. The alveoli are lined with a simple squamous epithelium, which provides the thinnest possible barrier to diffusion between the air within the alveoli and the blood within the pulmonary capillaries that surround them. In a slide of the lung, numerous alveoli and cross sections of some of the smaller airways (such as respiratory bronchioles) will be visible scattered throughout the slide. The transition from conducting zone structures to respiratory zone structures occurs deep within the lungs. The key feature in distinguishing respiratory zone structures from conducting zone structures is the presence of alveoli. Any structure that has at least one alveolus coming off it (as may be the case with a respiratory bronchiole) participates in gas exchange and, thus, is part of the respiratory zone.

INTEGRATE

CLINICAL VIEW

Tuberculosis

Pulmonary tuberculosis, typically caused by *Mycobacterium tuberculosis*, is a bacterial infection that impacts the lungs. While most people infected with *M. tuberculosis* lack symptoms, an active infection can cause severe complications, even death. Immune compromised individuals are particularly susceptible to developing an active tuberculosis infection. Typically, *M. tuberculosis* invades alveolar macrophages and begins replication, forming a tubercle (small nodule). Additional immune cells (i.e., B- and T-lymphocytes) surround the macrophages, forming a granuloma. Necrosis may occur at the center of the granuloma. A chest radiograph of a patient suffering from tuberculosis will typically show nodules near the apex of the lungs. Signs of an active infection include persistent cough, fever, and fatigue. Tuberculosis may be spread easily to others through contact with bodily fluids. While pulmonary tuberculosis is the most common type of tuberculosis, the disease may impact other organs. *M. tuberculosis* may enter the blood and/or lymph vessels and easily spread from the lungs to other organs. Treatment of an active infection often involves the use of multiple antibiotics. Health-care providers usually receive a screening for tuberculosis, called a TB test, before working with patients.

Table 24.8	Respiratory Volumes and Capacities			
Volumes				
Volume	**Definition**		**Normal Values (Male)**	**Normal Values (Female)**
Expiratory Reserve Volume (ERV)	The amount of air expelled from the lungs during a forced expiration, following a quiet expiration; ERV is a measure of lung and chest wall elasticity		1200 mL	700 mL
Inspiratory Reserve Volume (IRV)	The amount of air taken into the lungs during a forced inspiration, following a quiet inspiration; IRV is a measure of lung compliance		3100 mL	1900 mL
Residual Volume (RV)	Amount of air left in lungs following a forced expiration		1200 mL	1100 mL
Tidal Volume (TV)	Volume of air taken into or expelled out of lungs during a quiet breath		500 mL	500 mL
Capacities				
Capacity	**Formula**	**Definition**	**Normal Values (Male)**	**Normal Values (Female)**
Functional Residual Capacity	ERV + RV	Amount of air normally left (residual) in lungs after quiet expiration	2400 mL	1800 mL
Inspiratory Capacity	TV + IRV	Total ability to inspire	3600 mL	2400 mL
Total Lung Capacity	TV + IRV + ERV + RV	Total amount of air that can be in lungs	6000 mL	4200 mL
Vital Capacity	TV + IRV + ERV	Measure of the strength of respiration	4800 mL	3100 mL

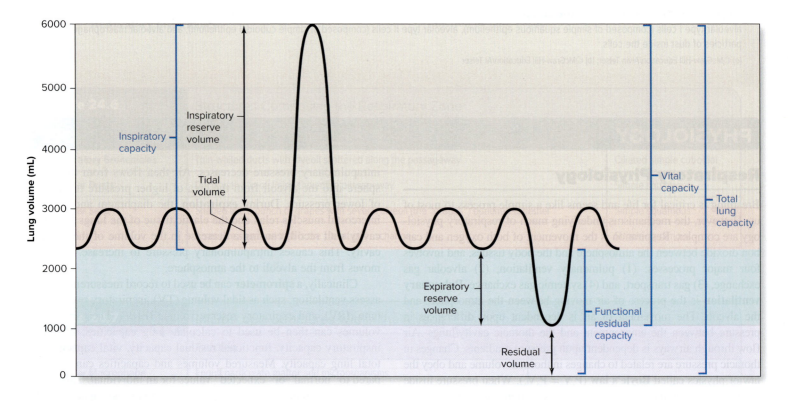

Figure 24.13   **Respiratory Volumes and Capacities.**

EXERCISE 24.10

MECHANICS OF VENTILATION

This exercise explores the mechanics of ventilation using a model that simulates the diaphragm, thoracic cavity, and lungs. Before begining, state the relationship between thoracic volume and intrapulmonary pressure, and how it relates to regulating the flow of air into and out of the lungs. _____

_____

_____

1. Obtain a model lung (**figure 24.14**). In this model, the balloons represent the lungs, the rubber sheeting represents the diaphragm, the Y tube represents the trachea and bronchi, and the bell jar represents the thoracic cavity.

2. Increase the volume of the thoracic cavity by pulling down on the "diaphragm" (rubber sheeting). Does the pressure inside the thoracic cavity increase or decrease?

   _____

   What happens to the balloons? _____

   _____

   Describe the direction of airflow. _____

   _____

3. Repeat step 2 by pushing up on the "diaphragm." Does the pressure inside the "thoracic cavity" increase or decrease? _____

   What happens to the balloons? _____

   _____

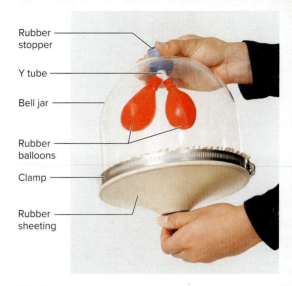

Figure 24.14   **Model Lung.**
©J and J Photography

Labels: Rubber stopper · Y tube · Bell jar · Rubber balloons · Clamp · Rubber sheeting

   Describe the direction of airflow. _____

   _____

4. Repeat step 2 and introduce air into the "thoracic cavity" by partially releasing the rubber stopper, thereby simulating a pneumothorax (air within pleural space). What happens to the balloons? _____

EXERCISE 24.11

AUSCULTATION OF RESPIRATORY SOUNDS

The sound of air flowing through respiratory structures can be detected by placing the diaphragm of a stethoscope on strategic locations of the chest and back, a process called **auscultation** (*auscultatio*, a listening). Physicians evaluating a patient with respiratory distress frequently perform this test. Physicians are listening for two main respiratory sounds: bronchial and vesicular. **Bronchial sounds**, caused by airflow through the bronchi, are often high-pitched and loud during both inspiration and expiration. **Vesicular sounds**, caused by air filling the alveolar sacs, are lower in pitch, quiet, and heard primarily during inspiration. When auscultation is performed between the first and second intercostal spaces, bronchovesicular sounds of intermediate pitch can be heard during inspiration and expiration. When inflammation or infection is present in the respiratory structures, these sounds take on different characteristics. Crackles, or popping noises, particularly during inspiration, may be indicative of fluid present in the lungs. Wheezing, or high-pitched sounds, particularly during expiration, may be indicative of increased resistance in the airway. The goal of this exercise is to auscultate these clinically relevant respiratory sounds.

   Before beginning, describe the optimal placement of the stethoscope diaphragm that will allow for auscultation of respiratory sounds in the space provided.

_____

_____

1. Obtain a stethoscope and alcohol swabs, and choose a study subject. First, clean the stethoscope by wiping the earpieces and diaphragm with an alcohol swab. Allow the stethoscope to dry. Insert the earpieces into your ears.

2. Locate the subject's larynx. Press the stethoscope diaphragm on the skin just inferior to the subject's larynx. Auscultate bronchial sounds as the subject takes several deep breaths through the mouth. While the subject is breathing, slowly move the stethoscope diaphragm inferiorly, following the anatomical course of the trachea, until the sounds are no longer audible.

3. To auscultate vesicular sounds, press the stethoscope diaphragm firmly on the subject's skin at the following locations on the right side of the body: (1) inferior to the clavicle, (2) in the intercostal spaces, and (3) along the medial border of the scapula. Be sure to keep the stethoscope diaphragm in one location during a complete cycle of inspiration and expiration.

4. Repeat the process described in step 3 on the left side of the body.

EXERCISE 24.12

PULMONARY FUNCTION TESTS

This laboratory exercise involves measuring respiratory volumes and capacities using a spirometer. Exercise 24.12A is designed for use with a wet spirometer (**figure 24.15**), whereas exercise 24.12B is designed for use with BIOPAC laboratory equipment. Before beginning, state a hypothesis regarding how vital capacity changes with a person's size, gender, and age.

_____

_____

_____

EXERCISE 24.12A Wet Spirometry

This exercise involves the use of a wet spirometer to measure respiratory volumes and capacities. A wet spirometer gets its name from the fact that the apparatus involves floating a chamber (the "bell") on top of a tank of water and measuring the volume of air that flows into and out of the bell (figure 24.15). A "dry" spirometer (as is used in the BIOPAC exercise in this chapter) measures air volumes by *calculating the volume of air that flows through a measuring device per unit time and comparing that to calibration values.*

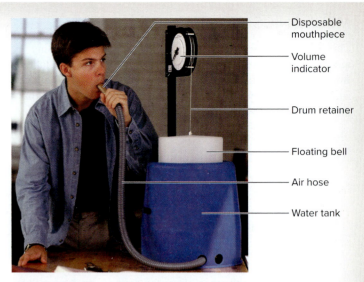

Figure 24.15   **Wet Spirometer.**   A bell floats on top of a tank of water. As air enters the bell, it displaces the water. This displacement is measured using the volume indicator, which is used to suspend the bell over the water.

©Phipps & Bird, Inc., Richmond, VA. Used with permission

1. Obtain a wet spirometer, a disposable mouthpiece, alcohol swabs, and a noseclip. Clean the noseclip by wiping it down with the alcohol swab and allowing it to air dry.

    Use a new (or clean) mouthpiece for each subject to ensure that potentially infectious microorganisms are not transmitted between subjects. Used mouthpieces should either be disinfected in an alcohol bath prior to reuse, or thrown away. Ask the laboratory instructor for instructions on proper disposal of used mouthpieces.

2. Become familiar with the parts of the spirometer by comparing it to figure 24.15 and identifying the following parts:

   ☐ **Air hose**   ☐ **Floating bell**
   ☐ **Disposable mouthpiece**   ☐ **Volume indicator**
   ☐ **Drum retainer**   ☐ **Water tank**

3. Prepare the wet spirometer by pushing the bell of the spirometer down as far as it will go. This will remove any excess air that lies inside the bell (and above the water). Note that air leaving the bell flows out of the air hose. When performing this exercise, as air flows into or out of the bell it will displace the bell above the water tank. Air volumes are recorded by noting the movement of the volume indicator that is attached to the bell by the drum retainer. Set the volume indicator to zero.

4. Place a disposable mouthpiece on the end of the air hose. Prepare the subject by placing a clean noseclip over the subject's nose. Have the subject sit quietly next to the wet spirometer. Instruct the subject to maintain an upright posture while performing the tests.

5. **Measure tidal volume (TV):** First, have the subject become relaxed with breathing into the mouthpiece by expiring several normal breaths through the mouthpiece. Be sure to push the bell of the spirometer down as far as it will go and reset the volume indicator to zero after each breath. Observe the movement of the volume indicator to determine tidal volume. When the subject is comfortable, have the subject expire a normal breath through the mouthpiece. Record the TV in **table 24.9**. Push the bell jar down as far as it will go and repeat for a total of five trials. Record each of the five TVs in table 24.9. Calculate the average TV from the five breaths and record it in table 24.9. Compare the results to expected values, which are listed in Table 24.8.

Table 24.9	Tidal Volume Measurements
Breath #	**Tidal Volume (mL)**
1	
2	
3	
4	
5	
Average TV	

6. Have the subject remove the mouthpiece and air hose from his or her mouth. Reset the spirometer by pushing the bell down to expel the air inside. Set the volume indicator to zero.

7. **Measure expiratory reserve volume (ERV):** Instruct the subject to start by taking five normal breaths through the mouthpiece. Then, have the subject perform a maximum exhalation following the fifth normal breath. Observe the movement of the volume indicator to determine the volume of air exhaled above and beyond the normal TV (= ERV). Record the ERV in **table 24.10**. Repeat this procedure for a total of five trials, recording the results from each trial in table 24.10. Calculate the average ERV from the five trials and record it as well. Compare the results to expected values listed in table 24.8. Repeat step 6 to reset the spirometer to zero.

Table 24.10	Expiratory Reserve Volume Measurements
Breath #	Expiratory Reserve Volume (mL)
1	
2	
3	
4	
5	
Average ERV	

8. **Measure vital capacity (VC):** Before beginning, explain to the subject that he/she must use as much effort as possible (including changing body position by bending over upon exhalation and leaning back upon inhalation) when performing a *maximum* exhalation/inhalation to obtain good results. To measure VC, instruct the subject to start by taking five normal breaths *without* the mouthpiece. Then, still without the mouthpiece in his/her mouth, have the subject perform a maximum exhalation, followed by a maximum inhalation following the fifth normal breath. Immediately after the maximum inhalation, have the study subject insert the mouthpiece into his/her mouth and exhale through the air hose as hard as possible to obtain a maximum exhalation.

Observe the movement of the volume indicator to determine the volume of air exhaled during the maximal exhalation that followed maximum inhalation (= VC). Record the VC in **table 24.11**. Reset both the bell jar down as far as it will go and the volume indicator to zero. Repeat this procedure for a total of five trials, recording the results from each trial in table 24.11. Calculate the average VC from the five trials and record it as well. Compare the results to expected values listed in table 24.8.

9. **Calculate inspiratory reserve volume (IRV):** IRV can be calculated using the formula: IRV = VC − (TV + ERV). Use the average values for ERV, TV, and VC measured in steps 6–8 to calculate IRV. Record the subject's IRV here: _____ mL. Compare the results to expected values listed in table 24.8.

Table 24.11	Vital Capacity Measurements
Trial #	Vital Capacity (mL)
1	
2	
3	
4	
5	
Average VC	

10. **Calculate remaining lung capacities.** Use the volumes measured in steps 6–9 to calculate the following lung capacities. Assume that residual volume (RV) is 1 liter.

☐ **Functional residual capacity (FRC) = RV + ERV**

☐ **Inspiratory capacity (IC) = TV + IRV**

☐ **Total lung capacity (TLC) = TV + IRV + ERV + RV**

☐ **Vital capacity (VC) = TV + IRV + ERV**

Record the calculated lung capacities in **table 24.12**. Compare the results to expected values listed in table 24.8.

Table 24.12	Calculation of Lung Volumes and Capacities
Lung Volume/ Capacity	Calculated Value (mL)
FRC	
IC	
TLC	

11. Upon completion of the exercise, dispose of the disposable mouthpiece, reset the spirometer to zero, and clean the workstation.

EXERCISE 24.12B BIOPAC Lesson 12: Pulmonary Function Tests

Obtain the Following:

- **BIOPAC airflow transducer (SS11LB)**
- **BIOPAC bacteriological filter (AFT1 for SS11LA or AFT36 for SS11LB)**
- **BIOPAC disposable mouthpiece (AFT2 for SS11LA or AFT36 for SS11LB)**
- **BIOPAC noseclip (AFT3)**
- **BIOPAC calibration syringe (AFT6/6A 0.6-L, AFT26 2-L, or AFT27 3-L)**
- **BIOPAC MP36/35 recording unit**
- **Laptop computer with BSL 4 software installed**

1. Prepare BIOPAC equipment by turning the computer ON and MP36/35 unit OFF. Plug the airflow transducer (SS11LA) into channel 1 (CH 1). Turn the MP36/35 unit ON.

(continued on next page)

(continued from previous page)

2. Start the BIOPAC Student Lab Program. Select L12-Pulmonary Function I from the Choose a Lesson menu, enter the filename, and click OK.

3. To calibrate, hold the airflow transducer horizontally such that it remains upright during the procedure. Click Calibrate. Ensure that data appear as a flat and centered line. If they do not, click Redo. If they do, click Continue.

4. For stage 2 of the calibration, place a disposable filter (AFT1 of AFT36) on the end of the calibration syringe and insert the entire assembly into the airflow transducer on the side labeled Inlet (**figure 24.16**). *Note: If the SS11LB airflow transducer is used, the AFT1 filter is not necessary during calibration.* Hold the airflow transducer/calibration syringe assembly horizontally such that the assembly remains upright during the entire procedure. Pull the calibration syringe plunger out to its maximal position. Click Calibrate. Pull the plunger in and out for five cycles, waiting 2 seconds between cycles. Click End Calibration. Compare the calibration data to **figure 24.17**. If there are differences, click Redo Calibration. Otherwise, click Continue.

5. OBTAIN THE DATA: Have the subject assume a relaxed, seated position. The subject should remain still and quiet, and the screen should not be visible to the subject during data collection. Prepare the airflow transducer by first removing the calibration syringe and then inserting a disposable mouthpiece on the side labeled Inlet. Have the subject put on a noseclip.

6. Click Record as the subject does the following: (1) takes five normal breaths, (2) inspires maximally once,

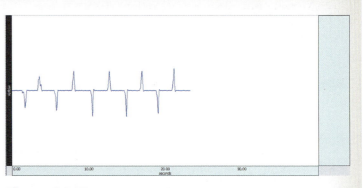

Figure 24.17 **Sample Calibration Data.**

Courtesy of and ©BIOPAC Systems, Inc. www.biopac.com

(3) expires maximally once, (4) takes five normal breaths. Click Stop. Review the data to ensure that positive spikes appear during inspiration and negative spikes appear during expiration. If the data do not show positive spikes during inspiration and negative spikes during expiration, click Redo. Otherwise, click Done.

7. ANALYZE THE DATA: To perform data analysis on data that was collected previously, click Review Saved Data. Note that CH 1 displays airflow and CH 2 displays volume. Review the airflow data and then turn this channel off by clicking on the channel number box and Alt + click (Windows) or Option + click (Mac). Note that data analysis is easier when airflow data are not visible.

8. To prepare to analyze the data, do the following:

   a. Use the zoom tool for optimal viewing of the recording.

   b. Set up measurement boxes at the top of the screen using the drop-down menus. Set CH 2 to each of the four boxes: P-P (max–min), Max, Min, Delta (last point–first point).

9. Locate the I-beam cursor. Select the point on the recording segment that corresponds to the vital capacity (VC; see figure 24.13) and record the P-P measurement in **table 24.13**.

10. To calculate the average tidal volume (TV), use the I-beam cursor to highlight the inspiration phase of cycle 3, which should correspond to the minimum and to the maximum of the third cycle. Record the P-P measurement as trial 1 in table 24.13. Repeat this procedure for the expiration phase of cycle 3, and record the P-P measurement as trial 2 in table 24.13. Calculate the average for these two trials, and record the calculation in table 24.13.

11. Measure the following respiratory volumes and capacities by using the I-beam cursor for the following values: IRV (Delta), ERV (Delta), RV (Min), IC (Delta), EC (Delta), TLC (Max). Consult figure 24.13 to review the portion of the recording that corresponds to each value. Record the data in table 24.13. Click Exit.

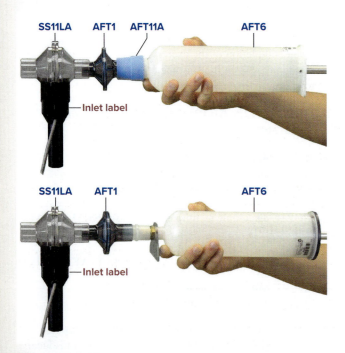

Figure 24.16 **Calibration Setup and Assembly.**

Courtesy of and ©BIOPAC Systems, Inc. www.biopac.com

12. Consult table 24.8 as a guide for the formulas to calculate each respiratory capacity. Calculate each capacity in table 24.13 and compare the measurements with expected values.

13. Make note of any pertinent observations here:

_____

_____

_____

Table 24.13	Measured and Calculated Respiratory Volumes and Capacities			
Abbreviation	**Volume/Capacity**	**Measurement**	**Calculated**	**Reference**
Volumes				
ERV	Expiratory Reserve Volume			700 – 1200 mL
IRV	Inspiratory Reserve Volume	Approximately 1 L		1900 – 3100 mL
RV	Residual Volume			1100 – 1200 mL
TV	Tidal Volume	(1) (2)	AVG =	500 mL
Capacities				
IC	Inspiratory Capacity			2400 – 3600 mL
FRC	Functional Residual Capacity			1800 – 2400 mL
TLC	Total Lung Capacity			4200 – 6000 mL
VC	Vital Capacity			3100 – 4800 mL

Source: BIOPAC Systems, Inc. Data from BIOPAC Table 12.2

INTEGRATE

CLINICAL VIEW

Obstructive versus Restrictive Respiratory Diseases

Spirometry tests can be used clinically to distinguish between obstructive and restrictive pulmonary diseases. Obstructive diseases, such as emphysema, involve a physical barrier (obstruction) in the airway, whereas restrictive diseases, such as pulmonary fibrosis (scarring), involve an inability to physically expand the lungs (restriction). Both types of diseases decrease the exchange of respiratory gases at the respiratory membrane, but for different reasons. If a patient presents with respiratory distress, doctors can measure respiratory volumes using a spirometer and calculate lung capacities to determine the disease classification. Two clinical measurements commonly used include

forced expiratory volume in 1 second (FEV)$_1$, which corresponds to the amount of air forcefully exhaled in 1 second, and **forced vital capacity** (FVC), the amount of air forcefully exhaled after a maximal inhalation. The ratio of these two values, or FEV$_1$%, is typically 75–80% in normal, healthy adults. Because patients with obstructive pulmonary disease have difficulty breathing out, their FEV$_1$% is typically reduced. Patients with an FEV$_1$% of less than 75% are diagnosed with an obstructive disease. In comparison, patients suffering from a restrictive disease have difficulty breathing in because of the limited ability to expand the lungs. These patients typically will have lower than normal inspiratory capacity and vital capacity. Once a patient's disease is classified as either obstructive or restrictive, further diagnostic testing can be performed and treatment plans may be devised to address the specific disease that caused the patient's respiratory distress.

Ph.I.L.S. LESSON 34: pH AND Hb-OXYGEN BINDING

Respiratory gases are exchanged between the air in the alveoli and the blood in the pulmonary capillaries during **alveolar gas exchange**. The epithelium of the alveoli, the endothelium of the alveolar capillaries, and the fused basement membranes between the two collectively form the **respiratory membrane**. For oxygen to enter the blood, or carbon dioxide to leave the blood, the gases must diffuse across this respiratory membrane. The rate at which each gas diffuses is governed by **Dalton's law of partial pressures**, which states that each gas in a mixture of gases exerts an individual, or partial, pressure in relation to its abundance in the mixture of gases. The partial pressure of each gas is calculated by multiplying the total pressure of the gas by the percentage of the individual gas in the mixture. For example, if the total pressure is 760 mm Hg, and oxygen is 21% of the mixture, then the partial pressure of oxygen (P_{O_2}) = 760 mm Hg $\times$ 0.21 = 159 mm Hg. Gases will diffuse along their individual partial pressure gradients from an area of high partial pressure to an area of low partial pressure.

Other factors that influence alveolar gas exchange include the total number of alveoli (surface area for gas exchange), the thickness of the respiratory membrane, and the solubility coefficient for the particular gas. The solubility coefficient dictates how readily a gas will dissolve in a liquid, and is governed by **Henry's law**. Carbon dioxide has a solubility coefficient that is roughly 24 times greater than that of oxygen; this means that carbon dioxide dissolves much more easily in liquid than does oxygen. Functionally, this means that carbon dioxide diffuses across the respiratory membrane much more readily than oxygen, or requires much smaller partial pressure differences to diffuse. A larger partial pressure gradient, increased surface area, thinner respiratory membrane, and greater solubility coefficient will all increase the rate of diffusion across the respiratory membrane.

Gas exchange also occurs between systemic arterial blood and tissue cells. Here the same concepts for gas exchange apply. However, in the tissues, the membrane the gases must cross consists of the capillary endothelium (and its basement membrane) and the plasma membrane of the cell.

In addition to **exchange** of gases between the air and the blood, and the blood and the tissues, gases must also be transported within the blood. The process of moving the gases through the cardiovascular system is known as **gas transport**. Because of the low solubility of oxygen in water, 98% of the oxygen transported in the blood is bound to the heme group of hemoglobin, the major component of erythrocytes (see chapter 20). The majority of carbon dioxide (~70%) is transported as bicarbonate ion. The conversion of carbon dioxide to bicarbonate ion occurs within erythrocytes, which contain the enzyme *carbonic anhydrase*. Carbonic anhydrase catalyzes the reaction between carbon dioxide (CO_2) and water (H_2O) to form the weak acid carbonic acid (H_2CO_3). In solution, carbonic acid easily dissociates into hydrogen ions (H^+) and bicarbonate ions (HCO_3^-). This reaction is represented by the following equation:

$$CO_2 + H_2O \leftrightarrow H_2CO_3 \leftrightarrow H^+ + HCO_3^-$$

This reaction is important in understanding not only carbon dioxide transport as bicarbonate ion in the blood, but also the relationship between the carbon dioxide levels in the blood and the pH of the blood. As more carbon dioxide accumulates in the blood, more of it is converted into hydrogen ions. As the concentration of hydrogen ions in the blood increases, the pH decreases.

Hemoglobin within red blood cells binds oxygen so oxygen can be transported to the tissues. Hemoglobin's affinity for oxygen increases when the partial pressure of oxygen (P_{O_2}) is high, as occurs in the lungs, and decreases when P_{O_2} is low, as occurs in the tissues. The affinity of hemoglobin for oxygen is also influenced by blood pH. Any shift in blood pH will dramatically affect the degree to which hemoglobin binds oxygen. In this exercise the partial pressure of oxygen in a blood sample will be altered using a vacuum pump to decrease the P_{O_2} of a blood sample. A spectrophotometer will then be used to evaluate the relationship between P_{O_2} of the blood sample and the percent saturation of hemoglobin. In addition, the percent saturation of hemoglobin at three different pH values (6.8, 7.4, and 8.0) will be analyzed to determine the effect of pH on hemoglobin saturation. Use the textbook as a guide while performing the following exercises. These exercises will guide your exploration of the major respiratory processes and the relationship between pH and respiration.

Before beginning, become familiar with the following concepts (use the textbook as a reference):

- Structure of hemoglobin
- Oxygen-hemoglobin dissociation curve
- Meaning of a right-shift or left-shift of the oxygen-hemoglobin dissociation curve

1. Open Ph.I.L.S. Lesson 34: pH and Hb-Oxygen Binding.

2. Read the objectives and introduction. Then take the pre-lab quiz. The laboratory exercise will open when the pre-lab quiz has been completed (**figure 24.18**).

Figure 24.18 Opening Screen for Ph.I.L.S. Lesson 34: pH and Hb-Oxygen Binding.

©McGraw-Hill Education

3. To calibrate the spectrophotometer, set the wavelength at 620 nm (wavelength at which intact hemoglobin absorbs light) and set the transmittance value to 0 using the arrows.

4. Open the lid of the spectrophotometer holder (if unsure, click on the words "spectrophotometer holder" in the instructions at the bottom of the screen).

5. Click on the globe and drag it to the top of one of the tubes (it will snap in place) to form the tonometer.

6. Click and drag the tonometer into the spectrophotometer. The Journal will open a table showing partial pressures of oxygen (Po_2) ranging from 160 mm Hg to 0 mm Hg for three different pH values—6.8, 7.4, and 8.0.

7. Set the transmittance to 100% using the arrows at Calibrate.

8. To record the transmittance through the sample, click Journal (orange rectangle at the bottom right of screen).

9. Close the Journal by clicking on the X. Click on the tonometer to remove it from the spectrophotometer. Then click and drag the tonometer back to the rack.

10. To change the partial pressure of oxygen in the sample, position the end of the vacuum tube on the top of the tonometer.

11. Click the down arrow on the vacuum once to reduce the barometric pressure (which simulates a decrease in partial pressure of oxygen). Click on the vacuum tube to return it to its original position.

12. Place the tonometer in the spectrophotometer. Click Journal to record.

13. Repeat steps 9–12 over the pressure range (160–0 mm Hg) for that sample until all values for the different partial pressures of oxygen have been recorded in the table on the computer.

14. When all values have been recorded in the table for a given sample, drag the tonometer to the biohazard waste container to dispose of the sample.

15. Repeat steps 5–14 with the remaining blood samples using different pH values.

16. Click Journal to see the complete table and graph. Complete **table 24.14** from the table created on the screen and graph the data in the space provided.

17. Open the post-lab quiz & lab report by clicking on it. Answer the post-lab questions that appear on the computer screen. Click Print Lab to print a hard copy of the report or click Save PDF to save an electronic copy of the report.

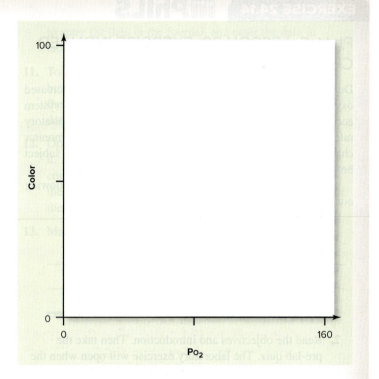

Table 24.14	Data Table for Partial Pressures of Oxygen and Blood pH		
Po_2	pH 6.8	pH 7.4	pH 8.0
160			
140			
120			
100			
80			
60			
40			
20			
0			

18. Make note of any pertinent observations here:

_____

_____

_____

GROSS ANATOMY

This laboratory session begins with a group of exercises that involve observing the gross anatomy of the kidneys, ureters, urinary bladder, and urethra. Particular attention will be focused on observing the location of these structures within the abdominopelvic cavity and their relationships with other organs.

The Kidney

The kidneys are located along the posterior body wall from vertebral levels T_{12}–L_3, with the right kidney slightly lower than the left kidney because of the location of the liver. The kidneys, like the ureters and urinary bladder, are retroperitoneal structures. The exercises in this section involve observing the gross anatomy of the kidney, the blood supply to the kidney, and the urine draining structures within the kidney.

EXERCISE 25.1

GROSS ANATOMY OF THE KIDNEY

1. Obtain a preserved animal kidney that has been sectioned along a coronal plane, or a classroom model of a coronal section of the kidney (**figure 25.1**). Whether viewing an actual kidney or a model of the kidney, the outermost structure that is visible is the **fibrous (renal) capsule**,

which is composed of dense irregular connective tissue. This is because the other protective layers (perinephric fat, renal fascia, and paranephric fat) have been removed.

2. Identify the structures listed in figure 25.1 on the kidney or on a classroom model of the kidney. Then label them in figure 25.1.

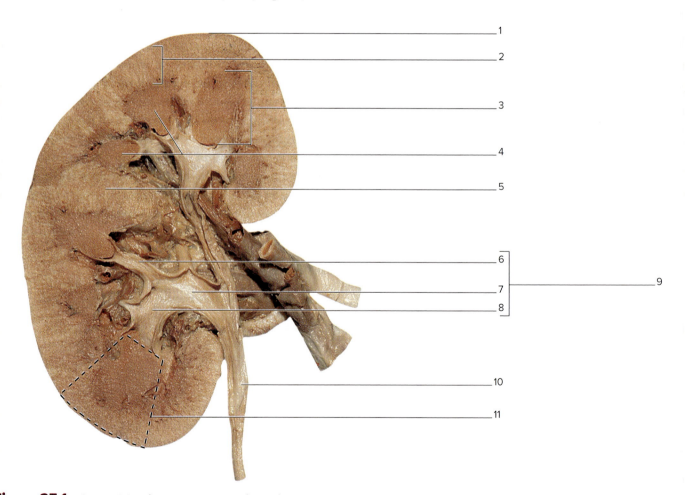

Figure 25.1   Coronal Section Through the Right Kidney. Use the terms listed to fill in the numbered labels in the figure.

☐ Fibrous (renal) capsule	☐ Renal column	☐ Renal medulla	☐ Renal sinus
☐ Major calyx	☐ Renal cortex	☐ Renal (medullary) Pyramid	☐ Ureter
☐ Minor calyx	☐ Renal lobe	☐ Renal pelvis	

©McGraw-Hill Education/Rebecca Gray

EXERCISE 25.2

BLOOD SUPPLY TO THE KIDNEY

The kidneys receive approximately 20 to 25% of the blood pumped by the heart each minute. This is not because they have a huge demand for oxygen or nutrients, but instead because their major function is to filter the blood to alter its volume and composition. The kidney accomplishes this through the processes of filtration, reabsorption, and secretion, which all involve transport of fluids and other substances between the functional units of the kidney (nephrons) and the blood. Thus, an understanding of blood flow into and out of the kidney is critical to understanding how the kidney functions.

1. Observe a classroom model of the kidney that demonstrates blood vessels of the kidney.

2. **Figure 25.2** diagrams the flow of blood through the kidney. Blood enters the kidney through the renal artery (a branch off the abdominal aorta). Blood is transported into the kidney by progressively smaller arteries until reaching the capillaries. The arrangement of capillaries around the nephrons of the kidney is unique. There are three capillary beds associated with each nephron: the glomerulus (within the renal corpuscle), the peritubular capillaries, and the vasa recta a specialized peritubular capillary bed surrounding the nephron loop in the medulla. The **glomerulus**, the first capillary bed, is the site of *filtration*, which is the movement of fluid from the blood across the filtration membrane into the capsular space of the renal corpuscle (figure 25.2*a,b*). Blood enters the glomerulus through an afferent arteriole and exits through an efferent arteriole. The efferent arteriole then leads into the second capillary bed, which is either the **peritubular capillaries** (within the cortex) or the **vasa recta** (within the medulla). The second capillary bed is the site of *exchange* of fluids, electrolytes, respiratory gases, and nutrients between the tubular portions of the nephron and the blood. Exchange

in these capillaries involves both *reabsorption* (movement of substances from the tubular lumen into the blood) and *secretion* (movement of substances from the blood into the tubular lumen). Finally, blood leaving the peritubular capillaries and vasa recta drains into progressively larger veins that carry the blood out of the kidney to the inferior vena cava.

3. Identify the blood vessels listed in **figure 25.3** on a classroom model of the kidney, using the textbook as a guide. Then label the structures in figure 25.3. What parts of the nephron do the peritubular capillaries surround?

   _____

   What parts of the nephron do the vasa recta surround?

   _____

4. In the space below, trace the flow of blood as it travels through the kidney. Start at the abdominal aorta and end at the inferior vena cava.

   _____

   _____

   _____

   _____

   _____

   _____

   _____

   _____

   _____

   _____

   _____

(continued on next page)

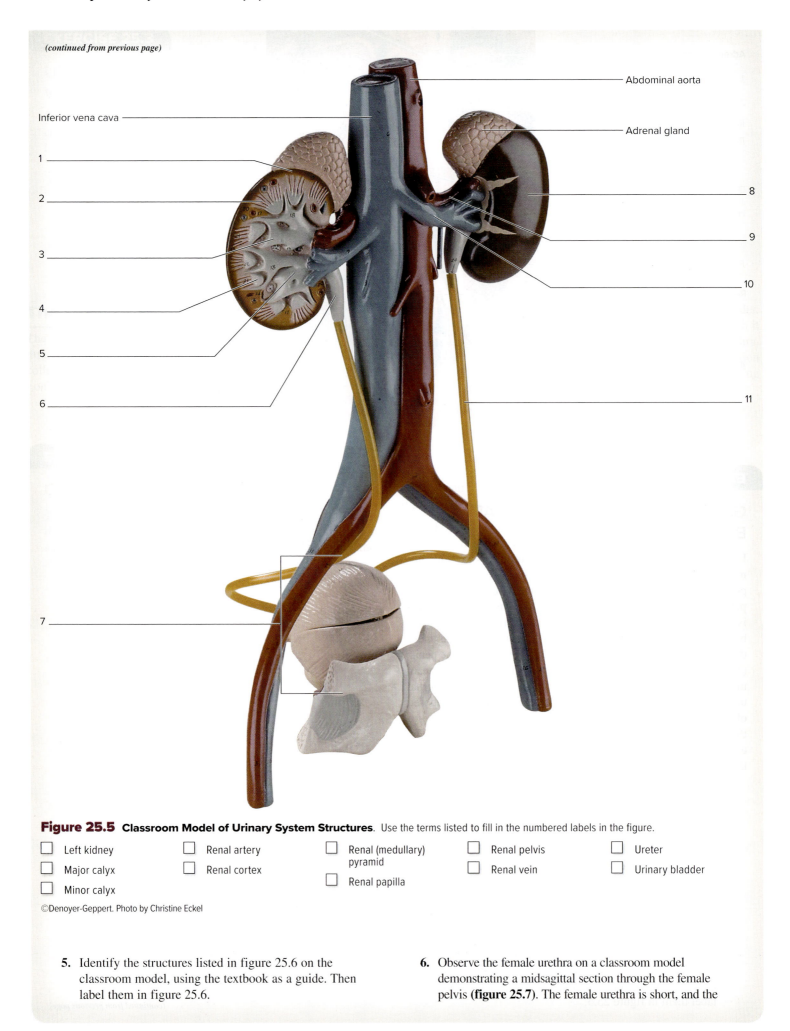

(continued from previous page)

Inferior vena cava

Abdominal aorta

Adrenal gland

1

2

3

4

5

6

7

8

9

10

11

Figure 25.5 **Classroom Model of Urinary System Structures**. Use the terms listed to fill in the numbered labels in the figure.

☐ Left kidney   ☐ Renal artery   ☐ Renal (medullary) pyramid   ☐ Renal pelvis   ☐ Ureter

☐ Major calyx   ☐ Renal cortex   ☐ Renal papilla   ☐ Renal vein   ☐ Urinary bladder

☐ Minor calyx

©Denoyer-Geppert. Photo by Christine Eckel

5. Identify the structures listed in figure 25.6 on the classroom model, using the textbook as a guide. Then label them in figure 25.6.

6. Observe the female urethra on a classroom model demonstrating a midsagittal section through the female pelvis **(figure 25.7)**. The female urethra is short, and the

external urethral orifice is located between the clitoris and the vagina. The entire urethra is surrounded by the internal urethral sphincter. The external urethral sphincter is skeletal muscle of the urogenital diaphragm, just as it is in the male.

7. Identify the structures listed in figure 25.7 on the classroom model, using the textbook as a guide. Then label them in figure 25.7.

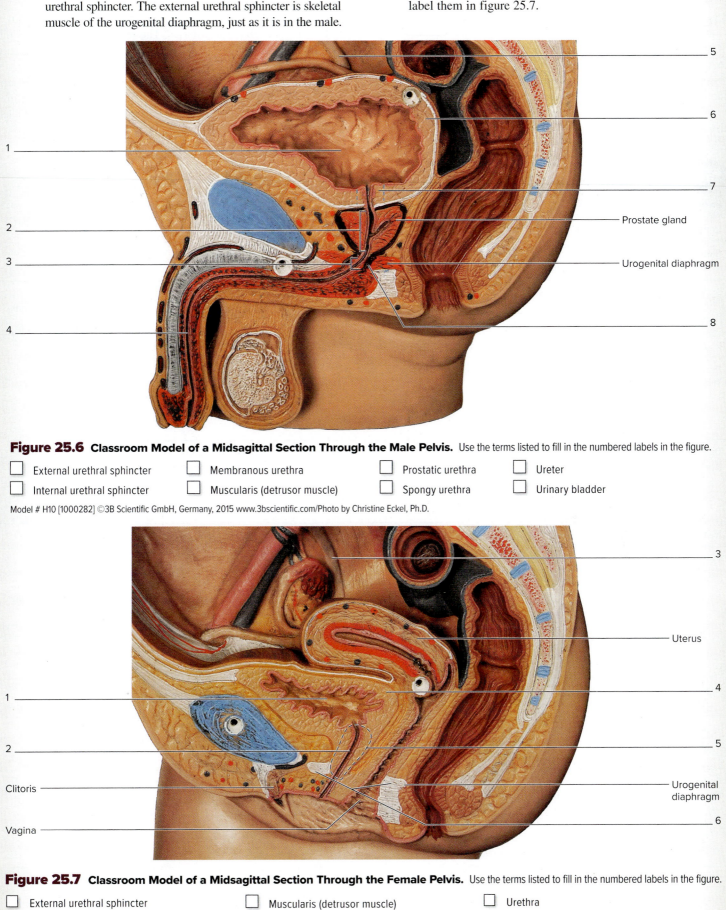

Figure 25.6 **Classroom Model of a Midsagittal Section Through the Male Pelvis.** Use the terms listed to fill in the numbered labels in the figure.

☐ External urethral sphincter   ☐ Membranous urethra   ☐ Prostatic urethra   ☐ Ureter
☐ Internal urethral sphincter   ☐ Muscularis (detrusor muscle)   ☐ Spongy urethra   ☐ Urinary bladder

Model # H10 [1000282] ©3B Scientific GmbH, Germany, 2015 www.3bscientific.com/Photo by Christine Eckel, Ph.D.

Figure 25.7 **Classroom Model of a Midsagittal Section Through the Female Pelvis.** Use the terms listed to fill in the numbered labels in the figure.

☐ External urethral sphincter   ☐ Muscularis (detrusor muscle)   ☐ Urethra
☐ Internal urethral sphincter   ☐ Ureter   ☐ Urinary bladder

Model # H10 [1000281] ©3B Scientific GmbH, Germany, 2015 www.3bscientific.com/Photo by Christine Eckel, Ph.D.

HISTOLOGY

The Kidney

Each kidney is composed of two major regions: an outer **renal cortex** and an inner **renal medulla (figure 25.8)**. The arrangement of nephrons along the **corticomedullary junction** means that some nephron components fall predominantly in the renal cortex and others in the renal medulla. Thus, the two regions exhibit distinct histological features. The renal cortex contains the renal corpuscles, proximal convoluted tubules (PCTs), distal convoluted tubules (DCTs), and peritubular capillaries. The renal medulla contains the nephron loops, collecting ducts (CDs), and vasa recta. Most structures within the kidney can be identified histologically by recognition of both the region (cortex or medulla) and the type of epithelium forming the structure. **Table 25.1** summarizes the type of epithelium that forms each of the structures and lists the major functions of each structure.

Table 25.1	Histological Features of the Kidney	
Structure	**Epithelium**	**Function**
Renal Cortex		
Glomerulus	Simple squamous epithelium with fenestrations.	Filtration
Visceral Layer	Specialized squamous epithelium called podocytes	Filtration: Extensions of podocytes, called pedicels contain actin filaments. The membrane-covered openings between the pedicles, called filtration slits, participate in the filtration process.
Parietal Layer	Simple squamous epithelium	Forms outer, impermeable wall of glomerular capsule
Proximal Convoluted Tubule (PCT)	Simple cuboidal epithelium with extensive microvilli and mitochondria. Lumen of tubule appears fuzzy and occluded. Nuclei are located at the basal surface but not always visible.	Reabsorbs glucose, amino acids, Ca^{2+}, PO_4, HCO_3, and 80% of the water and NaCl present in the filtrate; secretes substances like penicillin and toxins after they have undergone modification by the liver; also secretes organic acids and bases
Distal Convoluted Tubule (DCT)	Simple cuboidal epithelium with sparse microvilli. Lumen of tubule appears clear. Nuclei are located near the apical surface and usually visible. Cuboidal cells are smaller than those of the PCT.	Secretes H^+ and K^+; reabsorbs Na^+ and water; contains the macula densa of the juxtaglomerular apparatus, which is involved with the regulation of blood pressure
Renal Medulla		
Nephron Loop		
Descending Limb—Thick Segment	Simple cuboidal epithelium	Epithelial cells are impermeable to sodium, but water is drawn out into the interstitial spaces by osmosis. Thus, the filtrate becomes more concentrated as it moves down the descending nephron loop.
Descending Limb—Thin Segment	Simple squamous epithelium	Epithelial cells are impermeable to sodium, but water is drawn out into the interstitial spaces by osmosis. Thus, the filtrate becomes more concentrated as it moves down the descending nephron loop.
Ascending Limb—Thin Segment	Simple squamous epithelium	Epithelial cells are impermeable to water, but sodium passively diffuses out. Thus, the filtrate becomes less concentrated as it moves up the ascending nephron loop.
Ascending Limb—Thick Segment	Simple cuboidal epithelium	Epithelial cells are impermeable to water, and they actively transport sodium out of the tubule. Thus, the filtrate becomes less concentrated as it moves up the ascending nephron loop.
Collecting Tubule and Duct (CD)	Simple cuboidal or simple columnar epithelium. Cells have very precise boundaries. Overall tube diameter is the same as the PCT, but the CD has a larger lumen and no microvilli. Cells appear more pale than those of the thick segment of the nephron loop.	Concentrates urine under the influence of antidiuretic hormone (ADH). Decreases urine volume through the action of aldosterone. Both hormones act on principal cells. Intercalated cells participate in regulating acid-base balance.

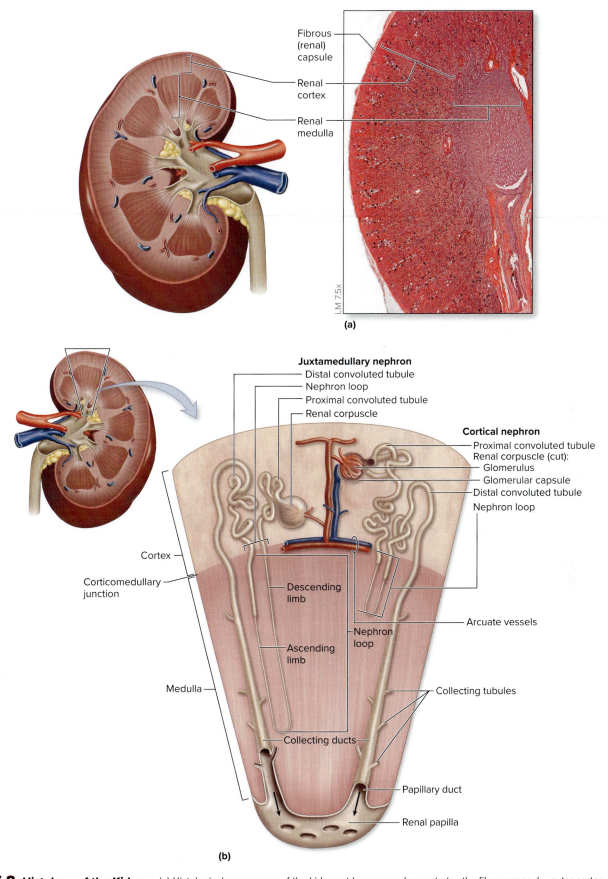

Figure 25.8 **Histology of the Kidney.** (*a*) Histological appearance of the kidney at low power demonstrates the fibrous capsule, outer cortex, and inner medulla. (*b*) Position of nephron structures within the renal cortex and renal medulla.

HISTOLOGY OF THE RENAL CORTEX

The renal cortex contains **renal corpuscles**, which are the site of filtration within the nephron. Each renal corpuscle is composed of a **glomerulus** (a tuft of capillaries), surrounded by a **glomerular capsule** (also known as Bowman's capsule). The glomerular capsule itself is composed of an inner **visceral layer**, which consists of modified simple squamous epithelial cells called **podocytes**, and an outer **parietal layer**, which consists of unmodified simple squamous epithelium. Unlike traditional simple squamous epithelium, which is specialized for diffusion and filtration, this parietal layer serves as an impermeable barrier surrounding the capsule. The renal cortex also contains **proximal convoluted tubules (PCTs)**, which are the site of most reabsorption and secretion, and **distal convoluted tubules (DCTs)**, which are involved in both reabsorption and secretion and compose part of the **juxtaglomerular apparatus**.

EXERCISE 25.6A The Renal Corpuscle

1. Obtain a compound microscope and a histology slide of the kidney and place the slide on the microscope stage.

2. Bring the tissue sample into focus at low power and scan the slide to distinguish between the outer cortex and inner medulla (figure 25.8). Next, move the microscope stage so the renal cortex is in the center of the field of view.

3. Scan the slide in the region of the cortex to locate a circular **renal corpuscle (figure 25.9)**. Bring the renal

corpuscle into the center of the field of view and then change to high power. Although the visceral layer of the glomerular capsule is indistinguishable from the glomerular capillaries, identify the tissue that contains both the visceral layer of the glomerular capsule and the glomerular capillaries (figure 25.9). Next, identify the parietal layer of the glomerular capsule. What type of epithelium composes the parietal layer of the glomerular capsule? _____

What is the name of the space located between the visceral and parietal layers of the glomerular capsule?

_____

The space just identified becomes continuous with the lumen of which of the following? (Circle one.)

PCT   **nephron loop**   **DCT**   **collecting duct**

4. Sketch the renal corpuscle as seen through the microscope in the space provided or on a separate sheet of paper. Be sure to label the glomerulus (and visceral layer of the glomerular capsule), the parietal layer of the glomerular capsule, and the capsular space on the drawing.

_____ ×

5. *Optional Activity*: **APR 13: Urinary System**—Review the "Kidney—Microscopic Anatomy" animation to visualize the parts of the nephron and their placement in the renal cortex and medulla.

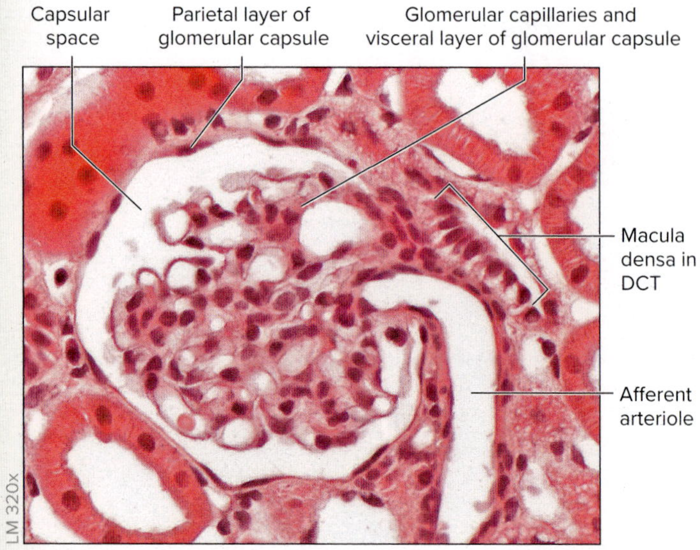

Capsular space | Parietal layer of glomerular capsule | Glomerular capillaries and visceral layer of glomerular capsule

Macula densa in DCT

Afferent arteriole

LM 320x

Figure 25.9 The Renal Corpuscle. The renal corpuscle consists of the glomerulus and the glomerular capsule.

©McGraw-Hill Education/Alvin Telser

EXERCISE 25.6B Proximal and Distal Convoluted Tubules

1. After completing exercise 25.6A, there should be a slide of the kidney on the microscope stage that has the outer cortex in the center of the field of view. If you did not perform exercise 25.6A, go through steps 1–2 of exercise 25.6A and then continue with this exercise.

2. Scan the slide in the region of the cortex to locate **proximal** and **distal convoluted tubules** (PCTs and DCTs) **(figure 25.10)**. When viewing a histology slide of the cortex of the kidney, cross sections of PCTs and DCTs, with a few renal corpuscles scattered throughout, should be visible. Although both PCTs and DCTs are lined with simple cuboidal epithelium, the proximal convoluted tubules have long, dense microvilli, which make the lumens of the PCTs appear "fuzzy." The distal convoluted tubules have very few, short microvilli, which make the lumens appear to be clear. What does the

presence of microvilli indicate about the function of an epithelium? _____

_____

3. Sketch cross sections of proximal and distal convoluted tubules as seen through the microscope in the space provided or on a separate sheet of paper.

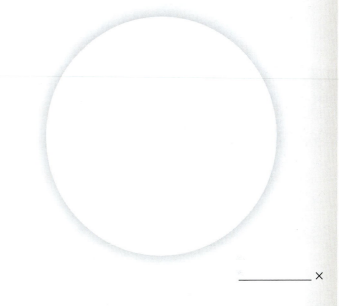

_____ ×

4. Identify the following structures on the slide of the cortex of the kidney, using figures 25.8 to 25.10 and table 25.1 as guides:

- [ ] **Capsular space**
- [ ] **Distal convoluted tubule**
- [ ] **Microvilli**
- [ ] **Parietal layer of glomerular capsule**
- [ ] **Proximal convoluted tubule**
- [ ] **Renal corpuscle**
- [ ] **Visceral layer of glomerular capsule and glomerular capillaries**

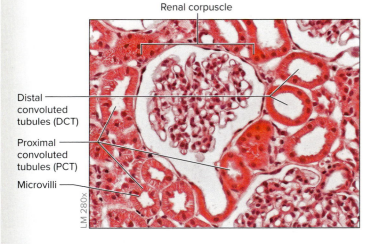

Renal corpuscle

Distal convoluted tubules (DCT)

Proximal convoluted tubules (PCT)

Microvilli

LM 280x

Figure 25.10 The Renal Cortex. The renal cortex contains renal corpuscles, proximal convoluted tubules (PCT), and distal convoluted tubules (DCT). PCTs have "fuzzy" lumens because of the presence of microvilli.
©McGraw-Hill Education/Al Telser

EXERCISE 25.7

HISTOLOGY OF THE RENAL MEDULLA

The renal medulla contains **nephron loops** (loops of Henle) and **collecting ducts**, with surrounding capillaries called the **vasa recta**. These structures are all elongated tubules that lie next to each other and function together to concentrate the urine formed by the nephrons.

1. Obtain a compound microscope and a histology slide of the kidney. Be sure to note the section type, whether a longitudinal or cross-section of the kidney. Place the slide on the microscope stage.

2. Bring the tissue sample into focus at low power and scan the slide to distinguish between the outer cortex and inner medulla. Next, move the microscope stage so the renal medulla is in the center of the field of view (figure 25.8*a*).

3. With the medulla in the center of the field of view, switch to high power and bring the tissue sample into focus once again. In this region of the kidney, thick and thin limbs of the nephron loops, collecting ducts, and possibly vasa recta should be visible (table 25.1, **figure 25.11**). Notice how, in longitudinal section, these structures appear as row after row of cells lined up next to each other. In cross sections, these structures appear as circular "tubes" with simple cuboidal cells surrounding the lumen. The main objective in viewing this part of the kidney is to gain an appreciation for the way these structures line up next to each other, which is critical to their function. An additional objective is to be able to distinguish thick limbs of the nephron loops from the collecting ducts based on the diameter of the lumens.

(continued on next page)

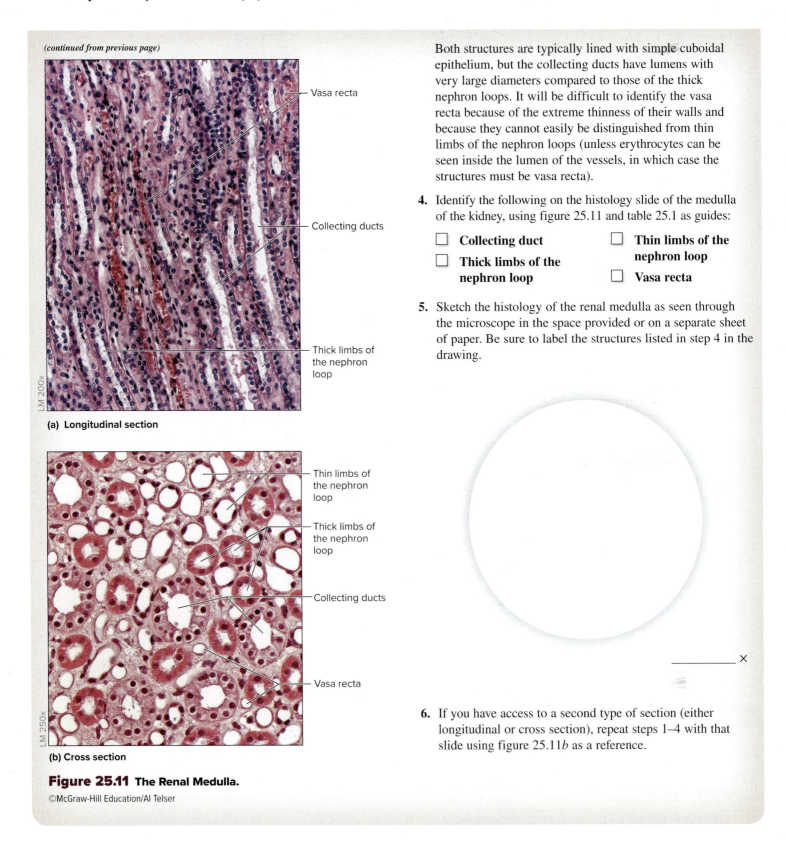

(continued from previous page)

LM 200x

Vasa recta

Collecting ducts

Thick limbs of the nephron loop

(a) Longitudinal section

LM 250x

Thin limbs of the nephron loop

Thick limbs of the nephron loop

Collecting ducts

Vasa recta

(b) Cross section

Figure 25.11 The Renal Medulla.
©McGraw-Hill Education/Al Telser

Both structures are typically lined with simple cuboidal epithelium, but the collecting ducts have lumens with very large diameters compared to those of the thick nephron loops. It will be difficult to identify the vasa recta because of the extreme thinness of their walls and because they cannot easily be distinguished from thin limbs of the nephron loops (unless erythrocytes can be seen inside the lumen of the vessels, in which case the structures must be vasa recta).

4. Identify the following on the histology slide of the medulla of the kidney, using figure 25.11 and table 25.1 as guides:

☐ **Collecting duct**   ☐ **Thin limbs of the nephron loop**

☐ **Thick limbs of the nephron loop**   ☐ **Vasa recta**

5. Sketch the histology of the renal medulla as seen through the microscope in the space provided or on a separate sheet of paper. Be sure to label the structures listed in step 4 in the drawing.

_____ ×

6. If you have access to a second type of section (either longitudinal or cross section), repeat steps 1–4 with that slide using figure 25.11*b* as a reference.

The Urinary Tract

The **urinary tract** consists of structures that function to transport or store urine: the ureters, urinary bladder, and urethra. This section involves making observations of the histological structure of the walls of the ureters and urinary bladder. **Table 25.2** lists the type of epithelium that lines the calyces of the kidney (the areas of kidney that drain urine from the kidney into the urinary tract) and each of the structures of the urinary tract.

Table 25.2	Urine-Draining Structures			
Structure	**Epithelium**	**Number of Cell Layers**	**Word Origin**	
Calyces	Transitional	2–3	*calyx,* cup of a flower	
Ureter	Transitional	4–5	*oureter,* urinary canal	
Urinary Bladder	Transitional	> 6	*urinary,* relating to urine	
Urethra	In males, the proximal portions are lined with transitional epithelium and the distal portions are lined with stratified squamous. Similarly in females, the tissue begins as transitional epithelium before changing to stratified squamous epithelium.	NA	*ourethra,* canal leading from the bladder	

EXERCISE 25.8

HISTOLOGY OF THE URETERS

The **ureters** are long, epithelial-lined, fibromuscular tubes that transport urine from the hilum of the kidney to the urinary bladder. They are lined with **transitional epithelium**, which has the ability to stretch when urine is being transported through the ureters. Like other tubular structures in the body, the wall of the ureter is composed of multiple layers. The walls of the ureters, however, do not have a submucosa. There are three layers to the walls of the ureters: the mucosa, muscularis, and adventitia (**figure 25.12**). Peristaltic contraction of smooth muscle in the muscularis layer of the ureters transports urine from the renal pelvis to the urinary bladder. The arrangement of smooth muscle layers in the muscularis of the ureters is just the opposite of that of the digestive tract organs. The inner layer is composed of **longitudinal** smooth muscle, whereas the outer layer is composed of **circular** smooth muscle. In addition, the outer layer is an **adventitia** (and not serosa) because the ureters are **retroperitoneal** (*retro-*, behind, + *peritoneal,* referring to the peritoneum) and are not covered by a visceral peritoneum.

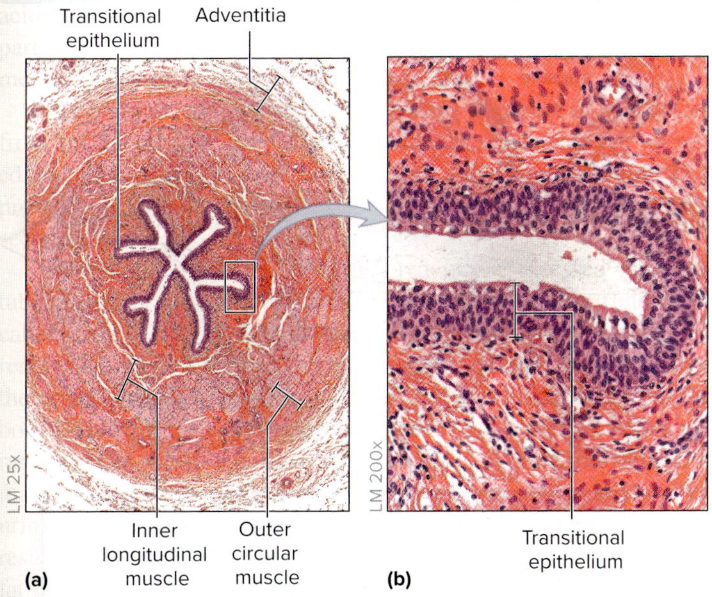

1. Obtain a histology slide demonstrating a cross-sectional view of a ureter. Place it on the microscope stage and bring the tissue sample into focus on low power.

2. Identify the following structures on the slide of the ureter, using figure 25.12 as a guide:

   ☐ **Adventitia**　　　　　　　☐ **Outer circular muscle**

   ☐ **Inner longitudinal**　　　☐ **Transitional**
   　　muscle　　　　　　　　　　**epithelium**

3. Sketch a cross section of a ureter as seen through the microscope in the space provided or on a separate sheet of paper. Be sure to label all the structures listed in step 2 in the drawing.

Figure 25.12 The Ureter. (*a*) Cross section through the entire ureter. (*b*) Close-up of the mucosa of the ureter.

(*a*) ©McGraw-Hill Education/Alvin Telser; (*b*) ©McGraw-Hill Education/Al Telser

(continued from previous page)

5. Test the following urine properties or constituents using individual test strips or combination test strips (such as Chemstrip® or Multistix®): pH, specific gravity, glucose, protein, ketones, bilirubin, leukocytes, and hemoglobin. Submerge a dipstick into the urine as directed on the strip container. Remove excess liquid on the strip by touching the strip to the inside rim of the urine container. Remove the strip and wait for the recommended time. Compare the color strip to the available color scale, as directed on the container. Record your observations in table 25.6.

6. Repeat steps 2–5 for an "unknown" sample provided by the instructor.

7. Make note of any pertinent observations here:

_____

_____

_____

EXERCISE 25.10B Urinalysis—Specific Gravity

1. Obtain a urine sample. If collecting your own urine, be sure to wash your hands prior to collecting the sample. Observe all laboratory safety procedures when handling urine samples. Obtain a clean, transparent collection cup. Be sure to collect a midstream sample (at least 50 mL) to avoid possible contamination by bacteria contained within the urethra. Refrigerate any unused urine immediately.

2. Pour the urine into a urinometer, filling it approximately two-thirds full.

3. Lower the specific gravity tube into the urinometer glass vial. Note that the specific gravity tube must float. If it does not float, continue to add urine until the urinometer glass vial floats.

4. Read the specific gravity by aligning the meniscus of the urine with the readings on the floating specific gravity tube (**figure 25.15**). Record the measurement in table 25.6 and compare this to normal values presented in the table.

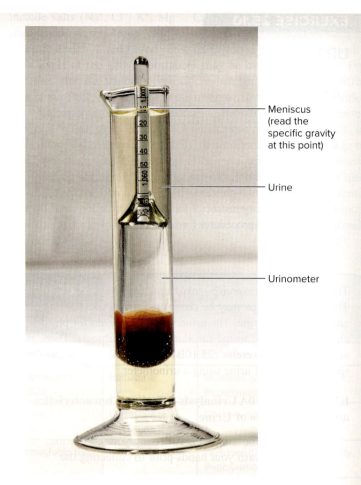

Meniscus (read the specific gravity at this point)

Urine

Urinometer

Figure 25.15 **Measuring Specific Gravity of Urine Using a Urinometer.** Fill the glass tube with enough urine to make the specific gravity tube float. Then observe the meniscus of the urine with the marks on the specific gravity tube to determine the specific gravity.

©Christine Eckel

5. Repeat steps 2–4 for an "unknown" urine sample provided by the instructor.

6. Make note of any pertinent observations here:

_____

_____

_____

Acid-Base Balance

Maintaining proper blood pH (7.35–7.45) is a critical homeostatic process that involves both short- and long-term regulation and employs multiple physiological processes. Because blood pH constantly changes, the first means of helping to prevent pH changes are the buffer systems. The most robust of the chemical buffers is the **carbonic acid** buffering system within the blood, which is described in detail in chapter 24. Additional chemical buffers in the blood include proteins and phosphates, which act in a similar manner to rapidly buffer against pH changes.

The respiratory system functions as a physiological buffering system. The respiratory system is tightly linked to the carbonic acid buffering system. As discussed in chapter 24, respiratory rates directly impact the amount of carbonic acid in the blood, thereby impacting blood pH. Accumulation of carbon dioxide, whether due to **hypoventilation** or resulting from reduced gas diffusion, leads to the increased formation of carbonic acid and the possible accumulation of hydrogen ions in the blood. This results in a drop in blood pH, or **respiratory acidosis**. Conversely, **hyperventilation** results in low levels of carbon dioxide (blowing off too much carbon dioxide), which may lead to **respiratory alkalosis**. When the source of acid does not involve changes in carbon dioxide levels due to alterations in the respiratory system, it is considered **metabolic**. Clinically, reviewing blood pH as well as carbon dioxide and bicarbonate levels reveals whether or not the patient is in acidosis (pH < 7.35) or in alkalosis (pH > 7.45), and if the source is respiratory (normal P_{CO_2} = 35–45 mm Hg) or metabolic (normal bicarbonate = 22–26 mEq/L blood). **Table 25.7** provides a quick reference guide for classifiying respiratory and metabolic acidosis and alkalosis.

The urinary system also functions as a physiological buffering system. For example, excessive fat metabolism and the loss of bicarbonate due to excessive diarrhea both significantly lower the blood pH. As the blood becomes more acidic, the kidneys secrete excess hydrogen ions and reabsorb more bicarbonate ions in the distal convoluted tubules and collecting tubules, thereby lowering urine pH. Conversely, as the blood becomes more basic, the rate of hydrogen secretion and bicarbonate reabsorption in the kidneys decreases.

Table 25.7	Classification of Respiratory versus Metabolic Acidosis/Alkalosis		
Condition		P_{CO_2}	**Bicarbonate**
Respiratory	Acidosis	> 45 mm Hg	22–26 mEq/L blood (normal range)
	Alkalosis	< 35 mm Hg	
Metabolic	Acidosis	35–45 mm Hg (normal range)	< 22 mEq/L blood
	Alkalosis		> 26 mEq/L blood

EXERCISE 25.11

A CLINICAL CASE STUDY IN ACID-BASE BALANCE

The purpose of this exercise is to apply the concept of acid-base balance to a real scenario, or clinical case. This exercise requires information presented in this chapter as well as previous chapters. Integrating multiple systems in a case study reinforces a concept that is so important in physiology: homeostasis.

Prior to beginning, state how both the respiratory system and the urinary system function to maintain homeostatic pH levels in the blood.

_____

_____

_____

_____

Consult with the laboratory instructor as to whether this exercise will be completed individually, in pairs, or in groups. Review the topics of diffusion and osmosis and the cardiovascular, lymphatic, respiratory, digestive, and urinary systems in the textbook and in previous chapters in this laboratory manual for assistance with this case study.

Case Study

Robert, now 60 years old, was diagnosed with type I diabetes mellitus at age 10 and has managed the disease with few complications by administering daily insulin injections. Fifteen years ago, Robert complained of extreme fatigue, and his health began to decline. Within a few years, Robert was diagnosed with a failing heart, chronic hypertension (high blood pressure), and peripheral artery disease (PAD), in which the peripheral vessels become narrowed or blocked.

Just 2 months ago, Robert exhibited signs of intermittent confusion followed by bouts of diarrhea and intense feelings of weakness. He was unable to regulate blood glucose levels, and he began experiencing respiratory symptoms. Eventually, his breathing became labored and difficult, particularly in the evenings as he would lie down to sleep. Sometimes, these symptoms would wake him up, sometimes after only 1 or 2 hours of sleep. Robert went to the hospital for evaluation.

Upon examination, doctors heard crackles (see chapter 24) bilaterally in his lungs. Doctors ordered a chest radiograph and a series of laboratory tests (**table 25.8**, 1st Admission). The results of these tests indicated that Robert was suffering from pneumonitis, or inflammation of the lung tissue. Robert was given a diuretic, which prevented the reabsorption of sodium in kidney tubule cells, and a broad-spectrum antibiotic to fight the infection. He was released from the hospital just two days later.

Robert's condition worsened during the following weeks. The frequency of diarrhea increased, and he was lethargic (lacking energy). Robert had used up his supply of insulin injections when his sister found him unresponsive in his home. Emergency technicians arrived on the scene and transported him to the hospital. They recorded a blood pressure of 90/50 mm Hg (reference = 110/70 mm Hg), a pulse of 90 beats per minute, and oxygen saturation of 90%. Soon, Robert's systolic blood pressure dropped to 70 mm Hg. Robert exhibited poor inspiratory effort, and crackles were heard bilaterally.

Robert was transported to the coronary care unit for further evaluation. Doctors performed an echocardiogram, which revealed some insufficiencies in the atrioventricular and semilunar valves, narrowing of the aorta, and hypertrophy of the left ventricle. Computed tomography (CT) of the chest revealed

(continued from previous page)

The corresponds to the Learning Objective(s) listed in the chapter opener outline.

Create and Evaluate

Analyze and Apply

Understand and Remember

Do You Know the Basics?

Exercise 25.1: Gross Anatomy of the Kidney

1. The layer of dense irregular connective tissue that directly covers and surrounds the entire kidney is the **1**

   _____ a. renal corpuscle

   _____ b. renal fascia

   _____ c. renal capsule

   _____ d. paranephric fat

2. The _____ (right/left) kidney is located more superiorly than the other. **2**

3. Place the following structures in order from innermost to outermost (1 = innermost; 4 = outermost). **3**

   _____ a. paranephric fat

   _____ b. perinephric fat

   _____ c. renal capsule

   _____ d. renal fascia

4. Structures that are located behind the peritoneum are called _____ (retroperitoneal/intraperitoneal). **4**

Exercise 25.2: Blood Supply to the Kidney

5. The renal _____ (artery/vein) supplies oxygen-rich blood to each kidney, whereas the renal _____ (artery/vein) drains blood from each kidney to the inferior vena cava. **5**

6. Place the following vessels in the correct order, from the entry into the kidney via the renal artery to the glomerulus (1 = first; 5 = last). **6**

   _____ a. afferent arteriole

   _____ b. arcuate artery

   _____ c. interlobar artery

   _____ d. interlobular artery

   _____ e. segmental artery

Exercise 25.3: Urine-Draining Structures Within the Kidney

7. Fluid that passes through nephron tubules, collecting tubules, and collecting ducts is called _____ (plasma/filtrate). **7**

8. Place the following structures in the correct order of urine flow (1 = first; 4 = last). **8**

   _____ a. major calyx

   _____ b. minor calyx

   _____ c. renal papilla

   _____ d. renal pelvis

Exercise 25.4: Gross Anatomy of the Ureters

9. Place the following structures in the correct order of urine flow from the kidney to the exterior of the body (1 = first; 3 = last). **9**

   _____ a. ureter

   _____ b. urethra

   _____ c. urinary bladder

10. The urinary bladder is located in the _____ (abdominal/pelvic) cavity. **10**

Exercise 25.5: Gross Anatomy of the Urinary Bladder and Urethra

11. Identify the three structures that form the boundaries of the urinary trigone. (Check three.) **11**

_____ a. right renal artery

_____ b. left renal artery

_____ c. right ureter

_____ d. left ureter

_____ e. urethra

12. Place the three sections of the male urethra in the correct order from proximal to distal (1 = first; 3 = last). **12**

_____ a. prostatic urethra

_____ b. membranous urethra

_____ c. spongy urethra

13. The internal urethral sphincter is composed of _____ (skeletal/smooth) muscle, whereas the external urethral sphincter is composed of _____ (skeletal/smooth) muscle. **13**

Exercise 25.6: Histology of the Renal Cortex

14. Which of the following are components of the nephron? (Check all that apply.) **14**

_____ a. collecting duct

_____ b. distal convoluted tubule

_____ c. glomerulus

_____ d. nephron loop

_____ e. proximal convoluted tubule

15. Fill in the following table with the type of epithelium lining each structure and a brief description of the function of each structure. **14** **15**

Structure	Epithelium	Function
Glomerulus		
Visceral Layer of Glomerular Capsule		
Parietal Layer of Glomerular Capsule		
PCT		
DCT		

16. The renal corpuscle consists of which of the following structures? (Circle all that apply.) **16**

_____ a. glomerular capillaries

_____ b. glomerular capsule

_____ c. macula densa

_____ d. podocytes

17. Filtration slits in the filtration membrane are formed by which of the following structures? (Circle one.) **17**

_____ a. basement membrane of glomerular capillaries

_____ b. endothelium of glomerular capillaries

_____ c. podocytes of the glomerular capsule

18. Which of the following structures are located in the renal cortex? (Check all that apply.) **18**

_____ a. collecting ducts _____ c. nephron loop _____ e. renal corpuscle

_____ b. distal convoluted tubule _____ d. proximal convoluted tubule

Exercise 25.7: Histology of the Renal Medulla

19. Which of the following structures are located in the renal medulla? (Check all that apply.) **19**

    _____ a. collecting ducts

    _____ b. distal convoluted tubule

    _____ c. nephron loop

    _____ d. proximal convoluted tubule

    _____ e. renal corpuscle

20. The thick limbs of the nephron loop are composed of simple _____ (cuboidal/squamous) epithelium, whereas the thin limbs of the nephron loop are composed of simple _____ (cuboidal/squamous) epithelium. **20**

Exercise 25.8: Histology of the Ureters

21. Place the layers of the wall of the ureter in order, from innermost to outermost (1 = innermost; 3 = outermost). **21**

    _____ a. adventitia

    _____ b. mucosa

    _____ c. muscularis

22. The muscularis layer of the ureter consists of an inner _____ (circular/longitudinal) layer of smooth muscle and an outer _____ (circular/longitudinal) layer of smooth muscle. **22**

Exercise 25.9: Histology of the Urinary Bladder

23. The urinary bladder has _____ (three/four) layers in its wall. **23**

24. The urinary bladder is lined with which type of epithelial tissue? (Circle one.) **24**

    a. simple cuboidal epithelium

    b. simple squamous epithelium

    c. stratified squamous epithelium

    d. transitional epithelium

Exercise 25.10: Urinalysis

25. Which of the following is the site of filtration in the nephron? (Circle one.) **25**

    a. distal convoluted tubule   c. nephron loop

    b. glomerulus   d. proximal convoluted tubule

26. Which of the following is an abnormal constituent of urine? (Circle one.) **26**

    a. glucose

    b. sodium ions

    c. trace amounts of protein

    d. urea

    e. water

27. Which of the following should normally not be found in urine? **27**

    a. sodium

    b. glucose

    c. potassium

    d. phosphates

28. A greater concentration of substances in urine (e.g., glucose or protein) would cause the urine to have a _____ (higher/lower) specific gravity than water. **28**

Exercise 25.11: A Clinical Case Study in Acid-Base Balance

29. The urinary system aids in _____ (short-/long-) term regulation of acid-base balance. **29**

30. Which of the following systems is/are employed to regulate acid-base balance in the blood? (Check all that apply.) **30**

    _____ a. cardiovascular system

    _____ b. lymphatic system

    _____ c. respiratory system

    _____ d. urinary system

31. Blood pH of 7.1 with elevated partial pressure of carbon dioxide and normal bicarbonate levels is indicative of _____ (metabolic/respiratory) _____ (acidosis/alkalosis). **31**

Can You Apply What You've Learned?

32. A surgeon has the job of removing a patient's kidney. Her approach to the kidney will be to cut through skin and adipose tissue of the lower back, remove portions of the most inferior ribs and a portion of the quadratus lumborum muscle, and then enter the retroperitoneal space. Once the surgeon has entered the retroperitoneal space, what are the layers of tissue (in order) that the surgeon must cut through to reach the kidney?

    _____

    _____

33. Describe why muscular contractions are needed to transport urine through the ureters. _____

    _____

    _____

34. Patients with chronic hypertension are at risk of suffering damage to the kidneys. Describe the specific structures of the nephron that might be damaged in a patient with hypertension. _____

    _____

    _____

    _____

35. Doctors will often first advise patients with hypertension to eat a diet low in sodium. Discuss why this may be beneficial in reducing systemic blood pressure. (Hint: Remember that an increased blood volume will correspond to an increase in systemic blood pressure.) _____

    _____

    _____

    _____

36. Patients in kidney failure require dialysis to perform the processes that are normally accomplished by the nephron. A dialysis membrane does this by mimicking the filtration membrane in the nephron. Describe the essential function typically accomplished by the kidneys that is replaced by the dialysis machine. _____

    _____

    _____

    _____

37. Describe why a person suffering from kidney failure would experience edema, which would be alleviated by dialysis. _____

    _____

    _____

    _____

38. Among its many effects, alcohol inhibits ADH secretion from the posterior pituitary gland. Describe the typical action of this hormone. Then discuss how ingestion of alcohol will affect urine formation. _____

    _____

    _____

    _____

39. A patient with suspected internal bleeding is admitted to the hospital. His systemic blood pressure is 80/50 mm Hg. Predict the patient's GFR given his current systemic blood pressure. Justify your answer. _____

    _____

    _____

Can You Synthesize What You've Learned?

40. In severe cases of hypertension, an angiotensin-converting enzyme (ACE) inhibitor may be prescribed for a patient. The enzyme ACE converts angiotensin I to angiotensin II. One target of angiotensin II is the adrenal cortex, where angiotensin II stimulates the release of the hormone aldosterone (see chapter 19). Aldosterone increases sodium reabsorption by kidney tubule cells. Describe how an ACE inhibitor will influence blood volume and blood pressure (as it relates to aldosterone). ___

_____

_____

_____

_____

41. In addition to stimulating the release of aldosterone, angiotensin II is a widespread vasoconstrictor. This means that smooth muscle in the walls of blood vessels will be stimulated to contract when angiotensin II is formed.

    a.   Describe what will happen to GFR when angiotensin II stimulates the afferent arterioles within the nephron to constrict. _____

    _____

    _____

    _____

    b.   Describe what effect widespread vasoconstriction will have on systemic blood pressure. _____

    _____

    _____

    _____

42. Susan, a 25-year-old female, contracted food poisoning likely due to eating undercooked eggs. She has been vomiting violently and cannot keep food or liquids down. Susan sought medical attention. Doctors measured her blood pH at 7.5, P_{CO_2} at 40 mm Hg, P_{O_2} at 95 mm Hg, O_2 saturation at 97%, and bicarbonate at 32 mEq/L.

    a.   Classify Susan's case as respiratory acidosis, respiratory alkalosis, metabolic acidosis, or metabolic alkalosis, and justify your answer. _____

    _____

    _____

    _____

    _____

    b.   Describe compensatory mechanisms that may have been initiated to adjust Susan's blood pH back toward normal homeostatic levels. _____

    _____

    _____

    _____

The Digestive System

OUTLINE AND LEARNING OBJECTIVES

Module 12: Digestive System

INTRODUCTION

The digestive (*digero*, to force apart, dissolve) system is responsible for the breakdown and absorption of molecules needed by the body for energy, maintenance, and ultimately survival. This is quite a complex feat! Most of the molecules we ingest in our food are large macromolecules like starches, proteins, and lipids. But in order to meet the specific needs of our cells, these macromolecules must first be broken down into their individual, absorbable building blocks, like monosaccharides, amino acids, and fatty acids. These processes occur within a long tube called the gastrointestinal (GI) tract, which consists of the mouth, pharynx, esophagus, stomach, small intestine, and large intestine. Foodstuffs moving through this tube are broken down by mechanical (physical) and chemical processes into their respective building blocks in an incredibly efficient manner. The organs within the GI tract do not work alone, however. Several important accessory structures, including the salivary glands, liver, gallbladder, and pancreas, add secretions to the GI lumen that are critical for digestion.

Each region of the GI tract is modified to suit the particular needs of the organ, and the structural composition of these organs reflects their functions. From the esophagus onward, these organs are composed of four layers (tunics): mucosa, submucosa, muscularis, and adventitia or serosa. These tunics share many similarities but have critical structural differences that allow each organ to perform its specific function. For example, the mucosa of the esophagus is composed of stratified squamous epithelial tissue and therefore is protective. In contrast, the mucosa of the small intestine is composed of simple columnar epithelial tissue and is important in absorption and secretion.

The laboratory exercises in this chapter explore how the various parts of the GI tract are modified to suit the particular needs of each organ. In addition, they explore the structure and function of accessory organs such as the liver and pancreas. As the structures of the digestive system are observed, various circulatory and lymphatic structures associated with these structures will also be reconsidered in the context of their roles in the digestive system.

The exercises in this chapter begin with an exploration of the gross anatomy of the GI tract and accessory organs. Next, the histological organization of these structures is examined. The Gross Anatomy and Histology sections are organized to study structures in order, beginning at the mouth and ending at the anus. In other words, the structures are studied in the order in which a bolus of food encounters them as it moves through the GI tract. It is not imperative that the structures be observed in this order. The advantage of studying them in this order is that it allows one to reflect on how the different parts of the GI tract work together to accomplish the process of digestion. Finally, the Physiology section of this chapter explores the processes of chemical digestion by digestive enzymes, including the effect of temperature on enzyme activity.

List of Reference Tables

Chapter 26: The Digestive System

Name: _____

Date: _____ Section: _____

These Pre-Laboratory Worksheet questions may be assigned by instructors through their ▦ connect course.

1. Humans have _____ (two/three) pairs of salivary glands.

2. The oral cavity opens into the _____ (nasopharynx/laryngopharynx/oropharynx).

3. The esophagus is located _____ (anterior/posterior) to the trachea.

4. The muscularis of the stomach contains _____ (two/three) layers of smooth muscle.

5. Which of the following cell types lines the mucosa of the small and large intestines? (Circle one.)

   a. simple cuboidal cells and goblet cells

   b. simple columnar cells and goblet cells

   c. ciliated simple cuboidal cells and goblet cells

   d. simple columnar cells with microvilli and goblet cells

   e. ciliated simple columnar cells and goblet cells

6. Which of the following places the three segments of the small intestine in order from proximal to distal? (Circle one.)

   a. jejunum, duodenum, ileum

   b. ileum, jejunum, duodenum

   c. duodenum, jejunum, ileum

   d. duodenum, ileum, jejunum

   e. jejunum, ileum, duodenum

7. The ducts from the liver, gallbladder, and pancreas empty into the _____ (pyloris/duodenum).

8. The liver consists of _____ (two/four) lobes. The largest lobe of the liver is the _____ (right/left) lobe.

9. The cecum is located at the beginning of the _____ (ascending/descending) colon.

10. Which of the following is the monomer that makes up proteins? (Circle one.)

    a. monosaccharides

    b. nucleic acids

    c. amino acids

    d. disaccharides

    e. fatty acids

11. Glucose is a _____ (monosaccharide/disaccharide), whereas sucrose is a _____ (monosaccharide/disaccharide).

12. Which of the following is the structural component of triglycerides to which fatty acids are attached? (Circle one.)

    a. glucose

    b. glycerol

    c. fatty acids

    d. amino acids

    e. glycogen

GROSS ANATOMY

The Gastrointestinal (GI) Tract

EXERCISE 26.1

OVERVIEW OF THE GI TRACT

Before covering the details of individual organs that compose the digestive system, it is useful to do a short review of the structures encountered in the GI tract as one moves from mouth to anus.

1. Identify the structures listed in **figure 26.1** on a torso model, using the textbook as a guide.

2. Label the organs of the GI tract in figure 26.1. When labeling each organ, identify at least one digestive function of that organ.

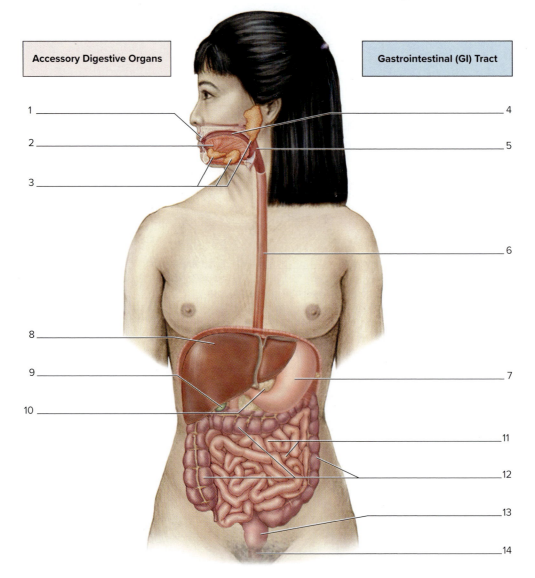

Accessory Digestive Organs

Gastrointestinal (GI) Tract

Figure 26.1   **Overview of the Digestive System.** Use the terms listed to fill in the numbered labels in the figure.

- ☐ Anal canal
- ☐ Esophagus
- ☐ Gallbladder
- ☐ Large intestine
- ☐ Liver
- ☐ Oral cavity
- ☐ Pancreas
- ☐ Pharynx
- ☐ Rectum
- ☐ Salivary glands
  Parotid
  Sublingual
  Submandibular
- ☐ Small intestine
- ☐ Stomach
- ☐ Teeth
- ☐ Tongue

The Oral Cavity, Pharynx, and Esophagus

The oral cavity contains a number of digestive system structures, including the teeth, salivary glands, lips, and tongue. These structures are important for wetting and manipulating food as it enters the GI tract.

The resulting **bolus** (*bolos*, lump; bolus = a wet mass of food) leaves the oral cavity, travels through the oropharynx (*oris*, mouth, + *pharynx*, throat), and enters the esophagus (*oisophagos*, gullet), which transports the bolus to the stomach. The esophagus meets up with the stomach just after it passes through the esophageal hiatus in the diaphragm.

GROSS ANATOMY OF THE ORAL CAVITY, PHARYNX, AND ESOPHAGUS

1. Observe a cadaver or a classroom model demonstrating the head and neck.

2. Using the textbook as a guide, identify the structures listed in **figure 26.2** on the cadaver or classroom model. Then label them in figure 26.2.

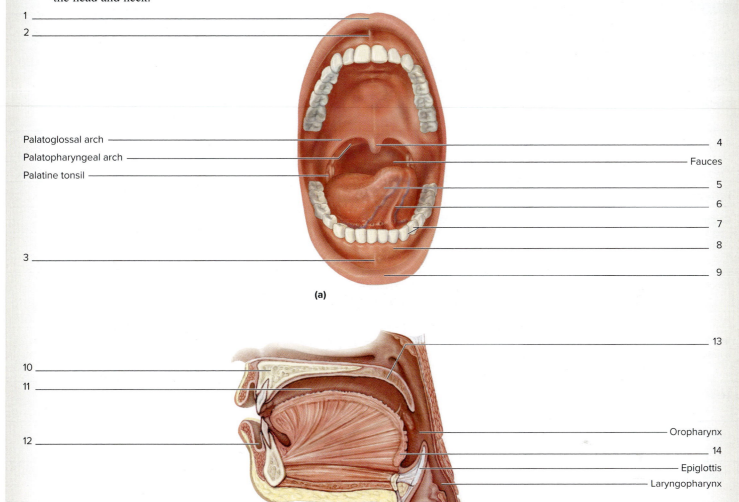

Palatoglossal arch

Palatopharyngeal arch

Palatine tonsil

4

Fauces

5

6

7

8

9

(a)

13

10

11

12

Oropharynx

14

Epiglottis

Laryngopharynx

15

Larynx

(b)

16

Parotid duct

Masseter muscle

Mucosa (cut)

Sublingual duct

Submandibular duct

17

Mylohyoid muscle (cut)

18

(c)

Figure 26.2 Oral Cavity. Use the terms listed to fill in the numbered labels in the figure.

(a) Anterior View of the Oral Cavity

☐ Gingivae
☐ Inferior labial frenulum
☐ Inferior lip
☐ Lingual frenulum
☐ Superior labial frenulum

☐ Superior lip
☐ Teeth
☐ Tongue
☐ Uvula

(b) Midsagittal View of the Oral Cavity and Pharynx

☐ Esophagus
☐ Hard palate
☐ Lingual tonsil

☐ Oral cavity
☐ Soft palate
☐ Vestibule

(c) Salivary Glands

☐ Parotid salivary gland
☐ Sublingual salivary gland

☐ Submandibular salivary gland

The Stomach

The stomach is a large, sac-like organ where both mechanical and chemical digestion continue on the bolus. The stomach is located in the epigastric abdominopelvic region, superficial to the pancreas, and deep to the anterior abdominal wall. **Table 26.1** lists the major features of the stomach and describes the function of each.

Table 26.1	Gross Anatomical Regions and Features Associated with the Stomach	
	Description	**Word Origin**
Regions		
Body	Main part of the stomach located between the fundus and the pylorus	*body*, the principal mass of a structure
Cardia	Small, narrow, superior portion of the stomach where esophagus enters	*kardia*, heart; relating to the part of the stomach nearest the heart
Fundus	The dome-shaped part of the stomach that lies superior to the cardiac notch	*fundus*, bottom
Pylorus	The distal region of the stomach that opens into the duodenum	*pyloros*, a gatekeeper
Features		
Gastric Folds (rugae)	Folds of the mucosal lining of the stomach	*ruga,* a wrinkle
Greater Curvature	The large, inferior convex portion of the stomach. It is one attachment point for the greater omentum.	*greater*, larger
Greater Omentum	A fold of four layers of peritoneum that attaches to the greater curvature of the stomach, drapes over the abdominal contents, and folds back upon itself to attach to the transverse colon	*omentum*, the membrane that encloses the bowels
Inferior Esophageal (cardiac) Sphincter	A physiological sphincter composed of the part of the diaphragm that surrounds the esophagus. When the diaphragm contracts, it closes off this opening, preventing reflux of stomach contents back into the esophagus.	*kardia*, heart; relating to the sphincter of the stomach nearest the heart
Lesser Curvature	The small, superior concave portion of the stomach. It is one attachment point of the lesser omentum.	*lesser*, smaller
Lesser Omentum	A fold of four layers of peritoneum that extends between the liver and the lesser curvature of the stomach	*omentum*, the membrane that encloses the bowels
Pyloric Sphincter	An anatomical sphincter composed of smooth muscle within the wall of the pylorus. It controls the passage of chyme from the stomach to the duodenum.	*pyloros*, a gatekeeper, + *sphinkter*, a band

EXERCISE 26.3

GROSS ANATOMY OF THE STOMACH

1. Observe a cadaver or a classroom model demonstrating the stomach.

2. Identify the structures listed in **figure 26.3** on the cadaver or classroom model, using the textbook as a guide. Then label them in figure 26.3.

3. Sketch the stomach in the space provided or on a separate sheet of paper. Be sure to label all of the structures listed below:

☐ **Body**   ☐ **Greater curvature**

☐ **Cardia**   ☐ **Lesser curvature**

☐ **Fundus**   ☐ **Pylorus**

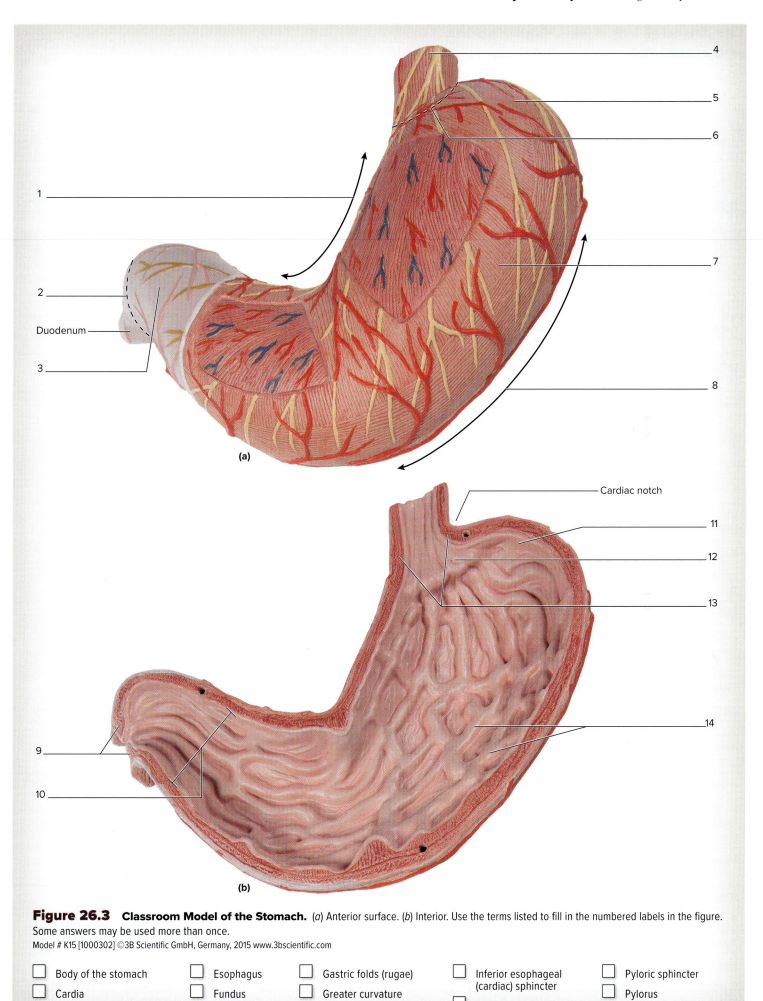

Figure 26.3 **Classroom Model of the Stomach.** (*a*) Anterior surface. (*b*) Interior. Use the terms listed to fill in the numbered labels in the figure. Some answers may be used more than once.

Model # K15 [1000302] ©3B Scientific GmbH, Germany, 2015 www.3bscientific.com

- ☐ Body of the stomach
- ☐ Cardia
- ☐ Esophagus
- ☐ Fundus
- ☐ Gastric folds (rugae)
- ☐ Greater curvature
- ☐ Inferior esophageal (cardiac) sphincter
- ☐ Lesser curvature
- ☐ Pyloric sphincter
- ☐ Pylorus

The Duodenum, Liver, Gallbladder, and Pancreas

The **duodenum** (*duodeno-*, breadth of 12 fingers) is the first part of the small intestine. It is C-shaped, and mostly retroperitoneal, which allows it to be anchored to the posterior abdominal wall. This is advantageous because a number of ducts coming from the liver, gallbladder, and pancreas empty their contents into the duodenum. The relationships between the duodenum and the liver, gallbladder, and pancreas are critically important for the process of digestion. The liver produces **bile**, a substance that emulsifies fats, which is temporarily stored and concentrated within the **gallbladder**. The pancreas produces **pancreatic juice**, which contains digestive enzymes and bicarbonate ion. When the liver, gallbladder, and pancreas release their secretions into the duodenum, the acidity of the chyme (*chymos*, juice) that has entered the duodenum from the stomach is neutralized and the digestion of proteins and carbohydrates continues. The digestion of fats and nucleic acids also occurs in the duodenum. **Table 26.2** lists the major features of the liver, pancreas, and duct system and describes the functions of each component.

Table 26.2	Gross Anatomical Features of the Liver, Gallbladder, Pancreas, and Their Associated Ducts	
	Description	**Word Origin**
Liver		
Caudate Lobe	A small lobe of the liver located between the right and left lobes and on the posterior, inferior part of the liver	*caudate*, possessing a tail, + *lobos*, lobe
Falciform Ligament	A fold of peritoneum that extends from the diaphragm and anterior abdominal wall to the liver. Its free inferior border contains the round ligament of the liver.	*falx*, sickle, + *forma*, form
Left Lobe	The second largest lobe of the liver. Extends from the falciform ligament toward the midline of the body.	*lobos*, lobe
Porta Hepatis	A depression on the inferomedial part of the liver that contains the hepatic artery, hepatic portal vein, and common bile duct	*porta*, gate, + *hepatikos*, liver
Quadrate Lobe	A small lobe of the liver located between the right and left lobes and on the anterior, inferior part of the liver between the gallbladder and the round ligament	*quadratus*, square, + *lobos*, lobe
Right Lobe	The largest lobe of the liver, it is on the right side of the abdomen and composes over half of the mass of the liver	*lobos*, lobe
Round Ligament of the Liver (ligamentum teres)	A remnant of the fetal umbilical vein, which connects to the umbilicus. Located within the free edge of the falciform ligament on the anterior abdominal wall.	*ligamentum*, a bandage, + *teres*, round
Gallbladder and Duct System		
Accessory Pancreatic Duct	Excretory duct located in the head of the pancreas. Empties into duodenum at the minor duodenal papilla.	*pankreas*, the sweetbread
Common Bile Duct	The bile duct formed from the union of the common hepatic duct and the cystic duct. Empties into the hepatopancreatic ampulla.	*bilis*, a yellow/green fluid produced by the liver
Common Hepatic Duct	Formed by the joining of the right and left hepatic ducts. Drains bile into the common bile duct.	*hepatikos*, liver
Cystic Duct	The bile duct that transports bile from the gallbladder to the common bile duct	*cystic*, relating to the gallbladder
Gallbladder	A sac-like appendage of the liver that stores and concentrates bile	*gealla*, bile, + *blaedre*, a distensible organ
Hepatopancreatic Ampulla	A duct formed by the joining of the common bile duct and the main pancreatic duct	*hepatikos*, relating to the liver, + *pancreatic*, relating to the pancreas, + *ampulla*, a two-handled bottle
Main Pancreatic Duct	The main excretory duct of the pancreas. Runs longitudinally in the center of the gland and empties into the duodenum at the major duodenal papilla.	*pankreas*, the sweetbread
Major Duodenal Papilla	A raised "nipple-like" bump located on the posterior wall of the descending part of the duodenum. The hepatopancreatic ampulla empties its contents here.	*major*, great, + *papilla*, a nipple
Minor Duodenal Papilla	A small, raised "nipple-like" bump located superior to the major duodenal papilla. Contains the opening of the accessory pancreatic duct.	*minor*, smaller, + *papilla*, a nipple
Right/Left Hepatic Duct	Drain bile from the right lobe and left lobe of the liver, respectively	*hepatikos*, liver
Pancreas		
Body	The main portion of the pancreas extending between the head and the tail	*pankreas*, the sweetbread
Head	The portion of the pancreas that sits in the depression formed by the curvature of the duodenum	*pankreas*, the sweetbread
Tail	The tapered, right end of the pancreas located near the hilum of the spleen	*pankreas*, the sweetbread

EXERCISE 26.4

GROSS ANATOMY OF THE DUODENUM, LIVER, GALLBLADDER, AND PANCREAS

1. Obtain a classroom model demonstrating the relationships among the duodenum, liver, gallbladder, and pancreas **(figure 26.4)**, or view these structures in the superior abdominal cavity of a prosected human cadaver.

2. Identify the structures listed in figure 26.4 on the classroom model or human cadaver, using table 26.2 and the textbook as guides. Then label them in figure 26.4.

3. Sketch the ducts coming from the liver, gallbladder, and pancreas in the space provided or on a separate sheet of paper. Use table 26.2 and the textbook as guides. Show how the ducts merge to eventually empty their contents into the duodenum. Label each duct and organ in the drawing.

4. Obtain a model of the liver **(figure 26.5)**, or view the liver from a human cadaver.

5. Identify the structures listed in figure 26.5 on the model of the liver or human cadaver, using table 26.2 and the textbook as guides. Then label them in figure 26.5.

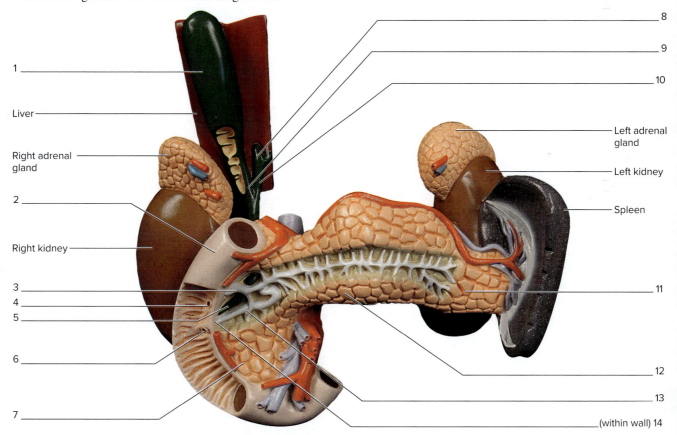

Figure 26.4 **Classroom Model of the Duodenum, Liver, Gallbladder, and Pancreas.** Anterior view. Use the terms listed to fill in the numbered labels in the figure.
©Denoyer-Geppert. Photo by Christine Eckel

☐ Accessory pancreatic duct
☐ Body of pancreas
☐ Common bile duct
☐ Common hepatic duct

☐ Cystic duct
☐ Duodenum
☐ Gallbladder
☐ Head of pancreas

☐ Hepatopancreatic ampulla
☐ Left and right hepatic ducts
☐ Main pancreatic duct
☐ Major duodenal papilla

☐ Minor duodenal papilla
☐ Tail of pancreas

(continued on next page)

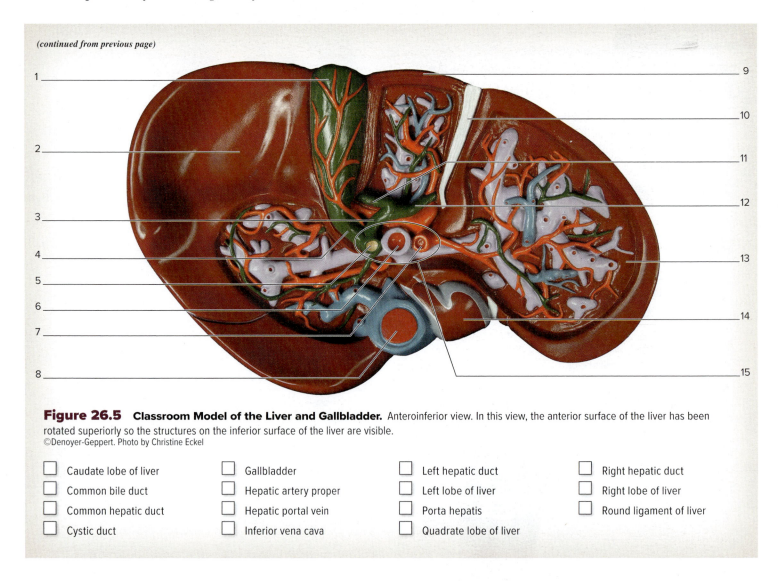

(continued from previous page)

Figure 26.5   **Classroom Model of the Liver and Gallbladder.** Anteroinferior view. In this view, the anterior surface of the liver has been rotated superiorly so the structures on the inferior surface of the liver are visible.
©Denoyer-Geppert. Photo by Christine Eckel

☐ Caudate lobe of liver   ☐ Gallbladder   ☐ Left hepatic duct   ☐ Right hepatic duct

☐ Common bile duct   ☐ Hepatic artery proper   ☐ Left lobe of liver   ☐ Right lobe of liver

☐ Common hepatic duct   ☐ Hepatic portal vein   ☐ Porta hepatis   ☐ Round ligament of liver

☐ Cystic duct   ☐ Inferior vena cava   ☐ Quadrate lobe of liver

The Jejunum and Ileum of the Small Intestine

The **jejunum** and **ileum** compose the majority of the small intestine. In general the jejunum is located in the upper left part of the abdominal cavity and the ileum is located in the lower right part of the abdominal cavity. The two segments can be distinguished from each other both histologically and grossly. Distinguishing anatomical features include: circular folds, encroaching fat, arterial arcades, and vasa recta (**figure 26.6**). **Circular folds** are the mucosal folds found within the lumen of the small intestine. **Encroaching fat** is mesenteric fat that "rides up" upon the wall of the intestine. **Arterial arcades** are arching branches of the mesenteric arteries, and **vasa recta** (*vasa*, vessel, + *rectus*, straight) are straight vessels that come off of the arterial arcades and enter the small intestine proper. **Table 26.3** summarizes the gross anatomical features that distinguish the jejunum from the ileum.

EXERCISE 26.5

GROSS ANATOMY OF THE JEJUNUM AND ILEUM OF THE SMALL INTESTINE

1. Observe a classroom model of the abdominal cavity or the abdominal cavity of a prosected human cadaver in which the small intestine is intact.

2. Identify the gross structures listed in figure 26.6 on the classroom model of the abdomen or in the abdominal cavity of the human cadaver. Use table 26.3 and the textbook as guides.

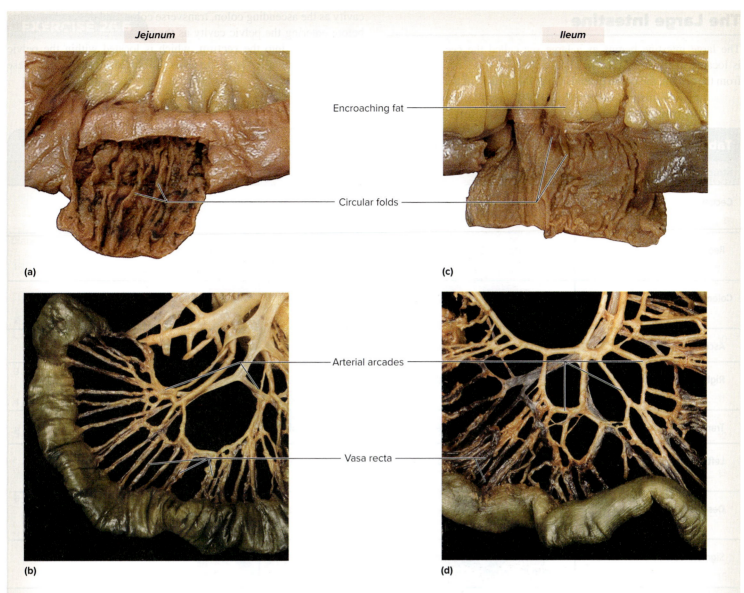

Figure 26.6 **The Jejunum and Ileum.** (*a*) The jejunum contains many deep circular folds and has no encroaching fat. (*b*) The blood vessels serving the jejunum consist of short arterial arcades and long vasa recta. (*c*) The ileum contains few shallow circular folds and has encroaching fat. (*d*) The blood vessels serving the ileum consist of large arterial arcades and short vasa recta.

(*a, c*) ©McGraw-Hill Education/Photo and Dissection by Christine Eckel; (*b, d*) Courtesy of David A. Morton and Chris Steadman, University of Utah School of Medicine

Table 26.3	Gross Anatomical Differences Between the Jejunum and the Ileum			
Part of the Small Intestine	**Circular Folds**	**Encroaching Fat**	**Arterial Arcades**	**Vasa Recta**
Jejunum	Deep, many	No	Fewer, larger	Longer
Ileum	Shallow, few	Yes	More, smaller, stacked upon each other	Shorter

Table 26.5	Salivary Gland Structures		
Structure	**Description and Location**	**Function**	**Word Origin**
Alveolus	The grape-shaped secretory portion of a gland	NA	*acinus*, grape
Mucous Cells	Cells have flattened nuclei that are located on the basal surface. Mainly located along the tubules of salivary glands.	Secrete mucin	*mucosus*, mucous
Myoepithelial Cells	Flattened cells are located around the alveoli and the long axes of the ducts. They are sometimes difficult to identify in light microscopy.	Contraction of these cells expels the secretions from salivary glands	*mys*, muscle, + *epithelial*, relating to epithelial tissues
Serous Cells	Cells have round nuclei and contain numerous secretory granules.	Secrete glycoproteins, electrolytes, and salivary amylase	*serous*, having a watery consistency

Table 26.6	Histological Characteristics of Salivary Glands			
Gland	**Secretory Cells**	**% of Saliva**	**Opening**	**Word Origin**
Parotid	**Serous:** mostly serous cells, with some adipose connective tissue	26–30%	Empties via the parotid duct opposite the second upper molar	*para*, beside, + *ous*, ear
Sublingual	**Mixed:** mostly mucous cells, with some serous cells	3–5%	Empties via multiple ducts into either the submandibular duct or directly into the oral cavity	*sub-*, under, + *lingual*, the tongue
Submandibular	**Mixed:** mostly serous cells (~80% of tissue), with some mucous cells (~20% of tissue)	60–70%	Empties via the submandibular ducts between the lingual frenulum and the mandible	*sub-*, under, + *mandible*, the mandible

EXERCISE 26.7

HISTOLOGY OF THE SALIVARY GLANDS

1. Obtain a compound microscope and histology slides of the parotid, submandibular, and sublingual salivary glands.

2. Place the slide of the **parotid gland** on the microscope stage. Bring the tissue sample into focus on low power and then switch to high power.

3. Identify the following structures on the slide of the parotid gland, using figure 26.9*b* and tables 26.5 and 26.6 as guides:

   ☐ **Adipocytes**   ☐ **Serous cells**

4. Place the slide of the **submandibular gland** on the microscope stage. Bring the tissue sample into focus on low power and then switch to high power.

5. The serous cells in the submandibular glands are located surrounding the mucous cells and are arranged in a half-moon shape. Scan the slide to locate these serous cells (figure 26.9*c*).

6. Identify the following structures on the slide of the submandibular gland, using figure 26.9*c* and tables 26.5 and 26.6 as guides:

   ☐ **Mucous cells**   ☐ **Serous cells**
   ☐ **Salivary duct**

7. Place the slide of the **sublingual gland** on the microscope stage. Bring the tissue sample into focus on low power and then switch to high power.

8. The sublingual gland is similar to the submandibular gland in that it contains serous cells. Two characteristics help distinguish the sublingual gland from the submandibular gland. The sublingual gland contains fewer adipocytes and greater numbers of mucous cells than the submandibular gland (table 26.6 and figure 26.9*d*).

9. Identify the following structures on the slide of the sublingual gland, using figure 26.9*d* and tables 26.5 and 26.6 as guides:

   ☐ **Mucous cells**   ☐ **Serous cells**
   ☐ **Salivary duct**

The Stomach

The **stomach** is an organ that mixes the bolus of food entering from the esophagus with gastric juices released from cells of the stomach mucosa. The bolus is changed to a liquid puree called **chyme** (*chymos*, juice). The stomach, like other organs along the GI tract, is composed of four tunics as illustrated in **figure 26.10**. Although these four tunics are present in the GI tract organs, modifications of each tunic may occur that are important for the function of each organ. For example, the mucosa of the stomach is modified to form **gastric pits**, invaginations of the epithelium that contain the openings of **gastric glands (figure 26.11** and **table 26.7)**. The cell types in the gastric pits and glands vary in different regions of the stomach. **Table 26.8** describes the structure of the pits and glands, and **table 26.9** describes the five cell types that compose the gastric pits and glands.

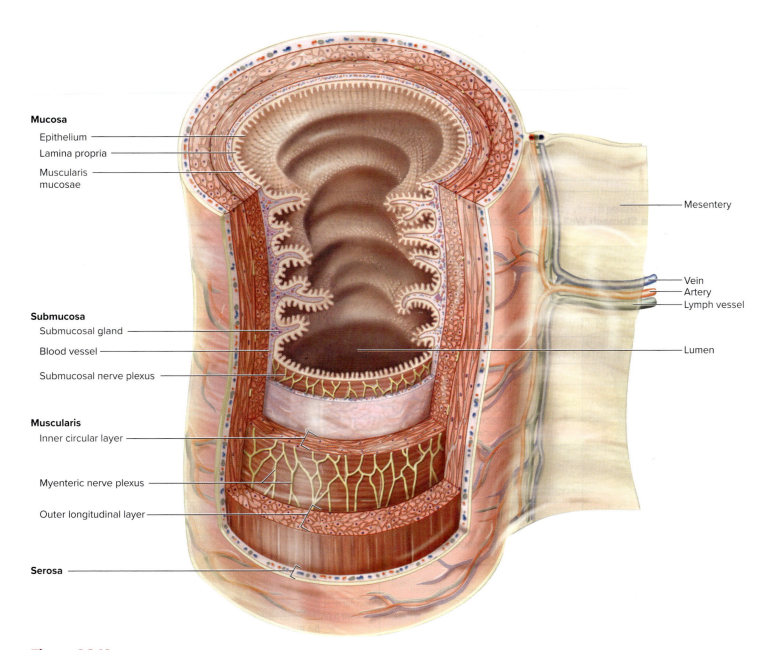

Mucosa
- Epithelium
- Lamina propria
- Muscularis mucosae

Submucosa
- Submucosal gland
- Blood vessel
- Submucosal nerve plexus

Muscularis
- Inner circular layer
- Myenteric nerve plexus
- Outer longitudinal layer

Serosa

Mesentery

Vein
Artery
Lymph vessel

Lumen

Figure 26.10 Wall Layers of the GI Tract. The four wall layers, or tunics, of the GI tract are the mucosa, submucosa, muscularis, and adventitia/serosa.

EXERCISE 26.9

HISTOLOGY OF THE SMALL INTESTINE

1. Obtain a histology slide of the small intestine and place it on the microscope stage. If the laboratory is equipped with slides of each section of the small intestine (duodenum, jejunum, and ileum), be sure to view all three. If not, use figure 26.13 to decide which part of the small intestine is on the slide.

2. Bring the tissue sample into focus at low power and move the microscope stage until the epithelium is at the center of the field of view. Then change to high power. What type of epithelium lines the mucosa of the small intestine?

_____

_____

What surface modifications are present? _____

_____

What is the purpose of these surface modifications?

_____

3. Identify the following structures on the slide(s), using figure 26.13 as a guide. Structures indicated with an asterisk (*) may not be present on all small intestine sections.

- ☐ Duodenal glands*
- ☐ Epithelium
- ☐ Lamina propria
- ☐ Mucosa
- ☐ Muscularis
- ☐ Muscularis mucosae
- ☐ Peyer patches*
- ☐ Serosa
- ☐ Submucosa
- ☐ Villi

4. Sketch the histology of the duodenum, jejunum, and ileum of the small intestine in the space provided or on a separate sheet of paper. Identify the histological features that differentiate these three portions of the small intestine from each other (refer to figure 26.13 for reference).

Duodenum

_____ ×

Jejunum

_____ ×

Ileum

_____ ×

The Large Intestine

Chyme leaving the small intestine enters the **large intestine** at the iliocecal junction. A valve, the **iliocecal valve**, controls the passage of chyme from the small intestine to the large intestine. Recall that the vast majority of nutrients are absorbed in the small intestine.

The function of the large intestine is mainly to absorb water (and some electrolytes) from the chyme that remains, and compact the waste products as **feces** for elimination from the body. The epithelium of the large intestine contains many goblet cells, which help lubricate the epithelium to ease the passage of feces through the large intestine.

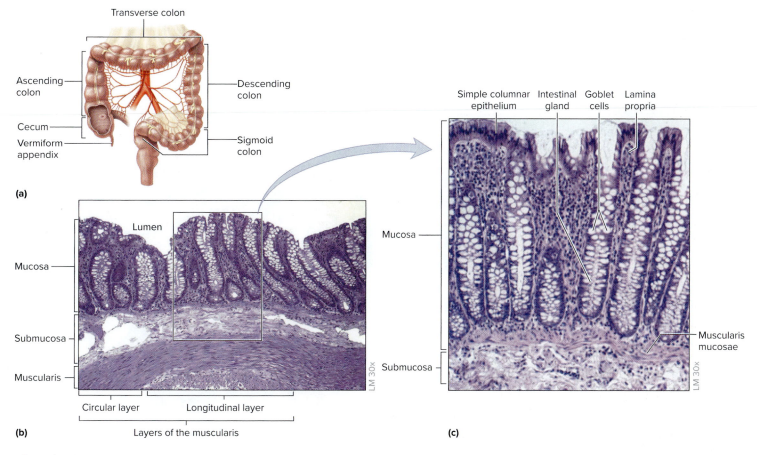

Figure 26.14 **The Large Intestine.** (*a*) Parts of the large intestine. (*b*) Histology of the wall of the large intestine. (*c*) Close-up of the epithelium lining the large intestine; numerous goblet cells are present.

(*b, c*) ©Victor P. Eroschenko

EXERCISE 26.10

HISTOLOGY OF THE LARGE INTESTINE

1. Obtain a histology slide of the large intestine (colon) and place it on the microscope stage.

2. Bring the tissue sample into focus on low power and identify the epithelial layer (**figure 26.14**). Move the microscope stage so the epithelium is at the center of the field of view and then change to high power.

3. Identify the following structures in the slide of the large intestine, using figure 26.14 and the textbook as guides:

   - ☐ **Epithelium**
   - ☐ **Goblet cells**
   - ☐ **Lamina propria**
   - ☐ **Mucosa**
   - ☐ **Muscularis**
   - ☐ **Muscularis: circular layer**
   - ☐ **Muscularis: longitudinal layer**
   - ☐ **Muscularis mucosae**
   - ☐ **Serosa**
   - ☐ **Submucosa**

4. Sketch the histology of the large intestine as seen through the microscope in the space provided or on a separate sheet of paper. Label the four major layers of the wall of the large intestine in the drawing.

_____ ✕

The Liver

The **liver** is the largest accessory organ in the digestive system. Indeed, it is the largest internal organ within the human body. The liver performs numerous vital functions that include detoxifying the blood, storing nutrients, and producing plasma proteins. However, its primary function in digestion is the production of bile. The structural and functional unit of the liver is a **hepatic lobule**, which is a hexagonally shaped structure consisting of strands of **hepatocytes** (the strands of hepatocytes are **hepatic cords**) radiating from a

central vein in the middle **(figure 26.15)**. In the areas where the outer edges of the hepatic lobules come together there are **portal triads**, which consist of a branch of the hepatic artery, a branch of the hepatic portal vein, and a bile duct. Within the hepatic lobules, in between the hepatic cords, are **hepatic sinusoids**: capillaries that carry blood from the branches of the hepatic artery and hepatic portal vein to the central veins. Along the sinusoids are several macrophage-like cells, **reticuloendothelial** (Kupffer) cells. These cells engulf microorganisms that enter the liver from the portal circulation.

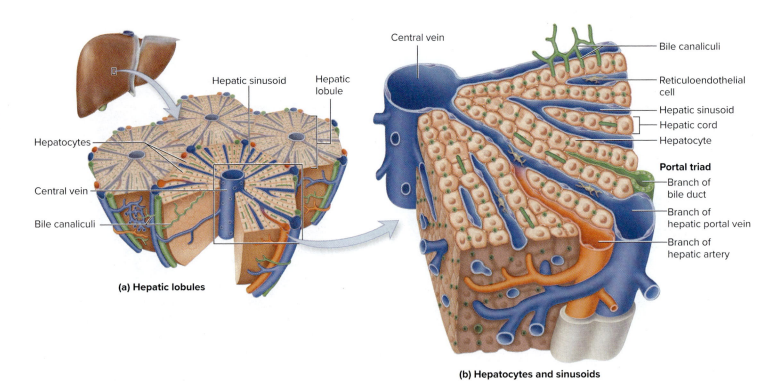

(a) Hepatic lobules

(b) Hepatocytes and sinusoids

Figure 26.15 **Structure of a Hepatic Lobule.**

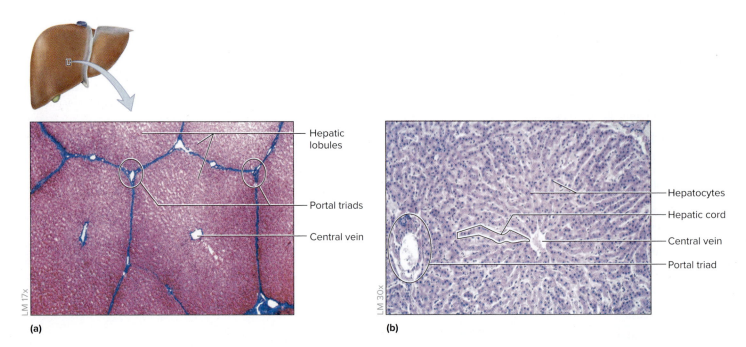

(a)

(b)

Figure 26.16 **The Liver.** (*a*) Low-magnification histology slide demonstrating multiple hepatic lobules with portal triads in the spaces between lobules. (*b*) Medium-magnification histology slide demonstrating a central vein, hepatocytes arranged into hepatic cords, and a portal triad.

©Victor P. Eroschenko

EXERCISE 26.11

HISTOLOGY OF THE LIVER

1. Obtain a histology slide of the liver and place it on the microscope stage. Bring the tissue sample into focus on low power and locate a hepatic lobule (**figure 26.16a**).

2. Move the microscope stage until the hepatic lobule is at the center of the field of view, and then change to high power (figure 26.16b). Identify the following structures on the slide of the liver, using figures 26.15 and 26.16 as guides:

   ☐ **Branch of bile duct**   ☐ **Hepatic cords**
   ☐ **Branch of hepatic**   ☐ **Hepatic lobule**
      artery   ☐ **Hepatic sinusoids**
   ☐ **Branch of hepatic**   ☐ **Hepatocyte**
      portal vein   ☐ **Portal triad**
   ☐ **Central vein**

_____ ×

3. Sketch the liver as seen through the microscope in the space provided or on a separate sheet of paper. Be sure to label all of the structures listed in step 2 in the drawing.

4. *Optional Activity:* **APR** **12: Digestive System**—Watch the "Liver" animation to visualize the organization and structure of a liver lobule.

The Pancreas

The **pancreas** is the second-largest accessory organ in the digestive system. It is both an endocrine and an exocrine gland. The histology of the endocrine part of the pancreas (the pancreatic islets) was covered in chapter 19 (The Endocrine System). The **exocrine** portion of the pancreas consists of grape-like bunches of cells called **acini** (s., *acinus*), which are similar in many ways to the cells that compose the salivary glands. The **acinar cells** composing acini produce many substances important for digestion, including digestive enzymes (e.g., pancreatic amylase). Cells of pancreatic ducts produce bicarbonate ion (HCO_3^-). Collectively, these secretions are called **pancreatic juice**. Pancreatic juice is transported from the acinar cells into small ducts that become larger ducts, and eventually dump the secretions into the duodenum via the **main pancreatic duct** (or accessory pancreatic duct) (**figure 26.17**).

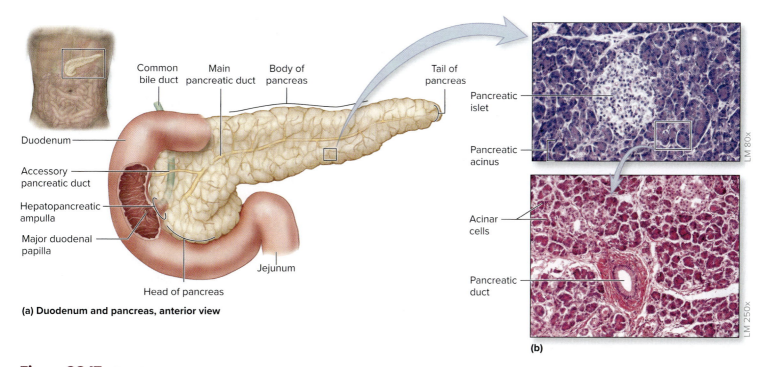

(a) Duodenum and pancreas, anterior view

Common bile duct | Main pancreatic duct | Body of pancreas | Tail of pancreas
Duodenum
Accessory pancreatic duct
Hepatopancreatic ampulla
Major duodenal papilla
Jejunum
Head of pancreas

Pancreatic islet
Pancreatic acinus
LM 80x

Acinar cells
Pancreatic duct
LM 250x

(b)

Figure 26.17   The Pancreas. (a) Location of the pancreas. (b) Histology of the pancreas demonstrating acinar cells, which secrete digestive enzymes, and pancreatic islets, which contain the hormone-secreting cells of the pancreas.

(b, top) ©Victor P. Eroschenko; (b, bottom) ©Alvin Telser/Science Source

HISTOLOGY OF THE PANCREAS

1. Obtain a histology slide of the pancreas and place it on the microscope stage. Bring the tissue sample into focus on low power. Scan the slide until a pancreatic islet surrounded by acinar cells is at the center of the field of view (figure 26.17*a*).

2. Move the microscope stage so the acinar cells that surround the pancreatic islet are in the center of the field of view, and then change to high power. Identify the following structures on the slide, using figure 26.17 as a guide:

   ☐ **Acinar cells**   ☐ **Pancreatic duct**
   ☐ **Pancreatic acinus**   ☐ **Pancreatic islet**

3. Sketch the pancreas as seen through the microscope in the space provided or on a separate sheet of paper. Be sure to label all of the structures listed in step 2 in the drawing.

_____ ×

LEARNING STRATEGY

After observing the histology of the pancreas, go back and review the histology of salivary glands. Note that the two tissues are extremely similar in structure. This is because they are also very similar in function. Acinar cells of the pancreas and serous cells of salivary glands both produce the enzyme *amylase*, which breaks down carbohydrates. When in doubt about which organ is under the microscope, look for pancreatic islets. The presence of pancreatic islets is the best way to confirm the organ being viewed is the pancreas. If they are absent, the organ is most likely a salivary gland. In addition, the presence of mucous cells also indicates the organ is a salivary gland.

PHYSIOLOGY

Digestive Physiology

Previous exercises in this chapter have detailed the anatomy of each of the digestive and accessory organs of the GI tract. The following laboratory exercises will explore some physiological functions of these organs. The overall function of the digestive system is to supply the body with nutrients, electrolytes, and water that are obtained from food and drink. Overall, the digestive system provides the means to **ingest** food, mechanically and chemically **digest** macromolecules within the ingested food **(table 26.10)**, **propel** food along the GI tract, **secrete** products that aid in digestion, **absorb** nutrients, and **eliminate** wastes. Specialized functions of each of the digestive and accessory organs allow these overall processes to take place. The following paragraph summarizes the fate of a food item from the moment it enters the mouth to the moment it exits the anus.

   As food enters the mouth, mechanical digestion of the entire food and chemical digestion of its carbohydrate components begins. Mechanical digestion (mastication) breaks down the food item into smaller pieces, which will allow enzymes to act on the macromolecules in the food and assist in their degradation (chemical digestion). The act of swallowing moves a bolus of moistened, partially digested food into the esophagus. This bolus is then transported through the esophagus by coordinated contractions of the longitudinal and circular layers of muscle that line the esophageal walls (peristalsis). From the esophagus, the bolus enters the stomach, where mechanical digestion (churning) continues and chemical digestion of proteins and fat begins. In addition, the highly acidic environment of the stomach promotes digestion of proteins. Digestion within the stomach converts the bolus that entered into a substance called chyme. From the stomach the acidic, liquid chyme is squeezed (3 mL at a time) into the duodenum of the small intestine. In the duodenum, secretions from accessory digestive organs—the liver, gallbladder, and pancreas—mix with the acidic chyme from the stomach to continue the chemical digestion of proteins, carbohydrates, fats, and nucleic acids. Pancreatic secretions (pancreatic "juice") contain enzymes that are required for protein, carbohydrate, fat, and nucleic acid digestion. Pancreatic secretions also contain bicarbonate ion, which is necessary to neutralize acids. The liver and gallbladder secrete bile into the duodenum. Bile is required for emulsification of fats. While all of this chemical digestion begins to take place, the small intestine also continues to

INTEGRATE

CLINICAL VIEW
Gallstones

Gallstones are hard stones that form from cholesterol and bile deposits in the gallbladder **(figure 26.18)**. The condition of having gallstones is called *chole-lithiasis* (*chole,* bile, + *lithos,* stone, + *-iasis,* condition). One of the most serious complications resulting from a gallstone occurs when the stone passes into the cystic duct and makes its way toward the duodenum, but becomes lodged somewhere en route to the duodenum. This blocks the flow of bile from the liver and gallbladder to the duodenum. If it lodges farther down, near the hepatopancreatic ampulla, it can also block the flow of pancreatic juice from the pancreas. Normally some of the pancreatic enzymes (proteolytic enzymes) are not activated until they enter the duodenum. However, when these enzymes build up within the pancreas because of a blockage of flow of pancreatic juice from the pancreas, these enzymes become activated and begin to digest the pancreas itself. This causes *pancreatitis* (inflammation of the pancreas), which can be life threatening.

Figure 26.18 **Gallbladder and Gallstone.** The gallstone is a mass of bile salts that have crystallized and condensed. Note the small crystals that have formed on the larger stone. A stone of this size might not cause problems for the patient if it were too large to enter into the cystic duct.

©McGraw-Hill Education/Photo and Dissection by Christine Eckel

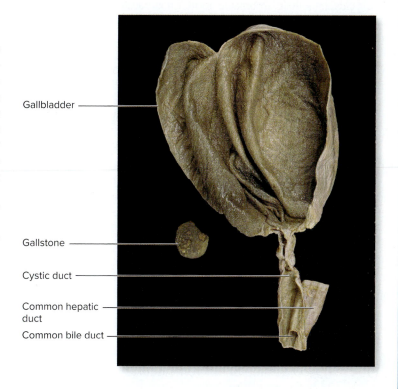

Gallbladder

Gallstone

Cystic duct

Common hepatic duct

Common bile duct

Table 26.10	Digestion and Absorption of Nutrients				
Substance	**Subunits**	**Site of Digestion**	**Site of Absorption**	**Mechanism of Absorption**	**Word Origin**
Carbohydrate	Monosaccharides (glucose, galactose, fructose)	Mouth, small intestine	Small intestine	Cotransport with sodium (glucose and galactose); facilitated diffusion (fructose); transport to hepatic portal vein	*carbo-, carbon* + *-hydro,* water
Lipid	Fatty acids, monoglycerides, glycerol	Stomach, small intestine	Small intestine	Diffuse into lacteals via chylomicrons; transported via lymph	*lipos, fat*
Nucleic Acid	Pentose sugars, nitrogenous bases, phosphates	Small intestine	Small intestine	Active transport; transport to hepatic portal vein	*nucleus, kernel*
Protein	Amino acids	Stomach, small intestine	Small intestine	Cotransport with sodium; transport to hepatic portal vein	*prote-, primary* + *-in, in*

mechanically digest the chyme using peristaltic contractions of its smooth muscle.

Following the digestion of carbohydrates, proteins, lipids, and nucleic acids into their absorbable form (monosaccharides, amino acids, fatty acids and monoglycerides, and the individual components of a nucleotide, respectively), each is transported from the GI tract into the blood or lymph. The surface area of the small intestine that is available for absorption is immense due to the presence of villi and microvilli. All nutrients (except lipids) are absorbed into the blood and then immediately enter the hepatic portal circulation (see chapter 20). In comparison, triglycerides (and other lipids) are packaged into specialized structures (chylomicrons) and absorbed into specialized lymph vessels called **lacteals** (see chapter 23). Water is reabsorbed along the length of the large intestine through the process of osmosis.

Any substances not absorbed remain within the lumen of the large intestine, are moved through the segments of the large intestine, and are eliminated as feces. **Table 26.11** summarizes the digestion of macromolecules, including the location and substances secreted that aid in the digestion of carbohydrates, proteins, lipids, and nucleic acids. Use the textbook as a guide while further exploring the processes of digestion, motility, and absorption in the following laboratory exercise.

INTEGRATE

CONCEPT CONNECTION

Motility in the GI tract is made possible by both an outer longitudinal and an inner circular layer of smooth muscle within the muscularis of the GI tract wall (see figure 26.10). Alternating waves of contraction of these two layers of muscle along the length of the tube moves substances along in a process termed peristalsis (*peri-*, around + *stalsis,* compression). Contraction of the outer, longitudinal layer decreases the length of the tube and increases its diameter; contraction of the inner, circular layer increases the length of the tube and decreases the diameter. This coordinated action is the same type of action that an earthworm uses to move about in the dirt. Contraction of the longitudinal layer of the worm's body wall musculature makes the worm short and fat, whereas contraction of the circular layer of the worm's body wall musculature makes the worm long and skinny (much like squeezing a tube of toothpaste).

These muscular contractions are coordinated by a network of nerves that lie adjacent to each muscle layer: the submucosal and myenteric plexuses. Collectively, these plexuses comprise the **enteric nervous system**, or "gut brain." While these nerves can operate independently, much like pacemaker cells in the heart, their activity is modulated by the autonomic nervous system.

Table 26.11	Digestion of Macromolecules			
Location	**Carbohydrates**	**Proteins**	**Lipids**	**Nucleic Acids**
Oral Cavity (saliva)	Starch—*salivary amylase*→ partially digested starch	No protein digestion	*Lingual lipase* added but activated in low pH of stomach	No nucleic acid digestion
Stomach (gastric juice)	No additional enzymes added	Protein—*pepsin*→ polypeptide and peptide fragments	Triglyceride—*lingual lipase*→monoglyceride and fatty acids (limited amounts) Triglyceride—*gastric lipase*→monoglyceride and fatty acids (limited amounts)	No nucleic acid digestion
Small Intestine (pancreatic juice secreted into duodenum)	Partially digested starch—*pancreatic amylase*→ oligosaccharides, maltose, and glucose	Protein—*trypsin*→ polypeptide and peptide fragments Protein—*chymotrypsin*→ polypeptide and peptide fragments Protein—*carboxypeptidase*→ amino acids from carboxy-end of peptides	Triglyceride—*pancreatic lipase*→monoglyceride and fatty acids (within micelles)	DNA—*pancreatic deoxyribonuclease*→ deoxyribonucleotides RNA—*pancreatic ribonuclease*→ ribonucleotides
Small Intestine (brush border enzymes)	Oligosaccharides—*dextrinase* and *glucoamylase*→ maltose, glucose Maltose—*maltase*→glucose Lactose—*lactase*→glucose, galactose Sucrose—*sucrase*→glucose, fructose	Dipeptides—*dipeptidase*→ amino acids Peptides—*aminopeptidase*→amino acids from amino-end of peptides	No lipid digestion completed by brush border enzymes	Nucleotides—*phosphatase*→ nucleosides and phosphate Nucleosides—*nucleosidase*→ nitrogenous base and sugar (ribose or deoxyribose)

INTEGRATE

CLINICAL VIEW
Cholera

Diarrhea (*dia-*, through + *-rrhoea*, to flow) is a relatively common condition in which water is either not absorbed or is "pulled" by osmosis into the lumen of the GI tract. When substances such as ions or undigested materials remain within the lumen of the GI tract, water may be attracted to them, thereby preventing the absorption of water. This is what happens when some pathogens invade the intestinal wall and cause diseases such as cholera, a disease that occurs in areas where sanitary conditions are poor. **Cholera** is caused by a bacterium, *Vibrio cholerae*, which establishes an extracellular infection within the mucosa of the small intestine. Once in place, the bacterium causes large quantities of ions, including sodium, potassium, and chloride, to be pumped into the lumen of the GI tract. The ions attract water via osmosis. The quantity of water within the GI tract becomes so great that the stretch receptors in the intestinal wall are stimulated, which increases gut motility. The result is production of massive amounts of watery diarrhea. Because dehydration is of great concern in a patient suffering from cholera, the "treatment" is administration of oral rehydration therapy (ORT). ORT promotes rehydration by taking advantage of the mechanisms of water absorption in the large intestine. That is, the solution contains sodium, glucose, and water. Sodium and glucose move across the epithelium of the intestinal wall via cotransport. Sodium is the body's greatest source of osmotic pressure. Thus, as sodium is reabsorbed from the lumen of the GI tract into the blood, glucose and, subsequently, water, will follow. Without treatment, a person suffering from cholera may lose up to 20 L of water per day, which can rapidly lead to death.

INTEGRATE

CONCEPT CONNECTION

Acidic and basic solutions play an important role in digestion. The stomach produces acidic chyme due to HCl produced by parietal cells. Cells of the pancreatic ducts produce bicarbonate ion to neutralize the acidic chyme once it enters the duodenum of the small intestine. Both processes (production of acidic and basic secretions) involve the same reaction, the *bicarbonate reaction,* which was introduced in chapter 24: $CO_2 + H_2O \leftrightarrow H_2CO_3 \leftrightarrow H^+ + HCO_3^-$. In the stomach, hydrogen ions move into the lumen of the stomach; in the pancreas, bicarbonate ions move into the pancreatic duct to be dumped into the small intestine. While this reaction is important for digestion, it is also involved in regulating blood pH. As hydrogen ions are pumped to the lumen of the stomach, the bicarbonate ions move into the blood. Similarly, as bicarbonate ions move to the pancreatic duct, the hydrogen ions move into the blood. When a person vomits excessively, that person's parietal cells must compensate for the loss of acidic chyme by moving more H^+ into the lumen of the stomach and releasing more HCO_3^- into the blood. The excess HCO_3^- in the blood may raise the blood pH, which can lead to **metabolic alkalosis**. Likewise, when a person suffers from excessive diarrhea, pancreatic acinar cells must compensate to replace the loss of bicarbonate ion by moving more HCO_3^- into the lumen of the small intestine and moving more H^+ into the blood. The excess H^+ in the blood may decrease blood pH, which can lead to **metabolic acidosis**. Recall from chapter 24 that changes in breathing rates can lead to respiratory acidosis or respiratory alkalosis.

EXERCISE 26.13

DIGESTIVE ENZYMES

The purpose of this laboratory exercise is to examine the properties of the enzyme amylase, which is involved in the chemical digestion of starch, and to observe the effect that temperature has on amylase function.

Before beginning, state a hypothesis regarding the effect of temperature on the rate of chemical digestion.

_____

_____

_____

Obtain the Following:
- **0.5% amylase solution**
- **0.5% starch solution**
- **37°C water bath**
- **6 test tubes**
- **Benedict's solution (or glucose test strips)**
- **hot plate**
- **ice**
- **Lugol's solution (iodine-potassium-iodide)**
- **porcelain spot plate**
- **test-tube clamp**
- **wax pencil**

1. Place three test tubes in a rack and label each (1, 2, and 3) with a wax pencil. Label the test tubes near the top of the test tube. The test tubes will be placed in boiling water. Thus, if the labels are below the water line they may come off. Fill test tube 1 with 6 mL of 0.5% amylase solution. Fill test tube 2 with 6 mL of 0.5% starch solution. Fill test tube 3 with 5 mL of 0.5% starch solution + 1 mL of 0.5% amylase solution. Swirl each tube gently until the contents are mixed well.

2. To heat the tubes, place the three test tubes in a warm water bath, 37°C (98.6°F), for approximately 10 minutes. Remove the test tubes from the warm water bath.

3. To test solution 1 for the presence of starch, remove 1 mL of the solution from test tube 1 and place it in one depression of a spot plate. Add 1 drop of Lugol's solution to the test solution and observe the color. A blue-black color indicates the presence of starch. Record the color observations and conclusions regarding the presence (+) or absence (−) of starch in **table 26.12**.

4. Repeat step 3 for the solutions in test tubes 2 and 3.

5. **Test each solution for the presence of sugars (monosaccharides and disaccharides):** to do this,

(continued on next page)

(continued from previous page)

remove 1 mL of the solution from test tube 1 and place it in a clean test tube. Add 1 mL of Benedict's solution to the test tube, place the tube in boiling water for 2 minutes, and observe the color. Remember to always use a test-tube clamp when lowering the test tube into boiling water. The color may vary depending on the amount of sugar present in the solution. The range in color follows that of the visible light spectrum (red, orange, yellow, green, and blue), where red indicates most and green indicates least sugar present. A solution that is blue in color indicates no sugar present. Record the color observations and conclusions regarding the presence (+) or absence (−) of sugar in table 26.12. If using glucose test strips to test for glucose, do not add Benedict's solution and do not boil the test tube. Instead, simply dip the glucose test strip into the test tube containing solution 1 and pull it back out. Shake the test strip gently to remove excess solution. Then, compare the color of the test strip to the chart provided with the glucose test strips to determine the presence or absence of glucose in the solution.

6. Repeat step 5 for the solutions in test tubes 2 and 3.

7. **Test the effect of temperature on enzyme activity:** To do this, place 1 mL of 0.5% amylase solution in each of three clean test tubes. Label test tube 1 "cold," test tube 2 "warm," and test tube 3 "hot." Vary the temperature conditions for the solutions by placing test tube 1 in a

beaker of ice water (~0°C), test tube 2 in a beaker with warm water (37°C), and test tube 3 in a beaker of boiling water (100°C). Use a test-tube clamp to lower test tube 3 into the boiling water.

8. Remove test tube 1 from the water bath. Add 5 mL of 0.5% starch solution to the test tube and mix thoroughly. Note that 0.5% starch solution added to test tube 1 should be chilled to approximately 0°C prior to being added to the amylase solution. Return the test tube to the respective water bath for 10 minutes.

9. Remove the test tube from the water bath and repeat steps 3 and 5 to test for the presence of starch and sugar, respectively. Record the color of each solution and conclusions regarding the presence (+) or absence (−) of starch and sugar in table 26.12.

10. Repeat steps 8 and 9 for the solutions in test tubes 2 and 3. Be careful when handling and mixing the contents in test tube 3 because the solution is very hot.

11. Describe the effect of increasing temperature on enzymatic activity. _____

_____

_____

_____

Table 26.12	Data for Chemical Digestion Test					
Test Tube #	Solution	Lugol's (Color)	Starch Present (+/−)	Benedict's (Color)	Sugar Present (+/−)
1	0.5% Amylase solution				
2	0.5% Starch solution				
3	0.5% Amylase and 0.5% Starch solution				
4	0.5% Amylase at 0°C				
5	0.5% Amylase at 37°C				
6	0.5% Amylase at 100°C				

A CLINICAL CASE STUDY IN DIGESTIVE PHYSIOLOGY

The purpose of this exercise is to apply concepts related to digestive physiology to a real scenario, or clinical case study. This exercise requires information presented in this chapter as well as in previous chapters. Integrating multiple systems in a case study reinforces understanding of homeostatic processes.

Prior to beginning the exercise, ask the laboratory instructor if this exercise is to be completed individually, in pairs, or in groups. Use the textbook and this laboratory manual as guides to review the following topics that relate to the case study: digestive system, membrane transport, autonomic nervous system, immune system, cardiovascular system, respiratory system, and acid-base balance.

Case Study

Beth, an 18-month-old female, was developing normally until she started having episodes of vomiting and diarrhea. Her symptoms worsened over the six months since the time they began. At times, she suffered from 10 to 15 bouts of diarrhea per day. A physical exam revealed that Beth's abdomen was soft and nontender, and had no palpable masses. Fecal examinations were negative for parasites, and blood tests were normal. However, Beth's weight steadily dropped. Her pediatrician noted that she had decreased from the fiftieth percentile in weight at her 12-month checkup to the tenth percentile at her 18-month checkup. Her parents shared with the physician that Beth was becoming increasingly irritable, and at times could not be consoled.

Beth was referred to a gastroenterologist to determine the cause of the gastrointestinal symptoms and weight loss. The gastroenterologist performed an upper gastrointestinal (GI) endoscopy and a colonoscopy to visualize the mucosal layers of the GI tract. The examination revealed normal mucosa in Beth's esophagus, stomach, and duodenum. Multiple nodules were observed in the descending colon, sigmoid colon, and rectum. However, the nodules were determined to be nonpathologic. These findings led the gastroenterologist to suspect that Beth was suffering from lactose intolerance, because milk had been introduced into Beth's diet around the same time that she began exhibiting gastrointestinal symptoms. However, blood tests revealed normal levels of lactase, the brush border enzyme required to digest lactose (the sugar found in dairy products). Gluten sensitivity (i.e., celiac disease) was also ruled out because Beth did not exhibit an abnormal immune response to gluten.

After ruling out the most common sources of diarrhea, the gastroenterologist next considered a more serious and rare diagnosis: presence of a hormonally active tumor, which could alter membrane transport across the intestinal epithelium. A computed tomography (CT) scan of Beth's abdomen revealed a 4.6 cm mass in the right paraspinal region, adjacent to the right kidney. A biopsy of the tumor demonstrated that the mass was a ganglioneuroma, a benign tumor that secretes the hormone vasoactive intestinal peptide (VIP). VIP is a hormone that resembles secretin, which is normally released by the small intestine. Secretin increases electrolyte secretion into the lumen of the GI tract and increases gut motility. The mass was removed surgically, and Beth was thriving just 1 month later. Her prognosis for a full recovery was excellent.

Answer the following questions concerning the case study just presented:

1. Failure to gain weight may be due to a low intake of nutrients, increased metabolic needs, or malabsorption. Which of these possibilities was the cause of Beth's weight loss? Justify the answer. _____

_____

_____

2. The gastroenterologist initially suspected either lactose intolerance or gluten sensitivity. Describe how each condition would affect the processes of digestion and absorption.

_____

_____

_____

3. Increased electrolyte secretion into the lumen of the gastrointestinal tract leads to secretory diarrhea. Describe how increasing the secretion of electrolytes into the lumen of the gastrointestinal tract could lead to excessive, watery stool and increased motility.

_____

_____

_____

4. Chronic diarrhea is a common cause of acid-base and fluid balance disturbances in the blood. Based on Beth's symptoms, predict her blood pH (acidic/basic), respiratory rate (high/low), heart rate (high/low), and urine pH prior to treatment. Justify these predictions.

_____

_____

_____

5. Which of the four conditions was she at most risk of: (a) metabolic acidosis, (b) metabolic alkalosis, (c) respiratory acidosis, or (d) respiratory alkalosis? Explain your answer. _____

_____

_____

6. Based on the name (ganglioneuroma) and location of the tumor, from what cells did the tumor likely originate?

_____

The ① corresponds to the Learning Objective(s) listed in the chapter opener outline.

Do You Know the Basics?

Exercise 26.1: Overview of the GI Tract

1. Place the organs of the GI tract in the correct order, from proximal to distal (1=most proximal; 6=most distal). **①**

    _____ a. anus

    _____ b. esophagus

    _____ c. large intestine

    _____ d. small intestine

    _____ e. rectum

    _____ f. stomach

Exercise 26.2: Gross Anatomy of the Oral Cavity, Pharynx, and Esophagus

2. The esophagus passes through the esophageal _____ (hiatus/sphincter) in the diaphragm prior to entering the stomach. **②**

Exercise 26.3: Gross Anatomy of the Stomach

3. A patient is suffering from gastroesophageal reflux disease (GERD). When she lies down, she feels a burning sensation in her esophagus caused by the reflux of

    stomach acids into the esophagus. This occurs because one of the sphincters in her stomach is not working properly. The sphincter that is not functioning

    properly is the _____ (cardiac/pyloric) sphincter. This sphincter is considered a(n) _____ (anatomical/physiological) sphincter.

    The type of muscle tissue that composes this sphincter is _____ (skeletal/smooth) muscle. **③**

Exercise 26.4: Gross Anatomy of the Duodenum, Liver, Gallbladder, and Pancreas

4. The ducts leading from the gallbladder and liver empty into the _____ (duodenum/pancreas). **④**

5. Which of the following is/are a lobe of the liver? (Check all that apply.) **⑤**

    _____ a. caudate

    _____ b. left

    _____ c. posterior

    _____ d. quadrate

    _____ e. right

6. Place the following terms in the correct order in which bile is transported from its site of production in the liver to its entry into the duodenum (1 = first; 4 = last).
   Assume that bile does not enter the gallbladder. **⑥**

    _____ a. common bile duct

    _____ b. common hepatic duct

    _____ c. left hepatic duct

    _____ d. bile canaliculus

Exercise 26.5: Gross Anatomy of the Jejunum and Ileum of the Small Intestine

7. Loops of arteries that supply blood to the small intestine are called _____ (arterial arcades/vasa recta), whereas straight vessels that supply blood to

    the small intestine are called _____ (arterial arcades/vasa recta). **⑦**

8. The ileum of the small intestine has _____ (few/many) arterial arcades and _____ (short/long) vasa recta, whereas the jejunum of the

    small intestine has _____ (few/many) arterial arcades and _____ (short/long) vasa recta. **⑧**

Exercise 26.6: Gross Anatomy of the Large Intestine

9. Which of the following places the four parts of the colon in order from proximal to distal? (Circle one.) **9**

    a. descending colon, transverse colon, ascending colon, sigmoid colon

    b. ascending colon, transverse colon, descending colon, sigmoid colon

    c. transverse colon, sigmoid colon, ascending colon, descending colon

    d. ascending colon, sigmoid colon, transverse colon, descending colon

10. The _____ (hepatic/splenic) flexure is located on the right side of the abdomen, whereas the _____ (hepatic/splenic) flexure is located on the left side of the abdomen. **10**

Exercise 26.7: Histology of the Salivary Glands

11. The cells within salivary glands that produce salivary amylase are _____ (mucous/serous) cells. **11**

12. Mary likes sour foods, and she decided to eat a slice of lemon. As she bit down on the lemon, she felt an uncomfortable squeezing-type sensation in her cheek as one of her salivary glands emptied its secretions into her mouth. Which of the three pairs of salivary glands did she feel? (Circle one.) **12**

    a. parotid

    b. sublingual

    c. submandibular

Exercise 26.8: Histology of the Stomach

13. Place the layers of the stomach wall in the correct order, from innermost to outermost (1 = innermost; 4 = outermost). **13**

    _____ a. mucosa

    _____ b. muscularis

    _____ c. serosa

    _____ d. submucosa

14. The stomach contains _____ (two/three) layers of smooth muscle in its muscularis layer. **14**

15. The mucosa of the stomach is lined with simple _____ (columnar/cuboidal) epithelial tissue. **15**

16. In the cardia of the stomach, gastric pits are _____ (short/long), whereas gastric glands are _____ (short/long). **16**

17. Mucous neck cells are found in gastric _____ (glands/pits). **17**

Exercise 26.9: Histology of the Small Intestine

18. Which of the following structure(s) increase(s) the total surface area of the small intestine? (Check all that apply.) **18**

    _____ a. cilia

    _____ b. circular folds

    _____ c. microvilli

    _____ d. villi

19. _____ (Goblet/Transitional) cells in the small intestine produce mucus, which protects and lubricates the epithelium of the small intestine. **19**

20. The presence of _____ (Peyer patches/intestinal crypts) is unique to the ileum. **20**

Exercise 26.10: Histology of the Large Intestine

21. Which of the following is the epithelial modification that is particularly abundant in the large intestine? (Circle one.) **21** **22**

    a. circular folds

    b. goblet cells

    c. microvilli

    d. submucosal glands

22. Place the wall layers of the large intestine in order from innermost to outermost (1 = innermost; 5 = outermost). **23**

    _____ a. mucosa

    _____ b. muscularis

    _____ c. muscularis mucosa

    _____ d. serosa

    _____ e. submucosa

Exercise 26.11: Histology of the Liver

23. The liver is composed of hexagonally shaped structures called _____ (hepatic lobules/portal triads) that consist of strands of _____ (hepatocytes/acinar cells). **24**

24. Hepatic lobules contain _____ (continuous, sinusoidal) capillaries, which receive both oxygen-rich blood from branches of the _____ (hepatic artery/hepatic portal vein) and nutrient-rich blood from branches of the _____ (hepatic artery/hepatic portal vein). This blood is transported to the central vein. **25** **26**

Exercise 26.12: Histology of the Pancreas

25. The acinar cells of the pancreas produce _____ (bicarbonate/digestive enzymes), whereas the cells that line the pancreatic ducts produce _____ (bicarbonate/digestive enzymes). **27**

26. An exocrine substance produced by _____ (acinar/islet) cells in the pancreas includes _____ (amylase/bicarbonate), which digests carbohydrates. **28**

27. The endocrine part of the pancreas consists of _____ (acinar/islet) cells that produce _____ (hormones/pancreatic juice). **29**

Exercise 26.13: Digestive Enzymes

28. Bile is produced by the _____ (liver/pancreas), pepsin is produced by the _____ (pancreas/stomach), and amylase is produced by both the _____ (salivary glands/stomach) and the _____ (pancreas/liver). **30**

29. Within normal physiological range, as temperature increases, enzyme activity _____ (increases/decreases), and as temperature decreases, enzyme activity _____ (increases/decreases). **31**

30. Initial digestion of carbohydrates occurs in the _____ (oral cavity/small intestine). **32**

Exercise 26.14: A Clinical Case Study in Digestive Physiology

31. Increased sodium secretion into the GI tract is likely to _____ (increase/decrease) motility and _____ (increase/decrease) absorption. **33**

Can You Apply What You've Learned?

32. Complete the following table, focusing on the specific histology of the tunics of each organ. Be sure to include any features that are "defining histological features" for that organ.

Tunic/Wall Layer	Stomach	Small Intestine Duodenum	Small Intestine Ileum	Large Intestine Colon
Mucosa				
Submucosa				
Muscularis				
Adventitia or Serosa (indicate which of these is present in the organ)				

33. The image to the right is a cross section through part of the small intestine.

   a. What part of the small intestine does this image represent (duodenum, jejunum, or ileum)?

     _____

   b. What characteristic(s) were used in determining which part of the small intestine this sample

     was taken from? _____

     _____

34. Using knowledge of the ducts draining bile from the liver and gallbladder and the pancreatic ducts, propose a location where a lodged gallstone might cause pancreatitis._____

   _____

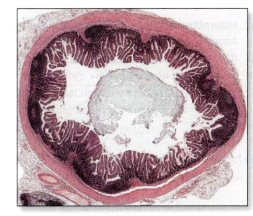

©McGraw-Hill Education/Christine Eckel

35. A patient presented to his physician with pain in the left lower quadrant of his abdomen. The source of the pain was an adhesion* between the visceral peritoneum covering part of the patient's colon and the parietal peritoneum lining his anterolateral abdominopelvic wall in that region. What part of the colon was most likely adhered to the abdominopelvic wall?

   _____

   _____

36. Describe why a person suffering from massive bouts of diarrhea is also at risk for electrolyte imbalance._____

   _____

   _____

37. Acidic chyme can denature and inactivate enzymes in the small intestine. Describe how an inability to produce bicarbonate in the pancreas can lead to incomplete digestion and issues with absorption. _____

   _____

   _____

*An adhesion (*adhaereo*, to stick to) in the abdominopelvic cavity is an area where two layers of peritoneum are stuck to each other with connective tissue. It usually is the result of some sort of injury or inflammation.

GROSS ANATOMY

Female Reproductive System

The internal detail, wall structures, and functions of most of the female reproductive organs will be covered in the histology section of this chapter. This section involves observing the gross structure of these organs to gain an appreciation of their location within the female pelvic cavity, and their relationships with each other. Note that there are numerous supporting ligaments that anchor the ovaries, uterine tubes, and uterus in place within the pelvic cavity. Primary anatomical features of the female breast will then be observed.

EXERCISE 27.1

GROSS ANATOMY OF THE OVARY, UTERINE TUBES, UTERUS, AND SUPPORTING LIGAMENTS

The ovary, uterine tubes, and uterus are all contained within the pelvic cavity of the female and are held in place by a number of supporting ligaments, many of which form from folds of the peritoneum as it drapes over these structures. **Table 27.1** summarizes the structure, function, and location of the supporting ligaments.

1. Observe a female human cadaver or classroom models of the female pelvis and female reproductive organs.

Also observe a classroom model demonstrating a sagittal section through a female pelvis.

2. Identify the structures listed in **figures 27.1** and **27.2** on the cadaver or on classroom models, using table 27.1 and the textbook as guides. Then label them in figures 27.1 and 27.2.

3. *Optional Activity*: **APR** 14: Reproductive System— In the Quiz section, select the "Structures: Female" option to test yourself on gross anatomy of female structures.

Table 27.1	Ligaments Supporting the Ovary, Uterine Tubes, and Uterus	
Supporting Ligament	**Description and Function**	**Word Origin**
Broad Ligament	Fold of peritoneum draped over the superior surface of the uterus. Portions of the broad ligament form the mesosalpinx and the mesovarium.	*broad*, wide, + *ligamentum*, a bandage
Mesosalpinx	The mesentery of the uterine tube, which is formed as a fold of the most superior part of the broad ligament	*meso-*, a mesentery-like structure, + *salpinx*, trumpet (tube)
Mesovarium	The mesentery of the ovary, which is formed as a posterior extension of the broad ligament	*meso-*, a mesentery-like structure, + *ovarium*, ovary
Ovarian Ligament	Ligament contained within folds of the broad ligament. It extends from the medial part of the ovary to the superolateral surface of the body of the uterus.	*ovarium*, ovary, + *ligamentum*, a bandage
Round Ligament of the Uterus	Ligament attached to the superolateral surface of the body of the uterus. It extends laterally to the deep inguinal ring, passes through the inguinal canal, and attaches to the skin of the labia majora.	*ligamentum*, a bandage, + *uterus*, uterus
Suspensory Ligament of the Ovary	A fold of peritoneum draping over the ovarian artery and vein superolateral to the ovary. Anchors the ovary to the lateral body wall.	*suspensio-*, to hang up, + *ligamentum*, a bandage
Transverse Cervical (Cardinal) Ligament	Ligament extending laterally from the cervix and vagina, connecting them to the pelvic wall	*transverse-*, across, + *cervix*, neck, + *ligamentum*, a bandage
Uterosacral Ligaments	Ligament connecting the inferior part of the uterus to the sacrum posteriorly	*utero-*, the uterus, + *sacral*, the sacrum

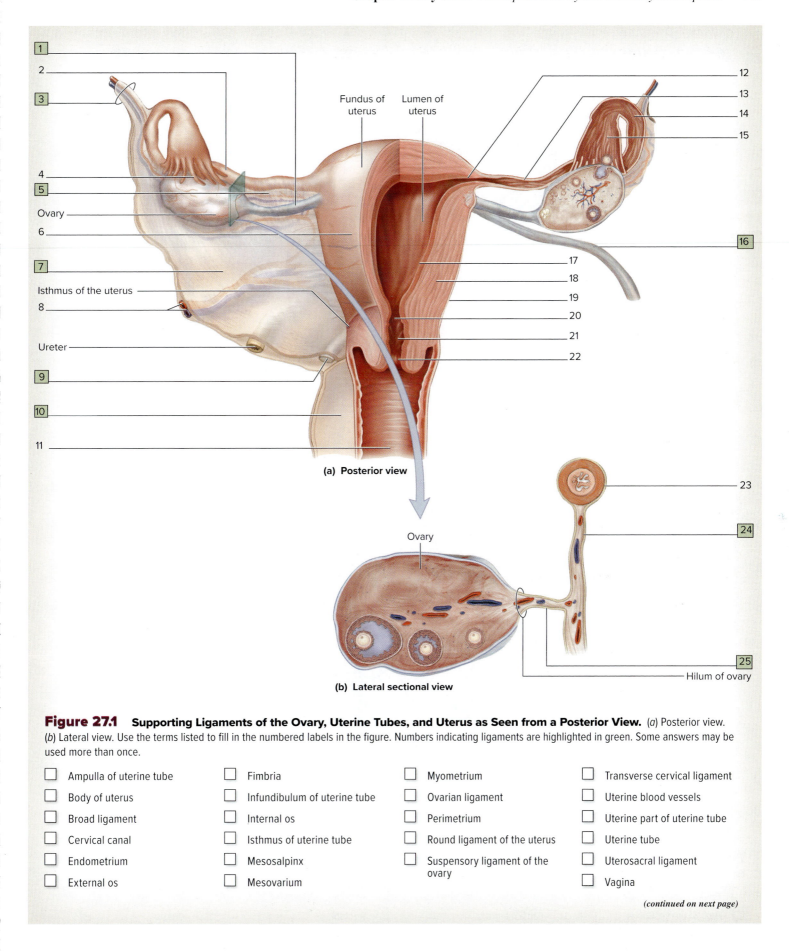

Figure 27.1 **Supporting Ligaments of the Ovary, Uterine Tubes, and Uterus as Seen from a Posterior View.** (*a*) Posterior view. (*b*) Lateral view. Use the terms listed to fill in the numbered labels in the figure. Numbers indicating ligaments are highlighted in green. Some answers may be used more than once.

- Ampulla of uterine tube
- Body of uterus
- Broad ligament
- Cervical canal
- Endometrium
- External os

- Fimbria
- Infundibulum of uterine tube
- Internal os
- Isthmus of uterine tube
- Mesosalpinx
- Mesovarium

- Myometrium
- Ovarian ligament
- Perimetrium
- Round ligament of the uterus
- Suspensory ligament of the ovary

- Transverse cervical ligament
- Uterine blood vessels
- Uterine part of uterine tube
- Uterine tube
- Uterosacral ligament
- Vagina

(*continued on next page*)

(continued from previous page)

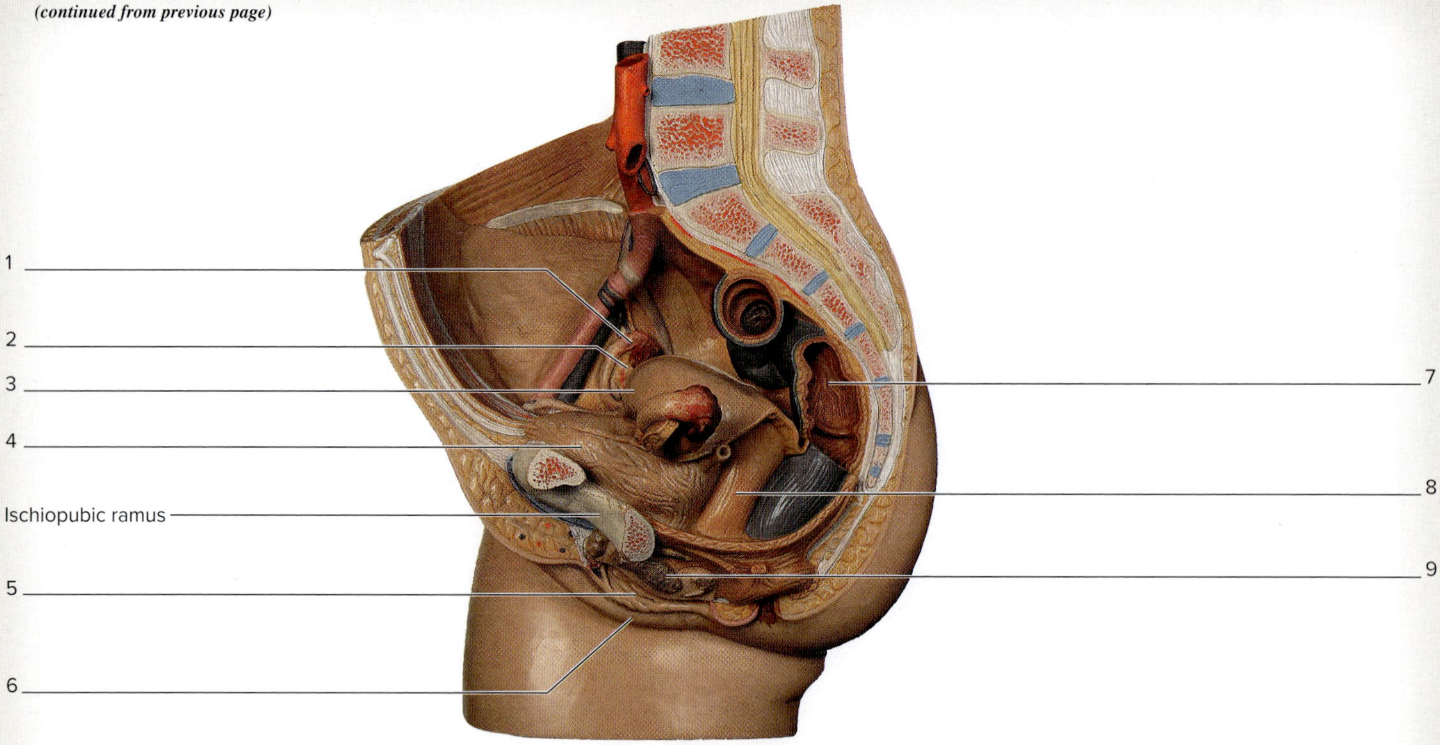

(a) Sagittal view

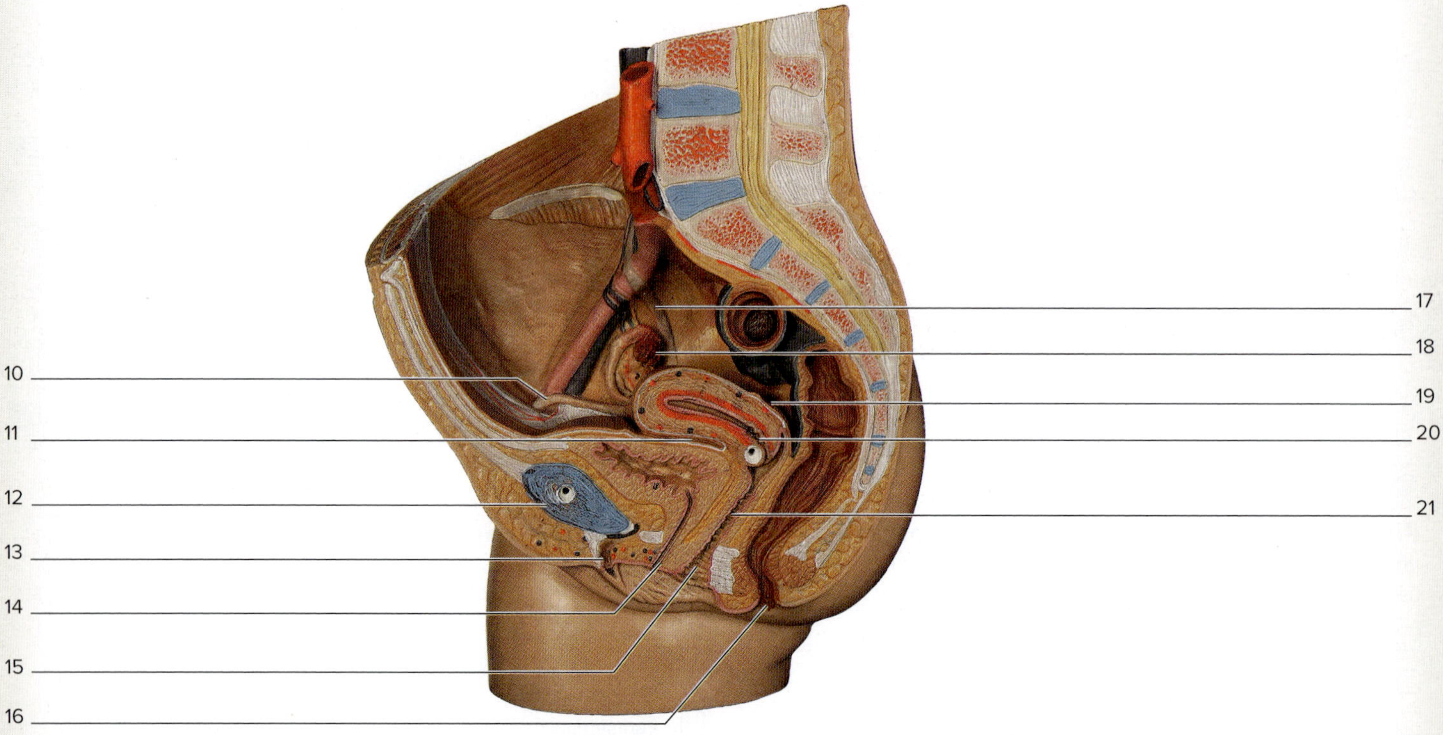

1
2
3
4
Ischiopubic ramus
5
6

7
8
9

(b) Midsagittal view

10
11
12
13
14
15
16

17
18
19
20
21

Figure 27.2   **Classroom Model of the Female Pelvic Cavity.** (*a*) Sagittal view with pelvic structures intact. (*b*) Midsagittal view. Use the terms listed to fill in the numbered labels in the figure. Some answers may be used more than once.

☐ Anus
☐ Bulb of the vestibule
☐ Cervix of uterus
☐ Clitoris
☐ External urethral orifice

☐ Fimbria of uterine tube
☐ Labia majora
☐ Labia minora
☐ Ovary
☐ Pubic symphysis

☐ Rectouterine pouch
☐ Rectum
☐ Round ligament of the uterus
☐ Ureter
☐ Urinary bladder

☐ Uterine tube
☐ Uterus
☐ Vagina
☐ Vaginal orifice
☐ Vesicouterine pouch

Model # H10 [1000281] ©3B Scientific GmbH, Germany, 2015 www.3bscientific.com/Photo by Christine Eckel, Ph.D.

GROSS ANATOMY OF THE FEMALE BREAST

The female breast consists largely of fatty tissue and suspensory ligaments. Imbedded within are numerous modified sweat glands, the **mammary glands** (*mamma*, breast), which are compound tubuloalveolar exocrine glands. These glands enlarge greatly during pregnancy, enabling them to produce milk to nourish the new baby. **Table 27.2** describes the structures that compose the female breast.

1. Observe the breast of a prosected female human cadaver or a classroom model of the female breast.

2. Identify the structures listed in **figure 27.3** on the cadaver or the classroom model of the female breast, using table 27.2 and the textbook as guides. Then label them in figure 27.3.

Table 27. 2	The Female Breast	
Structure	**Description and Function**	**Word Origin**
Alveoli	Secretory units of the mammary glands, which produce milk	*alveolus*, a concave vessel, a bowl
Areola	The pigmented area of skin surrounding the nipple	*areola*, area
Areolar Glands	Sebaceous glands deep to the skin of the areola; produce sebum, which keeps the skin of the areola moist, particularly during lactation	*areolar*, relating to the areola of the breast
Lactiferous Ducts	Ducts that form from the confluence of small ducts draining milk from the alveoli and lobules	*lacto-*, milk + *ductus*, to lead
Lactiferous Sinuses	10–20 large channels that form from the confluence of several lactiferous ducts; the spaces where milk is stored prior to release from the nipple	*lacto-*, milk + *sinus*, cavity
Lobes	Large subdivisions of the mammary glands	*lobos*, lobe
Lobules	Smaller subdivisions of the mammary glands, which contain the alveoli	*lobulus*, a small lobe
Nipple	A cylindrical projection in the center of the breast that contains the openings of the lactiferous ducts	*neb*, beak or nose
Suspensory Ligaments	Bands of connective tissue that anchor the breast skin and tissue to the deep fascia overlying the pectoralis major muscle	*suspensio*, to hang up, + *ligamentum*, a band

INTEGRATE

CONCEPT CONNECTION

Lactation, the process of milk production and release from the mammary glands, involves both positive and negative feedback mechanisms. Lactation is regulated by hormones that are present during pregnancy and after childbirth. During pregnancy, estrogen and progesterone levels are elevated, which stimulates proliferation of the acini and branching of the lactiferous ducts within mammary glands. Following birth, estrogen and progesterone levels drop, but prolactin levels are elevated, which stimulates milk production. The baby suckling on the mother's breast stimulates the release of oxytocin from the posterior pituitary gland, which further increases milk production. Oxytocin also stimulates myoepithelial cells within mammary glands to contract, thereby releasing milk from the breast. Milk moves through the lactiferous ducts and sinuses and exits the breast through openings in the nipple. Positive feedback mechanisms stimulate the production and ejection of milk. Negative feedback mechanisms may inhibit GnRH release by the hypothalamus, thereby inhibiting FSH and LH release from the anterior pituitary gland and preventing ovulation. Even so, women should use alternative methods of birth control while breastfeeding because some women continue to ovulate during this time.

(continued on next page)

(continued from previous page)

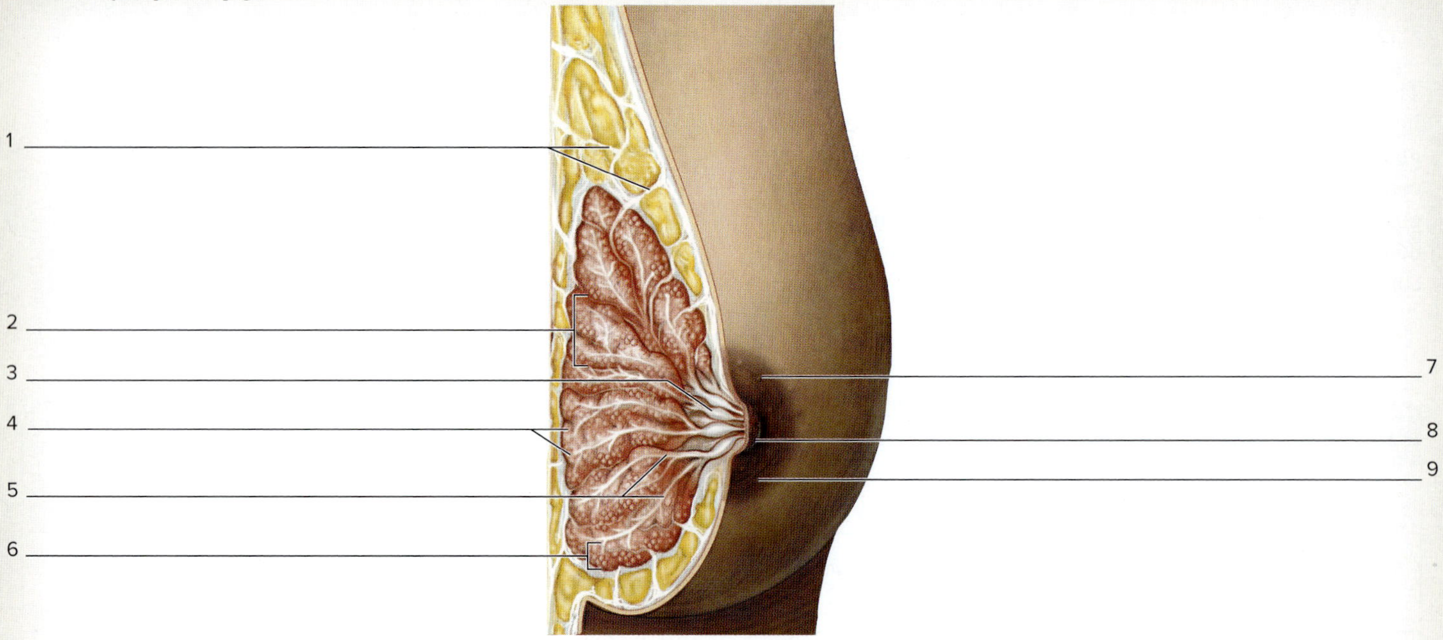

1

2

3

4

5

6

7

8

9

(a) Anteromedial view

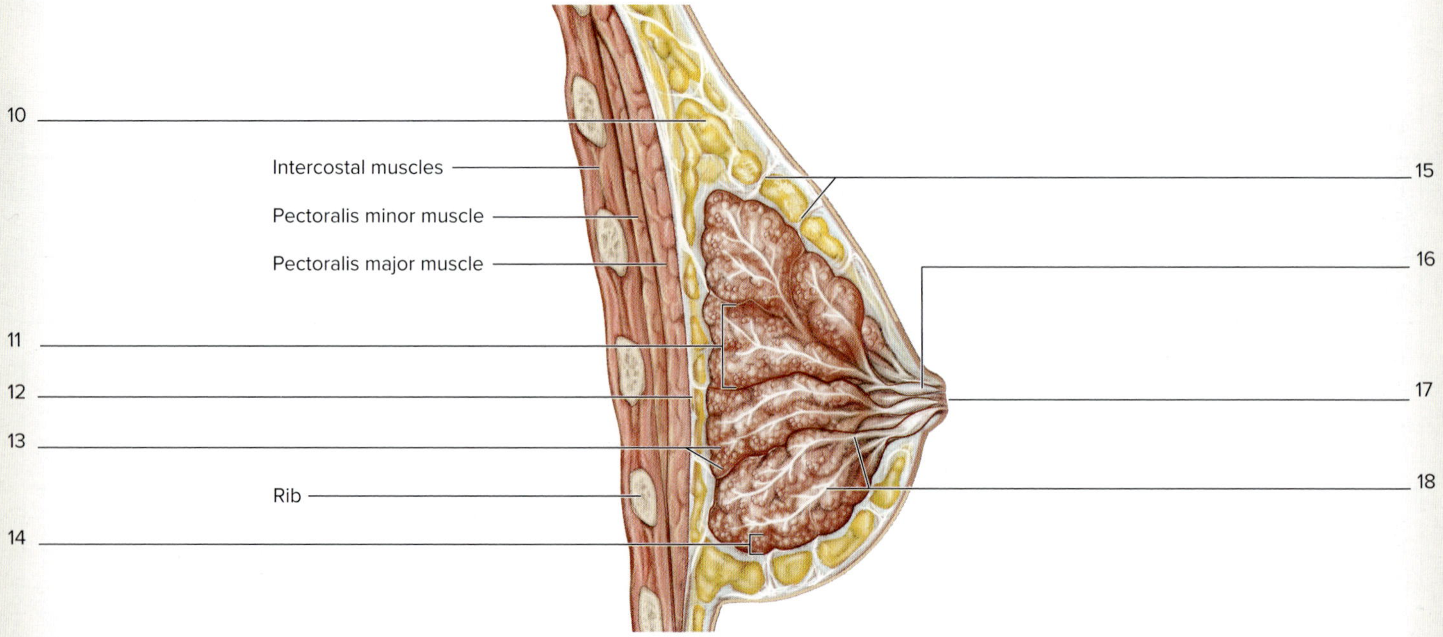

10

Intercostal muscles

Pectoralis minor muscle

Pectoralis major muscle

11

12

13

Rib

14

15

16

17

18

(b) Sagittal view

Figure 27.3 **The Female Breast.** (*a*) Anterior view. (*b*) Sagittal view. Use the terms listed to fill in the numbered labels in the figure. Some answers may be used more than once.

☐ Adipose tissue

☐ Alveoli

☐ Areola

☐ Areolar gland

☐ Deep fascia

☐ Lactiferous ducts

☐ Lactiferous sinus

☐ Lobe

☐ Lobule

☐ Nipple

☐ Suspensory ligaments

CLINICAL VIEW

Breast Cancer Screening and Breast Self-Examination (BSE)

Clinical Breast Exam (CBE)

All women of reproductive age are encouraged to make annual visits to their gynecologist for a clinical breast exam (CBE) and a pap smear to test for early cervical changes that are risk factors for cervical cancer. The physician performing the CBE is palpating for abnormal lumps in the breast. However, the physician can also help teach a woman how to perform breast self-exam (BSE) so she can monitor the condition of her breasts monthly instead of just during yearly visits. The vast majority of breast cancers are discovered by the patient herself.

Mammography

A mammogram is an x-ray of the breast that is used to look for areas of extra density or calcifications, which can be early signs of breast cancer. Most women are advised to start having mammograms at age 40, but those with a family history of the disease are often advised to have their first mammogram at age 35. Mammography is good at locating very small tumors (especially DCIS), which may not be palpable on self-examinations or clinical examinations.

Breast Self-Exam (BSE)

Women should perform a breast self-exam (BSE) at least once a month to look for any unusual bumps, lumps, or thickening of breast tissue. Upon palpation, a typical breast feels a bit lumpy, with some firmer areas and some softer areas. The firmer/harder areas can be thickenings of the suspensory ligaments, or fibrocystic changes to the breast. The softer areas are typically just adipose tissue. There are a number of benign (noncancerous) changes to the breast that can make breasts feel "lumpy." Specifically, fibrocystic changes in the breast consist of fluid-filled cysts and are often surrounded by dense fibrous tissue. This tissue, though often large and dense, is typically mobile (moves around easily), and often mirrors itself on the opposite breast. That is, if you find such changes in the lower-right quadrant of the right breast, you might also find it on the lower-right quadrant of the left breast. On the contrary, breast cancers tend to feel much firmer, like a kernel of unpopped popcorn, and they are immobile (they do not move when you touch them). They are generally painless and sometimes cause dimpling of the skin overlying the tumor.

The following are three procedures for performing a breast self-exam:

1. In the shower: Use soapy hands to palpate the breast while raising the arm on the same side as the breast you are palpating. Starting at the periphery of the breast, use two or three fingers to make small circles that progressively move toward the nipple. Squeeze the nipple to look for discharge. Be sure to palpate all the way up into the axilla, as the majority of tumors arise in the outer/upper quadrant toward the axillary region of the breast.

2. Lying down: Use the same procedure as in the shower. It is good to do an exam lying down because it makes the breasts lie flat, which may make it easier to feel certain lumps.

3. Standing in front of a mirror: Place your hands on your hips, tilt your elbows forward (anterior), and look for any indentations in the skin of the breast or "orange peel"–looking skin, which can be indicative of an underlying tumor.

If you discover anything that concerns you, make an appointment with your gynecologist as soon as possible to discuss your findings so the physician can help determine if what you felt was something benign or something that needs further exploration. Always remember that breast cancer is *highly curable* when caught early.

Male Reproductive System

The **testes**, the primary male reproductive organs, function in the production of sperm and testosterone. Sperm produced in the testes are stored in the **epididymis** and transported via the **ductus deferens** to the male urethra during ejaculation. The **seminal vesicles** and **prostate gland** are accessory reproductive glands that produce substances that nourish and protect the sperm and also compose the vast majority of semen. The **bulbourethral glands** are small glands located within the urogenital diaphragm that produce a mucus-like substance during sexual arousal that neutralizes the acidity of the male urethra and provides lubrication to ease the passage of semen during ejaculation.

The following exercise involves observing the male reproductive structures at the gross level.

EXERCISE 27.3

GROSS ANATOMY OF THE SCROTUM, TESTIS, SPERMATIC CORD, AND PENIS

1. Observe a prosected male human cadaver or classroom models of the male reproductive organs.

2. Identify the structures listed in **figures 27.4** and **27.5** on the cadaver or on classroom models, using the textbook as a guide. Then label them in figures 27.4 and 27.5.

3. *Optional Activity*: **APR** **14: Reproductive System**—In the Quiz section, select the "Structures: Male" option to test your knowledge of male reproductive gross anatomy.

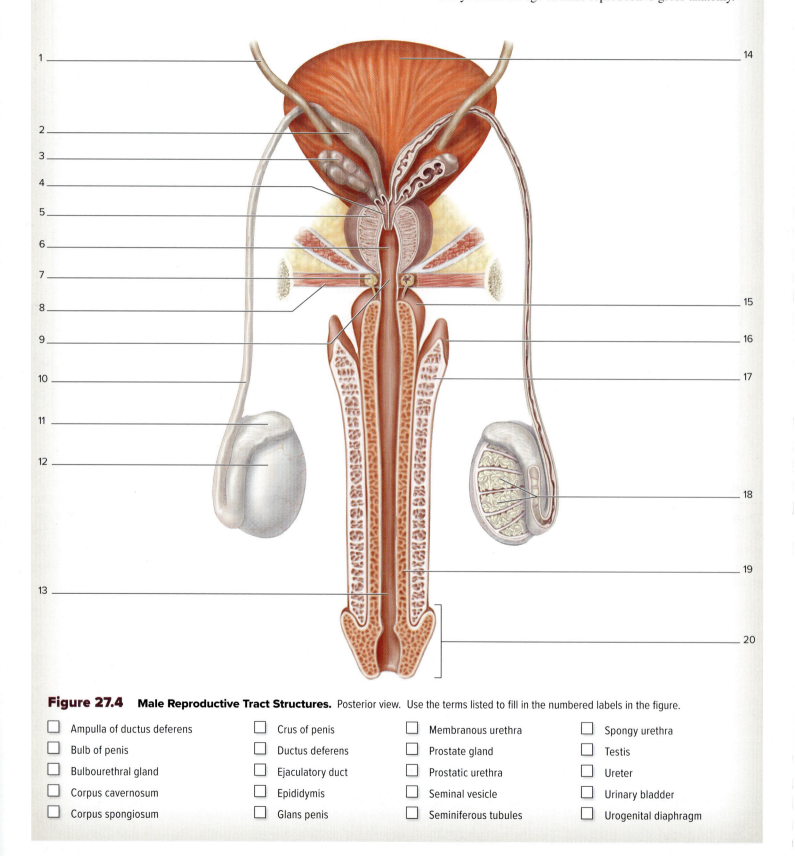

Figure 27.4 **Male Reproductive Tract Structures.** Posterior view. Use the terms listed to fill in the numbered labels in the figure.

☐ Ampulla of ductus deferens
☐ Bulb of penis
☐ Bulbourethral gland
☐ Corpus cavernosum
☐ Corpus spongiosum

☐ Crus of penis
☐ Ductus deferens
☐ Ejaculatory duct
☐ Epididymis
☐ Glans penis

☐ Membranous urethra
☐ Prostate gland
☐ Prostatic urethra
☐ Seminal vesicle
☐ Seminiferous tubules

☐ Spongy urethra
☐ Testis
☐ Ureter
☐ Urinary bladder
☐ Urogenital diaphragm

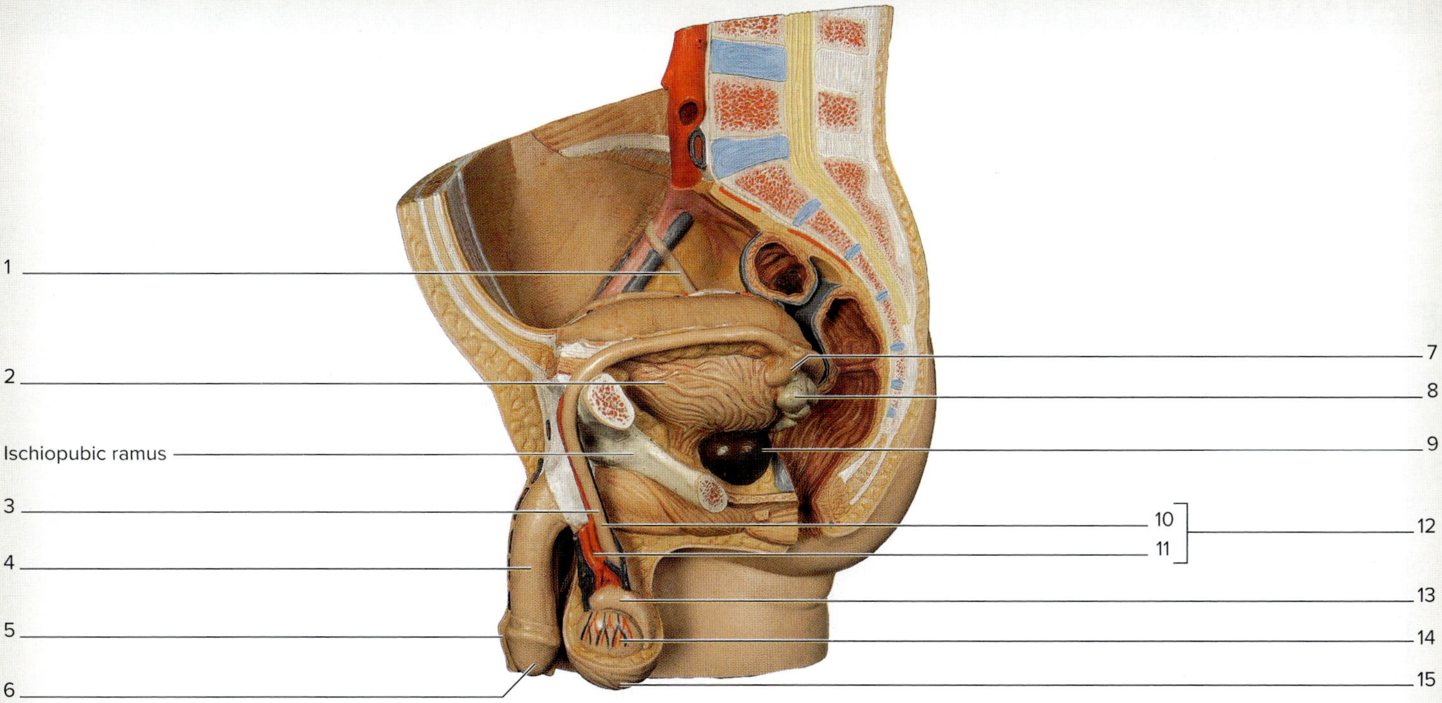

(a) Sagittal view

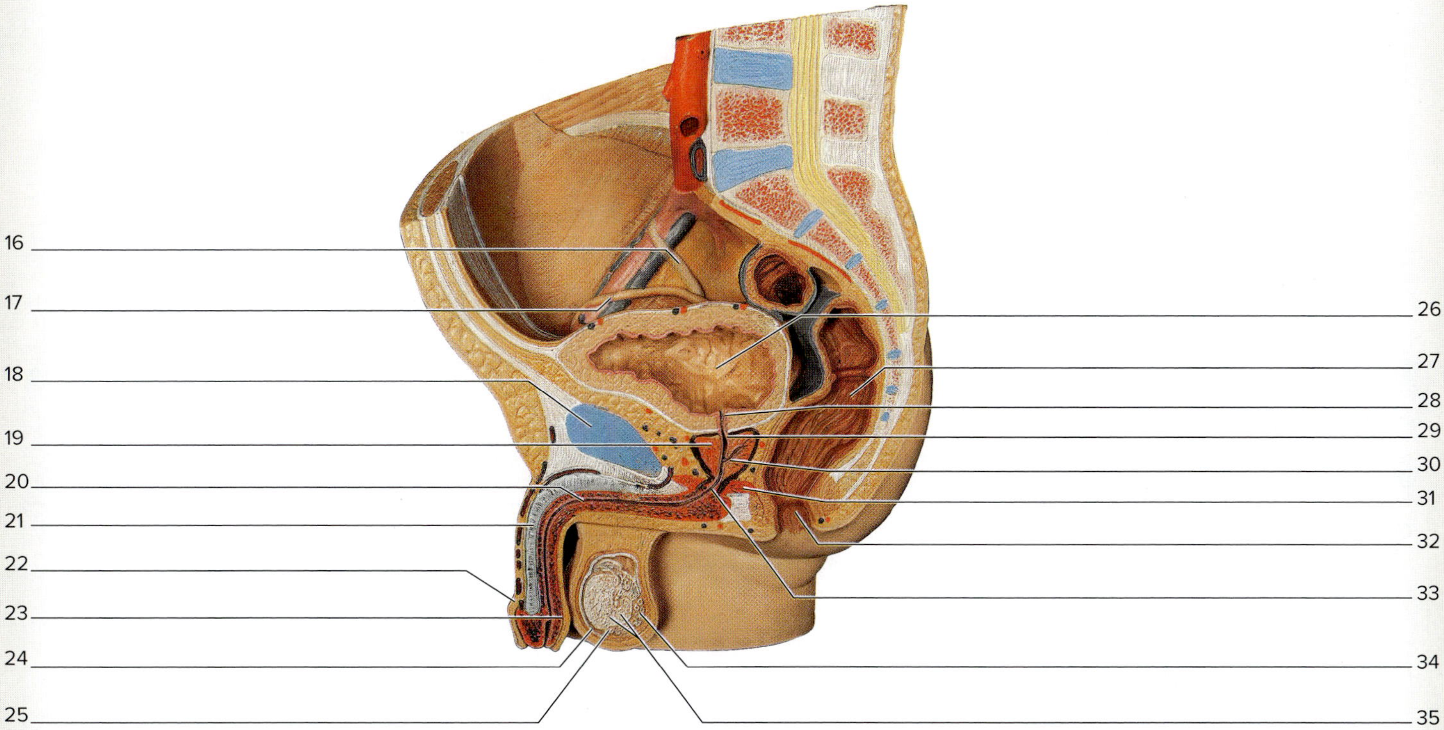

(b) Midsagittal view

Figure 27.5 **Classroom Model of the Male Pelvic Cavity.** (*a*) Sagittal view with pelvic structures intact. (*b*) Midsagittal view. Use the terms listed to fill in the numbered labels in the figure. Some answers may be used more than once.

☐ Anus

☐ Corpus cavernosum

☐ Corpus spongiosum

☐ Ductus deferens (ampulla)

☐ Ductus deferens

☐ Ejaculatory duct

☐ Epididymis

☐ Glans penis

☐ Internal urethral sphincter

☐ Membranous urethra

☐ Penis

☐ Prepuce

☐ Prostate gland

☐ Prostatic urethra

☐ Pubic symphysis

☐ Rectum

☐ Scrotum

☐ Seminal vesicle

☐ Seminiferous tubules

☐ Spermatic cord

☐ Spongy urethra

☐ Testicular artery and vein

☐ Testis

☐ Tunica albuginea of testis

☐ Tunica vaginalis of testis

☐ Ureter

☐ Urinary bladder

☐ Urogenital diaphragm

Model # H10 [1000282] ©3B Scientific GmbH, Germany, 2015 www.3bscientific.com/Photo by Christine Eckel, Ph.D.

HISTOLOGY

Female Reproductive System

The female reproductive system has two important functions: produce the female gametes (ova) and support, protect, and nourish the developing embryo/fetus that is formed when an ovum becomes fertilized by a sperm. Components of the female reproductive system include the ovaries, uterine tubes, uterus, vagina, and mammary glands. Exercises 27.4 to 27.7 explore the histological details of the ovary, uterine tube, uterus, and vagina.

EXERCISE 27.4

HISTOLOGY OF THE OVARY

The **ovary** is the primary reproductive organ in the female and functions in the production of the ova (eggs) and the female sex steroid hormones **estrogen** and **progesterone**. The ovaries contain several **ovarian follicles**, all at various stages of development. Each follicle contains a developing **oocyte** (**table 27.3**). The structure and function of the ovary are best viewed at the histological level because characteristics of ovarian follicles are evident at each stage of development.

1. Obtain a histology slide of an ovary and place it on the microscope stage.

2. Bring the tissue sample into focus on low power and identify the outer **cortex** and inner **medulla** of the ovary (**figure 27.6**). Next, move the microscope stage so the ovarian cortex is at the center of the field of view, and then change to high power.

3. Observe the outermost region of the ovarian cortex. Locate the two layers of tissue that compose the outer coverings of the ovary: the germinal epithelium and the tunica albuginea (**figure 27.7**). The **germinal epithelium**

(*germen*, a sprout) is a simple cuboidal epithelium that forms the outermost covering of the ovary. The name *germinal* refers to the fact that scientists once thought that the germ cells (oocytes) were formed from this epithelial layer. Most ovarian cancers arise in the germinal epithelium. Deep to the germinal epithelium is the **tunica albuginea** (*tunica*, a coat, + *albugineus*, white spot), which is composed of dense irregular connective tissue.

4. Focus on the cortex and identify follicles in each stage of follicular development listed in **table 27.4**.

5. Identify the following structures on the slide of the ovary, using tables 27.3 and 27.4 and figures 27.6 and 27.7 as guides:

- [ ] **Corpus albicans**
- [ ] **Corpus luteum**
- [ ] **Germinal epithelium**
- [ ] **Primary follicle**
- [ ] **Primary oocyte**
- [ ] **Primordial follicle**
- [ ] **Secondary follicle**
- [ ] **Secondary oocyte**
- [ ] **Tunica albuginea**
- [ ] **Vesicular (Graafian) follicle**

Table 27.3	Developmental Stages of an Oocyte			
Cell Name/ Oocyte Stage	**Description and Function**	**Stage of Mitosis/ Meiosis**	**Ploidy**	**Word Origin**
Oogonia	Primitive germ cells that undergo mitosis in the second to fifth months of embryonic life form oogonia that will subsequently develop into primary oocytes. Approximately 70% of the primary oocytes will degenerate in a process called atresia.	Formed by mitosis; undergo mitosis to form primary oocytes	Diploid (2n)	*oon*, egg, + *gonia*, generation
Primary Oocyte	Oogonia replicate DNA, begin meiosis, and arrest in prophase I of meiosis I, becoming primary oocytes. Primary oocytes will not undergo subsequent meiotic divisions unless the follicle is stimulated to mature by follicle-stimulating hormone (FSH) and luteinizing hormone (LH).	Arrested in prophase I of meiosis	Diploid (2n)	*primary*, first, + *oon*, egg, + *kytos*, cell
Secondary Oocyte	Under the influence of FSH and LH (after puberty), primary oocytes complete meiosis I, producing two cells of unequal sizes. The larger cell becomes a secondary oocyte and the smaller one becomes a polar body, which degenerates. The secondary oocyte will not undergo subsequent meiotic divisions unless fertilization occurs.	Arrested in metaphase II of meiosis	Haploid (1n)	*secondary*, second, + *oon*, egg, + *kytos*, cell
Ovum	Upon fertilization, the secondary oocyte undergoes the second meiotic division to become an ovum.	Formed after secondary oocyte is fertilized and completes meiosis	Haploid (1n)	*oon*, egg, + *kytos*, cell

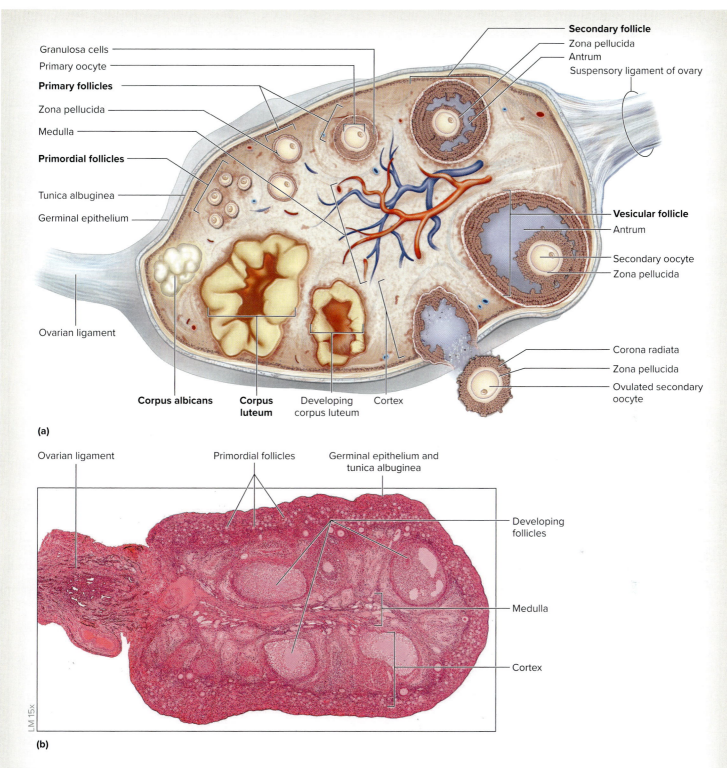

Granulosa cells
Primary oocyte
Primary follicles
Zona pellucida
Medulla
Primordial follicles
Tunica albuginea
Germinal epithelium
Ovarian ligament

Secondary follicle
Zona pellucida
Antrum
Suspensory ligament of ovary

Vesicular follicle
Antrum
Secondary oocyte
Zona pellucida

Corona radiata
Zona pellucida
Ovulated secondary oocyte

Corpus albicans   **Corpus luteum**   Developing corpus luteum   Cortex

(a)

Ovarian ligament   Primordial follicles   Germinal epithelium and tunica albuginea

Developing follicles

Medulla

Cortex

LM 15x

(b)

Figure 27.6   **The Ovary.** (*a*) The ovary is the primary reproductive organ of the female and produces the female gametes, the ova. (*b*) Histological structure of the ovary demonstrating the outer cortex containing follicles in various stages of development, and the inner medulla, which consists mainly of blood vessels and nerves.

(*b*) ©McGraw-Hill Education/Al Telser

6. Next move the microscope stage so a vesicular follicle is in the center of the field of view. A **vesicular (Graafian) follicle (figure 27.8)** is a mature follicle that is ready to be released during ovulation, after stimulation by a surge in luteinizing hormone (LH), which is secreted by the anterior pituitary gland. **Table 27.5** lists the histological features of a vesicular follicle. Identify the following parts of the vesicular follicle, using table 27.5 and figure 27.8 as guides:

☐ **Antrum**
☐ **Corona radiata**
☐ **Cumulus oophorus**
☐ **Granulosa cells**

☐ **Secondary oocyte**
☐ **Theca externa**
☐ **Theca interna**
☐ **Zona pellucida**

7. Identify a corpus luteum and corpus albicans on the slide, using table 27.4 as a guide. Then complete table 27.4, describing the histological features of each follicular stage.

(continued on next page)

(continued from previous page)

Table 27.4 — Developmental Stages of Ovarian Follicles

Follicle Stage	Primordial Follicle	Primary Follicle	Secondary (Maturing) Follicle
Photograph	LM 500x ©McGraw-Hill Education/Alvin Telser	LM 500x ©McGraw-Hill Education/Alvin Telser	LM 50x ©Ed Reschke/Stone/Getty Images
Follicle Description			
Oocyte Stage			

Developmental Stages of Ovarian Follicles

Follicle Stage	Vesicular (Graafian) Follicle	Corpus Luteum	Corpus Albicans
Photograph	LM 100x ©Ed Reschke	LM 25x ©McGraw-Hill Education/Alvin Telser	LM 80x ©McGraw-Hill Education/Alvin Telser
Follicle Description			
Oocyte Stage			

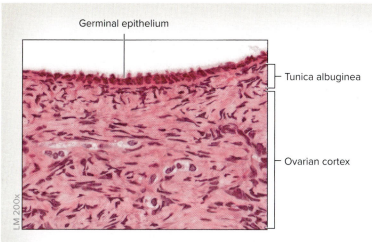

Figure 27.7   **Germinal Epithelium and Tunica Albuginea of the Ovary.** The germinal epithelium is the most common origin site for ovarian cancers. The tunica albuginea is a "white coat" of dense irregular connective tissue that surrounds the entire ovary internal to the germinal epithelium.

©McGraw-Hill Education/Al Telser

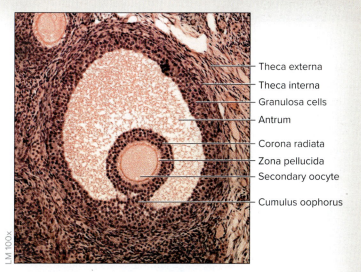

Figure 27.8   **Vesicular (Graafian) Follicle.** A vesicular follicle contains a secondary oocyte and a single antrum.

©McGraw-Hill Education/Christine Eckel, photographer

INTEGRATE

LEARNING STRATEGY

There are several distinguishing features used to identify follicles in various developmental stages. First, look for the number of cell layers and the shape of the cells that surround the oocyte. Primordial follicles contain a single layer of flattened cells, whereas primary follicles contain one or more layers of cuboidal cells. Next, look for the presence of fluid-filled antra. Secondary follicles contain one or more fluid-filled antra, whereas vesicular follicles are characterized by the presence of a single large antrum.

Table 27.5	Components of a Vesicular Follicle	
Structure	**Description and Function**	**Word Origin**
Antrum	The fluid-filled space in the center of the follicle	*antron,* a cave
Corona Radiata	A single layer of columnar cells derived from the cumulus oophorus that attach to the zona pellucida of the oocyte	*corona,* a crown, + *radiatus,* to shine
Cumulus Oophorus	A "mound" of granulosa cells that supports and surrounds the secondary oocyte within the follicle	*cumulus,* a heap, + *oophoron,* ovary
Granulosa Cells	Epithelial cells lining the follicle that will become the luteal cells of the corpus luteum after ovulation. They secrete the liquor folliculi, which is the fluid that fills the antrum.	*granulum,* a small grain
Theca Externa	The external fibrous layer of a well-developed vesicular follicle. The cells and fibers are arranged in concentric layers.	*theca,* a box, + *externus,* external
Theca Interna	The inner cellular layer of the vesicular follicle; these cells secrete androgen that is converted to estrogen by granulosa cells	*theca,* a box, + *internus,* internal
Zona Pellucida	A thick coat of glycoproteins that surrounds the oocyte	*zona,* zone, + *pellucidus,* clear

EXERCISE 27.5

HISTOLOGY OF THE UTERINE TUBES

The **uterine tubes** (fallopian tubes, or oviducts) are epithelia-lined fibromuscular tubes that extend from an ovary to the uterus. These tubes transport the ovum (or if fertilization occurs, the developing zygote) from the ovary to the uterus. The wall of the uterine tube is composed of three layers: mucosa, muscularis, and serosa.

1. Obtain a histology slide demonstrating a cross section of a uterine tube.

2. Place the slide on the microscope stage and bring the tissue sample into focus on low power. Note the highly folded **mucosa** of the tube **(figure 27.9)**.

3. Move the microscope stage until an area of folded mucosa appears at the center of the field of view. Then switch to high power. The epithelium of the uterine tube contains two cell types, **ciliated epithelial cells** and **secretory cells**. **Table 27.6** describes the histological components of the uterine tubes. Focus on the epithelial cells that line the uterine tube, and distinguish ciliated cells from secretory cells (figure 27.9*b*).

4. Identify the following structures on the slide of the uterine tube, using table 27.6 and figure 27.9 as guides:

   ☐ **Ciliated cells**   ☐ **Secretory cells**

   ☐ **Mucosal folds**   ☐ **Serosa**

   ☐ **Muscularis**

5. Sketch a cross section of the uterine tube, as seen through the microscope in the space provided or on a separate sheet of paper. Be sure to label the structures listed in step 4 in the drawing.

_____ ✕

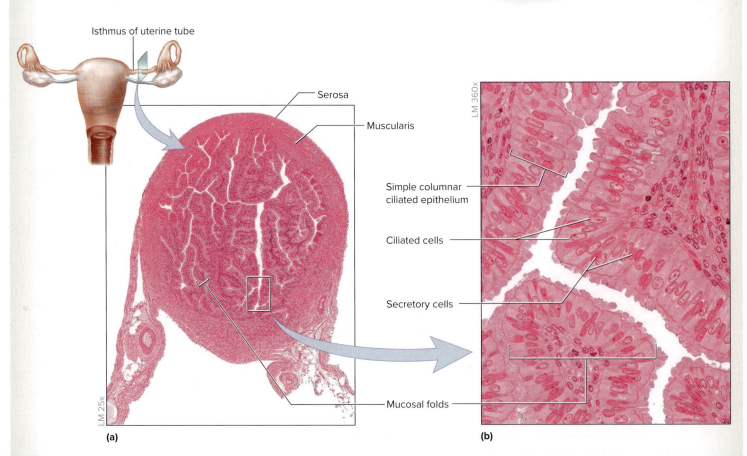

(a)   (b)

Figure 27.9 **Histology of the Uterine Tubes.** (*a*) Cross section through the isthmus of the uterine tubes demonstrating the many folds of the mucosa. (*b*) The epithelial cells lining the uterine tubes contain two cell types: ciliated cells and secretory cells. Ciliated cells have light nuclei, whereas secretory cells have dark nuclei.

©McGraw-Hill Education/Al Telser

Table 27.6	Components of the Uterine Tube	
Uterine Tube Structure	**Description and Function**	**Word Origin**
Regions		
Infundibulum	The funnel-like expansion of the ovarian end of the uterine tube. Finger-like extensions of the infundibulum are fimbriae (*fimbria*, fringe).	*infundibulum*, a funnel
Ampulla	The wide part of the uterine tube located medial to the infundibulum. Its mucosa is highly folded and lined with simple ciliated columnar epithelium and secretory cells. Fertilization most commonly occurs in the ampulla.	*ampulla*, a two-handled bottle
Isthmus	The narrow part of the uterine tube located right next to the uterus	*isthmos*, a constriction
Uterine Part (Interstitial Segment)	The part of the uterine tube that penetrates the wall of the uterus	*intra–*, within, + *muralis*, wall (*inter*, between, + *sisto*, to stand)
Layers		
Serosa	The outer layer of the uterine tube. It consists of simple squamous epithelium (a fold of peritoneum).	*serosus*, serous
Muscularis	The middle layer of the wall of the uterine tube. It consists of smooth muscle whose contraction assists in transporting the ovum toward the uterus.	*musculus*, mouse (referring to muscle)
Mucosa	The inner layer of the wall of the uterine tube. It is highly folded in the infundibulum and ampulla of the uterine tube. Contains ciliated columnar epithelium with a layer of areolar connective tissue underneath.	*mucosus*, mucous
Cells		
Ciliated Cells	Simple ciliated columnar epithelial cells; cilia beat toward the uterus to transport the ovum to the uterus. Nuclei stain lighter than those of the secretory cells.	*cilium*, eyelash
Secretory Cells	Nonciliated columnar epithelial cells that promote the activation of spermatozoa (capacitation) and provide nourishment for the ovum. Nuclei stain darker than those of the ciliated cells.	*secretus*, to separate

EXERCISE 27.6

HISTOLOGY OF THE UTERINE WALL

The **uterus** is a hollow, pear-shaped, muscular organ whose primary function is to support, protect, and nourish the developing embryo/fetus.

The uterine wall is composed of three layers: endometrium, myometrium, and perimetrium. The inner lining of the uterus, the **endometrium**, goes through a cycle of growth during the ovarian cycle. Each time a woman's ovary undergoes a single ovarian cycle, the endometrium becomes prepared for the possibility that a fertilized egg will become implanted. If no implantation occurs, the innermost layer of the endometrium, the **functional layer**, sloughs off in the process of **menstruation** (*menstruus*, monthly). **Table 27.7** describes the layers of the uterine wall and the components that make up each layer. **Table 27.8** shows histological images of the phases of the menstrual cycle.

1. Obtain a histology slide that includes a portion of the uterine wall. Place it on the microscope stage and bring the tissue sample into focus on low power.

2. Identify the following structures on the slide of the uterus, using tables 27.7 and 27.8 as guides:

   ☐ **Endometrium (basal layer)**   ☐ **Myometrium**
   ☐ **Endometrium (functional layer)**   ☐ **Perimetrium**

3. Move the microscope stage so the functional layer of the endometrium is in the center of the field of view. Switch to high power and identify **uterine glands**.

4. Complete table 27.8 and describe what occurs within the tissue during each phase.

(continued on next page)

(continued from previous page)

Table 27.7	Wall Layers of the Uterus	
Wall Layer	**Description and Function**	**Word Origin**
Endometrium	The mucous membrane composing the inner layer of the uterine wall. Consists of simple columnar epithelium and a lamina propria with simple tubular uterine glands. The structure, thickness, and state of the endometrium undergoes marked changes during the menstrual cycle.	*endon*, within, + *metra*, uterus
Functional Layer (Stratum Functionalis)	The apical layer of the endometrium. Most of this layer is shed during menstruation.	*stratum*, a layer, + *functus*, to perform
Basal Layer (Stratum Basalis)	The basal layer of the endometrium. It undergoes minimal changes during the menstrual cycle and serves as the basis for regrowth of the more apical stratum functionalis.	*stratum*, a layer, + *basalis*, basal
Myometrium	The muscular wall of the uterus composed of three layers of smooth muscle	*mys*, muscle, + *metra*, uterus
Perimetrium	The outermost covering of the uterus; a serous membrane formed from peritoneum	*peri-*, around, + *metra*, uterus

Table 27.8	Phases of the Menstrual Cycle		
Phase	**(a) Menstrual Phase**	**(b) Proliferative Phase**	**(c) Secretory Phase**
	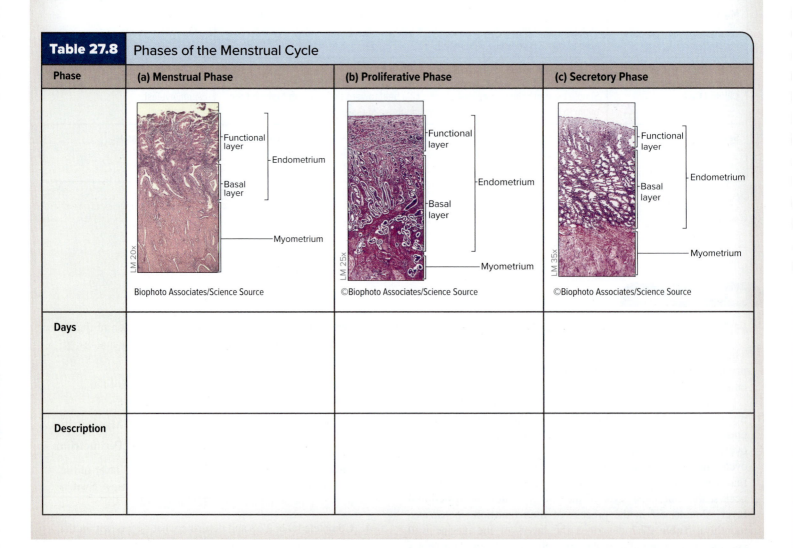		
Days			
Description			

LM 20x — Functional layer — Basal layer — Endometrium — Myometrium — Biophoto Associates/Science Source

LM 25x — Functional layer — Basal layer — Endometrium — Myometrium — ©Biophoto Associates/Science Source

LM 35x — Functional layer — Basal layer — Endometrium — Myometrium — ©Biophoto Associates/Science Source

HISTOLOGY OF THE VAGINAL WALL

The **vagina** (*vagina*, sheath) is a muscular tube that functions as the organ of copulation, serves as the passageway for menses, and serves as the birth canal during parturition (*parturio*, to be in labor).

1. Obtain a histology slide demonstrating a portion of the vaginal wall. Place it on the microscope stage and bring the tissue sample into focus on low power. The walls of the vagina are composed of a mucosa and muscularis; the mucosa is lined with a nonkeratinized stratified squamous epithelium (**figure 27.10**).

2. Identify the following structures on the slide of the vagina, using figure 27.10 as a guide:

   ☐ **Lamina propria**
   ☐ **Mucosa**
   ☐ **Muscularis**
   ☐ **Nonkeratinized stratified squamous epithelium**

3. *Optional Activity*: **APR** **14: Reproductive System—** Watch the "Female Reproductive System Overview" animation to review the female reproductive organs and their histological features.

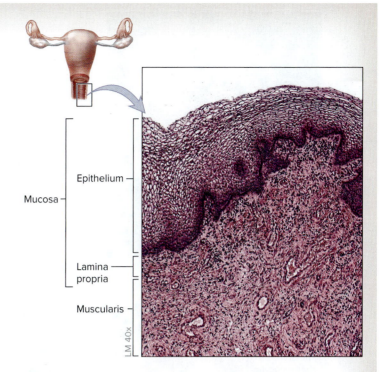

Figure 27.10 **Histology of the Vaginal Epithelium.** The epithelium is nonkeratinized stratified squamous.
©McGraw-Hill Education/Al Telser

Male Reproductive System

The male reproductive system consists of the testes, epididymis, ductus deferens, seminal vesicles, prostate gland, bulbourethral glands, and penis. The primary functions of the male reproductive system are to produce the male gamete (sperm), produce testosterone, and provide nourishment for the sperm and a mechanism for the sperm to be delivered to the female reproductive tract. Testes also produce testosterone hormone.

HISTOLOGY OF THE SEMINIFEROUS TUBULES

The **testes** are the primary reproductive organ in the male; they function in the production of sperm and the male sex steroid hormone **testosterone**. Within each testis are several hundred, coiled **seminiferous tubules (figure 27.11)**, which are the site of sperm production, or **spermatogenesis** (*sperma*, seed, + *genesis*, origin). Within each tubule, **spermatogonia** undergo successive meiotic divisions as they move from the basal to the apical surface of the tubule epithelium. **Table 27.9** describes the developmental stages of the sperm, and **table 27.10** describes the structure and function of the accessory cell types located within the testes.

1. Obtain a slide of the testes and place it on the microscope stage.

2. Observe the slide on low power and identify several cross sections of the seminiferous tubules. Move the microscope stage so that one tubule is at the center of the field of view, and change to high power.

(continued on next page)

(continued from previous page)

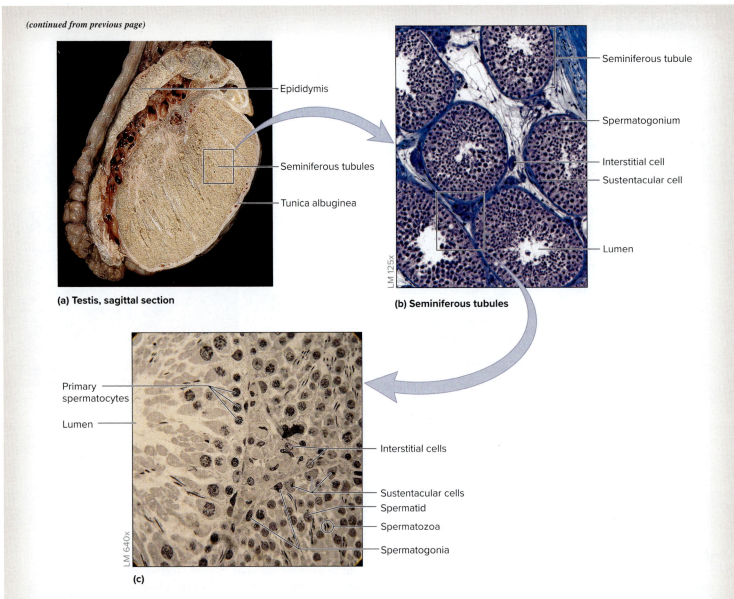

(a) Testis, sagittal section

Epididymis

Seminiferous tubules

Tunica albuginea

(b) Seminiferous tubules

Seminiferous tubule

Spermatogonium

Interstitial cell

Sustentacular cell

Lumen

LM 125x

(c)

Primary spermatocytes

Lumen

Interstitial cells

Sustentacular cells

Spermatid

Spermatozoa

Spermatogonia

LM 640x

Figure 27.11   Testes and Seminiferous Tubules. (*a*) Testis and epididymis showing seminiferous tubules within the testes and tunica albuginea surrounding the testis. (*b*) Medium-power histological appearance of seminiferous tubules. (*c*) High-power histological view of seminiferous tubules demonstrating stages of spermatogenesis and sustentacular cells.

(*a*) From Anatomy & Physiology Revealed, ©McGraw-Hill Education/The University of Toledo, photography and dissection; (*b*) ©McGraw-Hill Education/Al Telser; (*c*) ©McGraw-Hill Education/Christine Eckel, photographer

Table 27.9	Developmental Stages of Sperm			
Cell Name	**Description and Function**	**Stage of Mitosis/Meiosis**	**Ploidy**	**Word Origin**
Spermatogonia	Primitive male gametes that are formed by mitosis from male germline stem cells	Formed by mitotic division of germline stem cells; some of these spermatogonia become primary spermatocytes	Diploid (2n)	*sperma,* seed, + *gonia,* generation
Primary Spermatocyte	Male gamete that has replicated its DNA in preparation for meiosis. It will subsequently undergo meiosis I to form two secondary spermatocytes.	Formed after mitotic division of spermatogonia	Diploid (2n)	*primary,* first, + *sperma,* seed, + *kytos,* cell
Secondary Spermatocyte	Male gamete that has completed meiosis I and will subsequently undergo meiosis II to form two spermatids	Formed after the first meiotic division	Haploid (1n)	*secondary,* second, + *sperma,* seed, + *kytos,* cell
Spermatid	Male gamete that has completed meiosis, but has not undergone spermiogenesis, the process of becoming a mature spermatozoon	Formed after the second meiotic division	Haploid (1n)	*sperma,* seed, + *-id,* a young specimen
Spermatozoon (Sperm)	A mature male gamete that has undergone spermiogenesis	Formed by maturation of spermatids	Haploid (1n)	sperma, seed, + *zoon,* animal

Table 27.10	Accessory Cells of the Testis	
Cell Name	**Description and Function**	**Word Origin**
Interstitial (Leydig) Cells	Large cells located adjacent to the seminiferous tubule. Produce the steroid hormone testosterone, which is required for proper sperm development and for the development of male secondary sex characteristics.	*inter-*, between, + *sisto*, to stand
Sustentacular (Sertoli) Cells	Tall columnar cells that surround multiple developing spermatocytes to provide them with support and nourishment. Phagocytize excess cytoplasm from developing spermatocytes. Form the *blood-testis barrier*, which protects the developing spermatocytes from antigens that circulate in the blood. Produce androgen-binding protein (ABP), which concentrates testosterone around the developing spermatocytes.	*sustento*, to hold upright

3. Identify the following structures, using tables 27.9 and 27.10 and figure 27.11 as guides:

- ☐ **Interstitial cell**
- ☐ **Primary spermatocyte**
- ☐ **Seminiferous tubule**
- ☐ **Spermatids**
- ☐ **Spermatogonia**
- ☐ **Sustentacular cell**

4. Sketch a seminiferous tubule as seen through the microscope in the space provided or on a separate sheet of paper. Be sure to label the structures listed in step 3 in the drawing.

5. *Optional Activity*: **APR 14: Reproductive System—** Watch the "Spermatogenesis" animation to visualize the formation of sperm in the seminiferous tubules.

_____ ×

EXERCISE 27.9

HISTOLOGY OF THE EPIDIDYMIS

After spermatozoa are formed within the seminiferous tubules, they are transported through the straight tubules, rete testis, and efferent ductules to enter the long, coiled tube of the **epididymis** (*epi-*, upon, + *didymos*, a twin [related to *didymoi*, testes]). Spermatozoa undergo the process of maturation and are stored in the epididymis until ejaculation takes place. One of the most characteristic features of the epididymis is the presence of thousands of sperm in the lumen of the tube (**figure 27.12**). **Table 27.11** describes the structure and function of the male accessory reproductive structures, including the epididymis.

1. Obtain a slide of the epididymis and place it on the microscope stage. Bring the tissue sample into focus on low power and identify several cross sections of the epididymis. Move the microscope stage so that one part of the lumen of the epididymis is at the center of the field of view, and then change to high power. Notice the sperm inside the lumen of the tubule, and how they do not come right up against the apical surface of the columnar epithelial cells. This is because of the stereocilia on the apical surface of the epithelial cells. **Stereocilia** (*stereo*, solid, + *cilium*, eyelid) are single, long microvilli (ironically, they are not cilia at all!) that increase the surface area of the epithelial cells for the purpose of secreting substances that nourish the sperm and absorbing substances from the sperm as they undergo maturation.

2. Identify the following structures on the slide of the epididymis, using table 27.11 and figure 27.12 as guides:

- ☐ **Pseudostratified columnar epithelial cells with stereocilia**
- ☐ **Sperm cells**
- ☐ **Stereocilia**

3. Sketch the epididymis as seen through the microscope in the space provided or on a separate sheet of paper. Be sure to label the structures listed in step 2 in the drawing.

_____ ×

(continued on next page)

(continued from previous page)

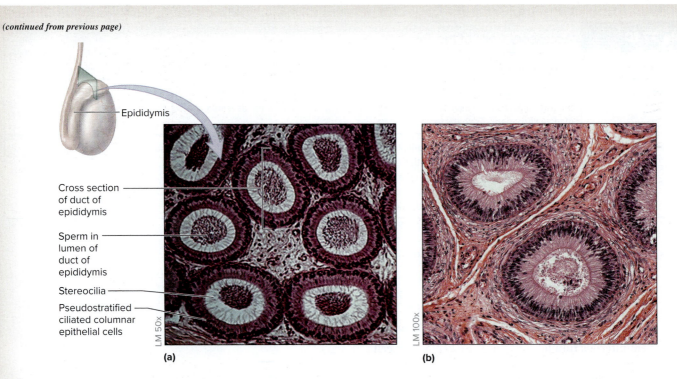

Figure 27.12 **Cross Section Through the Epididymis.** Several cross sections through the seminiferous tubules are visible. (*a*) Note the numerous sperm within the lumen. (*b*) A higher magnification view demonstrates pseudostratified columnar epithelium with stereocilia.

(*a*) ©Ed Reschke/Stone/Getty Images; (*b*) ©McGraw-Hill Education/Christine Eckel

Table 27.11	Male Accessory Reproductive Structures		
Structure	**Epithelium**	**Function**	**Word Origin**
Epididymis	Pseudostratified columnar epithelium with stereocilia	Site of storage and maturation of sperm. Walls contain some smooth muscle that will propel sperm into the ductus deferens during ejaculation.	*epi-*, upon, + *didymos*, a twin (related to *didymoi*, testes)
Ductus Deferens (Vas Deferens)	Pseudostratified columnar epithelium with stereocilia	A thick, muscular tube whose walls undergo peristaltic contractions during ejaculation to propel sperm into the urethra	*ductus*, to lead, + *defero*, to carry away
Prostate Gland	Glandular epithelium resembling pseudostratified or simple columnar epithelium	Accessory reproductive gland that contributes approximately 25% of the volume of semen. The fluid produced is rich in Vitamin C (citric acid) and enzymes (such as PSA, prostate-specific antigen) that are important for proper sperm function.	*prostates*, one standing before
Seminal Vesicles	Pseudostratified columnar epithelium	Accessory reproductive gland that contributes approximately 60% of the volume of semen. The fluid produced is rich in fructose, which is a source of energy for the sperm. They also produce prostaglandins, which are important in promoting sperm motility and may stimulate uterine contractions.	*seminal*, relating to sperm, + *vesicula*, a blister
Bulbourethral Glands	Glandular epithelium consisting of simple and pseudostratified columnar epithelium	Accessory reproductive gland that produces a mucus-like lubricating substance during sexual arousal (prior to ejaculation) that lubricates the urethra and neutralizes the acidity of the urine, thus preparing the way for the spermatozoa to pass	*bulbus*, bulb, + *urethral*, relating to the urethra

EXERCISE 27.10

HISTOLOGY OF THE DUCTUS DEFERENS

The **ductus deferens** (vas deferens) is a continuation of the epididymis and is the route by which sperm travel from the epididymis to the urethra during ejaculation (*ejaculo*, to shoot out). It is a highly muscular tube lined with pseudostratified columnar epithelium with stereocilia (**figure 27.13**).

1. Obtain a slide demonstrating the ductus deferens and place it on the microscope stage. Bring the tissue sample into focus on low power and identify the cross section of the ductus deferens (figure 27.13). Move the microscope stage so the lumen of the ductus deferens is at the center of the field of view, and then change to high power.

2. Identify the following structures on the slide of the ductus deferens, using table 27.11 and figure 27.13 as guides:

   ☐ **Lumen**

   ☐ **Mucosa**

   ☐ **Pseudostratified ciliated columnar epithelial cells**

   ☐ **Smooth muscle**

   ☐ **Stereocilia**

3. Sketch a cross section of the ductus deferens as seen through the microscope in the space provided or on a separate sheet of paper. Be sure to label the structures listed in step 2 in the drawing.

_____ ×

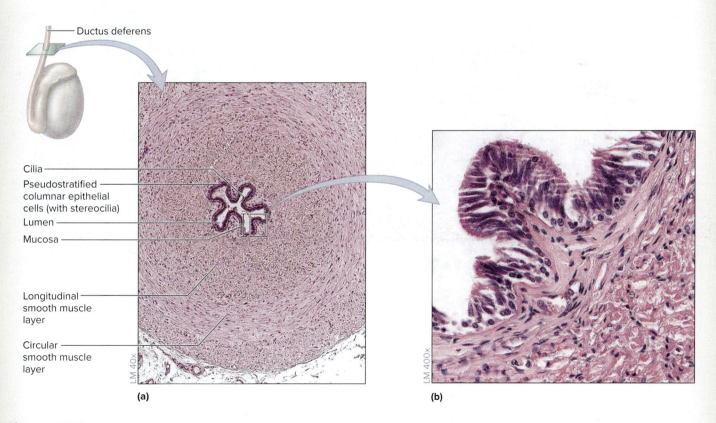

Figure 27.13 **Ductus Deferens.** (*a*) Cross section through the entire ductus deferens, demonstrating thick layers of smooth muscle surrounding the lumen. (*b*) Close-up view of the pseudostratified columnar epithelium with stereocilia.

(*a*) ©McGraw-Hill Education/Al Telser; (*b*) ©McGraw-Hill Education/Christine Eckel

EXERCISE 27.11

HISTOLOGY OF THE SEMINAL VESICLES

Near the posterior wall of the bladder, each ductus deferens comes together with the duct from a **seminal vesicle (figure 27.14)**. These accessory reproductive glands are lined with pseudostratified columnar epithelium and produce substances (including fructose, which is a source of energy for the sperm) that are an important component of **semen**. Semen is the fluid expelled from the penis during ejaculation (table 27.11).

1. Obtain a slide of the seminal vesicles and place it on the microscope stage. Bring the tissue sample into focus on low power. Identify cross sections of the seminal vesicles. Move the microscope stage so the lumen of a portion of the seminal vesicle is at the center of the field of view, and then change to high power.

2. Identify the following structures on the slide of the seminal vesicles, using table 27.11 and figure 27.14 as guides:

☐ **Lumen**

☐ **Mucosal folds**

☐ **Muscular wall**

3. Sketch the seminal vesicles as seen through the microscope in the space provided or on a separate sheet of paper. Be sure to label the structures listed in step 2 in the drawing.

_____ ×

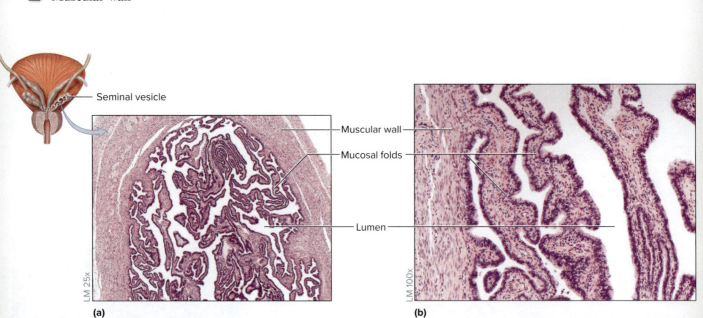

Seminal vesicle

Muscular wall

Mucosal folds

Lumen

LM 25x

(a)

LM 100x

(b)

Figure 27.14   The Seminal Vesicles. (*a*) Low magnification. (*b*) Medium magnification.

(*a*) ©McGraw-Hill Education/Alvin Telser; (*b*) ©McGraw-Hill Education/Al Telser

EXERCISE 27.12

HISTOLOGY OF THE PROSTATE GLAND

The **prostate** gland is an accessory reproductive gland located immediately inferior to the urinary bladder. Like the seminal vesicles, the prostate produces substances that will become part of semen and that are important for proper sperm function.

1. Obtain a slide demonstrating the prostate gland and place it on the microscope stage. Bring the tissue sample into focus on low power and identify the prostatic urethra and the prostate gland itself (**figure 27.15**). Move the microscope stage so a part of the gland relatively far away from the urethra is at the center of the field of view, and then change to a higher power.

2. Identify the following structure on the slide of the prostate, using table 27.11 and figure 27.15 as guides:

   ☐ **Tubuloalveolar glands**

3. As men age, calcifications often form in the prostate. Such calcifications are called **prostatic calculi** (corpora aranacea) (figure 27.15). Scan the slide and see if any of these are present in the specimen.

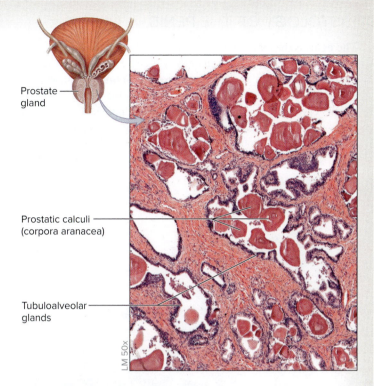

Prostate gland

Prostatic calculi (corpora aranacea)

Tubuloalveolar glands

LM 50x

Figure 27.15   **The Prostate Gland.** Note the many prostatic calculi (corpora aranacea).

©McGraw-Hill Education/Al Telser

INTEGRATE

CONCEPT CONNECTION

Recall from chapter 5 that smooth muscle is composed of spindle-shaped, nonstriated cells with a single centrally located nucleus. The function of this type of muscle is to move contents through the viscera and blood vessels. As you learned in chapter 26 (The Digestive System), the smooth muscle layers of the muscularis contract to churn and mix contents within the digestive organs as well as to move these contents forward with peristaltic contractions. Smooth muscle is also present in several male and female reproductive organs. In the male, smooth muscle in the ductus deferens undergoes peristaltic contraction to move sperm from the epididymis to the urethra during ejaculation. In the female, the smooth muscle that makes up the thick myometrial layer of the uterus contracts during labor, allowing the fetus to exit the birth canal.

INTEGRATE

CLINICAL VIEW

Testicular Cancer and Testicular Self Exams

Testicular cancer arises from an abnormal growth of cells within the testes. Most cases of testicular cancer are germ cell tumors and are classified based on their histological appearance and physiological behaviors. Although this form of cancer is rare, it has climbed over the past few decades, and treatment of this disease depends on the type of cancer diagnosed.

Often the first sign of testicular cancer is a noticeable bump felt on one of the testicles. Clinicians recommend that males perform monthly testicular self-examinations once they hit puberty. To perform a testicular self-exam, males should:

1. Perform the exam after a warm shower when the scrotal sac will be relaxed and easier to examine.

2. Examine the scrotum and testicles in the mirror, looking for any changes in size or swelling.

3. Hold a testicle between the thumb and index finger. Roll the testicle around between the fingers, feeling for any abnormal bumps. The testicular surface should be relatively smooth. However, be aware that there is a coiled mass of tissue behind each testicle (the epididymis).

4. Pay attention to any pain felt during the examination as this general process should be painless.

Any concerning findings should be reported to a clinician for follow-up examination and possible testing.

EXERCISE 27.13

HISTOLOGY OF THE PENIS

The **penis** is the male copulatory organ and is homologous to the female clitoris. It is composed of three erectile bodies, the paired **corpora cavernosa** and the single **corpus spongiosum**, which contains the male urethra (**figure 27.16**). In the center of each corpus cavernosum is a central artery that supplies blood to the penis. During an erection, the veins draining the penis become constricted, and venous blood fills up the erectile tissues.

1. Obtain a slide demonstrating a cross section of the penis and place it on the microscope stage. Observe at low power.

2. Identify the following structures on the slide of the penis, using figure 27.16 as a guide:

   ☐ **Central artery**
   ☐ **Corpus cavernosum**
   ☐ **Corpus spongiosum**
   ☐ **Dorsal surface**
   ☐ **Dorsal vein**
   ☐ **Superficial fascia (connective tissue)**
   ☐ **Urethra**
   ☐ **Ventral surface**

3. Sketch a cross section of the penis as seen through the microscope in the space provided or on a separate sheet of paper. Be sure to label the structures listed in step 2 in the drawing.

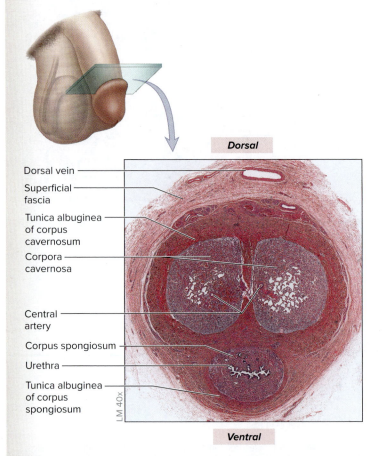

Dorsal

Dorsal vein
Superficial fascia
Tunica albuginea of corpus cavernosum
Corpora cavernosa
Central artery
Corpus spongiosum
Urethra
Tunica albuginea of corpus spongiosum
LM 40x

Ventral

_____ ×

Figure 27.16   Cross Section Through the Shaft of the Penis. The penis contains two paired corpora cavernosa dorsally, and a single corpus spongiosum ventrally. Note the male urethra in the center of the corpus spongiosum.

©McGraw-Hill Education/Christine Eckel

4. *Optional Activity*: **APR** 14: Reproductive System—Watch the "Male Reproductive System Overview" animation to review the male reproductive structures and their relationships.

INTEGRATE

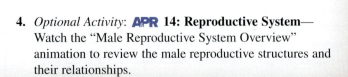

CLINICAL VIEW
Erectile Dysfunction

Erectile dysfunction (ED) occurs when a male is unable to achieve or maintain an erection. Stimulation of the parasympathetic nervous system causes central arteries within the corpora cavernosa to dilate. This dilation increases local blood flow to the erectile tissues. Concomitant constriction of veins at the base of the penis prevents the drainage of blood from the erectile tissues. The resulting engorging of the erectile tissues within the penis with blood causes an erection. The penis remains erect until ejaculation, a process stimulated by the sympathetic nervous system. Erectile dysfunction involves the interruption of the arousal response and is most commonly caused by neurological, cardiovascular, or hormonal disturbances. Interestingly, some treatments for those suffering from erectile dysfunction, including sildenafil (i.e., Viagra), were first introduced as antihypertensive drugs. Sildenafil reduces systemic blood pressure by dilating blood vessels and decreasing total peripheral resistance. As concerns an erection, sildenafil dilates the central arteries within the penis, which facilitates the process of achieving and maintaining an erection.

PHYSIOLOGY

Reproductive Physiology

Gametogenesis, the production of the **gametes** (sperm and secondary oocytes) is accomplished by meiosis. **Meiosis** is similar in many ways to mitosis, but there are two nuclear divisions (I and II), which result in a reduction of chromosome number from a diploid number (2n, or 46) to a haploid number (n, or 23). Another difference in meiosis is the process of **synapsis** (crossing over) that occurs when homologous chromosomes are adjacent to each other. Synapsis is a process that introduces variability in the genetic code, which ensures that daughter cells are genetically different from parent cells.

Table 27.12 depicts and describes each phase of meiosis and the defining features of each phase.

Human somatic cells contain 23 pairs of chromosomes, or **homologous pairs**. These include 22 pairs of autosomal chromosomes, and one pair of sex chromosomes. Human gametes (egg and sperm) contain 22 autosomal chromosomes and one sex chromosome, an X chromosome for oocytes and an X or Y chromosome for sperm. Meiosis yields four genetically different, haploid daughter cells. The process of **fertilization** brings haploid sperm together with a haploid secondary oocyte. When a sperm cell has fertilized an oocyte, the resulting cell (called a zygote) is once again diploid.

Table 27.12	**Phases of Meiosis**			
	Description			
Phase	**Meiosis I**		**Meiosis II**	**Word Origin**
Prophase	Homologous maternal and paternal chromosomes pair; crossover (genetic variability) occurs	Replicated chromosome / Sister chromatids / Tetrad	Nuclear envelope breaks down; chromosomes gather	*pro-*, before
Metaphase	Homologous chromosomes attach to spindle fibers and line the equator of the cell	Spindle fiber attached to centromere / Centromere / Equator	Chromosomes attach to spindle fibers and line the cell equator	*meta-*, middle

(continued on next page)

Table 27.12	Phases of Meiosis *(continued)*			
	Description			
Phase	**Meiosis I**		**Meiosis II**	**Word Origin**
Anaphase	Separation of chromosome pairs; reduction division	Homologous replicated chromosomes separate	Sister chromatids pulled apart (single-stranded chromosomes)	*ana-*, upward
			Sister chromatids separate / Sister chromatids separate	
Telophase	Completion of nuclear division; nuclear envelope reforms	Replicated chromosome	Completion of nuclear division; nuclear envelope reforms	*telo-*, end
Cytokinesis	Division of the cytoplasm; each cell contains 23 chromosomes	Cleavage furrow	Division of the cytoplasm; each of 4 cells contains 23 single-stranded chromosomes	*cyto-*, cell
			Cells separate into four haploid daughter cells / Single chromosomes	

Oogenesis

The process of producing a mature egg (ovum) in females requires both mitotic and meiotic division. Primordial germ cells, or oogonia (singular, *oogonium*), divide by mitosis. Primary oocytes start the process of meiosis. **Oogenesis** involves forming secondary oocytes from primary oocytes. However, rather than proceeding rapidly through the two phases of meiosis, primary and secondary oocytes experience periods of "arrest" where meiosis is stopped until hormones or events trigger continuation of meiosis. These "arrested" oocytes are housed within ovarian follicles. **Primordial follicles** contain primary oocytes, which are available to be selected each month once a female reaches puberty, the time at which monthly hormone fluctuations begin. Only a few primordial follicles will mature into primary follicles, and fewer still will become secondary follicles. At ovulation, a large vesicular follicle bursts and releases a secondary oocyte from the follicle's fluid-filled sac. The secondary oocyte will then be drawn into the uterine tubes to potentially be fertilized by sperm. The final phases of meiosis occur after fertilization (see table 27.12).

The result of fertilization is formation of a **zygote**, which will subsequently undergo mitosis to form a mass of cells that may implant in the uterine wall. Development *in utero* continues, as described in exercise 27.16. **Figure 27.17** depicts the overall process of oogenesis, including identification of the various phases and arrests of meiosis, corresponding stages of the ovarian follicles, and histological images of each stage.

Hormones are responsible for regulating the monthly changes that occur both in the ovaries (ovarian cycle) and in the uterus (uterine cycle). The overall purpose is to produce viable eggs (ova) for fertilization and to provide an attractive place for implantation of the fertilized egg. The release of gonadotropin-releasing hormone (GnRH) from the hypothalamus stimulates the release of luteinizing hormone (LH) and follicle-stimulating hormone (FSH) from the anterior pituitary gland. It is LH and FSH that stimulate the development of the follicles during the **follicular phase** of the ovarian cycle. As the follicles develop, the granulosa cells surrounding the follicle secrete estrogen. Estrogen stimulates the cells in the uterine endometrium to proliferate during the **proliferative phase** of the uterine cycle.

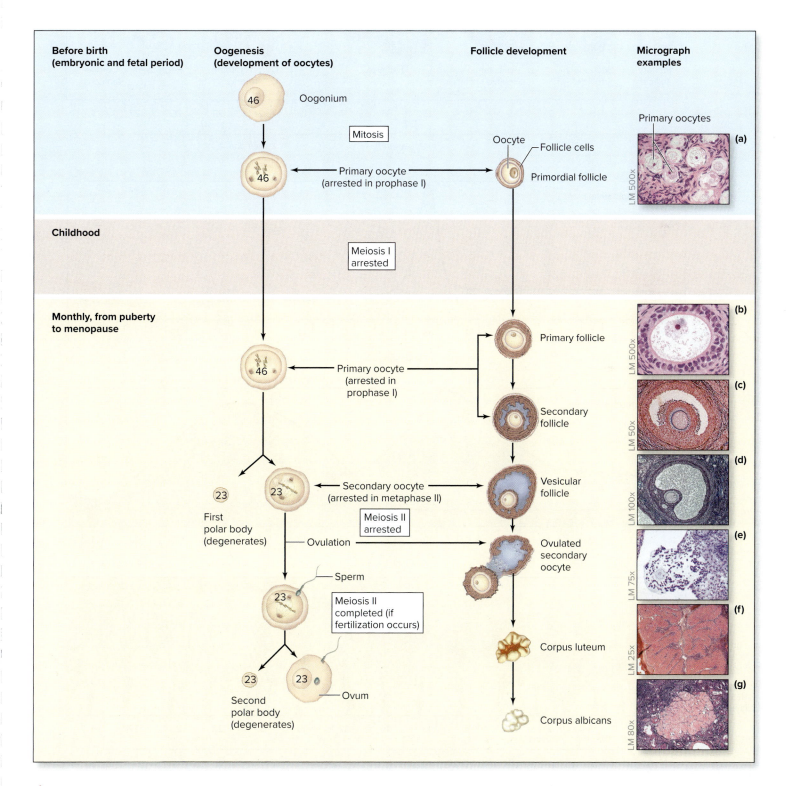

Figure 27.17 **Oogenesis.**

(*a* & *b*) & (*f* & *g*) ©McGraw-Hill Education/Alvin Telser; (*c*) ©Ed Reschke/Getty Images; (*d*) ©Ed Reschke; (*e*) ©Dr. Francisco Gaytan

Once ovulation occurs, the ruptured vesicular follicle becomes the corpus luteum. This structure begins producing progesterone and estrogen throughout the **luteal phase** of the ovarian cycle. Increasing levels of progesterone result in thickening and increased vascularization of the endometrium. In addition, the corpus luteum maintains elevated progesterone and estrogen levels in the event fertilization occurs, but prior to the formation of the placenta, the structure is responsible for maintaining a viable pregnancy. Human chorionic gonadotropin (hCG), the hormone secreted by the developing placenta, prevents the degradation of the corpus luteum in the event that fertilization and implantation occur. If fertilization and implantation do not occur, the corpus luteum degenerates into the corpus albicans. This degeneration also decreases levels of progesterone and estrogen. As the progesterone levels drop, blood vessels (spiral

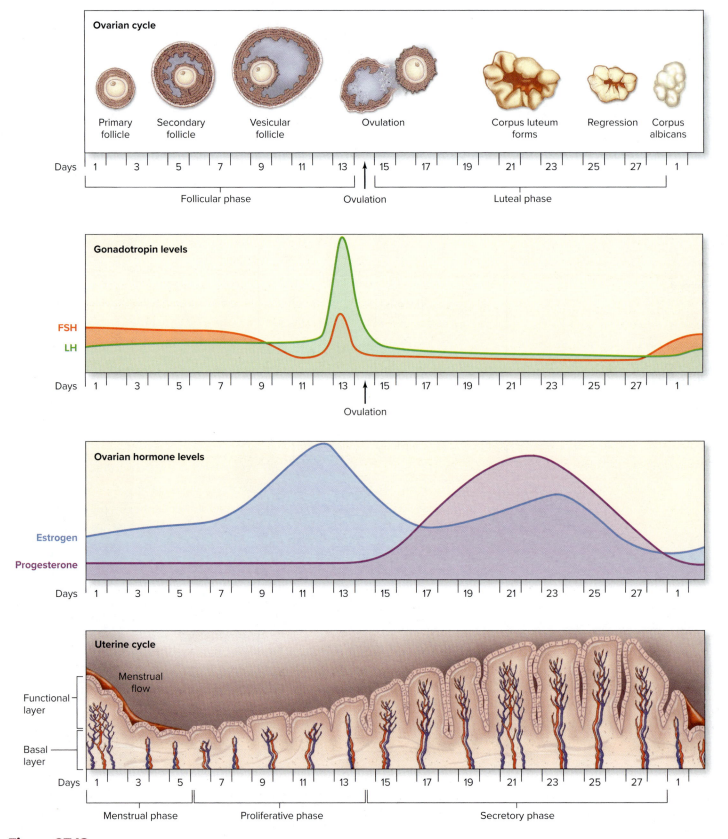

Figure 27.18 **Ovarian and Uterine Cycles.**

arterioles) in the uterine wall constrict, and the endometrial layer is shed for menses. Typically, these changes occur in 28-day cycles, as depicted in **figure 27.18**. However, cycles can vary in length. Research suggests that the duration of the luteal phase remains more consistent at roughly 14 days, whereas the duration of the follicular phase may vary widely among women.

INTEGRATE

CONCEPT CONNECTION

Estrogen released by follicular cells acts directly on the hypothalamus and the anterior pituitary gland to regulate the amount of gonadotropin-releasing hormone (GnRH) released from the hypothalamus, and of luteinizing hormone (LH) and follicle-stimulating hormone (FSH) released from the anterior pituitary gland. Prior to ovulation, the increasing estrogen is stimulatory (positive feedback) for the release of GnRH, LH, and FSH. However, once ovulation occurs, estrogen is inhibitory (negative feedback) to the release of these hormones. It is a peak in LH that allows ovulation to occur; therefore, this switch from positive to negative feedback limits the number of eggs released each month. Incidentally, fertility drugs will often override these feedback mechanisms, thereby releasing multiple secondary oocytes with the chance for multiples (twins, triplets, or more). The release of these hormones was first introduced in the chapter on the endocrine system, chapter 19. The concepts of negative and positive feedback appear throughout the book, and are critical to understanding homeostasis.

Spermatogenesis

The production of sperm, **spermatogenesis**, in males is similar to oogenesis in females. It is interesting that the overall process is regulated by the same hormones (GnRH, LH, and FSH) that regulate oogenesis in females, with the exception of males secreting testosterone in place of estrogen. GnRH from the hypothalamus stimulates the release of LH and FSH from the anterior pituitary gland. LH stimulates interstitial cells in the seminiferous tubules of the testes to release testosterone, and FSH stimulates sustentacular cells to release androgen binding protein (ABP) and inhibin. The elevated levels of testosterone are stimulatory for sperm production, yet are inhibitory for GnRH production by the hypothalamus. The process begins as spermatogonia, or primordial germ cells, actively undergo mitosis and remain at the basement membrane of the seminiferous tubules. These spermatogonia divide to form primary spermatocytes, which are diploid cells. Primary spermatocytes then undergo meiosis. Unlike females, males have the ability to continuously form new primary spermatocytes for gamete formation. After meosis I, the spermatocyte is considered secondary and haploid, with only 23 chromosomes. Once the secondary spermatocyte undergoes meiosis II, it is considered a spermatid and is located closer to the lumen of the seminferous tubule. Each of the four spermatids produced then undergoes the maturation process, or **spermiogenesis**. This process involves the development of a head, or chromosome-containing acrosome, a midpiece containing abundant mitochondria, and a tail (flagellum) of the functional spermatozoon (plural, *spermatozoa*). The spermatozoon continues the maturation process in the epididymis before becoming a viable sperm cell.

EXERCISE 27.14

A CLINICAL CASE IN REPRODUCTIVE PHYSIOLOGY

The purpose of this exercise is to apply the concept of reproductive physiology to a real scenario, or clinical case study. This exercise requires information presented in this chapter, as well as in previous chapters. Integrating multiple systems in a case study reinforces a concept that is very important in physiology: homeostasis.

Before beginning, state how the ovarian cycle may be influenced by increasing levels of testosterone.

_____

_____

_____

Consult with the laboratory instructor as to whether this exercise will be completed individually, in pairs, or in groups. Review the topics of the cardiovascular, lymphatic, respiratory, digestive, urinary, and reproductive systems in the textbook and in previous chapters in this laboratory manual as needed for assistance with the case study.

Case Study

Elizabeth, a 32-year-old newlywed, was eager to start a family. She was career focused, as was her husband, Frank, and both thrived under high-stress conditions. Their jobs required long hours, with little time to cook or clean at home. For that reason, they were constantly eating on the go, often getting takeout meals that were, admittedly, not the healthiest. Despite this hectic schedule, the couple was ready to have a child of their own. So, they began trying to conceive. Both Elizabeth and Frank were optimistic that they would be able to conceive despite having many friends that struggled with infertility. No women in Elizabeth's family had struggled before, so she had no real concerns.

Frank and Elizabeth's attempts to conceive were met with disappointment. They "tried" for 1 year with no luck, and ultimately sought medical advice, because Elizabeth was concerned that her age would soon make conception more difficult. Both Frank and Elizabeth visited a fertility clinic to investigate their case further. Doctors first tested Frank's sperm count and sperm viability, because this is the simplest test to conduct. There was no evidence of inadequate numbers of sperm or their inability to fertilize an egg. Frank and Elizabeth were relieved, yet still frustrated that they were unsure as to the cause of their infertility.

Doctors then made note of Elizabeth's physical health. At 5 feet, 0 inches, Elizabeth weighed 175 pounds. In addition to being overweight, she also suffered from chronic hypertension and an elevated heart rate. Doctors suggested a change in lifestyle to reduce stress, which they hoped would help Elizabeth lose weight and improve her cardiovascular health. Doctors also inquired about Elizabeth's reproductive health.

(continued on next page)

(continued from previous page)

Elizabeth admitted that she had been prescribed birth control pills at 15 years old to help regulate her irregular (unpredictable) periods. Since stopping the pill just 1 year ago, her periods were again irregular, with cycle durations ranging from 31 to 51 days. Menses would last 7 days with no remarkable pain or discomfort. She also noticed that she had an increased incidence of facial hair and acne, something that was quite unusual for her. Based on these symptoms, doctors ordered blood samples **(table 27.13)** to test for a condition called polycystic ovarian syndrome (PCOS).

PCOS is an endocrine disorder, characterized by elevated levels of testosterone, that is a major cause of infertility in females. Patients are often diabetic (type II) and exhibit signs of insulin resistance. Patients tend to be overweight and experience irregular periods due to a disruption in the positive and negative feedback mechanisms that regulate the ovarian and uterine cycles. Often, patients with PCOS do not ovulate, thereby preventing fertilization and implantation. Instead of ovulating, partially developed follicles may persist as "cysts" in the ovary. Clinically, ultrasound is used to observe ovarian cysts, and a diagnosis of PCOS requires that one ovary has more than 12 follicles ranging 2 to 9 mm in diameter. It is proposed that regulation of blood glucose levels and weight loss may allow ovulation to resume, although the exact mechanisms are unclear. In the event that dietary changes are unsuccessful, drugs can be administered to regulate blood glucose levels as well as stimulate ovulation.

Table 27.13	Elizabeth's Blood Test Results	
Blood Test	**Result**	**Reference Range**
Cholesterol (mg/dl)	218	< 200
FSH (mLU/L)	6	3–20
Glucose (mg/dl)	141	70–110
Insulin (µU/mL)	167	0–20
LH (mLU/L)	18	3–20
Testosterone (pg/mL)	15	1.1–6.3
Triglycerides (mg/dl)	291	40–150

Answer the following questions regarding Elizabeth's case:

a. Doctors often prescribe metformin, a drug that decreases glucose transport in the GI tract, for patients suffering from PCOS. Describe how this would be beneficial, given the symptoms of the disease.

_____

_____

b. Elizabeth's blood samples were taken on day 3 of the menstrual cycle. Describe the relative levels of estrogen, progesterone, LH, and FSH during this phase of the cycle.

_____

_____

c. How might these hormone levels be different if samples were collected on day 14 (assume a 28-day cycle)?

_____

_____

d. Describe the potential consequences of elevated LH when compared to FSH levels.

_____

_____

e. Discuss why elevated testosterone may lead to anovulation. (Hint: Testosterone exhibits similar feedback mechanisms in males and females).

_____

_____

f. Do you think that Elizabeth is suffering from PCOS?

_____

Fertilization and Development

The time period between **fertilization** and birth is considered the **prenatal period**. Fertilization yields a diploid cell, or **zygote**, that undergoes multiple mitotic divisions, known as **cleavage** events. Two cells become four cells, and then eight cells, until there is a mass of cells, called a **blastocyst**, which is ready for implantation in the uterine wall. This cleavage process takes approximately 2 weeks and constitutes the **pre-embryonic period**. Implantation marks the official start of gestation, a period that lasts 38 weeks in all. Weeks 3 through 8 constitute the **embryonic period**, a time in which the organism is called an **embryo**. During this time, major organs and organ systems develop. In the ninth week of development, the organism is classified as a **fetus**. The **fetal period** lasts for the remaining 30 weeks of gestation. During the fetal period, organs and systems continue to grow and develop in preparation for birth and to sustain life separate from the mother's womb. The following laboratory exercises explore the events of the pre-embryonic and embryonic periods.

EXERCISE 27.15

EARLY DEVELOPMENT: FERTILIZATION AND ZYGOTE FORMATION

The following laboratory exercise focuses on the changes that occur during the pre-embryonic period. Complete **table 27.14**, depicting the chronology of events in pre-embryonic development, using the textbook as a guide.

Table 27.14	Pre-Embryonic Period		
Appearance	**Developmental Stage**	**Location**	**Events**
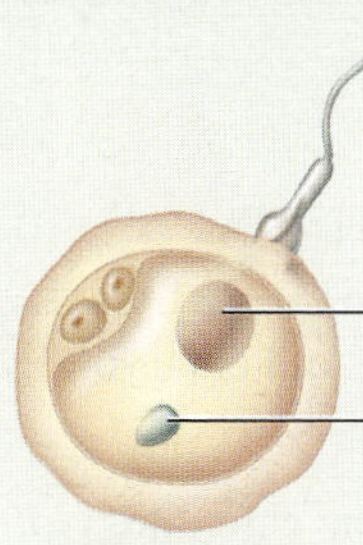 Ovum pronucleus — Sperm pronucleus ⊢ 120 µm ⊣			
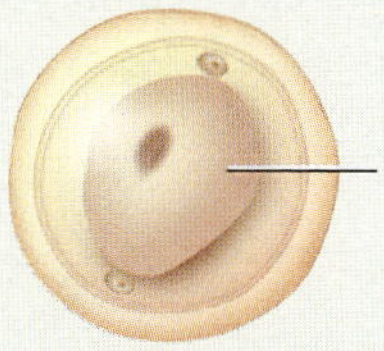 Nucleus ⊢ 120 µm ⊣			
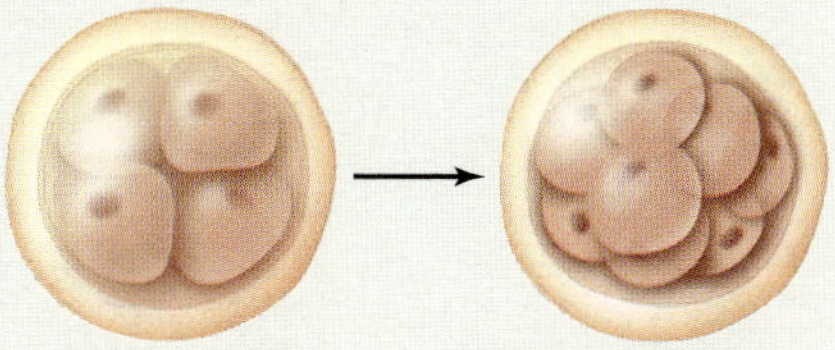 ⊢ 120 µm ⊣ ⊢ 120 µm ⊣ 4-cell stage 8-cell stage			
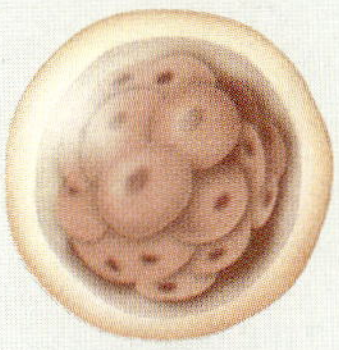 ⊢ 120 µm ⊣ Morula			

(continued on next page)

(continued from previous page)

Table 27.14	Pre-Embryonic Period *(continued)*		
Appearance	**Developmental Stage**	**Location**	**Events**
Embryoblast / Trophoblast — 120 μm			
Cytotrophoblast / Embryoblast / Syncytiotrophoblast			

EXERCISE 27.16

EARLY DEVELOPMENT: EMBRYONIC DEVELOPMENT

This laboratory exercise focuses on the changes that occur in an organism during the embryonic period. Complete the following using the textbook as a guide.

1. Label the structures of the 3-week-old embryo identified in **figure 27.19**.

2. Complete **table 27.15** depicting the chronology of events in embryonic development.

3. Label the structures of the 4-week-old embryo identified in **figure 27.20**.

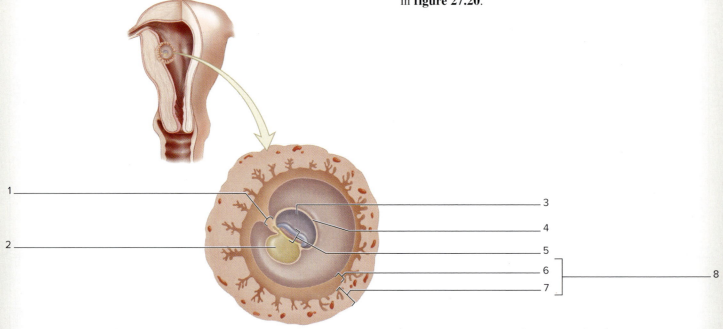

Figure 27.19 **Structures of the 3-Week-Old Embryo.** Use the terms listed to fill in the numbered labels in the figure.

☐ Amnion   ☐ Chorion   ☐ Embryo   ☐ Placenta

☐ Amniotic cavity   ☐ Connecting stalk   ☐ Functional layer of uterus   ☐ Yolk sac

Table 27.15	Stages of Embryonic Development	
Developmental Week		**Events**
Week 3	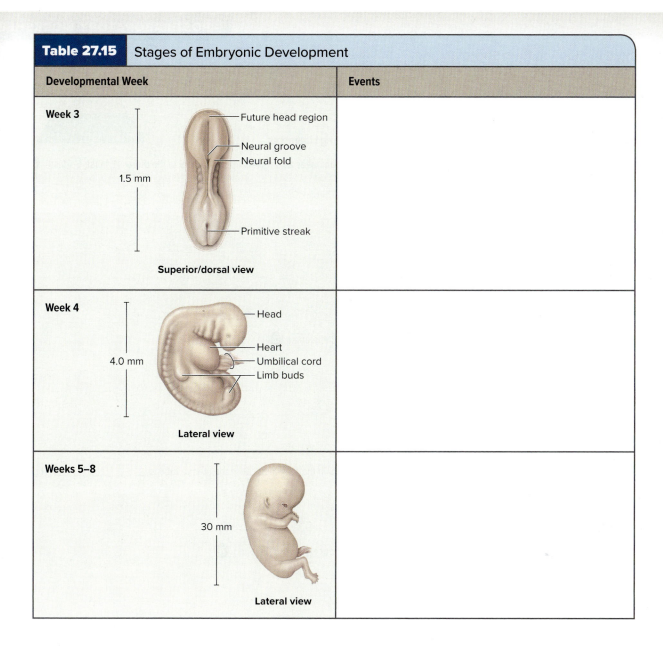	
Week 4		
Weeks 5–8		

Figure 27.20 **Structures of the 4-Week-Old (Late) Embryo.** Use the terms listed to fill in the numbered labels in the figure.

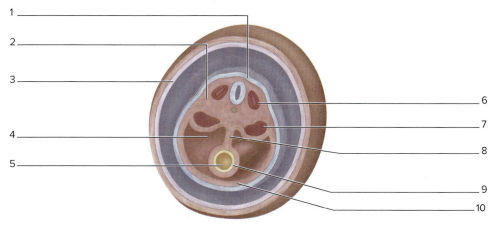

☐ Abdominal wall   ☐ Endoderm   ☐ Mesentery   ☐ Peritoneal cavity

☐ Amnion   ☐ Gut tube   ☐ Mesoderm   ☐ Somite

☐ Ectoderm   ☐ Intermediate mesoderm

The ❶ corresponds to the Learning Objective(s) listed in the chapter opener outline.

Do You Know the Basics?

Exercise 27.1: Gross Anatomy of the Ovary, Uterine Tubes, Uterus, and Supporting Ligaments

1. Place the following terms in the correct order to describe the pathway that an ovum takes as it travels from the ovary to the uterus (1 = first; 6 = last). ❶

   _____ a. ampulla of uterine tube

   _____ b. infundibulum of uterine tube

   _____ c. isthmus of uterine tube

   _____ d. ovarian cortex

   _____ e. uterine part of uterine tube

   _____ f. uterus

2. Which of the following structures travels through the inguinal canal in females? (Circle one.) ❷

   a. broad ligament

   b. ovarian ligament

   c. round ligament of the uterus

   d. suspensory ligament of the ovary

   e. uterosacral ligament

3. The peritoneal cavity is considered an _____ (open/closed) cavity in the female because the uterine tube is _____ (open/closed) at its distal end (the infundibulum). ❸

Exercise 27.2: Gross Anatomy of the Female Breast

4. Which of the following are the secretory units of the mammary glands, which produce milk? (Circle one.) ❹

   a. alveoli

   b. areolar glands

   c. lactiferous sinuses

   d. lobules

   e. nipples

Exercise 27.3: Gross Anatomy of the Scrotum, Testis, Spermatic Cord, and Penis

5. Match the description listed in column A with the associated structure listed in column B. ❺ ❽

   Column A

   _____ 1. located inferior to the urinary bladder

   _____ 2. produces fructose, an important component of semen

   _____ 3. produces a mucus-like substance that neutralizes the acidity of the urethra

   _____ 4. site of sperm maturation

   _____ 5. site of sperm production

   Column B

   a. bulbourethral gland

   b. epididymis

   c. prostate gland

   d. seminal vesicles

   e. testes

6. Which of the following is the portion of the male urethra that passes through erectile tissue? (Circle one.) ❻

   a. membranous urethra

   b. prostatic urethra

   c. spongy urethra

7. Which of the following are considered accessory reproductive structures that contribute substances to semen (other than sperm)?

   (Check all that apply.) **7** **8**

   _____ a. bulbourethral gland

   _____ b. epididymis

   _____ c. prostate gland

   _____ d. seminal vesicles

   _____ e. testes

8. The spermatic cord contains the testicular artery and a network of venous structures called the _____ (pampiniform plexus/rete testes). **9**

9. Place the following structures in the order in which a sperm travels through the male reproductive tract during ejaculation (1 = first; 5 = last). **10**

   _____ a. ductus deferens

   _____ b. epididymis

   _____ c. membranous urethra

   _____ d. prostatic urethra

   _____ e. spongy urethra

Exercise 27.4: Histology of the Ovary

10. The germinal epithelium is composed of which of the following types of epithelial tissue? (Circle one.) **11**

    a. simple columnar epithelium

    b. simple cuboidal epithelium

    c. simple squamous epithelium

    d. transitional epithelium

11. Identify the structure shown in the following photomicrograph. (Circle one.) **12**

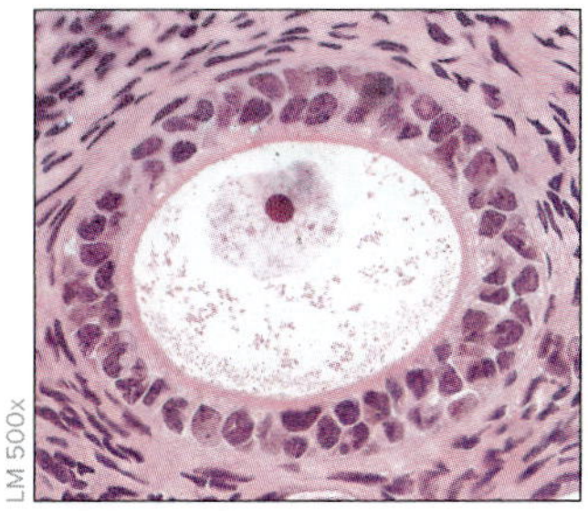

©McGraw-Hill Education/Alvin Telser

    a. primary follicle
    b. primordial follicle
    c. secondary follicle
    d. vesicular follicle

12. The arrow in the following figure points to a _____ (primary / secondary) follicle, which contains a _____ (primary/secondary) oocyte. **12**

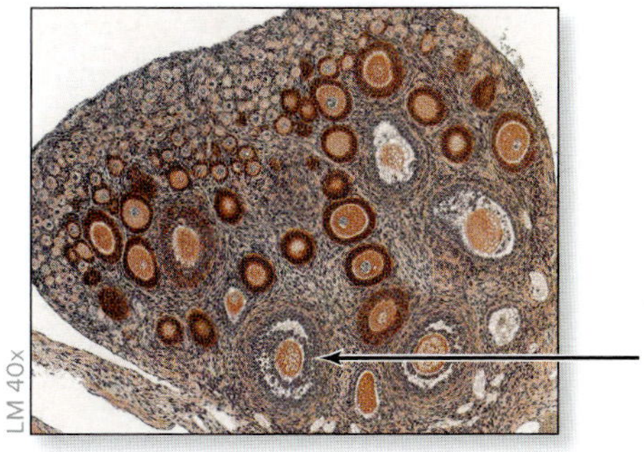

©McGraw-Hill Education/Christine Eckel

Exercise 27.5: Histology of the Uterine Tubes

13. Place the following layers that compose the wall of the uterine tube in order, from outermost to innermost (1 = outermost; 3 = innermost). **13** **14**

    _____ a. mucosa

    _____ b. muscularis

    _____ c. serosa

14. The uterine tube is lined with simple _____ (cuboidal/columnar) epithelium. **15**

Exercise 27.6: Histology of the Uterine Wall

15. The two layers of the uterine endometrium are the functional layer and the basal layer. The _____ (functional/basal) layer is shed during

    menstruation. **16**

16. The myometrium is composed of _____ (two/three) layers of smooth muscle. **17**

Exercise 27.7: Histology of the Vaginal Wall

17. The wall of the vagina is composed of which of the following layers? (Check all that apply.) **18**

    _____ a. adventitia

    _____ b. mucosa

    _____ c. muscularis

    _____ d. submucosa

18. The innermost lining of the vagina is lined with which of the following types of epithelial tissue? (Circle one.) **19**

    a. keratinized stratified squamous epithelium

    b. nonkeratinized stratified squamous epithelium

    c. simple columnar epithelium

    d. simple cuboidal epithelium

    e. simple squamous epithelium

Exercise 27.8: Histology of the Seminiferous Tubules

19. _____ (Sertoli/Spermatogonia) are tall columnar cells located in the seminiferous tubules that form the blood-testis barrier. **20**

20. Spermatogonia are located near the _____ (basement membrane/lumen) of the seminiferous tubules. **21**

21. Sustentacular cells, located within seminiferous tubules, produce _____ (androgen-binding protein/testosterone), whereas interstitial (Leydig) cells,

    located in between seminiferous tubules, produce _____ (androgen-binding protein/testosterone). **22**

Exercise 27.9: Histology of the Epididymis

22. Upon viewing the epididymis through a microscope, thousands of _____ (primary spermatocytes/spermatozoa) are visible in the lumen of the tube. **23**

23. The epididymis is lined with _____ (simple/pseudostratified) columnar epithelium with stereocilia. **24**

24. Which of the following is a function of stereocilia within the epidymis? (Check all that apply.) **25**

    _____ a. enhance sperm maturation

    _____ b. increase surface area

    _____ c. propel sperm

    _____ d. secretion and absorption

    _____ e. sperm production

Exercise 27.10: Histology of the Ductus Deferens

25. Identify the structure shown in the following photomicrograph. (Circle one.) **26**

    a. ductus deferens

    b. epididymis

    c. penis

    d. prostate gland

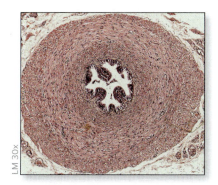

©Christine Eckel

26. The ductus deferens contains a mucosa and layers of smooth muscle. The innermost smooth muscle layer is _____(circular/longitudinal) muscle,

    whereas the outermost smooth muscle layer is _____(circular/longitudinal) muscle. **27**

27. The ductus deferens is lined with _____ (glandular epithelium/pseudostratified columnar epithelium with stereocilia) **28**

Exercise 27.11: Histology of the Seminal Vesicles

28. _____ (Bulbourethral glands/Seminal vesicles) release substances, including fructose, that nourish sperm. **29**

29. Seminal vesicles are lined with _____(glandular epithelium/pseudostratified columnar epithelium). **30**

Exercise 27.12: Histology of the Prostate Gland

30. Which of the following are distinguishing features of the prostate gland of an aging male when viewed through a microscope? (Check all that apply.) **31** **32**

    _____ a. cilia

    _____ b. mucosal folds

    _____ c. muscular wall

    _____ d. prostatic calculi

    _____ e. tubuloalveolar glands

Exercise 27.13: Histology of the Penis

31. The urethra is surrounded by which structure in the penis? (Circle one.) **33**

    a. central artery

    b. corpora cavernosa

    c. corpus spongiosum

    d. dorsal vein

32. The pair of erectile bodies located on the dorsal side of the penis are the _____ (corpora cavernosa/corpus spongiosum). **34**

Exercise 27.14: A Clinical Case in Reproductive Physiology

33. The product of mitosis is two _____ (diploid/haploid) daughter cells, whereas the product of meiosis is four _____ (diploid/haploid)

    daughter cells. **35**

34. Rising levels of _____ (estrogen/progesterone) correspond to the _____(proliferative/secretory) phase of the uterine cycle. **36**

35. Elevated levels of _____ (prolactin/testosterone) in females may inhibit ovulation. **37**

Exercise 27.15: Early Development: Fertilization and Zygote Formation

36. The diploid cell that forms as a result of fertilization is called a _____(blastocyst/zygote), whereas the structure that implants in the uterine wall is

    called a _____ (blastocyst/zygote). **38**

Exercise 27.16: Early Development: Embryonic Development

37. During embryonic development, limb buds appear in _____ (week 3/week 4). **39**

38. Which of the following are germ layers present in an embryo? (Check all that apply.) **40**

_____ a. amnion

_____ b. ectoderm

_____ c. endoderm

_____ d. mesoderm

_____ e. yolk sac

Can You Apply What You've Learned?

39. Unlike in the epididymis, it is highly unlikely that sperm cells are visible inside the lumen of the ductus deferens when viewed histologically. Why do you think this is?

_____

_____

_____

40. The homologous structure to the male penis is the female clitoris. The clitoris consists of two paired erectile tissues, the corpora cavernosa, but it does not contain a

corpus spongiosum. What structure is the clitoris "missing" as compared to the penis? _____

_____

41. A woman who is trying to conceive may purchase ovulation kits in order to test for the release of the secondary oocyte. An ovulation kit requires placing a dipstick in
a collected urine sample. A positive test indicates that ovulation will happen typically within 48 hours. For which hormone is this test sensitive?

_____

42. One cause of miscarriage, or loss of pregnancy, is low levels of progesterone in the initial stages of development.

a. Describe what mechanisms are in place to ensure adequate levels of progesterone in the event that fertilization occurs. _____

_____

b. What long-term mechanisms are in place to provide a continuous supply of progesterone (to maintain pregnancy)? _____

_____

Can You Synthesize What You've Learned?

43. Viagra®, a drug known for its ability to enhance erection in males, is a drug that was first prescribed to treat hypertension. Using the concepts of pressure, flow, and

resistance, describe why a drug used to counteract high blood pressure may also lead to an erection in males. _____

_____

_____

_____

44. Birth control pills work to override the hormone cycles in order to prevent a secondary oocyte from being released. Based on your knowledge of hormonal regulation
of the ovarian cycle, design a drug that could be used to prevent ovulation. Be specific about what hormones would be present in the female and why. Would you
expect the levels of these hormones to change throughout the 28-day cycle? _____

_____

_____

_____

Cat Dissection Exercises

OUTLINE AND LEARNING OBJECTIVES

Anatomy & Physiology Revealed® 4.0

Module 2: Cells & Chemistry

INTRODUCTION

Dissection is a powerful tool in learning anatomy, which it provides an unparalleled experience of discovering the beauty of the human/animal form. Dissection provides the opportunity to see the real size of structures and allows one to actually *feel* the texture of the tissues and organs! What's more, it allows investigation of the relationships between anatomical structures that, until now, were studied separately from each other. Although the process of dissection may cause apprehension at first, this feeling generally dissipates with experience.

The exercises in this chapter cover the dissection of a vertebrate mammal that has anatomy similar to the human: the domestic cat (*Felis domesticus*). The beauty of this comparative vertebrate anatomy is that it allows one to realize just how much all vertebrates have in common. For example, although a cat's posture is not the same as a human's, the structure and function of the cat's skeletal and muscular systems demonstrate remarkable similarities to those of a human. The exercises in this chapter require the dissector to take a regional approach and to focus on the relationships among the organs and the blood and nerve supplies to those organs at the same time. This regional approach stresses the interconnectedness of the systems in a much deeper way than learning each organ system individually.

List of Reference Tables

These Pre-Laboratory Worksheet questions may be assigned by instructors through their ▣ connect course.

1. The dorsal surface of a cat is the same as the _____ (posterior/superior) surface of a human, whereas the caudal surface of a cat is the same as the _____ (inferior/posterior) surface of a human.

2. The anterior surface of a cat is the same as the _____ (cranial/ventral) surface of a human, whereas the inferior surface of a cat is the same as the _____ (caudal/ventral) surface of a human.

3. In a cat, a transverse plane separates _____ (anterior/superior) portions from _____ (inferior/posterior) portions.

4. In a cat, a frontal plane separates _____ (anterior/dorsal) portions from _____ (posterior/ventral) portions.

5. Superficially, a male cat can be identified by the presence of a _____ (scrotum/vestibule), whereas the female cat is identified by the presence of a _____ (scrotum/vestibule).

6. When skinning the pudendal region of the male cat, which structures may be at risk for damage? (Check all that apply.)

   _____ a. epididymis

   _____ b. penis

   _____ c. spermatic cord

   _____ d. testis

   _____ e. vestibule

7. Which of the following dissection techniques is most likely to allow for skinning the cat while protecting underlying structures? (Circle one.)

   a. blunt dissection

   b. dissection with scissors

   c. dissection with a scalpel

   d. open scissors technique

8. When handling preserved organisms, one should wear _____. (Circle all that apply.)

   a. aprons or lab coats

   b. gloves

   c. safety glasses

   d. open-toed shoes

9. When storing dissected specimens between laboratory sessions, organisms like cats should always be _____ (wrapped in skin in the storage bag/ placed in a storage bag filled with embalming fluid).

10. During a dissection, nerves and arteries can usually be distinguished by feel; for example, nerves feel _____ (cord-like/tube-like) while arteries feel (cord-like/tube-like).

GROSS ANATOMY

Overview

INTEGRATE

CLINICAL VIEW

Proper Handling and Care of Preserved Cats

The dissection specimens under study in the laboratory are embalmed, or preserved, to enable safe dissection without risk of exposure to necrotic (dead) tissue or harmful microorganisms. Before beginning the dissection, review table 1.6 in chapter 1, which describes the composition of embalming fluids, safe handling of embalming fluids, and safety procedures for dissection. Some specimens may have been preserved with "biosafe" solutions, which generally do not contain the fixative formalin, a substance that can be harmful if handled incorrectly. The dissection specimens used in the lab may or may not have been preserved using formalin. If you are uncertain what preservative was used, it is safest to use the same precautions used for formalin-preserved specimens. That is, always wear gloves before handling the specimen, wear safety glasses when there is a risk of fluid splashing from the specimen, and never touch your face or eyes with your hands if they may have preservative fluid on them. You may also be required to wear dissecting aprons or lab coats and closed-toed shoes. If you get the solution in your eyes, rinse your eyes out thoroughly at the eyewash station. Thoroughly clean your work area and be sure to wash your hands and forearms thoroughly with soap and warm water before leaving the laboratory for the day.

INTEGRATE

LEARNING STRATEGY

Figures in this chapter are color-coded so that common structures can be easily identified. Refer back to this color key often while studying the anatomical structures of the cat.

Color Key for Common Structures

Arteries	Cartilage	Nerves	Pulmonary trunk	Tendons, Ligaments	Ureters
Bone	Muscle	Portal vessels	Salivary glands	Thymus and Adrenal glands	Veins

EXERCISE 28.1

DIRECTIONAL TERMS AND SURFACE ANATOMY

Before beginning dissection of the cat, be sure to read the Clinical View on the proper handling and care of preserved cats.

Figure 28.1 Directional Terms. Note how directional terms in the cat (*a*) differ from the same terms in a human (*b*) because the cat stands on four legs whereas the human stands on only two.
(*b*) ©Eric A. Wise

Directional Terms

One of the major differences encountered when comparing the anatomy of a cat to that of a human is the directional terminology used. This is because the normal anatomic position for the cat is with all four limbs on the ground (**figure 28.1*a***). Humans, on the other hand, have a normal anatomic position with only the lower limbs on the ground (figure 28.1*b*).

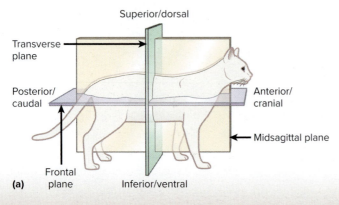

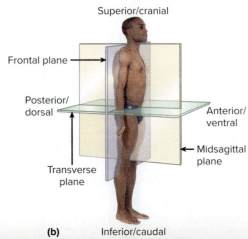

Obtain the Following:
- **dissecting tools**
- **dissecting tray**
- **gloves**
- **preserved cat**
- **storage bag**
- **paper towels**
- **wetting solution (for keeping the specimen moist)**

1. Lay the cat on the dissecting tray with its dorsal side up and start by observing the surface anatomy. On the head, make note of the **vibrissae** (whiskers), the **pinnae** (external ears), and the **nictitating membrane** of each eye (**figure 28.2**). The nictitating membrane is a membrane at the medial aspect of each eye that can brush across the anterior surface of the eye to keep it moist and protect it from debris when the eyes are open. Note that the fur superior to each eye is also elongated a bit to form "eyelashes." These, like human eyelashes, also help keep debris out of the cat's eyes.

2. Turn the cat over so it is ventral side up. Observe the neck and inguinal regions for any sign of incisions. Most commonly there will be an incision in the neck region. This is typically where injections of embalming fluid and any other substances have been introduced into the circulatory system. If there is an incision in the abdomen, the hepatic portal system may also have been injected.

3. Observe the skin adjacent to the midline of the ventral thorax and abdomen and look for the **mammary ridge**, which contains a series of **mammary glands** and **mammary papillae** (nipples). In a pregnant female cat, these ridges become raised because of the growth and development of mammary glands in the subcutaneous tissues. These structures come off with removal of the skin.

4. Determine the sex of the cat. This can be done by both visual inspection and palpation (*palpare*, to touch) of the inguinal and perineal regions. The external genitalia may not be easy to visualize because the cat is covered with fur. If the specimen is a male cat, a small, firm structure at the midline, which is the **penis**, can be palpated. Adjacent to the penis, two roundish structures can be palpated. These are the **testicles**, which are located within the **scrotum**. If the specimen is a female cat, there will be a small invagination anterior/cranial to the root of the tail, which is the **vestibule**.

Figure 28.2 Surface Anatomy of the Cat. This photo shows several terms related to the surface anatomy of a cat's head.
©McGraw-Hill Education/Christine Eckel

(continued on next page)

(continued from previous page)

5. In the space provided, record the sex of the specimen and make note of any pertinent incisions and/or surface anatomy observations that may be important. If the cat has a unique identification number, be sure to record that number as well.

6. Observe a specimen of the opposite sex so that you will be able to identify external genitalia of both a male and a female cat.

Sex of Cat: _____

Cat ID: _____

Back and Limbs

Skeletal System

The cat skeleton has almost all the same bones as the human skeleton, although they often vary in both size and shape, as well as in their orientation within the skeleton, compared with those in a human. Some of these differences reflect the cat's four-legged posture and locomotor capabilities.

EXERCISE 28.2

SKELETAL SYSTEM

1. Observe a cat skeleton (**figure 28.3**). Be sure to carefully observe all of the bones of the skeleton so that you are able to identify them later (e.g., on a laboratory exam).

2. Record in **table 28.1** the number of vertebrae in each region of the vertebral column in a human. Next, count and record the number of vertebrae in each region of the

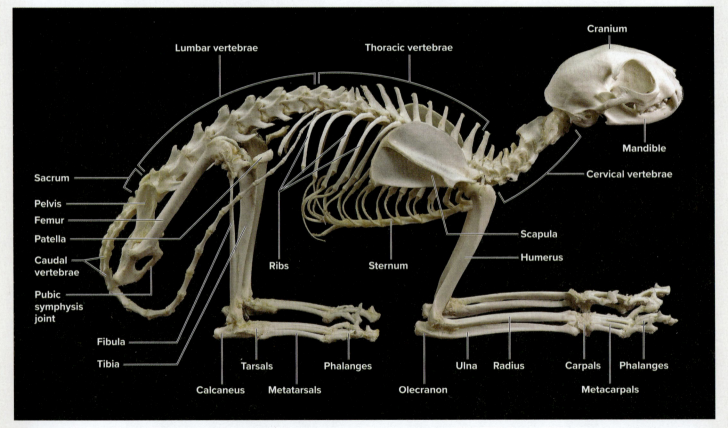

(a)

Figure 28.3 The Cat Skeleton. The cat skeleton is similar to the human skeleton in several ways. Several notable differences are the elongation of the pelvis, metacarpals, and metatarsals; the extra vertebrae; and a clavicle that does not articulate with any other bone. (*a*) Articulated cat skeleton in anatomic position.

©McGraw-Hill Education/Photo and Dissection by Christine Eckel

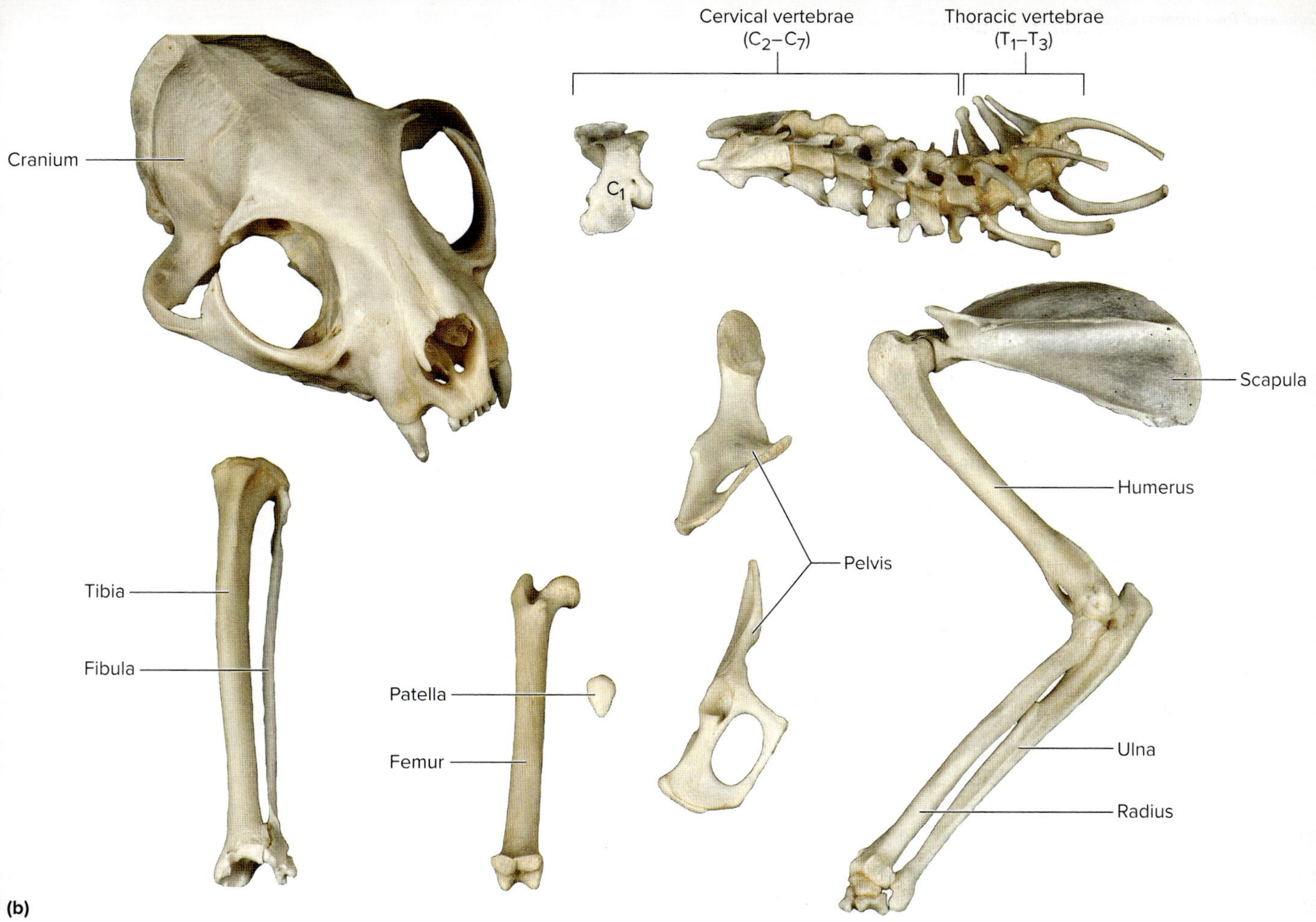

Cervical vertebrae
(C_2–C_7)

Thoracic vertebrae
(T_1–T_3)

Cranium

C_1

Scapula

Humerus

Pelvis

Tibia

Fibula

Patella

Femur

Ulna

Radius

(b)

Figure 28.3 **The Cat Skeleton** *(continued).* (*b*) Disarticulated (partial) cat skeleton demonstrating features of individual bones.
©McGraw-Hill Education/Christine Eckel

vertebral column of the cat. Is the number of vertebrae in each region of the cat skeleton the same as the number of vertebrae in a human skeleton? Explain:

_____

_____

_____

INTEGRATE

CONCEPT CONNECTION

When observing the cat skeleton, pay particular attention to the bones of the distal portions of the limbs. The bones that compose hands (metacarpals and phalanges) and feet (metatarsals and phalanges) of humans are very elongated in cats. Cats, in essence, walk on the tips of their phalanges, and their claws are homologous to human fingernails. When a human is standing, the metatarsals lie flat on the ground. When a cat is standing, the metatarsals are lifted off the ground. It is only when the cat lies down that the metacarpals and metatarsals are able to lie flat on the ground (as in the photo of the cat skeleton, figure 28.3*a*).

Table 28.1	Comparison of Human and Cat Vertebrae	
Number of Vertebrae	**Human**	**Cat**
Cervical		
Thoracic		
Lumbar		
Sacral (fused)		
Coccygeal / Caudal		

(continued on next page)

Muscular System

Just as with the skeletal system, the muscular system of the cat bears a remarkable resemblance to the human muscular system. Most of the differences we see in muscles are related to the adaption of the cat's muscular system for four-legged locomotion and the human muscular system for two-legged locomotion.

Before beginning the dissection, review the overall organization of the human muscular system presented in chapter 11, paying particular attention to major muscle groups. Similar muscle groups exist in the cat. Thus, taking a similar approach to learning the muscles of the cat will greatly assist the learning process. Information on the name, origin, insertion, and action of the muscles of the cat is contained in tables 28.2–28.4 and 28.6–28.7. These tables also make note of muscles that are different in the cat than in the human. Such muscles are different either because they are not found in the human or because they have a slightly different name in the cat.

Recall that one of the best ways to remember the action of a muscle is to learn its attachment points (do not worry about distinguishing which attachment point is origin and which is insertion). Then, consider the action that is performed when the muscle shortens, using the terms for joint movements and directional terms. In the cat, the terms **cranial/caudal** take the place of the terms *anterior/posterior*. To move a structure *cranial* is to move a structure anterior and toward the head of the cat. To move a structure *caudal* is to move a structure posterior and toward the tail of the cat.

The musculature of the cat is presented here as a series of exercises that look at muscles in one region of the body at a time. Each time a new muscle is identified, refer to the muscle tables to review that muscle's origin, insertion, and action. Also consider how the size, location, and function of the muscle compares with those of the homologous muscle(s) in a human. Most of the time, these will be very similar. If you have already started to master the anatomy of the human muscular system, then it will be most efficient to give extra attention only to muscles of the cat that are considerably different from those in the human.

In the following exercises, dissection and identification of muscles start on the ventral surface and move from cranial to caudal. Subsequent exercises follow a progression of dissection and identification of muscles of the forelimb and the hindlimb from both lateral and dorsal views. In each case, dissections begin with superficial observations and dissections and proceed to deep dissections.

EXERCISE 28.4

MUSCLES OF THE HEAD AND NECK

1. **Preparation:** Place the cat ventral side up on the dissecting tray. Begin by observing the muscles of the anterior neck **(figure 28.5)**. On one half of the neck you will do a superficial dissection, and on the other, you will do a deep dissection to observe underlying muscles. Some of the superficial muscles may have been damaged by the embalming/injection incision. If that is the case, use that side for the deep dissection and the unaffected side for the superficial dissection. A large vein may be visible on both sides of the neck, lying superficial to the muscles. This is the **external jugular vein**. Next to this vein there may be several small, kidney-bean-size (and -shaped) structures. These are **lymph nodes**. While dissecting and cleaning the muscles, take care not to cut the internal jugular veins or remove the lymph nodes, because these will be observed and discussed further in the context of the cardiovascular and lymphatic systems.

2. Superficial dissection: Although the word "dissect" literally means "to cut into two pieces" (*di,* two + *sectio,* to cut), in reality very little dissecting of muscles happens in these exercises. The process is more about cleaning muscles and separating them from each other so as to best view them and understand their location and functions. Often the best tools for this job are forceps and a pair of scissors. Use the forceps to pull on the fascia (connective tissue) overlying the muscles; then use the scissors to cut the fascia away. When cutting, always

cut in the direction of the muscle fibers, which prevents them from tearing. Before and during the dissection, preview the muscles in figure 28.5 to have an idea of what to look for. More important, this will also assist you in knowing where to be most careful, so as not to destroy muscles or related structures such as nerves and blood vessels (see the Learning Strategy on this page).

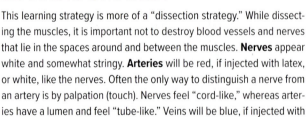

INTEGRATE

LEARNING STRATEGY

This learning strategy is more of a "dissection strategy." While dissecting the muscles, it is important not to destroy blood vessels and nerves that lie in the spaces around and between the muscles. **Nerves** appear white and somewhat stringy. **Arteries** will be red, if injected with latex, or white, like the nerves. Often the only way to distinguish a nerve from an artery is by palpation (touch). Nerves feel "cord-like," whereas arteries have a lumen and feel "tube-like." Veins will be blue, if injected with latex, or dark brownish-purple, if not. The dark brownish-purple color of veins comes from the dried blood that is still inside the vessels, which is visible due to the thin-walled nature of the veins.

Try to preserve (and clean) the arteries and nerves while dissecting. They are more likely to be kept intact if the dissection is performed from proximal to distal (rather than from distal to proximal). This is because the dissection begins where nerves and arteries are larger and follows their paths as they decrease in size.

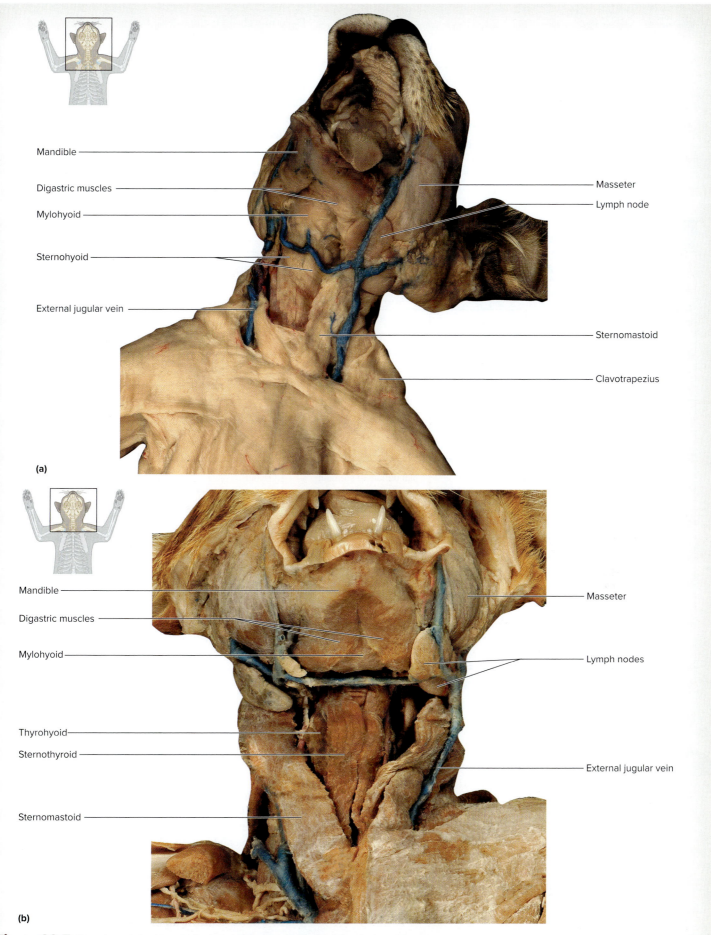

Mandible

Digastric muscles

Mylohyoid

Sternohyoid

External jugular vein

Masseter

Lymph node

Sternomastoid

Clavotrapezius

(a)

Mandible

Digastric muscles

Mylohyoid

Thyrohyoid

Sternothyroid

Sternomastoid

Masseter

Lymph nodes

External jugular vein

(b)

Figure 28.5 **Muscles of the Head, Neck, and Thorax (Anterior View).** Superficial muscles on the right side of the cat's neck are somewhat mangled from the embalming process, but the muscles can be seen clearly on the left side of the cat's neck (right side of the photo). (*a*) Superficial muscles. (*b*) Deep muscles.
©McGraw-Hill Education/Christine Eckel

(continued on next page)

(continued from previous page)

3. **Superficial muscles:** Clean, identify, and isolate the following muscles, using **table 28.2** and figure 28.5 as guides:

   - [ ] **Clavotrapezius**
   - [ ] **Digastric**
   - [ ] **Masseter**
   - [ ] **Mylohyoid**
   - [ ] **Sternohyoid**
   - [ ] **Sternomastoid**
   - [ ] **Sternothyroid**

4. **Deep muscles:** On the side of the neck that was previously determined to be most appropriate for a deep dissection, proceed to transect the sternomastoid and sternohyoid muscles. Clean, identify, and isolate the following two muscles, using table 28.2 and figure 28.5 as guides. Note that both of these muscles attach to the thyroid cartilage. The sternothyroid extends from the thyroid inferiorly to the sternum and the thyrohyoid extends from the thyroid superiorly to the hyoid bone.

   - [ ] **Sternothyroid**
   - [ ] **Thyrohyoid**

Table 28.2	Muscles of the Head and Neck			
Cat Muscle	**Human Muscle**	**Origin**	**Insertion**	**Action**
Muscles of the Neck (Anterior)				
Cleidomastoid	Sternocleidomastoid-clavicular head	Clavicle	Mastoid process	Rotates head
Digastric	Digastric	Mastoid and jugular processes of occipital bone	Mandible	Depresses mandible
Geniohyoid	Geniohyoid	Mandible	Hyoid	Moves hyoid cranially
Mylohyoid	Mylohyoid	Mandible	Median raphe	Elevates floor of oral cavity
Sternohyoid	Sternohyoid	Manubrium of sternum	Hyoid	Moves hyoid caudally
Sternomastoid	Sternocleidomastoid-sternal head	Manubrium of sternum	Lambdoidal ridge and mastoid process	Rotates head
Sternothyroid	Sternothyroid	Manubrium of sternum	Thyroid cartilage	Moves larynx caudally
Muscles of Mastication				
Masseter	Masseter	Zygomatic arch	Mandible	Elevates mandible
Temporalis	Temporalis	Temporal bone	Coronoid process of mandible	Elevates mandible

EXERCISE 28.5

MUSCLES OF THE THORAX

1. **Superficial muscles:** With the cat still ventral side up on the dissecting tray, prepare to dissect the muscles of the anterior thorax. There are four main muscles in this group. These muscles collectively mimic one action of the pectoralis major and minor muscles in humans: adduction of the forelimb. While cleaning the superficial fascia around these muscles, note the direction of the fibers of the **pectoantebrachialis** (transverse), **pectoralis major** (slightly oblique), **pectoralis minor**, and **xiphihumeralis** (both very oblique). Note also that the pectoantebrachialis and xiphihumeralis are unique to the cat.

While dissecting around the insertion of these muscles on the humerus, take care not to damage the latissimus dorsi and serratus ventralis muscles (**figures 28.6** and **28.7**), which will be studied with muscles of the back.

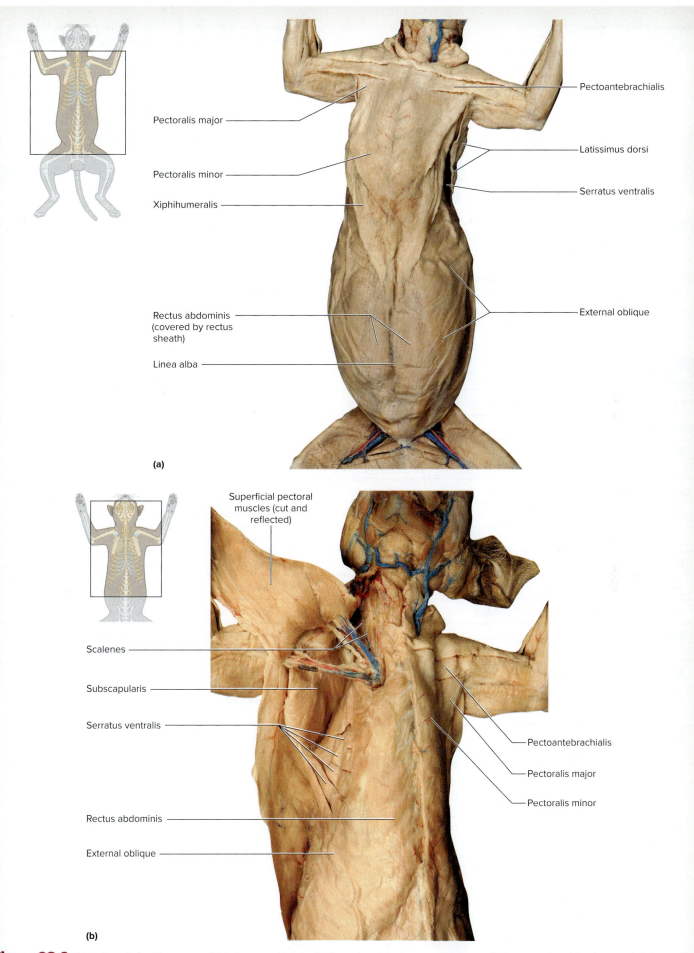

Pectoantebrachialis

Pectoralis major

Latissimus dorsi

Pectoralis minor

Serratus ventralis

Xiphihumeralis

Rectus abdominis
(covered by rectus
sheath)

External oblique

Linea alba

(a)

Superficial pectoral
muscles (cut and
reflected)

Scalenes

Subscapularis

Serratus ventralis

Pectoantebrachialis

Pectoralis major

Pectoralis minor

Rectus abdominis

External oblique

(b)

Figure 28.6 Muscles of the Thorax and Abdomen. (*a*) Superficial muscles of the thorax and abdomen. (*b*) Deep muscles of the thorax and abdomen. The superficial pectoral muscles (pectoantebrachialis, pectoralis major/minor, and xiphihumeralis) and the sternomastoid have been removed to show the deeper muscles.

©McGraw-Hill Education/Christine Eckel

(continued on next page)

(continued from previous page)

For what activity or activities would the pectoral group of muscles be most important in the cat?

_____

_____

2. Clean, identify, and isolate the following muscles, using **table 28.3** and figure 28.6 as guides:

   ☐ **Pectoantebrachialis**   ☐ **Pectoralis minor**

   ☐ **Pectoralis major**   ☐ **Xiphihumeralis**

3. **Deep muscles:** Use blunt dissection to gently separate the pectoral muscles from the underlying muscles. It may be helpful to work a probe deep to the pectoral muscles, taking care not to tear the deep muscles of the thorax. The pectoral muscles should be bisected close to the midline to prevent damage to the nerves of the brachial plexus, which are located in the axillary region. Keeping the probe deep to the pectoral muscles, use scissors to make a vertical incision through the pectoantebrachialis, pectoralis major and minor, and xiphihumeralis muscles close to the midline of the body on the side of the cat that was chosen for the deep dissection. Bisect and reflect these muscles to expose the deeper muscles. In this view the scalene muscle group will be visible in addition to muscles that connect the scapula to the axial skeleton (serratus ventralis) and one of the muscles composing the rotator cuff: the subscapularis.

4. Clean, identify, and isolate the following muscles, using **table 28.4** and figure 28.6 as guides:

   ☐ **Scalenes**

   ☐ **Serratus ventralis**

   ☐ **Subscapularis**

Table 28.3	Muscles of the Thorax and Abdomen			
Cat Muscle	**Human Muscle**	**Origin**	**Insertion**	**Action**
Superficial Muscles				
Pectoantebrachialis	NA	Manubrium	Fascia of forelimb near elbow	Adducts the forelimb
Pectoralis Major	Pectoralis major	Cranial sternebrae	Proximal humerus	Adducts the forelimb
Pectoralis Minor	Pectoralis minor	Caudal sternebrae	Proximal humerus	Adducts the forelimb
Xiphihumeralis	NA	Xiphoid process	Proximal humerus	Adducts the forelimb
Deep Muscles				
External Intercostals	External intercostals	Superior rib	Inferior rib	Draws ribs cranially
Internal Intercostals	Internal intercostals	Inferior rib	Superior rib	Draws ribs caudally
Serratus Ventralis	Serratus anterior	Ribs 1–10	Vertebral border of scapula	Depresses scapula and moves scapula cranially
Muscles of the Abdominal Wall				
External Oblique	External oblique	Lumbodorsal fascia and ribs	Linea alba	Compresses abdominal viscera
Internal Oblique	Internal oblique	Lumbodorsal fascia and pelvis	Linea alba	Compresses abdominal viscera
Rectus Abdominis	Rectus abdominis	Pubic symphysis	Sternum and costal cartilages	Flexes trunk and compresses abdominal viscera
Transversus Abdominis	Transversus abdominis	Ilium, lumbar vertebrae, and posterior ribs	Linea alba	Compresses abdominal viscera

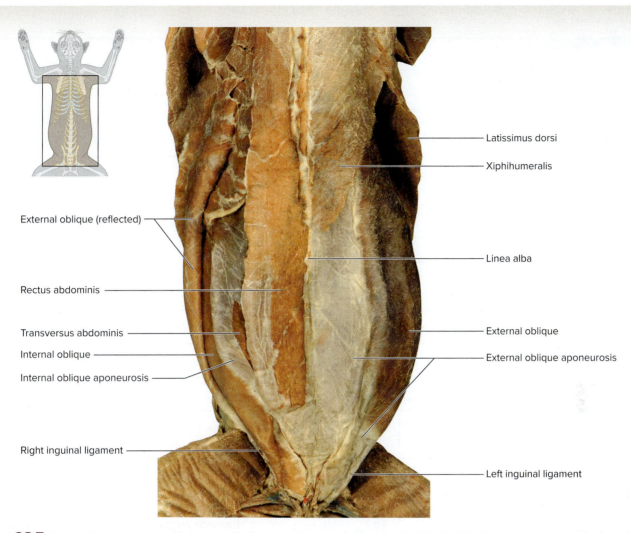

Latissimus dorsi

Xiphihumeralis

External oblique (reflected)

Rectus abdominis

Linea alba

Transversus abdominis

Internal oblique

External oblique

Internal oblique aponeurosis

External oblique aponeurosis

Right inguinal ligament

Left inguinal ligament

Figure 28.7 **Deep Muscles of the Abdomen.** In this photo the external oblique on the left side of the figure has been cut and reflected to show the rectus abdominis and internal oblique. The aponeurosis of the internal oblique has been removed to show the fibers of the deepest muscle, the transversus abdominis.

©McGraw-Hill Education/Christine Eckel

Table 28.4	Muscles of the Back and Shoulder			
Cat Muscle	**Human Muscle**	**Origin**	**Insertion**	**Action**
Muscles of the Back and Shoulder (superficial)				
Acromiodeltoid	Deltoid—anterior portion	Acromium of scapula	Humerus	Flexes and rotates humerus
Acromiotrapezius	Trapezius—superior fibers	Spines of vertebra C_1–T_1	Acromium process and spine of scapula	Adducts and elevates scapula
Clavobrachialis (clavodeltoid)	Deltoid—middle fibers	Clavicle	Proximal ulna	Abducts and rotates humerus
Clavotrapezius	Trapezius—middle fibers	Nuchal line of skull and nuchal ligament	Clavicle	Moves humerus cranially
Latissimus Dorsi	Latissimus dorsi	Lumbodorsal fascia and spines of lower thoracic and lumbar vertebrae	Medial humerus	Elevates and moves humerus caudally
Levator Scapulae Ventralis	NA	Occipital bone, transverse process of atlas	Acromial process	Moves scapula cranially

(continued on next page)

(continued from previous page)

Table 28.4	Muscles of the Back and Shoulder *(continued)*			
Cat Muscle	**Human Muscle**	**Origin**	**Insertion**	**Action**
Muscles of the Back and Shoulder (superficial) *(continued)*				
Spinodeltoid	Deltoid—posterior portion	Spine of scapula	Humerus	Flexes and rotates humerus
Spinotrapezius	Trapezius—inferior portion	Spines of thoracic vertebrae	Scapula	Elevates and moves scapula caudally
Teres Major	Teres major	Axillary border of scapula	Medial humerus	Adducts humerus
Muscles of the Back and Shoulder (deep)				
Infraspinatus	Infraspinatus	Infraspinous fossa	Greater tuberosity of humerus	Rotates humerus
Rhomboid	Rhomboid major and minor	Spines of thoracic vertebrae	Ventral border of scapula	Adducts scapula
Rhomboid Capitis	Levator scapulae	Superior nuchal line	Cranial angle of scapula	Moves scapula cranially
Scalenes	Scalenes	NA	NA	NA
Anterior	Anterior	Ribs 2–3	Transverse processes of cervical vertebrae	Flexes neck and moves ribs cranially
Medius	Middle	Ribs 6–9	Transverse processes of cervical vertebrae	Flexes neck and moves ribs cranially
Posterior	Posterior	Rib 3	Transverse processes of cervical vertebrae	Flexes neck and moves ribs cranially
Serratus Ventralis	Serratus anterior	Ribs 1–10	Vertebral border of scapula	Depresses and moves scapula cranially
Splenius	Splenius capitis and cervicis	Lambdoidal ridge of occipital bone	Nuchal line	Rotates and elevates/extends head
Subscapularis	Subscapularis	Subscapular fossa	Lesser tuberosity of humerus	Adducts humerus
Supraspinatus	Supraspinatus	Supraspinous fossa	Greater tuberosity of humerus	Extends humerus
Teres Minor	Teres minor	Axillary border of scapula	Greater tuberosity of humerus	Rotates humerus
Transverse Costarum	NA	Sternum	First rib	Moves sternum cranially

EXERCISE 28.6

MUSCLES OF THE ABDOMINAL WALL

1. **Superficial muscles:** With the cat still ventral side up on the dissecting tray, prepare to dissect the muscles of the abdominal wall. There are four main muscles in this group, just as there are in humans. Contraction of these muscles compresses the abdominal viscera, raising intra-abdominal pressure. Contraction of the rectus abdominis, as in humans, also flexes the vertebral column. These muscles have broad, flat **aponeuroses** for tendons and insert on the **linea alba**, just as in humans.

2. Clean, identify, and isolate the external oblique muscle and identify the following, using table 28.3 and figure 28.7 as guides:

   ☐ **External oblique muscle**

   ☐ **External oblique aponeurosis**

   ☐ **Linea alba**

3. **Deep muscles:** Using scissors and forceps, cut a small, vertical slit (a sagittal section) in the external oblique muscle close to its insertion on the linea alba. Carefully work a probe underneath the muscle so that it separates the external oblique from the underlying **rectus abdominis** and **internal oblique** muscle. Keeping the probe in place to protect the underlying muscles, continue the vertical incision of the external oblique from the inguinal ligament (figure 28.7) to the inferior border of the xiphihumeralis muscle. Reflect the external oblique to reveal the underlying rectus abdominis and internal oblique muscles.

4. Make an incision in the aponeurosis of the internal oblique muscle to reveal fibers of the **transversus abdominis** muscle. Either remove the aponeurosis of the internal oblique entirely or just cut a "window" through it to reveal the transversus abdominis (as shown in the left side of figure 28.7). When performing all of these dissections, take care not to cut too deep, which might result in exposing the abdominal cavity and possibly cutting into abdominal contents.

5. Clean, identify, and isolate the following muscles, using table 28.3 and figure 28.7 as guides:

   ☐ **Internal oblique**

   ☐ **Rectus abdominis**

   ☐ **Transverse abdominis**

6. Record the direction of fibers of the four abdominal muscles in **table 28.5**.

Table 28.5	Fiber Orientation of Cat Abdominal Muscles
Muscle	**Fiber Orientation**
External Oblique	
Internal Oblique	
Rectus Abdominis	
Transverse Abdominis	

EXERCISE 28.7

MUSCLES OF THE BACK AND SHOULDER

1. **Preparation:** Turn the cat over so that its ventral surface is on the dissecting tray and the dorsal surface is up. Begin at the head and progress toward the tail as was done in previous observations/dissections.

2. **Superficial muscles:** Recall that the most superficial "back" muscles in humans consist of the trapezius, deltoid, and latissimus dorsi muscles. These are not true "back" muscles because they do not move the vertebral column. Instead, their functions are to attach the scapula and upper limb to the axial skeleton, and to produce movements of the upper limb and scapula. Cats also have these muscles. The cat **deltoid** is divided into three separate muscles, each homologous to the three regions of the human deltoid muscle (table 28.4). The cat **trapezius** is similarly divided into three separate muscles, each homologous to the three regions of the human trapezius muscle. The cat also has a unique

   muscle, the **levator scapulae ventralis**, which has no human counterpart (table 28.4). The **latissimus dorsi** is similar in the cat and the human. While identifying and cleaning these muscles, think about the actions each performs and relate each muscle in the cat to its human homologue.

3. **Note about the clavicle:** In the cat, the clavicle is embedded within the two bellies of what appears to be a single muscle (at least superficially). These two muscle "bellies" are two different muscles: the **clavotrapezius** and the **clavodeltoid (figure 28.8)**. While identifying these two muscles, do not be confused if the names appear to relate to two parts of what appears to be a single muscle. Palpate the midpoint of the two muscles to feel the clavicle. Once the clavotrapezius and clavodeltoid have been reflected, the clavicle will be clearly visible.

(continued on next page)

(continued from previous page)

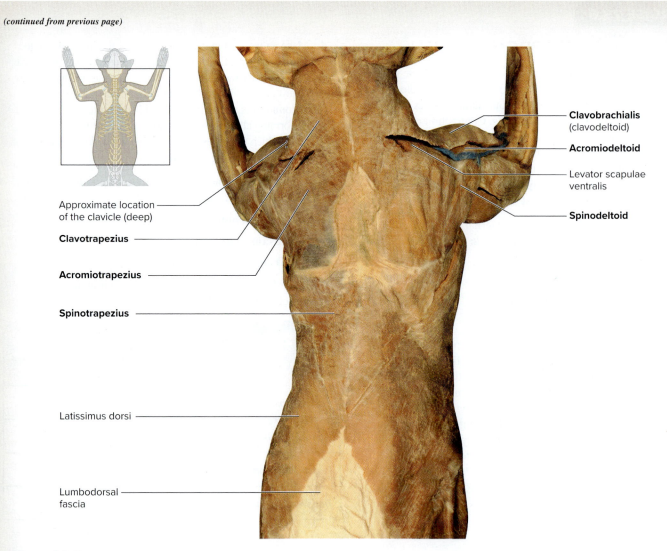

Approximate location of the clavicle (deep)

Clavotrapezius

Acromiotrapezius

Spinotrapezius

Latissimus dorsi

Lumbodorsal fascia

Clavobrachialis (clavodeltoid)

Acromiodeltoid

Levator scapulae ventralis

Spinodeltoid

Figure 28.8 **Superficial Back Muscles.** The three parts of the trapezius muscle are labeled on the left side of the figure. The three parts of the deltoid muscle are labeled on the right side of the figure.

©McGraw-Hill Education/Christine Eckel

4. Clean, identify, and isolate the following muscles, using table 28.4 and figure 28.8 as guides:

☐ **Acromiodeltoid**

☐ **Acromiotrapezius**

☐ **Clavobrachialis (clavodeltoid)**

☐ **Clavotrapezius**

☐ **Latissimus dorsi**

☐ **Levator scapulae ventralis**

☐ **Spinodeltoid**

☐ **Spinotrapezius**

5. **Deep muscles:** Using scissors and forceps, cut a small, vertical slit (a sagittal section) in the **acromiotrapezius** and **spinotrapezius** muscles on both sides. To do this, slip a probe deep to the muscles to protect the underlying rhomboid muscles. Then cut superficial to the probe. When the acromiotrapezius and spinotrapezius muscles are cut, the scapula will pull away from the midline and the rhomboids will be visible. The **rhomboid** in the cat is homologous to the rhomboid major and minor in humans, and the **rhomboid capitis** muscle is homologous to the

levator scapulae in humans. A third muscle that connects the scapula to the axial skeleton in the cat is the **serratus ventralis**, which is equivalent to the serratus anterior in humans. In the cat, however, this muscle is larger, with more attachments to the ribs, than the serratus anterior of humans.

6. Observe the muscles of the scapula that form part of the **rotator cuff** (see figure 13.1). Recall that these muscles form the major muscular support of the shoulder joint, and consist of the supraspinatus, infraspinatus, subscapularis, and teres minor. The **supraspinatus** and **subscapularis** are more easily visualized from a posterior view.

7. Clean, identify, and isolate the following muscles, using table 28.4 and **figure 28.9** as guides:

☐ **Rhomboid**

☐ **Rhomboid capitis**

☐ **Splenius**

☐ **Subscapularis**

☐ **Supraspinatus**

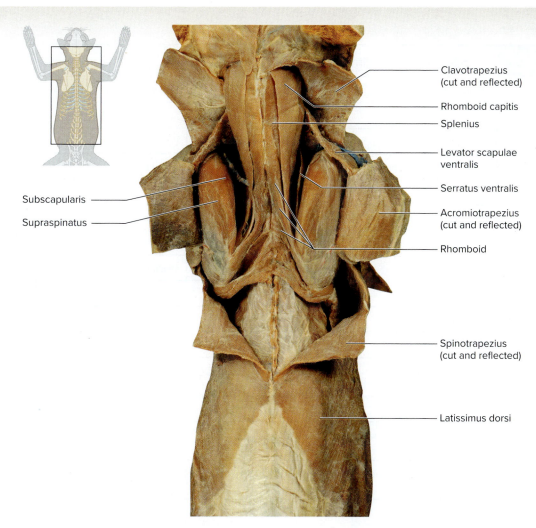

Clavotrapezius
(cut and reflected)

Rhomboid capitis

Splenius

Levator scapulae
ventralis

Serratus ventralis

Subscapularis

Supraspinatus

Acromiotrapezius
(cut and reflected)

Rhomboid

Spinotrapezius
(cut and reflected)

Latissimus dorsi

Figure 28.9 **Deep Back Muscles.** A better view of the deep muscles of the shoulder, back, and neck is possible after bisecting and reflecting the clavotrapezius, acromiotrapezius, and spinotrapezius muscles as shown here.

©McGraw-Hill Education/Christine Eckel

EXERCISE 28.8

MUSCLES OF THE FORELIMB

1. **Superficial muscles:** Recall that the human upper limb muscles are organized into four compartments: anterior and posterior *arm*, and anterior and posterior *forearm*. The muscles of the cat forelimb are similarly arranged. Thus, while dissecting out muscles in each region of the forelimb, always consider the common action of the muscles in that group before considering the specifics of each muscle. This will greatly facilitate the learning of these muscles. **Table 28.6** summarizes the human homologue, origin, insertion, and action of the forelimb muscles of the cat.

2. Observe the superficial muscles of the shoulder. The muscles of the deltoid group should already have been cleaned and identified. The dissection will now proceed distally along the forelimb, starting with the **triceps**

brachii (figure 28.10). The triceps brachii in the cat has the same three heads and insertion as the triceps brachii of the human. While cleaning the anterior border of the lateral head of the triceps brachii near the elbow, note the origin of several other muscles: the **brachialis**, the **brachioradialis**, and several extensors of the forearm.

3. This dissection exercise involves cleaning, identifying, and isolating all of the arm/forearm muscles as a group. The dissection also requires working in a small area. Therefore it is most efficient to pay attention to all muscles in that area at the same time. To effectively complete the dissection, reposition the cat occasionally to make it easier to get at the muscles on the lateral and medial sides of the forelimb. While dissecting the forelimb muscles, take particular care when cleaning

(continued on next page)

(continued from previous page)

Table 28.6	Muscles of the Forelimb			
Cat Muscle	**Human Muscle**	**Origin**	**Insertion**	**Action**
Muscles of the Arm				
Anconeus	Anconeus	Lateral humerus	Lateral ulna	Extends forearm
Biceps Brachii	Biceps brachii—long head	Supraglenoid tubercle	Radial tuberosity	Flexes forearm
Brachialis	Brachialis	Lateral humerus	Proximal ulna	Flexes forearm
Coracobrachialis	Coracobrachialis	Coracoid process of scapula	Proximal humerus	Adducts humerus
Triceps Brachii	Triceps brachii	Humerus	Olecranon	
Lateral head	Lateral head	Deltoid ridge of humerus	Olecranon of ulna	Extends forearm
Long head	Long head	Axillary border of scapula	Olecranon of ulna	Extends forearm
Medial head	Medial head	Humerus	Olecranon of ulna	Extends forearm
Muscles of the Forearm (dorsal)				
Abductor Pollicis Longus	Abductor pollicis longus	Lateral radius and ulna	First metacarpal	Abducts the thumb
Brachioradialis	Brachioradialis	Midhumerus	Distal radius	Supinates the manus
Extensor Carpi Radialis Brevis	Extensor carpi radialis brevis	Distal humerus	Third metacarpal	Extends the manus
Extensor Carpi Radialis Longus	Extensor carpi radialis longus	Distal humerus	Second metacarpal	Extends the manus
Extensor Digitorum Communis	Extensor digitorum	Distal humerus	Digits II–V	Extends digits II–V
Extensor Digitorum Lateralis	Extensor digiti minimi	Distal humerus	Digits II–V	Extend the digits
Extensor Carpi Ulnaris	Extensor carpi ulnaris	Lateral epicondyle of humerus	Fifth metacarpal	Extend the manus
Extensor Pollicis	Extensor pollicis brevis	Proximal ulna	Middle phalanx of first digit	Extend first digit
Supinator	Supinator	Lateral epicondyle of humerus	Radius	Supinate the manus
Muscles of the Forearm (ventral)				
Flexor Carpi Radialis	Flexor carpi radialis	Medial epicondyle of humerus	Metacarpals II and III	Flexes the manus
Flexor Carpi Ulnaris	Flexor carpi ulnaris	Medial epicondyle of humerus and ulna	Pisiform bone	Flexes the digits
Flexor Digitorum Profundus	Flexor digitorum profundus	Ulna, humerus, and radius	Digits I–V	Flexes the digits
Flexor Digitorum Superficialis	Flexor digitorum superficialis	Medial epicondyle of humerus	Digits II–V	Flexes the digits
Pronator Teres	Pronator teres	Medial epicondyle of humerus	Medial border of radius	Pronate the manus

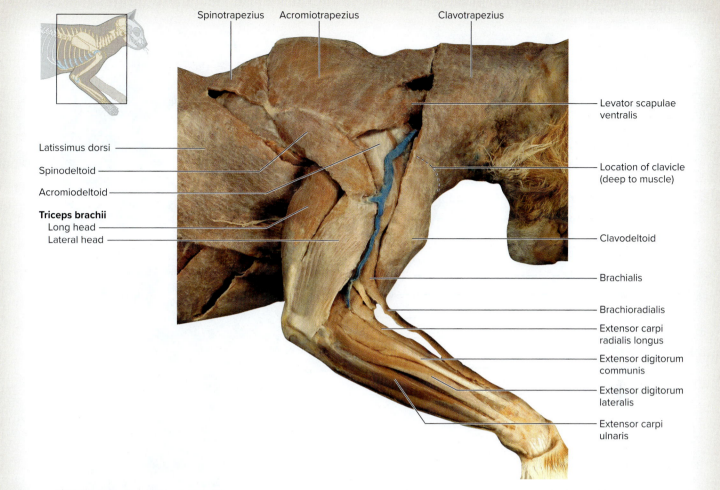

Spinotrapezius   Acromiotrapezius   Clavotrapezius

Levator scapulae ventralis

Latissimus dorsi

Spinodeltoid

Acromiodeltoid

Triceps brachii
Long head
Lateral head

Location of clavicle (deep to muscle)

Clavodeltoid

Brachialis

Brachioradialis

Extensor carpi radialis longus

Extensor digitorum communis

Extensor digitorum lateralis

Extensor carpi ulnaris

Figure 28.10 Muscles of the Lateral Forelimb. In a lateral view of the forelimb, the most visible muscles are the triceps brachii and the extensor muscles of the forearm.

©McGraw-Hill Education/Christine Eckel

the proximal attachments of the muscles that originate on the medial epicondyle of the humerus (**flexor carpi ulnaris, pronator teres, palmaris longus, flexor carpi radialis**). On the proximal end of these muscles, the fascia becomes embedded in the muscle, making it difficult to remove. Thus, do not remove the fascia when approaching that region. Instead, simply cut the fascia where it starts to embed the muscles so as not to tear the muscles. While dissecting these muscles, think about the homologous muscles in the human and note both similarities and differences.

4. The majority of muscles in the forelimb of the cat have counterparts in the human. The one major exception is the **epitrochlearis** muscle in the cat, which has no human counterpart. See more about this muscle in step 6.

5. Clean, identify, and isolate the following muscles of the arm and forearm, using table 28.6 and figures 28.10 and **28.11** as guides:

- ☐ **Brachialis**
- ☐ **Brachioradialis**
- ☐ **Extensor carpi radialis longus**
- ☐ **Extensor carpi ulnaris**
- ☐ **Extensor digitorum communis**
- ☐ **Extensor digitorum lateralis**
- ☐ **Flexor carpi radialis**
- ☐ **Flexor carpi ulnaris**
- ☐ **Triceps brachii— lateral head**
- ☐ **Triceps brachii— long head**

(continued on next page)

(continued from previous page)

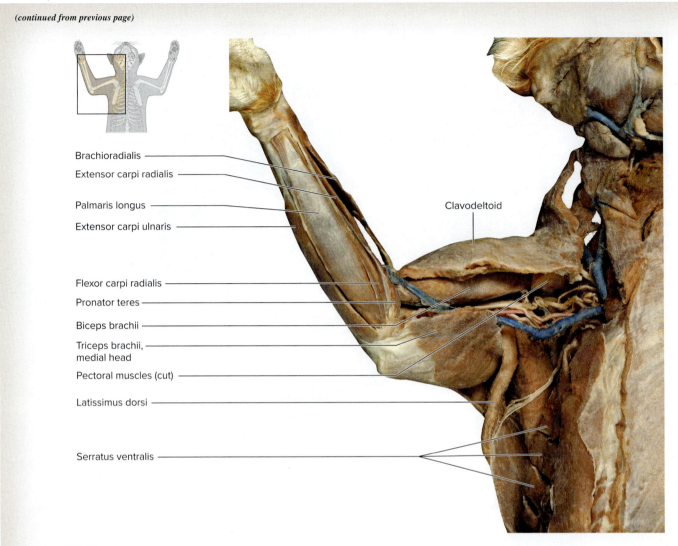

Brachioradialis
Extensor carpi radialis
Palmaris longus
Extensor carpi ulnaris
Flexor carpi radialis
Pronator teres
Biceps brachii
Triceps brachii, medial head
Pectoral muscles (cut)
Latissimus dorsi
Serratus ventralis
Clavodeltoid

Figure 28.11 Muscles of the Medial Forelimb. In a medial view of the forelimb, the biceps brachii and most of the flexors of the forearm can be viewed.

©McGraw-Hill Education/Christine Eckel

6. **Deep muscles:** Locate the **epitrochlearis** muscle in the medial forearm. This muscle originates near the insertion of the **latissimus dorsi**. While cleaning the muscle near that location, take care not to damage any of the nerves of the brachial plexus that run in the axilla. Once the epitrochlearis has been identified and cleaned, push a blunt probe deep to the muscle to isolate it from the deeper muscles. Next, use scissors to bisect and reflect the epitrochlearis. This will allow some of the deeper muscles of the medial forearm to be visualized.

7. Clean, identify, and isolate the following muscles of the forelimb, using table 28.6 and figure 28.11 as guides:

☐ **Biceps brachii** ☐ **Triceps brachii—medial head**
☐ **Pronator teres**

EXERCISE 28.9

MUSCLES OF THE HINDLIMB

EXERCISE 28.9A Muscles of the Hindlimb: Thigh

1. **Superficial muscles:** Begin the dissection of hindlimb with muscles of the posterior and lateral hip and thigh. Place the cat ventral side down on the dissection tray. Before starting the dissection, observe the hindlimb muscles in **figure 28.12**. Notice that while the cat has nearly the same gluteal and posterior thigh muscles as the human, the sizes of the muscles are quite different. These size differences have to do with the cat's four-legged posture as compared to the human's two-legged posture. That is, the gluteal muscles in the cat are comparatively small, and the posterior thigh, or "hamstring" muscles are comparatively large. This situation is largely reversed in humans, who tend to have larger gluteal muscles and smaller hamstring muscles (see Learning Strategy on hamstring muscles that appears at the end of this exercise).

Using figure 28.12 as a guide, begin cleaning the fascia off the gluteal muscles and proceed toward the distal thigh. Take special care around the insertion of the **gluteus maximus** and **caudofemoralis** muscles. The insertions of these muscles are on the **fascia lata** of the thigh, which is really just a thickening of the deep fascia that envelops the thigh muscles. The goal in this part of the dissection is to preserve the band of connective tissue that composes the fascia lata, or **iliotibial (IT) band**, while removing the remaining fascia.

2. Clean, identify, and isolate the following muscles of the hindlimb, using **table 28.7** and figure 28.12 as guides:

☐ **Biceps femoris**   ☐ **Gluteus medius**
☐ **Caudofemoralis**   ☐ **Sartorius**
☐ **Gastrocnemius**   ☐ **Tensor fasciae latae**
☐ **Gluteus maximus**

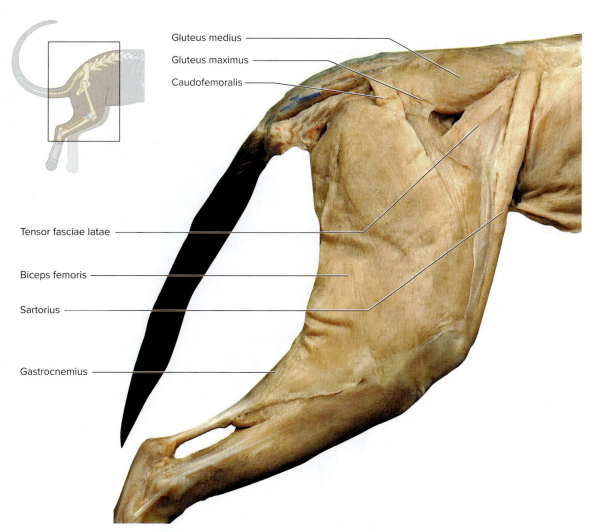

Figure 28.12 **Muscles of the Hindlimb: Thigh.** Superficial dissection, lateral view of right hindlimb of the cat.
©McGraw-Hill Education/Christine Eckel

(continued on next page)

(continued from previous page)

3. Turn the cat over so that the dorsal surface is down and the medial and anterior thigh muscles are visible **(figure 28.13a)**. While cleaning muscles on the anterior thigh, note the origin of the **sartorius** muscle. Observe the small, triangular region that lies medial to the origin of the sartorius. This is the **femoral triangle**, a region that contains the femoral artery, nerve, and vein. It may be necessary to clean some fat superficial to the triangle to observe the neurovascular structures. The **inguinal ligament**, the sartorius, and the **adductors** form the borders of the femoral triangle. While cleaning the muscles, take care not to destroy the nerves and vessels in this region. Carefully trace the sartorius muscle from origin to insertion (table 28.7), clean the muscle, and separate it from the underlying muscles. Note that the location and function of the sartorius is similar in cats and humans, with one major exception: The human sartorius *flexes* the leg (weakly), whereas the cat sartorius *extends* the leg.

4. **Deep muscles:** Once the sartorius is isolated from the surrounding muscles, slip a blunt probe deep to the muscle to protect the deeper muscles. Then *transect* the sartorius across the middle of the muscle belly and

reflect the two ends to expose underlying muscles. Next, isolate the **gracilis** from the deeper muscles. Using a probe to protect the underlying muscles, *transect* the gracilis, and *reflect* the two ends to expose the deeper muscles. Take care not to damage the femoral vessels and nerve while reflecting these two muscles.

5. **Anterior thigh:** Observe the **quadriceps femoris** group of muscles. These are homologous to the quadriceps femoris muscle group in humans. As in humans, all four muscles of the group come together to form a common tendon that encases the patella: the **patellar tendon**. This band of connective tissue extends distal to the patella to connect to the tibial tuberosity as the **patellar ligament**. Recall that the major action of the quadriceps femoris group of muscles is to extend the leg.

6. Clean, isolate, and identify the following structures, using figure 28.13 and table 28.7 as guides:

☐ **Patella**   ☐ **Vastus intermedius**
☐ **Patellar ligament**   ☐ **Vastus lateralis**
☐ **Patellar tendon**   ☐ **Vastus medialis**
☐ **Rectus femoris**

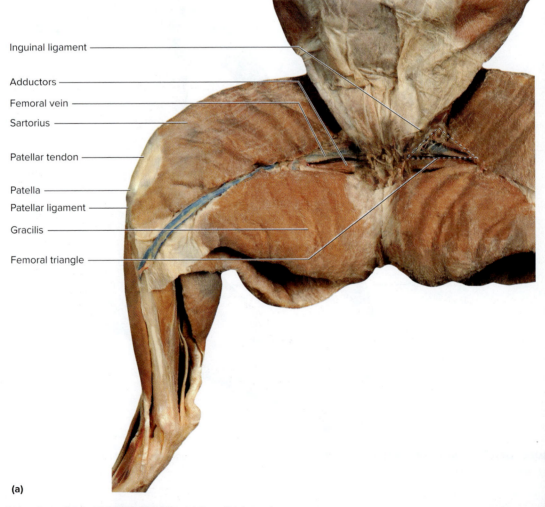

Inguinal ligament
Adductors
Femoral vein
Sartorius
Patellar tendon
Patella
Patellar ligament
Gracilis
Femoral triangle

(a)

Figure 28.13 Muscles of the Anterior and Medial Thigh. (*a*) Superficial muscles.
©McGraw-Hill Education/Christine Eckel

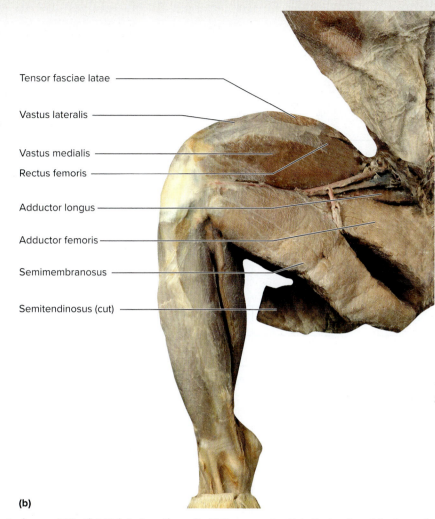

Tensor fasciae latae

Vastus lateralis

Vastus medialis

Rectus femoris

Adductor longus

Adductor femoris

Semimembranosus

Semitendinosus (cut)

(b)

Figure 28.13 **Muscles of the Anterior and Medial Thigh *(continued)*.** (*b*) Deep muscles. Note the location of the femoral artery and vein. Muscles lateral to the femoral vessels belong to the anterior compartment. Muscles medial to the femoral vessels belong to the medial compartment. Sartorius and gracilis have been cut and reflected, and thus are not visible in this photo. Vastus intermedius is deep to the vastus lateralis and vastus medialis and thus is not visible in this photo.

©McGraw-Hill Education/Christine Eckel

7. **Medial thigh:** Observe the muscles medial to the femoral vessels (figure 28.13*b*). The medial compartment muscles of the cat, the **adductor** muscles, are again very similar to the homologous muscles in the human. The major exceptions are that the **gracilis** is much larger in the cat, and the cat has one large adductor, the **adductor femoris**, in place of the adductor brevis and adductor magnus muscles of the human. In the medial view of the thigh, parts of the large **semimembranosus** and

semitendinosus muscle are also visible. These muscles are part of the posterior thigh, or "hamstring," muscles.

8. Clean, isolate, and identify the following muscles of the medial and posterior thigh, using figure 28.13*b* and table 28.7 as guides.

☐ **Adductor femoris**   ☐ **Semimembranosus**
☐ **Adductor longus**   ☐ **Semitendinosus**

Table 28.7	Muscles of the Hindlimb			
Cat Muscle	**Human Muscle**	**Origin**	**Insertion**	**Action**
Muscles of the Hip				
Caudofemoralis	NA	Proximal caudal vertebra	Patella	Abducts thigh
Gluteus Maximus	Gluteus maximus	Last sacral and first caudal vertebra	Fascia lata	Abducts thigh
Gluteus Medius	Gluteus medius	Iliac crest, sacral vertebrae, and first caudal vertebra	Greater trochanter of femur	Abducts thigh

(continued on next page)

(continued from previous page)

Table 28.7	Muscles of the Hindlimb *(continued)*			
Cat Muscle	**Human Muscle**	**Origin**	**Insertion**	**Action**
Gluteus Profundus/ Minimus	Gluteus minimus	NA	NA	NA
Gracilis	Gracilis	Pubic symphysis	Fascia of distal thigh	Adducts thigh
Iliopsoas	Iliopsoas	Lumbar vertebrae and ilium	Lesser trochanter of femur	Flexes and rotates thigh
Tensor Fasciae Latae	Tensor fasciae latae	Ilium	Fascia lata	Tightens fascia lata
Tenuissimus	NA (tenuissimus rare in humans)	Second caudal vertebra	Proximal tibia	Flexes the shank
Muscles of the Thigh (anterior and medial)				
Adductor Femoris	Adductor brevis and magnus	Pubic symphysis	Ventral femur	Adducts thigh
Adductor Longus	Adductor longus	Pubic symphysis	Femur	Adducts thigh
Pectineus	Pectineus	Pubis	Shaft of femur	Adducts thigh
Rectus Femoris	Rectus femoris	Ilium	Patella	Extends shank
Sartorius	Sartorius	Iliac crest	Patella and proximal tibia	Adducts and laterally rotates thigh, extend the shank
Vastus Intermedius	Vastus intermedius	Femur	Patella	Extends shank
Vastus Lateralis	Vastus lateralis	Femur	Patella	Extends shank
Vastus Medialis	Vastus medialis	Femur	Patella	Extends shank
Muscles of the Thigh (posterior)				
Biceps Femoris	Biceps femoris	Ischial tuberosity	Proximal tibia	Flexes shank, abducts thigh
Semimembranosus	Semimembranosus	Ischium	Medial epicondyle of femur, proximal tibia	Extends thigh
Semitendinosus	Semitendinosus	Ischium	Tibia	Flexes shank
Muscles of the Leg (anterior and lateral)				
Extensor Digitorum Longus	Extensor digitorum longus	Lateral epicondyle of femur	Digits II–V	Extends the digits
Fibularis Brevis	Fibularis brevis	Fibula	Metatarsals	NA
Fibularis Longus	Fibularis longus	Fibula	First metatarsal	NA
Tibialis Anterior	Tibialis anterior	Proximal tibia and fibula	First metatarsal	Flexes the foot
Muscles of the Leg (posterior)				
Flexor Digitorum Longus	Flexor digitorum longus	Tibia and head of fibula	Digits I–V	Flexes the digits
Flexor Hallucis Longus	Flexor hallucis longus	Tibia and fibula	Tendon of flexor digitorum longus	Flexes the digits

Table 28.7	Muscles of the Hindlimb *(continued)*			
Cat Muscle	**Human Muscle**	**Origin**	**Insertion**	**Action**
Muscles of the Leg (posterior) *(continued)*				
Gastrocnemius	Gastrocnemius	Distal, posterior femur, tendon of plantaris	Calcaneus	Extends the foot
Plantaris	Plantaris	Patella and femur	Calcaneus	Extends the foot
Soleus	Soleus	Proximal fibula	Calcaneus	Extends the foot
Tibialis Posterior	Tibialis posterior	Tibia and fibula	Tarsals	Extends the foot

EXERCISE 28.9B Muscles of the Hindlimb: Leg and Foot

1. **Superficial muscles:** Continue dissecting the hindlimb distally toward the leg. The muscles of the leg in cats are organized into compartments that are very similar to those of the human leg. These compartments are anterior, posterior, and lateral. Anterior compartment muscles dorsiflex the foot and extend the digits, posterior compartment muscles plantar flex the foot and flex the digits, and lateral compartment muscles evert the foot. The posterior compartment is the largest compartment of the leg, containing the triceps surae muscle group. The **triceps surae** consists of the gastrocnemius, soleus, and plantaris muscles, all of which insert on the calcaneus via

the calcaneal tendon. As in the human, the triceps surae group is the most important group of muscles acting about the ankle. They are the primary muscles used for plantar flexion of the foot, an action that is necessary for standing, jumping, walking, and running.

2. Clean, identify, and isolate the following muscles, using **figure 28.14** and table 28.7 as guides:

   ☐ **Extensor digitorum longus**
   ☐ **Fibularis brevis**
   ☐ **Fibularis longus**
   ☐ **Gastrocnemius**
   ☐ **Plantaris**
   ☐ **Soleus**
   ☐ **Tibialis anterior**

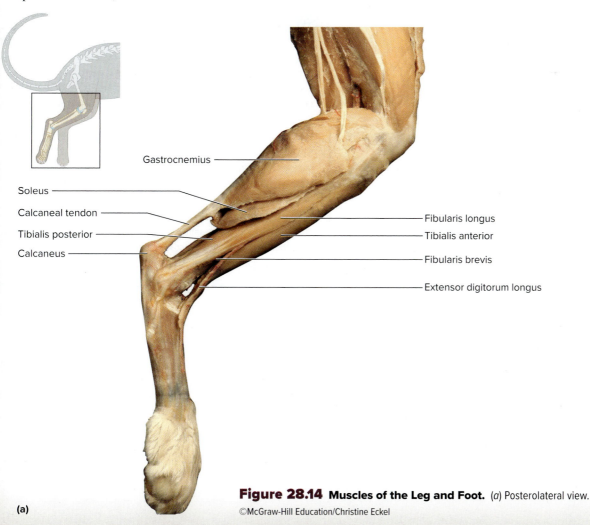

Gastrocnemius

Soleus

Calcaneal tendon

Tibialis posterior

Calcaneus

Fibularis longus

Tibialis anterior

Fibularis brevis

Extensor digitorum longus

Figure 28.14 Muscles of the Leg and Foot. (*a*) Posterolateral view.

(a)

©McGraw-Hill Education/Christine Eckel

(continued on next page)

(continued from previous page)

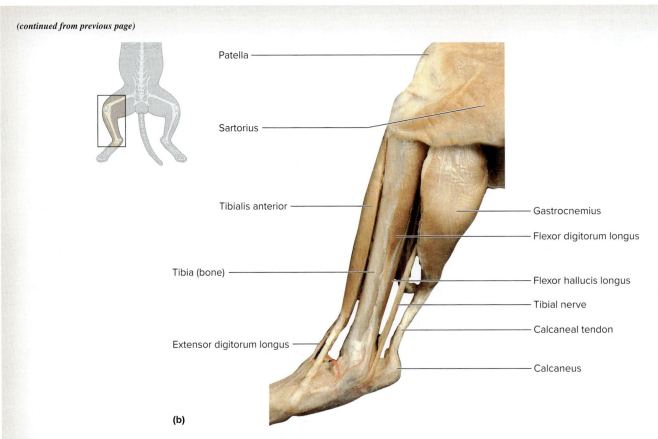

Patella

Sartorius

Tibialis anterior

Gastrocnemius

Flexor digitorum longus

Tibia (bone)

Flexor hallucis longus

Tibial nerve

Calcaneal tendon

Extensor digitorum longus

Calcaneus

(b)

Figure 28.14 **Muscles of the Leg and Foot** *(continued).* (*b*) Medial view.

©McGraw-Hill Education/Christine Eckel

3. **Deep muscles:** Viewed earlier, the largest, most superficial muscles of the posterior leg are the triceps surae muscles. One must reflect these to view the deep muscles of the posterior leg. The easiest way to do this is to detach the calcaneal tendon from the calcaneus, and then reflect the triceps surae muscles cranially.

4. Once the superficial muscles have been reflected, clean, identify, and isolate the following deep muscles, which are quite small. Use figure 28.14 and table 28.7 as guides. While dissecting these muscles, consider the homologous muscles in the human leg and note any similarities and differences.

☐ **Flexor digitorum longus**

☐ **Flexor hallucis longus**

☐ **Tibialis posterior**

INTEGRATE

LEARNING STRATEGY

The semitendinosus, semimembranosus, and biceps femoris muscles are commonly referred to as the **hamstring muscles** of the thigh. The term *hamstring* comes from the animal meat processing industry. When the hindlimbs of animals, particularly pigs, are hung in the slaughterhouse, they are typically hung by the tendons of these three muscles—thus the term *ham* (the back of a hog's knee) and *string*. So, literally, the pig is strung up by these muscles. The term *hamstrung* has, over time, come to take on several other meanings. Most commonly when people are said to be *hamstrung*, the term refers to the fact that the people are unable to make a decision or proceed with something because they are stuck or crippled (as if they were "hung up" by their hamstrings!).

EXERCISE 28.10

NERVOUS SYSTEM: PERIPHERAL NERVES

1. **Brachial plexus:** Place the cat ventral side up on the dissecting tray. Briefly review the muscles of the limbs. While doing so, note the whitish, cord-like structures within the limbs. These are the **peripheral nerves.** This dissection focuses on the axillary region and the major components of the brachial plexus. Start by reflecting the pectoral muscles laterally (they may have already been removed, depending on the order of dissection exercises). Note the mass of connective tissue and "stringy stuff" in the axillary region. To distinguish nerves from blood vessels, use the open scissors blunt dissection technique to gently loosen the connective tissue in this region. Using this technique, connective tissue is pulled apart, leaving blood vessels and nerves intact. However, be gentle with the technique so as not to tear blood vessels and nerves.

2. Try to locate the **subclavian artery** as it emerges under the clavicle and enters the axilla as the **axillary artery.** Recall that the cords of the brachial plexus are named for their location relative to the axillary artery. After identifying the axillary artery, gently trace the nerves that surround it both toward and away from their origin in the spinal cord. Identify the following components of the brachial plexus, using **figure 28.15** as a guide:

☐ **Axillary artery (reference)**   ☐ **Musculocutaneous nerve**

☐ **Axillary nerve**   ☐ **Posterior cord**

☐ **Lateral cord**   ☐ **Radial nerve**

☐ **Medial cord**   ☐ **Ulnar nerve**

☐ **Median nerve**

3. **Lumbosacral plexus:** This dissection focuses on the pelvic and inguinal region, which is the location of the lumbosacral plexus. Start by observing the internal anatomy of the lower posterior abdominal wall. To view the nerves of the lumbosacral plexus, first remove the urinary bladder, the ureters, the kidney, and any reproductive structures that lie ventral to the posterior abdominal wall. Identify the **psoas major muscle**, a major landmark in the pelvis **(figure 28.16)**. Trace the psoas major posteriorly toward its insertion on the femur. As it crosses the inguinal region, two nerves emerge adjacent to the muscle: the **femoral nerve** and the **obturator nerve.** Reflect one of the psoas major muscles superiorly to view the lumbar plexus from which these two nerves arise.

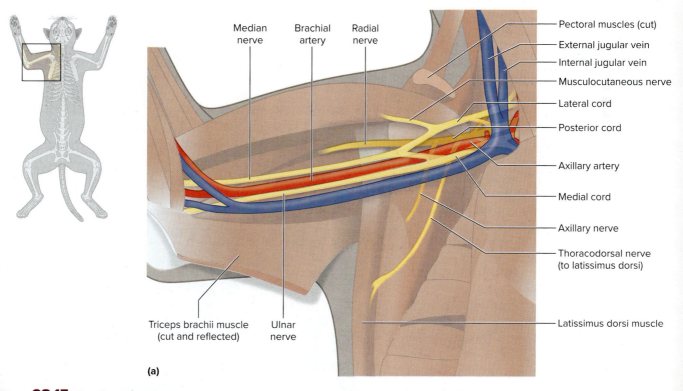

(a)

Figure 28.15 **The Brachial Plexus.** (*a*) Illustration.

(continued on next page)

(continued from previous page)

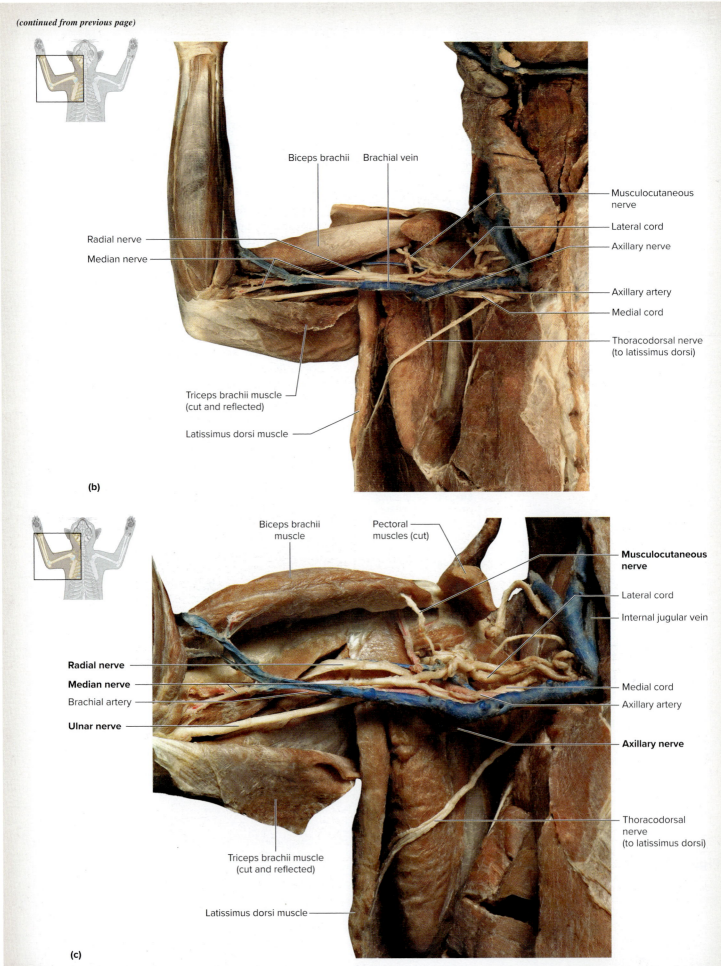

(b)

Biceps brachii Brachial vein

Radial nerve

Median nerve

Musculocutaneous nerve

Lateral cord

Axillary nerve

Axillary artery

Medial cord

Thoracodorsal nerve (to latissimus dorsi)

Triceps brachii muscle (cut and reflected)

Latissimus dorsi muscle

(c)

Biceps brachii muscle   Pectoral muscles (cut)

Musculocutaneous nerve

Lateral cord

Internal jugular vein

Radial nerve

Median nerve

Brachial artery

Ulnar nerve

Medial cord

Axillary artery

Axillary nerve

Thoracodorsal nerve (to latissimus dorsi)

Triceps brachii muscle (cut and reflected)

Latissimus dorsi muscle

Figure 28.15 **The Brachial Plexus (continued).** (b) Dissection. (c) Dissection in anatomic position.

©McGraw-Hill Education/Christine Eckel

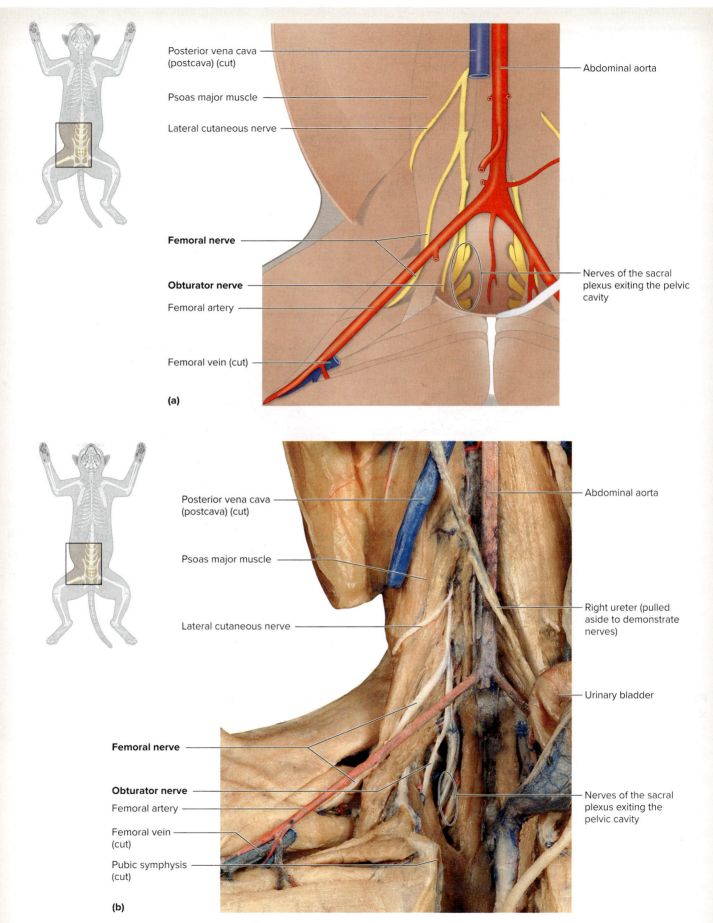

Figure 28.16 Lumbosacral Plexus: Lumbar Plexus. From an anterior view, the nerves of the lumbar portion of the lumbosacral plexus are best observed. (*a*) Illustration. (*b*) Dissection. Note the nerves of the sacral plexus that emerge within the pelvic cavity and run posterior to eventually exit within the gluteal region (figure 28.17).

(*b*) ©McGraw-Hill Education/Christine Eckel

(*continued on next page*)

(continued from previous page)

3. **Veins:** Identify the following veins of the cat, using **figure 28.19** as a guide:

- ☐ **Anterior facial**
- ☐ **Anterior tibial**
- ☐ **Anterior vena cava (precava)**
- ☐ **Axillary**
- ☐ **Azygos**
- ☐ **Brachial**
- ☐ **Brachiocephalic**
- ☐ **Cephalic**
- ☐ **Common iliac**
- ☐ **External iliac**
- ☐ **External jugular**
- ☐ **Femoral**
- ☐ **Gastrosplenic**
- ☐ **Gonadal (testicular or ovarian)**

- ☐ **Great saphenous**
- ☐ **Hepatic portal**
- ☐ **Iliolumbar**
- ☐ **Inferior (caudal) mesenteric**
- ☐ **Internal iliac**
- ☐ **Internal jugular**
- ☐ **Median cubital**
- ☐ **Popliteal**
- ☐ **Posterior tibial**
- ☐ **Posterior vena cava (postcava)**
- ☐ **Radial**
- ☐ **Renal**
- ☐ **Subclavian**
- ☐ **Superior (cranial) mesenteric**
- ☐ **Ulnar**

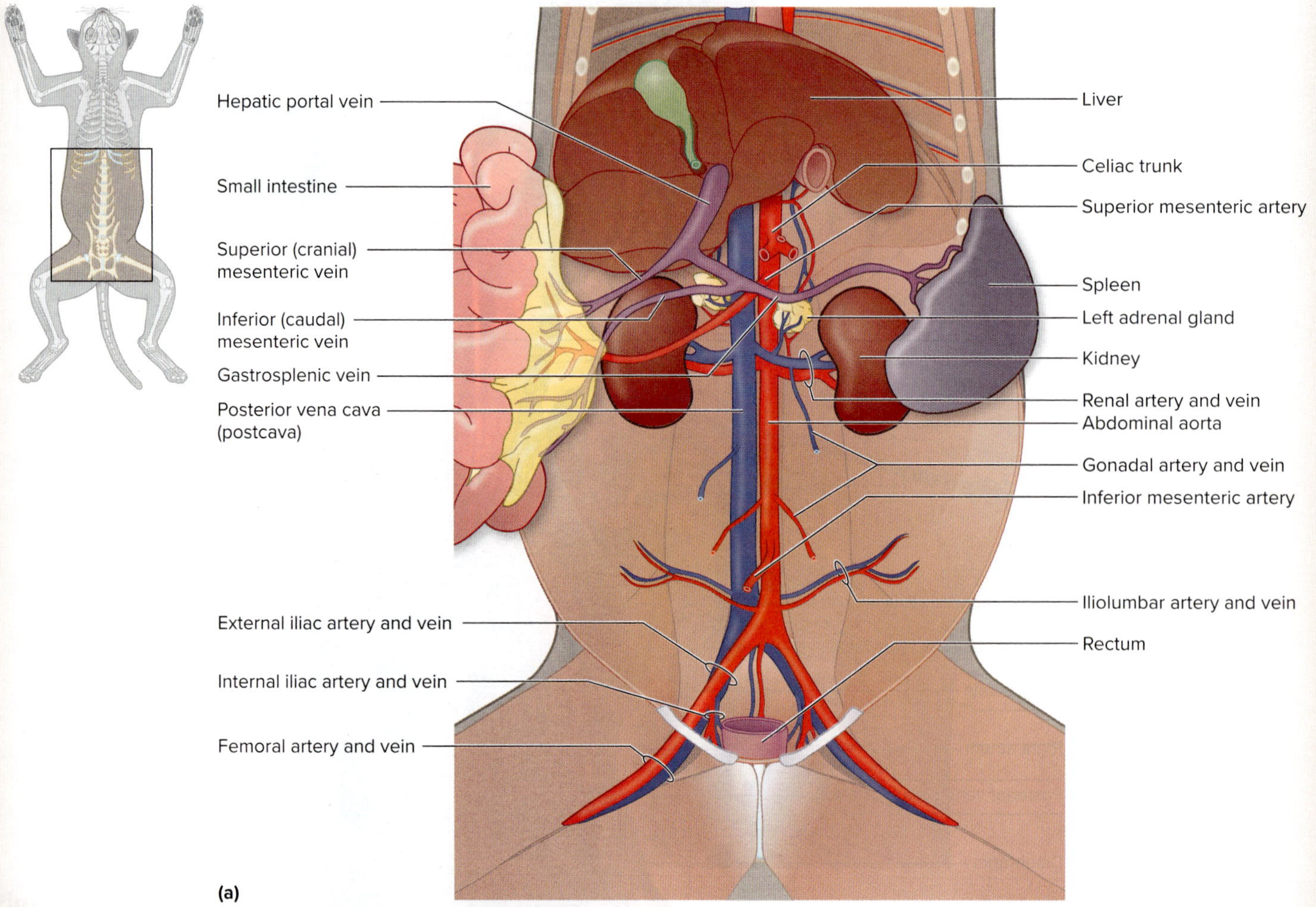

(a)

Figure 28.19 Abdominal and Lower Limb Vessels. (*a*) Illustration.

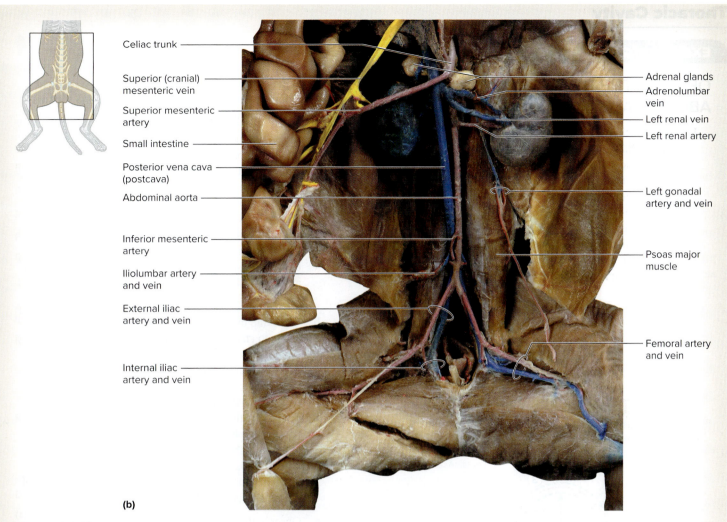

Celiac trunk

Superior (cranial) mesenteric vein

Superior mesenteric artery

Small intestine

Posterior vena cava (postcava)

Abdominal aorta

Inferior mesenteric artery

Iliolumbar artery and vein

External iliac artery and vein

Internal iliac artery and vein

Adrenal glands

Adrenolumbar vein

Left renal vein

Left renal artery

Left gonadal artery and vein

Psoas major muscle

Femoral artery and vein

(b)

Figure 28.19 **Abdominal and Lower Limb Vessels *(continued)*.** (*b*) Dissection demonstrating the major arteries and veins of the abdomen and lower limb of the cat in one figure. Observe the other dissection photos in this chapter to see individual vessels within specific regions of the body (e.g., upper limb, thorax, abdomen, etc.).

©McGraw-Hill Education/Christine Eckel

(continued from previous page)

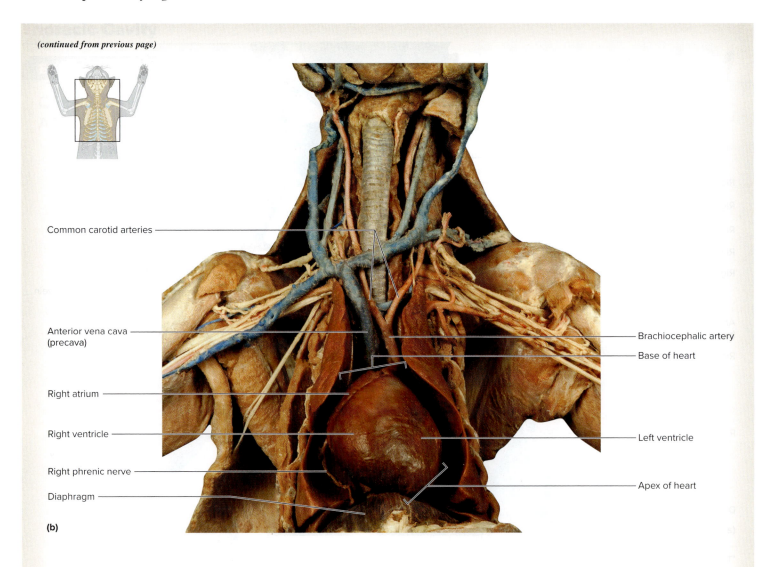

Common carotid arteries

Anterior vena cava (precava)

Right atrium

Right ventricle

Right phrenic nerve

Diaphragm

Brachiocephalic artery

Base of heart

Left ventricle

Apex of heart

(b)

Figure 28.21 The Thoracic Cavity (continued). (*b*) Deep dissection demonstrating the heart and lungs and structures of the superior mediastinum (trachea, esophagus, blood vessels, and nerves).
©McGraw-Hill Education/Christine Eckel

EXERCISE 28.13

THE HEART, LUNGS, AND MEDIASTINUM

1. **Superficial dissection:** When removing the sternum, the anterior part of the **parietal pericardium** (outer covering of the pericardial sac) may be adhered to the bone. The goal is to keep the pericardium with the heart, not the body wall. Similarly, the **parietal pleura** (outer covering of the **pleural cavities**) is adhered to the inside of the thoracic cage. The pleura is more difficult to separate from the thoracic cage than the pericardium. Therefore make sure to observe the inside of the thoracic cage and make note of the parietal pleura. Where is the visceral pleura located? _____

2. Begin to clean the structures in the superior mediastinum (e.g., trachea, carotid arteries, esophagus), taking care to pull on any "stringy" structures very gently so as not to destroy them. While cleaning the region superficial and lateral to the pericardium, gently separate the **parietal pericardium** from the **parietal pleura** on both sides and look for the right and left **phrenic nerves**, which lie in the space between these two layers. What structure do the phrenic nerves innervate? _____

3. Observe the area just cranial to the pericardial cavity, where the great vessels enter and leave the heart. Depending on the age of the cat, the thymus may or may not be visible in this location. In a young cat, the thymus often looks like a small, extra lobe of the lung, although it is not part of the lung. In an older cat, the thymus may not be visible because it largely has been replaced with adipose connective tissue. Recall the function of the thymus from chapter 23. The thymus in humans, just as with cats, is also large in adolescents, but generally absent in older individuals. In the space provided, explain why the thymus is not usually visible in an older cat or human.

_____

_____

4. Using scissors, carefully open the **pericardial sac** to reveal the heart within. Review the layers of the pericardium (exercise 21.1) while dissecting away the sac. At the base of the heart (remember, the base of the heart is the most cranial part), the pericardial sac is adhered to the great vessels. Of note, the major veins entering the cat heart are the *anterior* vena cava (precava) and *posterior* vena cava (postcava). These are the same as the *superior* vena cava and *inferior* vena cava in the human, respectively. As you detach the sac from the vessels, identify each of the great vessels (**figures** 28.21 and **28.22**).

☐ **Anterior vena**   ☐ **Aorta**
    cava (precava)   ☐ **Pulmonary trunk**

5. Identify the four chambers of the heart, using figure 28.22 as a guide.

☐ **Left atrium**   ☐ **Right atrium**

☐ **Left ventricle**   ☐ **Right ventricle**

6. **Deep dissection:** After identifying the great vessels and heart chambers, the next step is to remove the heart and lungs from the thoracic cavity. To do this, follow these steps:

a. Transect the **pulmonary arteries** and **veins** at the hilum of each lung. When doing this, also transect the **primary bronchi**, which enter the lungs adjacent to the pulmonary vessels.

b. Transect the **anterior vena cava (precava)** and **aorta** just superior to the heart. This means cutting through the *ascending* part of the aorta.

c. Bisect the **posterior vena cava (postcava)**, which lies between the inferior surface of the heart and the diaphragm.

7. **The lungs:** Because all of the pulmonary vessels have been transected between the heart and the lungs, the lungs can now be removed from the thoracic cavity (figure 28.22). Remove both the heart and the lungs. Place them next to the cat in the dissecting tray or in another receptacle, and observe their features. How many lobes does the right lung have? _____ How many lobes does the left lung have? _____ Are these the same or different from the number of lobes in the human lungs?

_____

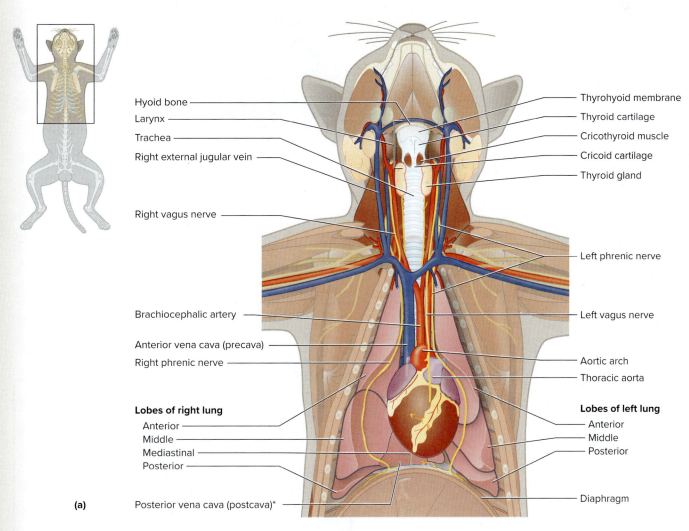

Hyoid bone
Larynx
Trachea
Right external jugular vein

Right vagus nerve

Brachiocephalic artery

Anterior vena cava (precava)
Right phrenic nerve

Lobes of right lung
Anterior
Middle
Mediastinal
Posterior

(a)   Posterior vena cava (postcava)*

Thyrohyoid membrane
Thyroid cartilage
Cricothyroid muscle
Cricoid cartilage
Thyroid gland

Left phrenic nerve

Left vagus nerve

Aortic arch
Thoracic aorta

Lobes of left lung
Anterior
Middle
Posterior

Diaphragm

*The posterior vena cava (postcava) in the cat is the same as the inferior vena cava in the human.

Figure 28.22 **The Respiratory System.** (*a*) Illustration.

(continued on next page)

(continued from previous page)

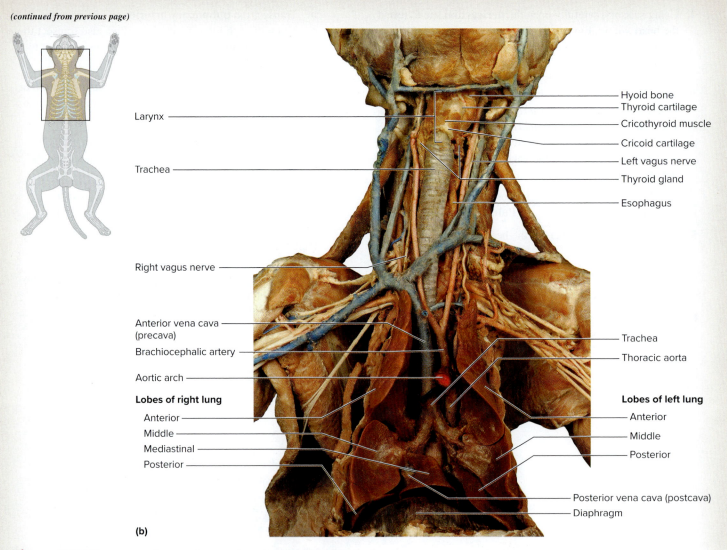

Larynx

Trachea

Right vagus nerve

Anterior vena cava (precava)

Brachiocephalic artery

Aortic arch

Lobes of right lung

Anterior

Middle

Mediastinal

Posterior

Hyoid bone

Thyroid cartilage

Cricothyroid muscle

Cricoid cartilage

Left vagus nerve

Thyroid gland

Esophagus

Trachea

Thoracic aorta

Lobes of left lung

Anterior

Middle

Posterior

Posterior vena cava (postcava)

Diaphragm

(b)

Figure 28.22 **The Respiratory System *(continued)*.** *(b)* Photo.

©McGraw-Hill Education/Christine Eckel

8. **The heart:** Observe the heart and identify the four chambers of the heart. Notice the relative size of the different chambers. Optional step: To observe internal structures of the heart, use a scalpel to make a coronal section of the heart, starting from the apex and ending at the base. Identify all of the valves of the heart and note the relative size of the chambers. Note any similarities or differences between the cat heart and human heart in the space provided.

_____

_____

_____

9. **Upper respiratory and digestive tracts:** This next step involves careful cleaning of structures that lie in the superior mediastinum: the **trachea**, the **esophagus**, the **thoracic duct**, and several blood vessels and nerves. Begin by locating the esophagus and the **azygos vein** on the posterior thoracic wall next to the vertebral column. The azygos vein is a vein that extends from caudal to cranial just to the right of the vertebral column. The azygos vein and thoracic duct are not visible in figure 28.22. Next, look for a thin, brown or colorless vessel located between the azygous vein and the

esophagus, which is the **thoracic duct** (see Learning Strategy on this page). Do not worry if the thoracic duct cannot be located; it is very small. The thoracic duct is the main lymphatic duct of the body. The thoracic duct begins as the **cisterna chyli** in the abdominal cavity near vertebral level L_1, and runs along the dorsal abdominal and thoracic walls before it turns and enters the left subclavian vein.

INTEGRATE

LEARNING STRATEGY

The thoracic duct is a very small vessel, particularly when compared to other structures within the mediastinum. One way of remembering how to find the thoracic duct is to look for what anatomists call the "duck between two gooses." This is a play on words resulting from pronouncing the names of the three structures slightly different than normal. The "duck" is the thoracic **duct**. The "two gooses" are the esopha**goose** (esophagus) and the azy**goose** (azygos vein). Thus, to find the thoracic duct, look for a small vessel that runs between the esophagus and the azygos vein.

10. Observe the most cranial part of the mediastinum. Clean the trachea as it enters the neck. Carefully clean the **larynx** and the **thyroid gland**. The thyroid gland wraps around the larynx, and portions of it extend between the larynx and the common carotid arteries. To perform most of this cleaning of connective tissue, use small scissors and forceps, taking care not to pull too hard on any structure so as not to tear it. While following the common carotid arteries into the neck, look for a small nerve that runs between the **common carotid artery** and the **internal jugular vein**. This is the vagus nerve (CN X), which was studied in chapter 15. Structures that are located cranial to the thyroid cartilage will be explored further in exercise 28.17, which covers structures of the head and neck. At this time, keep the rest of the head covered and hydrated so it will be easy to dissect later.

11. Clean, identify, and isolate the following structures in the mediastinum and anterior neck, using figures 28.22 and **28.23** as guides:

- ☐ **Anterior vena cava (precava)**
- ☐ **Aortic arch**
- ☐ **Common carotid arteries**
- ☐ **Esophagus**
- ☐ **External jugular veins**
- ☐ **Larynx**
- ☐ **Primary bronchi**
- ☐ **Subclavian arteries**
- ☐ **Thoracic aorta**
- ☐ **Thyroid gland**
- ☐ **Trachea**

12. When finished with the dissection of the thoracic cavity, place the heart and lungs back into the thoracic cavity, close the thoracic cage, sprinkle the cat with wetting solution, and then wrap the cat in its skin and place it in the storage bag.

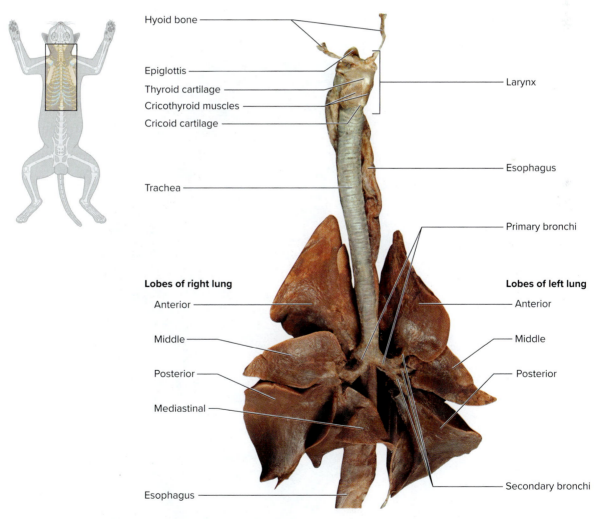

Figure 28.23 Respiratory System Structures. In this dissection the hyoid, larynx, trachea, esophagus, and lungs have been removed from the neck and thoracic cavity to make respiratory system structures more visible.

©McGraw-Hill Education/Christine Eckel

Abdomen and Pelvis

EXERCISE 28.14

THE ABDOMINAL CAVITY

The directions for making incisions to open both the thoracic and abdominal cavities are described in exercise 28.12, figure 28.20. While performing dissections on the abdominal cavity, make sure to keep the thoracic cavity closed so the heart and lungs don't fall out and also to keep them hydrated so they do not dry out. The contents of the abdominal cavity are more easily studied after removing the superficial peritoneum (**Figure 28.24**).

1. Observe the transverse cuts that were previously made below the diaphragm, and the vertical cut along the linea alba. Carefully reflect the anterior abdominal wall laterally along these lines.

2. **Superficial dissection:** One of the most prominent structures that becomes visible as soon as the abdominal cavity is opened is a huge "drape" of fat that overlies most of the abdominal contents. This is the **greater omentum (figure 28.25)**, which is a quadruple fold of peritoneum and contains relatively large amounts of fat in many cats. Recall from chapter 26 that the

greater omentum functions to help cushion and insulate abdominal organs. In addition, if an infection arises in the abdominal cavity, the greater omentum can move around until it "walls off" that area to prevent the spread of infection. What are the two structures to which the greater omentum is attached (see chapter 26 for review)?

_____ and

_____

Gently lift the greater omentum from the abdominal cavity using forceps. Then use scissors to detach it from its connections to the abdominal organs.

3. **Deep dissection:** With the greater omentum removed, the contents of the abdominal cavity can be explored more easily. This process requires much less work than dissection of the muscles of the body because there is not a lot of connective tissue that needs to be removed. Instead, the majority of the organs move around freely because of the peritoneum that covers them. Recall that the **peritoneum** is to the abdominal cavity what the pericardium is to the

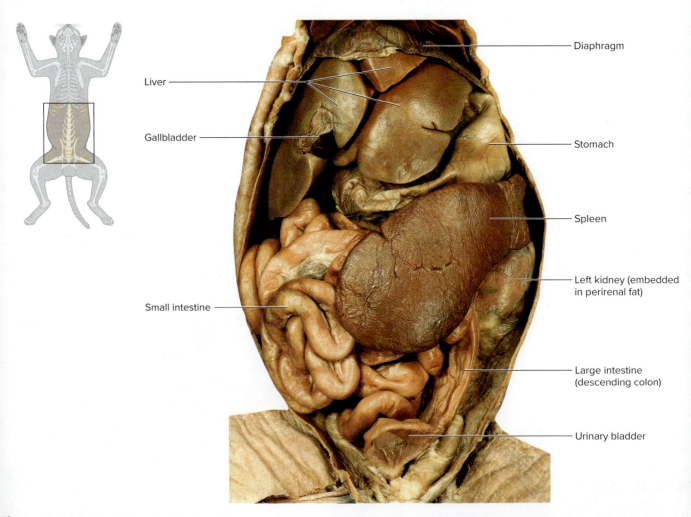

Figure 28.24 The Abdominal Cavity. The anterior abdominal wall and greater omentum have been removed to show a superficial view of the abdomen.

©McGraw-Hill Education/Christine Eckel

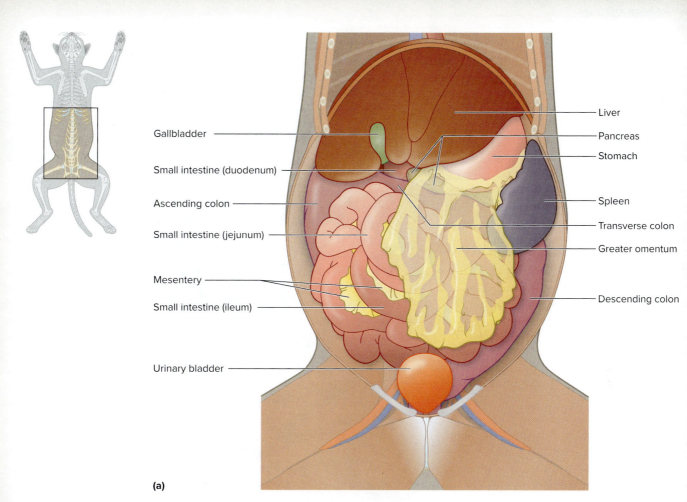

Gallbladder

Small intestine (duodenum)

Ascending colon

Small intestine (jejunum)

Mesentery

Small intestine (ileum)

Urinary bladder

Liver

Pancreas

Stomach

Spleen

Transverse colon

Greater omentum

Descending colon

(a)

Figure 28.25 **The Spleen, Jejunum, Ileum, and Large Intestine (Colon).** Trace the tail of the pancreas to see where it enters the hilum of the spleen. Then observe both the small and large intestines and the folds of mesentery that anchor each to the body wall. (*a*) Illustration.

pericardial cavity. That is, the **parietal peritoneum** is the inner lining of the abdominal wall, and the **visceral peritoneum** is the shiny tissue covering most abdominal organs, such as the intestines. It is the **visceral peritoneum** that forms both omenta and mesenteries (more later on mesenteries). The "dissection" of the abdominal cavity largely consists of gently separating the peritoneum that anchors organs to each other to view deeper structures and to appreciate the relationships among organs. Identify the following organs or parts of organs within the abdominal cavity, using figure 28.25 as a guide:

☐ **Large intestine**　　☐ **Spleen**

☐ **Liver**　　　　　　 ☐ **Stomach**

☐ **Small intestine**　　 ☐ **Urinary bladder**

4. **The liver and gallbladder:** Observe the liver. Note a band of peritoneum that connects the liver ventrally to the body wall. This is the **falciform ligament**. The posterior border of the falciform ligament contains another band of connective tissue, the **round ligament of the liver**. The round ligament is a remnant of the umbilical vein, which was present during fetal development. Observe the **lobes of the liver**. How

many are there? _____ Note also the connection between the cranial surface of the liver and the diaphragm. Here, the liver is connected to the diaphragm via a ligament called the **coronary ligament**. The term "coronary" means "crown" and applies in this case because the ligament forms a circle or "crown" of sorts atop the liver. Feel around for the coronary ligament; then cut the coronary ligament with scissors to detach the cranial surface of the liver from the diaphragm. After completing this step, look carefully on the cranial surface of the liver and note the **bare spot** of the liver. The bare spot is the only part of the liver that is not covered with peritoneum; it is characteristically dull in appearance as compared with the shiny appearance of parts of the liver that are covered with peritoneum. Note also the **gallbladder**, which has a characteristic green color. What substance is stored in the gallbladder, which is also the substance that gives it a green color? _____ What is the function of this substance? _____

_____

(*continued on next page*)

(continued from previous page)

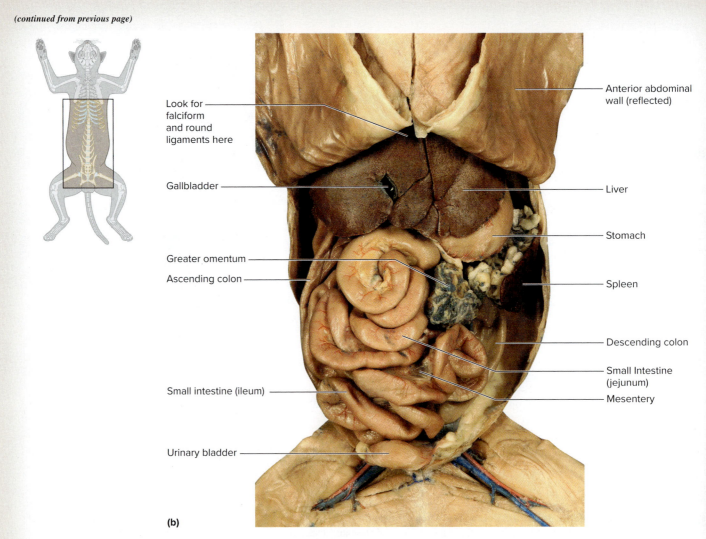

Look for falciform and round ligaments here

Gallbladder

Greater omentum

Ascending colon

Small intestine (ileum)

Urinary bladder

Anterior abdominal wall (reflected)

Liver

Stomach

Spleen

Descending colon

Small Intestine (jejunum)

Mesentery

(b)

Figure 28.25 **The Spleen, Jejunum, Ileum, and Large Intestine (Colon) *(continued).*** (*b*) Dissection.
©McGraw-Hill Education/Christine Eckel

5. **The stomach, pancreas, spleen, and duodenum:** The next step requires lifting the liver cranially to view the structures that lie deep to it. If moving it is too difficult, use a scalpel to carefully remove the left lateral lobe of the liver. However, be sure to check with the lab instructor before doing this. Deep to the caudal part of the liver is the **stomach**. Deep to the stomach is the **pancreas**, and near the *head* of the pancreas is the first part of the small intestine: the **duodenum (figure 28.26)**. Follow the *body* of the pancreas laterally to find the *tail* of the pancreas, which ends at the hilum of the **spleen**. Observe the area deep to the head of the pancreas and near the hilum of the liver to locate the following structures, using figure 28.26 as a guide:

☐ **Common bile duct**   ☐ **Head of pancreas**

☐ **Duodenum**   ☐ **Liver**

☐ **Gallbladder**   ☐ **Stomach**

6. **The jejunum, ileum, and large intestine:** Lay the stomach and pancreas back into place and observe the spleen and the remainder of the intestinal tract. To fully appreciate the length of the small intestine, pick it up at the end of the duodenum and "walk" your fingers along the length of the small intestine until they reach the *ileocecal valve*. Note that it takes some time to reach that location! While doing this, make note of the **mesenteries**, which anchor the small intestine to the dorsal body wall (figure 28.25). It may be difficult to distinguish the **jejunum** (middle part of the small intestine) from the **ileum** (last part of the small intestine) in the cat, but try to use the criteria listed in figure 26.6 to distinguish the two from each other. A fun exercise is to have one dissection group remove the small intestine in its entirety from the abdominal cavity, dissect away its mesentery so it can be extended fully, and measure its length. If this is done, record the length of the entire small intestine here: _____ cm

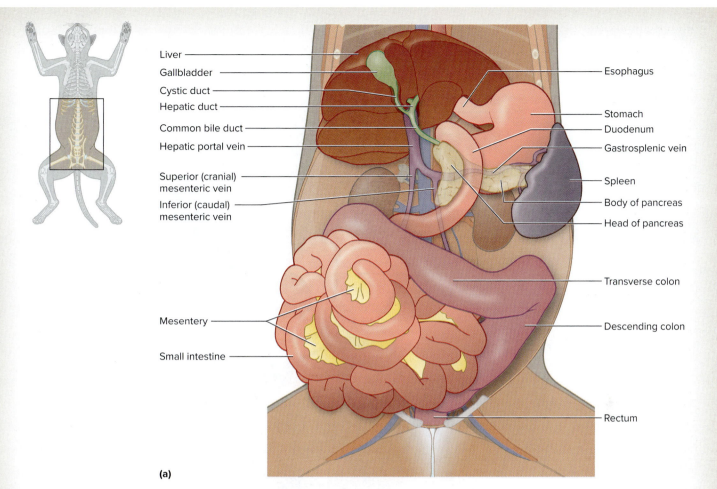

Liver
Gallbladder
Cystic duct
Hepatic duct
Common bile duct
Hepatic portal vein
Superior (cranial) mesenteric vein
Inferior (caudal) mesenteric vein
Mesentery
Small intestine

Esophagus
Stomach
Duodenum
Gastrosplenic vein
Spleen
Body of pancreas
Head of pancreas
Transverse colon
Descending colon
Rectum

(a)

Figure 28.26 The Hepatic Portal System. The hepatic portal system consists of the gastrosplenic vein, inferior (caudal) mesenteric vein, superior (cranial) mesenteric vein, and hepatic portal vein. (*a*) Illustration.

7. **The hepatic portal system:** Now that the organs of the gastrointestinal system within the abdominal cavity have been identified, focus on the blood supply to these organs. Before beginning, review the hepatic portal circulation of the human from chapter 22, figure 22.9. What are the three major veins that converge in the *human* to become the hepatic portal vein?

a. _____

b. _____

c. _____

In the cat, there are two major veins that come together to form the hepatic portal vein, the **gastrosplenic vein** and the **superior (cranial) mesenteric vein** (figure 28.26). The gastrosplenic vein receives blood largely from the spleen, stomach, and descending colon, thus making it homologous with both the splenic and inferior mesenteric veins in the human. The anterior mesenteric vein in the cat drains a region of the small and large intestines similar to the region drained by the superior mesenteric vein in the human. To locate these veins, use blunt dissection to gently tease away the mesentery of the small intestine. Both the arteries that supply arterial blood to the intestines (the **superior [cranial] mesenteric artery** and the **inferior [caudal] mesenteric artery**), as well as the veins that drain blood from the intestine, extend between the two layers of peritoneum that compose the mesenteries. Identify the following blood vessels in the cat, using figure 28.26 as a guide:

☐ **Abdominal aorta**

☐ **Celiac trunk**

☐ **Gastrosplenic vein**

☐ **Hepatic portal vein**

☐ **Inferior (caudal) mesenteric artery**

☐ **Superior (cranial) mesenteric artery**

☐ **Superior (cranial) mesenteric vein**

8. After completing the dissection of the abdominal cavity, place the organs gently back into the abdominal cavity, close the abdominal wall, sprinkle the cat with wetting solution, and then wrap the cat in its skin and place it in the storage bag.

(continued on next page)

(continued from previous page)

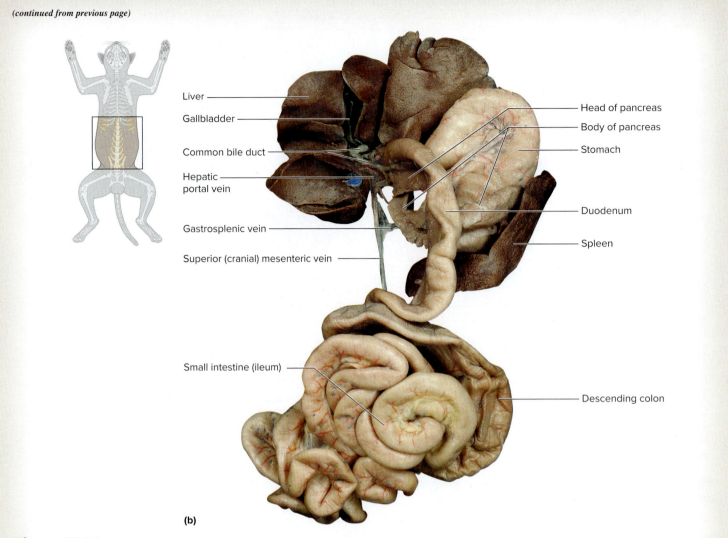

Liver

Gallbladder

Common bile duct

Hepatic portal vein

Gastrosplenic vein

Superior (cranial) mesenteric vein

Small intestine (ileum)

Head of pancreas

Body of pancreas

Stomach

Duodenum

Spleen

Descending colon

(b)

Figure 28.26 **The Hepatic Portal System** *(continued).* (*b*) Dissection. The inferior (caudal) mesenteric vein is not visible in the dissection photo.

©McGraw-Hill Education/Christine Eckel

THE DORSAL ABDOMINAL WALL

This exercise begins with removal of the gastrointestinal organs from the abdominal cavity. This involves detaching them from the vascular and mesenteric connections that anchor them to the dorsal abdominal wall. They will then be removed as a unit so they can be replaced within the abdominal cavity when the dissection of the dorsal abdominal wall is completed. Structures of the dorsal abdominal wall, most of which belong to the urinary and endocrine systems, will be observed in addition to the main blood vessels that either arise from the abdominal aorta or drain into the posterior vena cava (postcava).

1. Begin with the cat supine and with the abdominal wall open. Removing the GI organs will first require separating them from their vasculature using the following steps:

   a. Gently lift the pre-dissected (exercise 28.13) organs of the gastrointestinal system (liver, stomach, spleen,

small and large intestines) away from the dorsal body wall. Observe the midline region next to the hilum of the liver to locate the blood vessels that enter the liver here.

   b. Using scissors, detach the vessels where they enter the hilum of the liver while keeping a few millimeters of each attached to its vessel of origin for later identification.

   c. Next, follow the abdominal aorta caudally until the origins of the **superior** and **inferior mesenteric arteries** can be identified. Again, use scissors to detach these vessels from the aorta while keeping a few millimeters of each vessel attached to the aorta.

   d. Finally, carefully detach the liver from the **posterior vena cava** (postcava). This is the most difficult step, because the postcava is somewhat embedded in the dorsal wall of the liver. First place the scissors in the area between the cranial border of the liver and the caudal border of the diaphragm to cut the connection between the postcava and the right atrium of the heart (if this step was not already completed when the heart was removed). Next, gently lift the liver caudally and cut the **hepatic veins**, which drain blood from the liver into the postcava. This process may result in a postcava that ends up a little mangled. This is not something to worry about as it is somewhat unavoidable unless the dissector is very experienced.

e. With the vascular connections now severed, all of the GI organs can be removed from the abdominal cavity. While gently pulling them out, use blunt dissection and/or scissors (only when necessary) to detach any mesenteric connections that remain.

2. With the GI organs removed, observe the dorsal abdominal wall. The structures visible here are all **retroperitoneal** structures. Define retroperitoneal:

_____

_____

_____

Notice that the **kidneys** are embedded in a large amount of fat. The majority of this fat is the **perirenal fat** (see figure 28.24). This fat forms part of the protective layers of the kidney. Remove the fat to fully view the kidney. Before removing this fat, look carefully within the fat that lies cranial to each kidney. Embedded in the fat are the **adrenal glands** (see chapter 19, figure 19.7). To dig the adrenals out of the fat, the best technique to use is the open scissors technique (figure 1.15) because it allows the dissector to loosen up the fat without damaging underlying structures—as long as the dissector uses a *gentle* hand. The adrenal glands have a similar consistency to that of the pancreas and a very thin capsule. Thus, they can be destroyed easily if one is not careful.

3. After isolating and cleaning the adrenal glands (**figure 28.27**), remove the remainder of the perirenal fat from the kidney. Take care in the region of the hilum of

the kidney so as not to cut the **ureters, renal arteries,** or **renal veins**. While cleaning, trace these structures to their respective connections with the urinary bladder (ureters), aorta (renal arteries), and posterior vena cava (renal veins). On the left side, in particular, take care when cleaning the **renal vein**. Look for a small vein that comes off its caudal surface. This is the left **gonadal** (ovarian or testicular) **vein**.

4. **Internal anatomy of the kidney:** Choose one of the kidneys to bisect to view the internal anatomy. If completely removing one kidney from the dorsal abdominal wall, remove the right kidney. To do this, simply bisect the renal artery and vein midway between the hilum of the kidney and the aorta and posterior vena cava, respectively. Then bisect the ureter distal to the renal pelvis. Use a scalpel with a large blade or a dissecting knife to slice the kidney along the cranial/caudal plane (figure 28.27*b*). Identify the following structures, using figure 28.27 as a guide:

☐ **Cortex**   ☐ **Minor calyx**

☐ **Fibrous capsule**   ☐ **Renal artery**

☐ **Major calyx**   ☐ **Renal pelvis**

☐ **Medulla**   ☐ **Renal vein**

Are there observable differences between the cat kidney and the human kidney? _____ If so, describe them in the space provided:

_____

_____

_____

5. While tracing the ureters caudally toward the urinary bladder, identify the **psoas major muscle**, which lies adjacent to the vertebral column. What is the function of this muscle?

_____

_____

_____

(continued on next page)

(continued from previous page)

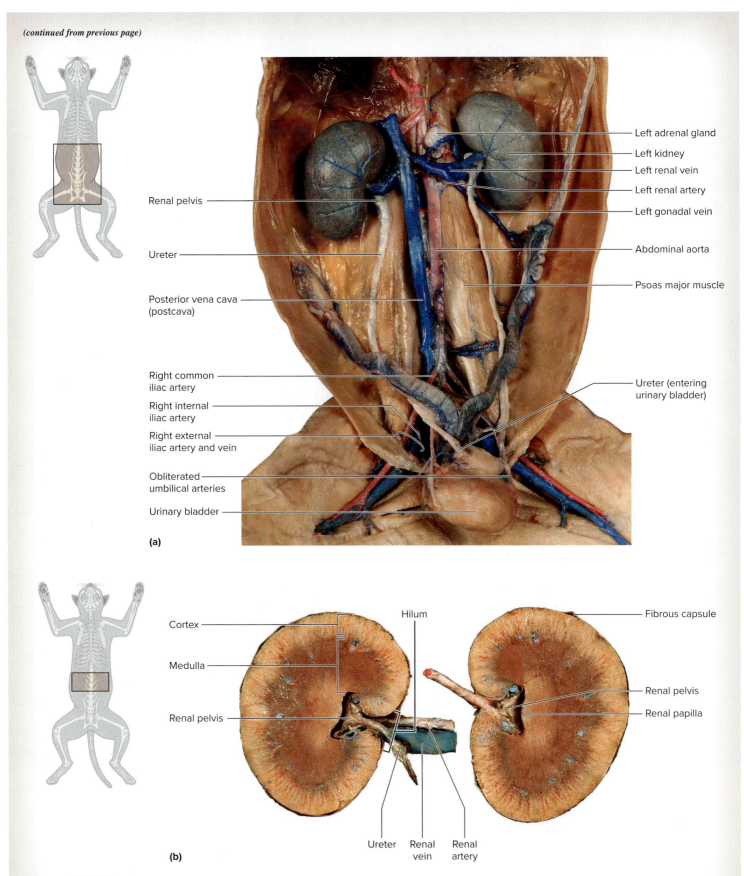

(a)

Left adrenal gland
Left kidney
Left renal vein
Left renal artery
Left gonadal vein
Abdominal aorta
Psoas major muscle
Ureter (entering urinary bladder)

Renal pelvis
Ureter
Posterior vena cava (postcava)
Right common iliac artery
Right internal iliac artery
Right external iliac artery and vein
Obliterated umbilical arteries
Urinary bladder

(b)

Cortex
Medulla
Renal pelvis
Hilum
Fibrous capsule
Renal pelvis
Renal papilla
Ureter
Renal vein
Renal artery

Figure 28.27 Dorsal Abdominal Wall. (*a*) Dorsal abdominal wall of a female cat. The urinary bladder is reflected posteriorly to demonstrate the course of the ureters superfical to the umbilical arteries. The ureters enter the dorsal, inferior surface of the urinary bladder. The umbilical arteries run alongside the urinary bladder and attach to the ventral abdominal wall internal to the umbilicus. (*b*) Bisected kidney.

©McGraw-Hill Education/Christine Eckel

6. **Identification of thoracic and abdominal vasculature:**
Before beginning, remove the heart and lungs from the thoracic cavity and the GI viscera from the abdominal cavity. Using **figure 28.28** as a guide, identify the following vessels in the abdominal and thoracic cavities:

☐ **Abdominal aorta**

☐ **Adrenal veins**

☐ **Anterior vena cava (precava)**

☐ **Aortic arch**

☐ **Azygos vein**

☐ **Brachiocephalic veins**

☐ **Celiac trunk**

☐ **Common iliac vein**

☐ **Descending thoracic aorta**

☐ **External iliac artery**

☐ **External iliac vein**

☐ **Hepatic veins**

☐ **Iliolumbar artery**

☐ **Iliolumbar vein**

☐ **Internal iliac artery**

☐ **Internal iliac vein**

☐ **Left gonadal vein**

☐ **Posterior vena cava (postcava)**

☐ **Renal arteries**

☐ **Renal veins**

☐ **Right gonadal vein**

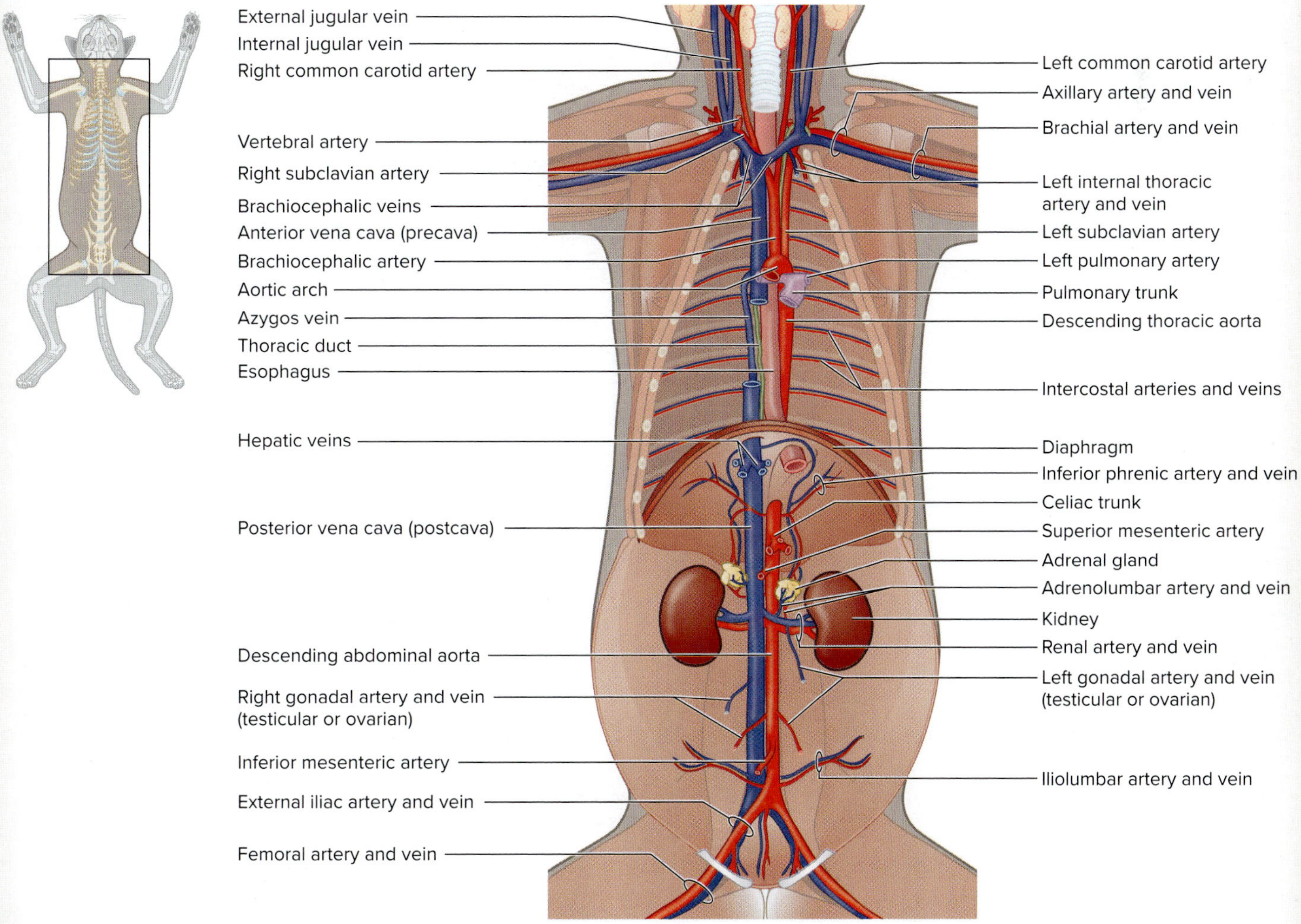

Figure 28.28 Vasculature of the Thoracic and Abdominal Cavities. This illustration demonstrates the arteries and veins of the thoracic and abdominal cavities (minus the hepatic portal system vessels).

UROGENITAL SYSTEMS

EXERCISE 28.14A Male Urogenital System

1. Follow the instructions here to dissect a male cat. If dissecting a female cat, be sure to identify all of the structures listed in this section that are pertinent to the male cat, because knowledge of both male and female reproductive structures is required.

2. **External genitalia:** Lay the cat ventral side up on the dissecting tray and observe the pelvic and perineal regions. When the muscles of the ventral hindlimb and abdominal wall were dissected in a previous exercise, the spermatic cord and external genitalia were avoided. At this time attention will be focused on revealing these structures. Begin by observing the illustration and dissection photo in **figure 28.29**, which demonstrate

the general location of male reproductive structures (figure 28.29*a*) and their appearance once the skin around them is removed. Next, refocus attention on the dissection specimen. Palpate the skin in the perineal region to feel for the penis and testes, then look for the **external urethral orifice** to definitively identify where the penis is located. The dissection in this area is somewhat difficult because the external genitalia are completely covered in thick, furry skin. Thus, before doing any cutting, be confident of what structures lie beneath the skin so they are not destroyed in the process of dissection. Also be sure to identify the spermatic cord extending from the inguinal region into the perineal region. The **spermatic cord** contains the testicular vessels, nerves, and the ductus deferens. Follow the path of the spermatic cord deep to the skin while dissecting caudally toward the testes.

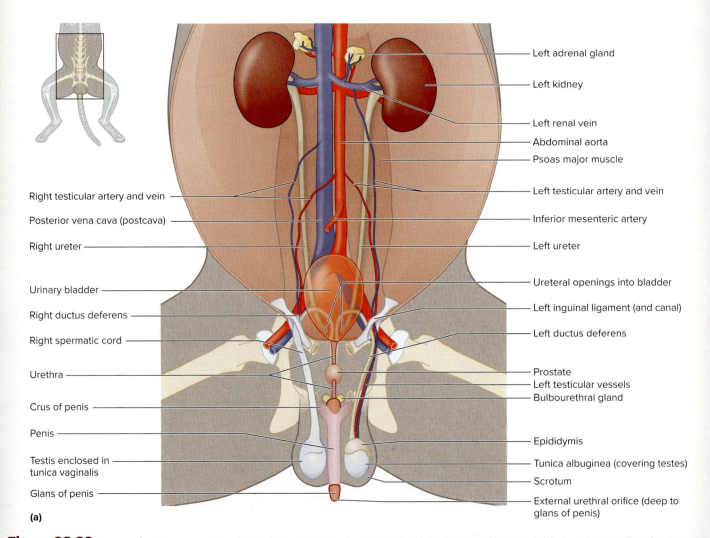

(a)

Figure 28.29 Urogenital System of the Male Cat. In both the illustration (*a*) and the dissection photograph (*b*), the pubic symphysis has been cut away to demonstrate the deep structures of the pelvis. Note in the dissection photo (*b*) that the hindlimb muscles have been removed for a better view of the male reproductive structures.

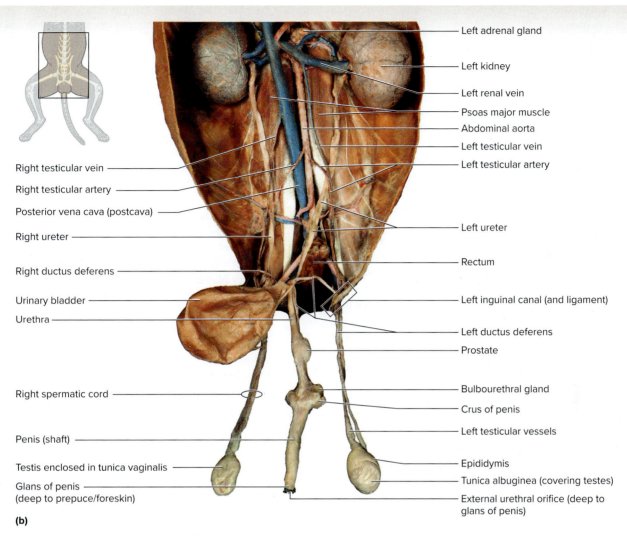

Right testicular vein
Right testicular artery
Posterior vena cava (postcava)
Right ureter
Right ductus deferens
Urinary bladder
Urethra
Right spermatic cord
Penis (shaft)
Testis enclosed in tunica vaginalis
Glans of penis
(deep to prepuce/foreskin)

Left adrenal gland
Left kidney
Left renal vein
Psoas major muscle
Abdominal aorta
Left testicular vein
Left testicular artery
Left ureter
Rectum
Left inguinal canal (and ligament)
Left ductus deferens
Prostate
Bulbourethral gland
Crus of penis
Left testicular vessels
Epididymis
Tunica albuginea (covering testes)
External urethral orifice (deep to glans of penis)

(b)

**Figure 28.29 Urogenital System of the Male Cat *(continued).* ** (*b*) Photo.
©McGraw-Hill Education/Christine Eckel

3. **Exposing the testes and penis:** Use tissue forceps to pull on the skin that remains in the perineal area. Then use a scalpel to *carefully* begin to cut the skin and the fascia that lies deep to it. Make the skin incisions next to the spermatic cord, not directly on top of it. Follow the spermatic cord inferiorly into the **scrotum** to the **testes**. Then carefully remove the skin around the testes. Superficially, the spermatic cord and scrotum are covered with connective tissue. Carefully dissect away this connective tissue around the testes and look for the "sac" that surrounds them. This is the **tunica vaginalis**, which is a remnant of the peritoneum. On one side only, use scissors to dissect away the tunica vaginalis to expose the **tunica albuginea** covering the testis and the **epididymis**. Look for a tube coming off of the epididymis. This is the **ductus deferens**. The vessels visible within the spermatic cord are the testicular artery and vein. In step 5 of the dissection we trace these structures back into the pelvic cavity, which also facilitates identification of each of these structures. Medial to the testes and slightly cranial to them is the **penis**. To dissect the penis, carefully carve the skin away from the erectile tissues,

because the skin and the penis are so tightly adhered to each other. Although a scalpel is the easiest tool to use for this task, it is sharp and easily cuts through important structures, so use extra caution with this step. Note that the tip of the penis contains the **glans**, which is covered with a **prepuce** (foreskin) just as in humans. Optional Activity: Once the penis has been identified, make a cross section through the shaft of the penis to identify the erectile tissues (corpora cavernosa and corpus spongiosum) and spongy urethra. Refer to figure 27.16 for a figure showing the homologous structures in the human for reference.

4. Identify the following structures, using figure 28.29 as a guide:

☐ **Ductus deferens**   ☐ **Spermatic cord**
☐ **Epididymis**   ☐ **Testes**
☐ **Glans of penis**   ☐ **Tunica albuginea**
☐ **Penis**   ☐ **Tunica vaginalis**
☐ **Prepuce**   ☐ **Urethra**

(continued on next page)

(continued from previous page)

5. **Internal urogenital structures:** Lay the cat ventral side up on the dissecting tray. Before opening the abdominopelvic cavity, identify the **spermatic cord** once again. Carefully tease apart the structures that compose the spermatic cord and locate the **ductus deferens**. Although it can be difficult to distinguish the testicular blood vessels from the ductus deferens, palpating the structures will demonstrate that the texture of the ductus deferens is hard compared to that of the more elastic vessels. Once the ductus deferens is identified, trace it as it extends toward the ventral abdominal wall and try to locate where it enters the *superficial inguinal ring*, extends through the *inguinal canal*, and emerges within the abdominopelvic cavity. Continue to trace the ductus deferens within the abdominopelvic cavity as it continues superficial to the **ureters** and dorsal to the **urinary bladder**, and then enters the dorsal, posterior surface of the urinary bladder. Next, locate the testicular artery and testicular vein as they arise from their respective origins on the **abdominal aorta** (testicular arteries), **posterior vena cava/postcava** (right testicular vein), and **left renal vein** (left testicular vein). Trace these vessels along the dorsal abdominal wall, through the inguinal canal, and within the spermatic cord toward the testes. It may seem odd that these vessels are so long. However, the testes begin their development in the embryo adjacent to the kidneys and then subsequently descend along the dorsal abdominal wall, through the inguinal canal, and out into the scrotum. Thus, tracing the path of the testicular vessels is similar to tracing the path the testes took during their embryonic/fetal descent.

6. Next, refocus attention on the abdominopelvic cavity and lift abdominal structures such as the small intestine out

of the way to obtain a better view of the pelvis and dorsal abdominal wall (or just remove them altogether if this was done with a previous dissection). Otherwise, pins or string may be used to hold the intestines out of the way while performing this part of the dissection. Locate the **kidneys**. Next, identify the hilum of the kidney and trace the ureters from there along the dorsal body wall to the dorsal aspect of the urinary bladder. To do this some of the **peritoneum** that covers the anterior surface of the urinary bladder and the ventral surface of the ureters may need to be removed (recall that urinary structures are all retroperitoneal). This is best accomplished by gently pulling on the peritoneum with tissue forceps. Once the peritoneum has been removed, clean the tissue around the urinary bladder and abdominopelvic wall. Using scissors, make a longitudinal incision in the ventral wall of the urinary bladder to view inside this organ. Within the urinary bladder, note the many folds of the walls, called **mucosal folds** (or *rugae*; similar to the folds within the stomach). Then, locate the openings of the two ureters, the opening of the urethra, and the triangular space delineated by the three openings: the **urinary trigone**. Does the wall of the urinary bladder look different in the region of the trigone?

_____

If so, describe its appearance. _____

_____

_____

7. Locate the **urethra** as it emerges posterior to the urinary bladder (figure 28.29*b*). Next, look for a small, round gland surrounding the urethra just caudal to the urinary bladder: the **prostate gland**. Then trace the urethra caudally until it enters the corpus spongiosum of the penis. Adjacent to this location, look for the **crus of the penis**. The crus ("legs") are extensions of the corpora cavernosa of the penis into the perineal region. They are covered with muscle, but palpating them will demonstrate that they feel a little "spongy" to the touch because of the erectile tissues within. Adjacent to the crus of the penis, locate the two small, paired **bulbourethral glands**. These can be difficult to locate, so do not worry if they cannot be located.

INTEGRATE

CLINICAL VIEW
Urinary Blockage in Male Cats

Note the relative length of the urethra in the male cat as compared to the urethra of a human male. In particular, note the relative length of urethra extending between the bladder and the prostate gland in the cat, which is different from that in a human. One problem some male cats develop over time is a blockage that prevents urine from passing through the urethra when uric acid crystals form. (Uric acid crystals are similar to kidney stones that form in humans.) The crystals must travel through a very long and relatively narrow urethra, which predisposes them to clogging the urethra and blocking the flow of urine if the crystals become large and numerous. Treatment for this condition involves catheterization to clear the blockage; a change in diet to prevent the formation of uric acid crystals; and in very severe cases (particularly when reblockage occurs), removal of the penis to shorten the overall length of the urethra so as to prevent recurrence of the problem.

INTEGRATE

LEARNING STRATEGY

As you trace the ductus deferens within the pelvic cavity, you will see that it travels over the ureters en route to its entry into the posterior, inferior surface of the urinary bladder. When you are observing both the ureters and the ductus deferens, you can remember which structure is which by thinking of the following: "Water flows under the bridge." That is, the tubes containing the "water" (the ureters) travel inferior to the "bridge" (the ductus deferens).

8. Identify the following structures, using figure 28.29 as a guide:

- ☐ **Bulbourethral glands**
- ☐ **Crus of penis**
- ☐ **Ductus deferens**
- ☐ **Glans of penis**
- ☐ **Prostate gland**
- ☐ **Shaft of penis**
- ☐ **Testicular arteries**
- ☐ **Testicular veins**
- ☐ **Ureters**
- ☐ **Urethra**
- ☐ **Urinary bladder**

EXERCISE 28.14B Female Urogenital System

1. Follow the instructions here to dissect a female cat. If dissecting a male cat, be sure to identify all of the structures listed in this section that are pertinent to the female cat, because knowledge of both male and female reproductive systems is required.

2. External genitalia: Lay the cat ventral side up on the dissecting tray and observe the perineal region of the female (**figure 28.30a**). Just inferior to the tail is the *vulva*, which surrounds the opening to the urogenital sinus. The **urogenital sinus** is unique in the cat as compared to the human vulva, in that both the urethra and the vagina open into the urogenital sinus in the cat.

3. **Internal urogenital structures:** To reach the internal urogenital structures of the female cat, first dissect through the **pubic symphysis**. Observe figure 28.3 (cat skeleton) before beginning the dissection to have an idea of how the pubic symphysis of the cat relates to the muscles of the perineal region. Obtain a pair of bone cutters, tissue forceps, and a scalpel. Using the tissue forceps and scalpel, cut through the muscles of the medial thigh ~0.5 cm to either side of the midline (where the pubic symphysis lies) to remove the muscular attachments adjacent to the pubic symphysis. Next, taking care not to go too deep and cut the urogenital sinus, use the bone cutters to cut through the pubic symphysis. Parts of the pubic bone and pubic symphysis will likely end up being removed in several small pieces. The main goal is to open up a wide enough "channel" through the mid-pubic region to view the urogenital sinus.

4. Once the urogenital sinus has been exposed, use scissors to carefully cut open the last centimeter of the urogenital

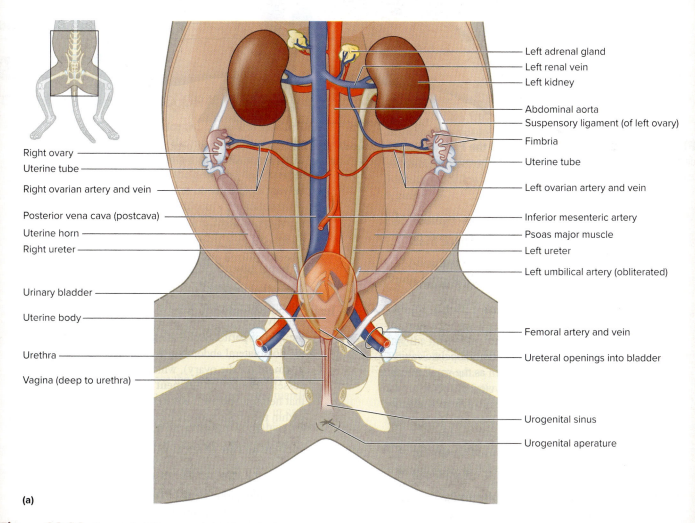

Left adrenal gland
Left renal vein
Left kidney
Abdominal aorta
Suspensory ligament (of left ovary)
Fimbria
Uterine tube
Left ovarian artery and vein
Inferior mesenteric artery
Psoas major muscle
Left ureter
Left umbilical artery (obliterated)
Femoral artery and vein
Ureteral openings into bladder
Urogenital sinus
Urogenital aperature

Right ovary
Uterine tube
Right ovarian artery and vein
Posterior vena cava (postcava)
Uterine horn
Right ureter
Urinary bladder
Uterine body
Urethra
Vagina (deep to urethra)

(a)

Figure 28.30 **Urogenital System of the Female Cat.** (*a*) Illustration.

(continued on next page)

(continued from previous page)

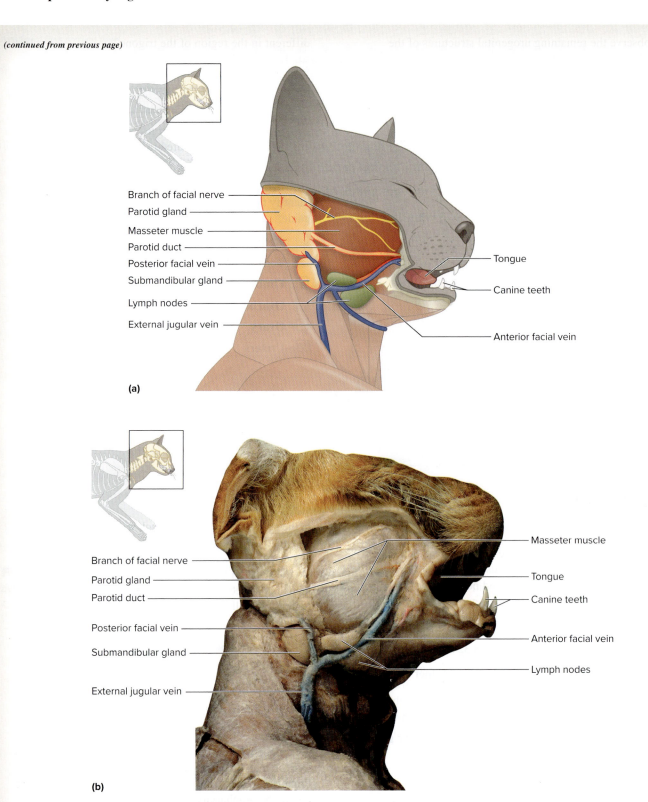

(a)

(b)

Figure 28.31 **Head, Neck, and Oral Cavity.** (*a*) Illustration showing structures of the head and neck, including the teeth. (*b*) Photo showing most of the structures in (*a*). All teeth are not visible.

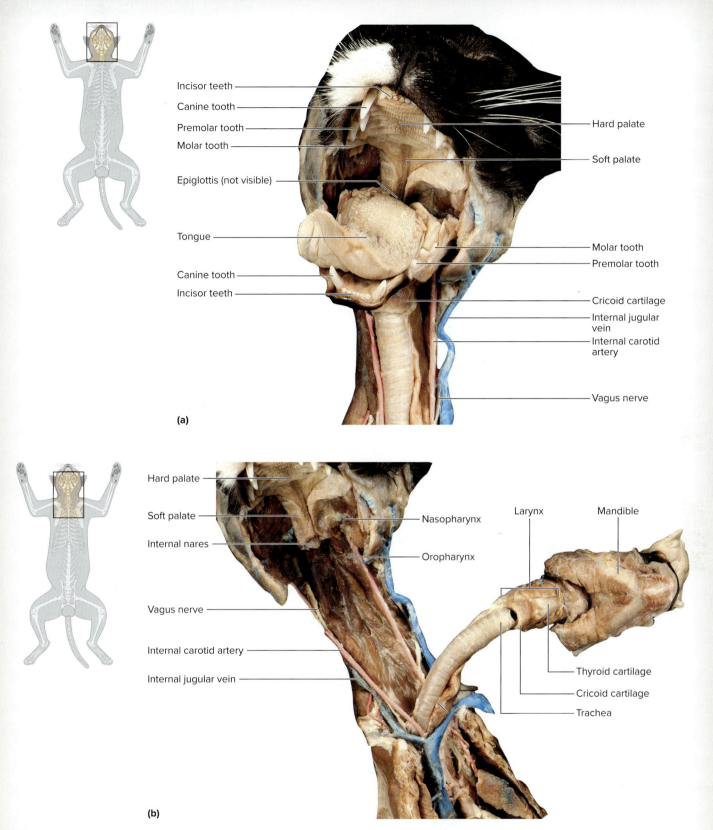

(a)

(b)

Figure 28.32 Deep Dissection of the Oral Cavity. Photos showing deep structures of the oral cavity of the cat. The thyroid gland has been removed. (*a*) Lower jaw in normal anatomic position. (*b*) Lower jaw, larynx, trachea, and esophagus displaced to the side for better view of deep structures.

©McGraw-Hill Education/Christine Eckel

The ❶ corresponds to the Learning Objective(s) listed in the chapter opener outline.

Do You Know the Basics?

Exercise 28.1: Directional Terms and Surface Anatomy

1. When handling a preserved specimen, which of the following is an *essential* piece of safety equipment that should be used every time you handle a specimen? (Check all that apply.) ❶

   _____ a. laboratory coat

   _____ b. nose plugs

   _____ c. protective gloves

   _____ d. protective mask

   _____ e. safety goggles

2. Match the appropriate description in column A with the surface anatomy structure of the cat listed in column B. ❷

   Column A

   _____ 1. ears

   _____ 2. eye membrane

   _____ 3. nasal openings

   _____ 4. whiskers

   Column B

   a. external nares

   b. nictitating membrane

   c. pinna

   d. vibrissae

3. Label the figure using appropriate directional terms. ❸

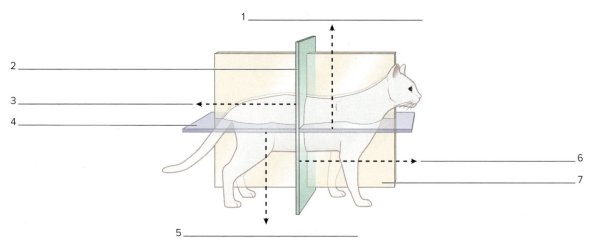

1 _____

2 _____

3 _____

4 _____

5 _____

6 _____

7 _____

4. Which of the following is the *surface anatomy* landmark that can be used to distinguish a female from a male cat? (Circle one.) ❹

   a. mammary glands   b. mammary ridge   c. vagina   d. vestibule

5. The integument structures present on a cat but not on a human are _____ (whiskers/mammary glands). ❺

Exercise 28.2: Skeletal System

6. The cat has _____ (7/12) vertebrae in the cervical region. ❻

7. In the cat, the _____ (clavicle/humerus) does not articulate with the scapula. ❼

Exercise 28.3: Skinning the Cat

8. The muscle with fibers within the skin of the anterior neck is the _____ (cutaneous maximus/platysma). ❽

9. Structures that appear as tight bands that run between the body wall and the skin and carry sensory information from receptors in the skin to the central nervous system are known as _____ (fascia/cutaneous nerves). ❾

Exercise 28.4: Muscles of the Head and Neck

10. Label the following photo of the head and neck muscles of the cat. **10**

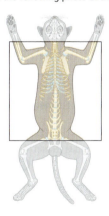

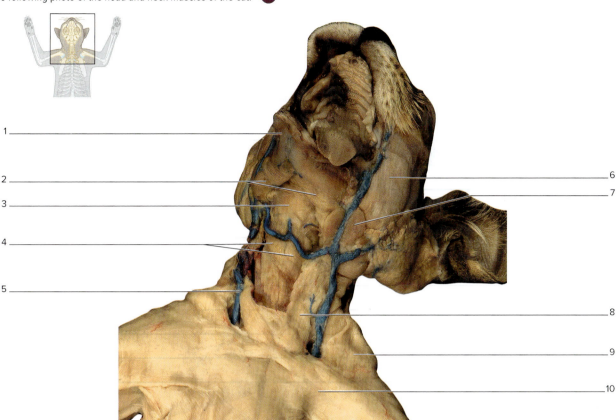

1 _____

2 _____

3 _____

4 _____

5 _____

6 _____

7 _____

8 _____

9 _____

10 _____

©McGraw-Hill Education/Christine Eckel

Exercise 28.5: Muscles of the Thorax

11. Label the following photo of the superficial thoracic muscles of the cat. **11**

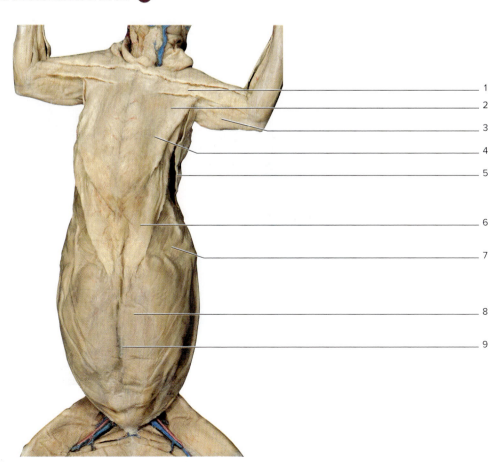

1 _____

2 _____

3 _____

4 _____

5 _____

6 _____

7 _____

8 _____

9 _____

©McGraw-Hill Education/Christine Eckel

24. Which of the following is the equivalent vessel to the superior vena cava in the human? (Circle one.) **24**

    a. aorta

    b. postcava

    c. precava

    d. pulmonary trunk

25. Similar to humans, the trachea in the cat is _____ (anterior/posterior) to the esophagus. **25**

Exercise 28.14: The Abdominal Cavity

26. Place the following digestive structures in sequential order beginning with the mouth (1 = first; 9 = last). **26**

    _____ a. duodenum   _____ f. large intestine

    _____ b. esophagus   _____ g. oral cavity

    _____ c. ileocecal valve   _____ h. rectum

    _____ d. ileum   _____ i. stomach

    _____ e. jejunum

27. The two major veins that form the hepatic portal vein in the cat are the gastrosplenic vein and the _____ (superior(cranial)/inferior(caudal)) mesenteric vein. **27**

Exercise 28.15: The Dorsal Abdominal Wall

28. Which of the following structures extend from the hilum of the kidney? (Check all that apply.) **28**

    _____ a. hepatic vein

    _____ b. renal artery

    _____ c. renal vein

    _____ d. ureter

Exercise 28.16: Urogenital Systems

29. Label the photo of the male urogenital system of the cat. **29** **30** **31** **32**

©McGraw-Hill Education/Christine Eckel

30. Label the photo of the female urogenital system of the cat.   29 30 31 32

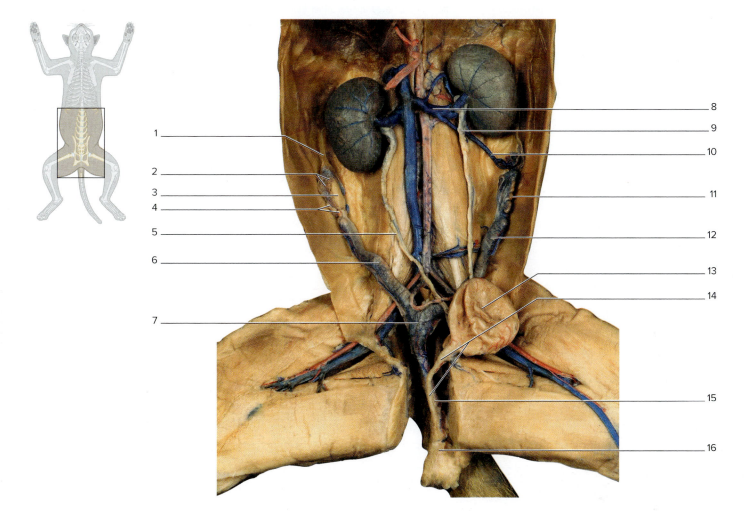

Exercise 28.17: The Head, Neck, and Oral Cavity

31. Which of the following covers the opening to the larynx in both humans and cats? (Circle one.) 33

    a. epiglottis

    b. glottis

    c. hyoid bone

    d. thyroid cartilage

Can You Apply What You've Learned?

32. Discuss two adaptations to the cat skeletal and muscular systems that allow for quadrupedal rather than bipedal locomotion.

   _____

   _____

33. Identify the nerves that branch from the brachial plexus to innervate the muscles of the cat's upper limb.

   _____

   _____

34. Identify the nerves that branch from the lumbosacral plexus to innervate the muscles of the cat's lower limb.

   _____

   _____

35. Compare hepatic portal circulation in humans and cats.

   _____

   _____

   _____

Can You Synthesize What You've Learned?

36. Trace the flow of urine from the ureters to the urethra in the male cat and in the female cat, identifying all structures along the path. Compare this path to that in humans.

   _____

   _____

   _____

   _____

37. Compare the size and structure of the uterine horn in the female cat with that of the fallopian tube in humans. Discuss why there is a dramatic difference between cats and humans.

   _____

   _____

   _____

   _____

Design Elements: Integrate: Clinical View icon (clipboard): ©Laia Design Studio/Shutterstock.com; Integrate Learning Strategy (pencil): ©Slavoljub Pantelic/Shutterstock.com

Fetal Pig Dissection Exercises

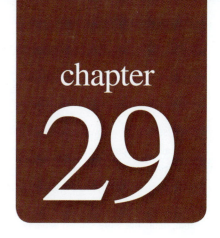

chapter
29

OUTLINE AND LEARNING OBJECTIVES

GROSS ANATOMY   852

Overview   852

EXERCISE 29.1: DIRECTIONAL TERMS AND SURFACE ANATOMY   852

1. Describe the proper handling and care of a preserved dissection specimen
2. Identify important surface anatomy landmarks in the pig
3. Note similarities and differences in directional terms as they concern the human and the pig
4. Determine the sex of the pig
5. Note modifications of the integument of a pig as compared to a human

Back and Limbs   854

EXERCISE 29.2: SKELETAL SYSTEM   857

6. Identify the major bones of the pig skeleton
7. Identify developing bones in an embryonic pig skeleton
8. Compare and contrast the skeletal system of the pig with that of the human

EXERCISE 29.3: SKINNING THE PIG   857

9. Observe the cutaneous nerves and vessels in the pig

EXERCISE 29.4: MUSCLES OF THE HEAD AND NECK   859

10. Identify the major head and neck muscles of the pig and compare them to those of a human

EXERCISE 29.5: MUSCLES OF THE THORAX   861

11. Identify the major thoracic muscles of the pig and compare them to those of a human

EXERCISE 29.6: MUSCLES OF THE ABDOMINAL WALL   863

12. Identify the major abdominal muscles of the pig and compare them to those of a human
13. Observe the fiber orientation of the abdominal muscles and note the similarities with human abdominal muscles

EXERCISE 29.7: MUSCLES OF THE BACK AND SHOULDER   864

14. Identify the major muscles located in the back and shoulder regions of the pig and compare them to those of a human

EXERCISE 29.8: MUSCLES OF THE FORELIMB   867

15. Identify the major forelimb muscles of the pig and compare them to those of a human

EXERCISE 29.9: MUSCLES OF THE HINDLIMB   870

16. Identify the major hindlimb muscles of the pig and compare them to those of a human
17. List the three muscles that compose the triceps surae muscle in both pigs and humans

EXERCISE 29.10: NERVOUS SYSTEM: SPINAL CORD AND PERIPHERAL NERVES   875

18. Identify nerves of the brachial plexus of the pig and compare them to those of a human

19. Identify nerves of the lumbosacral plexus of the pig and compare them to those of a human

EXERCISE 29.11: CARDIOVASCULAR SYSTEM: PERIPHERAL BLOOD VESSELS   879

20. Identify the major blood vessels of the forelimb and hindlimb of the pig and compare them to those of a human

Thoracic Cavity   884

EXERCISE 29.12: OPENING THE THORACIC AND ABDOMINAL CAVITIES   884

21. Prepare the thoracic and abdominal cavities for observation

EXERCISE 29.13: THE HEART, LUNGS, AND MEDIASTINUM   886

22. Identify the major structures of the pig heart and compare them to those of a human
23. Identify pulmonary structures of the pig and compare them to those of a human
24. Identify the major thoracic blood vessels of the pig and compare them to those of a human
25. Identify the thoracic duct, esophagus, and trachea/bronchi in the pig thorax and compare them to those of a human

Abdomen and Pelvis   889

EXERCISE 29.14: THE ABDOMINAL CAVITY   889

26. Identify digestive system structures of the pig and compare them to those of a human
27. Identify the major blood vessels of the abdominal cavity of the pig and compare them to those of a human

EXERCISE 29.15: THE DORSAL ABDOMINAL WALL   892

28. Identify structures of the dorsal abdominal wall (kidneys, adrenal glands, ureters) of the pig and compare them to those of a human

EXERCISE 29.16: UROGENITAL SYSTEMS   898

29. Identify the allantoic bladder and urethra on male and female pigs and compare them to those of a human
30. Identify external genitalia on both male and female pigs and compare them to those of a human
31. Identify accessory reproductive organs in both male and female pigs and compare them to those of a human
32. Identify the primary reproductive organs (testes and ovaries) on male and female pigs, respectively, and compare them to those of a human

Head and Neck   904

EXERCISE 29.17: THE HEAD, NECK, AND ORAL CAVITY   904

33. Identify structures of the oral cavity of the pig and compare them to those of a human

Anatomy & Physiology Revealed® 4.0

Module 2: Cells & Chemistry

849

INTRODUCTION

D issection is a powerful tool in learning anatomy and provides an unparalleled experience of discovering the beauty of the human/animal form. Dissection provides the opportunity to see the real size of structures and allows one to actually *feel* the texture of the tissues and organs! What's more, it allows investigation of the relationships between anatomical structures that, until now, were studied separately from each other. Although the process of dissection may cause apprehension at first, this feeling generally dissipates with experience.

The exercises in this chapter cover the dissection of a vertebrate mammal that has anatomy similar to the human: the fetal pig (*Sus scrofa domesticus*). The beauty of this comparative vertebrate anatomy is that it allows one to realize just how much all vertebrates have in common. For example, although a fetal pig has a different posture from that of a human, observation of the bones demonstrates that there is remarkable consistency in structure and function between the bones of the fetal pig and those of the human. The exercises in this chapter require the dissector to take a regional approach and to focus on the relationships among the organs and the blood and nerve supplies to those organs at the same time. This regional approach stresses the interconnectedness of the systems in a much deeper way than learning each organ system individually.

List of Reference Tables

These Pre-Laboratory Worksheet questions may be assigned by instructors through their ⬛ connect° course.

1. The dorsal surface of a fetal pig is the same as the _____ (posterior/superior) surface of a human, whereas the caudal surface of a fetal pig is the same as the _____ (inferior/posterior) surface of a human.

2. The anterior surface of a fetal pig is the same as the _____ (cranial/ventral) surface of a human, whereas the inferior surface of a fetal pig is the same as the _____ (caudal/ventral) surface of a human.

3. In a fetal pig, a transverse plane separates _____ (anterior/superior) portions from _____ (inferior/posterior) portions.

4. In a fetal pig, a frontal plane separates _____ (anterior/dorsal) portions from _____ (posterior/ventral) portions.

5. Superficially, a male fetal pig can be distinguished from a female fetal pig by the presence of a _____ (genital papilla/scrotum), whereas the female fetal pig is identified by the presence of a _____ (genital papilla/scrotum).

6. When skinning the pudendal region of the male fetal pig, which structures may be at risk for damage? (Check all that apply.)

    _____ a. epididymis

    _____ b. penis

    _____ c. spermatic cord

    _____ d. testis

    _____ e. vestibule

7. Which of the following dissection techniques is most likely to allow for skinning the fetal pig while protecting underlying structures? (Circle one.)

    a. blunt dissection

    b. dissection with scissors

    c. dissection with a scalpel

    d. open scissors technique

8. When handling preserved organisms, one should wear _____. (Circle all that apply.)

    a. aprons or lab coats

    b. gloves

    c. safety glasses

    d. open-toed shoes

9. In the fetal pig, which of the following circulatory structures are present? (Check all that apply.)

    _____ a. ductus arteriosus

    _____ b. round ligament of the liver

    _____ c. umbilical vein

    _____ d. ligamentum arteriosum

    _____ e. umbilical arteries

10. When storing dissected specimens between laboratory sessions, organisms like fetal pigs should always be _____ (wrapped in skin in the storage bag/placed in a storage bag filled with embalming fluid).

GROSS ANATOMY

Overview

CLINICAL VIEW
Proper Handling and Care of Preserved Fetal Pigs

⚠️ The dissection specimens under study in the laboratory are embalmed, or preserved, to enable safe dissection without risk of exposure to necrotic (dead) tissue or harmful microorganisms. Before beginning, review table 1.6 in chapter 1, which describes the composition of embalming fluids, safe handling of embalming fluids, and safety procedures for dissection. Some specimens may have been preserved with "biosafe" solutions, which generally do not contain the fixative formalin, a substance that can be harmful if handled incorrectly. The dissection specimens used in the lab may or may not have been preserved using formalin. If uncertain, it is safest to follow the same precautions you used for formalin-preserved specimens. That is, always wear gloves while handling the specimen, wear safety glasses when there is a risk of fluids splashing from the specimen, and never touch your face or eyes with your hands if they may have preservative fluid on them. You may also be required to wear a dissecting apron or lab coat and closed-toed shoes. If the solution gets in your eyes, rinse them out thoroughly at the eyewash station in the laboratory. Finally, before leaving the laboratory for the day, thoroughly clean the work area and be sure to wash your hands and forearms thoroughly with soap and warm water.

INTEGRATE

LEARNING STRATEGY

Figures in this chapter are color-coded so that common structures can be easily identified. Refer back to this color key often while studying the anatomical structures of the fetal pig.

Color Key for Common Structures

| Arteries | Cartilage | Nerves | Pulmonary trunk | Tendons, Ligaments | Ureters |
| Bone | Muscle | Portal vessels | Salivary glands | Thymus and Adrenal glands | Veins |

EXERCISE 29.1

DIRECTIONAL TERMS AND SURFACE ANATOMY

Before beginning dissection of the fetal pig, be sure to read the Clinical View on this page on the proper handling and care of preserved fetal pigs.

Figure 29.1 **Directional Terms.** Note how directional terms in the pig (a) differ from the same terms in a human (b) because the pig stands on four legs whereas the human stands on only two.
©Eric A. Wise

Directional Terms

One of the major differences between the anatomical nomenclature of a pig and a human has to do with directional terms. This is because the normal anatomical position for a pig has all four limbs on the ground (**figure 29.1a**). Humans, on the other hand, have a normal anatomical position with only the lower limbs on the ground (figure 29.1b).

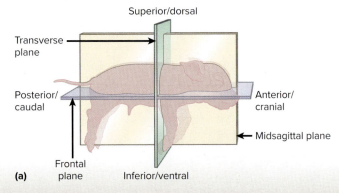

Superior/dorsal

Transverse plane

Posterior/caudal — Anterior/cranial

Midsagittal plane

Frontal plane — Inferior/ventral

(a)

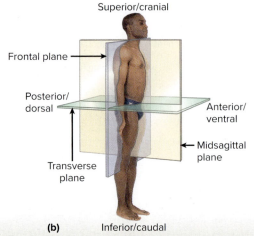

Superior/cranial

Frontal plane

Posterior/dorsal — Anterior/ventral

Transverse plane — Midsagittal plane

(b) — Inferior/caudal

Obtain the Following:
- dissecting tools
- dissecting tray
- gloves
- preserved fetal pig
- storage bag
- paper towels
- wetting solution (for keeping the specimen moist)

1. Lay the pig on the dissecting tray with its dorsal side up and start by observing its surface anatomy (**figure 29.2**). On the head, make note of the **rostrum** (snout), **pinnae** (earlobes), **hooves**, and **umbilical cord**. Because the pig is a fetal (unborn) pig, its circulatory system will demonstrate the fetal condition rather than the adult condition. That is, the umbilical cord has two umbilical arteries that send blood to the placenta to pick up oxygenated blood and one umbilical vein to return oxygenated blood to the fetal pig's circulatory system. This topic will be covered in more detail in the discussion of the circulatory system, but it is something to keep in mind while completing the dissections.

2. Turn the pig over so it is ventral side up. Observe the neck and inguinal regions for any sign of incisions. Unless a special injection was used, there should not be any skin incisions, because injections of embalming fluid and any other substances into the circulatory system are likely to have been made through the umbilical vessels. If there is an incision on the abdomen, the pig may have had its renal system injected with colored latex.

3. Observe the skin adjacent to the midline of the ventral thorax and abdomen and look for the mammary ridge, which contains a series of **mammary glands** (in a pregnant adult female, not in a fetal pig) and **mammary papillae** (nipples). In a pregnant pig, these ridges would be raised because of the growth and development of mammary glands in the subcutaneous tissues. These stuctures come off with removal of the skin.

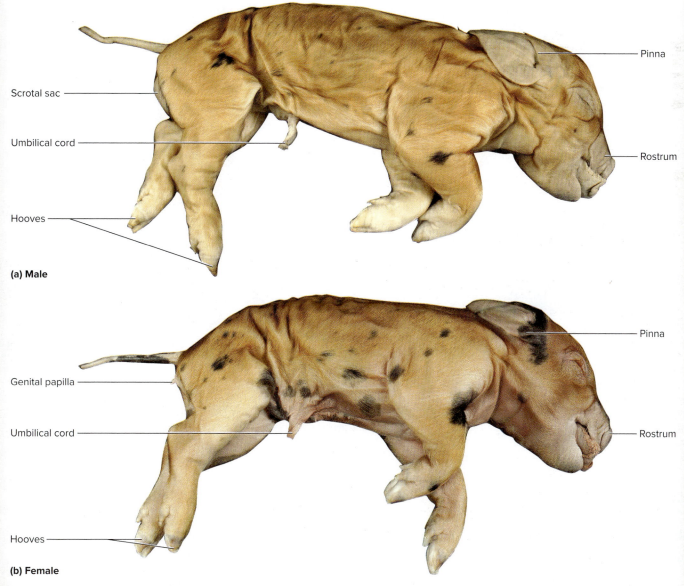

(a) Male

(b) Female

Figure 29.2 **Surface Anatomy of the Pig.** (*a*) Male pig, lateral view. (*b*) Female pig, lateral view.
©McGraw-Hill Education/Christine Eckel

(continued on next page)

(continued from previous page)

4. Determine the sex of the pig. To do this, observe the area just ventral to the tail of the pig. If the specimen is a **female pig**, there will be a small opening, the **urogenital opening**, which might be partially obscured by a **genital papilla** (figures 29.2 and **29.3**). The urogenital opening, as its name suggests, is an opening into both the urinary and the reproductive system structures of the female pig. If the specimen is a **male pig**, there may be the beginnings of a scrotum ventral to the tail (although if it is too young, the scrotum may not be visible because the testes have not yet descended). Male pigs also have a small raised area just caudal to the umbilical cord, the **urogenital opening**. The urogenital opening contains the male urethra, which will be the passageway for urine and sperm (although not at the same time) in the adult male pig. If the pig is a male, take extra care when removing the skin in the caudal region of the abdomen, so as not to destroy structures such as the **spermatic cord**, **testes**, or **penis**, which will be in early stages of development.

5. Record the sex of the pig in the space provided and make note of any pertinent incisions and/or surface anatomy observations that may be important. If the pig has a unique identification number, be sure to record that number as well.

6. Observe a specimen of the opposite sex so you are able to identify external genitalia of both a male and a female fetal pig.

Sex of Pig: _____

Pig ID: _____

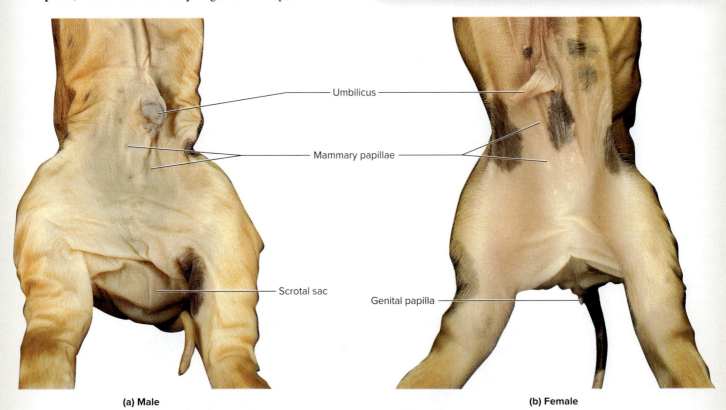

(a) Male   (b) Female

Figure 29.3   **External Genitalia.**   Ventral view of the pelvis of male and female fetal pigs. Note the scrotal sac in the male (*a*) and the genital papilla in the female (*b*). The opening to the female urogenital sinus lies anterior to the genital papilla.
©McGraw-Hill Education/Christine Eckel

Back and Limbs

Skeletal System

Because the skeleton of the fetal pig is still undergoing ossification, it is not possible to show a representative fetal pig skeleton in full. Instead, observe the diagram and x-ray of the fetal pig skeleton in **figure 29.4**, which illustrates the four-legged posture and relative locations of axial and appendicular bones. This figure will be particularly helpful when identifying the muscles of the fetal pig and drawing parallels between fetal pig and human structures. An embryonic pig skeleton, which is composed primarily of hyaline cartilage, will also be observed for comparison with the fetal pig.

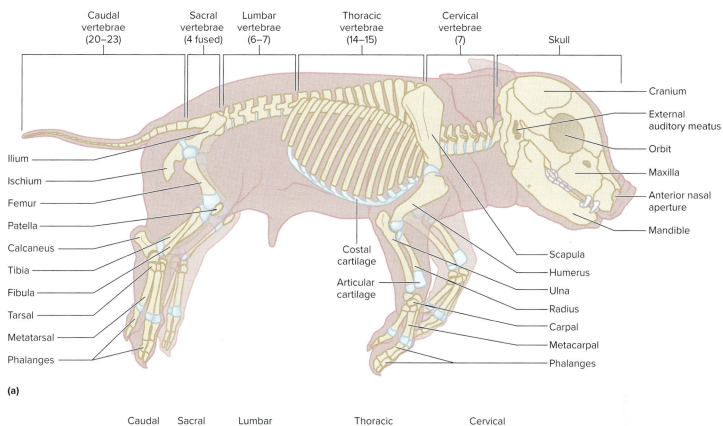

(a)

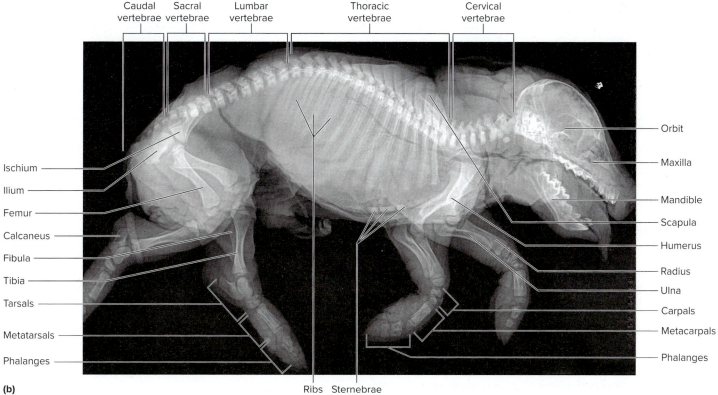

(b)

Figure 29.4 **The Pig Skeleton.** (*a*) Illustration of the skeleton of a fetal pig, lateral view. (*b*) X-ray image of a fetal pig, lateral view.

(*b*) ©McGraw-Hill Education/Christine Eckel

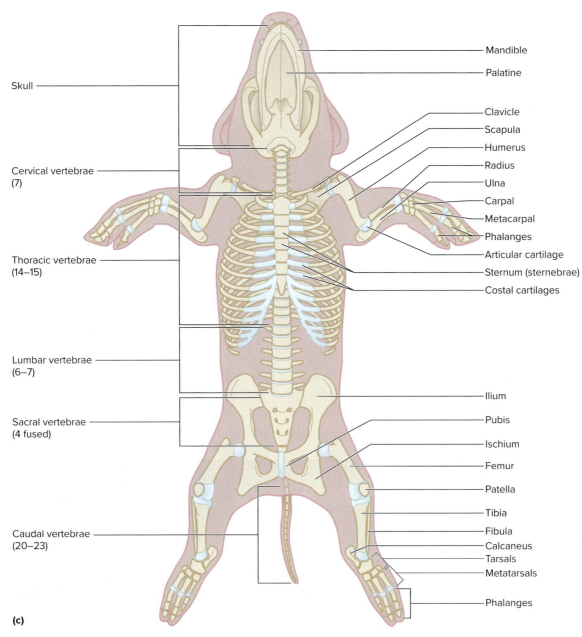

Skull

Cervical vertebrae
(7)

Thoracic vertebrae
(14–15)

Lumbar vertebrae
(6–7)

Sacral vertebrae
(4 fused)

Caudal vertebrae
(20–23)

Mandible

Palatine

Clavicle

Scapula

Humerus

Radius

Ulna

Carpal

Metacarpal

Phalanges

Articular cartilage

Sternum (sternebrae)

Costal cartilages

Ilium

Pubis

Ischium

Femur

Patella

Tibia

Fibula

Calcaneus

Tarsals

Metatarsals

Phalanges

(c)

Figure 29.4   The Pig Skeleton (*continued*). (c) Illustration of the fetal pig skeleton, ventral view.

INTEGRATE

CONCEPT CONNECTION

When observing the fetal pig skeleton, pay particular attention to the bones of the distal portions of the limbs. The bones that compose hands (metacarpals and phalanges) and feet (metatarsals and phalanges) of humans are elongated in pigs. Pigs in essence, walk on the tips of their phalanges, and their hooves are homologous to human fingernails. When a human is standing, the metatarsals lie flat on the ground. When a pig is standing, the metatarsals are lifted off the ground. It is only when the pig lies down that the metacarpals and metatarsals are able to lie flat on the ground.

EXERCISE 29.2

SKELETAL SYSTEM

Observe the diagrams and x-ray image of a fetal pig skeleton (figure 29.4a–c). Note the similarities and differences between the fetal pig skeleton and the human skeleton (figure 7.6) in the space provided.

_____

_____

_____

_____

_____

_____

Beginning the Dissection

The first step before dissection of the pig muscles is to remove the skin from the pig. Before begining this exercise, check with the instructor to determine how he or she wants the fetal pig skinned, because it may be different from the instructions provided here. When skinning the fetal pig, begin on one side, either dorsal or ventral, and proceed to remove the skin from the trunk, the neck, and all four limbs down to the hooves. The description given here is for skinning a fetal pig starting on the ventral surface. Remove the skin completely and dispose of it properly, most likely in a biohazard waste bag. When the dissection for the day is complete, wrap the pig in moist paper towels or cloth and put it in a plastic bag. Keeping the fetal pig wrapped and in an airtight bag during storage helps to keep the pig from drying out between laboratory sessions. When performing dissections with the pig supine (ventral side up), secure the forelimbs and hindlimbs to the dissecting pan with string.

EXERCISE 29.3

SKINNING THE PIG

1. **Midline incision:** Before begining, follow the proper incision guides in **figure 29.5** for the specimen, because the incisions are slightly different for male and female pigs. Place the pig supine on the dissecting tray (figure 29.5). If the pig has an embalming/injection incision, begin the incision there, because it will be easier. If there is no incision in the pig, then use forceps to gently lift some of the skin away from the sternum. Using scissors or a scalpel, make a small, vertical incision, taking care not to cut the underlying muscles (figure 29.5, step 1, and figure 1.11). While proceeding with this incision, take care to keep tension on the skin. This causes the underlying fascia to stretch and pull away from the muscles. The target of the dissection instrument is the skin and underlying fascia, not the muscles. After making the initial incision, use blunt dissection (see figure 1.16) or the open scissors technique (see figures 1.14 and 1.15) to disengage the skin from the underlying muscles.

2. **Perineal incision:** Next, extend the cut along the midline from the chin to the umbilicus (figure 29.5, step 2). When approaching the umbilical cord, take care to cut around the urogenital opening in male pigs (figure 29.5a). In female pigs, cut a circle around the umbilicus and then continue the vertical incision all the way to the genital papilla (figure 29.5b, step 3). In male pigs, extend the cuts around the umbilicus and urogenital opening up to the skin adjacent to the anus (figure 29.5a, step 3). Then carefully connect the two incision lines cranial to the anus.

(continued on next page)

(continued from previous page)

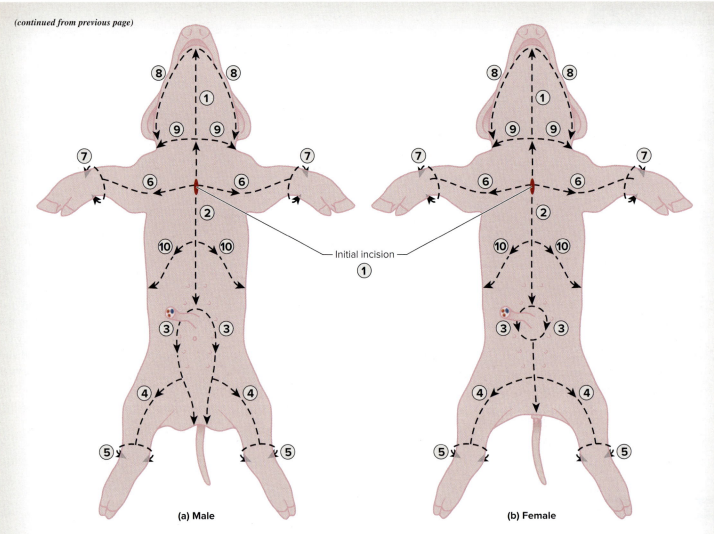

Figure 29.5   Ventral Surface Incisions of the (a) Male Pig and (b) Female Pig. If the pig has an incision in the throat from embalming/injecting, start the incision there and progress in the numbered sequence as shown.

3. **Hindlimb incisions:** Now begin an incision from the midline of the perineum to the ankle of the hindlimb (figure 29.5, step 4). The skin is both thinner and more tightly adhered to the underlying tissues here. To avoid damaging underlying structures, start by pushing a probe under the skin (see figure 1.12). Then make the incision superficial to the probe. When at the ankle, carefully extend the cut in a circle around the ankle to free it from the limb (figure 29.5, step 5). Beware: It is very easy to cut too deep in this area and damage important tendons, nerves, and blood vessels. Remove the skin from both hindlimbs.

4. **Forelimb incisions:** Begin an incision from the midline of the thorax to the wrist of the forelimb (figure 29.5, step 6). As with the skin on the hindlimbs, the skin is both thinner and more tightly adhered to the underlying tissues here. To avoid damaging underlying structures, start by pushing a probe under the skin (see figure 1.12). Then make the incision superficial to the probe. When at the wrist, carefully extend the cut in a circle around the wrist to free it from the limb (figure 29.5, step 7). Beware: It is very easy to cut too deep in this area and damage important tendons, nerves, and blood vessels. Remove the skin from both forelimbs.

5. **Neck in cisions:** Start with the midline incision below the chin. Extend the incision along the inferior border of the mandible toward the pinna of the ears (figure 29.5, step 8). Here again the skin is very tightly adhered to the underlying tissues. Thus, use a probe when necessary to avoid cutting into underlying structures. Continue the incision dorsally until it encircles the back of the neck (figure 29.5, step 9). The skin over the face and back of the head will be left intact.

6. **Removing the skin:** Once the neck incision is completed, the entirety of the skin can be removed in one piece. While dissecting the skin away from the trunk, look for several small, tight bands that run between the body wall and the skin. There may even be tiny blood vessels that run next to them if the pig has been injected with colored latex. The whitish structures are cutaneous nerves, which are branches of underlying mixed spinal nerves (these will be covered in more detail later). Cutaneous nerves carry sensory information from receptors in the skin to the central nervous system.

7. Once the skin is completely free from the body, dispose of it in the proper waste receptacle.

Muscular System

Just as with the skeletal system, the muscular system of the pig bears a remarkable resemblance to the human muscular system. Largely, the differences in muscles have resulted from adaptation of the pig's muscular system for four-legged locomotion and adaptation of the human muscular system for two-legged locomotion.

Before beginning the dissection, review the overall organization of the human muscular system presented in chapter 11, paying particular attention to major muscle groups. Note that there are muscle groups in the pig that are similar to those in humans. Thus, taking a similar approach to learning the muscles of the fetal pig will greatly assist the learning process. Information on the name, origin, insertion, and action of each muscle is contained in tables 29.1 through 29.6. These tables also make note of muscles that are different in the pig from those in the human. The muscles of the pig are different because either they are not found in the human or they are named slightly differently in the pig.

Recall that one of the best ways to remember the action of a muscle is to learn its attachment points (do not worry about distinguishing which is origin and which is insertion), and then to consider the action that is performed when the muscle shortens. In addition to the directional terms that are most commonly used in humans (see table 2.2), there are also references to movement that is **cranial** or **caudal** as concerns actions of pig muscles. To move a structure cranially is to move anteriorly and toward the head of the pig. To move a structure caudally is to move posteriorly and toward the tail of the pig.

The muscles of the fetal pig are presented here as a series of exercises that look at muscles in one particular region of the body at a time. After identifying a muscle, refer to the muscle tables to review that muscle's origin, insertion, and action. Also consider how the size, location, and function of the muscle compare with those qualities of the homologous muscle in a human. The majority of the time these will be the same. Those who already know the human muscular system can give extra attention to the muscles of the pig that are considerably different.

EXERCISE 29.4

MUSCLES OF THE HEAD AND NECK

1. **Preparation:** Place the pig ventral side up on the dissecting tray. Begin by observing the muscles of the anterior neck. A superficial dissection will be performed on one half of the neck, and a deep dissection will be peformed on the other to observe underlying muscles. Some of the superficial muscles may have been damaged by the embalming/injecting incision. If that is the case, use the injected side for the deep dissection and the other

for the superficial dissection **(figure 29.6)**. Look for a large vein on both sides of the neck that lies superficial to the muscles. This is the **internal jugular vein**. Next to this vein are several small structures the shape and size of kidney beans. These are **lymph nodes**. While dissecting and cleaning the muscles, take care to preserve the internal jugular vein and the lymph nodes for further study in the context of the cardiovascular and lymphatic systems.

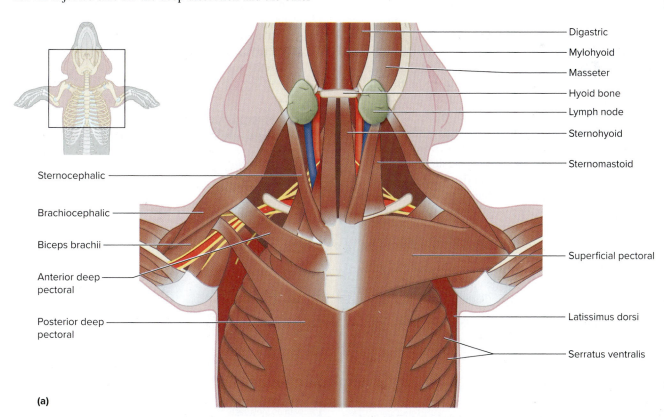

Sternocephalic

Brachiocephalic

Biceps brachii

Anterior deep pectoral

Posterior deep pectoral

Digastric

Mylohyoid

Masseter

Hyoid bone

Lymph node

Sternohyoid

Sternomastoid

Superficial pectoral

Latissimus dorsi

Serratus ventralis

(a)

Figure 29.6 **Muscles of the Head, Neck, and Thorax (Anterior View).** Superficial muscles are shown on the right side of the figure. The superficial pectoral and the sternomastoid have been removed on the left side of the figure to show the deeper muscles. (*a*) Illustration.

(continued on next page)

(continued from previous page)

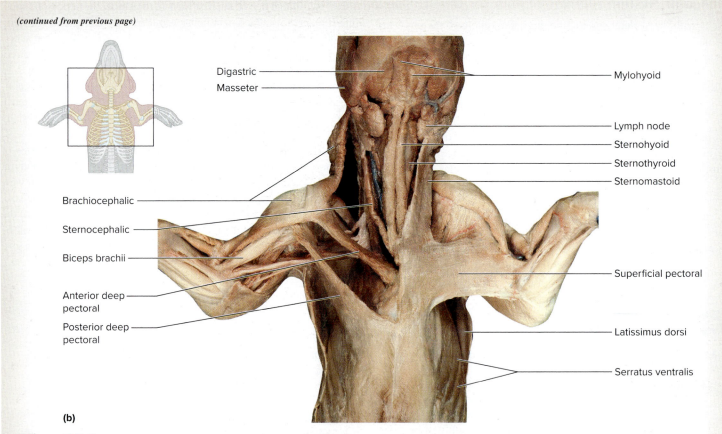

(b)

Figure 29.6   Muscles of the Head, Neck, and Thorax (Anterior View) (*continued*).   (*b*) Dissection.
(*b*) ©McGraw-Hill Education/Christine Eckel

2. **Superficial dissection:** Although the word "dissect" means "to cut apart" (*dis,* apart + *sectio,* to cut), in reality, muscles are rarely dissected. Rather, dissection is a process of *cleaning* muscles and separating them from one another so as to best view them and understand their location and functions. Often the best tools for this job are forceps and a pair of scissors. Use the forceps to pull on the fascia (connective tissue) overlying the muscles; then use the scissors to cut the fascia away. When cutting, always cut in the direction of the muscle fibers, which prevents them from tearing. Before and during the dissection, preview the muscles in figure 29.6 to get an idea of what to look for. More important, it is critical to know where to take extra care not to destroy muscles or related structures such as nerves and blood vessels (see the Learning Strategy on this page).

3. **Identifying superficial muscles:** Using **table 29.1** and figure 29.6 as guides, clean, identify, and isolate the following muscles from each other:

   ☐ **Digastric**   ☐ **Sternohyoid**

   ☐ **Masseter**   ☐ **Sternomastoid**

   ☐ **Mylohyoid**

4. **Deep dissection:** On the side of the neck that was chosen to be most appropriate for a deep dissection, transect the sternomastoid and sternohyoid muscles. Using table 29.1

and figure 29.6 as guides, clean, identify, and isolate the sternothyroid muscle, which attaches to the thyroid cartilage and is named for its origin and insertion. Note that the sternothyroid muscle lies deep to the sternohyoid muscle.

INTEGRATE

LEARNING STRATEGY

This learning strategy is more of a "dissection strategy." While dissecting the muscles, take care not to destroy blood vessels and nerves that lie in the spaces around and between the muscles. **Nerves** appear white and somewhat stringy. **Arteries** will be red, if injected with latex, or white, like the nerves. Often the only way to distinguish a nerve from an artery is by palpation (touch). Nerves feel "cord-like," whereas arteries have a lumen and feel "tube-like." Veins will be blue if injected with latex, or dark brownish-purple if not. The color of veins comes from the blood that is still inside the vessels. The blood is visible due to the thin-walled (and see-through) nature of the veins.

Try to preserve (and clean) the arteries and nerves while dissecting. They are more likely to be kept intact if the dissection is performed from proximal to distal (rather than from distal to proximal). This is because the dissection begins where nerves and arteries are larger and follows their path as they decrease in size.

Table 29.1	Muscles of the Head and Neck			
Pig Muscle	**Human Muscle**	**Origin**	**Insertion**	**Action**
Digastric	Digastric	Mastoid process and occipital bone	Mandible	Depresses mandible
Masseter	Masseter	Zygomatic arch	Mandible	Elevates mandible
Mylohyoid	Mylohyoid	Mandible	Hyoid bone	Elevates floor of oral cavity
Sternohyoid	Sternohyoid	Manubrium of sternum	Hyoid bone	Moves hyoid caudally
Sternomastoid	Sternocleidomastoid-clavicular head	Manubrium of sternum	Mastoid process	Flexes head
Sternothyroid	Sternothyroid	Manubrium of sternum	Thyroid cartilage	Moves larynx caudally

EXERCISE 29.5

MUSCLES OF THE THORAX

1. **Superficial muscles:** With the fetal pig still ventral side up on the dissecting tray, prepare to dissect the superficial muscles of the anterior thorax. The main muscle to dissect is the **superficial pectoral**, which adducts the forelimb. This muscle is homologous to the pectoralis major in the human. While dissecting around the insertion of this muscle on the humerus, take care not to damage the latissimus dorsi and serratus ventralis muscles (figure 29.6). These muscles arise from the dorsal surface of the fetal pig and will be studied with the back.

   For what activity of the pig would the pectoral group of muscles be most important?

   _____

   _____

2. Using **table 29.2** and figures 29.6 and **29.7** as guides, clean, identify, and isolate the superficial pectoral muscle.

3. **Deep muscles:** Use blunt dissection to gently separate the superficial pectoral muscle from the underlying muscles. It is helpful to work a probe deep to the pectoral muscles, taking care not to tear the deep muscles of the thorax. Be sure to bisect the superficial pectoral muscle close to the midline so as to avoid damaging the nerves of the brachial plexus, which are located in the axillary region. Keeping the probe deep to the pectoral muscles, use scissors to make a vertical incision through the superficial pectoral muscle. Bisect and reflect this muscle to expose the anterior and posterior deep pectoral muscles. This makes a muscle that connects the scapula to the axial skeleton visible—the **serratus ventralis** muscle.

4. Using table 29.2 and figure 29.7 as guides, clean, identify, and isolate the following muscles:

   ☐ **Anterior deep pectoral**

   ☐ **Posterior deep pectoral**

   ☐ **Serratus ventralis**

Table 29.2	Muscles of the Thorax			
Pig Muscle	**Human Muscle**	**Origin**	**Insertion**	**Action**
Superficial Muscles				
Superficial Pectoral	Pectoralis major	Sternum	Humerus	Adducts the forelimb
Deep Muscles				
Anterior Deep Pectoral	NA	Sternum	Scapula	Adducts and retracts the forelimb
Posterior Deep Pectoral	Pectoralis minor	Sternum and costal cartilages	Proximal humerus	Adducts the forelimb
Serratus Ventralis	Serratus anterior	Cervical vertebrae and ribs 1–9	Medial border of scapula	Depresses scapula and moves scapula cranially

(continued on next page)

EXERCISE 29.7

MUSCLES OF THE BACK AND SHOULDER

1. **Preparation:** Turn the pig over so that it is now lying on its side. If possible, start with its ventral surface down on the dissecting tray and the dorsal surface facing up. However, if the specimen does not want to cooperate, performing the dissection with the specimen lying on one side is fine. As with observations/dissection of the ventral muscles, begin observations/dissection of the dorsal muscles at the head and progress toward the tail.

2. **Superficial muscles:** Recall that the most superficial "back" muscles in humans consist of the trapezius, deltoid, and latissimus dorsi muscles. These are not true "back" muscles, because they do not move the vertebral column. Instead, their functions are to attach the scapula and upper limb to the axial skeleton, and to create large movements of the upper limb and scapula. Pigs also have

these muscles. In the pig the **deltoid** is smaller than the human deltoid muscle (**table 29.5**). The pig trapezius is divided into two separate muscles, the **acromiotrapezius** and the **spinotrapezius**; and the superior portion of the **brachiocephalic** muscle is homologous to the superior portion of the human trapezius muscle. The **latissimus dorsi** is similar in structure and function in the pig and the human. While identifying and cleaning these muscles, think about the actions each performs and relate each muscle in the pig to its human homologue.

3. Using table 29.5 and **figure 29.8** as guides, clean, identify, and isolate the following muscles:

- ☐ **Acromiotrapezius**
- ☐ **Brachiocephalic**
- ☐ **Deltoid**
- ☐ **Latissimus Dorsi**
- ☐ **Spinotrapezius**

Table 29.5	Muscles of the Back and Shoulder			
Pig Muscle	**Human Muscle**	**Origin**	**Insertion**	**Action**
Muscles of the Back and Shoulder (superficial)				
Acromiotrapezius	Trapezius	Spines of cervical vertebrae	Acromium process and spine of scapula	Elevates scapula
Brachiocephalic (superior: clavotrapezius; inferior: clavobrachialis)	Trapezius (clavotrapezius only)	Mastoid process of temporal bone and occipital bone	Distal end of humerus	Flexes forelimb
Deltoid	Deltoid—middle fibers	Spine of scapula	Proximal humerus	Abducts forelimb
Latissimus Dorsi	Latissimus dorsi	Spines of lower thoracic and lumbar vertebrae	Medial humerus	Elevates and moves humerus caudally
Spinotrapezius	Trapezius	Spines of thoracic vertebrae	Spine of scapula	Elevates and retracts scapula
Teres Major	Teres major	Axillary border of scapula	Medial humerus	Adducts humerus
Muscles of the Back and Shoulder (deep)				
Infraspinatus	Infraspinatus	Infraspinous fossa	Greater tuberosity of humerus	Rotates humerus
Rhomboid	Rhomboid major and minor	Spines of cervical and thoracic vertebrae	Ventral border of scapula	Moves scapula forward and cranially
Rhomboid Capitis	NA	Occipital bone	Ventral border of scapula	Moves scapula forward
Serratus Ventralis	Serratus anterior	Cervical vertebrae and ribs 1–9	Medial border of scapula	Depresses scapula and moves scapula cranially
Splenius	Splenius capitis and cervicis	Nuchal ligament	Occipital bone	Elevates head
Supraspinatus	Supraspinatus	Supraspinous fossa	Greater tuberosity of humerus	Extends humerus

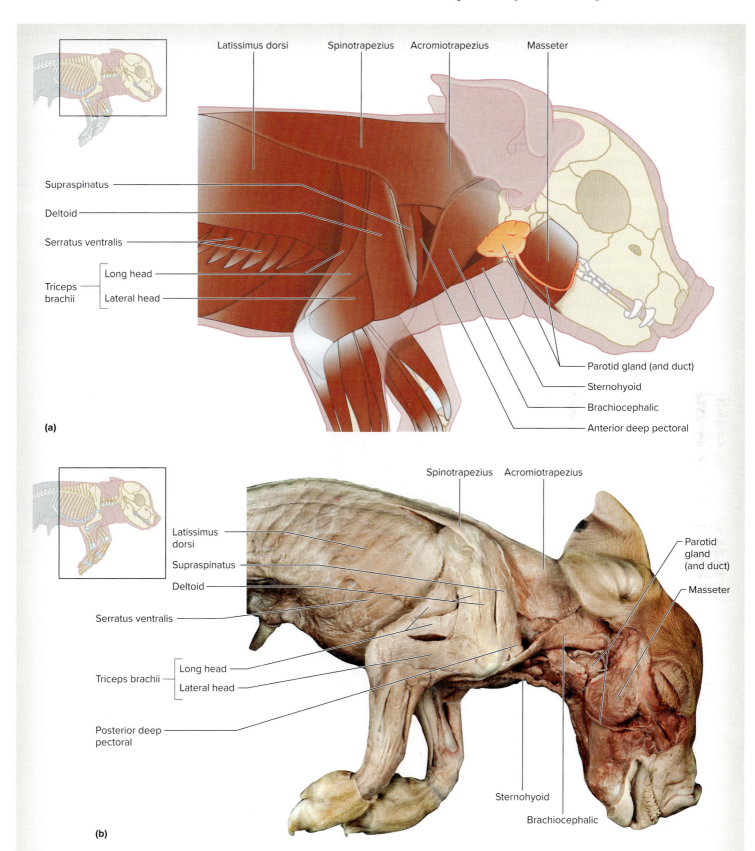

Figure 29.8 **Superficial Muscles of the Back and Shoulder.** Removal of the skin allows for visualization of the superficial muscles of the back and shoulder. (*a*) Illustration. (*b*) Dissection.

(*b*) ©McGraw-Hill Education/Christine Eckel

(*continued on next page*)

(continued from previous page)

4. **Deep muscles:** Using scissors and forceps, cut a small, vertical slit (a sagittal section) in the acromiotrapezius and spinotrapezius muscles. To do this, slip a probe deep to the muscles to protect the underlying rhomboid muscles. Then cut superficial to the probe. When the acromiotrapezius and spinotrapezius are cut, the scapula will pull away from the midline and the rhomboids will be visible. The **rhomboid** in the pig is homologous to the rhomboid major and minor in humans, and the **rhomboid capitis** muscle has no homologue in humans. A third muscle that connects the scapula to the axial skeleton is the **serratus ventralis**, which is equivalent to the serratus anterior of humans. In the pig, however, this muscle is larger, with more attachments to the ribs, than the serratus anterior of humans. This muscle was first identified in exercise 29.5; however, it appears more clearly in this view.

Observe to the muscles of the scapula that form part of the rotator cuff (see figure 13.1). Recall that these muscles form the major muscular support of the shoulder joint, and consist of the supraspinatus, infraspinatus, subscapularis, and teres minor. From a posterior view, the **infraspinatus** and **supraspinatus** muscles are visible. These two muscles are the rotator cuff muscles that are easiest to identify in the fetal pig; therefore, the remaining two muscles will not be identified in this chapter. Using table 29.5 and **figure 29.9** as guides, clean, identify, and isolate the following muscles:

- ☐ **Infraspinatus**
- ☐ **Rhomboid**
- ☐ **Rhomboid capitis**
- ☐ **Splenius**
- ☐ **Supraspinatus**
- ☐ **Teres major**

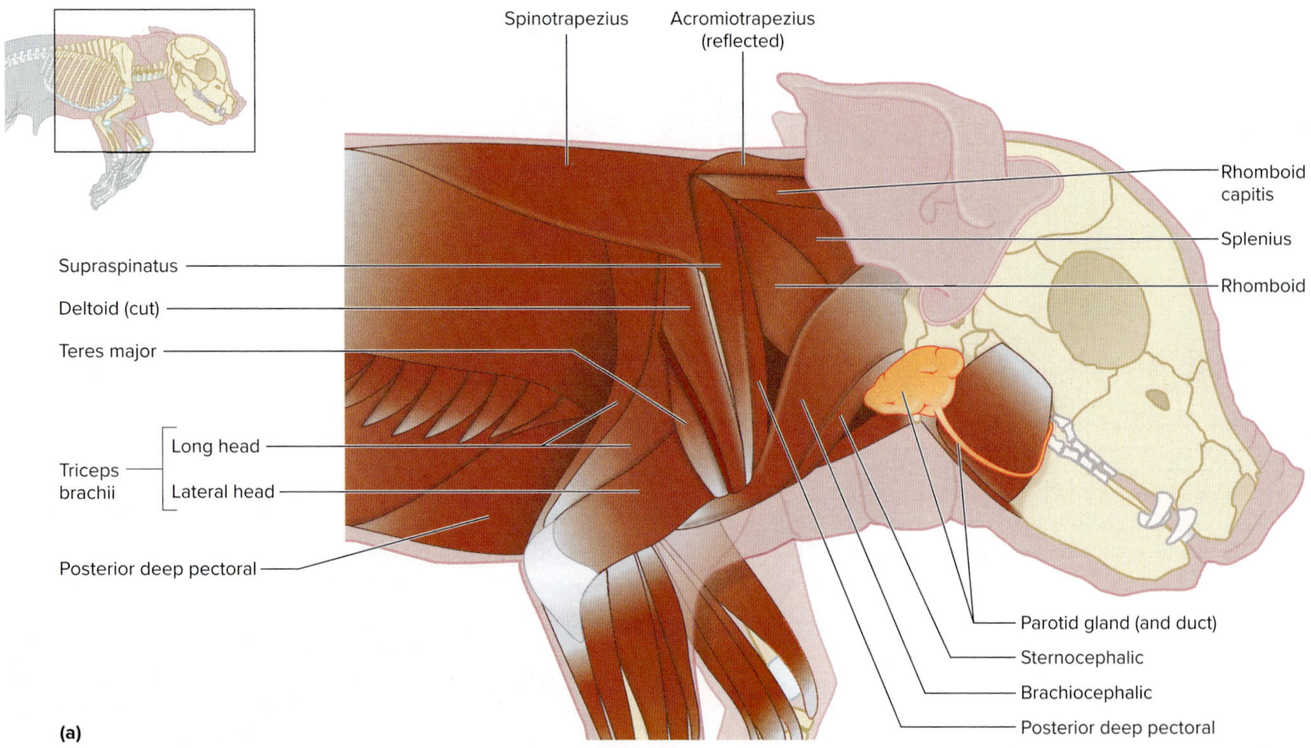

Figure 29.9   Deep Muscles of the Shoulder and Forelimb. After bisecting and reflecting the acromiotrapezius, deep muscles of the shoulder, back, and neck are better visualized. (*a*) Illustration.

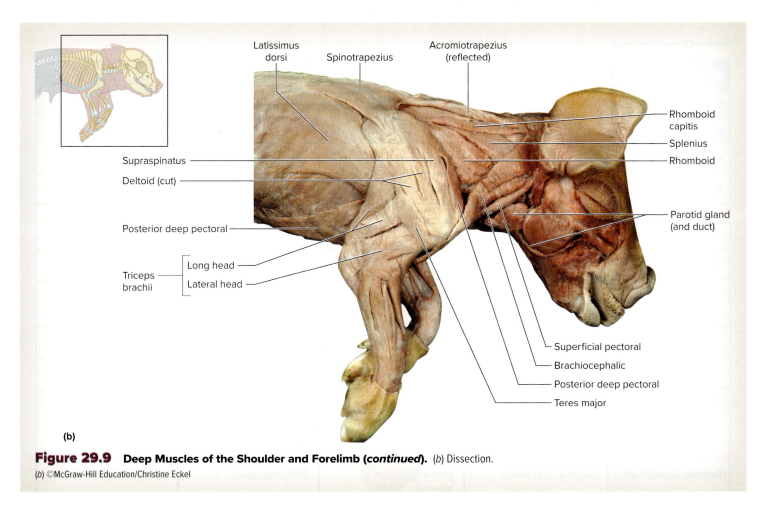

(b)

Figure 29.9 **Deep Muscles of the Shoulder and Forelimb (*continued*).** (*b*) Dissection.

(*b*) ©McGraw-Hill Education/Christine Eckel

EXERCISE 29.8

MUSCLES OF THE FORELIMB

1. **Superficial muscles:** Recall that the human upper limb muscles are organized into four compartments: anterior and posterior arm, and anterior and posterior forearm. The muscles of the pig forelimb are similarly arranged. Thus, while dissecting muscles in each region of the forelimb, always consider the common action of the muscles in that group before getting overly concerned with specifics. This will greatly facilitate the ease of learning these muscles. **Table 29.6** summarizes the human homologue, origin, insertion, and action of the forelimb muscles.

2. Observe the superficial muscles of the shoulder. At this point the muscles of the deltoid group should have been cleaned and identified already. The dissection will now proceed distally along the forelimb, starting with the **triceps brachii** (figure 29.9). The triceps brachii in the pig has the same three heads and the same insertion as the triceps brachii of the human. While cleaning the anterior border of the lateral head of the triceps brachii near the elbow, note the origin

of several other muscles: the brachialis and several extensors of the forearm.

3. In this dissection exercise all of the arm/forearm muscles will be cleaned, identified, and isolated as a group together because of the small region. It is best to pay attention to all muscles in the region at the same time. To effectively complete the dissection, the pig will need to be repositioned occasionally so as to make it easier to get at the muscles on the lateral and medial sides of the forelimb. While dissecting the forelimb muscles, take particular care when cleaning the proximal attachments of the muscles that originate on the medial epicondyle of the humerus (**flexor carpi ulnaris, flexor carpi radialis, flexor digitorum profundus,** and **flexor digitorum superficialis**). On the proximal end of these muscles, the fascia becomes embedded in the muscle, making it difficult to remove. Thus, do not try to remove the fascia when getting to that point. Instead, simply cut the fascia where it starts to embed the muscles, so as not to tear the muscles. While dissecting these muscles, think about the homologous muscles in the human and note both similarities and differences.

(*continued on next page*)

MUSCLES OF THE HINDLIMB

EXERCISE 29.9A Muscles of the Hindlimb: Thigh

1. **Lateral and posterior thigh: Superficial muscles:**
   Dissection of hindlimb muscles begins with a dissection of the muscles of the posterior and lateral hip and thigh. Place the fetal pig ventral side down on the dissection tray. As with the forelimb, the pig may end up placed on its side if it doesn't "cooperate" in staying on its ventral surface. Before beginning the dissection, observe the hindlimb muscles in **figure 29.11**. Notice that while the pig has nearly the same gluteal and posterior thigh muscles as the human, the sizes of the muscles are quite different. These size differences have to do with the pig's four-legged posture as compared to the human's two-legged posture. That is, the gluteal muscles in the pig are relatively small and the posterior thigh, or "hamstring," muscles are relatively large. This situation is largely reversed in humans, who tend to have larger gluteal muscles and smaller hamstrings (see Learning Strategy on hamstring muscles). Using **figure 29.12** as a guide, begin cleaning the fascia from the gluteal muscles and proceed toward the distal thigh.

2. Using **table 29.7** and figure 29.11 as guides, clean, identify, and isolate the following muscles of the hindlimb:

   ☐ **Biceps femoris**   ☐ **Iliacus**

   ☐ **Gluteus medius**   ☐ **Tensor fasciae latae**

   ☐ **Gracilis**   ☐ **Vastus lateralis**

Table 29.7	Muscles of the Hindlimb			
Pig Muscle	**Human Muscle**	**Origin**	**Insertion**	**Action**
Muscles of the Hip				
Gluteus Medius	Gluteus medius and minimus	Lumbodorsal fascia and lateral pelvis	Greater trochanter of femur	Abducts and extends thigh
Gracilis	Gracilis	Ventral pubis	Proximal tibia	Adducts thigh
Iliacus	Iliacus	Ventral ilium	Lesser trochanter of femur	Flexes hip and rotates thigh
Psoas	Psoas	Lumbar vertebrae	Lesser trochanter of femur	Flexes hip and rotates thigh
Tensor Fasciae Latae	Tensor fasciae latae	Ilium	Fasciae latae	Tighten fascia lata
Muscles of the Thigh (anterior and medial)				
Adductor Group	Adductor magnus, longus, and brevis	Pubis	Ventral femur	Adducts thigh
Pectineus	Pectineus	Pubis	Shaft of femur	Adducts thigh and flexes hip
Quadriceps Muscles				
Rectus Femoris	Rectus femoris	Ilium	Patella	Extends hindlimb and flexes hip
Sartorius	Sartorius	Iliac crest	Proximal tibia	Flexes hip and adducts thigh
Vastus Intermedius	Vastus intermedius	Femur	Patella	Extends hindlimb
Vastus Lateralis	Vastus lateralis	Femur	Patella	Extends hindlimb
Vastus Medialis	Vastus medialis	Femur	Patella	Extends hindlimb
Muscles of the Thigh (posterior)				
Hamstring Muscles				
Biceps Femoris	Biceps femoris	Sacrum and ischium	Proximal tibia	Flexes hindlimb and abducts thigh
Semimembranosus	Semimembranosus	Ischium	Proximal tibia and distal femur	Extends thigh and adducts hindlimb
Semitendinosus	Semitendinosus	Ischium	Proximal tibia	Extends thigh and flexes hindlimb
Muscles of the Leg (anterior and lateral)				
Extensor Digitorum Longus	Extensor digitorum longus	Proximal tibia and fibula	Digits	Extends the digits
Fibularis (peroneus) Muscles	Fibularis brevis and longus	Fibula	Metatarsals	Flexes foot
Tibialis Anterior	Tibialis anterior	Proximal tibia	Second tarsal and metatarsal	Flexes foot
Muscles of the Leg (posterior)				
Gastrocnemius	Gastrocnemius	Distal femur	Calcaneus	Extends foot
Soleus	Soleus	Proximal fibula	Calcaneus	Extends foot

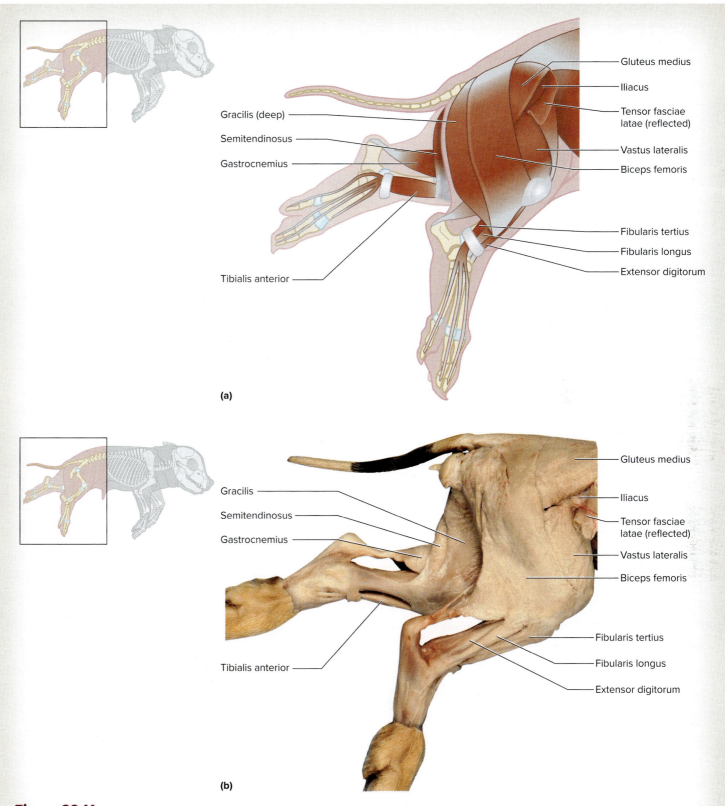

Figure 29.11 **Superficial Muscles of the Hindlimb.** In a lateral view of the hindlimb, the most visible muscles are the tensor fasciae latae, gluteus medius, and biceps femoris. (*a*) Illustration. (*b*) Dissection.
(*b*) ©McGraw-Hill Education/Christine Eckel

(continued on next page)

(continued from previous page)

3. **Lateral and posterior thigh: Deep muscles:** Locate the **biceps femoris** (figure 29.11) muscle. The part of the muscle visible in a superficial view is the *long head* of the biceps femoris. The distal end of this muscle becomes thin as it approaches its insertion. Carefully push a probe deep to the distal end of the long head of the biceps femoris muscle to protect the underlying muscles. Then transect the muscle and reflect it posteriorly to view the *short head* of the biceps femoris muscle (figure 29.12).

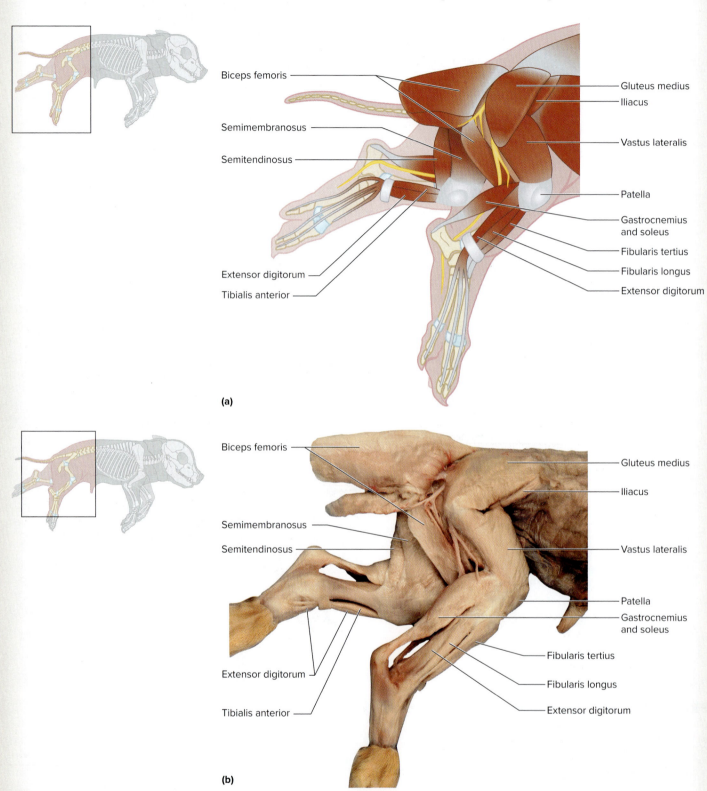

(a)

(b)

Figure 29.12 **Deep Muscles of the Hindlimb.** In a lateral view of the hindlimb, the most visible muscles are the vastus lateralis, iliacus, semitendinosus, semimembranosus, and gastrocnemius. (*a*) Illustration. (*b*) Dissection.

(b) ©McGraw-Hill Education/Christine Eckel

4. **Medial thigh: Superficial muscles:** Turn the pig over so that it is lying on its dorsal surface and the medial and anterior thigh muscles are visible **(figure 29.13)**. While proceeding to clean muscles on the medial and anterior thigh, carefully trace the **sartorius** and **gracilis** muscles from origin to insertion (table 29.7). Clean these muscles, and separate them from the underlying muscles.

5. **Medial thigh: Deep muscles:** Once the sartorius and gracilis muscles are separated from the surrounding muscles, slip a blunt probe deep to each muscle to protect the underlying muscles. Then transect each muscle across the middle of the muscle belly and reflect the two ends to expose the underlying muscles. Take care not to damage the femoral vessels and nerve while reflecting these two muscles.

6. **Anterior thigh:** Observe the **quadriceps femoris** muscle group. This is homologous to the quadriceps femoris muscle group in humans. As in humans, all four muscles of the group come together to form a common tendon that encases the patella, the **patellar tendon**. This band of connective tissue extends distal to the patella to connect to the tibial tuberosity as the **patellar ligament**. Recall that the major action of the quadriceps group of muscles is to extend the leg.

7. Using figures 29.12 and 29.13 and table 29.7 as guides, clean, isolate, and identify the following structures (Note: The vastus intermedius is deep to the rectus femoris and is not shown in figure 29.13.):

   ☐ **Patella**   ☐ **Vastus intermedius**

   ☐ **Patellar ligament**   ☐ **Vastus lateralis**

   ☐ **Patellar tendon**   ☐ **Vastus medialis**

   ☐ **Rectus femoris**

8. **Medial thigh:** Observe the medial compartment muscles of the pig (figure 29.13*b*). This group, the **adductor** group, is also very similar to the homologous muscles in the human. The major exception is that the **gracilis** is much larger in the pig than in the human. In the medial view of the thigh, the large **semimembranosus** and **semitendinosus** muscles are also visible. These are part of the posterior thigh, or "hamstring," muscle group. Medial to the semimembranosus muscle is the small **adductor group** of muscles. Just superior to the adductors, and deep to the gracilis, is a small muscle, the **pectineus**.

9. Using figure 29.13*b* and table 29.7 as guides, clean, isolate, and identify the following muscles of the medial and posterior thigh:

   ☐ **Adductor group**   ☐ **Pectineus**

   ☐ **Biceps femoris**   ☐ **Semimembranosus**

   ☐ **Gracilis**   ☐ **Semitendinosus**

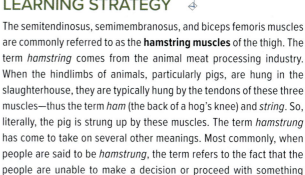

INTEGRATE

LEARNING STRATEGY

The semitendinosus, semimembranosus, and biceps femoris muscles are commonly referred to as the **hamstring muscles** of the thigh. The term *hamstring* comes from the animal meat processing industry. When the hindlimbs of animals, particularly pigs, are hung in the slaughterhouse, they are typically hung by the tendons of these three muscles—thus the term *ham* (the back of a hog's knee) and *string*. So, literally, the pig is strung up by these muscles. The term *hamstrung* has come to take on several other meanings. Most commonly, when people are said to be *hamstrung*, the term refers to the fact that the people are unable to make a decision or proceed with something because they are stuck or crippled (as if they were "hung up" by their hamstrings!).

EXERCISE 29.9 B Muscles of the Hindlimb: Leg and Foot

1. **Superficial muscles:** Continuing to dissect the hindlimb, move distally toward the leg. The muscles of the leg in pigs and humans are very similar. Most important, the muscles of the pig leg are organized into compartments that mimic those of the human leg. These compartments are anterior, posterior, and lateral. Anterior compartment muscles dorsiflex the foot and extend the digits, posterior compartment muscles plantar flex the foot and flex the digits, and lateral compartment muscles evert the foot. The triceps surae group of muscles is similar in the pig and the human. The **triceps surae** consists of the gastrocnemius, soleus, and plantaris muscles, all of which insert on the calcaneus via the calcaneal tendon. In pigs, as in humans, both the gastrocnemius and the soleus muscles are very large, and the plantaris is small, with very little function. The triceps surae group is the most important group of muscles acting about the ankle. These muscles are the primary muscles used for plantar flexion of the foot, which is necessary for standing, jumping, and the like.

2. Using figure 29.12 and table 29.7 as guides, clean, identify, and isolate the following muscles:

   ☐ **Extensor digitorum**   ☐ **Gastrocnemius**

   ☐ **Fibularis (peroneus)**   ☐ **Soleus**
       muscles
                                   ☐ **Tibialis anterior**
   ☐ **Flexor digitorum**

(continued on next page)

(continued from previous page)

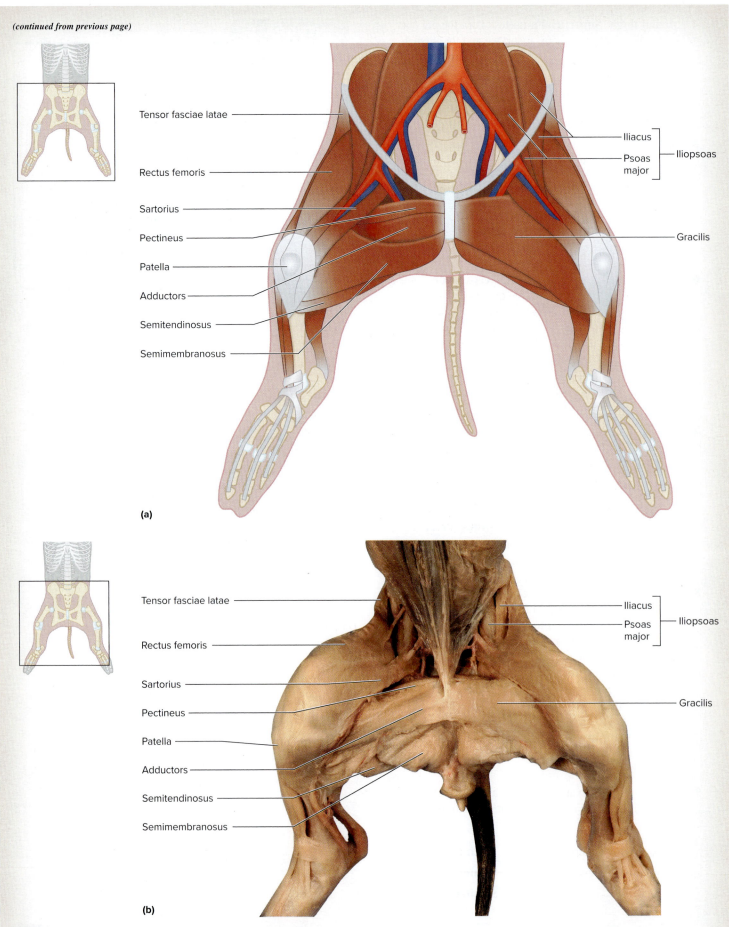

(a)

(b)

Figure 29.13 **Muscles of the Ventral Hindlimb.** The left side of the figure is a deep dissection, the right side is a superficial dissection. The gracilis has been removed from the deep dissection on the left side of the (*a*) illustration and (*b*) dissection.

(*b*) ©McGraw-Hill Education/Christine Eckel

EXERCISE 29.10

NERVOUS SYSTEM: SPINAL CORD AND PERIPHERAL NERVES

1. **Spinal cord:** Place the pig dorsal side up on the dissecting tray. Locate the scapula and the iliac crest; then locate a thoracic vertebra that lies about halfway between the two. It doesn't matter which thoracic vertebra is selected, just be sure to select one that it is in the mid to lower region of the vertebral column. Next, use scissors to perform a *laminectomy* to remove the vertebral arches and open the vertebral canal. To do this, place the point of the scissors between the laminae of two adjacent vertebrae (next to the midline) and cut through the most cranial of the

two laminae. Continue this cut cranially for about 6 to 8 vertebral segments; then perform the laminectomy on the other side of the vertebral column on the same vertebrae. Use forceps to pull the vertebral arches away from the vertebral bodies. Continue this process along the vertebral column until the vertebral arches have been removed from the entire vertebral canal and the entire spinal cord is visible **(figure 29.14).** If the process of getting the scissors in between the laminae proves to be difficult, particularly in the cervical region, then use scissors or bone cutters to remove the spinous processes first so they are out of the way. Once the laminectomy has been completed, identify

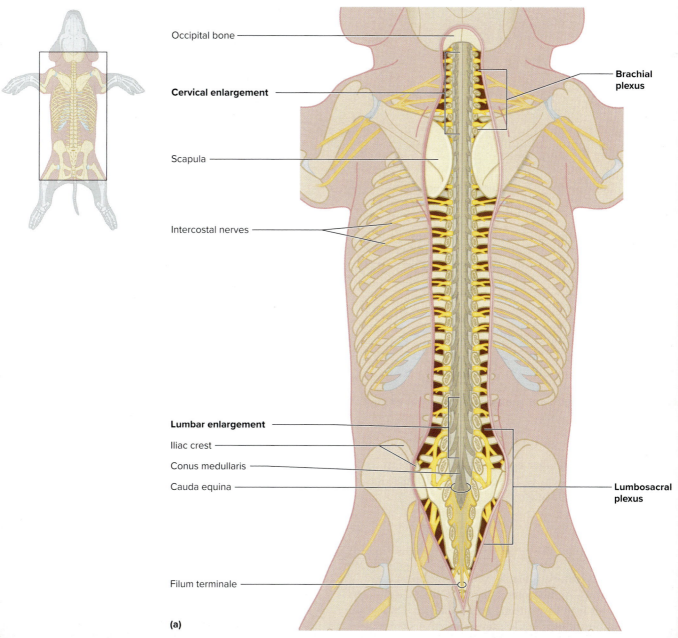

Occipital bone

Cervical enlargement

Scapula

Intercostal nerves

Brachial plexus

Lumbar enlargement

Iliac crest

Conus medullaris

Cauda equina

Lumbosacral plexus

Filum terminale

(a)

Figure 29.14   Spinal Cord and Peripheral Nerves of the Fetal Pig. Overall organization of the CNS and PNS of the fetal pig. Note the cervical and lumbar enlargements of the spinal cord where the nerves of the brachial and lumbosacral plexuses arise, respectively. (*a*) Illustration.

(continued on next page)

(continued from previous page)

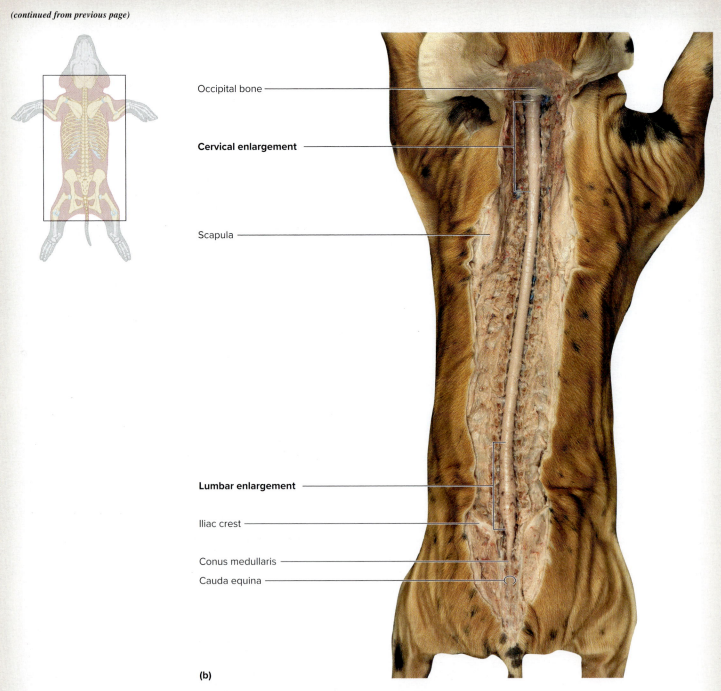

Occipital bone

Cervical enlargement

Scapula

Lumbar enlargement

Iliac crest

Conus medullaris

Cauda equina

(b)

Figure 29.14 **Spinal Cord and Peripheral Nerves of the Fetal Pig (*continued*).** (*b*) Dissection.
(*b*) ©McGraw-Hill Education/Christine Eckel

the **cervical** and **lumbar enlargements** of the spinal cord. Why are these regions of the spinal cord larger than the thoracic and sacral regions?

_____

_____

2. **Brachial plexus:** Place the pig ventral side up on the dissecting tray. At this point in the dissection, the muscles of the upper limb have already been identified. During the dissection of the muscles, several whitish, cord-like structures running within the limbs were likely observed. These are the peripheral nerves. This dissection focuses

on the axillary region to identify the major components of the brachial plexus. Start by reflecting the pectoral muscles laterally (they may have been removed already, depending on the order in which the dissection exercises have been performed). Note the mass of connective tissue and "stringy stuff" in the axillary region. To distinguish nerves from blood vessels, use the open scissors blunt dissection technique to gently loosen the connective tissue in this region (figure 1.16, p. 22). When using this technique, unimportant structures such as connective tissue tend to fall apart, whereas more robust structures such as blood vessels and nerves tend to remain intact. However, be gentle with the dissection to avoid damaging the blood vessels and nerves.

3. Locate the **subclavian artery** as it emerges under the clavicle and enters the axilla as the **axillary artery**. Recall that the cords of the brachial plexus are named for their location relative to the axillary artery. After identifying the axillary artery, gently trace the nerves that surround it, both toward and away from their origin in the spinal cord. Using **figure 29.15** as a guide, identify the following components of the brachial plexus:

- ☐ **Axillary nerve**
- ☐ **Median nerve**
- ☐ **Musculocutaneous nerve**
- ☐ **Radial nerve**
- ☐ **Thoracodorsal nerve (optional)**
- ☐ **Ulnar nerve**

4. **Lumbosacral plexus:** This dissection focuses on the pelvic and inguinal regions to identify the major components of the lumbosacral plexus (lumbar plexus + sacral plexus). Start by observing the internal anatomy of the lower dorsal body wall. To view the nerves of the lumbosacral plexus, remove the allantoic bladder, the ureters, the kidney, and any reproductive structures that lie ventral to the dorsal body wall. Identify the **psoas major muscle**, a major landmark in the pelvis. Trace the psoas major posteriorly toward its insertion on the femur. As it crosses the inguinal region, two nerves emerge adjacent to the muscle: the **femoral nerve** emerges lateral and the **obturator nerve** emerges medial **(figure 29.16a)**. Reflect one of the psoas major muscles superiorly to view the lumbar plexus from which these two nerves arise.

5. Turn the pig over so it is dorsal side up. Using a scalpel, cut through the **gluteus medius** muscle and reflect it laterally to expose the piriformis muscle. Note the large nerve that emerges posterior to the piriformis muscle. This is the largest nerve in the body, the **sciatic nerve**. To trace the sciatic nerve distally, you will need to bisect the biceps femoris muscle. Just cranial to the knee joint, you will see the sciatic nerve bifurcate into its two major branches in the pig, the **external popliteal nerve** and the **internal popliteal nerve**. These two nerves are homologous to the human common fibular and tibial nerves, respectively.

6. Using figure 29.16 as a guide, identify the following nerves of the lumbosacral plexus:

- ☐ **Common fibular nerve**
- ☐ **Femoral nerve**
- ☐ **Obturator nerve**
- ☐ **Sciatic nerve**
- ☐ **Tibial nerve**

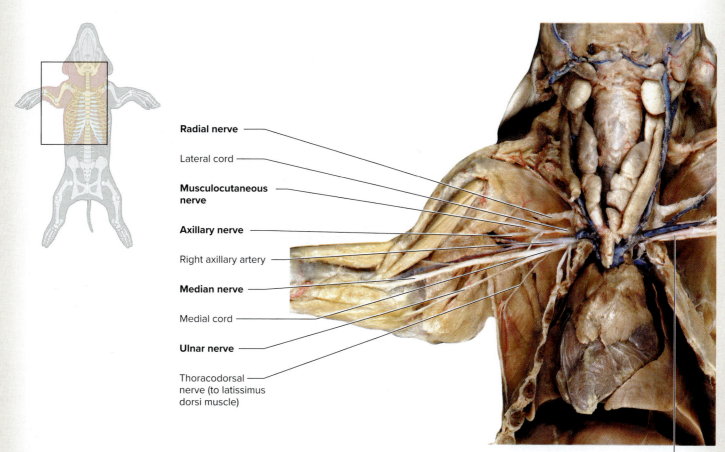

Radial nerve

Lateral cord

Musculocutaneous nerve

Axillary nerve

Right axillary artery

Median nerve

Medial cord

Ulnar nerve

Thoracodorsal nerve (to latissimus dorsi muscle)

Left axillary artery

Figure 29.15 **Nerves of the Brachial Plexus of the Fetal Pig.** The best way to locate the nerves of the brachial plexus is to first locate the axillary artery within the axilla. Note in this dissection specimen that the right axillary artery has some blue latex inside instead of red. This is a consequence of the injection process. It is, indeed, the artery in this dissection/photo, not the vein. Observe the left axillary artery (red) and vein (dark blue) for comparison.
©McGraw-Hill Education/Christine Eckel

(continued on next page)

(continued from previous page)

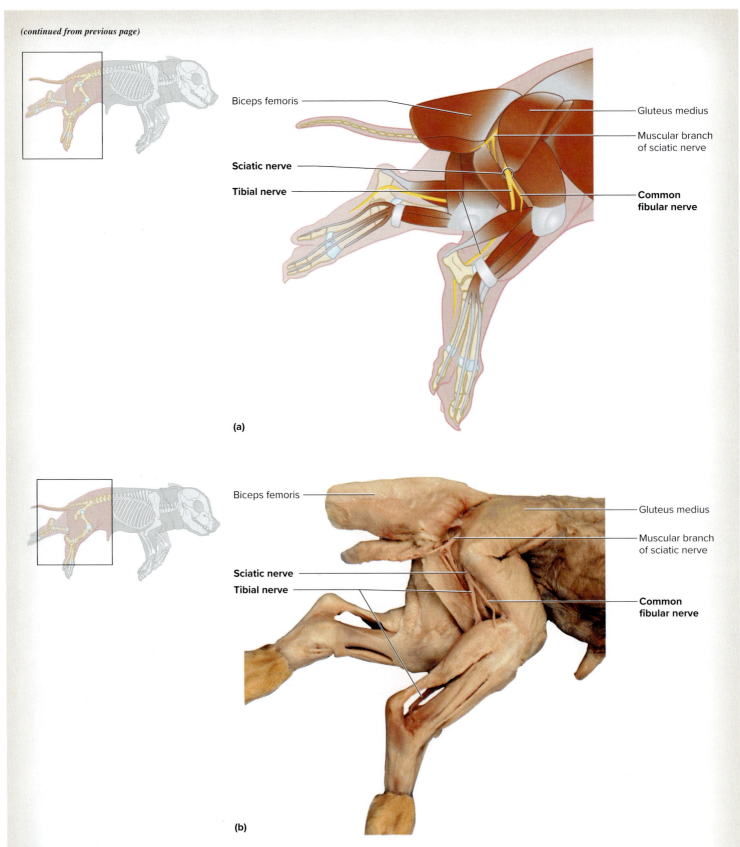

Biceps femoris

Gluteus medius

Muscular branch of sciatic nerve

Sciatic nerve

Tibial nerve

Common fibular nerve

(a)

Biceps femoris

Gluteus medius

Muscular branch of sciatic nerve

Sciatic nerve

Tibial nerve

Common fibular nerve

(b)

Figure 29.16 **Nerves of the Lumbosacral Plexus of the Fetal Pig.** In a lateral view of the hindlimb, the most visible nerves of the lumbosacral plexus are the sciatic nerve and its two branches: the tibial and common fibular nerves. The femoral and obturator nerves can be viewed from an anterior view of the medial hip and thigh (not shown here). (*a*) Illustration. (*b*) Dissection.

EXERCISE 29.11

CARDIOVASCULAR SYSTEM: PERIPHERAL BLOOD VESSELS

Although many of the blood vessels of the fetal pig have already been identified in the context of the structures they supply, this exercise is most useful as a succinct review of the major vessels and as a way to ensure that all of the major blood vessels of the pig have been identified. Place the pig ventral side up in the dissecting tray. While identifying the listed vessels, scissors and forceps may be used to clean the vessels as their routes through the body are traced.

Using **figures 29.17** to **29.19** as guides, identify the following arteries and veins of the pig:

- ☐ Anterior tibial
- ☐ Aortic arch
- ☐ Axillary
- ☐ Brachial
- ☐ Brachiocephalic
- ☐ Common carotid
- ☐ Descending abdominal aorta
- ☐ Descending thoracic aorta
- ☐ External carotid
- ☐ External iliac
- ☐ Femoral
- ☐ Gonadal (testicular or ovarian)
- ☐ Inferior (caudal) mesenteric
- ☐ Internal carotid
- ☐ Internal iliac
- ☐ Popliteal
- ☐ Posterior tibial
- ☐ Pulmonary
- ☐ Radial
- ☐ Renal
- ☐ Splenic
- ☐ Subclavian
- ☐ Superior (cranial) mesenteric
- ☐ Ulnar
- ☐ Umbilical

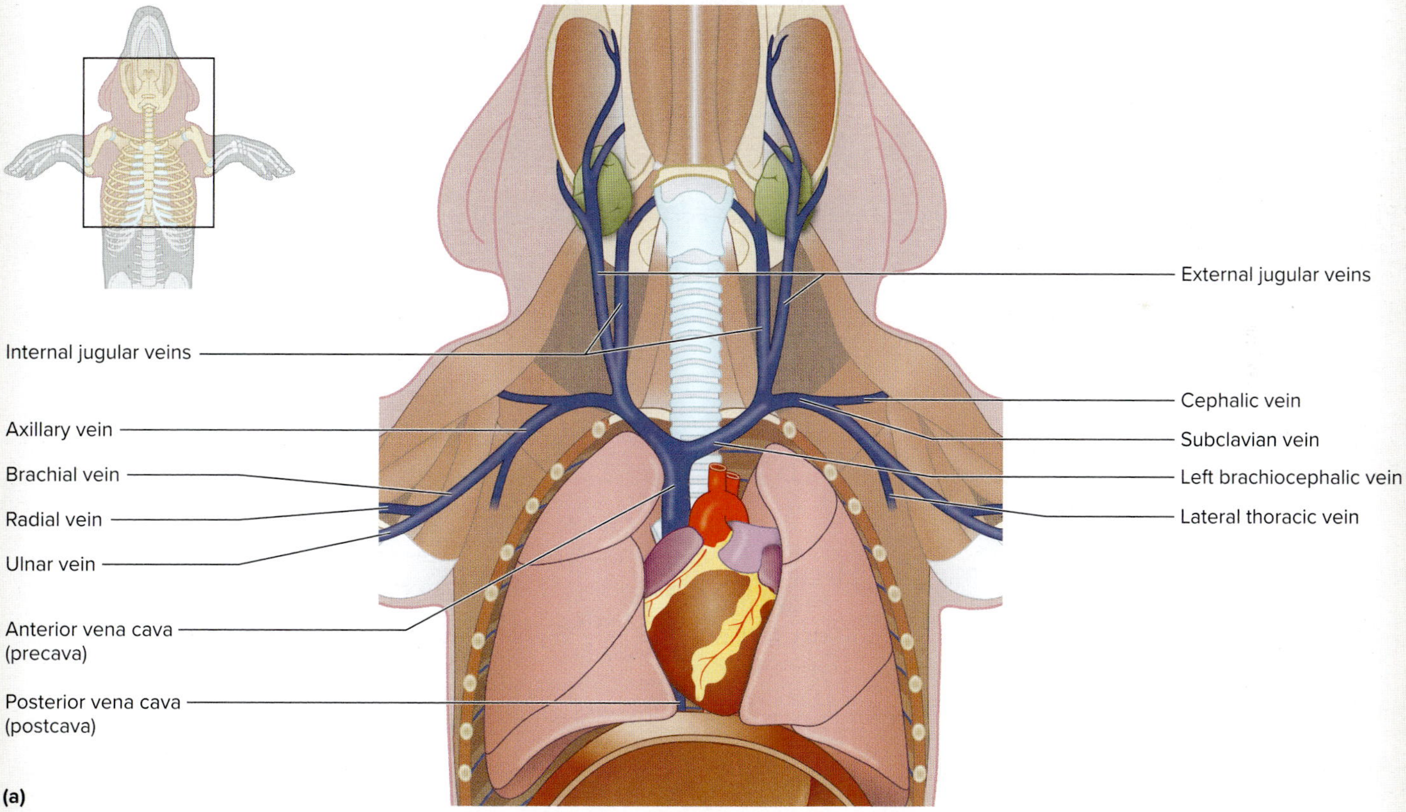

(a)

Figure 29.17 **Veins of the Neck, Thorax, and Forelimb of the Fetal Pig.** (*a*) Illustration.

Labels: External jugular veins; Internal jugular veins; Cephalic vein; Subclavian vein; Left brachiocephalic vein; Lateral thoracic vein; Axillary vein; Brachial vein; Radial vein; Ulnar vein; Anterior vena cava (precava); Posterior vena cava (postcava)

(continued on next page)

(continued from previous page)

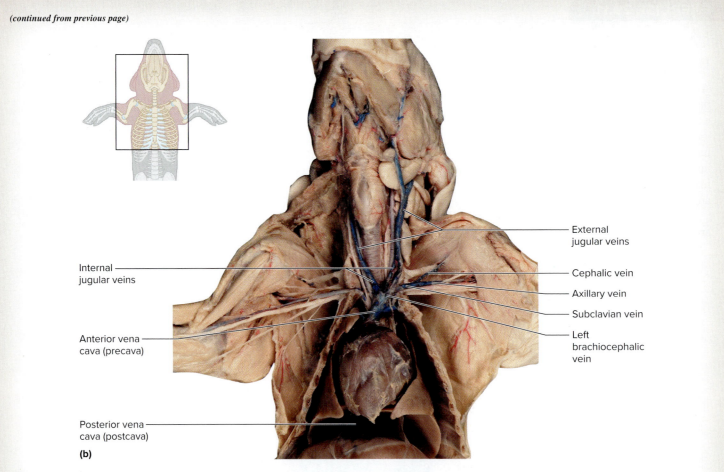

Internal jugular veins

Anterior vena cava (precava)

Posterior vena cava (postcava)

External jugular veins

Cephalic vein

Axillary vein

Subclavian vein

Left brachiocephalic vein

(b)

Figure 29.17 **Veins of the Neck, Thorax and Forelimb of the Fetal Pig (*continued*).** (*b*) Dissection (not all veins shown in the illustration can be seen in the dissection).

(*b*) ©McGraw-Hill Education/Christine Eckel

INTEGRATE

CONCEPT CONNECTION

Recall that the pig being dissected in these exercises is a *fetal* pig. That is, its circulatory and urinary systems are made to function within the context of the mother's uterus. In the fetus, the urinary bladder contains a stalk called the *allantoic stalk*, which exits the body within the umbilicus. The allantoic bladder drains urine into a structure called the *allantois*. Hence, the term *allantoic bladder*. When the fetus is born, the allantoic stalk closes off and urine then exits the urinary bladder via the urethra. When observing the allantoic bladder, note the area where its cranial portion converges with the two umbilical arteries at the umbilicus. In an adult human, a strand of tissue can be observed superior to the urinary bladder and running along the anterior abdominal wall toward the umbilicus. This is called the *median umbilical ligament*. The median umbilical ligament is a remnant of the allantoic bladder that existed during the human fetal developmental period.

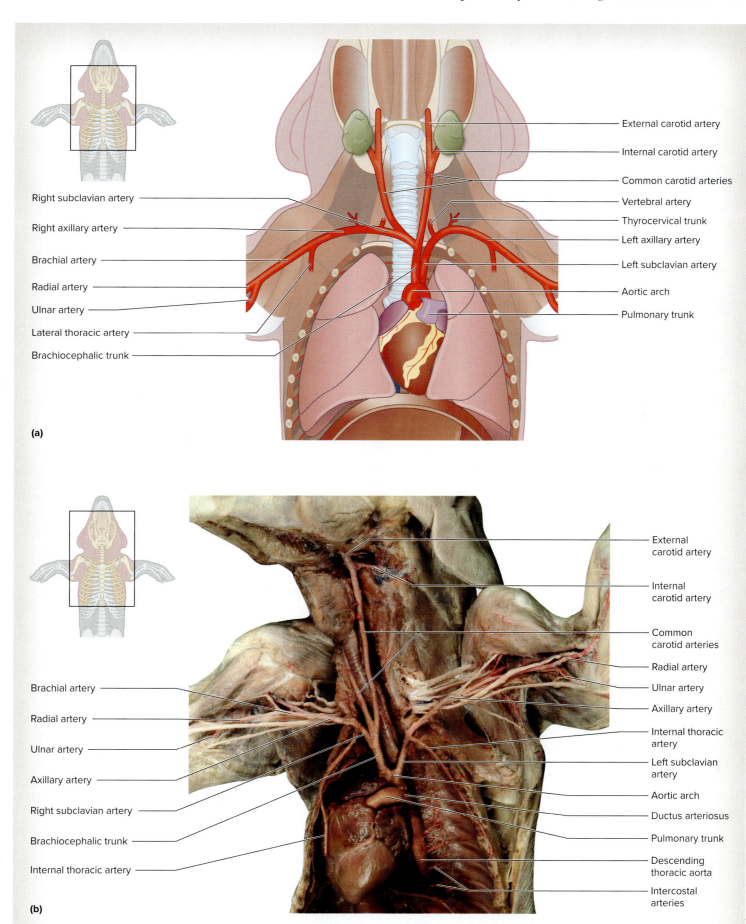

Figure 29.18 **Arteries of the Neck, Thorax, and Forelimb of the Fetal Pig.** (*a*) Illustration. (*b*) Dissection (not all arteries shown in the illustration can be seen in the dissection). Veins are not shown in this figure, but they may be found just adjacent to the corresponding arteries.

(*b*) ©McGraw-Hill Education/Christine Eckel

(*continued on next page*)

(continued from previous page)

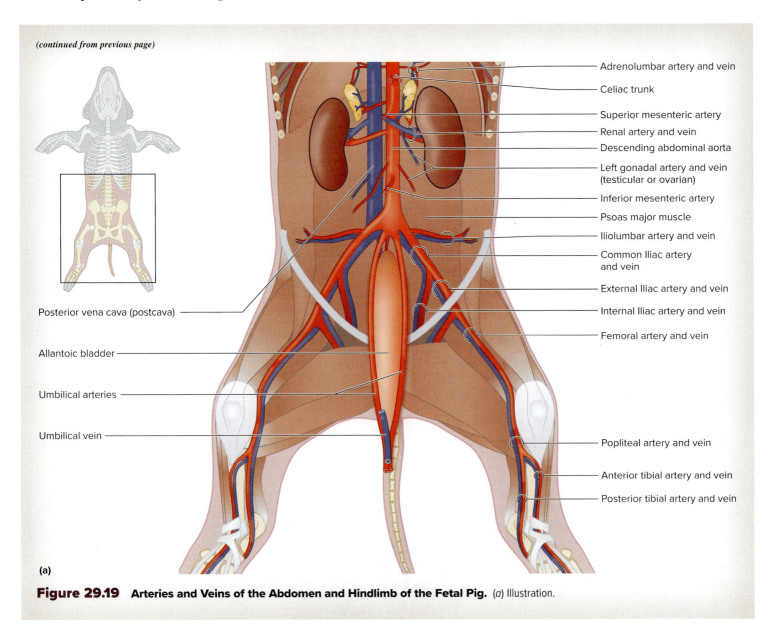

Adrenolumbar artery and vein

Celiac trunk

Superior mesenteric artery

Renal artery and vein

Descending abdominal aorta

Left gonadal artery and vein
(testicular or ovarian)

Inferior mesenteric artery

Psoas major muscle

Iliolumbar artery and vein

Common Iliac artery
and vein

External Iliac artery and vein

Internal Iliac artery and vein

Femoral artery and vein

Posterior vena cava (postcava)

Allantoic bladder

Umbilical arteries

Umbilical vein

Popliteal artery and vein

Anterior tibial artery and vein

Posterior tibial artery and vein

(a)

Figure 29.19   **Arteries and Veins of the Abdomen and Hindlimb of the Fetal Pig.** *(a)* Illustration.

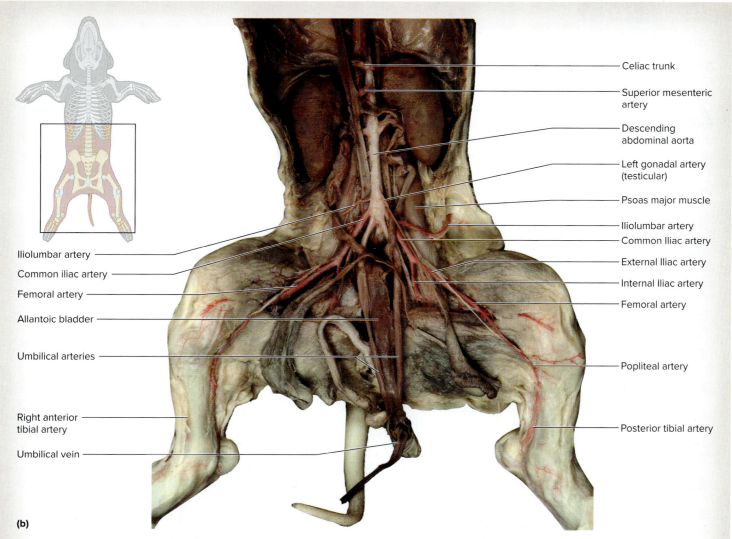

Celiac trunk

Superior mesenteric artery

Descending abdominal aorta

Left gonadal artery (testicular)

Psoas major muscle

Iliolumbar artery

Common Iliac artery

External Iliac artery

Internal Iliac artery

Femoral artery

Popliteal artery

Posterior tibial artery

Iliolumbar artery

Common iliac artery

Femoral artery

Allantoic bladder

Umbilical arteries

Right anterior tibial artery

Umbilical vein

(b)

Figure 29.19 **Arteries and Veins of the Abdomen and Hindlimb of the Fetal Pig (*continued*).** (*b*) Dissection (not all vessels shown in the illustration can be seen in the dissection). Most veins are not shown in this figure, but they may be found just adjacent to the corresponding arteries. The splenic artery is not shown in this figure.

(*b*) ©McGraw-Hill Education/Christine Eckel

OPENING THE THORACIC AND ABDOMINAL CAVITIES

⚠️ **Opening the thoracic cavity:** Opening the thoracic cavity of the pig requires first making a vertical incision just lateral to the midline in the anterior wall of both the abdominal and the thoracic cavities. Next, cut alongside the inferior border of the thoracic cage superior to the diaphragm to free the abdominal muscles from the thoracic cage. Finally, open the thoracic cavity by removing the sternum and reflecting the ribs laterally. Wear protective eyewear during all dissections, because excessive preservative fluid may be present in the thoracic and abdominal cavities. This fluid may be drained or removed using a sponge or paper towel. **Figure 29.20a** demonstrates the cuts that will be made to open up the thoracic and abdominal cavities.

1. Place the pig on the dissecting tray with the ventral side up. Obtain a pair of forceps, a blunt probe, and scissors. Using figure 29.20a as a guide, begin opening the thoracic and abdominal cavities by following these steps:

    a. Make a vertical incision lateral to the most caudal part of the anterior abdominal wall (figure 29.20a, step 1), taking care not to damage the spermatic cord or external genitalia if dissecting a male pig. Using forceps, lift the tissue composing the abdominal wall away from the abdominal viscera. Then, using scissors, continue the vertical incision cranially. While cutting, use the scissors to continue to lift the abdominal wall away from the contents of the abdominal cavity so as not to cut into abdominal organs. When the cut reaches the level of the umbilicus, stop. Start another, similar cut on the other side of the most inferior part of the anterior abdominal wall and extend the cut cranially until it meets up with the first cut. Continue this cut to the level of the xiphoid process.

    b. At the xiphoid, alter the vertical incision so it proceeds either to the left or to the right of the sternum (figure 29.20a, step 2). This cut must go through the costal cartilages. The cartilages generally are easy to cut through using scissors. As with the incision in the anterior abdominal wall, take care not to damage underlying structures as you make your incision. Continue the incision until it passes through the attachment of the most superior rib to the sternum.

    c. Observe the inferior border of the ribs and locate the **diaphragm**. To preserve the diaphragm and its relations with the thoracic and abdominal cavities, first separate its attachments to the anterior body wall both above and below the muscle. To do this, first make a lateral cut *cranial* to the diaphragm on both right and left sides to free the diaphragm from the anterior *thoracic* body wall (figure 29.20a, step 3). Make a lateral cut *caudal* to the diaphragm on both right and left sides to free the diaphragm from the anterior *abdominal* wall (figure 29.20a, step 4).

    d. Finally, make two longitudinal incisions in the lateral borders of the ribs to remove the anterior thoracic wall from the thoracic cavity (figure 29.20a, step 5).

After opening the thoracic and abdominal cavities, take the fetal pig to the sink and rinse out the thoracic and abdominal cavities with water to rid the cavities of debris from embalming and any dried blood that may remain. Be sure to ask the instructor which sink is appropriate for this task before going ahead with the rinsing of the cavities. Using a sink without proper disposal methods for dried blood and debris can result in clogged drains.

When finished, dry the pig with paper towels and return it to the the dissecting pan.

2. The contents of the abdominal cavity will not be observed at this time. Thus, keep the abdominal wall closed over the contents of the abdominal cavity while focusing attention on the thoracic cavity. Recall that inside the thoracic cavity there are three smaller cavities: two pleural cavities, which house the lungs, and a central pericardial cavity, which houses the heart. There is also a space, the mediastinum, located cranial to the heart, which contains the great vessels of the heart, the esophagus, the trachea, and the thymus.

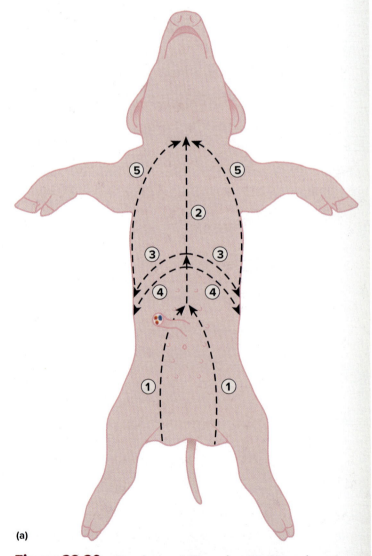

(a)

Figure 29.20 **Opening and Observing the Thoracic and Abdominal Cavities of the Fetal Pig.** (*a*) Location of incisions for opening the thoracic and abdominal cavities of the fetal pig.

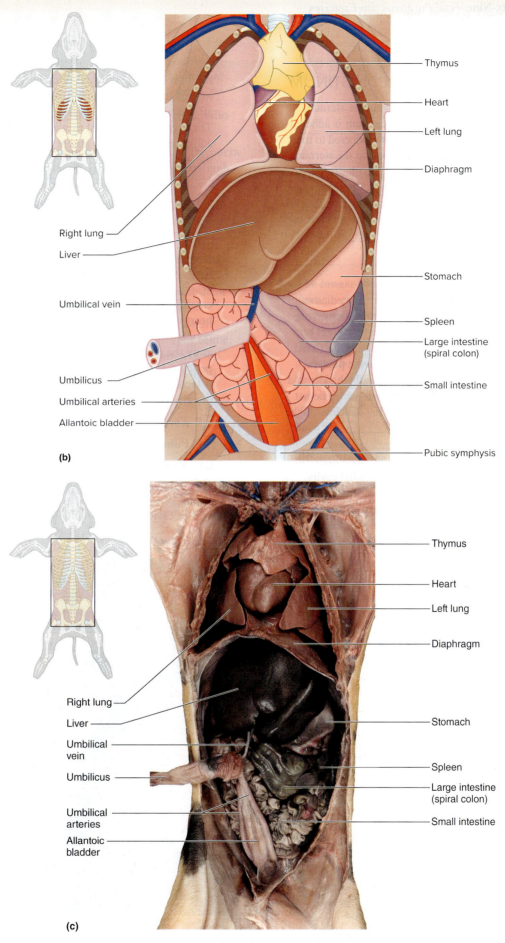

Figure 29.20 **Opening and Observing the Thoracic and Abdominal Cavities of the Fetal Pig (*continued*).** (*b*) Illustration demonstrating major organs of the ventral body cavity. (*c*) Dissection. Note in the dissection that the umbilicus retains its connections with the vasculature—specifically, the umbilical arteries and the umbilical vein.

(*c*) ©McGraw-Hill Education/Christine Eckel

(continued from previous page)

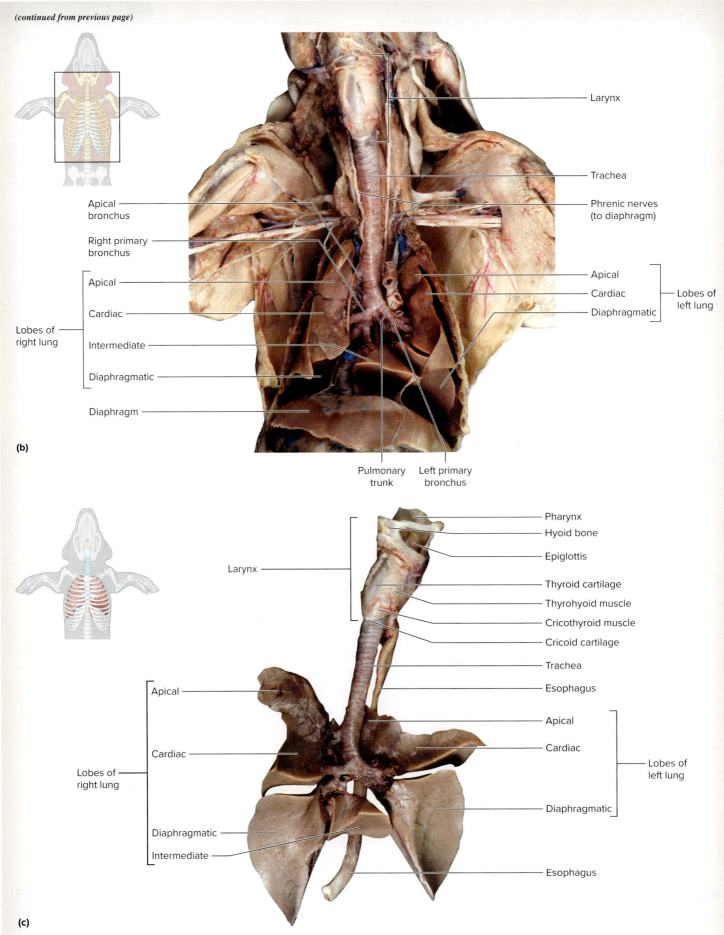

(b)

(c)

Figure 29.22   Respiratory System of the Fetal Pig (*continued*). (*b*) Dissection demonstrating the location of the lungs within the thoracic cavity. (*c*) Upper respiratory tract and lungs removed from the thoracic cavity to better visualize the lungs.
(*b, c*) ©McGraw-Hill Education/Christine Eckel

8. **Upper respiratory and digestive tracts:** The structures in the superior mediastinum, which consist of the **trachea**, the **esophagus**, the **thoracic duct**, and several blood vessels, will be dissected at this time. Begin by locating the esophagus and the **azygos vein** on the posterior thoracic wall next to the vertebral column (see figure 29.27). Next, look for a thin, brown or colorless vessel located between these two structures, which is the thoracic duct (see the Learning Strategy on this page). The thoracic duct is the main lymphatic duct of the body, which runs along the posterior abdominal and thoracic walls before it turns and enters the left subclavian vein. Identify the sympathetic ganglia just adjacent to the vertebral column. These ganglia contain cell bodies for postganglionic neurons of the sympathetic nervous system (see chapter 17 for more detail).

While approaching the most cranial part of the mediastinum, continue to clean the trachea as it enters the neck. There, carefully clean the larynx and the **thyroid gland**, a small, dark gland located anterior to the **larynx** between the common carotid arteries. To perform most of this cleaning of connective tissue, use small scissors and forceps, taking care not to pull too hard on any structure so as not to tear it. While following the common carotid arteries into the neck, identify a small nerve that runs between the **common carotid artery** and the **internal jugular vein** (figures 29.17 and 29.18).

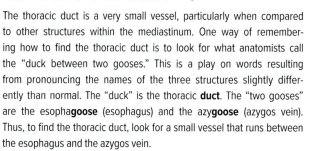

INTEGRATE

LEARNING STRATEGY

The thoracic duct is a very small vessel, particularly when compared to other structures within the mediastinum. One way of remembering how to find the thoracic duct is to look for what anatomists call the "duck between two gooses." This is a play on words resulting from pronouncing the names of the three structures slightly differently than normal. The "duck" is the thoracic **duct**. The "two gooses" are the esopha**goose** (esophagus) and the azy**goose** (azygos vein). Thus, to find the thoracic duct, look for a small vessel that runs between the esophagus and the azygos vein.

This is the **vagus nerve** (CN X). Dissection of structures that are located cranial from the thyroid cartilage is covered in exercise 29.17 (head and neck). Thus, for now, keep the rest of the head covered and hydrated so it will be easy to dissect later.

9. Upon completion of the dissection of the thoracic cavity, place the heart and lungs back into the thoracic cavity, close the thoracic cage, sprinkle the pig with wetting solution, and then wrap the pig in moist paper towels or a cloth and place it in the storage bag.

Abdomen and Pelvis

EXERCISE 29.14

THE ABDOMINAL CAVITY

While performing dissections on the abdominal cavity, make sure to keep the thoracic cavity closed so the heart and lungs don't fall out, and also keep them hydrated so they do not dry out.

1. Observe where the transverse cuts were previously made below the diaphragm, and the vertical cuts adjacent to the midline. Carefully reflect the anterior abdominal wall laterally. While doing this, note structures that run between the liver and the ventral abdominal wall. The structures are the *falciform ligament* and the **umbilical vein (figure 29.23)**. The falciform ligament anchors the liver to the ventral body wall. The umbilical vein runs in the caudal border of the falciform ligament. The umbilical arteries carry deoxygenated blood from the fetal pig to the placenta, and the umbilical vein carries oxygenated blood from the placenta to the fetal pig. Soon after birth, the umbilical vein degenerates and becomes the **round**

ligament of the liver, or ligamentum teres. Trace the umbilical vein from the area on the ventral body wall deep to the umbilicus to the location where it enters the liver. Note that the umbilical vein becomes the **ductus venosus** upon entering the liver. Locate the **umbilical arteries** (figure 29.23), which run from the umbilicus caudal alongside the developing urinary bladder, called the **allantoic bladder**, or urachus, in the fetal pig. The umbilical arteries arise from the internal iliac arteries and supply deoxygenated blood to the placenta via the umbilical cord. While reflecting the abdominal wall, cut the umbilical arteries, umbilical vein, and falciform ligament as close to the abdominal wall as possible so as to leave them "hanging" for later identification.

2. **Superficial dissection:** One superficial structure of great importance is the **greater omentum**, which overlies most of the contents of the abdominal cavity. In the fetal pig,

(continued on next page)

(continued from previous page)

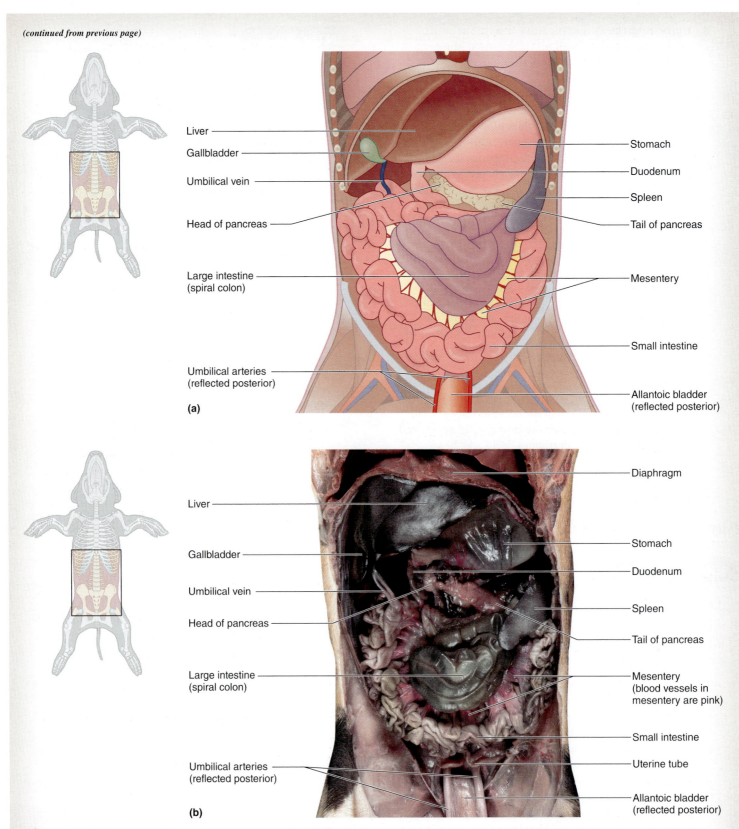

Liver
Gallbladder
Umbilical vein
Head of pancreas
Large intestine (spiral colon)
Umbilical arteries (reflected posterior)

(a)

Stomach
Duodenum
Spleen
Tail of pancreas
Mesentery
Small intestine
Allantoic bladder (reflected posterior)

Liver
Gallbladder
Umbilical vein
Head of pancreas
Large intestine (spiral colon)
Umbilical arteries (reflected posterior)

(b)

Diaphragm
Stomach
Duodenum
Spleen
Tail of pancreas
Mesentery (blood vessels in mesentery are pink)
Small intestine
Uterine tube
Allantoic bladder (reflected posterior)

Figure 29.23   Gastrointestinal Organs of the Fetal Pig. In this superficial dissection of the abdominal cavity of a female pig, the majority of the GI system organs can be seen clearly, with the exception of the falciform ligament, which was cut when the ventral abdominal wall was removed. The gallbladder can be found deep to the pointer tip that indicates its location. (*a*) Illustration. (*b*) Dissection.

arises in the abdominal cavity, the greater omentum can move around until it "walls off" that area of the cavity to prevent the spread of infection throughout the abdominal cavity. What are the two structures to which the greater omentum is attached (see chapter 24 for review)?

_____

and _____

3. **Deep dissection:** Exploration of the abdominal cavity requires much less work than dissection of the musculature because there is not a lot of connective tissue that needs to be removed. Instead, the majority of the organs move around freely because of the **peritoneum** that covers them. Thus, the "dissection" will largely consist of gently separating organs from one another to view underlying structures and to appreciate the relationships among these organs. Be sure to take extra care when dissecting each organ, because structures are delicate in the fetal pig and can be easily damaged. Using figure 29.23 as a guide, identify the following organs or parts of organs within the abdominal cavity:

☐ **Allantoic bladder**   ☐ **Pancreas**

☐ **Large intestine (spiral colon)**   ☐ **Small intestine**

☐ **Liver**   ☐ **Spleen**

      ☐ **Stomach**

4. **The liver and gallbladder:** Observe the liver. Note a band of peritoneum that connects the liver ventrally to the body wall. This is the **falciform ligament**. The posterior border of the falciform ligament contains the **umbilical vein**, which will become the **round ligament of the liver** in the adult. The falciform ligament of the liver was observed in step 1 of this dissection. Observe the **lobes of the liver**.

How many are there? _____
Note also the **gallbladder**, which has a characteristic green color. Also note the tubelike structure that connects the gallbladder to the **common hepatic duct**. This structure is the **cystic duct**. What substance is stored in the gallbladder, which is also the substance that gives it a green color? _____ What is the function of this substance?

_____

5. **The stomach, pancreas, spleen, and duodenum:** The liver will need to be lifted cranially to view the structures that lie deep to it. If keeping the liver out of the way is difficult, use a scalpel to carefully remove the left

lateral lobe of the liver. However, be sure to check with the instructor before doing this. Note the thin layer of peritoneum that connects the liver to the stomach. This is the **lesser omentum**. Deep to the caudal part of the liver is the **stomach**. Deep to the stomach is the **pancreas**, and near the **head of the pancreas** is the first part of the small intestine: the **duodenum** (figure 29.23). Follow the **body of the pancreas** laterally to find the **tail of the pancreas**, which ends at the hilum of the **spleen**. Using figure 29.23 as a guide, observe the area deep to the head of the pancreas and near the hilum of the liver to locate the following structures:

☐ **Cystic duct**   ☐ **Lesser omentum**

☐ **Duodenum**   ☐ **Spleen**

☐ **Gallbladder**   ☐ **Stomach**

☐ **Head of pancreas**   ☐ **Tail of pancreas**

☐ **Ileum**

6. **The jejunum, ileum, and large intestine:** Lay the stomach and pancreas back into place and observe the **spleen** and the remainder of the intestinal tract. To fully appreciate the length of the small intestine, pick it up at the end of the duodenum and "walk" your fingers along the length of the small intestine up to the **ileocecal valve**. It may take a long time to reach that location! While doing this, make note of the **mesenteries**, which anchor the small intestine to the dorsal body wall (figure 29.23). It may be difficult to determine jejunum from ileum in the pig, but try to use the criteria listed in figure 26.6 to distinguish the two from each other. It is often fun to have one dissection group remove the small intestine in its entirety from the abdominal cavity, dissect away its mesentery so it can be extended fully, and measure its length. Then record the length of the entire small intestine here: _____ cm

7. **The hepatic portal system:** After identifying the organs of the gastrointestinal system within the abdominal cavity, observe the blood supply to these organs. Before beginning, see figure 22.9 to review the hepatic portal circulation of the human. What are the three major veins that converge in the human to become the hepatic portal vein?

a. _____

b. _____

c. _____

(continued on next page)

(continued from previous page)

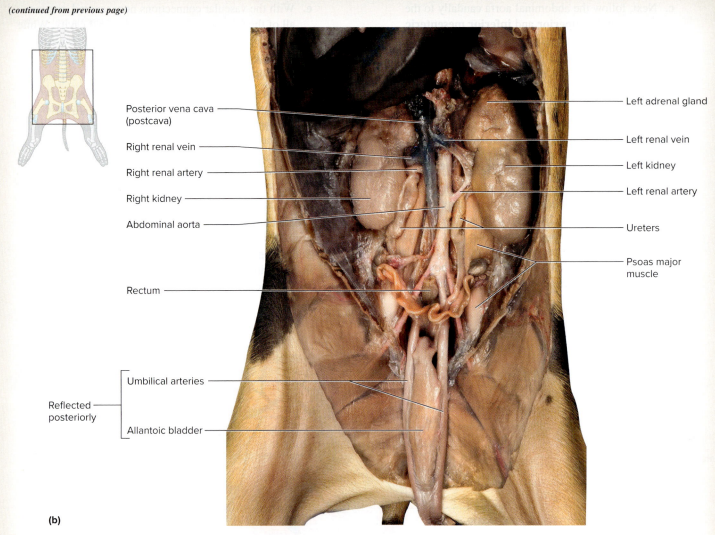

Posterior vena cava
(postcava)

Right renal vein

Right renal artery

Right kidney

Abdominal aorta

Rectum

Umbilical arteries

Reflected
posteriorly

Allantoic bladder

Left adrenal gland

Left renal vein

Left kidney

Left renal artery

Ureters

Psoas major
muscle

(b)

Figure 29.25   Dorsal Abdominal Wall of the Fetal Pig (*continued*). (*b*) Dissection.
(*b*) ©McGraw-Hill Education/Christine Eckel

4. Locate the hilum of the right and left kidneys. Take care in the hilum of the kidney so as not to cut the **ureters**, **renal arteries**, or **renal veins**. Gently tease each tube apart using blunt dissection, keeping in mind that each structure is quite delicate. While cleaning, trace these structures to their respective connections with the allantoic bladder (ureters), aorta (renal arteries), and posterior vena cava (renal veins). On the left side, in particular, take care when cleaning the **renal vein**. Look for a small vein that comes off of its caudal surface. This is the left **gonadal** (ovarian or testicular) **vein**.

5. **Internal anatomy of the kidney:** Choose one of the kidneys to bisect to view its internal anatomy. If deciding to completely remove one kidney, remove the right kidney. To do this, simply bisect the renal artery and vein midway between the hilum of the kidney and the aorta/posterior vena cava. Then bisect the ureter distal to the renal pelvis. Use a scalpel with a large blade or a dissecting knife to slice the kidney along the

cranial/caudal plane (**figure 29.26**). Using figure 29.26 as a guide, identify the following structures:

☐ **Cortex**   ☐ **Renal artery**
☐ **Fibrous capsule**   ☐ **Renal papilla**
☐ **Major calyx**   ☐ **Renal pelvis**
☐ **Medulla**   ☐ **Renal vein**
☐ **Minor calyx**   ☐ **Ureter**

Are there visible differences between the pig kidney and the human kidney? _____ If so, describe them in the space provided:

_____

_____

_____

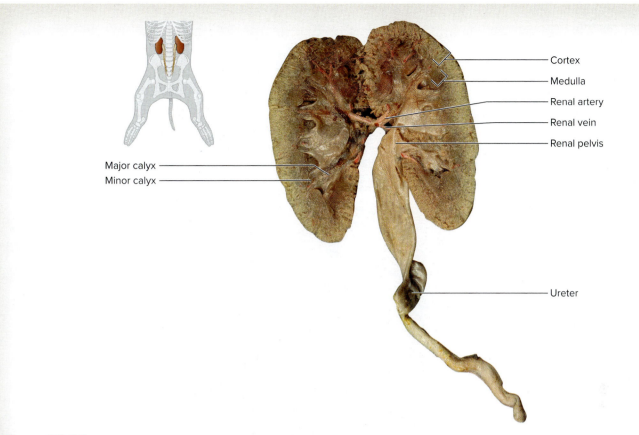

Figure 29.26 **Internal Anatomy of the Fetal Pig Kidney.** Horizontal section through the right kidney of the fetal pig. The kidney is encased in a fibrous capsule (not shown).
©McGraw-Hill Education/Christine Eckel

6. While tracing the ureters caudally toward the allantoic bladder, identify the **psoas major muscle**, which lies adjacent to the vertebral column.

7. **Identification of thoracic and abdominal vasculature:** Before beginning, remove the heart and lungs from the thoracic cavity and the GI viscera from the abdominal cavity. Using **figure 29.27** as a guide, identify the following vessels in the abdominal and thoracic cavities:

- ☐ **Abdominal aorta**
- ☐ **Adrenal veins**
- ☐ **Anterior vena cava (precava)**
- ☐ **Aortic arch**
- ☐ **Azygos vein**
- ☐ **Brachiocephalic veins**
- ☐ **Celiac trunk**

- ☐ **Common iliac vein**
- ☐ **Descending thoracic aorta**
- ☐ **External iliac artery**
- ☐ **External iliac vein**
- ☐ **Hepatic veins**
- ☐ **Iliolumbar artery**
- ☐ **Iliolumbar vein**
- ☐ **Internal iliac artery**
- ☐ **Internal iliac vein**
- ☐ **Left gonadal vein**
- ☐ **Posterior vena cava (postcava)**
- ☐ **Renal arteries**
- ☐ **Renal veins**
- ☐ **Right gonadal vein**

8. Upon completion of the dissection of the dorsal abdominal wall, place the organs gently back into the abdominal cavity, close the abdominal wall, sprinkle the pig with wetting solution, and then wrap the pig in wet paper towels and place it in the storage bag.

(continued on next page)

(continued from previous page)

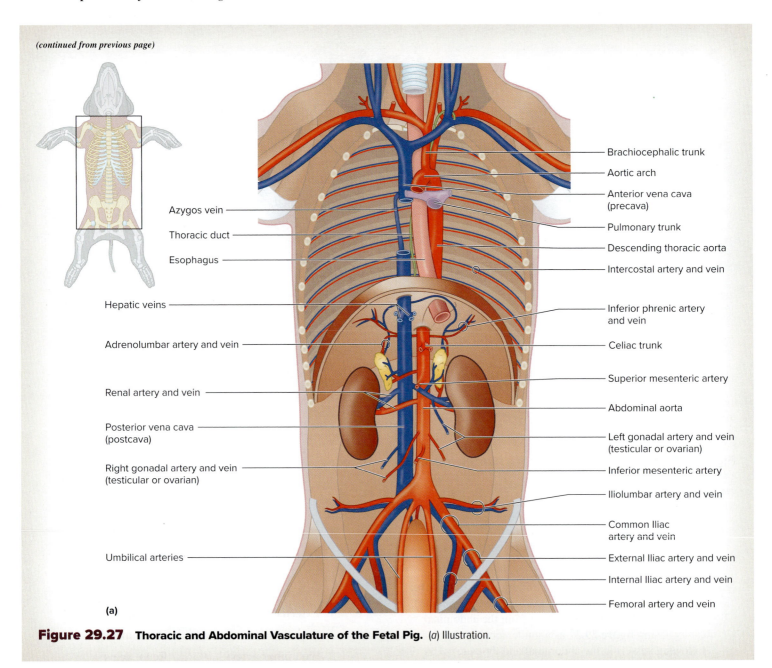

Figure 29.27 **Thoracic and Abdominal Vasculature of the Fetal Pig.** (*a*) Illustration.

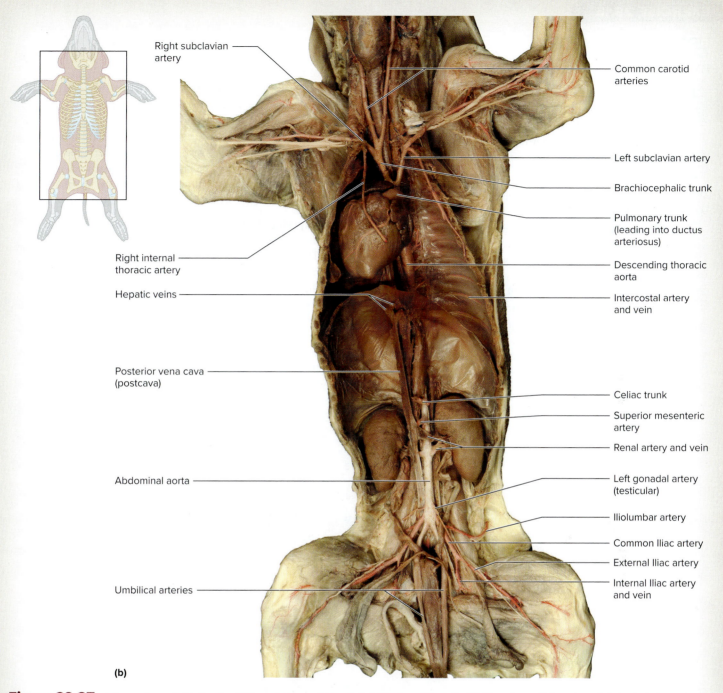

Right subclavian artery

Common carotid arteries

Left subclavian artery

Brachiocephalic trunk

Pulmonary trunk (leading into ductus arteriosus)

Right internal thoracic artery

Descending thoracic aorta

Hepatic veins

Intercostal artery and vein

Posterior vena cava (postcava)

Celiac trunk

Superior mesenteric artery

Renal artery and vein

Abdominal aorta

Left gonadal artery (testicular)

Iliolumbar artery

Common Iliac artery

External Iliac artery

Umbilical arteries

Internal Iliac artery and vein

(b)

Figure 29.27 **Thoracic and Abdominal Vasculature of the Fetal Pig (*continued*).** (*b*) Dissection—all thoracic and abdominal viscera have been removed. Most veins are not shown in this figure, yet may be found just adjacent to the corresponding arteries. The azygos and adrenal veins are not shown in this figure.

(*b*) ©McGraw-Hill Education/Christine Eckel

EXERCISE 29.16

UROGENITAL SYSTEMS

EXERCISE 29.16 A Male Urogenital System

1. Follow the instructions here to dissect a male pig. If dissecting a female pig, be sure to come back and identify all of the structures listed in this section that are pertinent to the male pig, because knowledge of both male and female reproductive structures is required.

2. **External genitalia:** Lay the pig ventral side up on the dissecting tray and observe the ventral abdomen caudal/ posterior to the umbilicus. When the muscles of the ventral hindlimb and the ventral abdominal wall was dissected, the area around the spermatic cord and external genitalia was preserved. These structures will now be the focus of the dissection. Begin by observing the illustration and dissection photo in **figure 29.28**, which demonstrate the general location of male reproductive structures (figure 29.28*a*) and their appearance once the skin around them is removed. Next, refocus attention on the dissection specimen. Palpate the skin in the ventral abdominopelvic region to feel for the penis and testes. Next, look for the

external urethral orifice (preputial orifice), which lies just caudal to the umbilicus. This is where the tip (glans) of the penis is located. Unlike in the human, the penis of the fetal pig is located almost entirely beneath the surface of the skin on the ventral abdomen. Thus, before doing any cutting, palpate the penis as it travels from the external urethral orifice caudal/posterior until it reaches the pubic symphysis. Also identify the spermatic cord. The **spermatic cord** contains the testicular vessels and the ductus deferens. Follow the path of the spermatic cord deep to the skin while dissecting toward the testes.

3. **Exposing the testes and penis:** Use tissue forceps to pull on the skin that remains in the ventral body wall. Then use a scalpel to *carefully* begin to cut the skin and the fascia that lies deep to it. Make skin incisions next to the spermatic cord, not directly on top of it. Follow the spermatic cord inferiorly into the **scrotum** until it reaches the **testes**. Then carefully remove the skin around the testes. The spermatic cord and scrotum are covered with connective tissue superficially. Carefully dissect away this connective tissue around the testes and look for the

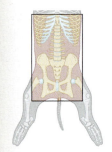

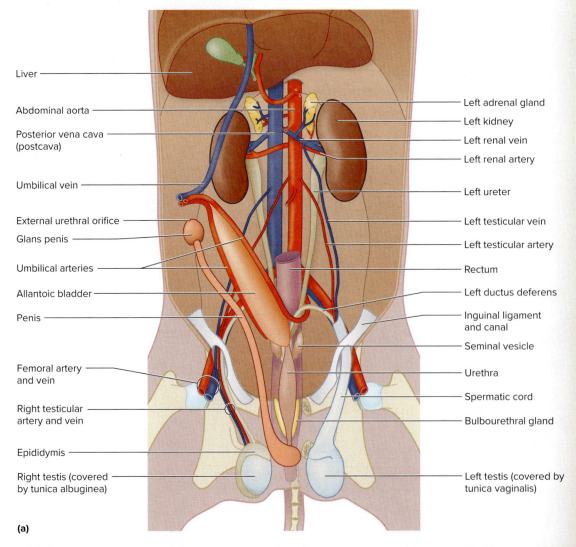

Liver

Abdominal aorta

Posterior vena cava (postcava)

Umbilical vein

External urethral orifice

Glans penis

Umbilical arteries

Allantoic bladder

Penis

Femoral artery and vein

Right testicular artery and vein

Epididymis

Right testis (covered by tunica albuginea)

Left adrenal gland

Left kidney

Left renal vein

Left renal artery

Left ureter

Left testicular vein

Left testicular artery

Rectum

Left ductus deferens

Inguinal ligament and canal

Seminal vesicle

Urethra

Spermatic cord

Bulbourethral gland

Left testis (covered by tunica vaginalis)

(a)

Figure 29.28   Male Urogenital System. In this figure the allantoic bladder, vessels, and penis are shown in near anatomical position. (*a*) Illustration.

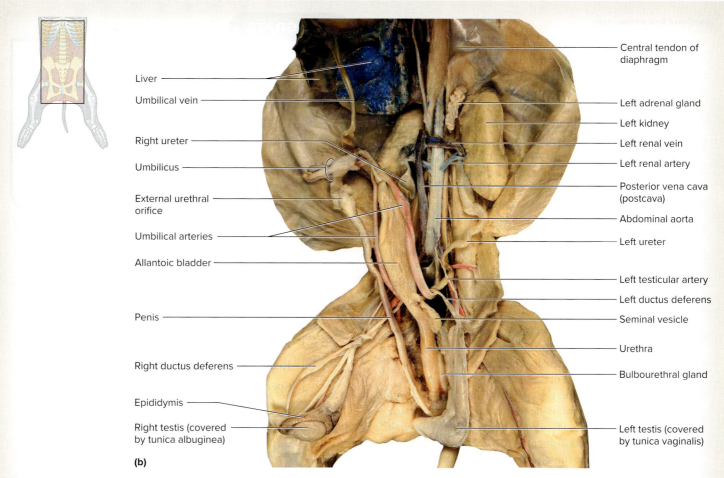

(b)

Figure 29.28 **Male Urogenital System (*continued*).** (*b*) Dissection.

(*b*) ©McGraw-Hill Education/Christine Eckel

"sac" that surrounds them. This is the **tunica vaginalis**, which is a remnant of the peritoneum. On one side only, use scissors to dissect away the tunica vaginalis to expose the **tunica albuginea**, which directly covers the testis, and the **epididymis**, which lies next to it. Look for a "vessel" coming off of the epididymis. This is the **ductus deferens**. The other vessels contained within the spermatic cord are the testicular artery and vein. In step 4 of the dissection, these structures will be traced as they travel back into the pelvic cavity. The process of tracing the path of these structures back into the pelvic cavity assists with identification. Medial to the testes and slightly superior to them is the **penis**, which extends from the urethra to the area immediately caudal to the umbilicus. Note that the tip of the penis expands to form the **glans**. Optional Activity: After identifying the penis, make a cross section through the shaft of the penis and identify the erectile tissues and spongy urethra. Refer to figure 27.16 for a figure

demonstrating the homologous structures in the human for reference.

Using figure 29.28 as a guide, identify the following structures:

- ☐ **Adrenal glands**
- ☐ **Allantoic bladder**
- ☐ **Bulbourethral gland**
- ☐ **Ductus deferens**
- ☐ **Epididymis**
- ☐ **External urethral orifice**
- ☐ **Glans penis**
- ☐ **Kidneys**
- ☐ **Penis**
- ☐ **Renal artery**
- ☐ **Renal vein**
- ☐ **Seminal vesicle**
- ☐ **Spermatic cord**
- ☐ **Testicular artery**
- ☐ **Testis (right and left)**
- ☐ **Tunica albuginea**
- ☐ **Tunica vaginalis**
- ☐ **Umbilical arteries**
- ☐ **Umbilical vein**
- ☐ **Umbilicus**
- ☐ **Ureters**

(*continued on next page*)

(continued from previous page)

and cutting toward the **vagina**. After opening the urogenital sinus, identify the following structures:

- ☐ **Allantoic bladder**
- ☐ **Body of uterus**
- ☐ **External urethral orifice**
- ☐ **Genital papilla**
- ☐ **Kidney**
- ☐ **Ovary**
- ☐ **Renal artery**
- ☐ **Renal vein**
- ☐ **Umbilical arteries**
- ☐ **Umbilical vein**
- ☐ **Ureter**
- ☐ **Urogenital sinus**
- ☐ **Uterine horn**
- ☐ **Vagina**

5. Next, lift abdominal structures such as the small intestine out of the way to obtain a better view of the pelvis and dorsal abdominal wall (or just remove them altogether if this was done in a previous dissection). Otherwise, pins or string may be used to hold the intestines out of the way while performing this part of the dissection. Once again locate the **kidneys**. Next, trace the ureters from the renal pelvis along the dorsal body wall until they pass deep to the **uterine horns**. Unlike the human uterus, which has a single **fundus** superior to the **body**, the pig uterus has two long **uterine horns**, which converge at the midline to form the **body** of the uterus. These horns are where fertilized ova implant and where developing fetal pigs reside during pregnancy.

The presence of these horns is what allows a female pig to produce a litter of several pigs, whereas the human uterus is designed to support only a single developing fetus. The muscular structure separating the uterus from the vagina is the **cervix** (not visible in figure 29.30c because it is an internal structure).

6. Note the fold of peritoneum that overlies the uterine horns. This is the **broad ligament**, which is a fold of the peritoneum that helps support the uterus, uterine tubes, and ovaries within the abdominopelvic cavity. In the fetal pig, the broad ligament is barely perceptible. While separating reproductive structures, note the thin membrane that covers them; this is the broad ligament. Trace the uterine horns laterally until they narrow where they converge with the **uterine tubes**. Continue to follow the uterine tubes until the openings into the tubes are visible. The openings into the uterine tubes are surrounded by small, delicate, finger-like projections called **fimbria(e)** (not visible in figure 29.30c). Medial to the fimbriae is a small, hard, oval structure, the **ovary**.

7. Next, lift abdominal GI system structures such as the small intestine out of the way to obtain a good view of the pelvis and dorsal body wall. Use dissecting pins or string to hold the intestines out of the way while performing this part of the dissection. Once again locate the **kidneys**. Next, trace the ureters from the renal pelvis along the dorsal body wall to the posterior aspect of the allantoic bladder. To do this, it may be necessary to remove some of the

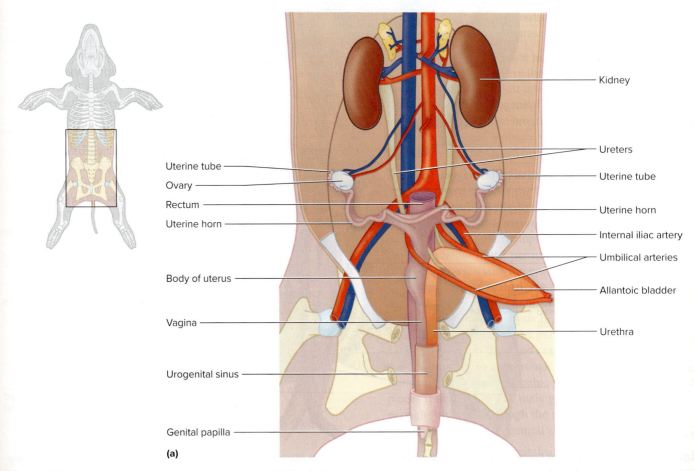

(a)

Figure 29.30 **Urogenital System of the Female Fetal Pig.** (a) Illustration.

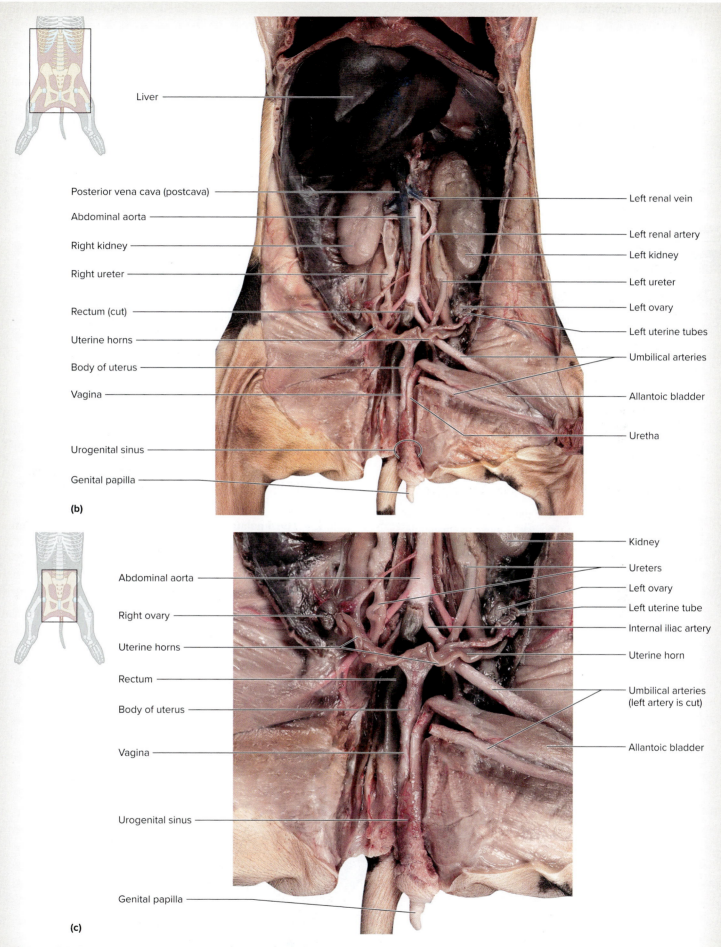

Liver

Posterior vena cava (postcava) —
Abdominal aorta —
Right kidney —
Right ureter —

Rectum (cut) —
Uterine horns —
Body of uterus —
Vagina —

Urogenital sinus —
Genital papilla —

Left renal vein
Left renal artery
Left kidney
Left ureter
Left ovary
Left uterine tubes
Umbilical arteries
Allantoic bladder
Uretha

(b)

Abdominal aorta —
Right ovary —
Uterine horns —
Rectum —
Body of uterus —
Vagina —
Urogenital sinus —
Genital papilla —

Kidney
Ureters
Left ovary
Left uterine tube
Internal iliac artery
Uterine horn
Umbilical arteries (left artery is cut)
Allantoic bladder

(c)

Figure 29.30 **Urogenital System of the Female Fetal Pig (*continued*).** (*b*) Superficial dissection showing internal urogenital system structures with the allantoic bladder and umbilical arteries reflected. (*c*) Deep dissection/close-up view of lower female urogenital system structures. The pubic symphysis has been removed, and the allantoic bladder and umbilical arteries have been reflected laterally in this dissection.

(*b, c*) ©McGraw-Hill Education/Christine Eckel

(*continued on next page*)

(continued from previous page)

peritoneum that covers the superior surface of the **allantoic bladder**. This is best accomplished by gently pulling on the peritoneum with tissue forceps. Once the peritoneum is removed, clean the tissue around the allantoic bladder and pelvic wall. Locate the **urethra** as it emerges inferior to the allantoic bladder. Then note how the urethra and vagina come together in the urogenital sinus.

8. Using figure 29.30 as a guide, identify the following structures:

☐ **Allantoic bladder** ☐ **Urogenital sinus**

☐ **Body of uterus** ☐ **Uterine horns**

☐ **Ovary** ☐ **Vagina**

☐ **Ureters**

Head and Neck

EXERCISE 29.17

THE HEAD, NECK, AND ORAL CAVITY

1. **Superficial dissection:** Place the pig ventral side up on the dissecting tray. Identify the **masseter muscle (figure 29.31)**, as well as the **lymph nodes** of the neck and the **external jugular veins**. Rotate the pig to get a clear view of the lateral aspect of the head. Using figure 29.31 as a guide, identify the following structures:

☐ **External jugular veins** ☐ **Parotid gland**

☐ **Lymph nodes** ☐ **Teeth**

☐ **Masseter muscle** ☐ **Tongue**

☐ **Parotid duct**

2. **Deep dissection:** To expose the deeper respiratory and digestive structures of the neck, remove the mandible first. Check with the instructor before proceeding in case the instructor wishes some groups to keep this region undissected. Obtain tissue forceps, a scalpel, and bone cutters. Use the scalpel to cut through the masseter muscle just inferior to the zygomatic arch. It is easier to cut through the bone after cutting through the soft tissues attached to it (i.e., the masseter muscle in this case). Obtain the bone cutters and place them in the corner of the mouth (the angle of the mouth) so they lie in the groove that was cut through the masseter muscle. Cut the mandible and any remaining associated tissues to free the **temporomandibular joint**. Do the same on the other side of the mouth. Now that the mandible is free, open the mouth as wide as it will go. Using **figure 29.32** as a guide, identify the following structures:

☐ **Hard palate** ☐ **Oropharynx** ☐ **Soft palate**

☐ **Nasopharynx** ☐ **Rostrum** ☐ **Tongue**

Parotid gland

External jugular vein

Submandibular gland

Sublingual gland (posterior part)

Mandible

Facial nerve (CN VII)

Masseter muscle

Parotid duct

Tongue

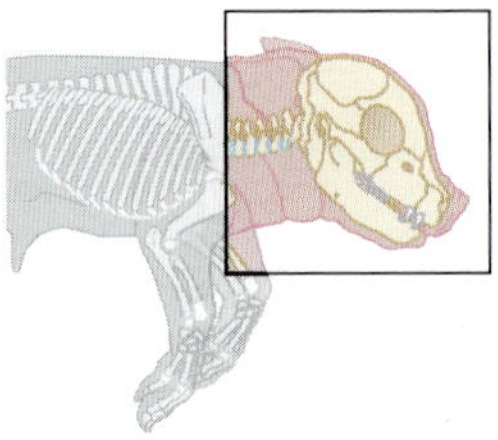

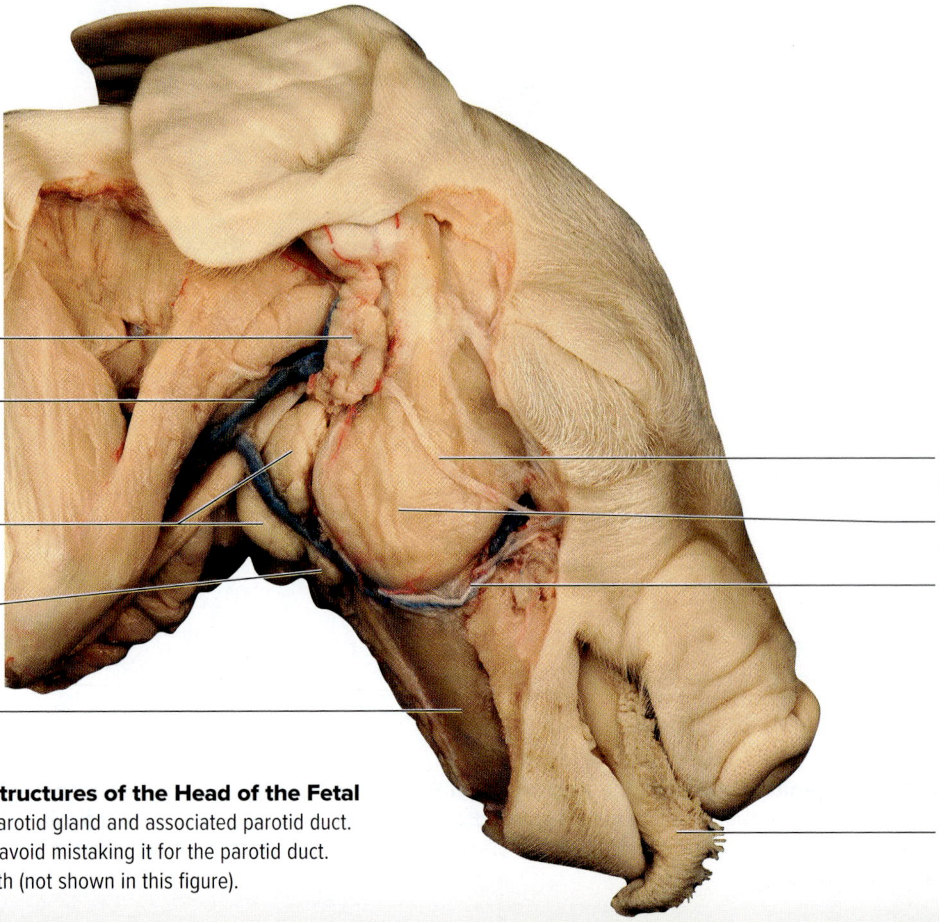

Figure 29.31 **Superficial Structures of the Head of the Fetal Pig.** Dissection demonstrating the parotid gland and associated parotid duct. The facial nerve (CN VII) is labeled to avoid mistaking it for the parotid duct. Open the mouth wide to view the teeth (not shown in this figure).
©McGraw-Hill Education/Christine Eckel

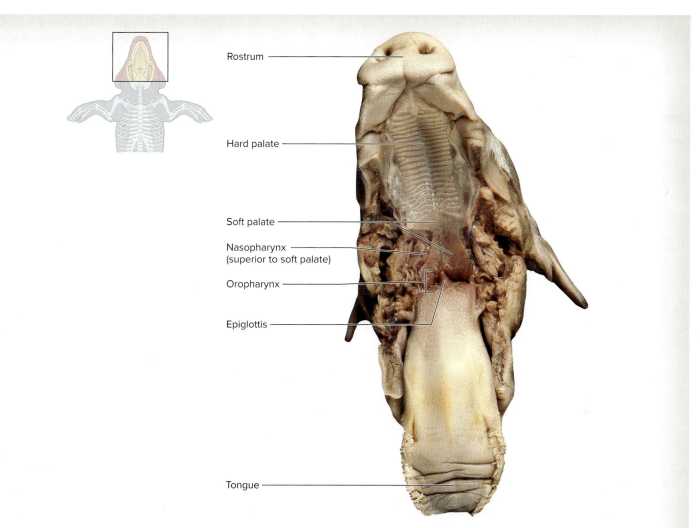

Rostrum

Hard palate

Soft palate

Nasopharynx
(superior to soft palate)

Oropharynx

Epiglottis

Tongue

Figure 29.32 **Deep Structures of the Fetal Pig Head.** The temporomandibular joint has been cut, allowing the mouth to open far enough to view the opening into the oropharynx. The nasopharynx is located beneath the soft palate.
©McGraw-Hill Education/Christine Eckel

The ❶ corresponds to the Learning Objective(s) listed in the chapter opener outline.

 Do You Know the Basics?

Exercise 29.1: Directional Terms and Surface Anatomy

1. When handling a preserved specimen, which of the following is an *essential* piece of safety equipment (i.e., it should be used every time you handle a specimen)? (Circle all that apply.) ❶

   a. protective mask

   b. safety goggles

   c. protective gloves

   d. nose plugs

   e. laboratory coat

2. Label the following figure of the surface anatomy of the pig. ❷

1 _____

2 _____

3 _____

4 _____

©McGraw-Hill Education/Christine Eckel

3. Match the human directional terms listed in column A with the corresponding fetal pig directional term listed in column B. ❸

Column A	Column B
_____ anterior	a. caudal
_____ inferior	b. cranial
_____ posterior	c. dorsal
_____ superior	d. ventral

4. The *surface anatomy* landmark that can be used to distinguish a female from a male fetal pig is _____. (Circle one.) ❹

   a. genital papilla

   b. mammary ridge

   c. mammary glands

   d. vestibule

   e. vagina

5. The integumentary structure that is unique to a fetal pig (as compared with a human) is _____ (hooves/hair). ❺

Exercise 29.2: Skeletal System

6. The fetal pig has _____ (7/12) cervical vertebrae. **6**

7. The embryonic fetal pig skeleton is primarily composed of _____ (hyaline cartilage/fibrocartilage). **7**

8. When a pig is standing the metatarsals are _____ (lifted off/lying flat on) the ground whereas when a human is standing the metatarsals are _____ (lifted off/lying flat on) the ground. **8**

Exercise 29.3: Skinning the Pig

9. Structures that appear as tight bands that run between the body wall and the skin and carry sensory information from receptors in the skin to the central nervous system are known as _____ (fascia/cutaneous nerves) **9**

Exercise 29.4: Muscles of the Head and Neck

10. Label the following figure of the head and neck muscles of the pig. **10**

©McGraw-Hill Education/Christine Eckel

Exercise 29.5: Muscles of the Thorax

11. Which of the following is a muscle of the pig for which there is no human analogue? (Circle one.) **11**

   a. anterior deep pectoral

   b. biceps brachii

   c. biceps femoris

   d. plantaris

   e. superficial pectoral

Exercise 29.6: Muscles of the Abdominal Wall

12. The fetal pig abdominal wall muscles are all the same as human abdominal wall muscles with one exception: There are no _____ (muscle fascicles/tendinous intersections) in the rectus abdominis muscle. **12**

13. The external oblique muscle in the pig has fibers that are directed _____ (medial/lateral) and _____ (cranial/caudal). **13**

Exercise 29.7: Muscles of the Back and Shoulder

14. Complete the following table of the superficial muscles of the back and shoulder regions of the pig. For each muscle, identify the human analogue and the action. **14**

Pig Muscle	Human Muscle	Action
Acromiotrapezius		
Clavotrapezius		
Deltoid		
Latissimus Dorsi		
Spinotrapezius		
Teres Major		

Exercise 29.8: Muscles of the Forelimb

15. Label the following diagram of the forelimb muscles of the pig. **15**

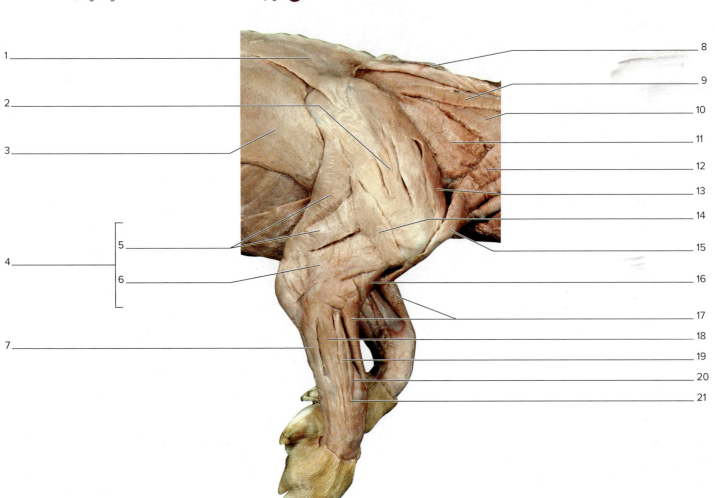

©McGraw-Hill Education/Christine Eckel

Exercise 29.9: Muscles of the Hindlimb

16. Label the following diagram of the hindlimb muscles of the pig. **16**

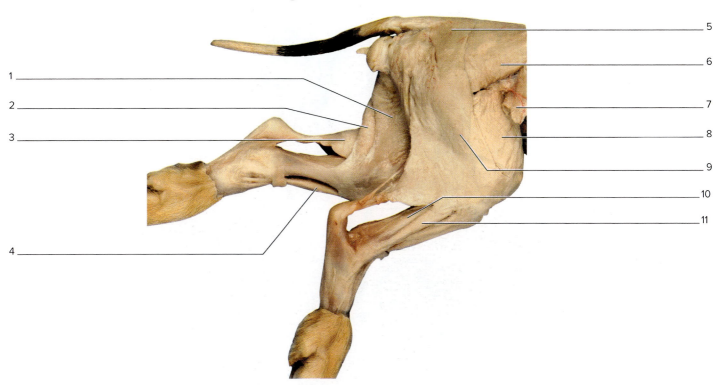

©McGraw-Hill Education/Christine Eckel

17. Which of the following muscles is *not* a muscle that composes the triceps surae in both pigs and humans? (Circle one.) **17**

a. gastrocnemius

b. plantaris

c. popliteus

d. soleus

Exercise 29.10: Nervous System: Spinal Cord and Peripheral Nerves

18. The group of nerves supplying the upper limb, which is found in the axillary region of the pig, is known as the

_____ (brachial/lumbosacral) plexus. **18**

19. The group of nerves supplying the lower limb, which is found in the pelvic region of the pig, is known as the

_____ (brachial/lumbosacral) plexus. **19**

Can You Apply What You've Learned?

35. Discuss adaptations to the pig skeletal and muscular systems that allow for quadrupedal (four-legged) rather than bipedal (two-legged) locomotion.

_____

_____

_____

36. Discuss the purpose of the thymus gland, and compare its size and structure in the fetal pig and the human.

_____

_____

_____

37. Identify the nerves that branch from the brachial plexus to innervate the muscles of the pig upper limb.

_____

_____

_____

38. Identify the nerves that branch from the lumbosacral plexus to innervate the muscles of the pig lower limb.

_____

_____

_____

39. Compare hepatic portal circulation in humans and pigs.

_____

_____

_____

Can You Synthesize What You've Learned?

40. Trace the flow of urine from the ureters to the urethra in the male and female pig, identifying all structures encountered along the path. Compare this path to that in humans.

_____

_____

_____

_____

41. Compare the size and structure of the uterine horn in the fetal pig with that of the uterine tube in humans. Discuss why there would be a dramatic difference between pigs and humans.

_____

_____

_____

_____

42. Describe the structure and purpose of the umbilical cord in fetuses of both pigs and humans, and discuss how the structures that compose the umbilical cord change upon birth.

_____

_____

_____

_____

Design Elements: Integrate: Clinical View icon (clipboard): ©Laia Design Studio/Shutterstock.com; Integrate Learning Strategy (pencil): ©Slavoljub Pantelic/Shutterstock.com

This appendix includes answers to labeling exercises, table completion activities, and calculation questions found in each chapter. Answers to in-exercise questions prompting students to record their personal observations and drawings are not provided due to their variability.

Chapter 1

Exercise 1.3, #1
Practice:
a. 0.5 cm
b. 0.74 ounces
c. 0.5 L
d. 12.7 cm
e. 9.9 ft
f. 11.4 L
g. 3000 mL

Figure 1.2   Identification of Common Dissection Instruments
1. Forceps
2. Scalpel blades
3. Scalpel (disposable)
4. Scalpel blade handle (#3)
5. Scalpel blade handle (#4)
6. Scissors (curved)
7. Scissors (pointed)
8. Blunt probe
9. Dissecting pins
10. Dissecting needle
11. Blunt probe
12. Hemostat

Exercise 1.5
1. A
2. B
3. A
4. A
5. B
6. B
7. A
8. C
9. A

Chapter 2

Figure 2.2   Sections Through a Human Brain
1. Coronal
2. Midsagittal
3. Transverse
4. Sagittal

Figure 2.4   Regional Terms
1. Oral
2. Cervical
3. Axillary
4. Brachial
5. Antebrachial
6. Carpal
7. Digital
8. Femoral
9. Crural
10. Frontal
11. Orbital
12. Mammary
13. Pelvic
14. Inguinal
15. Tarsal
16. Vertebral
17. Sacral
18. Calcaneal
19. Occipital
20. Lumbar
21. Perineal
22. Popliteal

Figure 2.5   Posterior View of an Individual with Three Reference Locations Marked
1. Right posterior thoracic region; lateral to vertebral region
2. Antebrachial; posterior forearm distal to the elbow
3. Femoral posterior thigh, distal to the gluteal region

Figure 2.7   Body Cavities
1. Posterior aspect
2. Ventral cavity
3. Cranial cavity
4. Vertebral canal
5. Thoracic cavity
6. Abdominopelvic cavity
7. Thoracic diaphragm
8. Abdominal cavity
9. Pelvic cavity
10. Thoracic cavity
11. Abdominopelvic cavity
12. Mediastinum
13. Pleural cavity
14. Pericardial cavity
15. Abdominal cavity
16. Pelvic cavity

Exercise 2.4, # 3
Abdominal cavity: Peritoneum
Thoracic cavity: Pleura
Pericardial cavity: Pericardium

Exercise 2.5, #2

Organ	Quadrant(s)	Region(s)
Left Kidney	Left upper	Left hypochondriac, left lumbar
Liver	Right upper, left upper	Right hypochondriac, right lumbar, epigastric
Pancreas	Right and left upper	Epigastric, left hypochondriac
Small Intestine	Right and left lower	Umbilical, hypogastric, right and left lumbar and right and left iliac
Spleen	Left upper	Left hypochondriac
Stomach	Right and left upper	Epigastric, left hypochondriac, umbilical
Urinary Bladder	Right and left lower	Hypogastric

Chapter 3

No Activities

Chapter 4

No Activities

Chapter 5

No Activities

Chapter 6

Table 6.5   Parts of a Nail

Structure	Description	Word Origins
Body	The main portion of the nail	
Eponychium	The fold of skin at the root of the nail; folds over the body of the nail	*epi*, upon, + *onyx*, nail
Hyponychium	The skin underneath the free border of the nail	*hypo*, under, + *onyx*, nail
Lunula	A white, curved area at the base of the nail	*luna*, moon
Nail Root	The portion of the nail embedded in the skin	

Chapter 7

Figure 7.2   Identifying Classes of Bones Based on Shape
1. Flat
2. Short
3. Long
4. Irregular
5. Long
6. Sesamoid (or short)
7. Irregular

Figure 7.6   The Human Skeleton
1. Skull
2. Mandible
3. Scapula
4. Humerus
5. Rib
6. Vertebra
7. Ilium
8. Sacrum
9. Coccyx
10. Ischium
11. Radius
12. Ulna
13. Pubis
14. Carpals
15. Metacarpals
16. Phalanges (of the hand)
17. Femur
18. Fibula
19. Tibia
20. Tarsals
21. Metatarsals
22. Phalanges (of the foot)

9. Head
10. Greater trochanter
11. Intertrochanteric crest
12. Lesser trochanter
13. Linea aspera

14. Popliteal surface
15. Adductor tubercle
16. Medial epicondyle
17. Medial condyle
18. Intercondylar fossa

Figure 9.16   The Tibia
1. Lateral condyle
2. Medial condyle
3. Tibial tuberosity
4. Anterior border
5. Medial malleolus

6. Medial condyle
7. Intercondylar eminence
8. Lateral condyle
9. Fibular articular facet
10. Medial malleolus

Figure 9.17   The Right Fibula
1. Head
2. Neck

3. Shaft
4. Lateral malleolus

Figure 9.18   The Tarsals
1. Phalanges
2. Metatarsals
3. Medial cuneiform
4. Navicular
5. Intermediate cuneiform

6. Lateral cuneiform
7. Cuboid
8. Talus
9. Calcaneus

Exercise 9.5D, #5

Tarsal Bone	Bone Shape/Appearance	Word Origin
Talus	Convex, triangular	*talus,* ankle
Calcaneus	Elongated	*calcaneous,* heel
Navicular	Shaped like a boat	*navis,* ship
Medial Cuneiform	Wedge-shaped	*cuneus,* wedge
Intermediate Cuneiform	Wedge-shaped	*cuneus,* wedge
Lateral Cuneiform	Wedge-shaped	*cuneus,* wedge
Cuboid	Cube-shaped	*kybos,* cube

Figure 9.19   The Metatarsals and Phalanges
1. Metatarsal II
2. Metatarsal III
3. Metatarsal IV
4. Metatarsal V
5. Lateral cuneiform
6. Cuboid
7. Distal phalanx
8. Middle phalanx

9. Proximal phalanx
10. Metatarsal I
11. Medial cuneiform
12. Intermediate cuneiform
13. Navicular
14. Talus
15. Calcaneus

Figure 9.20   Pelvis Radiograph, Anterior View
1. Ilium
2. Acetabulum
3. Head of femur
4. Neck of femur

5. Greater trochanter
6. Obturator foramen
7. Lesser trochanter

Figure 9.21   Radiograph of Knee, Medial View
1. Femur
2. Medial condyle
3. Neck of fibula
4. Fibula

5. Patella
6. Tibial tuberosity
7. Tibia

Figure 9.22   Surface Anatomy of the Lower Limb
1. Iliac crest
2. Anterior superior iliac spine
3. Sacrum
4. Coccyx
5. Greater trochanter of femur
6. Ischial tuberosity
7. Lateral epicondyle of femur
8. Head of fibula
9. Lateral malleolus

10. Medial epicondyle of femur
11. Patella
12. Tibial tuberosity
13. Shaft of tibia
14. Medial malleolus
15. Calcaneus
16. Metatarsals
17. Phalanges

Chapter 10

Figure 10.1   Fibrous Joints
1. Syndesmosis
2. Gomphosis
3. Suture

Figure 10.2   Cartilaginous Joints
1. Synchondrosis
2. Symphysis

3. Synchondrosis
4. Symphysis*

Figure 10.3   Diagram of a Representative Synovial Joint
1. Fibrous layer of articular capsule
2. Synovial membrane
3. Joint (articular) cavity

4. Articular capsule
5. Articular cartilage
6. Ligament

Figure 10.4   Classifications of Synovial Joints
1a. Saddle
1b. Biaxial
2a. Hinge
2b. Uniaxial
3a. Plane
3b. Uniaxial

4a. Ball and socket
4b. Multiaxial
5a. Condylar
5b. Biaxial
6a. Pivot
6b. Uniaxial

Figure 10.6   A Representative Synovial Joint: The Right Knee Joint
1. Posterior cruciate ligament
2. Lateral meniscus
3. Fibular collateral ligament
4. Anterior cruciate ligament
5. Medial meniscus
6. Tibial collateral ligament
7. Anterior cruciate ligament
8. Medial meniscus

9. Posterior cruciate ligament
10. Tibial collateral ligament
11. Fibular collateral ligament
12. Lateral meniscus
13. Suprapatellar bursa
14. Prepatellar bursa
15. Patellar ligament
16. Infrapatellar bursa

Chapter 11

Exercise 11.2 (Figure 11.1)   Muscles of the Human Body

	Muscle Name	Architecture
1	Orbicularis oculi	Circular
2	Deltoid	Multipennate
3	Pectoralis major	Convergent
4	Sartorius	Parallel
5	Rectus femoris	Bipennate
6	Trapezius	Convergent
7	Triceps brachii	Bipennate
8	Extensor digitorum	Parallel
9	Gastrocnemius	Bipennate (individual heads) Multipennate (entire muscle)

Table 11.4   Fascial Compartments of the Limbs and Their General Muscle Actions

Compartment	General Description of Muscle Actions
Compartments of the Arm	
Anterior	Flexion of shoulder and elbow
Posterior	Extension of shoulder and elbow
Compartments of the Forearm	
Anterior	Flexion of the wrist and digits (fingers)
Posterior	Extension of the wrist and digits (fingers)
Compartments of the Thigh	
Anterior	Extension of the knee, flexion of the hip
Posterior	Extension of the hip, flexion of the knee
Medial	Adduction of the thigh
Compartments of the Leg	
Anterior	Dorsiflexion and inversion of the ankle, extension of the digits (toes)
Posterior	Plantarflexion and inversion of the ankle, flexion of the digits (toes)
Lateral	Eversion of the ankle

*Note: Figure 10.2(4) illustrates a cross-section of the cauda equina in the lumbar region of the vertebral column.

Figure 11.12   The Crossbridge Cycle
1. Calcium binds to troponin, causing a conformational change in troponin and tropomyosin. Myosin binding sites on actin are exposed.
2. Myosin head attaches to the binding site on actin.
3. ADP and P_i are released, resulting in a power stroke.
4. ATP binds to the ATP binding site on myosin, resulting in the detachment of the myosin head from actin.
5. The myosin ATPase hydrolyzes ATP, forming ADP and P_i. The myosin head remains in a "reset" position.

Chapter 12

Figure 12.1   Muscles of Facial Expression
1. Epicranius (occipitofrontalis)
2. Procerus
3. Levator labii superioris
4. Zygomaticus minor
5. Zygomaticus major
6. Depressor labii inferioris
7. Corrugator supercilii
8. Nasalis
9. Levator anguli oris
10. Orbicularis oris
11. Mentalis
12. Epicranius (occipitofrontalis)
13. Buccinator
14. Orbicularis oculi
15. Zygomaticus minor
16. Zygomaticus major
17. Orbicularis oris
18. Depressor labii inferioris
19. Depressor anguli oris
20. Platysma

Figure 12.2   Muscles of Mastication
1. Temporalis
2. Masseter
3. Temporalis
4. Lateral pterygoid
5. Medial pterygoid

Figure 12.3   Muscles That Move the Tongue
1. Palatoglossus
2. Styloglossus
3. Hyoglossus
4. Genioglossus

Figure 12.4   Muscles of the Pharynx
1. Tensor veli palatini
2. Levator veli palatini
3. Superior constrictor
4. Stylopharyngeus
5. Middle constrictor
6. Inferior constrictor

Figure 12.5   Muscles of the Head and Neck
1. Mylohyoid
2. Digastric (anterior belly)
3. Digastric (posterior belly)
4. Omohyoid (superior belly)
5. Sternohyoid
6. Sternocleidomastoid
7. Thyrohyoid
8. Sternothyroid
9. Scalenes
10. Splenius capitis
11. Sternocleidomastoid
12. Scalenes
13. Omohyoid (inferior belly)
14. Digastric (posterior belly)
15. Mylohyoid
16. Digastric (anterior belly)
17. Sternothyroid
18. Omohyoid (superior belly)
19. Sternohyoid
20. Semispinalis capitis
21. Sternocleidomastoid
22. Splenius capitis

Concept Connection (p. 293)
The right sternocleidomastoid rotates the neck to the left. The right splenius capitis rotates the neck to the right. In summary: To rotate your neck to the right, you use the sternal head of the sternocleidomastoid on the left side of the neck, and the splenius capitis muscle on the right side of the neck.

Figure 12.6   Muscles of the Vertebral Column
1. Splenius capitis
2. Splenius cervicis
3. Iliocostalis
4. Longissimus
5. Spinalis
6. Semispinalis capitis
7. Semispinalis cervicis
8. Semispinalis thoracis
9. Multifidus
10. Quadratus lumborum

Figure 12.7   Muscles of Respiration
1. External intercostals
2. Internal intercostals
3. Transversus thoracis
4. Diaphragm
5. Internal intercostals
6. External intercostals
7. Diaphragm

Figure 12.9   Diaphragm
1. Caval opening (for inferior vena cava)
2. Aortic opening (hiatus)
3. Esophageal opening (hiatus)

Figure 12.10   Muscles of the Abdominal Wall
1. Tendinous intersections
2. Rectus abdominis
3. Transversus abdominis
4. Internal oblique
5. External oblique

Chapter 13

Figure 13.1   Muscles That Act About the Pectoral Girdle and Glenohumeral Joint
1. Trapezius
2. Deltoid
3. Pectoralis major
4. Biceps brachii (long head)
5. Biceps brachii (short head)
6. Subscapularis
7. Coracobrachialis
8. Pectoralis minor
9. Serratus anterior
10. Trapezius
11. Rhomboid minor
12. Rhomboid major
13. Deltoid
14. Rhomboid major
15. Latissimus dorsi
16. Levator scapulae
17. Supraspinatus
18. Infraspinatus
19. Teres minor
20. Teres major
21. Serratus anterior

Figure 13.2   Anterior (Flexor) Compartment of the Arm
1. Coracobrachialis
2. Biceps brachii (short head)
3. Biceps brachii (long head)
4. Tendon of the long head of biceps brachii
5. Coracobrachialis
6. Brachialis

Figure 13.3   Posterior (Extensor) Compartment of the Arm
1. Lateral head of triceps brachii
2. Long head of triceps brachii

Figure 13.4   Anterior (Flexor) Compartment of the Forearm
1. Pronator teres
2. Brachioradialis
3. Flexor retinaculum
4. Flexor carpi radialis
5. Palmaris longus
6. Flexor carpi ulnaris
7. Flexor digitorum superficialis
8. Palmar aponeurosis
9. Flexor pollicis longus
10. Pronator quadratus
11. Flexor digitorum profundus

Figure 13.6   Posterior (Extensor) Compartment of the Forearm
1. Extensor carpi ulnaris
2. Extensor digiti minimi
3. Extensor digitorum tendons
4. Extensor carpi radialis longus
5. Extensor carpi radialis brevis
6. Extensor digitorum
7. Abductor pollicis longus
8. Extensor pollicis brevis
9. Extensor pollicis longus
10. Extensor indicis
11. Supinator
12. Abductor pollicis longus
13. Extensor pollicis brevis

Figure 13.7   Intrinsic Muscles of the Hand
1. Medial lumbrical
2. Flexor digiti minimi brevis
3. Abductor digiti minimi
4. First dorsal interosseous
5. Lateral lumbricals
6. Adductor pollicis
7. Flexor pollicis brevis
8. Abductor pollicis brevis
9. Thenar group
10. Hypothenar group

Figure 13.9   Muscles That Act About the Hip Joint/Thigh
1. Gluteus maximus
2. Piriformis
3. Superior gemellus
4. Obturator internus
5. Inferior gemellus
6. Gluteus medius
7. Gluteus minimus
8. Quadratus femoris
9. Iliacus
10. Psoas major
11. Iliopsoas
12. Tensor fasciae latae

Figure 13.10   Medial Compartment of the Thigh
1. Pectineus
2. Adductor brevis
3. Adductor longus
4. Gracilis

Figure 13.11   Anterior Compartment of the Thigh
1. Inguinal ligament
2. Tensor fasciae latae
3. Iliotibial tract
4. Rectus femoris
5. Vastus lateralis
6. Pectineus
7. Adductor longus
8. Gracilis
9. Sartorius
10. Vastus medialis
11. Quadriceps tendon

Figure 13.12   Muscles That Flex the Knee Joint/Leg
1. Semimembranosus
2. Semitendinosus
3. Biceps femoris (long head)
4. Biceps femoris (short head)

Figure 13.13   Anterior Compartment of the Leg
1. Extensor digitorum longus
2. Extensor hallucis longus
3. Fibularis tertius
4. Tibialis anterior

Figure 13.14   Posterior Compartment of the Leg
1. Plantaris
2. Popliteus
3. Tibialis posterior
4. Flexor digitorum longus
5. Flexor hallucis longus
6. Calcaneal tendon

Figure 13.15   Lateral View of the Leg
1. Gastrocnemius
2. Soleus
3. Fibularis longus
4. Fibularis brevis
5. Extensor digitorum brevis
6. Tibialis anterior
7. Extensor digitorum longus
8. Extensor hallucis longus
9. Fibularis tertius
10. Extensor hallucis brevis

Figure 13.16   Intrinsic Muscles of the Foot
1. Flexor digitorum brevis
2. Abductor hallucis
3. Abductor digiti minimi
4. Lumbricals
5. Tendon of flexor hallucis longus
6. Tendons of flexor digitorum longus
7. Quadratus plantae
8. Adductor hallucis
9. Flexor hallucis brevis
10. Flexor digiti minimi brevis
11. Plantar interossei
12. Dorsal interossei

Chapter 14

No Activities

Chapter 15

Figure 15.1   Superior View of the Brain
1. Frontal lobe
2. Precentral gyrus
3. Central sulcus
4. Postcentral gyrus
5. Longitudinal fissure
6. Parietal lobe
7. Occipital lobe

Figure 15.2   Lateral View of the Brain
1. Frontal lobe
2. Parietal lobe
3. Central sulcus
4. Precentral gyrus
5. Postcentral gyrus
6. Lateral sulcus
7. Temporal lobe
8. Pons
9. Medulla oblongata
10. Occipital lobe
11. Transverse fissure
12. Cerebellum

Figure 15.3   Inferior View of the Brain
1. Frontal lobe
2. Infundibulum
3. Mammillary bodies
4. Temporal lobe
5. Pons
6. Occipital lobe
7. Olfactory bulb
8. Olfactory tract
9. Optic chiasm
10. Optic nerve
11. Optic tract
12. Midbrain
13. Cerebellum
14. Medulla oblongata

Figure 15.4   Midsagittal View of the Brain
1. Cingulate gyrus
2. Corpus callosum
3. Septum pellucidum
4. Hypothalamus
5. Mamillary body
6. Infundibulum
7. Cerebral peduncle
8. Pons
9. Medulla oblongata
10. Central sulcus
11. Thalamus
12. Parieto-occipital sulcus
13. Pineal body (gland)
14. Tectal plate (corpora quadrigemina)
15. Midbrain
16. Cerebral aqueduct
17. Fourth ventricle
18. Cerebellum

Figure 15.6   Meningeal Structures

(a) Coronal Section
1. Superior sagittal sinus
2. Falx cerebri
3. Pia mater
4. Subarachnoid space
5. Arachnoid mater
6. Falx cerebri
7. Dura mater (periosteal layer)
8. Dura mater
9. Superior sagittal sinus
10. Arachnoid villi
11. Dura mater (meningeal layer)

(b) Midsagittal Section
12. Falx cerebri
13. Superior sagittal sinus
14. Inferior sagittal sinus
15. Diaphragma sellae
16. Straight sinus
17. Tentorium cerebelli
18. Transverse sinus
19. Confluence of sinuses
20. Falx cerebelli
21. Occipital sinus

Figure 15.7   Cast of the Ventricles of the Brain
1. Lateral ventricles
2. Interventricular foramen
3. Third ventricle
4. Cerebral aqueduct
5. Fourth ventricle

Figure 15.8   Cerebrospinal Fluid (CSF) Production and Circulation
1. Periosteal dura
2. Arachnoid villus
3. Superior sagittal sinus
4. Meningeal dura
5. Arachnoid mater
6. Subarachnoid space
7. Pia mater
8. Cerebral cortex
9. Subarachnoid space
10. Arachnoid villi
11. Dura mater
12. Superior sagittal sinus
13. Pia mater
14. Choroid plexus in third ventricle
15. Choroid plexus in lateral ventricle
16. Interventricular foramen
17. Cerebral aqueduct
18. Fourth ventricle
19. Choroid plexus in fourth ventricle
20. Median or lateral apertures
21. Subarachnoid space
22. Central canal of spinal cord

Exercise 15.7, #2
1. Lateral ventricles
2. Interventricular foramen
3. Third ventricle
4. Cerebral aqueduct
5. Fourth ventricle
6. Median and lateral apertures
7. Subarachnoid space
8. Arachnoid villi

Exercise 15.8, #4

Point of Exit	Cranial Nerve
Midbrain	III, IV
Pons	V, VI, VII, part of VIII
Medulla Oblongata	part of VIII, IX, X, XI, XII

Chapter 16

Figure 16.1   Regional Gross Anatomy of the Spinal Cord
1. Cervical plexus
2. Cervical enlargement
3. Brachial plexus
4. Lumbosacral enlargement
5. Conus medullaris
6. Lumbar plexus
7. Sacral plexus
8. Posterior median sulcus
9. Posterior roots
10. Anterior roots
11. Pia mater
12. Denticulate ligament
13. Conus medullaris
14. Dura mater
15. Cauda equina
16. Filum terminale
17. Posterior root ganglion

Figure 16.3   The Phrenic Nerves in the Thoracic Cavity
1. Right phrenic nerve
2. Diaphragm
3. Left phrenic nerve

Figure 16.5   Major Nerves of the Brachial Plexus
1. Lateral cord
2. Posterior cord
3. Medial cord
4. Musculocutaneous nerve
5. Axillary nerve
6. Radial nerve
7. Median nerve
8. Ulnar nerve
9. Superior trunk
10. Middle trunk
11. Inferior trunk

Figure 16.6   Nerves of the Lumbar Plexus
1. Obturator nerve
2. Femoral nerve

Figure 16.7   Nerves of the Sacral Plexus Within the Gluteal and Popliteal Regions
1. Inferior gluteal nerve
2. Posterior femoral cutaneous nerve
3. Pudendal nerve
4. Superior gluteal nerve
5. Sciatic nerve
6. Tibial nerve
7. Common fibular nerve

Figure 16.8   Model of a Spinal Cord in Cross Section
1. Posterior funiculus
2. Posterior horn
3. Lateral horn
4. Posterior rootlets
5. Posterior root ganglion
6. Spinal nerve
7. Anterior root
8. Anterior horn
9. Posterior median sulcus
10. Anterior rootlets
11. Anterior median fissure

Figure 16.10   Components of a Reflex
1. Control center/interneuron
2. Sensory neuron
3. Motor neuron
4. Effector
5. Receptor

Exercise 16.5, #7

Region of Spinal Cord	Relative Size and Shape of the Spinal Cord	Predominantly White or Gray Matter?	Other Distinguishing Characteristics
Cervical	Relatively large; Oval shape	White matter	Large ventral horns
Thoracic	Relatively small; Circular	White matter	Presence of lateral horns
Lumbar	Largest cross-sectional area; Circular	Gray matter	Large broad ventral horn
Sacral	Smallest cross-sectional area; Circular	Gray matter	Numerous nerve roots surrounding the spinal cord

Chapter 17

Figure 17.3   Overview of the Parasympathetic Division of the ANS
1. Oculomotor nerve (CN III)
2. Facial nerve (CN VII)
3. Glossopharyngeal nerve (CN IX)
4. Vagus nerve (CN X)
5. Cardiac plexus
6. Abdominal aortic plexus
7. Pelvic splanchnic nerves

Table 17.2   Parasympathetic Division Outflow

Nerve(s)	Origin of Preganglionic Neurons	Autonomic Ganglia	Effectors Innervated	Example Effector Response
Oculomotor (CN III)	Midbrain	Ciliary	Eye ciliary muscles; sphincter pupillae muscle	Alters shape of the lens for close vision; constricts pupil
Facial (CN VII)	Pons	Pterygopalatine; submandibular	Lacrimal glands; glands of nasal cavity, palate, oral cavity; submandibular and sublingual salivary glands	Increases saliva production
Glossopharyngeal (CN IX)	Medulla oblongata	Otic	Parotid salivary glands	Increases saliva production
Vagus (CN X)	Medulla oblongata	Terminal and intramural	Thoracic viscera and most abdominal viscera	Decreases heart rate; constricts bronchi/bronchioles; stimulate gastrointestinal secretions and motility; stimulates glycogenesis
Pelvic Splanchnic Nerves	S2 – S4 (spinal cord)	Terminal and intramural	Some abdominal viscera and most pelvic viscera	Increase gastrointestinal secretions and motility; contract smooth muscle in the urinary bladder and relax the internal urethral sphincter; erection of the female clitoris and the male penis

Figure 17.4   Overview of the Sympathetic Division of the ANS
1. Sympathetic trunk ganglia (paravertebral)
2. Adrenal medulla
3. Prevertebral ganglia
4. Sympathetic trunk

Table 17.3   Sympathetic Division Pathways

Pathway	Origin (Spinal Segment)	Destination	Effectors Innervated	Effector Response
Spinal Nerve	T1 – L2	Integumentary structures	Sweat glands, arrector pili muscles, blood vessels in skin of neck, torso, and limbs	Increases sweat production
Postganglionic Sympathetic Nerve	T1 – T5	Head and neck viscera; thoracic organs	Sweat glands; arrector pili muscles, and blood vessels in skin of head; dilator pupillae muscle of eye; superior tarsal muscle of eye; neck viscera; esophagus, heart, lungs, blood vessels within thoracic cavity	Increases sweat production; dilates pupil; dilates bronchi/bronchioles; constricts blood vessels
Splanchnic Nerve	T5 – L2	Abdominal and pelvic organs	Abdominal portion of esophagus, stomach, liver, gallbladder, spleen, pancreas, small intestine, most of large intestine, kidneys, ureters, adrenal glands, blood vessels within abdominopelvic cavity; distal portion of large intestine, anal canal, and rectum; distal part of ureters; urinary bladder; reproductive organs	Decreases gastrointestinal secretion and motility; constricts blood vessels; relaxes smooth muscle in the urinary bladder and contracts the internal urethral sphincter
Adrenal Medulla	T8 – T12	Adrenal gland	Neurosecretory cells of adrenal medulla	Increases epinephrine release

Chapter 18

Figure 18.2b   Accessory Structures of the Eye
1. Eyelashes
2. Medial canthus
3. Eyebrow
4. Superior eyelid
5. Pupil
6. Sclera
7. Iris
8. Inferior eyelid

Figure 18.4   Extrinsic Eye Muscles
1. Superior oblique
2. Superior rectus
3. Lateral rectus
4. Inferior rectus
5. Inferior oblique
6. Superior rectus
7. Superior oblique
8. Medial rectus
9. Inferior rectus
10. Inferior oblique

Figure 18.8a   Classroom Model of the Ear
1. Tensor tympani
2. Semicircular canals
3. Stapes
4. Malleus
5. Incus
6. Auricle (pinna)
7. External acoustic meatus
8. Tympanic membrane
9. Auditory tube

Chapter 19

Figure 19.1   Labeling Major Endocrine Glands of the Body
1. Parathyroid glands
2. Thyroid gland
3. Thymus
4. Adrenal cortex
5. Adrenal medulla
6. Testes
7. Pineal gland
8. Hypothalamus
9. Pituitary gland
10. Pancreas
11. Ovaries
12. Hypothalamus
13. Anterior pituitary gland
14. Pineal gland
15. Posterior pituitary gland

Table 19.1   The Hormone-Secreting Cells of the Pituitary Gland

Pituitary Gland	Hormone Released	Hormone Action	Cell Type	Hypothalamic Control of Release	Word Origin
Anterior Pituitary (Adenohypophysis)	Growth hormone (GH)	Stimulates liver to produce IGF-1 (insulin-like growth factor-1); IGF-1 promotes bone and muscle growth	Somatotropes (acidophil)	Growth hormone releasing hormone (GHRH) Growth hormone inhibiting hormone (GHIH)	*grothr*, growth, + *hormon*, to set in motion
	Prolactin (PRL)	Stimulates development of mammary glands; milk production	Lactotropes (acidophil)	Prolactin-releasing hormone (PRH) Prolactin-inhibiting hormone (PIH)	*pro-*, before, + *lac*, milk
	Follicle-stimulating hormone (FSH)	Stimulates growth and maturation of ovarian follicles (females); stimulates spermatogenesis (males)	Gonadotropes (basophil)	Gonadotropin-releasing hormone (GnRH)	*folliculus*, a small sac (referring to the ovarian follicle)
	Luteinizing hormone (LH)	Induces ovulation and stimulates estrogen production (females); stimulates testosterone production (males)			*luteus*, yellow (referring to the corpus luteum of the female ovary)
	Adrenocorticotropic hormone (ACTH)	Stimulates growth, development, and secretion of steroid hormones from adrenal cortex	Corticotropes (basophil)	Corticotropin-releasing hormone (CRH)	*adrenocortico*, referring to the adrenal cortex, + *trophe*, nourishment
	Thyroid-stimulating hormone (TSH)	Stimulates thyroid hormone secretion from the thyroid gland	Thyrotropes (basophil)	Thyrotropin-releasing hormone (TRH)	*thyroid*, shaped like an oblong shield
Pituitary Gland	**Hormone Released**	**Hormone Action**	**Cell Type**	**Hypothalamic Nuclei Primarily Responsible for Production**	**Word Origin**
Posterior Pituitary (Neurohypophysis)	Antidiuretic hormone (ADH)	Increases water retention by the kidneys (maintains blood volume and pressure)	Axon terminals	Supraoptic nuclei	*anti*, against, + *diuresis*, excretion of urine
	Oxytocin	Stimulates uterine contractions; stimulates milk ejection from mammary glands	Axon terminals	Paraventricular nuclei	*okytckos*, swift birth

Chapter 20

No Activities

Chapter 21

Figure 21.1   Location of the Heart Within the Thoracic Cavity
1. Right lung/pleural cavity
2. Diaphragm
3. Mediastinum
4. Left lung/pleural cavity
5. Heart

Figure 21.2   Pericardium
1. Fibrous pericardium
2. Parietal layer of serous pericardium
3. Pericardial cavity
4. Visceral layer of serous pericardium
5. Myocardium
6. Endocardium
7. Pericardial cavity
8. Myocardium
9. Endocardium
10. Visceral layer of serous pericardium
11. Fibrous pericardium
12. Diaphragm

Figure 21.9   Circulation to and from the Heart Wall
1. Right coronary artery
2. Marginal artery
3. Small cardiac vein
4. Left coronary artery
5. Circumflex artery
6. Great cardiac vein
7. Anterior interventricular artery
8. Coronary sinus
9. Right coronary artery
10. Posterior interventricular artery
11. Middle cardiac vein

Table 21.3   Comparisons Between Cardiac and Skeletal Muscle Tissues

Muscle Tissue	Cardiac Muscle	Skeletal Muscle
Nervous Control	Autonomic	Somatic
Appearance of Cells	Short, branched, striated	Long, cylindrical, striated
Number of Nuclei	Uninucleate	Multinucleate
Location of Nuclei	Central	Peripheral

Figure 21.14   Cardiac Muscle Tissue
1. Striations
2. Intercalated disc
3. Nucleus
4. Cardiac muscle cell

Chapter 22

Exercise 22.1, #4

right ventricle → pulmonary semilunar valve → pulmonary trunk → pulmonary arteries → lungs → pulmonary veins → left atrium

Figure 22.1   Pulmonary Circuit
1. Right atrium
2. Right AV valve
3. Right ventricle
4. Pulmonary semilunar valve
5. Pulmonary trunk
6. Pulmonary arteries
7. Branch of pulmonary artery
8. Pulmonary capillaries
9. Branch of pulmonary vein
10. Pulmonary veins
11. Left atrium
12. Left AV valve
13. Left ventricle
14. Aortic semilunar valve
15. Aorta

Figure 22.2   Great Vessels of the Heart
1. Right subclavian artery
2. Right common carotid artery
3. Left common carotid artery
4. Left subclavian artery
5. Brachiocephalic trunk
6. Aortic arch
7. Right brachiocephalic vein
8. Left brachiocephalic vein
9. Superior vena cava
10. Inferior vena cava

Figure 22.3   Circulation to the Head and Neck

(a) Arterial Supply
1. Internal carotid artery
2. External carotid artery
3. Common carotid artery
4. Vertebral artery
5. Thyrocervical trunk
6. Subclavian artery
7. Superficial temporal artery
8. Occipital artery
9. Facial artery
10. Brachiocephalic trunk (artery)

(b) Venous Drainage

1. Vertebral vein
2. External jugular vein
3. Internal jugular vein
4. Subclavian vein
5. Right brachiocephalic vein
6. Superficial temporal vein
7. Facial vein
8. Superior thyroid vein

Figure 22.4   Circulation from the Aortic Arch to the Anterior Part of the Right Parietal Bone and Back to the Superior Vena Cava

1. Brachiocephalic trunk
2. Right common carotid artery
3. Right external carotid artery
4. Right superficial temporal artery
5. Right superficial temporal vein
6. Right internal jugular vein
7. Right brachiocephalic vein

Figure 22.5   Circulation to the Brain

(a) Arterial Supply

1. Middle cerebral artery
2. Internal carotid artery
3. Posterior cerebral artery
4. Anterior communicating artery
5. Anterior cerebral artery
6. Internal carotid artery
7. Posterior communicating artery
8. Posterior cerebral artery
9. Basilar artery
10. Vertebral artery

(b) Venous Drainage

1. Straight sinus
2. Transverse sinus
3. Sigmoid sinus
4. Internal jugular vein
5. Superior sagittal sinus
6. Inferior sagittal sinus
7. Cavernous sinus

Figure 22.6   Circulation from the Aortic Arch to the Right Parietal Lobe of the Brain and Back to the Right Brachiocephalic Vein

1. Brachiocephalic trunk (artery)
2. Right common carotid artery
a. Alternate 2: right subclavian artery
3. Right internal carotid artery
b. Alternate 3: right vertebral artery
4. Middle cerebral artery
c. Alternate 4: basilar artery
d. Alternate 5: posterior cerebral artery
e. Alternate 6: posterior communicating artery
5. Superior sagittal sinus
6. Transverse sinus
7. Sigmoid sinus
8. Right internal jugular vein
9. Right brachiocephalic vein

Figure 22.7   Circulation to the Thoracic and Abdominal Walls

(a) Arterial Supply

1. Right subclavian artery
2. Brachiocephalic trunk
3. Internal thoracic artery
4. Anterior intercostal arteries
5. Superior epigastric artery
6. Descending abdominal aorta
7. Inferior epigastric artery
8. Aortic arch
9. Left subclavian artery
10. Posterior intercostal arteries
11. Descending thoracic aorta
12. Common iliac artery

(b) Venous Drainage

1. Right subclavian vein
2. Right brachiocephalic vein
3. Superior vena cava
4. Anterior intercostal veins
5. Azygos vein
6. Internal thoracic vein
7. Inferior vena cava
8. Superior epigastric vein
9. Inferior epigastric vein
10. Left subclavian vein
11. Left brachiocephalic vein
12. Accessory hemiazygos vein
13. Posterior intercostal vein
14. Hemiazygos vein
15. Inferior vena cava
16. Left common iliac vein

Figure 22.8   Arterial Supply to Abdominal Organs

(a) Arterial Supply to the Stomach, Spleen, Pancreas, Duodenum, and Liver

1. Celiac trunk
2. Common hepatic artery
3. Hepatic artery proper
4. Left hepatic artery
5. Right hepatic artery
6. Gastroduodenal artery
7. Right gastric artery
8. Left gastric artery
9. Splenic artery
10. Descending abdominal aorta

(b) Arterial Supply to the Small and Large Intestines

1. Middle colic artery
2. Right colic artery
3. Ileocolic artery
4. Celiac trunk

5. Superior mesenteric artery
6. Left colic artery
7. Descending abdominal aorta
8. Inferior mesenteric artery
9. Sigmoid arteries
10. Superior rectal artery

Figure 22.9   The Hepatic Portal System

1. Inferior vena cava
2. Hepatic veins
3. Hepatic portal vein
4. Superior mesenteric vein
5. Gastric vein
6. Splenic vein
7. Inferior mesenteric vein

Figure 22.10   Circulation from the Left Ventricle of the Heart to the Right Kidney and Back to the Right Atrium of the Heart

1. Ascending aorta
2. Aortic arch
3. Descending thoracic aorta
4. Descending abdominal aorta
5. Right renal artery
6. Right renal vein
7. Inferior vena cava

Figure 22.11   Circulation from the Abdominal Aorta to the Spleen and Back to the Right Atrium of the Heart

1. Abdominal aorta
2. Celiac trunk
3. Splenic artery
4. Splenic vein
5. Hepatic portal vein
6. Hepatic veins
7. Inferior vena cava

Exercise 22.6, #8

Left ventricle → ascending aorta → aortic arch → descending aorta → abdominal aorta → inferior mesenteric a. → sigmoid a. → sigmoid colon → sigmoid v. → inferior mesenteric v. → splenic v. → hepatic portal v. → liver → hepatic v. → inferior vena cava → right atrium

Figure 22.13   Circulation to the Upper Limb

(a) Arterial Supply

1. Subclavian artery
2. Axillary artery
3. Deep brachial artery
4. Brachial artery
5. Ulnar artery
6. Radial artery
7. Deep palmar arch
8. Superficial palmar arch
9. Digital arteries

(b) Venous Drainage

1. Brachiocephalic vein
2. Subclavian vein
3. Axillary vein
4. Cephalic vein
5. Basilic vein
6. Brachial veins
7. Median cubital vein
8. Radial veins
9. Cephalic vein
10. Deep palmar venous arch
11. Superficial palmar venous arch
12. Dorsal venous network
13. Digital veins
14. Ulnar veins

Figure 22.14   Circulation from the Aortic Arch to the Anterior Surface of the Index Finger and Back Along a Superficial Route to the Superior Vena Cava

1. Brachiocephalic trunk
2. Subclavian artery
3. Axillary artery
4. Brachial artery
5. Radial artery
6. Superficial palmar arch
7. Digital artery
8. Digital vein
9. Superficial palmar venous arch
10. Cephalic vein
a. Alternate 11: median cubital vein
b. Alternate 12: basilic vein
c. Alternate 13: axillary vein
11. Subclavian vein
12. Brachiocephalic vein
13. Superior vena cava

Figure 22.15   Circulation from the Aortic Arch to the Capitate Bone of the Wrist and Back Along a Deep Route to the Superior Vena Cava

1. Brachiocephalic trunk
2. Subclavian artery
3. Axillary artery
4. Brachial artery
5. Ulnar artery
6. Deep palmar arch
7. Deep palmar venous arch
8. Ulnar veins
9. Brachial vein
10. Axillary vein
11. Subclavian vein
12. Brachiocephalic vein
13. Superior vena cava

Figure 26.7   Classroom Model of the Abdominal Cavity and Large Intestine

1. Right lobe of liver
2. Falciform ligament
3. Pylorus of stomach
4. Gallbladder
5. Greater curvature of stomach
6. Ascending colon
7. Ileum of small intestine
8. Esophagus
9. Left lobe of liver
10. Body of stomach
11. Left colic (splenic) flexure
12. Transverse colon
13. Taenia coli
14. Jejunum of small intestine
15. Descending colon
16. Esophagus
17. Duodenum
18. Right colic (hepatic) flexure of colon
19. Ascending colon
20. Taenia coli
21. Cecum
22. Tail of pancreas
23. Body of pancreas
24. Left colic (splenic) flexure of colon
25. Transverse colon
26. Descending colon
27. Rectum
28. Sigmoid colon

Chapter 27

Figure 27.1   Supporting Ligaments of the Ovary, Uterine Tubes, and Uterus as Seen from a Posterior View

1. Ovarian ligament
2. Uterine tube
3. Suspensory ligament of the ovary
4. Fimbria
5. Mesosalpinx
6. Body of uterus
7. Broad ligament
8. Uterine blood vessels
9. Uterosacral ligament
10. Transverse cervical ligament
11. Vagina
12. Uterine part of uterine tube
13. Isthmus of uterine tube
14. Ampulla of uterine tube
15. Infundibulum of uterine tube
16. Round ligament of the uterus
17. Endometrium
18. Myometrium
19. Perimetrium
20. Internal os
21. Cervical canal
22. External os
23. Uterine tube
24. Mesosalpinx
25. Mesovarium

Figure 27.2   Classroom Model of the Female Pelvic Cavity

1. Uterine tube
2. Ovary
3. Uterus
4. Urinary bladder
5. Labia minora
6. Labia majora
7. Rectum
8. Vagina
9. Bulb of the vestibule
10. Round ligament of the uterus
11. Vesicouterine pouch
12. Pubic symphysis
13. Clitoris
14. External urethral orifice
15. Vaginal orifice
16. Anus
17. Ureter
18. Fimbria of uterine tube
19. Rectouterine pouch
20. Cervix of uterus
21. Vagina

Figure 27.3   The Female Breast

1. Suspensory ligaments
2. Lobe
3. Lactiferous sinus
4. Alveoli
5. Lactiferous ducts
6. Lobule
7. Areolar gland
8. Nipple
9. Areola
10. Adipose tissue
11. Lobe
12. Deep fascia
13. Alveoli
14. Lobule
15. Suspensory ligaments
16. Lactiferous sinus
17. Nipple
18. Lactiferous ducts

Figure 27.4   Male Reproductive Tract Structures

1. Ureter
2. Ampulla of ductus deferens
3. Seminal vesicle
4. Ejaculatory duct
5. Prostate gland
6. Prostatic urethra
7. Bulbourethral gland
8. Urogenital diaphragm
9. Membranous urethra
10. Ductus deferens
11. Epididymis
12. Testis
13. Spongy urethra
14. Urinary bladder
15. Bulb of penis
16. Crus of penis
17. Corpus cavernosum
18. Seminiferous tubules
19. Corpus spongiosum
20. Glans penis

Figure 27.5   Classroom Model of the Male Pelvic Cavity

1. Ureter
2. Urinary bladder
3. Ductus deferens
4. Penis
5. Prepuce
6. Glans penis
7. Ductus deferens (ampulla)
8. Seminal vesicle
9. Prostate gland
10. Ductus deferens
11. Testicular artery and vein
12. Spermatic cord
13. Epididymis
14. Testis
15. Scrotum
16. Ureter
17. Ductus deferens
18. Pubic symphysis
19. Prostate gland
20. Spongy urethra
21. Corpus cavernosum
22. Prepuce
23. Corpus spongiosum
24. Tunica albuginea of testis
25. Tunica vaginalis of testis
26. Urinary bladder
27. Rectum
28. Internal urethral sphincter
29. Prostatic urethra
30. Ejaculatory duct
31. Urogenital diaphragm
32. Anus
33. Membranous urethra
34. Epididymis
35. Seminiferous tubules

Table 27.4   Developmental Stages of Ovarian Follicles

Follicle Stage	Primordial Follicle	Primary Follicle	Secondary (Maturing) Follicle
Photograph	LM 500x	LM 500x	LM 50x
Follicle Description	Single layer of flattened follicular cells surround the oocyte.	One or more layers of cuboidal cells surround the oocyte. Glycoprotein coat (zona pellucida) forms around the oocyte.	Multiple layers of cuboidal cells surround the oocyte. Fluid-filled space or spaces begin to form (antrum/antra). Thecal cells develop into internal and external layers.
Oocyte Stage	Primary oocyte	Primary oocyte	Primary oocyte

Developmental Stages of Ovarian Follicles

Follicle Stage	Vesicular (Graafian) Follicle	Corpus Luteum	Corpus Albicans
Photograph	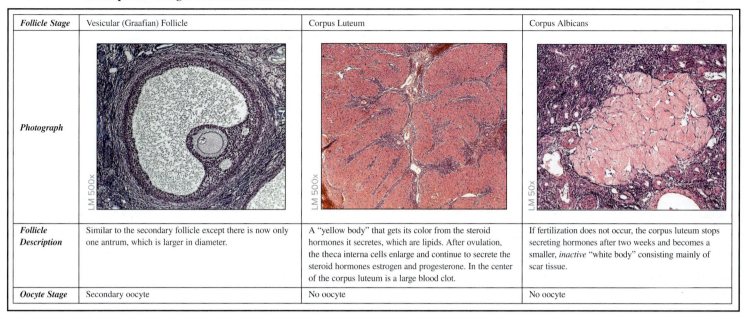		
Follicle Description	Similar to the secondary follicle except there is now only one antrum, which is larger in diameter.	A "yellow body" that gets its color from the steroid hormones it secretes, which are lipids. After ovulation, the theca interna cells enlarge and continue to secrete the steroid hormones estrogen and progesterone. In the center of the corpus luteum is a large blood clot.	If fertilization does not occur, the corpus luteum stops secreting hormones after two weeks and becomes a smaller, *inactive* "white body" consisting mainly of scar tissue.
Oocyte Stage	Secondary oocyte	No oocyte	No oocyte

Table 27.8   Phases of the Menstrual Cycle

Phase	(a) Menstrual Phase	(b) Proliferative Phase	(c) Secretory Phase
	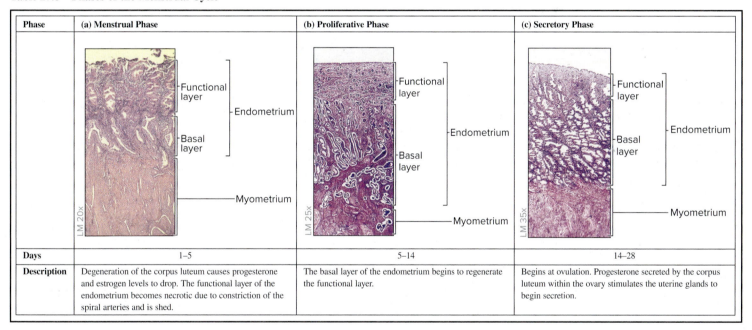		
Days	1–5	5–14	14–28
Description	Degeneration of the corpus luteum causes progesterone and estrogen levels to drop. The functional layer of the endometrium becomes necrotic due to constriction of the spiral arteries and is shed.	The basal layer of the endometrium begins to regenerate the functional layer.	Begins at ovulation. Progesterone secreted by the corpus luteum within the ovary stimulates the uterine glands to begin secretion.

Table 27.14   Pre-Embryonic Period

Developmental Stage	Location	Events
Fertilization	Ampulla of uterine tube	Sperm penetrates secondary oocyte; secondary oocyte completes meiosis and becomes an ovum
Zygote	Ampulla of uterine tube	Ovum and sperm pronuclei fuse to produce diploid cell
Cleavage	Uterine tube	Zygote undergoes cell division
Morula	Uterine tube	Structure formed resembles a solid ball of cells (16 or more cells present)
Blastocyst	Uterus	Hollow ball of cells; outer ring formed by trophoblast; embryoblast is an inner cell cluster
Implantation	Functional layer of endometrium of uterus	Blastocyst adheres to functional layer of uterus; placenta forms (trophoblast cells and functional layer)

Figure 27.19   Structures of the 3-Week-Old Embryo

1. Connecting stalk
2. Yolk sac
3. Amniotic cavity
4. Amnion
5. Embryo
6. Chorion
7. Functional layer of uterus
8. Placenta

Table 27.15   Stages of Embryonic Development

Developmental Week	Events
Week 3	Three primary germ layers form; notochord develops
Week 4	Derivatives of three germ layers form; limb buds appear
Weeks 5–8	Head enlarges; eyes, ears, and nose appear; major organs form

Figure 27.20   Structures of the 4-Week-Old (Late) Embryo

1. Ectoderm
2. Mesoderm
3. Amnion
4. Peritoneal cavity
5. Gut tube
6. Somite
7. Intermediate mesoderm
8. Mesentery
9. Endoderm
10. Abdominal wall

Chapter 28

Table 28.1   Comparison of Human and Cat Vertebrae

Number of Vertebrae	Human	Cat
Cervical	7	7
Thoracic	12	13
Lumbar	5	7
Sacral (fused)	5	3
Coccygeal/Caudal	1–2	~20

Table 28.5   Fiber Orientation of Cat Abdominal Muscles

Muscle	Fiber Orientation
External Oblique	Medial and caudal
Internal Oblique	Medial and cranial
Rectus Abdominis	Straight along the sagittal plane
Transverse Abdominis	Straight along the transverse plane

Chapter 29

Table 29.4   Fiber Orientation of Pig Abdominal Muscles

Muscle	Fiber Orientation
External Oblique	Medial and caudal
Internal Oblique	Medial and cranial
Rectus Abdominis	Straight along the sagittal plane
Transverse Abdominis	Straight along the transverse plane